# Lecture Notes in Computer Science 5117

*Commenced Publication in 1973*
Founding and Former Series Editors:
Gerhard Goos, Juris Hartmanis, and Jan van Leeuwen

Andrei Voronkov (Ed.)

# Rewriting Techniques and Applications

19th International Conference, RTA 2008
Hagenberg, Austria, July 15-17, 2008
Proceedings

Springer

Volume Editor

Andrei Voronkov
The University of Manchester
School of Computer Science
Kilburn Building, Oxford Road, Manchester M13 9PL, UK

Library of Congress Control Number: 2008930434

CR Subject Classification (1998): F.4, F.3.2, D.3, I.2.2-3, I.1

LNCS Sublibrary: SL 1 – Theoretical Computer Science and General Issues

ISSN 0302-9743
ISBN-10 3-540-70588-0 Springer Berlin Heidelberg New York
ISBN-13 978-3-540-70588-8 Springer Berlin Heidelberg New York



Springer is a part of Springer Science+Business Media

springer.com


Printed in Germany

Typesetting: Camera-ready by author, data conversion by Scientific Publishing Services, Chennai, India
Printed on acid-free paper SPIN: 12323005 06/3180 5 4 3 2 1 0

# Preface

This volume contains the papers presented at RTA 2008: 19th International Conference on Rewriting Techniques and Applications held July 15–17th in Hagenberg, Austria and organised by the Research Institute on Symbolic Computation.

There were 57 submissions. Each submission was reviewed by at least four Programme Committee members. The committee decided to accept 30 papers. The submission and Programme Committee work was organised through the EasyChair system.

I thank the Programme Committee members for their very efficient work. My special thanks to Temur Kutsia, Aart Middeldorp, Robert Nieuwenhuis and Maribel Fernandez for their help and advice on many aspects of the RTA organisation and traditions. I would also like to thank RTA General Chair Bruno Buchberger.

April 2008 Andrei Voronkov

# Conference Organisation

## General Chair

Bruno Buchberger (Johannes Kepler University Linz)

## Programme Chair

Andrei Voronkov (University of Manchester)

## Organising Committee Chair

Temur Kutsia (Johannes Kepler University Linz)

## Programme Committee

Maribel Fernández (King's College London)
Neil Ghani (University of Nottingham)
Jürgen Giesl (RWTH Aachen)
Guillem Godoy (Universidad Politécnica de Cataluña)
Jean Goubault-Larrecq (ENS Cachan)
Tetsuo Ida (University of Tsukuba)
Claude Kirchner (INRIA)
Konstantin Korovin (University of Manchester)
Temur Kutsia (Johannes Kepler University Linz)
Aart Middeldorp (University of Innsbruck)
Paliath Narendran (University at Albany - SUNY)
Robert Nieuwenhuis (Universidad Politécnica de Cataluña)
Michaël Rusinowitch (LORIA-INRIA)
Aaron Stump (Washington University in St. Louis)
Jean-Marc Talbot (Université de Provence)
Yoshihito Toyama (Tohoku University)
Ralf Treinen (Université Paris Diderot)
Hans Zantema (Technische Universiteit Eindhoven)

## External Reviewers

Markus Aderhold
Maria Alpuente
Thorsten Altenkirch
Takahito Aoto
Emilie Balland
Richard Banach
Peter Baumgartner
Clara Bertolissi

Hariolf Betz
Gavin Bierman
Stefan Blom
Yohan Boichut
Iovka Boneva
Guillaume Bonfante
Thierry Boy de la Tour
Sander Bruggink
Serge Burckel
Sergiu Bursuc
Christophe Calves
Benjamin Carle
Jacques Chabin
Jeremie Chalopin
Witold Charatonik
James Cheney
Yannick Chevalier
Yuki Chiba
Olaf Chitil
Horatiu Cirstea
Evelyne Contejean
Veronique Cortier
Vincent Danos
Jeremy Dawson
Philippe de Groote
Fer-Jan de Vries
Stephanie Delaune
Kevin Donnelly
Dan Dougherty
Daniel Dougherty
Rachid Echahed
Steven Eker
Jörg Endrullis
Emmanuel Filiot
Severine Fratani
Laurent Fribourg
Carsten Fuhs
Thomas Genet
Clemens Grabmayer
Bernhard Gramlich
Yves Guiraud
Andrew Haas
Makoto Hamana
Michael Hanus
Hugo Herbelin
Nao Hirokawa
Clement Houtmann
Florent Jacquemard
Hélène Kirchner
Stefan Kahrs
Yukiyoshi Kameyama
Yevgeny Kazakov
Jeroen Ketema
Zurab Khasidashvili
Kentaro Kikuchi
Vladimir Klebanov
Adam Koprowski
Martin Korp
Sébastien Limet
Luigi Liquori
Francisco Javier Lopez-Fraguas
Salvador Lucas
Denis Lugiez
Wolfgang Lux
Ian Mackie
Michael Maher
Yitzhak Mandelbaum
Claude Marché
Mircea Marin
Narciso Marti-Oliet
Robert McNaughton
Catherine Meadows
Yasuhiko Minamide
Jean-François Monin
Pierre-Etienne Moreau
Georg Moser
Frantisek Mraz
Masaki Nakamura
Enrica Nicolini
Joachim Niehren
Naoki Nishida
Kazuhiro Ogata
Luke Ong
David Pichardie
Brigitte Pientka
Detlef Plump
Romain Péchoux
Yves RooS
Matthias Raffelsieper
Pierre Rety

Christophe Ringeissen
Enric Rodriguez-Carbonell
Grigore Rosu
Grigore Rosu
David Rydeheard
Didier Rémy
Masahiko Sakai
John Sasso
Alexis Saurin
Manfred Schmidt-Schauss
Peter Schneider-Kamp
Helmut Seidl
Paula Severi
Francois-Regis Sinot
Christian Sternagel
Christoph Sticksel
Stephan Swiderski
Luc Ségoufin
Géraud Sénizergues
Toshinori Takai
Carolyn Talcott
Hendrik Tews
René Thiemann
Rene Thiemann
Wolfgang Thomas
Sophie Tison
Ashish Tiwari
Christian Urban
Camille Vacher
Steffen van Bakel
Vincent van Oostrom
Femke van Raamsdonk
Lionel Vaux
Alicia Villanueva
Eelco Visser
Uwe Waldmann
Johannes Waldmann
Christoph Weidenbach
Edwin Westbrook
Freek Wiedijk
Toshiyuki Yamada

# Table of Contents

# Modular Termination of Basic Narrowing*

María Alpuente, Santiago Escobar, and José Iborra

Universidad Politécnica de Valencia, Spain
{alpuente,sescobar,jiborra}@dsic.upv.es

**Abstract.** Basic narrowing is a restricted form of narrowing which constrains narrowing steps to a set of non-blocked (or basic) positions. Basic narrowing has a number of important applications including equational unification in canonical theories. Another application is analyzing termination of narrowing by checking the termination of basic narrowing, as done in pioneering work by Hullot. In this work, we study the modularity of termination of basic narrowing in hierarchical combinations of TRSs, including a generalization of proper extensions with shared subsystem. This provides new algorithmic criteria to prove termination of basic narrowing.

## 1 Introduction

Narrowing [12] is a generalization of term rewriting that allows free variables in terms (as in logic programming) and replaces pattern matching with syntactic unification. Narrowing was originally introduced as a mechanism for solving equational unification problems [15], hence termination results for narrowing have been traditionally achieved as a by–product of addressing the decidability of equational unification. Basic narrowing [15] is a refinement of narrowing which restricts narrowing steps to a set of non-blocked (or basic) positions, and is still complete for equational unification in canonical TRSs. Termination of basic narrowing was first studied by Hullot in [15], where a faulty termination result for narrowing was enunciated, namely the termination of all narrowing derivations in canonical theories when all basic narrowing derivations issuing from the right–hand sides (rhs's) of the rules terminate. This result was implicitly corrected in [16], downgrading it to the more limited result of basic narrowing termination (instead of ordinary narrowing) under the basic narrowing termination requirement for the rhs's of the rules. The missing condition to recover narrowing termination in [15] is to require that the TRS satisfies Réty's maximal commutation condition for narrowing sequences [23], as we proved[1] in [2]. From this result, we also distilled in [2] a syntactic characterization of TRSs where

* This work has been partially supported by the EU (FEDER) and Spanish MEC project TIN2007-68093-C02-02, Integrated Action Hispano-Alemana A2006-0007, the UPV grant 3249 PAID0607 and the UPV grant FPI-UPV 2006-01.

[1] We also explicitly dropped in [2] the superfluous requirement of canonicity from Hullot's termination result, as cognoscenti tacitly do.

A. Voronkov (Ed.): RTA 2008, LNCS 5117, pp. 1–16, 2008.

termination of basic narrowing implies termination of narrowing, namely right-linear TRSs that are either left-linear or regular and where narrowing computes[2] only normalized substitutions.

The main motivation for this paper is proving termination of narrowing via termination of basic narrowing. We present several criteria for modular termination of basic narrowing in hierarchical combinations of TRSs, including generalized proper extensions with shared subsystem. By adopting the divide-and-conquer principle, this allows us to prove (basic) narrowing termination in a modular way, thus extending the class of TRSs for which termination of basic narrowing (and hence termination of narrowing) can be proved. We assume a standard notion of modularity, where a property $\varphi$ of TRSs is called *modular* if, whenever $\mathcal{R}_1$ and $\mathcal{R}_2$ satisfy $\varphi$, then their combination $\mathcal{R}_1 \cup \mathcal{R}_2$ also satisfies $\varphi$. Our modularity results for basic narrowing rely on a commutation result for basic narrowing sequences that has not been identified in the related literature before.

In [21], a modularity result for decidability of unification (via termination of narrowing) in canonical TRSs is given. However, this result does not imply the modularity of narrowing termination for a particular class of TRSs but rather the possibility to define a terminating, modular narrowing procedure. Namely, the result in [21] is as follows: given a canonical TRS $\mathcal{R}$ such that narrowing terminates for $\mathcal{R}_1$ and $\mathcal{R}_2$ and $\mathcal{R}{\downarrow} \subseteq \mathcal{R}_1{\downarrow}\mathcal{R}_2{\downarrow}$ (i.e. normalization with $\mathcal{R} = \mathcal{R}_1 \cup \mathcal{R}_2$ can be obtained by first normalizing with $\mathcal{R}_1$ followed by a normalization with $\mathcal{R}_2$), then there is a terminating and complete, modular narrowing strategy for $\mathcal{R}$. Any complete strategy can be used within the modular procedure given in [21], including the basic narrowing strategy. As far as we know this is the only previous modularity result in the literature that concerns the modular termination of basic narrowing.

After some preliminaries in Section 2, we study commutation properties of basic narrowing derivations in Section 3. Section 4 recalls some standard notions for modularity of rewriting and presents our main modularity results for the termination of basic narrowing. In order to prove in Section 4.3 that termination of basic narrowing is modular for proper extensions [22], we first prove an intermediate result: in Section 4.2 we prove that basic narrowing termination is modular for a restriction of proper extensions called *nice* extensions [22]. In Section 5 we generalize our results and prove modularity for a wider class of TRSs called *relaxed proper extensions*. We conclude in Section 6. Proofs of all results in this paper are included in [1].

## 2 Preliminaries

In this section, we briefly recall the essential notions and terminology of term rewriting [9,20,24].

$\mathcal{V}$ denotes a countably infinite set of variables, and $\Sigma$ denotes a set of function symbols, or signature, each of which has a fixed associated arity. Terms are

[2] This includes some popular classes of TRSs, including linear constructor systems.

viewed as labelled trees in the usual way, where $\mathcal{T}(\Sigma, \mathcal{V})$ and $\mathcal{T}(\Sigma)$ denote the non-ground term algebra and the ground algebra built on $\Sigma \cup \mathcal{V}$ and $\Sigma$, respectively. Positions are defined as sequences of positive natural numbers used to address subterms of a term, with $\epsilon$ as the root (or top) position (i.e., the empty sequence). Concatenation of positions $p$ and $q$ is denoted by $p.q$, and $p < q$ is the usual prefix ordering. Two positions $p, q$ are disjoint, denoted by $p \parallel q$, if neither $p < q$, $p > q$, nor $p = q$. Given $S \subseteq \Sigma \cup \mathcal{V}$, $\mathcal{P}os_S(t)$ denotes the set of positions of a term $t$ that are rooted by function symbols or variables in $S$. $\mathcal{P}os_{\{f\}}(t)$ with $f \in \Sigma \cup \mathcal{V}$ will be simply denoted by $\mathcal{P}os_f(t)$, and $\mathcal{P}os_{\Sigma\cup\mathcal{V}}(t)$ will be simply denoted by $\mathcal{P}os(t)$. $t|_p$ is the subterm at the position $p$ of $t$. $t[s]_p$ is the term $t$ with the subterm at the position $p$ replaced with term $s$. By $Var(s)$, we denote the set of variables occurring in the syntactic object $s$. By $\bar{x}$, we denote a tuple of pairwise distinct variables. A *fresh* variable is a variable that appears nowhere else. A *linear* term is one where every variable occurs only once.

A *substitution* $\sigma$ is a mapping from the set of variables $\mathcal{V}$ into the set of terms $\mathcal{T}(\Sigma, \mathcal{V})$, with a finite domain $D(\sigma)$ and image $I(\sigma)$. A substitution is represented as $\{x_1/t_1, \ldots, x_n/t_n\}$ for variables $x_1, \ldots, x_n$ and terms $t_1, \ldots, t_n$. The application of substitution $\theta$ to term $t$ is denoted by $t\theta$, using postfix notation. Composition of substitutions is denoted by juxtaposition, i.e., the substitution $\sigma\theta$ denotes $(\theta \circ \sigma)$. We write $\theta_{\restriction Var(s)}$ to denote the restriction of the substitution $\theta$ to the set of variables in $s$; by abuse of notation, we often simply write $\theta_{\restriction s}$. Given a term $t$, $\theta = \nu\ [t]$ iff $\theta_{\restriction Var(t)} = \nu_{\restriction Var(t)}$, that is, $\forall x \in Var(t)$, $x\theta = x\nu$. A substitution $\theta$ is more general than $\sigma$, denoted by $\theta \leq \sigma$, if there is a substitution $\gamma$ such that $\theta\gamma = \sigma$. A *unifier* of terms $s$ and $t$ is a substitution $\vartheta$ such that $s\vartheta = t\vartheta$. The *most general unifier* of terms $s$ and $t$, denoted by $mgu(s,t)$, is a unifier $\theta$ such that for any other unifier $\theta'$, $\theta \leq \theta'$.

A *term rewriting system* (TRS) $\mathcal{R}$ is a pair $(\Sigma, R)$, where $R$ is a finite set of rewrite rules of the form $l \to r$ such that $l, r \in \mathcal{T}(\Sigma, \mathcal{V})$, $l \notin \mathcal{V}$, and $Var(r) \subseteq Var(l)$. We will often write just $\mathcal{R}$ or $(\Sigma, R)$ instead of $\mathcal{R} = (\Sigma, R)$. Given a TRS $\mathcal{R} = (\Sigma, R)$, the signature $\Sigma$ is often partitioned into two disjoint sets $\Sigma = \mathcal{C} \uplus \mathcal{D}$, where $\mathcal{D} = \{f \mid f(t_1, \ldots, t_n) \to r \in R\}$ and $\mathcal{C} = \Sigma \setminus \mathcal{D}$. Symbols in $\mathcal{C}$ are called *constructors*, and symbols in $\mathcal{D}$ are called *defined functions*. The elements of $\mathcal{T}(\mathcal{C}, \mathcal{V})$ are called *constructor terms*. We let $Def(\mathcal{R})$ denote the set of defined symbols in $\mathcal{R}$. A rewrite step is the application of a rewrite rule to an expression. A term $s \in \mathcal{T}(\Sigma, \mathcal{V})$ *rewrites* to a term $t \in \mathcal{T}(\Sigma, \mathcal{V})$, denoted by $s \xrightarrow{p}_{\mathcal{R}} t$, if there exist $p \in \mathcal{P}os_\Sigma(s)$, $l \to r \in \mathcal{R}$, and substitution $\sigma$ such that $s|_p = l\sigma$ and $t = s[r\sigma]_p$. When no confusion can arise, we omit the subscript in $\to_{\mathcal{R}}$. We also omit the reduced position p when it is not relevant. A term $s$ is a *normal form* w.r.t. the relation $\to_{\mathcal{R}}$ (or simply a normal form), if there is no term $t$ such that $s \to_{\mathcal{R}} t$. A term is a reducible expression or *redex* if it is an instance of the left hand side of a rule in $\mathcal{R}$. A term $s$ is a *head normal form* if there are no terms $t, t' s.t.\ s \to^*_{\mathcal{R}} t' \xrightarrow{\epsilon}_{\mathcal{R}} t$. A TRS $\mathcal{R}$ is *($\to$)-terminating* (also called strongly normalizing or noetherian) if there are no infinite reduction sequences $t_1 \to_{\mathcal{R}} t_2 \to_{\mathcal{R}} \ldots$.

Narrowing is a symbolic computation mechanism that generalizes rewriting by replacing pattern matching with syntactic unification. Many redundancies in the narrowing algorithm can be eliminated by restricting narrowing steps to a distinguished set of *basic* positions, which was proposed by Hullot in [15].

### 2.1 Basic Narrowing

Basic narrowing is the restriction of narrowing introduced by Hullot [15] which is essentially based on forbidding narrowing steps on terms brought in by instantiation. We use the definition of basic narrowing given in [14], where the expression to be narrowed is split into a *skeleton* $t$ and an *environment* part $\theta$, i.e., $\langle t, \theta \rangle$. The environment part keeps track of the accumulated substitution so that, at each step, substitutions are composed in the environment part, but are not applied to the expression in the skeleton part, as opposed to ordinary narrowing. For TRS $\mathcal{R}$, $l \to r \ll \mathcal{R}$ denotes that $l \to r$ is a *fresh* variant of a rule in $\mathcal{R}$, i.e., all the variables are *fresh*.

**Definition 1 (Basic narrowing).** *[14] Given a term $s \in \mathcal{T}(\Sigma, \mathcal{V})$ and a substitution $\sigma$, a basic narrowing step for $\langle s, \sigma \rangle$ is defined by $\langle s, \sigma \rangle \overset{b}{\leadsto}_{p,\mathcal{R},\theta} \langle t, \sigma' \rangle$ if there exist $p \in \mathcal{P}os_\Sigma(s)$, $l \to r \ll \mathcal{R}$, and substitution $\theta$ such that $\theta = mgu(s|_p\sigma, l)$, $t = (s[r]_p)$, and $\sigma' = \sigma\theta$.*

Along a basic narrowing derivation, the set of *basic* occurrences of $\langle t, \theta \rangle$ is $\mathcal{P}os_\Sigma(t)$, and the non–basic occurrences are $\mathcal{P}os_\Sigma(t\theta) - \mathcal{P}os_\Sigma(t)$. When $p$ is not relevant, we simply denote the basic narrowing relation by $\overset{b}{\leadsto}_{\mathcal{R},\theta}$. By abuse of notation, we often relax the skeleton-environment notation for basic narrowing steps, i.e., $\langle s, \sigma \rangle \overset{b}{\leadsto}_{\mathcal{R},\theta} \langle t, \sigma' \rangle$, and use the more compact notation $s\sigma \overset{b}{\leadsto}_{\mathcal{R},\theta} t\sigma\theta$ instead; but then suitable track of the basic positions along the narrowing sequences is implicitly done.

We say that $\mathcal{R}$ is $(\overset{b}{\leadsto})$-terminating when every basic narrowing derivation issuing from any term terminates. All modular termination results in this paper are based on the following termination result for basic narrowing. It is essentially Hullot's basic narrowing termination result, where we have explicitly dropped the superfluous requirement of canonicity [2].

**Theorem 1 (Termination of Basic Narrowing).** [15,2] *Let $\mathcal{R}$ be a TRS. If for every $l \to r \in \mathcal{R}$, all basic narrowing derivations issuing from $r$ terminate, then $\mathcal{R}$ is $(\overset{b}{\leadsto})$-terminating.*

In the literature, this condition has been approximated by requiring that every rhs of a rewrite rule is a variable, in [15], or a constructor term, in [21]. This approximation has been generalized in [2] by requiring the rhs's to be a *rigid normal form* (rnf), i.e., unnarrowable.

## 3 A Commutation Result for Basic Narrowing Derivations

The commutation properties of ordinary narrowing were extensively studied by Rety in [23]. We analyze here those of basic narrowing, in Rety's style. First let us recall the notion of *antecedent* of a position in a rewriting sequence [23].

**Definition 2 (Antecedent of a position).** [23] *Let $t \xrightarrow{p}_{l\to r} t'$ be a rewriting step, $v \in \mathcal{P}os(t)$, and $v' \in \mathcal{P}os(t')$. Position $v$ is an* antecedent *of $v'$ iff either*

1. *$v \parallel p$, i.e., $v$ and $p$ are disjoint, and $v = v'$, or*
2. *there exists an occurrence $u' \in \mathcal{P}os_x(r)$ of a variable $x$ in $r$ s.t. $v' = p.u'.w$ and $v = p.u.w$, where $u \in \mathcal{P}os_x(l)$ is an occurrence of $x$ in $l$.*

With the notations of the previous definition, we have:

1. $t|_v = t'|_{v'}$,
2. $v'$ may have no antecedent if $v' = p.u'$ with $u' \in \mathcal{P}os_\Sigma(r)$, or if $v' < p$,

This notion extends to a rewrite sequence by transitive closure of the rewriting relation in the usual way. The notion of antecedent can also be extended to narrowing sequences as follows.

**Definition 3 (Narrowing antecedent of a position).** [23] *Let $t \overset{b}{\leadsto}{}^*_{\mathcal{R},\sigma} t'$, $v \in \mathcal{P}os(t)$, and $v' \in \mathcal{P}os(t')$. We say $v$ is an antecedent of $v'$ iff $v$ is an antecedent of $v'$ in the rewrite sequence $t\sigma \to^*_{\mathcal{R}} t'$.*

And note that now we have $t|_v\sigma = t'|_{v'}$. In the following, we consider basic narrowing derivations of the form

$$s \overset{b}{\leadsto}_{p,g\to d,\sigma} t \overset{b}{\leadsto}_{q,l\to r,\theta} u \tag{1}$$

and we are interested in the conditions that allow us to commute the first two steps by first applying to $s$ the rule $l \to r$ and then the rule $g \to d$ to the resulting term. If the subterm $t|_q$ already exists in $s$, i.e., if $q$ admits at least one antecedent in $s$, the idea essentially consists in applying $l \to r$ to all the antecedents of $q$, and then applying $g \to d$ to the resulting term. Let us give an example for motivation.

*Example 1.* [23] Let us consider the following TRS $\mathcal{R}_4$:

$$\mathcal{R}_4 = \{\ \mathtt{f}(x,x) \ \to\ x\ (\mathrm{r1}) \qquad \mathtt{g}(x,\mathtt{h}(x)) \ \to\ x\ (\mathrm{r2})\ \}$$

and the basic narrowing derivation:

$$\langle \mathtt{h}(\mathtt{f}(0,x),\mathtt{g}(x,y)),\{\}\rangle \overset{b}{\leadsto}_{p=1,r1,\{x/0\}} \langle \mathtt{h}(x,\mathtt{g}(x,y)),\{x/0\}\rangle$$
$$\overset{b}{\leadsto}_{q=2,r2,\{y/\mathtt{h}(0)\}} \langle \mathtt{h}(x,x),\{x/0\}\rangle.$$

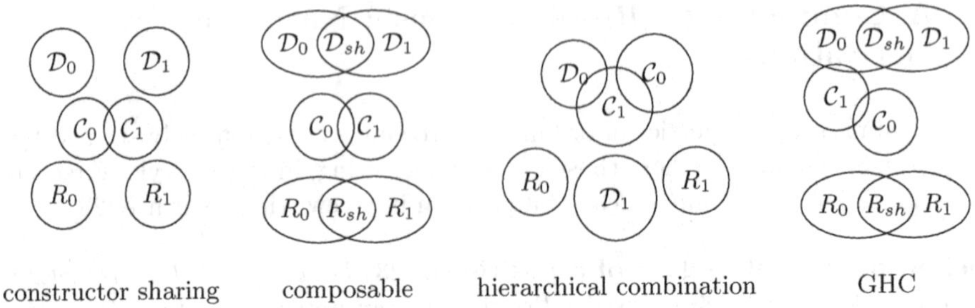


**Fig. 1.** Standard modular combinations

The occurrences $p$ and $q$ are disjoint, therefore $q$ has an antecedent in $s$ at $q' = 2$. By first applying $r2$ at $q' = 2$, and then $r1$ at $p$ we get:

$$\langle \mathtt{h}(\mathtt{f}(0,x),\mathtt{g}(x,y)),\{\}\rangle \overset{b}{\leadsto}_{q'=2,r2,\{y/\mathtt{h}(x)\}} \langle \mathtt{h}(\mathtt{f}(0,x),x),\{\}\rangle$$
$$\overset{b}{\leadsto}_{p=1,r1,\{x/0\}} \langle \mathtt{h}(x,x),\{x/0\}\rangle$$

The following result establishes that in a basic narrowing derivation, the antecedent of a position is always in the skeleton part, and case 2 of Definition 2 cannot happen.

**Lemma 1.** *Given a basic narrowing derivation* $t \overset{b}{\leadsto}_{p,l\to r,\sigma} t' \overset{b}{\leadsto}_{q',g\to d,\theta} u$, *if* $q \in \mathcal{P}os(t)$ *is an antecedent of* $q'$, *then* $q$ *and* $p$ *are necessarily disjoint,* $q'$ *is in the skeleton, and* $q = q'$.

Now we show that basic narrowing steps can be commuted under certain conditions. This result is the basis for the modularity results of Section 5.

**Proposition 1 (Commutation of Basic Narrowing).** *Let* $\mathcal{R}$ *be a TRS and*

$$\langle s,\theta\rangle \overset{b}{\leadsto}_{p,g\to d,\sigma_1} \langle s[d]_p,\theta\sigma_1\rangle \overset{b}{\leadsto}_{q,l\to r,\sigma_2} \langle s[d]_p[r]_q,\theta\sigma_1\sigma_2\rangle \tag{2}$$

*be a sequence of two basic narrowing steps s.t.* $q$ *admits an antecedent in* $s$. *Then (2) can be commuted to the following equivalent basic narrowing derivation:*

$$\langle s,\theta\rangle \overset{b}{\leadsto}_{q,l\to r,\sigma_3} \langle s[r]_q,\theta\sigma_3\rangle \overset{b}{\leadsto}_{p,g\to d,\sigma_4} \langle s[r]_q[d]_p,\theta\sigma_3\sigma_4\rangle \tag{3}$$

*where* $\sigma_1\sigma_2 = \sigma_3\sigma_4[s]$.

## 4 Modular Termination of Basic Narrowing

Let us recall some standard notions regarding modularity of rewriting, as defined in [20], that will be used throughout the paper. Figure 1 shows diagramatic renditions of these definitions.

**disjoint.** $(\Sigma_0, \mathcal{R}_0)$ and $(\Sigma_1, \mathcal{R}_1)$ are *disjoint* if they do not share symbols, that is, $\Sigma_0 \cap \Sigma_1 = \varnothing$. Their union, called *direct sum*, is denoted $\mathcal{R} = \mathcal{R}_0 \uplus \mathcal{R}_1$.

**constructor sharing.** $(\mathcal{D}_0 \uplus \mathcal{C}_0, \mathcal{R}_0)$ and $(\mathcal{D}_1 \uplus \mathcal{C}_1, \mathcal{R}_1)$ are *constructor sharing* if they do not share defined symbols, i.e., $\mathcal{D}_0 \cap \mathcal{D}_1 = \varnothing$.

**composable.** Two systems $(\mathcal{D}_0 \uplus \mathcal{D}_{sh} \uplus \mathcal{C}_0, \mathcal{R}_0)$ and $(\mathcal{D}_1 \uplus \mathcal{D}_{sh} \uplus \mathcal{C}_1, \mathcal{R}_1)$ are *composable* if $\mathcal{D}_0 \cap \mathcal{C}_1 = \mathcal{D}_1 \cap \mathcal{C}_0 = \varnothing$ and both systems share all the rewrite rules that define every shared defined symbol, i.e., $\mathcal{R}_{sh} \subseteq \mathcal{R}_0 \cap \mathcal{R}_1$ where $\mathcal{R}_{sh} = \{l \to r \in \mathcal{R}_0 \cup \mathcal{R}_1 \mid root(l) \in \mathcal{D}_{sh}\}$.

**hierarchical combination.** A system $\mathcal{R} = \mathcal{R}_0 \cup \mathcal{R}_1$ is the *hierarchical combination* (HC) of a base system $(\mathcal{D}_0 \uplus \mathcal{C}_0, \mathcal{R}_0)$ and an extension system $(\mathcal{D}_1 \uplus \mathcal{C}_1, \mathcal{R}_1)$ iff $\mathcal{D}_0 \cap \mathcal{D}_1 = \varnothing$ and $\mathcal{C}_0 \cap \mathcal{D}_1 = \varnothing$.

**generalized hierarchical combination.** A system $\mathcal{R} = \mathcal{R}_0 \cup \mathcal{R}_1$ is the *generalized hierarchical combination* (GHC) of a base system $(\mathcal{D}_0 \uplus \mathcal{D}_{sh} \uplus \mathcal{C}_0, \mathcal{R}_0)$ and an extension $(\mathcal{D}_1 \uplus \mathcal{D}_{sh} \uplus \mathcal{C}_1, \mathcal{R}_1)$ with shared subsystem $(\mathcal{F}, R_{sh})$ iff $\mathcal{D}_0 \cap \mathcal{D}_1 = \varnothing$, $\mathcal{C}_0 \cap \mathcal{D}_1 = \varnothing$, $R_{sh} = R_0 \cap R_1$ where $R_{sh} = \{l \to r \in \mathcal{R}_0 \cup \mathcal{R}_1 \mid root(l) \in \mathcal{D}_{sh}\}$, and $\mathcal{F} = \{f \in \mathcal{F} \mid f \text{ occurs in } R_{sh}\}$.

Roughly speaking, in a hierarchical combination $\mathcal{R} = \mathcal{R}_0 \cup \mathcal{R}_1$ the sets of function symbols defined in $\mathcal{R}_0$ and $\mathcal{R}_1$ are disjoint, and the defined function symbols of the base ($\mathcal{R}_0$) can occur in rules of the extension, but not viceversa. GHCs generalize both HCs and composable systems.

As noted by [22], this classification of combinations of TRSs is straightforwardly applicable to programming languages and incremental program development. The modularity results of direct–sums can be used when two subsystems are defined over different domains, e.g. the natural numbers and the Boolean domain. The modularity results of constructor sharing unions can be used when two subsystems define independent functions (none of the two systems use the procedures defined in the other) over a common domain. HCs model the notion of *modules* in programming languages. The following example borrowed from [20] illustrates these notions.

*Example 2.* Consider the following TRSs:

$$\mathcal{R}_+ = \begin{cases} 0 + y \to y \\ \mathtt{s}(x) + y \to \mathtt{s}(x + y) \end{cases} \qquad \mathcal{R}_- = \begin{cases} 0 - \mathtt{s}(y) \to 0 \\ x - 0 \to x \\ \mathtt{s}(x) - \mathtt{s}(y) \to x - y \end{cases}$$

$$\mathcal{R}_* = \begin{cases} 0 * y \to 0 \\ \mathtt{s}(x) * y \to (x * y) + y \end{cases} \qquad \mathcal{R}_{pow} = \begin{cases} \mathtt{pow}(x, 0) \to \mathtt{s}(0) \\ \mathtt{pow}(x, \mathtt{s}(y)) \to x * \mathtt{pow}(x, y) \end{cases}$$

$$\mathcal{R}_{app} = \begin{cases} \mathtt{nil}\texttt{++}ys \to ys \\ (x : xs)\texttt{++}ys \to x : (xs\texttt{++}ys) \end{cases}$$

$\mathcal{R}_+$ and $\mathcal{R}_{app}$ are disjoint, $\mathcal{R}_+$ and $\mathcal{R}_-$ are constructor–sharing, $\mathcal{R}_+ \cup \mathcal{R}_*$ is composable with $\mathcal{R}_+ \cup \mathcal{R}_{app}$, and $\mathcal{R}_* \cup \mathcal{R}_+$ is a HC where $\mathcal{R}_*$ extends $\mathcal{R}_+$. Lastly, the system $\mathcal{R}_1 = \mathcal{R}_{pow} \cup \mathcal{R}_+$ extends $\mathcal{R}_0 = \mathcal{R}_* \cup \mathcal{R}_+$ in a GHC with shared subsystem $\mathcal{R}_{sh} = \mathcal{R}_+$.

Note that constructor sharing systems generalize disjoint unions, and are themselves generalized by both composable and HCs. Finally, these last two notions are subsumed by GHCs.

### 4.1 Constructor–Sharing and Composable Unions

The following result is a direct consequence of Theorem 1.

**Theorem 2 (Modularity of Constructor–Sharing Unions).** *Termination of basic narrowing is modular for constructor–sharing systems.*

This implies modularity for disjoint unions too, as in the following well-known example.

*Example 3 (Toyama).* Let us consider Toyama's example [25]:

$$\mathcal{R}_0 : \mathtt{f}(0,1,x) \to \mathtt{f}(x,x,x) \qquad \mathcal{R}_1 : \mathtt{g}(x,y) \to x \quad \mathtt{g}(x,y) \to y$$

Basic narrowing trivially terminates on each system, since every rhs is clearly unnarrowable. By Theorem 2, it also terminates for $\mathcal{R}_0 \cup \mathcal{R}_1$.

It is well known that Toyama's example is not $(\to)$-terminating. However, it is innermost terminating. This shows that $(\overset{b}{\leadsto})$-termination does not entail $(\to)$-termination, which suggests that the modularity requirements for $(\overset{b}{\leadsto})$-termination are less restrictive than those of $(\to)$-termination. Actually, the modularity properties of $(\overset{b}{\leadsto})$-termination are comparable to those of innermost $(\to)$-termination (see e.g. [20]). The next theorem extends the modularity of $(\overset{b}{\leadsto})$-termination to composable systems.

**Theorem 3 (Modularity of Composable Unions).** *Termination of basic narrowing is a modular property of composable systems.*

In Section 4.3 we further extend this result up to generalized proper extensions (GPE) [22], a fairly general restriction of HCs. To achieve this result, we proceed as follows. First we prove in Section 4.2 the modularity for generalized *nice* extensions (GNE) [22], a restriction of GPEs. Then, we apply a result from [22] that relates GPEs to GNEs, which delivers the desired result.

In the remaining of this section we make use of the following notion.

**Definition 4 (Dependency Relation $\trianglerighteq_{\mathcal{R}}$).** [22] *For a TRS $(\mathcal{D} \uplus \mathcal{C}, \mathcal{R})$ the dependency relation $\trianglerighteq_{\mathcal{R}}$ is the smallest preorder satisfying the condition $f \trianglerighteq_{\mathcal{R}} g$ whenever there is a rewrite rule $f(s_1, \ldots, s_m) \to r \in \mathcal{R}$ and $g(t_1, \ldots, t_n)$ is a subterm of $r$, with $g \in \mathcal{D}$.*

We often omit $\mathcal{R}$ from $\trianglerighteq_{\mathcal{R}}$ when it is clear from the context. We say that a symbol $f \in \mathcal{D}$ depends on a symbol $g \in \mathcal{D}$ if $f \trianglerighteq g$. Intuitively, $f \trianglerighteq g$ if the evaluation of $f$ involves a call to $g$ after one or more rewrite steps.

### 4.2 Nice Extensions

Nice extensions (NE) are a restriction of PEs introduced by Krishna Rao [22]. NEs are a useful intermediate notion, because it can be shown that every PE can be modelled as a *pyramid* of NEs, which we do in Section 4.3.

**Definition 5 (Split).** *Let $(\mathcal{D} \uplus \mathcal{C}, \mathcal{R})$ be a GHC of a base system $(\mathcal{D}_0 \uplus \mathcal{D}_{sh} \uplus \mathcal{C}_0, \mathcal{R}_0)$ and the extension $(\mathcal{D}_1 \uplus \mathcal{D}_{sh} \uplus \mathcal{C}_1, \mathcal{R}_1)$. The set $\mathcal{D}_1$ of defined symbols of $\mathcal{R}_1$ is split in two sets $\mathcal{D}_1^0$ and $\mathcal{D}_1^1$ where $\mathcal{D}_1^0$ contains all the symbols that depend on function symbols from $\mathcal{R}_0$, i.e., $\mathcal{D}_1^0 = \{f \in \mathcal{D}_1 \mid \exists g \in \mathcal{D}_0, f \trianglerighteq_{\mathcal{R}} g\}$ and $\mathcal{D}_1^1 = \mathcal{D}_1 \setminus \mathcal{D}_1^0$. We can then split $\mathcal{R}_1$ in two subsystems $\mathcal{R}_1^0$ and $\mathcal{R}_1^1$ as $\mathcal{R}_1^0 = \{l \to r \in \mathcal{R}_1 \mid root(l) \in \mathcal{D}_1^0\}$ and $\mathcal{R}_1^1 = \{l \to r \in \mathcal{R}_1 \mid root(l) \in \mathcal{D}_1^1\}$.*

**Definition 6 (Generalized Nice Extension).** *[22] Let $\mathcal{R} = \mathcal{R}_0 \cup \mathcal{R}_1$ be the GHC of the extension $(\mathcal{D}_1 \uplus \mathcal{D}_{sh} \uplus \mathcal{C}_1, \mathcal{R}_1)$ over the base $(\mathcal{D}_0 \uplus \mathcal{D}_{sh} \uplus \mathcal{C}_0, \mathcal{R}_0)$. $\mathcal{R}_1$ is a generalized* nice *extension (GNE) of $\mathcal{R}_0$ if, for every rewrite rule $l \to r \in \mathcal{R}_1$, and for every subterm $s$ of $r$ such that $root(s) \in \mathcal{D}_1^0$, $s$ contains no function symbol of $\mathcal{D}_0 \cup \mathcal{D}_1^0$ strictly below its root.*

Figure 4 shows a diagramatic rendition of NEs, and an example can be found in Example 4 later.

We identify a special set $\mathcal{S}_{\mathcal{R}_0 \cup \mathcal{R}_1}$ of terms that represent the right hand sides of rules of the TRSs that can be obtained as GNEs. This allows us to prove that basic narrowing w.r.t. $\mathcal{R} = \mathcal{R}_0 \cup \mathcal{R}_1$ terminates only if it terminates for the terms in $\mathcal{S}_{\mathcal{R}_0 \cup \mathcal{R}_1}$. Let us introduce the standard notion of context here. A context is a term $C$ with zero or more 'holes', i.e., the fresh constant symbol $\square$. If $C$ is a context and $\bar{t}$ a list of terms, $C[\bar{t}]$ denotes the result of replacing the holes in $C$ by the terms in $\bar{t}$.

**Definition 7 ($\mathcal{S}_{\mathcal{R}_0 \cup \mathcal{R}_1}$ terms).** *Let $(\mathcal{D} \uplus \mathcal{C}, \mathcal{R})$ be the union of a base system $(\mathcal{D}_0 \uplus \mathcal{D}_{sh} \uplus \mathcal{C}_0, \mathcal{R}_0)$ and a GNE $(\mathcal{D}_1 \uplus \mathcal{D}_{sh} \uplus \mathcal{C}_1, \mathcal{R}_1)$. Define the sets $\mathcal{D}_1^0$, $\mathcal{D}_1^1$, $\mathcal{R}_1^0$ and $\mathcal{R}_1^1$ as in Definition 5. Let $\mathcal{CC}_{01}$ be the set of contexts of $(\mathcal{C} \cup \mathcal{D}_0 \cup \mathcal{D}_{sh} \cup \mathcal{D}_1^1)$. We define $\mathcal{S}_{\mathcal{R}_0 \cup \mathcal{R}_1}$ as the set of all terms of the form $C[s_1, \ldots, s_n]$, where $C \in \mathcal{CC}_{01}$ and the following conditions hold:*

1. *for all $i \in \{1 \cdots n\}$, $root(s_i) \in \mathcal{D}_1^0$, and*
2. *$s_i$ contains no function symbol of $\mathcal{D}_0 \cup \mathcal{D}_1^0$ strictly below its root.*

By definition, $\mathcal{S}_{\mathcal{R}_0 \cup \mathcal{R}_1}$ terms have the property that no $\mathcal{R}_1^0$ reduction step is possible within the context $C$. Also, the set $\mathcal{S}_{\mathcal{R}_0 \cup \mathcal{R}_1}$ is closed under $\stackrel{b}{\leadsto}_{\mathcal{R}_0 \cup \mathcal{R}_1}$ if $\mathcal{R}_1$ is a GNE of $\mathcal{R}_0$.

The reader can check that the right hand sides of the rules in a GNE fulfill the conditions above. In order to prove the $(\stackrel{b}{\leadsto})$-termination of a system, by Theorem 1 it suffices to prove that derivations starting from the right hand sides of the rules are finite. We prove in Section 5 the more general result that derivations starting from $\mathcal{S}_{\mathcal{R}_0 \cup \mathcal{R}_1}$ terms are finite.

**Corollary 1.** *Let $\mathcal{R}_1$ be a GNE over $\mathcal{R}_0$. Every basic narrowing derivation in $\mathcal{R}_0 \cup \mathcal{R}_1$ starting from a term of $\mathcal{S}_{\mathcal{R}_0 \cup \mathcal{R}_1}$ terminates.*

Now we can easily generalize this result to any term by applying Theorem 1.

**Corollary 2.** *Termination of basic narrowing is modular for generalized nice extensions.*

### 4.3 Proper Extensions

In this section, we extend our previous modularity results from NEs to PEs, by reusing a result from Krishna Rao that relates proper and generalized nice extensions.

**Definition 8 (Generalized Proper Extension).** *[22] Let $\mathcal{R} = \mathcal{R}_0 \cup \mathcal{R}_1$ be the GHC of a base system $(\mathcal{D}_0 \uplus \mathcal{D}_{sh} \uplus \mathcal{C}_0, \mathcal{R}_0)$ and an extension $(\mathcal{D}_1 \uplus \mathcal{D}_{sh} \uplus \mathcal{C}_1, \mathcal{R}_1)$. Define the sets $\mathcal{D}_1^0$, $\mathcal{D}_1^1$, $\mathcal{R}_1^0$ and $\mathcal{R}_1^1$ as in Definition 5. $\mathcal{R}_1$ is a generalized* proper *extension (GPE) of $\mathcal{R}_0$ if each rewrite rule $l \to r \in \mathcal{R}_1^0$ satisfies that, for every subterm $t$ of $r$ such that $root(t) \in \mathcal{D}_1^0$ and $root(t) \trianglerighteq_{\mathcal{R}} root(l)$, $t$ contains no function symbol of $\mathcal{D}_0 \cup \mathcal{D}_1^0$ strictly below its root.*

Figure 4 shows a diagramatic rendition of GPEs.

*Example 4.* Consider computing the factorial of a number in tail recursive style.

$$\mathcal{R}_! = \begin{cases} \mathtt{fact}(x) \to \mathtt{factacc}(x, 1) \\ \mathtt{factacc}(0, x) \to x \\ \mathtt{factacc}(\mathtt{s}(y), x) \to \mathtt{factacc}(y, x * \mathtt{s}(y)) \end{cases} \qquad \mathcal{R}_* = \begin{cases} 0 * y \to 0 \\ \mathtt{s}(x) * y \to (x * y) + y \end{cases}$$

$\mathcal{R}_!$ is a hierarchical extension of $\mathcal{R}_*$, but it is not a PE (because of the 3rd rule). On the other hand, the standard, non tail recursive presentation of factorial is a PE, and moreover a NE.

To understand why non proper extensions can be troublesome for termination, consider the following example.

*Example 5.* Consider the following TRSs, whose combination is hierarchical but not proper:

$$\mathcal{R}_1 : \{\mathtt{f(a)} \to \mathtt{f(b)}\} \qquad \mathcal{R}_0 : \{\mathtt{b} \to \mathtt{a}\}$$

There exists the following infinite basic narrowing derivation

$$\mathtt{f(a)} \overset{b}{\leadsto} \mathtt{f(b)} \overset{b}{\leadsto} \mathtt{f(a)} \overset{b}{\leadsto} \cdots$$

produced by the nesting of a redex w.r.t. $\mathcal{R}_0$ inside the recursive call to `f` in the rhs of the rule of $\mathcal{R}_1$.

PEs are less restrictive than NEs because they allow nesting of $\mathcal{R}_0$ functions only as long as they do not occur inside a recursive definition, whereas NEs forbid any function nesting. That is, every NE is also a PE, but not the other way around. As stated before, we can model any GPE as a finite pyramid of one or more GNEs. Essentially, the idea is similar to the modular decomposition of a

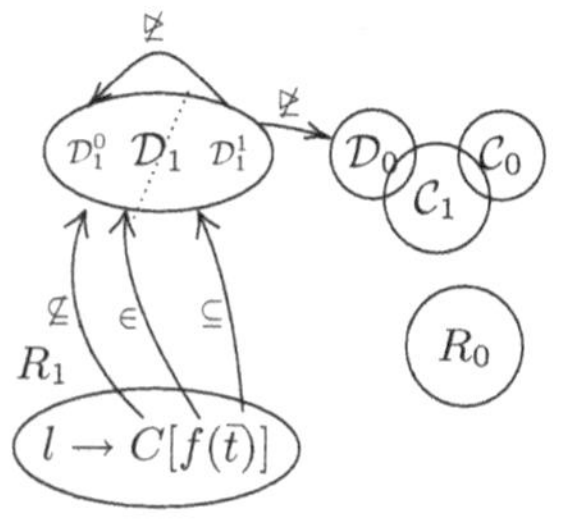

**Fig. 2.** Nice extension

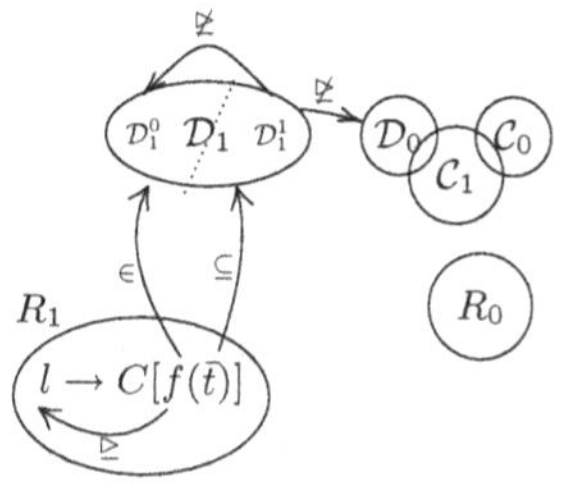

**Fig. 3.** Proper extension

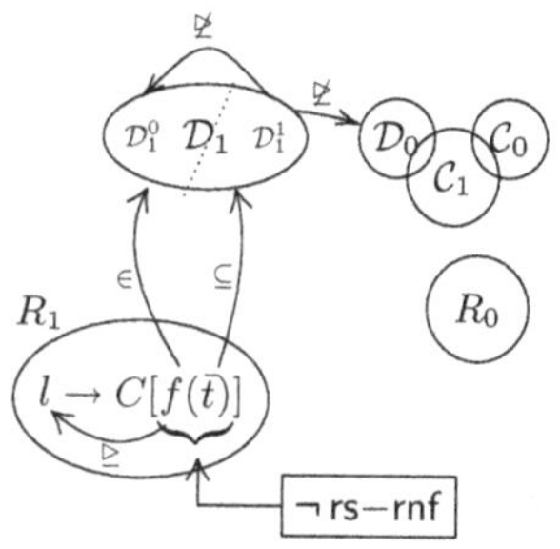


**Fig. 4.** Relaxed Proper extension

TRS in [26]. What we do is to reduce a given PE to the canonical modular form, a modular partition such that each of the individual modules cannot be split up. In order to achieve this we employ the graph induced by the dependency relation $\trianglerighteq$ on defined function symbols, and the rules corresponding to the symbols of every strongly connected component become a module (i.e., a GNE).

**Lemma 2.** *[22] Let $\mathcal{R}_1$ be a finite TRS such that it is a GPE of $\mathcal{R}_0$. Then $\mathcal{R}_1$ can be seen as a finite pyramid of GNEs.*

We are now ready to give our final modularity result for termination of basic narrowing in GPEs, which follows from the previous lemma and Corollary 2.

**Corollary 3.** *Termination of basic narrowing is modular for generalized proper extensions.*

In the following section, by weakening some conditions of GPEs, we provide a novel class of composition of TRSs called *relaxed proper extensions* for which the modularity of basic narrowing termination still holds.

## 5 Relaxed Proper Extensions

Let us introduce the main idea behind our generalization of GPEs by means of the following example.

*Example 6.* Consider the following TRS, an encoding[3] of the exponentiation $x^y$ and the exclusive or operators that are commonly used in the specification of many cryptographic protocols [7,8], where the constructor symbol $\mathsf{g}$ is used as a generator for the exponentiation.

$$\mathcal{R}_1 : \mathtt{exp}(\mathtt{exp}(\mathtt{g}, X), Y) \to \mathtt{exp}(\mathtt{g}, X{*}Y)$$
$$\mathcal{R}_0 : X{*}X^{-1} \to \mathtt{1} \qquad X{*}\mathtt{1} \to X \qquad \mathtt{1}{*}X \to X$$

[3] We are aware that this encoding is not complete since the exclusive or operator is associative and commutative; nevertheless, the example is useful for motivation.

Basic narrowing trivially terminates on each system separately, since every rhs is clearly unnarrowable. However, their combination $\mathcal{R} = \mathcal{R}_0 \cup \mathcal{R}_1$ is not a PE, since the base defined symbol $*$ appears below the extension defined symbol exp in a recursive call. It is easy to see that basic narrowing indeed terminates in $\mathcal{R}$, because the outer function symbol exp in the recursive invocation occurring in the right hand side of the rule of $\mathcal{R}_1$ is blocked forever. The following novel notion of *relaxed* proper extension (RPE) captures this idea.

We introduce the notion of *root-stable rigid normal form*, which lifts to narrowing the standard concept of head normal form. By abuse of notation, we apply this notion, with no change, to basic narrowing.

**Definition 9 (Root-Stable Rigid Normal Form).** [2] *A term $s$ is a* root-stable rigid normal form *(*rs−rnf*) w.r.t. $\mathcal{R}$ if either $s$ is a variable or there are no substitutions $\theta$ and $\theta'$ and terms $s'$ and $s''$ s.t. $s\theta \stackrel{>\epsilon *}{\to}_{\mathcal{R}} s' \stackrel{b}{\leadsto}_{\epsilon,\mathcal{R},\theta'} s''$.*

**Definition 10 (Generalized Relaxed Proper Extension).** *Let $(\mathcal{D} \uplus \mathcal{C}, \mathcal{R})$ be a GHC of a base system $(\mathcal{D}_0 \uplus \mathcal{D}_{sh} \uplus \mathcal{C}_0, \mathcal{R}_0)$ and the extension $(\mathcal{D}_1 \uplus \mathcal{D}_{sh} \uplus \mathcal{C}_1, \mathcal{R}_1)$. Define the sets $\mathcal{D}_1^0$, $\mathcal{D}_1^1$, $\mathcal{R}_1^0$ and $\mathcal{R}_1^1$ as in Definition 5. $\mathcal{R}_1$ is a generalized* relaxed proper *extension (GRPE) of $\mathcal{R}_0$ if every rule in $\mathcal{R}_1^0$ satisfies the following condition:*

**(H1)** *for each subterm $t$ of $r$ such that (a) $root(t) \in \mathcal{D}_1^0$, (b) $t$ is not a* rs−rnf*, and (c) $root(t) \trianglerighteq_{\mathcal{R}} root(l)$, $t$ does not contain a function symbol of $\mathcal{D}_0 \cup \mathcal{D}_1^0$ strictly below its root.*

Figure 4 shows a diagramatic rendition of GRPEs, and the reader can check that the TRS of Example 6 is indeed a GRPE. In the following, we show that $(\stackrel{b}{\leadsto})$-termination is modular for RPEs by showing first its modularity for GRNEs, and then establishing a relation between GRPEs an GRNEs. The reasoning is similar to the one followed in Section 4.3.

**Definition 11 (Generalized Relaxed Nice Extension).** *Let $(\mathcal{D} \uplus \mathcal{C}, \mathcal{R})$ be a GHC of a base system $(\mathcal{D}_0 \uplus \mathcal{D}_{sh} \uplus \mathcal{C}_0, \mathcal{R}_0)$ and an extension $(\mathcal{D}_1 \uplus \mathcal{D}_{sh} \uplus \mathcal{C}_1, \mathcal{R}_1)$. Define the sets $\mathcal{D}_1^0$, $\mathcal{D}_1^1$, $\mathcal{R}_1^0$ and $\mathcal{R}_1^1$ as in Definition 5. $\mathcal{R}_1$ is a generalized relaxed* nice *extension (GRNE) of $\mathcal{R}_0$ if it is a GRPE, and for every rewrite rule $l \to r \in \mathcal{R}_1$ the following condition holds:*

**(N1)** *for each subterm $t$ of $r$ such that $t$ is not a* rs−rnf *and $root(t) \in \mathcal{D}_1^0$, $t$ contains no function symbol of $\mathcal{D}_0 \cup \mathcal{D}_1^0$ strictly below its root.*

We can extend $\mathcal{S}_{\mathcal{R}_0 \cup \mathcal{R}_1}$ to precisely capture the right hand sides of GRNEs.

**Definition 12 ($\mathcal{S}^{\mathsf{rs-rnf}}_{\mathcal{R}_0 \cup \mathcal{R}_1}$ Terms).** *Let $(\mathcal{D} \uplus \mathcal{C}, \mathcal{R})$ be a GRNE of a base system $(\mathcal{D}_0 \uplus \mathcal{D}_{sh} \uplus \mathcal{C}_0, \mathcal{R}_0)$ and an extension $(\mathcal{D}_1 \uplus \mathcal{D}_{sh} \uplus \mathcal{C}_1, \mathcal{R}_1)$. Define the sets $\mathcal{D}_1^0$, $\mathcal{D}_1^1$, $\mathcal{R}_1^0$ and $\mathcal{R}_1^1$ as in Definition 5. We define $\mathcal{S}^{\mathsf{rs-rnf}}_{\mathcal{R}_0 \cup \mathcal{R}_1}$ as the set of all terms of the form $C[s_1, \ldots, s_n]$, where $C$ is a context in $(\mathcal{D} \cup \mathcal{C})$ and the following conditions hold:*

1. *no reduction is possible in $\mathcal{R}_1^0$ at a position within the context $C$,*
2. *for all $i \in \{1 \cdots n\}$, $root(s_i) \in \mathcal{D}_1^0$, $s_i$ is not a $\mathsf{rs{-}rnf}$, and*
3. *$s_i$ contains no function symbol of $\mathcal{D}_0 \cup \mathcal{D}_1^0$ strictly below its root.*

Note that $\mathcal{S}_{\mathcal{R}_0 \cup \mathcal{R}_1} \subseteq \mathcal{S}^{\mathsf{rs-rnf}}_{\mathcal{R}_0 \cup \mathcal{R}_1}$. The set $\mathcal{S}^{\mathsf{rs-rnf}}_{\mathcal{R}_0 \cup \mathcal{R}_1}$ also enjoys the property of being closed under $\overset{b}{\leadsto}_{\mathcal{R}_0 \cup \mathcal{R}_1}$ in a GRNE.

**Lemma 3.** *If $t \in \mathcal{S}^{\mathsf{rs-rnf}}_{\mathcal{R}_0 \cup \mathcal{R}_1}$ and $t \overset{b}{\leadsto}_{\mathcal{R}_0 \cup \mathcal{R}_1, \theta} t'$, then $t' \in \mathcal{S}^{\mathsf{rs-rnf}}_{\mathcal{R}_0 \cup \mathcal{R}_1}$.*

The rest of this section is devoted to extending Corollary 1 to the set $\mathcal{S}^{\mathsf{rs-rnf}}_{\mathcal{R}_0 \cup \mathcal{R}_1}$. First, let us recall some general results on quasi–commutation of abstract relations.

**Definition 13 (Abstract Reduction System).** *An abstract reduction system (ARS) is a structure $\mathcal{A} = (A, \{\to_\alpha \mid \alpha \in I\})$ consisting of a set $A$ and a set of binary relations $\to_\alpha$ on $A$, indexed by a set $I$. We write $(A, \to_1, \to_2)$ instead of $(A, \{\to_\alpha \mid \alpha \in \{1, 2\}\})$.*

**Definition 14 (Quasi-commutation).** *[6] Let $\to_0$ and $\to_1$ be two relations on a set $S$. The relation $\to_1$ quasi-commutes over $\to_0$ if, for all $s, u, t \in S$ s.t. $s \to_0 u \to_1 t$, there exists $v \in S$ s.t. $s \to_1 v \to_{01}^* t$, where $\to_{01}^*$ is the transitive-reflexive closure of $\to_0 \cup \to_1$.*

**Theorem 4.** *[6] If the relations $\to_0$ and $\to_1$ in the ARS$(S, \to_0, \to_1)$ are strongly normalizing and $\to_1$ quasi-commutes over $\to_0$, the relation $\to_0 \cup \to_1$ is strongly normalizing too.*

We now define an ARS with skeleton–environment pairs as elements, where the skeletons come from the set $\mathcal{S}^{\mathsf{rs-rnf}}_{\mathcal{R}_0 \cup \mathcal{R}_1}$ of terms, and the relationships $\to_0$ and $\to_1$ of the ARS are restrictions of basic narrowing.

**Definition 15.** *Let $\mathcal{R} = \mathcal{R}_0 \cup \mathcal{R}_1$ be a generalized relaxed nice combination. We define the ARS $\mathcal{A}(\mathcal{R}_0, \mathcal{R}_1) = (\mathcal{S}^{\mathsf{rs-rnf}}_{\mathcal{R}_0 \cup \mathcal{R}_1} \times Subst, \to_0, \to_1)$, where the relations $\to_0$ and $\to_1$ are defined as follows. Let $s = C[s_0, \ldots, s_n]$ be a term in $\mathcal{S}^{\mathsf{rs-rnf}}_{\mathcal{R}_0 \cup \mathcal{R}_1}$. Then*

1. *$\langle C[s_0, \ldots, s_n], \sigma\rangle \to_0 \langle C'[s_0, \ldots, s_n], \theta\sigma\rangle$ if $\langle C[s_0, \ldots, s_n], \sigma\rangle \overset{b}{\leadsto}_{\mathcal{R}_0 \cup \mathcal{R}_1^1, \theta}$ $\langle C'[s_0, \ldots, s_n], \sigma\theta\rangle$ is a basic narrowing step given within the context $C$*
2. *$\langle C[s_0, \ldots, s_n], \sigma\rangle \to_1 \langle C[s_0, \ldots, s_{i-1}, s_i', s_{i+1}, \ldots, s_n], \theta\sigma\rangle$ if $\langle C[s_0, \ldots, s_n], \sigma\rangle$ $\overset{b}{\leadsto}_{\mathcal{R}_1, \theta} \langle C[s_0, \ldots, s_{i-1}, s_i', s_{i+1}, \ldots, s_n], \theta\sigma\rangle$ is a basic narrowing step given at a subterm $s_i$, with $i \in \{0, \ldots, n\}$.*

The relation $\to_1 \cup \to_0$ is exactly the basic narrowing relation over $\mathcal{S}^{\mathsf{rs-rnf}}_{\mathcal{R}_0 \cup \mathcal{R}_1}$. In the following we establish that both $\to_0$ and $\to_1$ are terminating relations.

**Lemma 4.** *Given the ARS $\mathcal{A}(\mathcal{R}_0, \mathcal{R}_1)$ of Definition 15, the relations $\to_0$ and $\to_1$ are terminating if $\mathcal{R}_0$ and $\mathcal{R}_1$ are $(\overset{b}{\leadsto})$-terminating.*

We are now in a position to prove the quasi-commutation of the relation $\rightarrow_1$ over the relation $\rightarrow_0$ in the ARS $\mathcal{A}(\mathcal{R}_0,\mathcal{R}_1)$. The proof of this result relies on Proposition 1.

**Theorem 5.** *Given the ARS $\mathcal{A}(\mathcal{R}_0,\mathcal{R}_1)$ of Definition 15, the relation $\rightarrow_1$ quasi-commutes over the relation $\rightarrow_0$.*

Then, as a straightforward consequence of Theorem 4 and Theorem 5, we derive the relaxed version of Corollary 1.

**Corollary 4.** *Let $\mathcal{R}_1$ be a GRNE over $\mathcal{R}_0$. Every basic narrowing derivation in $\mathcal{R}_0 \cup \mathcal{R}_1$ starting from a term of $\mathcal{S}^{\mathsf{rs-rnf}}_{\mathcal{R}_0 \cup \mathcal{R}_1}$ terminates.*

By Theorem 1, we obtain the desired modularity result for basic narrowing in our generalization of GNEs.

**Corollary 5.** *Termination of basic narrowing is modular for generalized relaxed nice extensions.*

We now study the connection between GRNEs and GRPEs, and extend the results and proofs from [22] extending them to our generalized relaxed nice extensions.

**Lemma 5.** *Let $\mathcal{R}_1$ be a finite TRS such that it is a GRPE of $\mathcal{R}_0$. $\mathcal{R}_1$ can be seen as a finite pyramid of GRNEs.*

Finally, we are able to establish the most general result of the paper, which follows directly from Corollary 5 and Lemma 5.

**Corollary 6.** *Termination of basic narrowing is modular for generalized relaxed proper extensions.*

## 6 Conclusions

The completeness and termination properties of basic narrowing have been studied previously in landmark work [15,23,19]. Recently we have contributed to the study of narrowing termination based on the termination of basic narrowing in [2]. In this paper, we improve our characterization of basic narrowing termination by proving modular termination in several hierarchical combinations of TRSs, including generalized proper extensions with shared subsystem.

Our main motivation for this work is proving termination of narrowing [15,2]. Narrowing has received much attention due to the different applications, such as automated proofs of termination [5], execution of multiparadigm programming languages [13,17], symbolic reachability [18], verification of cryptographic protocols [10], equational unification [15], equational constraint solving [3,4], and model checking [11], among others. Termination of narrowing is, therefore, of much interest to these applications.

## References

1. Alpuente, M., Escobar, S., Iborra, J.: Modular termination of basic narrowing. Technical Report DSIC-II/04/08, Universidad Politécnica de Valencia (2007)
2. Alpuente, M., Escobar, S., Iborra, J.: Termination of Narrowing revisited. Theoretical Computer Science (to appear, 2008)
3. Alpuente, M., Falaschi, M., Gabbrielli, M., Levi, G.: The semantics of equational logic programming as an instance of CLP. In: Logic Programming Languages, pp. 49–81. The MIT Press, Cambridge (1993)
4. Alpuente, M., Falaschi, M., Levi, G.: Incremental Constraint Satisfaction for Equational Logic Programming. Theoretical Computer Science 142(1), 27–57 (1995)
5. Arts, T., Giesl, J.: Termination of term rewriting using dependency pairs. Theor. Comput. Sci. 236(1-2), 133–178 (2000)
6. Bachmair, L., Dershowitz, N.: Commutation, transformation, and termination. In: Proc. of the 8th Int'l Conf. on Automated Deduction, January 1986, pp. 5–20 (1986)
7. Comon-Lundh, H.: Intruder Theories (Ongoing Work). In: Walukiewicz, I. (ed.) FOSSACS 2004. LNCS, vol. 2987, pp. 1–4. Springer, Heidelberg (2004)
8. Cortier, V., Delaune, S., Lafourcade, P.: A Survey of Algebraic Properties used in Cryptographic Protocols. Journal of Computer Security 14(1), 1–43 (2006)
9. Dershowitz, N., Jouannaud, J.-P.: Rewrite systems. In: Handbook of Theoretical Computer Science, vol. B, pp. 244–320. Elsevier Science, Amsterdam (1990)
10. Escobar, S., Meadows, C., Meseguer, J.: A Rewriting-Based Inference System for the NRL Protocol Analyzer and its Meta-Logical Properties. TCS 367 (2006)
11. Escobar, S., Meseguer, J.: Symbolic model checking of infinite-state systems using narrowing. In: Baader, F. (ed.) RTA 2007. LNCS, vol. 4533, pp. 153–168. Springer, Heidelberg (2007)
12. Fay, M.: First-Order Unification in an Equational Theory. In: Fourth Int'l Conf. on Automated Deduction, pp. 161–167 (1979)
13. Hanus, M.: The Integration of Functions into Logic Programming: From Theory to Practice. Journal of Logic Programming 19&20, 583–628 (1994)
14. Hölldobler, S. (ed.): Foundations of Equational Logic Programming. LNCS, vol. 353. Springer, Heidelberg (1989)
15. Hullot, J.-M.: Canonical Forms and Unification. In: Bibel, W. (ed.) CADE 1980. LNCS, vol. 87, pp. 318–334. Springer, Heidelberg (1980)
16. Hullot, J.-M.: Compilation de Formes Canoniques dans les Théories ́quationelles. Thése de Doctorat de Troisième Cycle. PhD thesis, Université de Paris Sud, Orsay (France) (1981)
17. Meseguer, J.: Multiparadigm logic programming. In: Kirchner, H., Levi, G. (eds.) ALP 1992. LNCS, vol. 632, pp. 158–200. Springer, Heidelberg (1992)
18. Meseguer, J., Thati, P.: Symbolic reachability analysis using narrowing and its application to verification of cryptographic protocols. HOSC, 123–160 (2007)
19. Middeldorp, A., Hamoen, E.: Completeness Results for Basic Narrowing. J. of Applicable Algebra in Engineering, Comm. and Computing 5, 313–353 (1994)
20. Ohlebusch, E.: Advanced Topics in Term Rewriting. Springer, Heidelberg (2002)
21. Prehofer, C.: On Modularity in Term Rewriting and Narrowing. In: Jouannaud, J.-P. (ed.) CCL 1994. LNCS, vol. 845, pp. 253–268. Springer, Heidelberg (1994)
22. Krishna Rao, M.R.K.: Modular proofs for completeness of hierarchical term rewriting systems. Theoretical Computer Science (January 1995)

23. Réty, P.: Improving Basic Narrowing Techniques. In: Lescanne, P. (ed.) RTA 1987. LNCS, vol. 256. Springer, Heidelberg (1987)
24. TeReSe (ed.): Term Rewriting Systems. Cambridge University Press, Cambridge (2003)
25. Toyama, Y.: Counterexamples to termination for the direct sum of term rewriting systems. Inf. Process. Lett. 25(3), 141–143 (1987)
26. Urbain, X.: Modular & incremental automated termination proofs. Int. J. Approx. Reasoning 32(4), 315–355 (2004)

# Linear-Algebraic λ-Calculus: Higher-Order, Encodings, and Confluence

Pablo Arrighi and Gilles Dowek

Université de Grenoble and IMAG Laboratories,
46 Avenue Félix Viallet, 38031 Grenoble Cedex, France
Pablo.Arrighi@imag.fr
École polytechnique and INRIA,
LIX, 91128 Palaiseau Cedex, France
Gilles.Dowek@polytechnique.edu

**Abstract.** We introduce a minimal language combining higher-order computation and linear algebra. This language extends the λ-calculus with the possibility to make arbitrary linear combinations of terms $\alpha.\mathbf{t} + \beta.\mathbf{u}$. We describe how to "execute" this language in terms of a few rewrite rules, and justify them through the two fundamental requirements that the language be a language of linear operators, and that it be higher-order. We mention the perspectives of this work in the field of quantum computation, whose circuits we show can be easily encoded in the calculus. Finally we prove the confluence of the calculus, this is our main result.

## 1 Motivations

The objective of this paper is to merge higher-order computation, be it terminating or not, in its simplest and most general form (namely the untyped λ-calculus) together with linear algebra in its simplest and most general form also (we take just an oriented version of the axioms of vectorial spaces). We see this as a platform for various applications, including quantum computation, each of them probably requiring its own type systems. Next we develop the various contexts in which this calculus may bring some decisive advances.

*Quantum programming languages.* There are two ways a quantum mechanical system may evolve: according to a unitary transformation or under a measurement. The former is often thought of as "purely quantum": it is deterministic and will typically be used to obtain quantum superpositions of base vectors. The latter is probabilistic in the classical sense, and will typically be used to obtain some classical information about a quantum mechanical system, whilst collapsing the system to a mere base vector.

One may say that measurement-based models of quantum computation – whether reliant upon teleportation[21], state transfer [23] or more astonishingly graph states [24] – lie on one extreme, as they keep the "quantumness" to a minimum. A more balanced approach is to allow for both unitary transformations

A. Voronkov (Ed.): RTA 2008, LNCS 5117, pp. 17–31, 2008.

and quantum measurements. Such models can be said to formalize the existing algorithm description methods to a strong extent: they exhibit quantum registers upon which quantum circuits may be applied, together with classical registers and programming structures in order to store measurements results and control the computation [25]. For this reason they are the more practical route to quantum programming. Whilst this juxtaposition of "quantum data, classical control" has appeared ad-hoc and heterogeneous at first, functional-style approaches together with linear type systems [26,2] have ended up producing elegant quantum programming languages.

Finally we may evacuate measures altogether – leaving them till the end of the computation and outside the formalism. This was the case for instance in [28], but here the control structure remained classical. In our view, such a language becomes even more interesting once we have also overcome the need for any additional classical registers and programming structures, and aim to draw the full consequence of quantum mechanics: "quantum data, quantum control". Quantum Turing Machines [6], for instance, lie on this other extreme, since the entire machine can be in a superposition of base vectors. Unfortunately they are a rather oblivious way to describe an algorithm. Functional-style control structure, on the other hand, seem to merge with quantum evolution descriptions in a unifying manner. The functional language we describe may give rise to a "purely quantum" programming language, i.e. one which has no classical registers, no classical control structure, no measurement and that allows arbitrary quantum superpositions of base vectors – once settled the question of restricting to unitary operators.

*General computable linear operators, and the restriction to unitary.* In our view, the problem of formulating a simple algebra of higher order computable operators upon infinite dimensional vector spaces was the first challenge that needed to be met, before even aiming to have a physically executable language. In the current state of affairs computability in vector spaces is dealt with matrices and compositions, and hence restricted to finite-dimensional systems – although this limitation is sometimes circumvented by introducing an extra classical control structure e.g. via the notions of uniform circuits or linear types. The language we provide achieves this goal of a minimal calculus for describing higher-order computable linear operators in a wide sense. Therefore this work may serve as a basis for studying wider notions computability upon abstract vector spaces.

The downside of this generality as far as the previously mentioned application to quantum computation are concerned is that our operators are not restricted to being unitary. A further step towards specializing our language to quantum computation would be to restrict to unitary operators, as required by quantum physics. There may be several ways to do so. A first lead would be to design an a posteriori static analysis that enforces unitarity – exactly like typability is not wired in pure lambda-calculus, but may be enforced a posteriori. A second one would be to require a formal unitarity proof from the programmer. With a term and a unitarity proof, we could derive a more standard representation of the operator, for instance in terms of an universal set of quantum gates [9]. This transformation may be seen as part of a compilation process.

In its current state our language can be seen as a specification language for quantum programs, as it possesses several desirable features of such a language: it allows a high level description of algorithms without any commitment to a particular architecture, it allows the expression of black-box algorithms through the use of higher order functionals, its notation remains close to both linear algebra and functional languages. The game is then be to prove that some quantum program expressed in a standard way (as a composition of universal quantum gates, say) is observationally equivalent to such a specification (a term of our language) under the operational semantics given next.

*Type theory, logics, models.* In this article linearity is understood in the sense of linear algebra, but a further aim to this research would be to investigate connections with linear λ-calculus, i.e. a calculus which types are formulae of linear logic [16]. The paper may also be viewed as part of a wave of probabilistic extensions of calculi, e.g. [18,8]. Type theories for probabilistic extensions of the λ-calculus such as ours or the recent [15] may lead to interesting forms of quantitative logics. The idea of superposing λ-terms is also reminiscent of several other works in λ-calculus, in particular Boudol's parallel λ-calculus [7], Ehrhard and Regnier's differential λ-calculus [14,29] (although our scalars may be arbitrary), Dougherty's algebraic extension [13] for normalizing terms of the λ-calculus.

The functions expressed in our language are linear operators upon the space constituted by its terms. It is strongly inspired from [4] where terms clearly form a vector space. However because it is higher-order, as functions may be passed as arguments to other functions, we get forms of infinity coming into the game. Thus, the underlying algebraic structure is not as obvious as in [4]. In this paper we provide the rules for executing the language in a consistent fashion (confluence), but we leave open the precise nature of the model which lies beneath.

*Confluence techniques.* A standard way to describe how a program is executed is to give a small step operational semantic for it, in the form of a finite set rewrite rules which gradually transform a program into a value. The main theorem proved in this paper is the confluence of our language. What this means is that the order in which those transformations are applied does not affect the end result of the computation. Confluence results are milestones in the study of programming languages and more generally in the theory of rewriting. Our proof uses many of the theoretical tools that have been developed for confluence proofs in a variety of fields (local confluence and Newman's lemma; strong confluence and the Hindley-Rosen lemma) as well as the avatar lemma for parametric rewriting as introduced in [3]. These are fitted together in an elaborate architecture which may have its own interest whenever one seeks to merge a non-terminating conditional confluent rewrite system together with a a terminating conditional confluent rewrite system.

*Outline.* Section 2 presents the designing principles of the language, Section 3 formally describes the linear-algebraic $\lambda$-calculus and its semantics. Section 4 shows that the language is expressive enough for classical and quantum computations. These are the more qualitative sections of the paper. We chose to postpone till Section 5 the proof of the confluence of the calculus, which is more technical. This is our main result. Due to page number constraints, some proofs are omitted in this extended abstract, we refer to the long version of the paper [5], for the full proofs.

## 2 Main Features of the Language

We introduce a minimal language combining higher-order computation and linear algebra, i.e. we extend the $\lambda$-calculus with the possibility to make linear combinations of terms $\alpha.\mathbf{t} + \beta.\mathbf{u}$.

*Higher-order.* In quantum computing, many algorithms fall into the category of "black-box" algorithms. I.e. some mysterious implementation of a function $f$ is provided to us which we call "oracle" – and we wish to evaluate some property of $f$, after a limited number of queries to its oracle. For instance in the Deutsch-Josza quantum algorithm, $f$ is a function $f : \{\mathbf{false}, \mathbf{true}\}^n \longrightarrow \{\mathbf{false}, \mathbf{true}\}$ which is either constant (i.e. $\exists c \forall x[f(x) = c]$) or balanced (i.e. $|\{x \,\text{such that}\, f(x) = \mathbf{false}\}| = |\{x \,\text{such that}\, f(x) = \mathbf{true}\}|$), whose corresponding oracle is a unitary transformation $U_f : \mathcal{H}_2^{n+1} \longrightarrow \mathcal{H}_2^{n+1}$ such that $U_f : \mathbf{x} \otimes \mathbf{b} \mapsto \mathbf{x} \otimes (\mathbf{b} \oplus f(\mathbf{x}))$, where $\mathcal{H}_2^{n+1}$ stands for a tensor product of $n+1$ two-dimensional Hilbert spaces, $\otimes$ is the tensor product and $\oplus$ just the addition modulo two. Our aim is to determine whether $f$ is constant or balanced, and it turns out we can do so in one single query to its oracle. The algorithm works by applying $H^{\otimes^{n+1}}$ upon $(\mathbf{false}^{\otimes^n} \otimes \mathbf{true})$, then $U_f$, and then $H^{\otimes^{n+1}}$ again, where $H^{\otimes^{n+1}}$ means applying the Hadamard gate on each of the $n+1$ qubits. It is clear from this example that a desirable feature for a linear-algebraic functional language is to be able to express algorithms as a function of an oracle. E.g. we may want to define

$$\mathbf{Dj}_1 \equiv \lambda\mathbf{x}\,((H \otimes H)\,(\mathbf{x}\,((H \otimes H)\,(\mathbf{false} \otimes \mathbf{true})))$$

so that $\mathbf{Dj}_1\, U_f$ reduces to $(H \otimes H)\,(U_f\,((H \otimes H)\,(\mathbf{false} \otimes \mathbf{true})))$. More importantly even, one must be able to express algorithms, whether they are "black-box" or not, independent of the size of their input. This is what differentiates programs from fixed-size circuits acting upon finite dimensional vector spaces, and demonstrates the ability to have control flow. The way to achieve this in functional languages involves duplicating basic components of the algorithm an appropriate number of times. E.g. we may want to define some $\mathbf{Dj}$ operator so that $(\mathbf{Dj}\,\mathbf{n})\,U_f$ reduces to the appropriate $(\mathbf{Dj}_n)\,U_f$, where $\mathbf{n}$ is a natural number.

Clearly the languages of matrices and circuits do not offer an elegant presentation for this issue. Higher-order appears to be a desirable feature to have for black-box computing, but also for expressing recursion and for high-level programming.

*Copying.* We seek to design a λ-calculus, i.e. have the possibility to introduce and abstract upon variables, as a mean to express functions of these variables. In doing so, we must allow functions such as $\lambda\mathbf{x}\,(\mathbf{x}\otimes\mathbf{x})$, which duplicate their argument. This is necessary for expressiveness, for instance in order to obtain the fixed point operator or any other form of iteration/recursion.

Now problems come up when functions such as $\lambda\mathbf{x}\,(\mathbf{x}\otimes\mathbf{x})$ are applied to superpositions (i.e. sums of vectors). Linear-algebra brings a strong constraint: we know that cloning is not allowed, i.e. that the operator which maps any vector $\psi$ onto the vector $\psi\otimes\psi$ is not linear. In quantum computation this impossibility is referred to as the "no-cloning theorem" [30]. Most quantum programming language proposals so far consist in some quantum registers undergoing unitary transforms and measures on the one hand, together with classical registers and programming structures ensuring control flow on the other, precisely in order to avoid such problems. But as we seek to reach beyond this duality and obtain a purely quantum programming language, we need to face it in a different manner.

This problem may be seen as a confluence problem. Faced with the term $(\lambda\mathbf{x}\,(\mathbf{x}\otimes\mathbf{x}))\,(\mathbf{false}+\mathbf{true})$, one could either start by substituting $\mathbf{false}+\mathbf{true}$ for $\mathbf{x}$ and get the normal form $(\mathbf{false}+\mathbf{true})\otimes(\mathbf{false}+\mathbf{true})$, or start by using the fact that all the functions defined in our language must be linear and get $((\lambda\mathbf{x}\,(\mathbf{x}\otimes\mathbf{x}))\,\mathbf{false})+((\lambda\mathbf{x}\,(\mathbf{x}\otimes\mathbf{x}))\,\mathbf{true})$ and finally the normal form $(\mathbf{false}\otimes\mathbf{false})+(\mathbf{true}\otimes\mathbf{true})$, leading to two different results. More generally, faced with a term of the form $(\lambda\mathbf{x}\,\mathbf{t})\,(\mathbf{u}+\mathbf{v})$, one could either start by substituting $\mathbf{u}+\mathbf{v}$ for $\mathbf{x}$, or start by applying the right-hand-side linearity of the application, breaking the confluence of the calculus. So that operations remain linear, it is clear that we must start by developing over the $+$ first, until we reach a base vector and then apply $\beta$-reduction. By base vector we mean a term which does not reduce to a superposition. Therefore we restrict the $\beta$-reduction rules to cases where the argument is a base vector, as formalized later.

With this restriction, we say that our language allows *copying* but not *cloning* [4,2]. It is clear that copying has all the expressiveness required in order to express control flow, since it behaves exactly like the standard $\beta$-reduction as long as the argument passed is not in a superposition. This is the appropriate linear extension of the $\beta$-reduction, philosophically it comprehends classical computation as a (non-superposed) sub-case of linear-algebraic/quantum computation.

The same applies to erasing: the term $\lambda\mathbf{x}\lambda\mathbf{y}\ \mathbf{x}$ expresses the linear operator mapping the base vector $\mathbf{b}_i\otimes\mathbf{b}_j$ to $\mathbf{b}_i$. Again this is in contrast with other programming languages where erasing is treated in a particular fashion whether for the purpose of linearity of bound variables or the introduction of quantum measurement.

*Higher-order & copying.* The main conceptual difficulty when seeking to let our calculus be higher-order is to understand how it combines with this idea of "copying", i.e. duplicating only base vectors. Terms of the form $(\lambda\mathbf{x}\,(\mathbf{x}\,\mathbf{x}))\ \ (\lambda\mathbf{x}\ \ \mathbf{v})$

raise the important question of whether the $\lambda$-term $\lambda \mathbf{x}\, \mathbf{v}$ must be considered to be a base vector or not. As we want higher-orderness in the traditional sense, i.e. $(\lambda \mathbf{x}\, \mathbf{t})\, (\lambda \mathbf{y}\, \mathbf{u}) \longrightarrow \mathbf{t}[\lambda \mathbf{y}\, \mathbf{u}/\mathbf{x}]$, abstractions must be the base vectors.

The eventual algebraic consequences of this notion of a privileged basis arising only because of the higher-order level are left as a topic for further investigations. An important intuition is that $(\lambda \mathbf{x}\, \mathbf{v})$ is not the vector itself, but its classical description, i.e. the machine constructing it – hence it is acceptable to be able to copy $(\lambda \mathbf{x}\, \mathbf{v})$ so long as we cannot clone $\mathbf{v}$. The calculus does exactly this distinction.

*Infinities & confluence.* It is possible, in our calculus, to define fixed point operators. For instance for each term $\mathbf{b}$ we can define the term

$$\mathbf{Y_b} = ((\lambda \mathbf{x}\, (\mathbf{b} + (\mathbf{x}\, \mathbf{x})))\, (\lambda \mathbf{x}\, (\mathbf{b} + (\mathbf{x}\, \mathbf{x}))))$$

Then the term $\mathbf{Y_b}$ reduces to $\mathbf{b} + \mathbf{Y_b}$, i.e. the term reductions generate a computable series of vectors $(n.\mathbf{b} + \mathbf{Y_b})_n$ whose "norm" grows towards infinity. This was expected in the presence of both fixed points and linear algebra, but the appearance of such infinities entails the appearance of indefinite forms, which we must handle with great caution. Marrying the full power of untyped $\lambda$-calculus, including fixed point operators etc., with linear-algebra therefore jeopardizes the confluence of the calculus, unless we introduce some further restrictions.

**Example 1.** *If we took an unrestricted factorization rule* $\alpha.\mathbf{t} + \beta.\mathbf{t} \longrightarrow (\alpha + \beta).\mathbf{t}$, *then the term* $\mathbf{Y_b} - \mathbf{Y_b}$ *would reduce to* $(1 + (-1)).\mathbf{Y_b}$ *and then* $\mathbf{0}$. *It would also be reduce to* $\mathbf{b} + \mathbf{Y_b} - \mathbf{Y_b}$ *and then to* $\mathbf{b}$, *breaking the confluence.*

Thus, exactly like in elementary calculus $\infty - \infty$ cannot be simplified to 0, we need to introduce a restriction to the rule allowing to factor $\alpha.\mathbf{t} + \beta.\mathbf{t}$ into $(\alpha + \beta).\mathbf{t}$ to the cases where $\mathbf{t}$ is finite. But what do we mean by finite? Notions of norm in the usual mathematical sense seem difficult to import here. In order to avoid infinities we would like to ask that $\mathbf{t}$ be normalizable, but this is impossible to test in general. Hence, we restrict further this rule to the case where the term $\mathbf{t}$ is normal. It is quite striking to see how this restriction equates the algebraic notion of "being normalized" with the rewriting notion of "being normal". The next two examples show that this indefinite form may pop up in some other, more hidden, ways.

**Example 2.** *Consider the term* $(\lambda \mathbf{x}\, ((\mathbf{x}\, \_) - (\mathbf{x}\, \_)))\, (\lambda \mathbf{y}\, \mathbf{Y_b})$ *where* $\_$ *is any base vector, for instance* **false**. *If the term* $(\mathbf{x}\, \_) - (\mathbf{x}\, \_)$ *reduced to* $\mathbf{0}$ *then this term would both reduce to* $\mathbf{0}$ *and also to* $\mathbf{Y_b} - \mathbf{Y_b}$, *breaking confluence.*

Thus, the term $\mathbf{t}$ we wish to factor must also be closed, so that it does not contain any hidden infinity.

**Example 3.** *If we took an unrestricted rule* $(\mathbf{t} + \mathbf{u})\, \mathbf{v} \longrightarrow (\mathbf{t}\, \mathbf{v}) + (\mathbf{u}\, \mathbf{v})$ *the term* $(\lambda \mathbf{x}\, (\mathbf{x}\, \_) - \lambda \mathbf{x}\, (\mathbf{x}\, \_))\, (\lambda \mathbf{y}\, \mathbf{Y_b})$ *would reduce to* $\mathbf{Y_b} - \mathbf{Y_b}$ *and also to* $\mathbf{0}$, *breaking confluence.*

Thus we have to restrict this rule to the case where $\mathbf{t} + \mathbf{u}$ is normal and closed.

**Example 4.** *If we took an unrestricted rule* $(\alpha.\mathbf{u})\,\mathbf{v} \longrightarrow \alpha.(\mathbf{u}\,\mathbf{v})$ *then the term* $(\alpha.(\mathbf{x}+\mathbf{y}))\,\mathbf{Y_b}$ *would reduce both to* $(\alpha.\mathbf{x}+\alpha.\mathbf{y})\,\mathbf{Y_b}$ *and to* $\alpha.((\mathbf{x}+\mathbf{y})\,\mathbf{Y_b})$*, breaking confluence due to the previous restriction.*

Thus we have to restrict this rule to the case where **u** is normal and closed.

This discussion motivates each of the restrictions $(*)$ – $(****)$ in the rules below. These restrictions are not just a fix: they are a way to formalize vectorial spaces in the presence of limits/infinities. It may come as a surprise, moreover, that we are able to tame these infinities with this small added set of restrictions, and without any need for context-sensitive conditions, as we shall prove in Section 5.

## 3 The Language

We consider a first-order language, called *the language of scalars*, containing at least constants 0 and 1 and binary function symbols + and ×. The *language of vectors* is a two-sorted language, with a sort for vectors and a sort for scalars, described by the following term grammar:

$$\mathbf{t} ::= \quad \mathbf{x} \quad | \quad \lambda\mathbf{x}\,\mathbf{t} \quad | \quad (\mathbf{t}\ \mathbf{t}) \quad | \quad \mathbf{0} \quad | \quad \alpha.\mathbf{t} \quad | \quad \mathbf{t}+\mathbf{t}$$

where $\alpha$ has the sort of scalars.

In this paper we consider only semi-open terms, i.e. terms containing vector variables but no scalar variables. In particular all scalar terms will be closed.

As usual we write $(\mathbf{t}\,\mathbf{u}_1 \ldots \mathbf{u}_n)$ for $(\ldots(\mathbf{t}\,\mathbf{u}_1)\ldots\mathbf{u}_n)$. Vectors appear in bold.

**Definition 1 (The system $S$ – scalar rewrite system).** *A scalar rewrite system $S$ is an arbitrary rewrite system defined on scalar terms and such that*

- *$S$ is terminating and confluent on closed terms,*
- *for all closed terms $\alpha$, $\beta$ and $\gamma$, the pair of terms*
  - $0+\alpha$ *and* $\alpha$, $0\times\alpha$ *and* $0$, $1\times\alpha$ *and* $\alpha$,
  - $\alpha\times(\beta+\gamma)$ *and* $(\alpha\times\beta)+(\alpha\times\gamma)$,
  - $(\alpha+\beta)+\gamma$ *and* $\alpha+(\beta+\gamma)$, $\alpha+\beta$ *and* $\beta+\alpha$,
  - $(\alpha\times\beta)\times\gamma$ *and* $\alpha\times(\beta\times\gamma)$, $\alpha\times\beta$ *and* $\beta\times\alpha$

  *have the same normal forms,*
- $0$ *and* $1$ *are normal terms.*

Examples of scalar rewrite systems for $\mathbb{D}$ and $\mathbb{D}[i,\sqrt{2}]$ are given in [4], where $\mathbb{D}$ is the set of rational numbers whose denominators is a power of two, as this is enough to express quantum computations. The same thing could be done for $\mathbb{Q}$ or any finite extension of $\mathbb{Q}$. Basically the notion of a scalar rewrite systems lists the few basic properties that scalars are usually expected to have: neutral elements, associativity of + etc. The following two definitions are standard for rewriting modulo associativity and commutativity.

**Definition 2 (The relation $\longrightarrow_{XAC}$).** *We define the relation $=_{AC}$ as the congruence generated by the associativity and commutativity axioms of the symbol $+$. Let $X$ be a rewrite system, we define the relation $\longrightarrow_{XAC}$ as follows $t \longrightarrow_{XAC} u$ if there exists a term $t'$ such that $t =_{AC} t'$, an occurrence $p$ in $t'$, a rewrite rule $l \longrightarrow r$ in $X$ and a substitution $\sigma$ such that $t'_{|p} = \sigma l$ and $u =_{AC} t'[p \leftarrow \sigma r]$).*

**Definition 3 (The system $L$ – vector spaces).** *Our small-step semantics is defined by the relation $\longrightarrow L\ AC$ where $L$ is the system formed with the rules of $S$ and the union of four groups of rules $E$, $F$, $A$ and $B$:*

- *Group $E$ – elementary rules*

$\mathbf{u}+\mathbf{0} \longrightarrow \mathbf{u}, \quad 0.\mathbf{u} \longrightarrow \mathbf{0}, \quad 1.\mathbf{u} \longrightarrow \mathbf{u}, \quad \alpha.\mathbf{0} \longrightarrow \mathbf{0},$
$\alpha.(\beta.\mathbf{u}) \longrightarrow (\alpha \times \beta).\mathbf{u}, \quad \alpha.(\mathbf{u}+\mathbf{v}) \longrightarrow \alpha.\mathbf{u}+\alpha.\mathbf{v},$

- *Group $F$ – factorisation*

$\alpha.\mathbf{u}+\beta.\mathbf{u} \longrightarrow (\alpha+\beta).\mathbf{u}, \quad \alpha.\mathbf{u}+\mathbf{u} \longrightarrow (\alpha+1).\mathbf{u}, \quad \mathbf{u}+\mathbf{u} \longrightarrow (1+1).\mathbf{u}, \quad (*)$

- *Group $A$ – application*

$(\mathbf{u}+\mathbf{v})\ \mathbf{w} \longrightarrow (\mathbf{u}\ \mathbf{w})+(\mathbf{v}\ \mathbf{w}), \quad \mathbf{w}\ (\mathbf{u}+\mathbf{v}) \longrightarrow (\mathbf{w}\ \mathbf{u})+(\mathbf{w}\ \mathbf{v}), \quad (**)$
$(\alpha.\mathbf{u})\ \mathbf{v} \longrightarrow \alpha.(\mathbf{u}\ \mathbf{v}), \quad \mathbf{v}\ (\alpha.\mathbf{u}) \longrightarrow \alpha.(\mathbf{v}\ \mathbf{u}), \quad (***)$
$\mathbf{0}\ \mathbf{u} \longrightarrow \mathbf{0}, \quad \mathbf{u}\ \mathbf{0} \longrightarrow \mathbf{0},$

- *Group $B$ – beta reduction*

$(\lambda\mathbf{x}\ \mathbf{t})\ \mathbf{b} \longrightarrow \mathbf{t}[\mathbf{b}/\mathbf{x}] \quad (****)$

*where $+$ is an $AC$ symbol. And:*

$(*)$ *the three rules apply only if* $\mathbf{u}$ *is a closed $L$-normal term.*
$(**)$ *the two rules apply only if* $\mathbf{u}+\mathbf{v}$ *is a closed $L$-normal term.*
$(***)$ *the two rules apply only if* $\mathbf{u}$ *is a closed $L$-normal term.*
$(****)$ *the rule applies only when* $\mathbf{b}$ *is a "base vector" term, i.e. an abstraction or a variable.*

Notice that the restriction $(*)$, $(**)$ and $(***)$ are well-defined as the terms to which the restrictions apply are smaller than the left-hand side of the rule.

Notice also that the restrictions are stable by substitution. Hence these conditional rules could be replaced by an infinite number of non conditional rules, i.e. by replacing the restricted variables by all the closed normal terms verifying the conditions.

Finally notice how the rewrite system $R = S \cup E \cup F \cup A$, taken without restrictions, is really just an oriented version of the axioms of vectorial spaces, as is further explained in [3]. Intuitively the restricted systems defines a notion of vectorial space with infinities.

*Normal forms.* We have explained why abstractions ought to be considered as "base vectors" in our calculus. We have highlighted the presence of non-terminating terms and infinities, which make it impossible to interpret the calculus in your usual vector space structure. The following result shows that terminating closed terms on the other hand can really be viewed as superposition of abstractions.

**Proposition 1 (Form of closed normal forms).** *A L-closed normal form is either the null vector or of the form $\sum_i \alpha_i.\lambda\mathbf{x}\,\mathbf{t}_i + \sum_i \lambda\mathbf{x}\,\mathbf{u}_i$ where the scalars are different from* 0 *and* 1 *and the abstractions are all distinct.*

## 4 Encoding Classical and Quantum Computation

The restrictions we have placed upon our language are still more permissive than those of the call-by-value λ-calculus, hence any classical computation can be expressed in the linear-algebraic λ-calculus just as it can in can in the call-by-value λ-calculus. For expressing quantum computation we need a specific language of scalars, together with its scalar rewrite system. This bit is not difficult, as was shown in [4]. It then suffices to express the three universal quantum gates **H**, **Phase**, **Cnot**, which we will do next.

*Encoding booleans.* We encode the booleans as the first and second projections, as usual in the classical λ-calculus: $\mathbf{true} \equiv \lambda\mathbf{x}\,\lambda\mathbf{y}\,\mathbf{x}$, $\mathbf{false} \equiv \lambda\mathbf{x}\,\lambda\mathbf{y}\,\mathbf{y}$. Again, note that these are conceived as linear functions, the fact we erase the second/first argument does not mean that the term should be interpreted as a trace out or a measurement.

*Encoding unary quantum gates.* For the Phase gate the naive encoding will not work, i.e.

$$\mathbf{Phase} \not\equiv \lambda\mathbf{y}\left(\mathbf{y}\,(e^{i\frac{\pi}{4}}.\mathbf{true})\,\mathbf{false}\right)$$

since by bilinearity this would give $\mathbf{Phase\ false} \longrightarrow^* e^{i\frac{\pi}{4}}.\mathbf{false}$, whereas the Phase gate is supposed to place an $e^{i\frac{\pi}{4}}$ only on **true**. The trick is to use abstraction in order to retain the $e^{i\frac{\pi}{4}}$ phase on **true** only (where _ is any base vector, for instance **false**).

$$\mathbf{Phase} \equiv \lambda\mathbf{y}\left(\left(\mathbf{y}\,\lambda\mathbf{x}\,(e^{i\frac{\pi}{4}}.\mathbf{true})\,\lambda\mathbf{x}\,\mathbf{false}\right)\,\_\right)$$

Now **Phase true** yields $e^{i\frac{\pi}{4}}.\mathbf{true}$ whereas **Phase false** yields **false**. This idea of using a dummy abstraction to restrict linearity can be generalized with the following construct: $[\mathbf{t}] \equiv \lambda\mathbf{x}\,\mathbf{t}$, whose effect is to associate a base vector $[\mathbf{t}]$ to any state, and its converse: $\{\mathbf{t}\} \equiv \mathbf{t}\,\_$ where _ is any base vector, for instance **false**. We then have the derived rule $\{[\mathbf{t}]\} \longrightarrow \mathbf{t}$, thus $\{.\}$ is a "left-inverse" of [.], but not a "right inverse", just like eval and ′ (quote) in LISP. Note that these hooks do not add anymore power to the calculus, in particular they do not enable cloning. We cannot clone a given state $\alpha.\mathbf{t} + \beta.\mathbf{u}$, but we can copy its classical description $[\alpha.\mathbf{t} + \beta.\mathbf{u}]$. For instance the function $\lambda\mathbf{x}\,[\mathbf{x}]$ will never "canonize" anything else than a base vector, because of restriction $(****)$. The phase gate can then be written

$$\mathbf{Phase} \equiv \lambda\mathbf{y}\,\{(\mathbf{y}\,[e^{i\frac{\pi}{4}}.\mathbf{true}])\,[\mathbf{false}]\}$$

For the Hadamard gate the game is just the same:

$$\mathbf{H} \equiv \lambda\mathbf{y}\,\{\mathbf{y}\,[\frac{\sqrt{2}}{2}.(\mathbf{false}+\mathbf{true})]\,[\frac{\sqrt{2}}{2}.(\mathbf{false}-\mathbf{true})]\}$$

*Encoding tensors.* In quantum mechanics, vectors are put together via the bilinear symbol $\otimes$. But because in our calculus application is bilinear, the usual encoding of pairs does just what is needed.

$$\otimes \equiv \lambda\mathbf{x}\,\lambda\mathbf{y}\,\lambda\mathbf{f}\,(\mathbf{f}\,\mathbf{x}\,\mathbf{y}),\quad \pi_1 \equiv \lambda\mathbf{x}\,\lambda\mathbf{y}\,\mathbf{x},\quad \pi_2 \equiv \lambda\mathbf{x}\,\lambda\mathbf{y}\,\mathbf{y},$$

$$\bigotimes \equiv \lambda\mathbf{f}\,\lambda\mathbf{g}\,\lambda\mathbf{x}\left(\otimes\ (\mathbf{f}\,(\pi_1\,\mathbf{x}))\,(\mathbf{g}\,(\pi_2\,\mathbf{x}))\right)$$

E.g. $\mathbf{H}^{\otimes 2} \equiv (\bigotimes\mathbf{H}\,\mathbf{H})$. From there on the infix notation for tensors will be used, i.e. $\mathbf{t}\otimes\mathbf{u} \equiv \otimes\,\mathbf{t}\,\mathbf{u}$, $\mathbf{t}\bigotimes\mathbf{u} \equiv \bigotimes\,\mathbf{t}\,\mathbf{u}$.

The Cnot gate can be defined in a similar way.

*Expressing the Deutsch-Josza algorithm parametrically.* We can now express algorithms parametrically. Here is the well-known simple example of the Deutsch algorithm.

$$\mathbf{Dj}_1 \equiv \lambda\mathbf{x}\left(\mathbf{H}^{\otimes 2}\left(\mathbf{x}\left(\mathbf{H}^{\otimes 2}\,(\mathbf{false}\otimes\mathbf{true})\right)\right)\right)$$

But we can also express control structure and use them to express the dependence of the Deutsch-Josza algorithm with respect to the size of the input. Encoding the natural number $n$ as the Church numeral $\mathbf{n} \equiv \lambda\mathbf{x}\,\lambda\mathbf{f}\,(\mathbf{f}^n\,\mathbf{x})$ the term $(\mathbf{n}\,\mathbf{H}\,\lambda\mathbf{y}\,(\mathbf{H}\bigotimes\mathbf{y}))$ reduces to $\mathbf{H}^{\otimes^{n+1}}$ and similarly the term $(\mathbf{n}\,\mathbf{true}\,\lambda\mathbf{y}\,(\mathbf{false}\otimes\mathbf{y}))$ reduces to $\mathbf{false}^{\otimes^n}\otimes\mathbf{true}$. Thus the expression of the Deutsch-Josza algorithm term of the introduction is now straightforward.

## 5 Confluence

The main theorem of this paper is the confluence of the system $L$. We shall proceed in two steps and prove first the confluence of the system $R = S \cup E \cup F \cup A$, i.e. the system $L$ minus the rule $B$. To prove the confluence of $R$ we prove its termination and local confluence. To be able to use a critical pair lemma, we shall use a well-known technique, detailed in the Section 5.1, and introduce an extension $R_{ext} = S \cup E \cup F_{ext} \cup A$ of the system $R$ as well as a more restricted form of $AC$-rewriting. This proof will proceed step by step as we shall prove first the local confluence of the system $S \cup E$ (Section 5.2) then that of $S \cup E \cup F_{ext}$ (Section 5.2) and finally that of $S \cup E \cup F_{ext} \cup A$ (Section 5.3). The last step towards our main goal is to show that the $B^{\parallel}$ rule is strongly confluent on the term algebra, and commutes with $R^*$, hence giving the confluence of $L$ (Section 5.4).

To prove the local confluence of the system $S \cup E$ we shall prove that of the system $S_0 \cup E$ where $S_0$ is a small avatar of $S$. Then we use a novel proof technique in order to extend from $S_0$ to $S$, hereby obtaining the confluence of $S \cup E$. As the system $R$ does not deal at all with lambda abstractions and bound variables, we have, throughout this first part of the proof, considered $\lambda\mathbf{x}$ as a unary function symbol and the bound occurrences of $\mathbf{x}$ as constants. This way we can safely apply known theorems about first-order rewriting.

### 5.1 Extensions and the Critical Pairs Lemma

The term $((\mathbf{a}+\mathbf{b})+\mathbf{a})+\mathbf{c}$ is $AC$-equivalent to $((\mathbf{a}+\mathbf{a})+\mathbf{b})+\mathbf{c}$ and thus reduces to $((1+1).\mathbf{a}+\mathbf{b})+\mathbf{c}$. However, no subterm of $((\mathbf{a}+\mathbf{b})+\mathbf{a})+\mathbf{c}$ matches $\mathbf{u}+\mathbf{u}$. Thus we cannot restrict the application of a rewrite rule to a subterm of the term to be reduced, and we have to consider all the $AC$-equivalents of this term first. This problem has been solved by [22,19] that consider a simpler form of application (denoted $\longrightarrow_{X,AC}$) and an extra rule $(\mathbf{u}+\mathbf{u})+\mathbf{x} \longrightarrow (1+1).\mathbf{u}+\mathbf{x}$. Notice that now the term $((\mathbf{a}+\mathbf{b})+\mathbf{a})+\mathbf{c}$ has a subterm $(\mathbf{a}+\mathbf{b})+\mathbf{a}$ that is $AC$-equivalent to an instance of the left-hand-side of the new rewrite rule.

**Definition 4 (The relation $\longrightarrow_{X,AC}$).** *Let $X$ be a rewrite system, we define the relation $\longrightarrow_{X,AC}$ as follows $t \longrightarrow_{X,AC} u$ if there exists an occurrence $p$ in $t$, a rewrite rule $l \longrightarrow r$ in $X$ and a substitution $\sigma$ such that $t_{|p} =_{AC} \sigma l$ and $u =_{AC} t[p \leftarrow \sigma r]$.*

**Definition 5 (The extension rules).** $(\alpha.\mathbf{u} + \beta.\mathbf{u}) + \mathbf{x} \longrightarrow (\alpha+\beta).\mathbf{u}+\mathbf{x}$, $(\alpha.\mathbf{u}+\mathbf{u})+\mathbf{x} \longrightarrow (\alpha+1).\mathbf{u}+\mathbf{x}$, $(\mathbf{u}+\mathbf{u})+\mathbf{x} \longrightarrow (1+1).\mathbf{u}+\mathbf{x}$ $(*)$ *We call $F_{ext}$ the system formed by the rules of $F$ and these three rules and $R_{ext}$ the system $S \cup E \cup F_{ext} \cup A$.*

As we shall see the confluence of $\longrightarrow_{R\ AC}$ is a consequence of that of $\longrightarrow_{(R_{ext}),AC}$. As usual we write $\mathbf{t} \longrightarrow^* \mathbf{u}$ if and only if $\mathbf{t} = \mathbf{u}$ or $\mathbf{t} \longrightarrow \ldots \longrightarrow \mathbf{u}$. We also write $\mathbf{t} \longrightarrow^? \mathbf{u}$ if and only if $\mathbf{t} = \mathbf{u}$ or $\mathbf{t} \longrightarrow \mathbf{u}$.

The notions of confluence, strong confluence, local confluence and critical pair are as usual. We use the critical pair lemma and the following lemma that is a consequence of the Theorems 8.9, 9.3 and 10.5 of [22].

**Proposition 2.** *If $\longrightarrow_{R_{ext},AC}$ is locally confluent and $\longrightarrow_{R\ AC}$ terminates then $\longrightarrow_{R\ AC}$ is confluent.*

Thus to prove the confluence of $\longrightarrow_{R\ AC}$ we shall prove its termination and the local confluence of $\longrightarrow_{R_{ext},AC}$.

### 5.2 Local Confluence of $S \cup E$

**Definition 6 (The rewrite system $S_0$).** *The system $S_0$ is formed by the rules* $0+\alpha \longrightarrow \alpha$, $0 \times \alpha \longrightarrow 0$, $1 \times \alpha \longrightarrow \alpha$, $\alpha \times (\beta+\gamma) \longrightarrow (\alpha \times \beta) + (\alpha \times \gamma)$ *where $+$ and $\times$ are $AC$ symbols.*

**Proposition 3.** *The system $S_0 \cup E$ is locally confluent.*

*Proof.* We check that all the critical pair close. This can be automatically done using, for instance, the system CIME [10].

**Definition 7 (Subsumption).** *A terminating and confluent relation $S$ subsumes a relation $S_0$ if whenever $t \longrightarrow_{S_0} u$, $t$ and $u$ have the same $S$-normal form.*

**Definition 8 (Commuting relations).** *Two relations $X$ and $Y$ are said to be commuting if whenever $t \longrightarrow_X u$ and $t \longrightarrow_Y v$, there exists a term $w$ such that $u \longrightarrow_Y w$ and $v \longrightarrow_X w$.*

**Proposition 4 (The avatar lemma).** *[4] Let $E$, $S$ and $S_0$ be three relations defined on a set such that:*

- *$S$ is terminating and confluent;*
- *$S$ subsumes $S_0$;*
- *$S_0 \cup E$ is locally confluent;*
- *$E$ commutes with $S^*$.*

*Then, the relation $S \cup E$ is locally confluent.*

**Proposition 5.** *For any scalar rewrite system $S$ the system $S \cup E$ is locally confluent.*

### 5.3 Local Confluence and Confluence of *R*

**Proposition 6.** *The system $S \cup E \cup F_{ext}$ is locally confluent.*

*Proof.* This system is made of two subsystems : $S \cup E$ and $F_{ext}$. To prove that it is locally confluent, we prove that all critical pairs close. We used an *AC*-unification algorithm to compute these critical pairs. If both rules used are rules of the system $S \cup E$, then the critical pair closes by Proposition 5. We check the 43 other critical pairs by hand. The detail can be found in the long version of the paper [5].

**Proposition 7.** *The system $R = S \cup E \cup F_{ext} \cup A$ is locally confluent.*

*Proof.* Similar to above. See [5].

**Proposition 8.** *The system $R$ terminates.*

**Proposition 9.** *The system $R$ is confluent.*

### 5.4 The System *L*

We now want to prove that the system $L$ is confluent. With the introduction of the rule $B$, we lose termination, hence we cannot use Newman's lemma [20] anymore. Thus we shall use for this last part techniques coming from the proof of confluence of the $\lambda$-calculus and prove that the relation $\longrightarrow_B^{\|}$ is strongly confluent. In our case as we have to mix the rule $B$ with $R$ we shall also prove that it commutes with $\longrightarrow_R^*$.

**Definition 9 (The relation $\longrightarrow^{||}_{B}$).** *The relation $\longrightarrow^{||}_{B}$ is the smallest reflexive congruence such that if* $\mathbf{u}$ *is a base vector,* $\mathbf{t} \longrightarrow^{||}_{B} \mathbf{t}'$ *and* $\mathbf{u} \longrightarrow^{||}_{B} \mathbf{u}'$ *then*

$$(\lambda \mathbf{x}\ \mathbf{t})\ \mathbf{u} \longrightarrow^{||}_{B} \mathbf{t}'[\mathbf{u}'/\mathbf{x}]$$

**Proposition 10 ($\longrightarrow^{*}_{R}$ commutes with $\longrightarrow^{||}_{B}$).**
*If* $\mathbf{t} \longrightarrow^{*}_{R} \mathbf{u}$ *and* $\mathbf{t} \longrightarrow^{||}_{B} \mathbf{v}$ *then there exists* $\mathbf{w}$ *such that* $\mathbf{u} \longrightarrow^{||}_{B} \mathbf{w}$ *and* $\mathbf{v} \longrightarrow^{*}_{R} \mathbf{w}$.

**Proposition 11 (Strong confluence of $B^{||}$).**
*If* $\mathbf{t} \longrightarrow^{||}_{B} \mathbf{u}$ *and* $\mathbf{t} \longrightarrow^{||}_{B} \mathbf{v}$ *then there exists* $\mathbf{w}$ *such that* $\mathbf{u} \longrightarrow^{||}_{B} \mathbf{w}$ *and* $\mathbf{v} \longrightarrow^{||}_{B} \mathbf{w}$.

**Proposition 12 (Hindley-Rosen lemma).** *If the relations $X$ and $Y$ are strongly confluent and commute then the relation $X \cup Y$ is confluent.*

**Theorem 1.** *The system $L$ is confluent.*

*Proof.* By Proposition 9, the relation $\longrightarrow_{R}$ is confluent, hence $\longrightarrow^{*}_{R}$ is strongly confluent. By Proposition 11, the relation $\longrightarrow^{||}_{B}$ is is strongly confluent. By Proposition 10, the relations $\longrightarrow^{*}_{R}$ and $\longrightarrow^{||}_{B}$ commute. Hence, by proposition 12 the relation $\longrightarrow^{*}_{R} \cup \longrightarrow^{||}_{B}$ is confluent. Hence, the relation $\longrightarrow_{L}$ is confluent.

## 6 Conclusion

*Summary.* When merging the untyped λ-calculus with linear algebra one faces two different problems. First of all simple-minded duplication of a vector is a non-linear operation ("cloning") unless it is restricted to base vectors and later extended linearly ("copying"). Second of all because we can express computable but nonetheless infinite series of vectors, hence yielding some infinities and the troublesome indefinite forms. Here again this is fixed by restricting the evaluation of these indefinite forms, this time to normal vectors. Both problems show up when looking at the confluence of the linear-algebraic λ-calculus.

The architecture of the proof of confluence seems well-suited to any non-trivial rewrite systems having both some linear algebra and some infinities as its key ingredients. Moreover the proof of confluence entails the following no-cloning result: there is no term CLONE such that for all term $\mathbf{v}$, (CLONE $\mathbf{v}$) $\longrightarrow^{*}$ $(\mathbf{v} \otimes \mathbf{v})$. Note that $\lambda \mathbf{x}\, \mathbf{v}$ on the other hand can be duplicated, because it is thought as the (plans of) the classical machine for building $\mathbf{v}$ – in other words it stands for potential parallelism rather than actual parallelism. As expected there is no way to transform $\mathbf{v}$ into $\lambda \mathbf{x}\, \mathbf{v}$ in general; confluence ensures that the calculus handles this distinction in a consistent manner.

*Perspectives.* The linear-algebraic λ-calculus merges higher-order computation with linear algebra in a minimalistic manner. Such a foundational approach is also taking place in [1] via some categorical formulations of quantum theory exhibiting nice composition laws and normal forms, but no explicit states, fixed point or the possibility to replicate gate descriptions yet. As for [1] although we have shown that quantum computation can be encoded in our language, the linear-algebraic $\lambda$-calculus remains some way apart from a model of quantum computation, because it allows evolutions which are not unitary. Establishing formal connections with this categorical approach does not seem an easy matter but is part of our objectives.

These connections might arise through typing. Typing is not only our next step on the list in order to enforce the unitary constraint, it is actually the principal aim and motivation for this work: we wish to extend the Curry-Howard isomorphism between proofs/propositions and programs/types to a linear-algebraic, quantum setting. Having merged higher-order computation with linear-algebra in a minimalistic manner, which does not depend on any particular type systems, grants us a complete liberty to now explore different forms of this isomorphism. For instance we may expect different type systems to have different fields of application, ranging from fine-grained entanglement-analysis for quantum computation, to opening connections with linear logic or even giving rise to some novel, quantitative logics.

We leave as an entirely open problem the search for a model of the linear-algebraic $\lambda$-calculus. One can notice already that the non-trivial models of the untyped $\lambda$-calculus are all uncountable, and hence the setting cannot be that of Hilbert spaces. This is also the reason why we have no provided a formal semantics in terms of linear operators in this paper. We suspect that models of the linear-algebraic $\lambda$-calculus will have to do with a C*-algebra endowed with some added higher-order structure – and may have a mathematical interest of its own.

## Acknowledgments

The authors would like to thank Evelyne Contejean, Jean-Pierre Jouannaud, Philippe Jorrand, Claude Marché and Simon Perdrix for some enlightening discussions.

## References

1. Abramsky, S., Coecke, B.: A categorical semantics of quantum protocols LICS. IEEE Computer Society, 415–425 (2004)
2. Altenkirch, T., Grattage, J., Vizzotto, J.K., Sabry, A.: An Algebra of Pure Quantum Programming. In: Third International Workshop on Quantum Programming Languages. Electronic Notes of Theoretical Computer Science, vol. 170C, pp. 23–47 (2007)
3. Arrighi, P., Dowek, G.: A computational definition of the notion of vectorial space. ENTCS 117, 249–261 (2005)
4. Arrighi, P., Dowek, G.: Linear-algebraic lambda-calculus. In: Selinger, P. (ed.) International workshop on quantum programming languages, Turku Centre for Computer Science General Publication, vol. 33, pp. 21–38 (2004)
5. Arrighi, P., Dowek, G.: Linear-algebraic lambda-calculus: higher-order, encodings, confluence, arXiv:quant-ph/0612199.
6. Bernstein, E., Vazirani, U.: Quantum Complexity Theory. In: Annual ACM symposium on Theory of Computing, vol. 25 (1993)

7. Boudol, G.: Lambda-calculi for (strict) parallel functions. Information and Computation 108(1), 51–127 (1994)
8. Bournez, O., Hoyrup, M.: Rewriting Logic and Probabilities. In: Nieuwenhuis, R. (ed.) RTA 2003. LNCS, vol. 2706. Springer, Heidelberg (2003)
9. Boykin, P., Mor, T., Pulver, M., Roychowdhury, V., Vatan, F.: On universal and fault-taulerant quantum computing, arxiv:quant-ph/9906054
10. `http://cime.lri.fr/`
11. Dershowitz, N., Jouannaud, J.-P.: Rewrite systems. In: Handbook of theoretical computer science. formal models and semantics, vol. B. MIT press, Cambridge (1991)
12. Deutsch, D., Josza, R.: Rapid solution of problems by quantum computation. Proc. of the Roy. Soc. of London A 439, 553–558 (1992)
13. Dougherty, D.: Adding Algebraic Rewriting to the Untyped Lambda Calculus. In: Proc. of the Fourth International Conference on Rewriting Techniques and Applications (1992)
14. Ehrhard, T., Regnier, L.: The differential lambda-calculus. Theoretical Computer Science 309, 1–41 (2003)
15. Di Pierro, A., Hankin, C., Wiklicky, H.: Probabilistic λ-calculus and quantitative program analysis. J. of Logic and Computation 15(2), 159–179 (2005)
16. Girard, J.-Y.: Linear logic. Theoretical Computer Science 50, 1–102 (1987)
17. Grover, L.K.: Quantum Mechanics Helps in Searching for a Needle in a Haystack. Phys. Rev. Lett. 79(2), 325–328 (1997)
18. Herescu, O.M., Palamidessi, C.: Probabilistic asynchronous pi-calculus. In: Tiuryn, J. (ed.) ETAPS 2000 and FOSSACS 2000. LNCS, vol. 1784, pp. 146–160. Springer, Heidelberg (2000)
19. Jouannaud, J.-P., Kirchner, H.: Completion of a Set of Rules Modulo a Set of Equations. SIAM J. of Computing 15(14), 1155–1194
20. Newman, M.H.A.: On theories with a combinatorial definition of "equivalence". Annals of Mathematics 432, 223–243 (1942)
21. Nielsen, M.A.: Universal quantum computation using only projective measurement, quantum memory, and preparation of the 0 state. Phys. Rev. A 308, 96–100 (2003)
22. Peterson, G.E., Stickel, M.E.: Complete Sets of Reductions for Some Equational Theories. J. ACM 28(2), 233–264 (1981)
23. Perdrix, S.: State transfer instead of teleportation in measurement-based quantum computation. Int. J. of Quantum Information 1(1), 219–223 (2005)
24. Raussendorf, R., Browne, D.E., Briegel, H.J.: The one-way quantum computer - a non-network model of quantum computation. Journal of Modern Optics 49, 1299 (2002)
25. Selinger, P.: Towards a quantum programming language. Math. Struc. in Computer Science 14(4), 527–586 (2004)
26. Selinger, P., Valiron, B.: A lambda calculus for quantum computation with classical control. Math. Struc. in Computer Science 16(3), 527–552 (2006)
27. Shor, P.W.: Polynomial-Time Algorithms for Prime Factorization and Discrete Logarithms on a Quantum Computer. SIAM J. on Computing 26, 1484–1509 (1997)
28. Van Tonder, A.: A Lambda Calculus for Quantum Computation, arXiv:quant-ph/0307150 (July 2003)
29. Vaux, L.: On linear combinations of lambda-terms. In: Baader, F. (ed.) RTA 2007. LNCS, vol. 4533, pp. 374–388. Springer, Heidelberg (2007)
30. Wooters, W.K., Zurek, W.H.: A single quantum cannot be cloned. Nature 299, 802–803 (1982)

# Term-Graph Rewriting Via Explicit Paths

Emilie Balland and Pierre-Etienne Moreau

UHP & LORIA, and INRIA & LORIA,
BP 101, 54602 Villers-lès-Nancy Cedex France
{Emilie.Balland,Pierre-Etienne.Moreau}@loria.fr

**Abstract.** The notion of path is classical in graph theory but not directly used in the term rewriting community. The main idea of this work is to raise the notion of path to the level of first-order terms, i.e. paths become *part* of the terms and not just meta-information about them. These paths are represented by words of integers (positive or negative) and are interpreted as relative addresses in terms. In this way, paths can also be seen as a generalization of the classical notion of position for the first-order terms and are inspired by de Bruijn indexes.

In this paper, we define an original framework called Referenced Term Rewriting where paths are used to represent pointers between subterms. Using this approach, any term-graph rewriting systems can be simulated using a term rewrite-based environment.

## 1 Introduction

The notion of position is of course central as soon as one deals with data structures. Absolute as well as relative positions are at the heart of algorithms manipulating data structures and their appropriate use and representation can lead to very important differences in the complexity behavior and more generally in the expression of the algorithms themselves. A typical example is the notion of de Bruijn indexes for lambda-terms [10] that not only allows for an easier expression of data-structure manipulations, typically substitution, but also completely changes the way the algorithm is designed, because in this case alpha-conversion is useless.

The main idea of this paper is to make the notion of path first-class, i.e. paths become *part* of the terms and not just meta-information about them. Paths are defined as a generalization of positions and denote a relation from a source position to a target one. A main difference with classical positions that specify a subterm with respect to a global term is that the source position is not necessarily the root.

The first contribution of the paper is to introduce the notion of referenced terms to ground an extension of term rewriting where paths are used to express references and thus to provide a natural way to add pointers in classical terms. For instance, the term $f(a, g(a))$ where we want to make explicit the fact that the subterm $a$ is shared, will be represented as the referenced term $f(a, g(\text{-1.-2.1}))$, where -1 and -2 denote backward move from respectively the first subterm of $g$

A. Voronkov (Ed.): RTA 2008, LNCS 5117, pp. 32–47, 2008.

and the second subterm of $f$. The rational term $g(g(g(\ldots)))$ (a rational term is a possibly infinite term with finitely many subterms, see [9]) is represented by the referenced term $g(\text{-}1)$.

Based on the formalization of paths and a notion of rewrite relation for referenced terms, a strong contribution of this paper is to establish a simulation of term-graph rewriting by referenced term rewriting. Since this simulation is completely based on standard first-order terms, another main interest of the approach is to provide a safe and efficient way to represent and transform term-graphs in a purely term-rewriting based language. Beside the theoretical interest, this is very useful to implement program analysis tools where the representation, the analysis, and the transformation of control-flow and data-flow graphs are crucial. This new representation of terms generalizes standard first-order terms. It requires us to carefully design this new notion of terms and its use to get syntactic correctness. For instance, $g(\text{-}1 \,\bm{.}\, \text{-}1)$ makes no sense as such. Completeness with respect to the standard notions of term and term-graph rewritings have also to be established. This leads to an original and clean way for representing and transforming graphs in a maximally shared rewrite-based environment, making in particular possible the use of term rewriting strategies [15] for term-graph rewriting.

The paper is organized as follows. In Section 2, we formalize the notion of paths and referenced terms where paths are interpreted as pointers. Section 3 shows the relation between referenced terms and cyclic term-graphs as well as the implementation that has been done in the Tom system. Section 4 presents related work and Section 5 concludes. We refer to the long version of the paper [5] for the omitted proofs.

## 2 Paths and Referenced Terms

We assume the reader to be familiar with the basic definitions of first-order terms given, in particular, in [3]. We briefly recall or introduce notations for a few concepts that will be used throughout this paper.

A signature $\mathcal{F}$ is a set of function symbols, each one associated to a natural number by an arity function. $\mathcal{T}(\mathcal{F}, \mathcal{X})$ is the set of *terms* built from a given finite set $\mathcal{F}$ of function symbols and a denumerable set $\mathcal{X}$ of variables. $\mathtt{symb}(t)$ is a partial function from $\mathcal{T}(\mathcal{F}, \mathcal{X})$ to $\mathcal{F}$, which associates to each term $t$ its top-symbol $f \in \mathcal{F}$. The set of variables occurring in a term $t$ is denoted by $\mathcal{V}ar(t)$. If $\mathcal{V}ar(t)$ is empty, $t$ is called a *ground term* and we note $\mathcal{T}(\mathcal{F}) = \mathcal{T}(\mathcal{F}, \emptyset)$ the set of ground terms. Given a set of terms $\mathcal{T}(\mathcal{F}, \mathcal{X})$, a *substitution* $\sigma$ is a function from $\mathcal{X}$ to $\mathcal{T}(\mathcal{F})$, denoted $\sigma = \{x_1 \mapsto t_1, \ldots, x_k \mapsto t_k\}$ when its domain is finite. $\mathcal{R}an(\sigma)$ denotes its codomain. By abuse of notation, we mix up the term $a \in \mathcal{T}(\mathcal{F})$ and the function symbol $a \in \mathcal{F}$ when the arity of $a$ is 0.

In term rewriting, the concept of positions is used to denote a subterm in a global term (i.e. the path from the root to this subterm). In this section, we define the notion of *path*, which generalizes the notion of position by denoting a path from a subterm to another one, and not only the path from the root to a given subterm. Negative numbers designate bottom-up displacements

**Definition 1 (Path).** *We denote $\mathcal{P}$ the set of words on $\mathbb{Z} \setminus \{0\}$. We denote $\epsilon$ the empty word, $p_1 \centerdot p_2$ the concatenation of two words $p_1$ and $p_2$ and $|p|$ the length of a word p. We denote $\mathcal{P}^*$ the set $\mathcal{P} \setminus \{\epsilon\}$.*

*Example 1.* $\epsilon$, $1 \centerdot 3$, and $-2 \centerdot 1 \centerdot 3$ are paths, elements of $\mathcal{P}$.

The notion of path is oriented and corresponds to the route from a subterm to another one. For example, considering the term $f(a, b)$, the path $-1 \centerdot 2$ describes how to reach the subterm $b$ starting from $a$. The negative integer -1 denotes a backward move from $a$ to $f$, whereas 2 goes from $f$ to $b$. The *inverse* of a path $p$ is denoted by $\overline{p}$ and can be calculated using the equations $\overline{\epsilon} = \epsilon$ and $\overline{i \centerdot p} = \overline{p} \centerdot -i$. For example, $\overline{-1 \centerdot 2} = -2 \centerdot 1$.

Note that positions are a subset of paths (paths only composed of positive integers). In the rest of the paper, they will be denoted by Greek letters $\omega, \delta$. The empty word $\epsilon$ is also a position and is generally called the top-position because in term-rewriting, positions are only interpreted from the root of the term. Given two positions, we will denote by $\sqsubseteq$ the classical relation of prefixation between two words ($\omega_1 \sqsubseteq \omega_2$ if there exists a position $p$ such that $\omega_1 \centerdot p = \omega_2$). The subterm of $t$ at position $\omega$ is denoted $t_{|\omega}$. The replacement at position $\omega$ of the subterm $t_{|\omega}$ by $t'$ is written $t[t']_\omega$. The set of positions of $t$ is denoted $\mathcal{P}os(t)$. Given two positions $\omega_{src}$ and $\omega_{dest}$ (for *source* and *destination*), note that the path $\overline{\omega_{src}} \centerdot \omega_{dest}$ corresponds to a path connecting $t_{|\omega_{src}}$ to $t_{|\omega_{dest}}$. We denote $<_\mathcal{P}$ the lexicographic order on positions. For example $\epsilon <_\mathcal{P} 1$ and $1 \centerdot 2 <_\mathcal{P} 1 \centerdot 3$.

If we want to use these paths to define term-graphs, it is necessary to consider equivalence classes. Informally, two paths are equivalent if for every source position, their target positions are equal. For example, paths $1 \centerdot 2 \centerdot -2$ and 1 are equivalent.

In the following we define the notion of *canonical form* as the smallest path of this equivalence class. Moreover, when interpreting a negative integer as a backward move from the $i$-th child to the father, we must ensure that if the previous move in the word is positive, it leads to the same $i$-th child. For example, the path $1 \centerdot -2$ cannot be considered as valid because a move downward to its first child is followed by a move backward from its second child. These observations lead us to introduce the notion of well-formed paths, as well as a constant $\bot$ for representing ill-formed paths.

**Definition 2 (Canonical path and path equivalence).** *The canonical form of a path $p \in \mathcal{P}$, denoted $(\!|p|\!)$, is obtained by maximal application of the rule $i \centerdot -i \rightarrow \epsilon$ if $i \in \mathbb{Z}^*$. Two paths $p_1$ and $p_2$ are said equivalent if $(\!|p_1|\!) = (\!|p_2|\!)$.*

It is easy to show that the rule using to obtain canonical paths is confluent and terminating.

**Definition 3 (Well-formed path).** *We introduce a constant $\bot$ for denoting ill-formed paths. A path $p \in \mathcal{P}$ is well-formed if $(\!|p|\!) \not\rightarrow_R^* \bot$ with $R$ is defined by the rule $p \centerdot i \centerdot -j \centerdot p' \rightarrow \bot$ if $i > 0$, $j > 0$ and $i \neq j$*

*Example 2.* $1 \centerdot 2 \centerdot -2$ and $2 \centerdot 3 \centerdot -3 \centerdot -2 \centerdot 1$ are well-formed paths, but $1 \centerdot -2$ is not.

Note that positions can be seen as a subset of well-formed paths because they correspond to paths only composed of positive integers. Note also that the inverse preserves the well-formedness. On the other hand, the concatenation of two well-formed paths does not always lead to a well-formed path: 1 is well-formed, -2 is well-formed, but $1 \mathbin{.} \text{-}2$ is not.

From these definitions, we show how the notion of paths can be used to extend an algebraic signature in order to represent referenced terms.

## 2.1 Referenced Terms

A referenced term is a term whose leaves may be a path, which denotes a reference to another subterm.

**Definition 4 (Referenced terms).** *Given a set of symbols $\mathcal{F}$ and a set of variables $\mathcal{X}$ such that $\mathcal{F}$, $\mathcal{X}$, $\mathcal{P}$ are disjoint, we denote by $\mathcal{T}_r(\mathcal{F},\mathcal{X})$ the set of* referenced terms *$\mathcal{T}(\mathcal{F} \cup \mathcal{P},\mathcal{X})$, where the elements of $\mathcal{P}$ are symbols of arity* 0.

*Example 3.* For $\mathcal{F} = \{f, g, a\}$, $a$, $g(\text{-}1)$, and $g(f(a, \text{-}2 \mathbin{.} 1))$ are *referenced terms*, elements of $\mathcal{T}_r(\mathcal{F})$. For any $\mathcal{F}$ and $\mathcal{X}$, we have $\mathcal{T}(\mathcal{F},\mathcal{X}) \subset \mathcal{T}_r(\mathcal{F},\mathcal{X})$.

**Definition 5 (Dereferencing).** *Given $t \in \mathcal{T}_r(\mathcal{F},\mathcal{X})$ and $\omega \in \mathcal{P}os(t)$:*

$$\mathtt{deref}(t,\omega) = \begin{cases} (\!|\omega \mathbin{.} \mathtt{symb}(t_{|\omega})|\!) & \text{if } \mathtt{symb}(t_{|\omega}) \in \mathcal{P} \\ \omega & \text{otherwise.} \end{cases}$$

We recall that $\mathtt{symb}(t)$ is a partial function from $\mathcal{T}(\mathcal{F},\mathcal{X})$ to $\mathcal{F}$, which associates to each term $t$ its top-symbol $f \in \mathcal{F}$. The operation $\mathtt{deref}(t,\omega)$ returns the position pointed by $t_{|\omega}$ when $\mathtt{symb}(t_{|\omega}) \in \mathcal{P}$, and $\omega$ otherwise. For example, $\mathtt{deref}(g(\text{-}1), 1) = \epsilon$, but $\mathtt{deref}(g(a), 1) = 1$. Note that when $\omega \mathbin{.} \mathtt{symb}(t_{|\omega})$ is ill-formed, the result of $\mathtt{deref}(t,\omega)$ is meaningless.

We now introduce a notion of *valid referenced terms*. The first condition ensures that the value returned by $\mathtt{deref}(t,\omega)$ is a position of $\mathcal{P}os(t)$, and thus is a well-formed path. For example, $\mathtt{deref}(g(-2), 1)$ is not a valid term. The second condition forbids pointers of pointers like in $f(-1 \mathbin{.} 2, -2 \mathbin{.} 1)$. This last requirement is introduced only for simplicity but is not mandatory in term-graph simulation. Indeed, it could be interesting to consider such terms for modeling imperative languages for example.

**Definition 6 (Valid referenced terms).** *A term $t \in \mathcal{T}_r(\mathcal{F},\mathcal{X})$ is a* valid referenced term *if $\forall \omega \in \mathcal{P}os(t)$ such that $\mathtt{symb}(t_{|\omega}) \in \mathcal{P}$ we have:*

- $\mathtt{deref}(t,\omega) \in \mathcal{P}os(t)$,
- $\mathtt{symb}(t_{|\mathtt{deref}(t,\omega)}) \notin \mathcal{P}^*$.

*We denote by $\mathcal{T}_{vr}(\mathcal{F},\mathcal{X})$ the set of* valid referenced terms *and $\mathcal{T}_{vr}(\mathcal{F})$ the set of ground* valid referenced terms.

Empty paths (denoted $\epsilon$) are allowed in valid referenced terms in order to deal with degenerated cycles that can appear when applying a collapsing rule, e.g. a rule of the form $I(x) \to x$. In the term-graph rewriting formalism, a fresh constant called *black hole* and denoted by $\bullet$ is generally introduced [2]. In our context, it is not necessary since $\epsilon$ corresponds intuitively to this constant.

*Example 4.* The terms $\epsilon$, $g(\text{-}1)$, $f(\text{-}1 \,.\, 2, a)$ and $f(\text{-}1 \,.\, 2, \epsilon)$ are valid, but $g(3)$ and $f(\text{-}1 \,.\, 2, \text{-}2)$ are not. Terms corresponding to non-empty paths (1, -1 . 2, etc.) are elements of $\mathcal{T}_r(\mathcal{F}, \mathcal{X})$ but are not valid (i.e. $\notin \mathcal{T}_{vr}(\mathcal{F}, \mathcal{X})$). The term $t = f(\text{-}1 \,.\, 1 \,.\, \text{-}1, \text{-}2 \,.\, 3)$ is invalid because $\mathtt{deref}(t, 2) = (\!|2 \,.\, \text{-}2 \,.\, 3|\!) = 3$ which is not in $\mathcal{P}os(t) = \{\epsilon, 1, 2\}$.

## 3 Term-Graph Rewriting

There exist different formalisations of term-graph rewriting, category-theory oriented [14], equationally oriented [2,17] or implementation oriented [6]. The difference between terms and term-graphs is the notion of horizontal and vertical sharing. In this section, we base our work on the equational framework introduced in [2]. This well established framework allows the definition of possibly cyclic term-graphs, using systems of recursion equations.

**Definition 7 (System of recursion equations from [2]).** *Given a finite set $\mathcal{F}$ of function symbols and a denumerable set $\mathcal{X}$ of variables, a system of recursion equations is of the form $\{\alpha_1 \mid \alpha_1 = t_1, \ldots, \alpha_n = t_n\}$, $\forall i, j \in [1, n]$ $\alpha_i \in \mathcal{X}$, $\alpha_i \neq \alpha_j$, $t_i \in \mathcal{T}(\mathcal{F}, \mathcal{X})$ is of the form $f(\beta_1, \ldots, \beta_m)$, $f \in \mathcal{F}$, $\forall j \in [1, m]$ $\beta_j \in \mathcal{X}$. Moreover, $\forall i \in [1, n]$ $\alpha_i$ must be reachable from $\alpha_1$.*

Given a system of recursion equations $L$, the root is denoted $\mathtt{root}(L)$ and corresponds to the recursion variable $\alpha_1$. The set of equations is denoted $\mathtt{set}(L)$. A variable $\alpha$ is said *bound* when it appears in the left-hand side of an equation. Otherwise, $\alpha$ is *free*. Note that systems of recursion equations are considered modulo renaming of the recursion variables. In Definition 7, the systems of recursion equations have been presented in *flattened form* ($t_i$ of the form $f(\beta_1, \ldots, \beta_m)$) which ensures the unicity of the representation of a term-graph (modulo renaming of recursion variables). An example of term-graph is given in Figure 1.

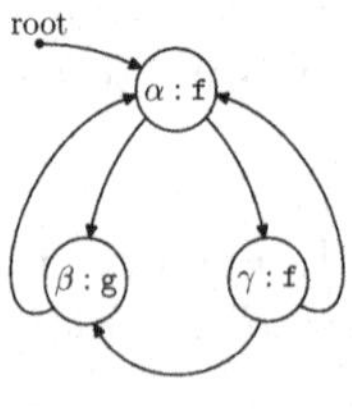


**Fig. 1.** This cyclic term-graph corresponds to the system of recursion equations $\{\alpha \mid \alpha = f(\beta, \gamma), \beta = g(\alpha), \gamma = f(\beta, \alpha)\}$. It contains horizontal sharing: $\alpha$ and $\gamma$ share the same subterm $\beta$; as well as vertical sharing: $\alpha$ is a subterm of both $\beta$ and $\gamma$.

**Definition 8 (Equational term-graph rewriting [2]).** *Given a rewrite rule composed of two systems of recursion equations $L_1$ and $L_2$ with the same root ($L_1$ and $L_2$ are not necessarily in flattened form) and where the free variables of $L_2$ are included in the set of free variables of $L_1$, a system of recursion equations $L$ is rewritten into $L'$ by the rule $L_1 \to L_2$ if there exists a variable substitution $\sigma$ (Definition 4.1 of [2]) and a recursion equation $\alpha = t$ in $L$ such that $\mathtt{set}(\sigma(L_1)) \subseteq \mathtt{set}(L)$ and $\alpha = \mathtt{root}(\sigma(L_1))$. $\mathtt{root}(L') = \mathtt{root}(L)$ and $\mathtt{set}(L') = \mathtt{set}(L) \setminus \{\alpha = t\} \cup \mathtt{set}(\sigma'(L_2))$ where $\sigma'(L_2)$ denotes $\sigma(L_2)$ in which every bound variable (except the root) has been renamed using a fresh name. To obtain from $L'$ a system as defined in Definition 7, equations corresponding to unreachable bound recursion variables are removed and equations are flattened (see [2] for more details). Degenerated cycles, i.e. equations of the form $\alpha = \alpha$ are replaced by $\alpha = \bullet$. In case of equations of type $\alpha = \beta$, each occurrence of $\alpha$ is substituted by $\beta$ and the equation $\alpha = \beta$ is removed.*

*Example 5.* Suppose we want to apply the rule $\{\beta_1 \mid \beta_1 = f(\beta_2, \beta_3)\} \to \{\beta_1 \mid \beta_1 = \beta_2\}$, which corresponds to $f(x, y) \to x$, on the following term-graph:

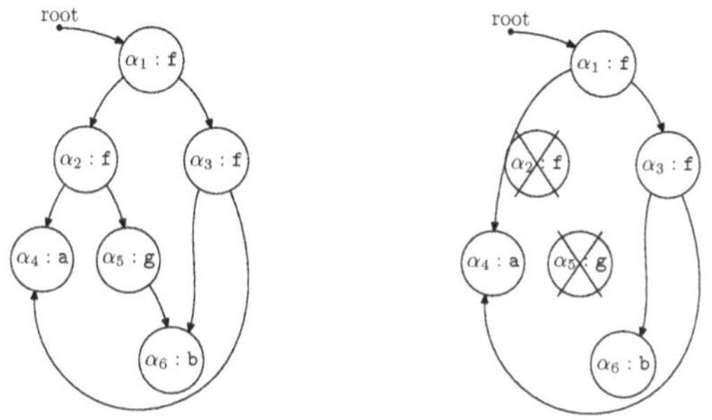


initial subject $s$ final result $s'$

The initial term-graph is $L = \{\alpha_1 \mid \alpha_1 = f(\alpha_2, \alpha_3), \alpha_2 = f(\alpha_4, \alpha_5), \alpha_3 = f(\alpha_6, \alpha_4), \alpha_4 = a, \alpha_5 = g(\alpha_6), \alpha_6 = b\}$. When applying the rule at position 1 (i.e. on $\alpha_2$), we have $\sigma = \{\beta_1 \mapsto \alpha_2, \beta_2 \mapsto \alpha_4, \beta_3 \mapsto \alpha_5\}$ and $\alpha_2$ is the selected bound variable. $\sigma(L_2) = \{\alpha_2 = \alpha_4\}$ (note that renaming with fresh variables is not necessary in this case) and we get $L' = L \setminus \{\alpha_2 = f(\alpha_4, \alpha_5)\} \cup \sigma(L_2)$ as final result. This corresponds to the system $\{\alpha_1 \mid \alpha_1 = f(\alpha_4, \alpha_3), \alpha_3 = f(\alpha_6, \alpha_4), \alpha_4 = a, \alpha_6 = b\}$ after cleanup.

### 3.1 Referenced Term Equivalence

In order to simulate term-graphs with referenced terms, we need to establish equivalence classes between valid referenced terms. For example, $f(\text{-}1.2, a)$ and $f(a, \text{-}2.1)$ should be equivalent. They both correspond to the term-graph rooted by $f$ whose two children correspond to the shared subterm $a$. To define the equivalence, we introduce three intermediate functions that characterize relocation, expansion and sharing.

The first one, called *subterm relocation*, is essential in the following. Given a term $t$ and two positions $\omega_1$, $\omega_2$, as illustrated in Figure 2, the relocation in $t$

from $\omega_1$ to $\omega_2$, denoted $t_{[\omega_1 \twoheadrightarrow \omega_2]}$, returns the subterm $t_{|\omega_1}$ where the references contained in $t_{|\omega_1}$ that point outside $t_{|\omega_1}$ have been updated as if $t_{|\omega_1}$ was moved to the position $\omega_2$.

**Definition 9 (Subterm relocation).** *Given $t \in \mathcal{T}_{vr}(\mathcal{F}, \mathcal{X})$ and two positions $\omega_1$, $\omega_2$, we consider the* subterm relocation $t_{[\omega_1 \twoheadrightarrow \omega_2]}$ *defined such that $\mathcal{P}os(t_{[\omega_1 \twoheadrightarrow \omega_2]}) = \mathcal{P}os(t_{|\omega_1})$ and $\forall \delta \in \mathcal{P}os(t_{|\omega_1})$:*

$$symb(t_{[\omega_1 \twoheadrightarrow \omega_2]|\delta}) = \begin{cases} (\!|\overline{\omega_2 \centerdot \delta} \centerdot \mathtt{deref}(t, \omega_1 \centerdot \delta)|\!) & \text{if } symb(t_{|\omega_1 . \delta}) \in \mathcal{P}^* \\ & \text{and } \omega_1 \not\sqsubseteq \mathtt{deref}(t, \omega_1 \centerdot \delta) \\ symb(t_{|\omega_1 . \delta}) & \text{otherwise.} \end{cases}$$

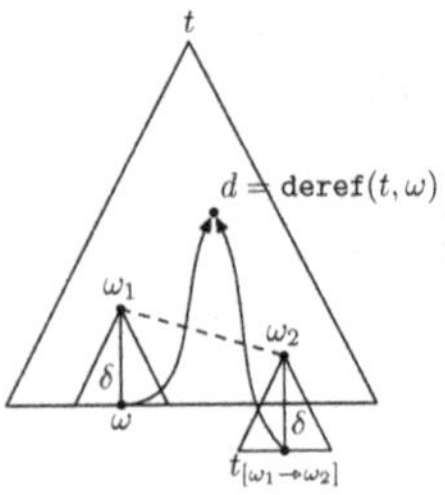


**Fig. 2.** Given $t$ and $\omega_1$, let us suppose that at position $\omega$ the subterm $t_{|\omega}$ contains a reference to the subterm $d$ (i.e. $d = \mathtt{deref}(t, \omega)$). The relocation from $\omega_1$ to $\omega_2$ corresponds to an update of $t_{|\omega_1}$ as if it was moved to $\omega_2$. To maintain the pointers to the referenced terms, the paths stored in $t_{|\omega_1}$ are updated. The result of this operation is the updated subterm $t_{|\omega_1}$. For example, $f(g(\text{-}1 \centerdot \text{-}1), a)_{[1 \twoheadrightarrow 2]} = g(\text{-}1 \centerdot \text{-}2)$.

The operation *expansion* noted **exp** consists in replacing all the sharing by duplication. Given a set of function symbols $\mathcal{F}$, **exp** is a function from $\mathcal{T}_{vr}(\mathcal{F})$ to $\mathcal{T}^\infty(\mathcal{F} \cup \{\epsilon\})$ where $\mathcal{T}^\infty(\mathcal{F} \cup \{\epsilon\})$ is the set of infinite terms over $\mathcal{F} \cup \{\epsilon\}$ defined as partial functions from the infinite set of positions to $\mathcal{F} \cup \{\epsilon\}$. We denote $\bot$ the undefined term represented by the empty function $\emptyset \to \mathcal{F} \cup \{\epsilon\}$. See [8] for more details.

**Definition 10 (Expansion).** *Given $t \in \mathcal{T}_{vr}(\mathcal{F})$, we consider the chain $\{t_i\}_{i \in \mathbb{N}}$ of terms of $\mathcal{T}^\infty(\mathcal{F} \cup \mathcal{P})$ defined as follows:*

- $t_0 = t$
- $t_{n+1} = t_n[t_{n[\mathtt{deref}(t_n, \omega) \twoheadrightarrow \omega]}]_\omega$
  *where $\omega$ is the smallest position of $\mathcal{P}os(t_n)$ such that $symb(t_{n|\omega}) \in \mathcal{P}^*$ for the order defined as $\omega < \omega'$ if $|\omega| < |\omega'| \vee (|\omega| = |\omega'| \wedge \omega <_{\mathcal{P}} \omega')$.*

*$exp(t) \in \mathcal{T}^\infty(\mathcal{F})$ is defined as $\bigcup_{i=0}^{\infty} t'_i$ where $t'_i$ corresponds to $t_i$ where every path $p \in \mathcal{P}^*$ has been replaced by $\bot$.*

**Proposition 1.** *Given $t \in \mathcal{T}_{vr}(\mathcal{F})$, $exp(t)$ is a total function (i.e. total w.r.t. to the arities, not totally defined over of the set of all positions. See [8] for details.*

*Proof.* By definition of the chain $\{t_i\}_{i\in\mathbb{N}}$, $\texttt{exp}(t)$ is a function. Moreover, as every path is replaced by a subterm, $\texttt{exp}(t)$ does not contain $\perp$ and the symbol arities are respected. □

*Example 6.* The function exp replaces in a valid referenced term every reference by the corresponding expanded subterm. $\texttt{exp}(f(\text{-}1 \,.\, 2, a)) = \texttt{exp}(f(a, \text{-}2 \,.\, 1)) = f(a, a)$ and $\texttt{exp}(g(\text{-}1)) = g(g(g(\ldots)))$. Note that in case of a cycle, the expanded term corresponds to an infinite term. A non-trivial example is $f(g(\text{-}1.\text{-}1.2), h(\text{-}1. \text{-}2 \,.\, 1))$. In this case, we need to update paths at every application of the rule. As the shared subterms are dependent, the result is infinite. We finally obtain $f(g(h(g(h(\ldots)))), h(g(h(g(\ldots)))))$.

Thus two equivalent referenced terms have the same expansion. However, this condition is necessary but not sufficient because the two terms to compare must also have a similar sharing. For example, $f(a, a)$ is not equivalent to $f(a, \text{-}2 \,.\, 1)$ because $a$ is not explicitly shared in the first one. For this, we introduce a third relation called *sharing* which computes the set of shared positions.

**Definition 11 (Sharing).** *Given $t \in \mathcal{T}_{vr}(\mathcal{F})$, we consider:*

- $\textit{share}_0(t) = \{\{\omega, \textit{deref}(t, \omega)\} \mid \omega \in \mathcal{P}os(t) \text{ and } \textit{symb}(t_{|\omega}) \in \mathcal{P}^*\}$
- $\textit{share}_{n+1}(t) = \{\{\omega' \,.\, q, q'\} \mid \{\omega, \omega'\}, \{\omega \,.\, q, q'\} \in \textit{share}_n(t)\}$

*The function $\textit{share}(t)$ is defined as* $\bigcup_{n=0}^{\infty} \textit{share}_n(t)$.

*Example 7.* The function share computes the set of unordered pairs of positions. For example, $\texttt{share}(f(\text{-}1 \,.\, 2, a)) = \{\{1, 2\}\}$ and $\texttt{share}(g(\text{-}1)) = \{\{1, \epsilon\}\}$. A non-trivial example is $f(g(\text{-}1 \,.\, \text{-}1 \,.\, 2), h(\text{-}1 \,.\, \text{-}2 \,.\, 1))$. At the first step, $\texttt{share}_0(t) = \{\{1, 2 \,.\, 1\}, \{2, 1 \,.\, 1\}\}$. In this case, it is necessary to close the relation of sharing with prefixes. We finally obtain the infinite set of related positions $\texttt{share}(t) = \{\{1, 2.1\}, \{2, 1.1\}, \ldots, \{1.(1.1)^*, 2.1.(1.1)^*\}, \{2.(1.1)^*, 1.1.(1.1)^*\}\}$ ($^*$ denotes the sublist repetition). This infinite result is due to the inter-dependency of the two references.

**Definition 12 (Equivalence).** *Two valid referenced terms $t_1, t_2$ are equivalent (denoted by $t_1 \sim t_2$) if $\textit{share}(t_1) = \textit{share}(t_2)$ and $\textit{exp}(t_1) = \textit{exp}(t_2)$.*

## 3.2 Canonical Referenced Terms

For every equivalence class, we define a canonical form using $<_{\mathcal{P}}$, the lexicographic order on positions.

**Definition 13 (Canonical referenced terms).** *A valid referenced term $t \in \mathcal{T}_{vr}(\mathcal{F}, \mathcal{X})$ is canonical if for every position $\omega \in \mathcal{P}os(t)$ such that $\textit{symb}(t_{|\omega}) \in \mathcal{P}^*$, $\textit{symb}(t_{|\omega})$ is a canonical path and $\textit{deref}(t, \omega) <_{\mathcal{P}} \omega$.*
*We denote by $\mathcal{T}_g(\mathcal{F}, \mathcal{X})$ the set of canonical referenced terms and $\mathcal{T}_g(\mathcal{F})$ the set of ground canonical referenced terms.*

*Example 8.* The term $f(a, \text{-}2 \centerdot 1)$ is canonical but $f(\text{-}1 \centerdot 2, a)$ is not because $\mathtt{deref}(f(\text{-}1 \centerdot 2, a), 1) = 2$ (the position of the pointed subterm $a$) is not smaller than 1.

To define the normalization function that returns the canonical form of any valid referenced term, we introduce a swapping function that permutes two subterms and updates all the paths contained in the global term in order to preserve the sharing. First, we translate the two subterms. This relocation updates the pointers from the subterms to the external context. To obtain a valid referenced term, we still have to update every pointer from the outside to the subterms.

**Definition 14 (Swapping).** *Given $t \in \mathcal{T}_{vr}(\mathcal{F}, \mathcal{X})$ and two disjoint positions $\omega_1, \omega_2$ ($\omega_1 \not\sqsubseteq \omega_2$ and $\omega_2 \not\sqsubseteq \omega_1$), we consider $u = t_{[\omega_1 \twoheadrightarrow \omega_2]}$, $v = t_{[\omega_2 \twoheadrightarrow \omega_1]}$, $t' = t[v]_{\omega_1}[u]_{\omega_2}$. The swapping in $t$ of the subterms at position $\omega_1$ and $\omega_2$ is denoted by $t_{[\omega_1 \leftrightarrow \omega_2]}$, and is defined such that $\forall \omega \in \mathcal{P}os(t')$, we have:*

$$\mathit{symb}(t_{[\omega_1 \leftrightarrow \omega_2]|\omega}) = \begin{cases} (\!|\overline{\omega} \centerdot \omega_2 \centerdot \delta|\!) \text{ if } \mathit{symb}(t'_{|\omega}) \in \mathcal{P}^*,\ \omega \not\sqsubseteq \omega_1 \\ \qquad \text{and } \exists \delta \text{ s.t. } \mathtt{deref}(t', \omega) = \omega_1 \centerdot \delta \\ (\!|\overline{\omega} \centerdot \omega_1 \centerdot \delta|\!) \text{ if } \mathit{symb}(t'_{|\omega}) \in \mathcal{P}^*,\ \omega \not\sqsubseteq \omega_2 \\ \qquad \text{and } \exists \delta \text{ s.t. } \mathtt{deref}(t', \omega) = \omega_2 \centerdot \delta \\ \mathit{symb}(t'_{|\omega}) \text{ otherwise.} \end{cases}$$

*Example 9.* $f(a,b)_{[1\leftrightarrow 2]} = f(b,a)$, $f(g(\text{-}1 \centerdot \text{-}1 \centerdot 2), h(\text{-}1 \centerdot \text{-}2 \centerdot 1))_{[1\leftrightarrow 2]} = f(h(\text{-}1 \centerdot \text{-}1 \centerdot 2), g(\text{-}1 \centerdot \text{-}2 \centerdot 1))$. A more complex example is the swap of $a$ and $b$ in $t = f(f(a,b), \text{-}2 \centerdot 1 \centerdot 2)$. In this case, the reference $\text{-}2 \centerdot 1 \centerdot 2$ has to be updated because it has to reference $b$. Thus, the result is $f(f(b,a), \text{-}2 \centerdot 1 \centerdot 1)$.

The swapping function preserves the notion of validity (Definition 6). Its complexity is linear on the size of the term since in the worst case, all the references in the term must be updated. In the following, we introduce a *normalization function* that associates to every valid referenced term its canonical form.

**Definition 15 (Normalization).** *We define $[\![\ ]\!] : \mathcal{T}_{vr}(\mathcal{F}, \mathcal{X}) \to \mathcal{T}_g(\mathcal{F}, \mathcal{X})$ such that given $t \in \mathcal{T}_{vr}(\mathcal{F}, \mathcal{X})$, $[\![t]\!]$ is the normal form of $t'$ ($t$ where every path is in canonical form) with respect to the conditional rule: $t' \to t'_{[\omega \leftrightarrow \mathtt{deref}(t', \omega)]}$ if $\omega <_{\mathcal{P}} \mathtt{deref}(t', \omega)$. The proof of this rule convergence can be found in [5].*

*Example 10.* $[\![a]\!] = a$, $[\![f(\text{-}1 \centerdot 2, a)]\!] = f(a, \text{-}2 \centerdot 1)$, and $[\![f(g(\text{-}1 \centerdot \text{-}1 \centerdot 2), h(\text{-}1 \centerdot \text{-}2 \centerdot 1))]\!] = f(g(h(\text{-}1 \centerdot \text{-}1)), \text{-}2 \centerdot 1 \centerdot 1)$

Note that the normalization is linear in the size of the term when the swapping is applied in a leftmost-innermost way.

**Proposition 2.** *$\forall t \in \mathcal{T}_{vr}(\mathcal{F}, \mathcal{X})$, we have $t \sim [\![t]\!]$.*

*Proof.* Given $t \in \mathcal{T}_{vr}(\mathcal{F}, \mathcal{X})$, $t'$ is trivially equivalent to $t$ and every rewriting step of normalization preserves the two functions $\mathtt{share}(t)$ and $\mathtt{exp}(t)$. In fact, the swapping between a pointer and its corresponding pointed subterm preserves the sharing and as only references are updated, the expansion is the same. □

**Proposition 3.** *$\forall t_1, t_2 \in \mathcal{T}_{vr}(\mathcal{F}, \mathcal{X})$, we have: $[\![t_1]\!] \sim [\![t_2]\!] \Leftrightarrow [\![t_1]\!] = [\![t_2]\!]$.*

*Proof.* First, the proof that $[\![t_1]\!] = [\![t_2]\!] \Rightarrow [\![t_1]\!] \sim [\![t_2]\!]$ is trivial because as $\sim$ is an equivalence relation and thus is reflexive. Secondly, we prove that $[\![t_1]\!] \sim [\![t_2]\!] \Rightarrow [\![t_1]\!] = [\![t_2]\!]$. Suppose they are not equal, it means that there exists a shared subterm at a position $\omega$ referenced at a position $\omega'$ in $[\![t_1]\!]$ and the contrary in $[\![t_2]\!]$. As $[\![t_1]\!], [\![t_2]\!]$ are canonical, it implies that $\omega <_{\mathcal{P}} \omega'$ and $\omega' <_{\mathcal{P}} \omega$ which is impossible due to the total order on positions. So $[\![t_1]\!] = [\![t_2]\!]$. □

**Theorem 1.** *$\forall t_1, t_2 \in \mathcal{T}_{vr}(\mathcal{F}, \mathcal{X})$, we have: $t_1 \sim t_2 \Leftrightarrow [\![t_1]\!] = [\![t_2]\!]$*

*Proof.* Direct consequence of the propositions 2 and 3. □

In practice, to verify that two terms are equivalent we compare their canonical forms. It is simpler and more realistic than computing $\texttt{exp}(t)$ and $\texttt{share}(t)$, which can be infinite.

### 3.3 Term-Graph Rewriting Using Canonical Referenced Terms

We introduce an original algorithm for implementing term-graph rewriting using canonical referenced terms. By manipulating only canonical referenced terms, we obtain a mapping one-to-one with term-graphs which makes easier the encoding of matching and rewriting. In Figure 3 we present a set of rules ($\mathcal{T}_g$-Matching), which is a specialization of the syntactic pattern matching algorithm presented in [16].

| | | |
|---|---|---|
| Decompose | $E \wedge f(p_1, \ldots, p_n) \ll_{\delta}^{s,\omega} f(t_1, \ldots, t_n) \parallel \Delta$ | $\twoheadrightarrow E \wedge \bigwedge_{i=1}^{n} p_i \ll_{\delta.i}^{s,\omega.i} t_i \parallel \Delta \cup \{(\delta, \omega)\}$ |
| Variable | $E \wedge x \ll_{\delta}^{s,\omega} t \parallel \Delta$ | $\twoheadrightarrow E \parallel \Delta \cup \{(\delta, \omega)\}$ |
| Stability | $E \wedge \pi \ll_{\delta}^{s,\omega} f(t_1, \ldots, t_n) \parallel \Delta$ | $\twoheadrightarrow E \parallel \Delta \cup \{(\delta, \omega)\}$ if $((\!\vert \delta \,.\, \texttt{symb}(\pi) \vert\!), \omega) \in \Delta$ |
| Dereferencing | $E \wedge p \ll_{\delta}^{s,\omega} \pi \parallel \Delta$ | $\twoheadrightarrow E \wedge p \ll_{\delta}^{s,\omega'} s_{\vert\omega'} \parallel \Delta \cup \{(\delta, \omega)\}$ where $\omega' = (\!\vert \omega \,.\, \texttt{symb}(\pi) \vert\!)$ |

**Fig. 3.** We consider the set of rules $\mathcal{T}_g$-Matching where $E$ is a conjunction of constraints, $\Delta$ is a set of pairs, $x$ is a variable ($\in \mathcal{X}$), $f$ is a symbol, element of $\in \mathcal{F} \cup \{\epsilon\}$, (remember that $\epsilon$ corresponds to $\bullet$), $s$, $t$, $t_1, \ldots, t_n$ are ground referenced term ($\in \mathcal{T}_g(\mathcal{F})$), $\pi$ is a non-empty path ($\texttt{symb}(\pi) \in \mathcal{P}^*$), $p$ is a pattern not reduced to a variable ($\in \mathcal{T}_g(\mathcal{F}, \mathcal{X}) \setminus \mathcal{X}$), $p_i$ are patterns ($\in \mathcal{T}_g(\mathcal{F}, \mathcal{X})$), $\wedge$ is the classical boolean connector, which is associative and commutative. Starting from a constraint $l \ll_{\epsilon}^{s,\omega} s_{\vert\omega} \parallel \emptyset$, the reduction leads either to $\top$ (the neutral element of $\wedge$) or a conjunction of matching constraints of the form $p \ll_{\delta}^{s,\omega} t$, where $\delta$ and $\omega$ correspond to the positions of $p$ and $t$ with respect to $l$ and $s$ (i.e. $p = l_{\vert\delta}$ and $t = s_{\vert\omega}$). The context $\Delta$ corresponds to the set of positions already visited. This set is necessary to correctly handle the case of cyclic terms.

**Definition 16 (Rule application).** *Given $s \in \mathcal{T}_g(\mathcal{F})$ and $l, r \in \mathcal{T}_g(\mathcal{F}, \mathcal{X})$, the rule $l \to r$ can be* applied *to subject $s$ at position $\omega$ if $l \ll_{\epsilon}^{s,\omega} s_{|\omega} \parallel \emptyset$ reduces to $\top \parallel \Delta$ by application of $\mathcal{T}_g$-*Matching.

Note that the algorithm given Figure 3 does not compute a substitution. Moreover, contrary to syntactic term matching algorithms, there is no rule for handling variables that have multiple occurrences. The notion of non-linearity in *term rewriting* should not be confused with the notion of non-linearity in *term-graph rewriting.* The latter one corresponds to subterms sharing. For example, $\{\alpha \mid \alpha = f(\beta, \beta)\}$ denotes a term-graph where the two subterms of $f$ are *shared.* This does not match $\{\alpha \mid \alpha = f(\beta, \gamma), \beta = a, \gamma = a\}$. In our formalism, *linear* term-rewriting is sufficient to simulate *non-linear* term-graph rewriting. For example, the system of recursion equations $\{\alpha \mid \alpha = f(\beta, \beta)\}$ can be encoded by $f(x, -2 \centerdot 1)$, where $x$ appears only once.

**Proposition 4.** *Given a subject $s \in \mathcal{T}_g(\mathcal{F})$, a pattern $l \in \mathcal{T}_g(\mathcal{F}, \mathcal{X})$, a position $\omega \in \mathcal{P}os(s)$, the reduction of $l \ll_{\epsilon}^{s,\omega} s_{|\omega} \parallel \emptyset$ by $\mathcal{T}_g$-*Matching *is convergent.*

*Proof.* First, we prove the termination. We consider the lexicographic combination of prefix ordering on positions ($\sqsubset$) and $<_{\mathcal{P}}$. This strict order is well-founded because $\mathcal{P}os(l) \times \mathcal{P}os(s)$ is finite. Its multiset extension to $\biguplus_{(p \ll_{\delta}^{s,\omega} t) \in E}(\delta, \omega)$ decreases at each application of $\mathcal{T}_g$-Matching rules.

Secondly, proving the local confluence is trivial because there is no interference between the rules. When two rules $r_1$ and $r_2$ can be applied on a subject $t$, we obtain the same result $t'$ when applying $r_1$ followed by $r_2$ or $r_2$ followed by $r_1$. As $\mathcal{T}_g$-Matching terminates, local confluence implies convergence. □

**Definition 17 (Rewriting algorithm).** *Given $t \in \mathcal{T}_g(\mathcal{F})$ and $l, r \in \mathcal{T}_g(\mathcal{F}, \mathcal{X})$:*

- *$l$ and $r$ are both linear.*
- *we denote by $\omega_{xl}$ the position of the variable $x$ in $l$.*
- *$t$ is rewritten into $t'$ by the rule $l \to r$ if:*
  1. *there exists a position $\omega$ such that the rule can be applied to $t$ (following the Definition 16),*
  2. *$t' = [\![\langle \dot{t}, \dot{r} \rangle_{[1.\omega \leftrightarrow 2]}]\!]_{|1}$, where*
     *$\langle \_, \_ \rangle$ is a fresh binary symbol,*
     *$\dot{t}$ corresponds to $t$ where every path towards the position $1.\omega$ has been replaced by a path towards the position 2,*
     *$\dot{r}$ is the ground term corresponding to $r$ in which the occurrence of a variable $x$ (whose position is denoted by $\omega_{xr}$) is replaced by the path $(\!|\overline{\omega_{xr}} \centerdot \text{-}2 \centerdot \mathit{deref}(\langle \dot{t}, r \rangle, 1 \centerdot \omega \centerdot \omega_{xl})|\!)$.*

In this algorithm, no substitution is explicitly computed. Instead, the right-hand side of the rule is instantiated by replacing every variable by paths to their corresponding subterm in $t_{|\omega}$. The binary symbol $\langle \_, \_ \rangle$ enables to connect it with the global subject in a valid referenced term $\langle \dot{t}, \dot{r} \rangle$. By swapping the redex and the right-hand side of the rule in $\langle \dot{t}, \dot{r} \rangle_{[1.\omega \leftrightarrow 2]}$, the substitution is automatically applied. The main subtlety of the algorithm is to rebalance the whole term using normalization. All the shared subterms in $t_{|\omega}$ that must be conserved are moved to the term and the unused part is left in the right part of $\langle \dot{t}, \dot{r} \rangle_{[1 \centerdot \omega \leftrightarrow 2]}$. At the end, the result of the rewrite step corresponds exactly to the left child of $\llbracket \langle \dot{t}, \dot{r} \rangle_{[1 \centerdot \omega \leftrightarrow 2]} \rrbracket$.

The complexity of the rewriting step is linear in the size of the global subject because the complexity of the swapping and the normalization are linear.

*Example 11.* Suppose we want to apply the rule $f(x, y) \rightarrow f(y, x)$ on the subject $t = f(a, b)$. The rule is applied at top-position, therefore we have $\omega_{xl} = 1$, $\omega_{yl} = 2$, $\omega_{xr} = 2$, $\omega_{yr} = 1$, $\dot{t} = f(a, b)$ and $\dot{r} = f(\llparenthesis \overline{\omega_{yr}} \centerdot \text{-}2 \centerdot (1 \centerdot \epsilon \centerdot \omega_{yl}) \rrparenthesis, \llparenthesis \overline{\omega_{xr}} \centerdot \text{-}2 \centerdot (1 \centerdot \epsilon \centerdot \omega_{xl}) \rrparenthesis) = f(\text{-}1 \centerdot \text{-}2 \centerdot 1 \centerdot 2, \text{-}2 \centerdot \text{-}2 \centerdot 1 \centerdot 1)$. Starting from $\langle f(a, b), \dot{r} \rangle$, we get $\langle f(a, b), \dot{r} \rangle_{[1 \leftrightarrow 2]} = \langle f(\text{-}1 \centerdot \text{-}1 \centerdot 2 \centerdot 2, \text{-}2 \centerdot \text{-}1 \centerdot 2 \centerdot 1), f(a, b) \rangle$. When computing the canonical form, we obtain $\llbracket \langle f(a, b), \dot{r} \rangle_{[1 \leftrightarrow 2]} \rrbracket = \langle f(b, a), f(\text{-}1 \centerdot \text{-}2 \centerdot 1 \centerdot 1, \text{-}2 \centerdot \text{-}2 \centerdot 1 \centerdot 2) \rangle$. Finally, the result is $\llbracket \langle f(a, b), \dot{r} \rangle_{[1 \leftrightarrow 2]} \rrbracket_{|1} = f(b, a)$ as expected.

The following example illustrates how collapsing rules are handled. Suppose we want to apply the rule $f(x) \rightarrow x$ to the subject $t = f(\text{-}1)$. Since -1 is a path to the top of the redex, we have $\dot{t} = f(\text{-}1 \centerdot \text{-}1 \centerdot 2)$. In a second step $\dot{r}$ is evaluated to $\dot{r} = \llparenthesis \text{-}2 \centerdot \mathtt{deref}(\langle f(\text{-}1 \centerdot \text{-}1 \centerdot 2), x \rangle, 1 \centerdot 1) \rrparenthesis = \llparenthesis \text{-}2 \centerdot 2 \rrparenthesis = \epsilon$ and we get $\langle f(\text{-}1 \centerdot \text{-}1 \centerdot 2), \epsilon \rangle_{[1 \leftrightarrow 2]} = \langle \epsilon, f(\text{-}1 \centerdot \text{-}2 \centerdot 1) \rangle$ which is already normalized. The result of the rewrite step is $\langle \epsilon, f(\text{-}1 \centerdot \text{-}2 \centerdot 1) \rangle_{|1} = \epsilon$ in accordance with [2].

By applying the rule $f(x, y) \rightarrow x$ to the first subterm, this last example shows what happens to pointers to subterms that disappear:

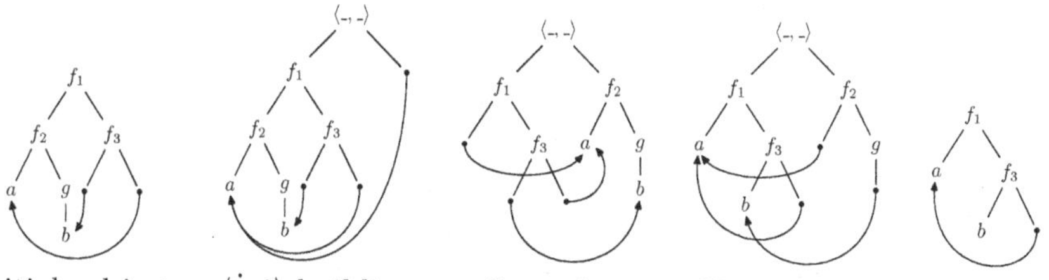

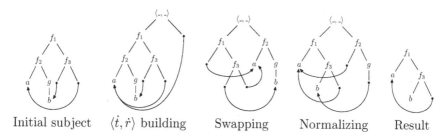

Initial subject $\langle \dot{t}, \dot{r} \rangle$ building Swapping Normalizing Result

In this example, the subterm $g(b)$ is not preserved by the rule, therefore the reference to $b$ is replaced by a copy of the subterm.

**Theorem 2.** *The set of ground canonical terms $\mathcal{T}_g(\mathcal{F})$ is closed under rewriting.*

The proof of the Theorem 2 can be found in [5]. As the main result of the paper we show in the next section that every term-graph can be represented by a canonical referenced term and that term-graph rewriting (Definition 8) can be simulated by the algorithm introduced in Definition 17.

### 3.4 Simulation of Term-Graph Rewriting

We introduce a function $\phi$ that translates any valid referenced term in $\mathcal{T}_{vr}(\mathcal{F}, \mathcal{X})$ into a system of recursion equations (under the same set $\mathcal{F}$ and $\mathcal{X}$). For this purpose, we associate to every term $t \in \mathcal{T}_{vr}(\mathcal{F}, \mathcal{X})$ a total function $\psi_t$ from $\mathcal{P}os(t)$ to $\mathcal{X}$ defined as follows:

$$\psi_t(\omega) = \begin{cases} x & \text{if } \mathtt{symb}(t_{|\omega}) \in \mathcal{F} \cup \{\epsilon\} \\ & \text{where } x \text{ is a fresh variable} \\ t_{|\omega} & \text{if } t_{|\omega} \in \mathcal{X} \\ \psi_t(\omega') & \text{if } t_{|\omega} \in \mathcal{P}^* \\ & \text{where } \omega' = \mathtt{deref}(t, \omega) \end{cases}$$

**Definition 18.** *Given a valid referenced term $t \in \mathcal{T}_{vr}(\mathcal{F}, \mathcal{X})$, we define its representation in systems of recursion equations by $\phi(t) = \{\alpha \mid \Delta\}$ where $\alpha = \psi_t(\epsilon)$ is the root, and $\Delta$ is a set of equation defined by:*

$$\begin{aligned} \Delta = \{\beta = f(\beta_1, \ldots, \beta_n) \mid & \ \omega \in \mathcal{P}os(t), \beta = \psi_t(\omega), \beta_i = \psi_t(\omega \centerdot i), \\ & \ symb(t_{|\omega}) = f \in \mathcal{F}, arity(f) = n\} \\ \cup \{\gamma = \bullet \qquad\quad & \mid \omega \in \mathcal{P}os(t), \gamma = \psi_t(\omega), symb(t_{|\omega}) = \epsilon\} \end{aligned}$$

*$\phi$ can be naturally extended to rules: $\phi(l \to r) = \phi(l) \to \phi(r)$.*

Note that the equational representations of two equivalent valid referenced terms are equal modulo renaming. Moreover, the $\phi$ function is surjective so every system of recursion equations has an unique representation in $\mathcal{T}_g(\mathcal{F}, \mathcal{X})$.

**Theorem 3 (Rewrite step simulation).** *Given a canonical referenced term $t \in \mathcal{T}_g(\mathcal{F})$ and a rule $R$, we can show that $t \to_R t' \Leftrightarrow \phi(t) \to_{\phi(R)} \phi(t')$.*

Theorem 3 shows how to simulate term-graph rewriting using term-rewriting and provides a technique for easily extending rule-based languages with term-graphs and more generally with a notion of pointers. The proof can be found in [5].

Acyclic term graphs are widely used to obtain efficient term rewriting implementation [13]. On the contrary, our goal is not to improve the efficiency of term-rewriting engines but to offer a support for graph structure manipulation. The main objective of such extensions is to perform static analysis by rewriting control flow graphs or data-structures with pointers.

### 3.5 Integration in the Tom Language

As canonical referenced terms are terms, it is possible to extend in a non-intrusive way any rule-based language in order to support term-graph rewriting. The presented formalism is implemented in Tom[1]. For now, it is possible to automatically generate from a signature the extended version for referenced terms where the normalization is integrated. As the Tom terms are implemented with maximal sharing, so are the term-graphs. This part of the implementation is presented in [4]. All the operations on paths have been also implemented. Thus users can define a system of term-graph rules and it is automatically compiled in a basic

[1] Available at `http://tom.loria.fr`

TOM strategy based on the rewriting algorithm (Definition 17). These term-graph rewriting rules can then be integrated in a more complex strategy using TOM strategy combinators. As a consequence, all TOM features are available for term-graphs.

## 4 Related Work

The notion of term graph has been intensively studied in the literature, in particular in [6], where a restricted form of *acyclic* term-graph has been used to represent terms with sharing. There is also a rich literature on modeling functional languages [7,1]. This approach leads to efficient implementations of functional languages or term rewriting engines, as in Clean, Elan, or Maude.

We are in a dual situation: we already have a very efficient term rewriting engine. This implementation is based on the notion of *maximally shared terms*, internally represented by acyclic graphs. Such a representation is purely functional and does not allow any side effect. This constraint makes the implementation of graphs and term-graphs difficult. The contribution of the paper is to provide a solution to represent and transform graphs in a functional environment. In addition to reuse efficient term structures, a main advantage of using a classical underlying term representation is to make possible the reuse of the notion of term rewriting strategy [15] which allows control over how rules are applied.

The extension to cyclic term-graph rewriting has been studied and in particular linked up with rational term rewriting [9]. Especially, a mapping from cyclic term-graph rewriting to rational parallel term rewriting can be defined. In this context, it is often difficult to deal with graph homomorphism. In this work, we choose to simulate term-graph rewriting as defined in [2] and to favor practical aspects. In this formalism, the matching corresponds to functional bisimulation. As a consequence, the pattern $g(\text{-}1)$ cannot be matched with the subject $g(g(\text{-}1 \centerdot \text{-}1))$ at the root position even if they both represent the same infinite term $g(g(g(\ldots)))$.

Moreover, the set of valid referenced terms where references are not interpreted as sharing but as oriented pointers (Definition 6) is to our knowledge a new approach that can be interesting to study and simulate object-oriented languages [11,12]. For example, it could be used to model garbage collection algorithms. In the context of term-graph rewriting, an original approach is the formalism presented in [12] where the right-hand side of the rules consists in a set of actions on the pointers. The work presented in this paper is a first step towards an implementation of such a formalism.

## 5 Conclusion

We have generalized the notion of term positions with *term paths* that are closely related to the notion of path in graphs and to the concept of de Bruijn indices [10]. By extending a signature with paths we obtained a new kind of terms called *referenced terms* which can contain pointers. This representation of pointers can be useful to model the semantics of imperative programming languages for instance. In the second part of the paper, we introduced *canonical referenced terms* to represent term-graphs. As in the case of de Bruijn indices, the interest of this representation is to avoid problems of alpha conversion, compared to representations with labels or variables. Another advantage is that contrary to a recursion equation, the hierarchical structure of the term-graph is explicit because of its term representation. The last contribution of the paper is an original algorithm that simulates cyclic term-graph rewriting using canonical referenced terms. Thanks to pointers, the substitution can be applied in an unusual way, using swapping between the redex and the right-hand side of the rule.

To conclude, this formalism opens promising perspectives in terms of program transformation and code analysis. To the best of our knowledge, the integration in the TOM language constitutes actually one of the most active and maintained implementation of term-graph rewriting and thus provides a solid platform to experiment graph transformations in a concise and expressive way.

**Acknowledgments.** We are extremely grateful to Claude Kirchner who has been strongly involved in this work. He greatly contributed to the foundations of the paper. We also sincerely thank Rachid Echahed and Joe Wells for fruitful exchanges and remarks about preliminary versions of this work. We also kindly thank the anonymous reviewers for their valuable and helpful comments.

## References

1. Ariola, Z.M., Blom, S.: Cyclic lambda calculi. In: Ito, T., Abadi, M. (eds.) TACS 1997. LNCS, vol. 1281, pp. 77–106. Springer, Heidelberg (1997)
2. Ariola, Z.M., Klop, J.W.: Equational term graph rewriting. Fundamenta Informaticae 26(3/4), 207–240 (1996)
3. Baader, F., Nipkow, T.: Term Rewriting and all That. Cambridge University Press, Cambridge (1998)
4. Balland, E., Brauner, P.: Term-graph rewriting in tom using relative positions. In: TermGraph 2007: International Workshop on Computing with Terms and Graphs (2007)
5. Balland, E., Moreau, P.-E.: Term-graph rewriting via explicit paths. Technical report (2008), `http://hal.inria.fr/inria-00173535/fr/`
6. Barendregt, H.P., van Eekelen, M., Glauert, J., Kennaway, J., Plasmeijer, M., Sleep, M.: Term graph rewriting. In: de Bakker, J.W., Nijman, A.J., Treleaven, P.C. (eds.) PARLE 1987. LNCS, vol. 259, pp. 141–158. Springer, Heidelberg (1987)
7. Benaissa, Z.-E.-A., Lescanne, P., Rose, K.H.: Modeling sharing and recursion for weak reduction strategies using explicit substitution. In: Kuchen, H., Swierstra, S.D. (eds.) PLILP 1996. LNCS, vol. 1140, pp. 393–407. Springer, Heidelberg (1996)
8. Corradini, A., Gadducci, F.: Cpo models for infinite term rewriting. In: Alagar, V.S., Nivat, M. (eds.) AMAST 1995. LNCS, vol. 936, pp. 368–384. Springer, Heidelberg (1995)
9. Corradini, A., Gadducci, F.: Rational term rewriting. In: Nivat, M. (ed.) ETAPS 1998 and FOSSACS 1998. LNCS, vol. 1378, pp. 156–171. Springer, Heidelberg (1998)

10. de Bruijn, N.G.: Lambda calculus notation with nameless dummies. a tool for automatic formula manipulation with application to the church-rosser theorem. Indagationes Mathematicae 34, 381–392 (1972)
11. Dougherty, D.J., Lescanne, P., Liquori, L.: Addressed term rewriting systems: Application to a typed object calculus. Mathematical Structures in Computer Science 16, 667–709 (2006)
12. Echahed, R., Peltier, N.: Narrowing data-structures with pointers. In: Corradini, A., Ehrig, H., Montanari, U., Ribeiro, L., Rozenberg, G. (eds.) ICGT 2006. LNCS, vol. 4178, pp. 92–106. Springer, Heidelberg (2006)
13. Hoffmann, B., Plump, D.: Implementing term rewriting by jungle evaluation. RAIRO: Theoretical Informatics and Applications 25 (1991)
14. Kennaway, R.: On graph rewritings. TCS 52(1-2), 37–58 (1987)
15. Kirchner, C.: Strategic rewriting. In: International Workshop on Reduction Strategies in Rewriting and Programming - WRS. ENTCS, vol. 124 (2004)
16. Kirchner, C., Kirchner, H.: Rewriting, solving, proving. A preliminary version of a book (1999), `www.loria.fr/~ckirchne/=rsp/rsp.pdf`
17. Plump, D.: Term graph rewriting. In: Handbook of Graph Grammars and Computing by Graph Transformation, pp. 3–61. World Scientific Publishing, Singapore (1999)

# Finer Is Better: Abstraction Refinement for Rewriting Approximations*

Y. Boichut[1], R. Courbis[2], P.-C. Héam[2], and O. Kouchnarenko[2]

[1] INRIA/PAREO
615 rue du Jardin Botanique
BP-101 F-54602 Villers-Lès Nancy Cedex
boichut@loria.fr
[2] INRIA/CASSIS
LIFC / University of Franche-Comté
16 route de Gray
F-25030 Besançon Cedex
{lastname.firstname}@lifc.univ-fcomte.fr

**Abstract.** Term rewriting systems are now commonly used as a modeling language for programs or systems. On those rewriting based models, reachability analysis, i.e. proving or disproving that a given term is reachable from a set of input terms, provides an efficient verification technique. For disproving reachability (i.e. proving non reachability of a term) on non terminating and non confluent rewriting models, Knuth-Bendix completion and other usual rewriting techniques do not apply. Using the tree automaton completion technique, it has been shown that the non reachability of a term $t$ can be shown by computing an over-approximation of the set of reachable terms and prove that $t$ is not in the over-approximation. However, when the term $t$ is in the approximation, nothing can be said.

In this paper, we improve this approach as follows: given a term $t$, we try to compute an over-approximation which does not contain $t$ by using an approximation refinement that we propose. If the approximation refinement fails then $t$ is a reachable term. This semi-algorithm has been prototyped in the Timbuk tool. We present some experiments with this prototype showing the interest of such an approach w.r.t. verification on rewriting models.

## 1 Introduction

In the rewriting theory, the reachability problem is the following: given a term rewriting system (TRS) $\mathcal{R}$ and two terms $s$ and $t$, can we decide whether $s \to_{\mathcal{R}}^{*} t$ or not? This problem, which can easily be solved on strongly terminating TRSs (by rewriting $s$ into all its possible reduced forms and compare them to $t$), is undecidable on non terminating TRSs. There exists several syntactic classes of TRSs for which this problem becomes decidable: some are surveyed in [12],

* This work has been funded by the French ANR-06-SETI-014 RAVAJ project.

A. Voronkov (Ed.): RTA 2008, LNCS 5117, pp. 48–62, 2008.

more recent ones are [16,20]. In general, the decision procedures for those classes compute a finite tree automaton recognising the possibly infinite set of terms reachable from a set $E \subseteq \mathcal{T}(\mathcal{F})$ of initial terms, by $\mathcal{R}$, denoted by $\mathcal{R}^*(E)$. Then, provided that $s \in E$, those procedures check whether $t \in \mathcal{R}^*(E)$ or not. On the other hand, outside of those decidable classes, one can prove $s \not\to_{\mathcal{R}}^* t$ using over-approximations of $\mathcal{R}^*(E)$ [17,12] and proving that $t$ does not belong to this approximation.

Recently, reachability analysis turned out to be a very efficient verification technique for proving properties on infinite systems modeled by TRSs. Some of the most successful experiments, using proofs of $s \not\to_{\mathcal{R}}^* t$, were done on cryptographic protocols [19,13,4] where protocols and intruders are described using a TRS $\mathcal{R}$, $E$ represents the set of initial configurations of the protocol and $t$ a possible flaw. Some other have been carried out on Java byte code programs [2] and in this context, $\mathcal{R}$ encodes the byte code instructions and the evolution of the Java Virtual Machine (JVM), $E$ specifies the set of initial configurations of the JVM and $t$ a possible flaw.

Then reachability analysis can prove the absence of flaws (if $\forall s \in E : s \not\to_{\mathcal{R}}^* t$). In [12], the method we propose to improve, given a TRS $\mathcal{R}$, a set of terms $E$ and an abstraction function $\gamma$, a sequence of sets of terms $App_0^\gamma, App_1^\gamma, \ldots, App_k^\gamma$ is built such that $App_0^\gamma = E$ and $\mathcal{R}(App_i^\gamma) \subseteq App_{i+1}^\gamma$. The role of the abstraction $\gamma$ is to define equivalence classes of terms and to allot each term to an equivalence class. The computation stops when on the one hand, the number of equivalence classes introduced by the abstraction function is bounded, and on the other hand, each equivalence class is $\mathcal{R}$−closed, i.e. when there exists $N \in \mathbb{N}$ such that $\mathcal{R}(App_N^\gamma) = App_N^\gamma$. Then, $App_N^\gamma$ represents an over-approximation of terms reachable by $\mathcal{R}$ from $E$. The abstraction function $\gamma$ should be well designed in such a way that on one hand $App_N^\gamma$ exists, and on the other hand $t \notin App_N^\gamma$. However, the main drawback of this technique based on tree automata, is that if $t \notin \mathcal{R}^*(E)$ then it is not trivial (when it is possible) to compute a such fix-point over-approximation $App_N^\gamma$. Indeed, a high-level expertise in this technique is required for defining a pertinent abstraction function. At the same time, it is easy to define simple abstraction functions leading to inconclusive analyses. So, the question is : *Is it possible to obtain conclusive analyses starting from simple abstraction functions?* This problem becomes crucial when approximations are used to prove security and safety properties and when a large community of users is targeted.

This paper addresses this question and proposes a solution that automatically attempts to show that a term $t$ is not a term of $\mathcal{R}^*(E)$. We proceed as follows. For a simple abstraction function $\gamma$, we compute a sequence $App_1^\gamma, \ldots, App_k^\gamma$ such that: either $App_k^\gamma$ is a fix-point automaton whose language is an over-approximation of reachable terms and $t \notin App_k^\gamma$, or $App_k^\gamma$ recognises $t$. For the former, everything is fine and we are done. For the latter, we first detect in the sequence $App_1^\gamma, \ldots, App_k^\gamma$ where the abstraction function has been too coarse. Second, we automatically refine $\gamma$, i.e., we fix $\gamma$ in order to remove $t$ from the over-approximation. The construction of the sequence restarts from the problematic set $App_i^\gamma$ with the refined abstraction function and so on. Moreover, if the algorithm fails for finding the reason concerning the abstraction function which makes $t$ be in $App_k^\gamma$ then the term $t$ is a term in $\mathcal{R}^*(E)$.

Note that this solution is a semi-algorithm and it has been prototyped in the Timbuk tool [14]. For a lack of space, the proofs of this paper are available at `http://lifc.univ-fcomte.fr/~kouchna/Kouchnarenko-english.html`.

**Layout of the paper** The paper is organised as follows. After giving preliminary notions on tree-automata and TRSs, we introduce in Section 2 the completion technique we want to improve. Section 3 presents the main contributions concerning the refinement of abstraction functions, the backward completion for the computation of the ancestors of a set of terms by rewriting. A semi-algorithm including those both processes is also given. Finally, before concluding, Section 4 reports on experimental results showing the feasibility and the interest of the proposed approach.

## 2 Preliminaries

### 2.1 Terms and TRSs

Comprehensive surveys can be found in [11,1] for term rewriting systems, and in [9,15] for tree automata and tree language theory.

Let $\mathcal{F}$ be a finite set of symbols, associated with an arity function $ar : \mathcal{F} \to \mathbb{N}$, and let $\mathcal{X}$ be a countable set of variables. $\mathcal{T}(\mathcal{F}, \mathcal{X})$ denotes the set of terms, and $\mathcal{T}(\mathcal{F})$ denotes the set of ground terms (terms without variables). The set of variables of a term $t$ is denoted by $\mathcal{V}ar(t)$. A substitution is a function $\sigma$ from $\mathcal{X}$ into $\mathcal{T}(\mathcal{F}, \mathcal{X})$, which can be extended uniquely to an endomorphism of $\mathcal{T}(\mathcal{F}, \mathcal{X})$. A position $p$ for a term $t$ is a word over $\mathbb{N}$. The empty sequence $\epsilon$ denotes the top-most position. The set $\mathcal{P}os(t)$ of positions of a term $t$ is inductively defined by $\mathcal{P}os(t) = \{\epsilon\}$ if $t \in \mathcal{X}$ and by $\mathcal{P}os(f(t_1, \ldots, t_n)) = \{\epsilon\} \cup \{i.p \mid 1 \leq i \leq n \text{ and } p \in \mathcal{P}os(t_i)\}$ otherwise. If $p \in \mathcal{P}os(t)$, then $t|_p$ denotes the subterm of $t$ at position $p$ and $t[s]_p$ denotes the term obtained by replacement of the subterm $t|_p$ at position $p$ by the term $s$. We also denote by $t(p)$ the symbol occurring in $t$ at position $p$. Given a term $t \in \mathcal{T}(\mathcal{F}, \mathcal{X})$, we denote $\mathcal{P}os_A(t) \subseteq \mathcal{P}os(t)$ the set of positions of $t$ such that $\mathcal{P}os_A(t) = \{p \in \mathcal{P}os(t) \mid t(p) \in A\}$. Thus $\mathcal{P}os_{\mathcal{F}}(t)$ is the set of functional positions of $t$. A TRS $\mathcal{R}$ is a set of *rewrite rules* $l \to r$, where $l, r \in \mathcal{T}(\mathcal{F}, \mathcal{X})$ and $l \notin \mathcal{X}$. A rewrite rule $l \to r$ is *left-linear* (resp. right-linear) if each variable of $l$ (resp. $r$) occurs only once within $l$ (resp. $r$). A TRS $\mathcal{R}$ is left-linear (resp. right-linear) if every rewrite rule $l \to r$ of $\mathcal{R}$ is left-linear (resp. right-linear). A TRS $\mathcal{R}$ is linear if it is right and left-linear. The TRS $\mathcal{R}$ induces a rewriting relation $\to_{\mathcal{R}}$ on terms whose reflexive transitive closure is written $\to_{\mathcal{R}}^\star$. The set of $\mathcal{R}$-descendants of a set of ground terms $E$ is $\mathcal{R}^*(E) = \{t \in \mathcal{T}(\mathcal{F}) \mid \exists s \in E \text{ s.t. } s \to_{\mathcal{R}}^\star t\}$. Symmetrically, the set of $\mathcal{R}$-ancestors of a set of ground terms $E$ is $\mathcal{R}^{-1\star}(E) = \{s \in \mathcal{T}(\mathcal{F}) \mid \exists t \in E \text{ s.t. } s \to_{\mathcal{R}}^\star t\}$.

## 2.2 Tree Automata Completion

Note that $\mathcal{R}^*(E)$ is possibly infinite: $\mathcal{R}$ may not terminate and/or $E$ may be infinite. The set $\mathcal{R}^*(E)$ is generally not computable [15]. However, it is possible to over-approximate it [12] using tree automata, i.e. a finite representation of infinite (regular) sets of terms. We next define tree automata.

Let $\mathcal{Q}$ be a finite set of symbols, of arity 0, called *states* such that $\mathcal{Q} \cap \mathcal{F} = \emptyset$. $\mathcal{T}(\mathcal{F} \cup \mathcal{Q})$ is called the set of *configurations.*

**Definition 1 (Transition and normalised transition).** *A* transition *is a rewrite rule $c \to q$, where $c \in \mathcal{T}(\mathcal{F} \cup \mathcal{Q})$ is a configuration and $q \in \mathcal{Q}$. A* normalised transition *is a transition $c \to q$ where $c = f(q_1, \ldots, q_n)$, $f \in \mathcal{F}$, $ar(f) = n$, and $q_1, \ldots, q_n \in \mathcal{Q}$.*

**Definition 2 (Bottom-up non-deterministic finite tree automaton).** *A bottom-up non-deterministic finite tree automaton (tree automaton for short) is a quadruple $\mathcal{A} = \langle \mathcal{F}, \mathcal{Q}, \mathcal{Q}_f, \Delta \rangle$, $\mathcal{Q}_f \subseteq \mathcal{Q}$ and $\Delta$ is a finite set of normalised transitions.*

The *rewriting relation* on $\mathcal{T}(\mathcal{F} \cup \mathcal{Q})$ induced by the transition set $\Delta$ of $\mathcal{A}$ is denoted $\to_\Delta$. When $\Delta$ is clear from the context, $\to_\Delta$ is also written $\to_\mathcal{A}$.

**Definition 3 (Recognised language).** *The tree language recognised by $\mathcal{A}$ in a state $q$ is $\mathcal{L}(\mathcal{A}, q) = \{t \in \mathcal{T}(\mathcal{F}) \mid t \to_\mathcal{A}^\star q\}$. The language recognised by $\mathcal{A}$ is $\mathcal{L}(\mathcal{A}) = \bigcup_{q \in \mathcal{Q}_f} \mathcal{L}(\mathcal{A}, q)$. A tree language is regular if and only if it is recognised by a tree automaton.*

Let us now recall how tree automata and TRSs can be used for term reachability analysis. Given a tree automaton $\mathcal{A}$ and a TRS $\mathcal{R}$, the tree automata completion algorithm proposed in [12] computes a tree automaton $\mathcal{A}_\mathcal{R}^k$ such that $\mathcal{L}(\mathcal{A}_\mathcal{R}^k) = \mathcal{R}^*(\mathcal{L}(\mathcal{A}))$ when it is possible (for the classes of TRSs where an exact computation is possible, see [12]), and such that $\mathcal{L}(\mathcal{A}_\mathcal{R}^k) \supseteq \mathcal{R}^*(\mathcal{L}(\mathcal{A}))$ otherwise.

The tree automata completion works as follows. From $\mathcal{A} = \mathcal{A}_\mathcal{R}^0$ the completion builds a sequence $\mathcal{A}_\mathcal{R}^0, \mathcal{A}_\mathcal{R}^1 \ldots \mathcal{A}_\mathcal{R}^k$ of automata such that if $s \in \mathcal{L}(\mathcal{A}_\mathcal{R}^i)$ and $s \to_\mathcal{R} t$ then $t \in \mathcal{L}(\mathcal{A}_\mathcal{R}^{i+1})$. If there is a fix-point automaton $\mathcal{A}_\mathcal{R}^k$ such that $\mathcal{R}^*(\mathcal{L}(\mathcal{A}_\mathcal{R}^k)) = \mathcal{L}(\mathcal{A}_\mathcal{R}^k)$, then $\mathcal{L}(\mathcal{A}_\mathcal{R}^k) = \mathcal{R}^*(\mathcal{L}(\mathcal{A}_\mathcal{R}^0))$ (or $\mathcal{L}(\mathcal{A}_\mathcal{R}^k) \supseteq \mathcal{R}^*(\mathcal{L}(\mathcal{A}))$ if $\mathcal{R}$ is in no class of [12]). To build $\mathcal{A}_\mathcal{R}^{i+1}$ from $\mathcal{A}_\mathcal{R}^i$, a *completion step* is achieved. It consists of finding *critical pairs* between $\to_\mathcal{R}$ and $\to_{\mathcal{A}_\mathcal{R}^i}$. To define the notion of critical pair, the substitution definition is extended to terms in $\mathcal{T}(\mathcal{F} \cup \mathcal{Q})$. For a substitution $\sigma : \mathcal{X} \mapsto \mathcal{Q}$ and a rule $l \to r \in \mathcal{R}$ such that $\mathcal{V}ar(r) \subseteq \mathcal{V}ar(l)$, if there exists $q \in \mathcal{Q}$ satisfying $l\sigma \to_{\mathcal{A}_\mathcal{R}^i}^* q$ then $l\sigma \to_{\mathcal{A}_\mathcal{R}^i}^* q$ and $l\sigma \to_\mathcal{R} r\sigma$ is a critical pair. Note that since $\mathcal{R}$ and $\mathcal{A}_\mathcal{R}^i$ are finite, there is only a finite number of critical pairs. Thus, for every critical pair detected between $\mathcal{R}$ and $\mathcal{A}_\mathcal{R}^i$ such that $r\sigma \not\to_{\mathcal{A}_\mathcal{R}^i}^* q$, the tree automaton $\mathcal{A}_\mathcal{R}^{i+1}$ is constructed by adding a new transition $r\sigma \to q$ to $\mathcal{A}_\mathcal{R}^i$. Consequently, $\mathcal{A}_\mathcal{R}^{i+1}$ recognises $r\sigma$ in $q$, i.e. $r\sigma \to_{\mathcal{A}_\mathcal{R}^{i+1}} q$.

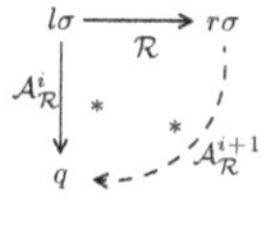

However, the transition $r\sigma \to q$ is not necessarily normalised. Then, we use abstraction functions whose goal is to define a set of normalised transitions $Norm$ such that $r\sigma \to^*_{Norm} q$. Thus, instead of adding the transition $r\sigma \to q$ which is not normalised, the set of transitions $Norm$ is added to $\Delta$, i.e., the transition set of the current automaton $\mathcal{A}^i_{\mathcal{R}}$. Notice that the completion process introduces new states. We give below a very general definition of abstraction functions which allot a state in $\mathcal{Q}$ to each functional position of $r\sigma$. The role of an abstraction function is to define equivalence classes of terms where one class corresponds to one state in $\mathcal{Q}$.

**Definition 4 (Abstraction Function).** *An abstraction function $\gamma$ is a function $\gamma : ((\mathcal{R} \times (\mathcal{X} \to \mathcal{Q}) \times \mathcal{Q}) \mapsto \mathbb{N}^*) \mapsto \mathcal{Q}$ such that $\gamma(l \to r, \sigma, q)(\epsilon) = q$.*

Thus, given an abstraction function $\gamma$, the normalisation of a transition $r\sigma \to q$ is defined as follows.

**Definition 5 ($\gamma$–normalisation).** *Let $\gamma$ be an abstraction function, $\Delta$ be a transition set, $l \to r \in R$ with $\mathcal{V}ar(r) \subseteq \mathcal{V}ar(l)$ and $\sigma : \mathcal{X} \to \mathcal{Q}$ such that $l\sigma \to^*_{\Delta} q$. The $\gamma$–normalisation of the transition $r\sigma \to q$, written $Norm_\gamma(l \to r, \sigma, q)$, is defined by:*

$$
\begin{aligned}
Norm_\gamma(l \to r, \sigma, q) = \{ & r(p)(\beta_{p.1}, \ldots, \beta_{p.n}) \to \beta \mid \\
& p \in \mathcal{P}os_{\mathcal{F}}(r), \\
& \beta = \gamma(l \to r, \sigma, q)(p) \\
& \beta_{p.i} = \begin{cases} \sigma(r(p.i)) \ if \ r(p.i) \in \mathcal{X} \\ \gamma(l \to r, \sigma, q)(p.i) \ otherwise. \end{cases}
\end{aligned}
$$

*Example 1 (Normalisation of a transition using an abstraction function).* Let $\mathcal{A} = \langle \mathcal{F}, \mathcal{Q}, \mathcal{Q}_f, \Delta \rangle$ be the tree automaton such that $\mathcal{F} = \{a, b, c, d, e, f, \omega\}$ with $ar(s) = 1$ with $s \in \{a, b, c, d, e, f\}$ and $ar(\omega) = 0$, $\mathcal{Q} = \{q_b, q_f, q_\omega\}$, $\mathcal{Q}_f = \{q_f\}$ and $\Delta = \{\omega \to q_\omega, b(q_\omega) \to q_b, a(q_b) \to q_f\}$. Thus, $\mathcal{L}(\mathcal{A}) = \{a(b(\omega))\}$. Given the TRS $\mathcal{R} = \{a(x) \to c(d(x)),\ b(x) \to e(f(x))\}$, two critical pairs are computed: $a(q_b) \to^*_{\mathcal{A}} q_f$, $a(q_b) \to_{\mathcal{R}} c(d(q_b))$ and $b(q_\omega) \to^*_{\mathcal{A}} b(q_\omega) \to_{\mathcal{R}} e(f(q_\omega))$. Let $\gamma$ be the abstraction function such that $\gamma(a(x) \to c(d(x)), \{x \to q_b\}, q_f)(\epsilon) = q_f$, $\gamma(a(x) \to c(d(x)), \{x \to q_b\}, q_f)(1) = q_f$, $\gamma(b(x) \to e(f(x)), \{x \to q_\omega\}, q_b)(\epsilon) = q_b$ and $\gamma(b(x) \to e(f(x)), \{x \to q_\omega\}, q_b)(1) = q_b$.

So, $Norm_\gamma(a(x) \to c(d(x)), \{x \to q_b\}, q_f) = \{d(q_b) \to q_f, c(q_f) \to q_f\}$ and $Norm_\gamma(b(x) \to e(f(x)), \{x \to q_\omega\}, q_b) = \{f(q_\omega) \to q_b, e(q_b) \to q_b\}$.

Now we formally define what a completion step is.

**Definition 6 (One Completion Step).** *Let $\mathcal{A} = \langle \mathcal{F}, \mathcal{Q}, \mathcal{Q}_f, \Delta \rangle$ be a tree automaton, $\gamma$ an abstraction function and $\mathcal{R}$ a left-linear TRS. We define a tree automaton $C^{\mathcal{R}}_\gamma(\mathcal{A}) = \langle \mathcal{F}, \mathcal{Q}', \mathcal{Q}'_f, \Delta' \rangle$ with $\mathcal{Q}' = \{q \mid c \to q \in \Delta'\}$ (and $\mathcal{Q} \subseteq \mathcal{Q}'$), $\mathcal{Q}'_f = \mathcal{Q}_f$ and $\Delta' = \Delta \cup \bigcup_{l \to r \in \mathcal{R},\ \sigma : \mathcal{X} \mapsto \mathcal{Q},\ l\sigma \to^*_{\mathcal{A}} q, r\sigma \not\to^*_{\mathcal{A}} q} Norm_\gamma(l \to r, \sigma, q)$.*

*Example 2.* Given $\mathcal{A}$, $\mathcal{R}$ and $\gamma$ of Example 1, performing one completion step on $\mathcal{A}$ gives the automaton $C^{\mathcal{R}}_\gamma(\mathcal{A})$ such that $C^{\mathcal{R}}_\gamma(\mathcal{A}) = \langle \mathcal{F}, \mathcal{Q}, \mathcal{Q}_f, \Delta' \rangle$ where

$\Delta' = \Delta \cup Norm_\gamma(a(x) \rightarrow c(d(x)), \{x \rightarrow q_b\}, q_f) \cup Norm_\gamma(b(x) \rightarrow e(f(x)), \{x \rightarrow q_\omega\}, q_b) = \{\omega \rightarrow q_\omega, b(q_\omega) \rightarrow q_b, a(q_b) \rightarrow q_f, d(q_b) \rightarrow q_f, c(q_f) \rightarrow q_f, f(q_\omega) \rightarrow q_b, e(q_b) \rightarrow q_b\}$. Notice that $C_\gamma^{\mathcal{R}}(\mathcal{A})$ is $\mathcal{R}$-close, and in fact an over-approximation of $\mathcal{R}^*(\mathcal{L}(\mathcal{A}))$ is computed. Indeed, the tree automaton $C_\gamma^{\mathcal{R}}(\mathcal{A})$ recognises the term $a(e(e(f(\omega))))$ when $\mathcal{R}^*(\mathcal{L}(\mathcal{A}))=\{a(b(\omega)), a(e(f(\omega))), c(d(b(\omega))), c(d(e(f(\omega))))\}$.

**Proposition 1 (Adaptation of [12, Theorem 1]).** *Let $\mathcal{A}$ be a tree automaton and $\mathcal{R}$ be a TRS such that $\mathcal{A}$ is deterministic or $\mathcal{R}$ is left-linear, and for every $l \rightarrow r \in \mathcal{R}$, $\mathcal{V}ar(r) \subseteq \mathcal{V}ar(l)$. One has $\mathcal{L}(\mathcal{A}) \cup \mathcal{R}(\mathcal{L}(\mathcal{A})) \subseteq C_\gamma^{\mathcal{R}}(\mathcal{A})$, for any abstraction function $\gamma$.*

However, abstraction functions can be defined in such a way that only actually reachable terms are computed. We call this class of abstraction functions $(\mathcal{A}, \mathcal{R})-$exact abstraction functions.

**Definition 7 ($(\mathcal{A}, \mathcal{R})-$exact abstraction function).** *Let $\mathcal{A} = \langle \mathcal{F}, \mathcal{Q}, \mathcal{Q}_f, \Delta \rangle$ be a tree automaton and $\mathcal{R}$ be a TRS. Let $Im(\gamma) = \{q' \mid \exists l \rightarrow r \in \mathcal{R},\ \exists p \in \mathcal{P}os_{\mathcal{F}}(r),\ p \neq \epsilon,\ \exists \sigma : \mathcal{X} \rightarrow \mathcal{Q}\ .\ \gamma(l \rightarrow r, \sigma, q)(p) = q'\}$. An abstraction function $\gamma$ is $(\mathcal{A}, \mathcal{R})-$exact if $\gamma$ is injective and $Im(\gamma) \cap \mathcal{Q} = \emptyset$.*

By adapting the proof of Theorem 2 in [12] to the new class of abstractions, we show that with such abstraction functions, only reachable terms are computed.

**Proposition 2 ([12, Theorem 2]).** *Let $\mathcal{A}$ be a tree automaton and $\mathcal{R}$ be a left-linear TRS such that $\mathcal{A}$ is deterministic or $\mathcal{R}$ is also right-linear. Let $\alpha$ be an $(\mathcal{A}, \mathcal{R})-$exact abstraction function. One has: $C_\alpha^{\mathcal{R}}(\mathcal{A}) \subseteq \mathcal{R}^*(\mathcal{L}(\mathcal{A}))$.*

*Example 3 (Exact automaton with $(\mathcal{A}, \mathcal{R})-$exact abstraction functions).* Let $\mathcal{A}$, $\mathcal{R}$ be the tree automaton and the TRS from Example 1. Let $\alpha$ be the $(\mathcal{A}, \mathcal{R})-$abstraction function such that $Norm_\alpha(a(x) \rightarrow c(d(x)), \{x \rightarrow q_b\}, q_f) = \{c(q_1) \rightarrow q_f,\ d(q_b) \rightarrow q_1\}$ and $Norm_\alpha(b(x) \rightarrow e(f(x)), \{x \rightarrow q_\omega\}, q_b) = \{e(q_2) \rightarrow q_b, f(q_\omega) \rightarrow q_2\}$. Note that $q_1$ and $q_2$ are not states of $\mathcal{A}$. Then, $C_\alpha^{\mathcal{R}}(\mathcal{A})$ is the tree automaton $\langle \mathcal{F}, \mathcal{Q} \cup \{q_1, q_2\}, \mathcal{Q}_f, \Delta' \rangle$ where $\Delta' = \{\omega \rightarrow q_\omega, b(q_\omega) \rightarrow q_b, a(q_b) \rightarrow q_f, d(q_b) \rightarrow q_1, c(q_1) \rightarrow q_f, f(q_\omega) \rightarrow q_2, e(q_2) \rightarrow q_b\}$.

Figure 1 gives a graphical representation of Example 3 and Example 2 using word automata. Indeed, considering the symbol $\omega$ as the empty word, the term $a(b(\omega))$ can be read as the word $ab$. The state $q_f$ of $\mathcal{A}$ becomes the initial state of the word automaton, and $q_\omega$ is its final state. Note that we do not consider the empty word $\omega$ in our representation. So, $C_\gamma^{\mathcal{R}}(\mathcal{A})$ from Example 2 is represented by the word automaton without non dashed transitions implying $q_1$ and $q_2$. It recognises, for example, the word $aeef$ because of the abstraction. Using $(\mathcal{A}, \mathcal{R})-$exact approximation function $\alpha$ gives $C_\alpha^{\mathcal{R}}(\mathcal{A})$, the word automaton in Fig. 1 without dashed transitions.

We now give the general result in [12] saying that, if there exists a fix-point automaton, then its language contains all the terms actually reachable by rewriting, at least.

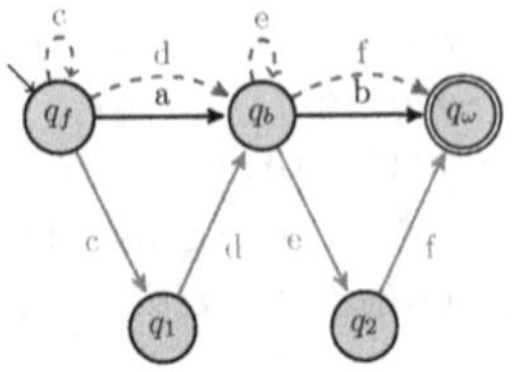


**Fig. 1.** $C_\gamma^{\mathcal{R}}(\mathcal{A})$ and $C_\alpha^{\mathcal{R}}(\mathcal{A})$ as simple word automata

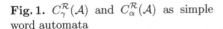

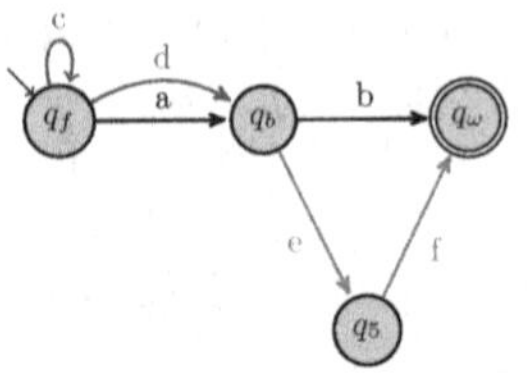


**Fig. 2.** Word automaton for $C_{\gamma\text{Disc}}^{\mathcal{R}}$

**Theorem 1 ([12, Theorem 1]).** *Let $\mathcal{A}$ and $\mathcal{R}$ be respectively a tree automaton and a left-linear TRS. For any abstraction function, if there exists $N \in \mathbb{N}$ and $N \geq 0$ such that $(C_\gamma^{\mathcal{R}})^{(N)}(\mathcal{A}) = (C_\gamma^{\mathcal{R}})^{(N+1)}(\mathcal{A})$, then $\mathcal{R}^*(\mathcal{L}(\mathcal{A})) \subseteq \mathcal{L}((C_\gamma^{\mathcal{R}})^{(N)}(\mathcal{A}))$.*

# 3 Approximation Refinement

Let consider a tree automaton $\mathcal{A}$, a TRS $\mathcal{R}$ and an abstraction function $\gamma$ such that there exists $N \in \mathbb{N}$ for which $(C_\gamma^{\mathcal{R}})^{(N)}(\mathcal{A}) = (C_\gamma^{\mathcal{R}})^{(N+1)}(\mathcal{A})$. Let $\mathcal{A}_p$ be a tree automaton recognising a set of unwanted terms, i.e., we want that terms to be unreachable. Let suppose that the intersection between the languages of $(C_\gamma^{\mathcal{R}})^{(N)}(\mathcal{A})$ and $\mathcal{A}_p$ is not empty; the method in [12] cannot conclude. In this section we refine $\gamma$ to show that: either a term recognised by $\mathcal{A}_p$ is actually reachable by rewriting from a term in $\mathcal{L}(\mathcal{A})$, either all terms recognised by $\mathcal{A}_p$ are actually unreachable. For the former, we present in Section 3.1 how to refine an abstraction function when that function gives rise to the non-empty intersection. For the latter, in Section 3.2, we describe the computation of the ancestors of a set of terms by rewriting, using a completion on TRSs whose rewrite rules are reversed. Then, in Section 3.3, the backward analysis is used for refining abstraction functions when the assumptions in Section 3.1 are not satisfied.

## 3.1 Abstraction Refinement

In this section, $\mathcal{A} = \langle \mathcal{F}, \mathcal{Q}, \mathcal{Q}_f, \Delta \rangle$ and $\mathcal{A}_P$ are two tree automata, $\mathcal{R}$ is a TRS, $\gamma$ is an abstraction function and $\alpha$ is a $(\mathcal{A}, \mathcal{R})-$exact abstraction function. The tree automaton $\mathcal{A}_p$ recognises a set of forbidden terms. We assume that: (1) $\mathcal{L}(C_\gamma^{\mathcal{R}}(\mathcal{A})) \cap \mathcal{L}(\mathcal{A}_p) \neq \emptyset$, $\mathcal{L}(\mathcal{A}) \cap \mathcal{L}(\mathcal{A}_p) = \emptyset$ and (2) $\mathcal{L}(C_\alpha^{\mathcal{R}}(\mathcal{A})) \cap \mathcal{L}(\mathcal{A}_p) = \emptyset$. In other words, these assumptions mean that the abstraction function is too coarse since a term of $\mathcal{L}(\mathcal{A}_p)$ is reachable using $\gamma$, while it is not with an exact computation. So, it means that some transitions introduced by $\gamma$ are problematic, i.e., lead to the non-empty intersection. The following definition allows the refinement of a given abstraction function according to a given set of problematic transitions.

**Definition 8 (Refined abstraction function).** *Let $l\sigma \rightarrow_{\mathcal{R}} r\sigma$, $l\sigma \rightarrow_{\mathcal{A}}^* q$ be a critical pair where $q \in \mathcal{Q}$, $l \rightarrow r$ is a rewrite rule and $\sigma$ a substitution from $\mathcal{X}$*

*into $Q$. Let $\Delta_0$ be a set of transitions. For any functional position $p$ of $r$, the refined abstraction function $\gamma_{\Delta_0}$ is built as follows:*

$$\gamma_{\Delta_0}(l \to r, \sigma, q)(p) \stackrel{\text{def}}{=} \begin{cases} \alpha(l \to r, \sigma, q)(p) & \textit{if } Norm_\gamma(l \to r, \sigma, q) \cap \Delta_0 \neq \emptyset \\ \gamma(l \to r, \sigma, q)(p) & \textit{otherwise.} \end{cases}$$

For the abstraction function $\gamma$ (resp. $\alpha$), we denote by $\Delta_\gamma$ (resp. $\Delta_\alpha$) the set of transitions occurring in $C^{\mathcal{R}}_\gamma(\mathcal{A})$ (resp.$C^{\mathcal{R}}_\alpha(\mathcal{A})$) but not in $\mathcal{A}$, and by $Q_\gamma$ (resp. $Q_\alpha$) the set of states occurring in the transitions of $\Delta_\gamma$ (resp. $\Delta_\alpha$). The following proposition claims that, given $\gamma$ and $\mathcal{A}$, we are able to refine $\gamma$ in such a way that no unwanted term is recognised by the tree automaton resulting from one completion step on $\mathcal{A}$ when using $\mathcal{R}$ and the refined abstraction function.

**Proposition 3 (Refined automaton existence).** *Considering assumptions (1) and (2), there exists* Disc $\subseteq \Delta_\gamma$ *such that* $\mathcal{L}(C^{\mathcal{R}}_{\gamma_{\text{Disc}}}(\mathcal{A})) \cap \mathcal{L}(\mathcal{A}_P) = \emptyset$.

*Example 4 (Refined Automaton).* Let $\mathcal{A}$, $\gamma$ and $\alpha$ be respectively the tree automaton, the two abstraction functions defined in Example 3. In Sect. 2.2, the word $aeef$ is recognised by the word automaton representing $C^{\mathcal{R}}_\gamma(\mathcal{A})$. Let Disc $\subseteq \Delta_\gamma$ be a set of transitions such that Disc $= \{e(q_b) \to q_b\}$. Indeed, this transition gives rise to infinite terms of the form $e^*(b(\omega))$ or $e^*(f(\omega))$ (infinite words of the form $e^*b$ or $e^*f$). So, $\gamma_{\text{Disc}}$ uses $\alpha$ and $\gamma$ for normalising respectively the transitions $e(f(q_\omega)) \to q_b$ and $c(d(q_b)) \to q_f$. A word automaton representing $C^{\mathcal{R}}_{\gamma_{\text{Disc}}}$ is given in Fig 2. Note that the word $aeef$ is not recognised anymore.

For applying Proposition 3 in practice, it could be more efficient to fix only problematic transitions instead of performing an exact completion step by taking Disc $= \Delta_\gamma$.

### 3.2 Backward Reachability Analysis by Completion

The backward analysis we expose in this section can be viewed as an exact completion performed on an automaton $\mathcal{A}$, using a TRS $\mathcal{R}$ whose rules have been reversed and an $(\mathcal{A}, \mathcal{R})$-exact abstraction function. Let $\mathcal{R}^t$ be a TRS built from $\mathcal{R}$ in the following way: $\mathcal{R}^t = \{r \to l | l \to r \in \mathcal{R}\}$. So, for $\mathcal{R}^t$, some rules may not satisfy the conditions of Propositions 1 and 2, in particular that for each $l \to r \in \mathcal{R}^t$, $\mathcal{V}ar(r) \subseteq \mathcal{V}ar(l)$. This is why we extend the completion definition in Section 2.2. Before, we establish the relation between $\mathcal{R}^t(E)$ and $\mathcal{R}^{-1}(E)$ for a given set of terms $E \subseteq \mathcal{T}(\mathcal{F})$.

**Proposition 4.** *For every set of terms* $E \subseteq \mathcal{T}(\mathcal{F})$, $\mathcal{R}^t(E) = \mathcal{R}^{-1}(E)$.

A consequence of this proposition is that $(\mathcal{R}^t)^*(E) = (\mathcal{R}^{-1})^*(E)$. So, we can now define the completion algorithm in order to analyse $(\mathcal{R}^{-1})^*(E)$. For a given set of functional symbols $\mathcal{F}$, we introduce a set of transitions $T(q)$ reducing each term of $\mathcal{T}(\mathcal{F})$ to the single state $q$. $T(q)$ is built as follows: $T(q) = \{s \to q \mid s \in \mathcal{T}(\mathcal{F}), s = f(q_1, \ldots, q_n), f \in \mathcal{F},\ ar(f) = n,\ n \geq 0,\ \textit{and}\ q_i = q\ \textit{for}\ i = 1 \ldots n\}$.

This set is useful for handling variables occurring in the right-hand side of a rule but not in its left-hand side.

In the following definitions, a substitution is transformed according to a given set of variables and a given state.

**Definition 9.** *Let $q$ be a state and $Y \subseteq \mathcal{X}$. Given a substitution $\sigma : \mathcal{X} \to \mathcal{Q}$ and $x \in \mathcal{X}$,*

$$Chg_Y^q(\sigma)(x) = \begin{cases} \sigma(x) & \textit{if } x \in Y \\ q & \textit{otherwise.} \end{cases}$$

In the remainder of this section we do not assume anymore that for each rule $l \to r \in \mathcal{R}$, $\mathcal{V}ar(r) \subseteq \mathcal{V}ar(l)$. Using Definition 9, Definition 10 extends the completion algorithm for handling such TRSs.

**Definition 10 (Backward Completion).** *Let $\mathcal{A} = \langle \mathcal{F}, \mathcal{Q}, \mathcal{Q}_f, \Delta \rangle$ be a tree automaton, $\gamma$ an abstraction function and $q_{all}$ a state such that $q_{all} \notin \mathcal{Q}$. Let $\mathcal{R}$ be a left-linear TRS. We define a tree automaton $K_\alpha^{\mathcal{R}}(\mathcal{A}) = \langle \mathcal{F}, \mathcal{Q}', \mathcal{Q}'_f, \Delta' \rangle$ where:*

$$\Delta' = \Delta \cup \bigcup_{l \to r \in \mathcal{R}, l\sigma \to_{\mathcal{A}}^* q, r(Chg_{\mathcal{V}ar(l)}^{q_{all}}(\sigma)) \not\to_{\mathcal{A}}^* q} (Norm_\gamma(l \to r, Chg_{\mathcal{V}ar(l)}^{q_{all}}(\sigma), q)) \cup T(q_{all}),$$

$\mathcal{Q}' = \{q \mid c \to q \in \Delta'\}$ *and* $\mathcal{Q}'_f = \mathcal{Q}_f$.

Note that there is a particular processing for variables occurring in the right-hand side of a rule which do not appear in its left-hand side. Indeed, each of these variables is substituted by the special state $q_{all}$ because it is impossible to determine terms substituting these variables. Actually, considering the set $T(q_{all})$ of transitions, $q_{all}$ is such that for each term $t \in \mathcal{T}(\mathcal{F})$, $t \to_{T(q_{all})}^* q_{all}$. So, for the rule $f(x) \to g(x, y)$, $y$ is replaced by $q_{all}$, i.e., $f(x) \to g(x, q_{all})$. Roughly, we can say that a term of the form $f(x)$ can be rewritten into $g(x,t)$ where $t$ is any term of $\mathcal{T}(\mathcal{F})$.

Finally, using an $(\mathcal{A}, \mathcal{R})$–exact abstraction function, we can perform a backward analysis of the set of reachable terms thanks to the new backward completion given above.

**Proposition 5 (Extension of Propositions 1 and 2).** *Let $\mathcal{A}$ be a tree automaton, $\mathcal{R}$ be TRS and $\alpha$ be an abstraction function. Thus, one has:*

1) *if $\mathcal{A}$ is deterministic or $\mathcal{R}$ is left-linear: $\mathcal{L}(\mathcal{A}) \cup \mathcal{R}(\mathcal{L}(\mathcal{A})) \subseteq \mathcal{L}(K_\alpha^{\mathcal{R}}(\mathcal{A}))$;*
2) *if $\mathcal{R}$ is linear and $\alpha$ be an $(\mathcal{A}, \mathcal{R})$–exact abstraction function: $\mathcal{L}(K_\alpha^{\mathcal{R}}(\mathcal{A})) \subseteq \mathcal{R}^*(\mathcal{L}(\mathcal{A}))$.*

### 3.3 Semi-algorithm

In Sect. 2.2, reachability analysis can be performed by computing either an over-approximation of reachable terms with a fine tuned abstraction function or an under-approximation using an $(\mathcal{A}, \mathcal{R})$–exact abstraction function. The former may allow the proof of unreachability of terms and the latter may show that terms are reachable. Nevertheless, using the completion algorithm of [12], a choice must be done according to the kind of analysis we want to perform.

We propose in this section a new semi-algorithm which attempts to perform both analyses automatically using abstraction refinement. More precisely, let $\mathcal{A}$, $\mathcal{R}$, $\gamma$, $\alpha$ and $\mathcal{A}_p$ be respectively a tree automaton, a linear TRS, an abstraction function, an $(\mathcal{A}, \mathcal{R})-$exact abstraction function and a tree automaton recognising a set of unwanted terms. A sequence of automaton $\mathcal{A}_0, \ldots, \mathcal{A}_k$, where $\mathcal{A}_0 = \mathcal{A}$, is computed by completion until $\mathcal{A}_k$ where $\mathcal{A}_k$ is either a fix-point automaton such that $\mathcal{L}(\mathcal{A}_k) \cap \mathcal{L}(\mathcal{A}_P) = \emptyset$, or $\mathcal{L}(\mathcal{A}_k) \cap \mathcal{L}(\mathcal{A}_P) \neq \emptyset$. For the former, by Proposition 1 each term of $\mathcal{L}(\mathcal{A}_P)$ is unreachable. For the latter, an exact completion step on $\mathcal{A}_{k-1}$ is performed using $\alpha$ ($C_\alpha^{\mathcal{R}}(\mathcal{A}_{k-1})$). If $\mathcal{L}(C_\alpha^{\mathcal{R}}(\mathcal{A}_{k-1})) \cap \mathcal{L}(\mathcal{A}_P) = \emptyset$ then the abstraction function $\gamma$ has been too coarse at $k^{th}$ completion step. So, according to Proposition 3, a new abstraction function $\gamma'$ is obtained by refining $\gamma$ to ensure that $\mathcal{L}(C_{\gamma'}^{\mathcal{R}}(\mathcal{A}_{k-1})) \cap \mathcal{L}(\mathcal{A}_P) = \emptyset$. Otherwise, the backward analysis following Proposition 5 is performed from $\mathcal{A}_P$ in order to detect the completion step which is guilty of this non-empty intersection. The completion step $i$ is guilty if $\mathcal{L}(\mathcal{A}_i) \cap \mathcal{L}((K_\alpha^{\mathcal{R}})^{(k-i)}(\mathcal{A}_P)) \neq \emptyset$. As soon as the incriminated completion step is detected, the abstraction function $\gamma$ is refined and the completion restarts from this completion step and using the new abstraction function, and so on. If no completion step is guilty then $\mathcal{R}^*(\mathcal{L}(\mathcal{A})) \cap \mathcal{L}(\mathcal{A}_P) \neq \emptyset$.

In Algorithm 1 and for the remainder of this section, $\mathcal{A}^e$ denotes $C_\alpha^{\mathcal{R}}(\mathcal{A})$. $\mathcal{A}_i$ is the $i^{th}$ element of the list *aut_list*. Let *aut_list* be a list of $n$ elements $[e_1, \ldots, e_n]$, *aut_list[i]* denotes the sublist $[e_1, \ldots, e_i]$ with $i \leq n$. The function *aut_list::x* adds $x$ at the end of the list *aut_list*.

**Algorithm 1 (Refinement semi-algorithm).** *Given $\mathcal{R}$ a linear TRS, $\mathcal{A}$ a tree automaton, $\mathcal{A}_P$ a tree automaton recognising a set of unwanted terms, $\gamma$ an abstraction function and $\alpha$ an $(\mathcal{A}, \mathcal{R})-$exact abstraction function, $Comp_{Ref}(\mathcal{A}, \mathcal{R}, \mathcal{A}_P, \gamma, \alpha)$ is defined as follows:*

***Variables***
$\mathcal{A}_P^{temp} := \mathcal{A}_P$;
*aut_list* := $[\mathcal{A}; C_\gamma^{\mathcal{R}}(\mathcal{A})]$; (* list of automata *)
$i$ := 1 ; (* completion step number *)
*result* := *true*;
**Begin**
**While** $(\mathcal{A}_i \neq \mathcal{A}_{i-1})$ *and* (*result* = *true*) **do**
  **If** $\mathcal{L}(\mathcal{A}_i) \cap \mathcal{L}(\mathcal{A}_P) = \emptyset$ **then** — *If the intersection is empty between*
    *aut_list* := *aut_list*:: $C_\gamma^{\mathcal{R}}(\mathcal{A}_i)$; — $\mathcal{L}(\mathcal{A}_i)$ *and* $\mathcal{L}(\mathcal{A}_P)$ *then a normal*
    $i := i + 1$; — *completion step is performed.*
  **Else**
    *status* := $\mathcal{L}(C_\alpha^{\mathcal{R}}(\mathcal{A}_{i-1})) \cap \mathcal{L}(\mathcal{A}_P)$; — *While the intersection between* $\mathcal{L}(\mathcal{A}_i^e)$
    **While** (*status* $\neq \emptyset$) *and* $(i > 0)$ **do** — *and* $\mathcal{L}(\mathcal{A}_P)$ *is not empty, and while*
      $\mathcal{A}_P^{temp} := K_\alpha^{\mathcal{R}^t}(\mathcal{A}_P^{temp})$; — $i > 0$*, Definition 10 is used*
      $i := i - 1$; — *to compute a new automaton* $A_P$.
    **If** $i > 0$ **then**
      *status* := $\mathcal{L}(C_\alpha^{\mathcal{R}}(\mathcal{A}_{i-1})) \cap \mathcal{L}(\mathcal{A}_P^{temp})$;

12 **EndIf**
13 **Done**
14 $\mathcal{A}_P^{temp} := \mathcal{A}_P$;
15 **If** *(i = 0)* **then**
16 *result := false ;*
17 **Else**
18 *Find Disc* $\subseteq \Delta_i \setminus \Delta_{i-1}$;
19 $\gamma := \gamma_{Disc}$
20 *aut_list := aut_list[i-1]::*$C_\gamma^{\mathcal{R}}(\mathcal{A}_i)$;
21 **EndIf**
22 **EndIf**
23 **Done**
24 *return result;*
25 **End**

*There are 2 cases to make* **while** *stop:*
- $i = 0$ *(and* $\mathcal{L}(\mathcal{A}_i^e) \cap \mathcal{L}(\mathcal{A}_P) \neq \emptyset$*).* *In this case we can conclude that* $\mathcal{R}^*(\mathcal{L}(\mathcal{A}_0)) \cap \mathcal{L}(\mathcal{A}_P) \neq \emptyset$;
- $\mathcal{L}(\mathcal{A}_i^e) \cap \mathcal{L}(\mathcal{A}_P) = \emptyset$ *(and* $i \geq 0$*).*

*In this case* $\gamma$ *is refined and a completion step is performed.*

Theorems 2 shows that the semi-algorithm above is sound.

**Theorem 2 (Soundness of Algorithm 1).** *If* $Comp_{Ref}(\mathcal{A}, \mathcal{R}, \mathcal{A}_P, \gamma, \alpha) =$ *true then* $\mathcal{R}^*(\mathcal{L}(\mathcal{A})) \cap \mathcal{L}(\mathcal{A}_P) = \emptyset$ *else if* $Comp_{Ref}(\mathcal{A}, \mathcal{R}, \mathcal{A}_P, \gamma, \alpha) =$ *false then* $\mathcal{R}^*(\mathcal{L}(\mathcal{A})) \cap \mathcal{L}(\mathcal{A}_P) \neq \emptyset$.

*Proof.* Algorithm 1 terminates when either a fix-point automaton is computed or *result = false*. If *result = false* then line 16 has been executed. Moreover, according to lines 7, 8, 9 and 11, $\mathcal{R}^*(\mathcal{L}(\mathcal{A})) \cap \mathcal{L}(\mathcal{A}_P) \neq \emptyset$. Indeed, if $i = 0$ at line 9 then $i$ has been equal to 1 either at line 6 (before entering in the **while**) or at line 9 (at the previous iteration). Consequently, $\mathcal{L}(C_\alpha^{\mathcal{R}}(\mathcal{A})) \cap \mathcal{L}(\mathcal{A}_P^{temp}) \neq \emptyset$ since $i$ becomes 0 next, at line 9. Consequently, as $\mathcal{A}_P^{temp}$ represents $(K_\alpha^{\mathcal{R}^t})^k(\mathcal{A}_P)$, according to Propositions 5 and 4, one deduces that $\mathcal{R}^*(\mathcal{L}(\mathcal{A})) \cap \mathcal{L}(\mathcal{A}_P) \neq \emptyset$.

If *result = true* and $\mathcal{A}_i$ is a fix-point automaton then line 15 has never been executed. So, to break **while** at line 7, *status* needs to be $\emptyset$. So, there exist $n \in \mathbb{N}$ and an abstraction function $\gamma'$ built from $\gamma$ such that $(C_{\gamma'}^{\mathcal{R}})^{(n)}(\mathcal{A}) = (C_{\gamma'}^{\mathcal{R}})^{(n+1)}(\mathcal{A})$. According to Proposition 1, $\mathcal{L}((C_{\gamma'}^{\mathcal{R}})^{(n)}(\mathcal{A})) \supseteq \mathcal{R}^*(\mathcal{L}(\mathcal{A}))$. Consequently, $\mathcal{R}^*(\mathcal{L}(\mathcal{A})) \cap \mathcal{L}(\mathcal{A}_P) = \emptyset$.

The case when *result = false* and $\mathcal{A}_i$ is a fix-point automaton, reduces to the first case handled in this proof.

Finally, Theorem 3 claims that our semi-algorithm is complete in the sense that if an unwanted term is reachable then Algorithm 1 returns false.

**Theorem 3 (Partial Completeness of Algorithm 1).** *If* $\mathcal{R}^*(\mathcal{L}(\mathcal{A})) \cap \mathcal{L}(\mathcal{A}_P) \neq \emptyset$ *then* $Comp_{Ref}(\mathcal{A}, \mathcal{R}, \mathcal{A}_P, \gamma, \alpha) =$ *false.*

*Proof.* Suppose each time, at line 18 *Disc* is set such that $Disc = \Delta_i \setminus \Delta_{i-1}$. Thus $\gamma$ tends to behave as $\alpha$. So, let $\gamma$ be $\alpha$. Since $\mathcal{R}^*(\mathcal{L}(\mathcal{A})) \cap \mathcal{L}(\mathcal{A}_P) \neq \emptyset$, there exists $n \in \mathbb{N}$ and $n > 0$ such that $\mathcal{L}((C_\gamma^{\mathcal{R}})^{(n)}(\mathcal{A})) \cap \mathcal{L}(\mathcal{A}_P) \neq \emptyset$ and $\mathcal{L}((C_\gamma^{\mathcal{R}})^{(n-1)}(\mathcal{A})) \cap \mathcal{L}(\mathcal{A}_P) = \emptyset$. Consequently, in this setting, *status* at line 6 is different from $\emptyset$ and $i = n$. Consequently, $K_\alpha^{\mathcal{R}^t}(\mathcal{A}_P)$ is computed. According to

Propositions 2 and 5, one can deduce that $\mathcal{L}((C_\gamma^\mathcal{R})^{(n-1)}(\mathcal{A})) \cap \mathcal{L}(K_\alpha^{\mathcal{R}^t}(\mathcal{A}_P)) \neq \emptyset$. So, by a simple induction, one trivially obtains $\mathcal{L}((C_\gamma^\mathcal{R})^{(0)}(\mathcal{A})) \cap \mathcal{L}((K_\alpha^{\mathcal{R}^t})^{(n)}(\mathcal{A}_P)) \neq \emptyset$ that corresponds to the value stored in *status* at line 11 after $n-1$ iteration in the **while** at line 7. Since $i = 1$ and *status*$\neq \emptyset$, a new iteration is performed. Finally, $i = 0$ and *status*$\neq \emptyset$. Thus, *result* is set to false, the **while** at line 1 is broken and Alg. 1 returns false.

An advantage of our approach is that the abstraction function $\gamma$ given as input to $Comp_{Ref}$ does not require to be very pertinent. As explained at the very beginning of this paper, it is very easy to generate abstraction functions leading to inconclusive analyses. Our algorithm attempts to fix this kind of abstraction functions in order to perform an unreachability analysis. If the inconclusive analysis is not of the abstraction function concern then our algorithm states that some of the unwanted terms are actually reachable. This algorithm has been prototyped in the Timbuk tool [14].

# 4 Experiments

Our abstraction refinement technique for completion has been applied for the verification of a simple two processes counting system. The following TRS describes the behaviour of two processes each one equipped with an input list and a FIFO. Each process receives a list of symbols '+' and '−' to count, as an input. One of the processes, say $P_+$, is counting the '+' symbols and the other one, say $P_-$ is counting the '−' symbols. When $P_+$ receives a '+', it counts it and when it receives a '−', it adds the symbol to $P_-$'s FIFO. The behaviour of $P_-$ is symmetric. When a process' input list and FIFO is empty then it stops and gives the value of its counter.

Here is a possible rewrite specification of this system, given in the Timbuk language, where `S(_,_,_,_)` represents a configuration with a process $P_+$, a process $P_-$, $P_+$'s FIFO and $P_-$'s FIFO. The term `Proc(_,_)` represents a process with an input list and a counter, `add(_,_)` implements adding of an element in a FIFO, and `cons`, `nil`, `s`, `o` are the usual constructors for lists and natural numbers.

```
Ops
       S:4 Proc:2 Stop:1 cons:2 nil:0 plus:0 minus:0 s:1 o:0 end:0 add:2
Vars   x y z u c m n
TRS R1
 add(x, nil) -> cons(x, nil)
 add(x, cons(y, z)) -> cons(y, add(x, z))
 S(Proc(cons(plus, y), c), z, m, n) -> S(Proc(y, s(c)), z, m, n)
 S(Proc(cons(minus, y), c), u, m, n) -> S(Proc(y, c), u, m, add(minus, n))
 S(x, Proc(cons(minus, y), c), m, n) -> S(x, Proc(y, s(c)), m, n)
 S(x, Proc(cons(plus, y), c), m, n) -> S(x, Proc(y, c), add(plus, m), n)
 S(Proc(x, c), z, cons(plus,m), n) -> S(Proc(x, s(c)), z, m, n)
 S(x, Proc(z, c), m, cons(minus,n)) -> S(x, Proc(z, s(c)), m ,n)
 S(Proc(nil, c), z, nil, n)  -> S(Stop(c), z, nil, n)
 S(x, Proc(nil, c), m, nil) -> S(x, Stop(c), m, nil)
```

The set of initial configurations of the system is described by the following tree automaton, where each process has a counter initialised to 0 and has an unbounded input list (with both '+' and '−') and with at least one symbol.

```
Automaton A1
States q0 qinit qzero qnil qlist qsymb
Final States q0
Transitions
 cons(qsymb, qnil) ->qlist    cons(qsymb, qlist) -> qlist        o -> qzero
 Proc(qlist, qzero) -> qinit  S(qinit, qinit, qnil, qnil) -> q0  nil -> qnil
 plus -> qsymb                minus -> qsymb
```

On this specification, we aim at proving that, for any input lists, there is no possible deadlock. In this example, a deadlock is a configuration where a process has stopped but there are still symbols to count in its FIFO. This property is specified by a tree automaton `Bad_state` recognising a system for which one of the two processes has stopped and whose FIFO is not empty.

Before computing an over-approximation of the reachable configurations of the two processes and according to Def. 4, we give the simple abstraction function $\gamma$ that satisfies the following property: Let $\sigma_1, \sigma_2$ be two substitutions from $\mathcal{X}$ into $\mathcal{Q}$ such that $\sigma_1 \neq \sigma_2$. Let $l \to r$ be a rule of the TRS R of the Timbuk specification. Then, for any state $q$ and for each functional position $p$ of $r$, $\gamma(l \to r, \sigma_1, q)(p) = \gamma(l \to r, \sigma_2, q)(p)$. With such an abstraction function, the completion converges to a fix-point automaton $\mathcal{A}$ within 8 completion steps. Unfortunately, the intersection between the tree automata $\mathcal{A}$ and `Bad_state` is non-empty. No conclusion can be drawn since $\mathcal{A}$ is an over-approximation.

Using the abstraction refinement technique, the following scenario sets: Three completion steps are performed and a non-empty intersection is found. So, the backward analysis is run and finally reaches the initial tree automaton. In conclusion, there exists a deadlock for this system.

The problem found here can be fixed by adding an additional symbol: 'end' which has to be added by process $P_+$ to $P_-$ FIFO when $P_+$ has reached the end of its list, and symmetrically for $P_-$. Then, a process can stop if and only if it has reached the end of its list and if it has read the 'end' symbol in its FIFO. Then, the TRS of the previous specification is modified a little. For the new specification, we use the same kind of abstraction function as in the first experiment. Timbuk finds a fix-point automaton $\mathcal{A}'$ within 9 completion steps. Unfortunately, the intersection between the tree automata $\mathcal{A}'$ and `Bad_state` is non-empty anew. Using the abstraction refinement, Timbuk finds another fix-point automaton $\mathcal{A}''$ whose language contains no term recognised by the tree automaton `Bad_state`. Consequently, we have managed to prove that the patched system is actually deadlock free.

## 5 Conclusion

The paper describes a new approach for automatically generating abstraction-based over-approximations guided by a set of unwanted terms. In the infinite state system verification framework, our work can be considered as an abstraction-based approach guided by a safety/security property which either can conclude that a safety property is satisfied or can detect a violation of the given property or may not terminate. The last point is not surprising since the reachability problem is undecidable on non terminating TRSs. Furthermore, in [3], we show that in some cases, unreachable terms are in all computable over-approximations. So, refinement may be unfruitful in the rewriting approximation framework. However, our first experimental results are promising. Moreover, the positive results obtained in the framework of security protocols and Java programs analysis, let us think that our refinement approach can work in practice and then, can make the reachability analysis detailed in [12] available to a larger community of users. More experimentations are needed to compare the technique in this paper and the abstraction technique in [5], and to determine our interactive backward reachability analysis efficiency and pertinence on, e.g., Java programs. In many system analyses, backward analysis provides better results than forward analysis.

**Related works.** The notion of abstraction function presented in [12] is not so far from the basic definition used in the framework of the abstraction-based verification of infinite-state systems.

Abstraction refinement is already used to make the definition of a good abstraction function easier and, consequently, to make the system verification easier. In [6], the CEGAR (*Counterexample-Guided Abstraction Refinement*) paradigm has been summarised and a general algorithm consisting in refining an abstraction function by analysing a spurious counterexample has been given. Our work fits almost exactly with this framework.

In [8,7], the authors use abstraction refinement on Kripke structure and ACTL specification. When an abstract counterexample is found, the corresponding concrete counterexample is computed. If it does not correspond to a path in the concrete model, it means that it is a spurious counterexample, and the abstraction function is then refined to make the concrete model correspond to the given specification. The spurious counterexample analysis is done with a forward method.

In [18], an abstraction refinement method is used on transition systems for verifying invariants with a technique combining model checking, abstraction and deductive verification. Contrary to the three previous articles, the authors do not consider liveness properties, and the spurious counterexample analysis is done with a backward method. In [10], as part of predicate abstraction, predicates are automatically discovered by analysing spurious counterexamples. The method exposed in this paper is close to the above methods, but it works on different data structures.

In the field of tree automata, [5] computes an over-approximation with the help of an initial tree automata, tree transducers, and by merging states which either recognise the same language for a given depth or satisfy a given predicate. This merging is the origin of the over-approximation. A refinement can be done either by increasing the depth or by extending the predicate with a spurious counterexample. In our case, term rewriting systems are used instead of transducers. Moreover, the states fusion is guided by a safety/security property together with an abstraction function.

## References


1. Baader, F., Nipkow, T.: Term Rewriting and All That. Cambridge University Press, Cambridge (1998)
2. Boichut, Y., Genet, T., Jensen, T., Le Roux, L.: Rewriting approximations for fast prototyping of static analyzers. In: Baader, F. (ed.) RTA 2007. LNCS, vol. 4533, pp. 48–62. Springer, Heidelberg (2007)
3. Boichut, Y., Héam, P.-C.: A Theoretical Limit for Safety Verification Techniques with Regular Fix-point Computations. Information Processing Letters ( to appear, 2008)
4. Boichut, Y., Héam, P.-C., Kouchnarenko, O.: Automatic Verification of Security Protocols Using Approximations. Technical Report RR-5727, INRIA (2005)
5. Bouajjani, A., Habermehl, P., Rogalewicz, A., Vojnar, T.: Abstract regular tree model checking. In: proceedings of INFINITY. BRICS Notes Series, vol. 4, pp. 15–24 (2005)
6. Clarke, E.: Counterexample-guided abstraction refinement. In: proceedings of TIME (2003)
7. Clarke, E.M., Grumberg, O., Jha, S., Lu, Y., Veith, H.: Counterexample-Guided Abstraction Refinement. In: Emerson, E.A., Sistla, A.P. (eds.) CAV 2000. LNCS, vol. 1855, pp. 154–169. Springer, Heidelberg (2000)
8. Clarke, E.M., Lu, Y., Jha, S., Veith, H.: Tree-like counterexamples in model checking. In: proceedings of LICS, pp. 19–29 (2002)
9. Comon, H., Dauchet, M., Gilleron, R., Jacquemard, F., Lugiez, D., Tison, S., Tommasi, M.: Tree Automata Techniques and Applications (2002)
10. Das, S., Dill, D.L.: Counter-example based predicate discovery in predicate abstraction. In: Aagaard, M.D., O'Leary, J.W. (eds.) FMCAD 2002. LNCS, vol. 2517. Springer, Heidelberg (2002)
11. Dershowitz, N., Jouannaud, J.-P.: Rewrite Systems. In: Handbook of Theoretical Computer Science, vol. B, ch.6, pp. 244–320. Elsevier Science Publishers B. V (1990)
12. Feuillade, G., Genet, T., VietTriemTong, V.: Reachability analysis over term rewriting systems. Journal of Automated Reasonning 33(3-4) (2004)
13. Genet, T., Klay, F.: Rewriting for Cryptographic Protocol Verification. In: McAllester, D. (ed.) CADE 2000. LNCS, vol. 1831, pp. 271–290. Springer, Heidelberg (2000)
14. Genet, T., Tong, V.V.T.: Reachability Analysis of Term Rewriting Systems with timbuk. In: Nieuwenhuis, R., Voronkov, A. (eds.) LPAR 2001. LNCS (LNAI), vol. 2250, pp. 691–702. Springer, Heidelberg (2001)
15. Gilleron, R., Tison, S.: Regular tree languages and rewrite systems. Fundamenta Informatica 24(1/2), 157–174 (1995)
16. Gyenizse, P., Vágvölgyi, S.: Linear Generalized Semi-Monadic Rewrite Systems Effectively Preserve Recognizability. Theoretical Computer Science 194(1-2), 87–122 (1998)
17. Jacquemard, F.: Decidable approximations of term rewriting systems. In: Ganzinger, H. (ed.) RTA 1996. LNCS, vol. 1103, pp. 362–376. Springer, Heidelberg (1996)
18. Lakhnech, Y., Bensalem, S., Berezin, S., Owre, S.: Incremental Verification by Abstraction. In: Margaria, T., Yi, W. (eds.) ETAPS 2001 and TACAS 2001. LNCS, vol. 2031. Springer, Heidelberg (2001)
19. Monniaux, D.: Abstracting cryptographic protocols with tree automata. In: Cortesi, A., Filé, G. (eds.) SAS 1999. LNCS, vol. 1694. Springer, Heidelberg (1999)
20. Takai, T., Kaji, Y., Seki, H.: Right-linear finite-path overlapping term rewriting systems effectively preserve recognizability. In: Bachmair, L. (ed.) RTA 2000. LNCS, vol. 1833. Springer, Heidelberg (2000)

# A Needed Rewriting Strategy for Data-Structures with Pointers*

Rachid Echahed and Nicolas Peltier

LIG, CNRS
46, avenue Félix Viallet
38031 Grenoble Cedex, France
Rachid.Echahed@imag.fr, Nicolas.Peltier@imag.fr

**Abstract.** We propose a reduction strategy for systems of rewrite rules operating on term-graphs. These term-graphs are intended to encode pointer-based data-structures that are commonly used in programming, with cycles and sharing. We show that this reduction strategy is optimal w.r.t. a given dependency schema, which intuitively encodes the "interferences" among the nodes in the term-graph. We provide a new way of computing such dependency schemata.

## 1 Introduction

It is well-known that term-graph rewrite systems are non confluent in general, even if we restrict ourselves to standard orthogonal systems. A system as simple as $f(x) \to x, g(x) \to x$ is non confluent when applied on a cyclic term-graph $\alpha = f(g(\alpha))$ (two distinct normal forms exist: $\alpha = f(\alpha)$ or $\alpha = g(\alpha)$). Things get even worse if the rules are allowed to "physically" affect the term-graph by relabeling some of its nodes or by redirecting some edges occurring in it, because in this case two distinct functions may modify or access to the same nodes, making the result obviously dependent on the evaluation ordering. Assume for instance that we write a function *insert* that physically inserts an element at the end of a list. We can call this function on the list $\alpha = [1,2,3]$ and on the element *length*$(\alpha)$, where $\alpha$ points to the same (physical) list $[1,2,3]$. According to the order in which the two functions *length* and *insert* are evaluated, we can obtain either $[1,2,3,3]$ (if *length* is evaluated first, then the result inserted in the list) or $[1,2,3,4]$ if we proceed in a "lazy" way, by inserting the element before computing its value.

Obviously, functions of this kind, operating on pointer-based data-structures, are ubiquous in programming, because they allow the programmer to avoid duplication of information, hence to reduce the amount of needed memory. For instance, redirections are needed if one wants to define in-place algorithms for reversing or sorting a list. Another well-known example is the Shorr-Waite algorithm [11] which uses a link reversal technique to avoid the need for a stack during the exploration of a graph.

* This work has been partly funded by the project ARROWS of the French *Agence Nationale de la Recherche.*

A. Voronkov (Ed.): RTA 2008, LNCS 5117, pp. 63–78, 2008.

In [7] we proposed a solution to this problem (for constructor-based rewrite rules). The idea is to fix a specific evaluation ordering, by assuming a given *priority ordering* (denoted by $\preceq$) among the nodes. Then the "strict" rewriting strategy merely consists in reducing systematically the maximal node (according to $\preceq$). This strategy is deterministic (if the rewrite system is orthogonal) thus trivially confluent. Of course it is not satisfactory since it may be inefficient (it can be compared to purely imperative programs, or to a kind of "innermost" rewriting where the priority ordering replaces the subterm relation). It is only useful as a way to define the *semantics* of the term-graph rewrite systems (namely the expected normal forms), but should not be used in practice to compute the values. Therefore we defined more flexible reduction strategies that are allowed to reduce nodes that are non maximal w.r.t. $\succeq$ but only under some particular conditions which are strong enough to ensure that confluence is preserved.

In order to define these conditions, we introduced the notion of *dependency schema*, which is a set of relations specifying, in some sense, which nodes interfere with a given node in a term-graph (i.e., given a node $\alpha$ in a term-graph, which nodes affect $\alpha$ and which nodes depend on $\alpha$). We proved confluence of the flexible strategy for a class of rewrite systems. A dependency schema needs not to be provided by the user: we presented an algorithm to automatically compute a dependency schema having the desired properties.

In the present paper we extend our results on two aspects. First, we introduce a refined dependency schema, which is more powerful than the previous one, in the sense that it provides a more precise approximation of the "ideal" interference relation (which is non computable). Second, we propose a particular rewrite strategy consisting in computing only the nodes that are – in some sense – *relevant* for the considered term-graph. We show that this strategy is *normalizing* (i.e. all the normal forms can be reached). Moreover, for a restricted class of term-graph rewrite systems, this strategy is *optimal*, in the sense that it only rewrites nodes that are really needed to obtain the normal form (i.e. they are reduced in any derivation leading to the considered normal form). It should be mentioned that optimality is defined here w.r.t. a particular dependency schema (optimality in general is trivially impossible). The class of rules we consider is not too restrictive because every inductively/strongly sequential rewrite system (in the standard sense [9,10,8]) can be reduced to it.

Our work extends existing results for term rewrite systems [1] and graph rewrite systems [4,5] for which optimal rewriting strategies are known since a long time ago (for inductively/strongly sequential systems). We hope that it provides a good theoretical basis for defining a programming language based on term-graphs, that will offer similar features as rewrite-based languages such as Haskell [12] (namely efficient, lazy reduction strategies) and in the same time will be sufficiently expressive to allow the programmer to fully control the allocation of memory when (and if) needed.

The rest of the paper is organized as follows. Section 2 defines the notion of term-graphs and actions operating on them. Section 3 introduces the notion of *graph rewrite rules*, *dependency schemata*, and defines *rewriting relations* based on them. We prove that the flexible rewriting relation is *confluent* for orthogonal systems. The notions we use are slightly different (and much simpler than) from the ones in [7], but the previous results are essentially the same[1]. Section 4 provides an example of a dependency schema, distinct from (and strictly more powerful than) the one already proposed in [7]. Section 5 is the heart of the paper: it contains the main new result, namely a reduction strategy which is optimal for a particular class of rules (called *elementary*). Section 6 provides simple examples of applications.

The reader can refer to [3] for additional references and comparisons with existing works in the field.

## 2 Basic Definitions

### 2.1 Term-Graphs

We use the word "term-graph" to denote data-structures defined by a set of *nodes* connected by *labeled directed edges*. Edges are assumed to be unique, i.e. given a node $\alpha$ and a label $a$, there can be at most one edge starting from $\alpha$ and labeled by $a$.

Formally, we assume given a set of *nodes* $\mathcal{N}$, denoted by Greek letters, and a set of *features* $\mathcal{F}$, denoted by $a, b, \ldots$ Features may be seen as (partial) functions from $\mathcal{N}$ to $\mathcal{N}$, or as *edge labels*. $\mathcal{F}$ contains at least a special element $l$, which will be used to denote the label (or head symbol) of the node $\alpha$. We also assume given a total ordering $\succeq$ on $\mathcal{N}$, called the *priority ordering*, and specifying the order in which the nodes should be reduced (as explained in the Introduction).

Let $\mathcal{P}$ be a subset of $\mathcal{N}$, called the set of *predefined nodes*. We use predefined nodes mainly to encode function symbols: $\mathcal{P}$ contains a set of *functions* $\Sigma$ (denoted by $f, g, h, \ldots$) divided into two disjoint sets of symbols: a set $\mathcal{D}$ of *defined symbols* and a set $\mathcal{C}$ of *constructors*. Predefined nodes could also be used to denote built-in values such as reals or integers.

**Definition 1.** *A* term-graph $t$ *is defined by a set of* nodes $\mathcal{N}(t) \subseteq \mathcal{N} \setminus \mathcal{P}$ *and a function mapping each symbol* $a$ *in* $\mathcal{F}$ *to a partial function* $a_t$ *from* $\mathcal{N}(t)$ *to* $\mathcal{N}(t) \cup \mathcal{P}$. *If* $a_t(\alpha) = \beta$ *then we say that* $t$ contains an edge from $\alpha$ to $\beta$, labeled by $a$, *or that* the feature $a$ of $\alpha$ is $\beta$. *We assume that* $l_t(\alpha) \in \Sigma$ *if* $l_t(\alpha)$ *is defined.*

*A* rooted term-graph *is a term-graph associated with a distinguished node* $\alpha$, *called the* root *of* $t$ *and denoted by* $root(t)$.

For instance, a term $f(a, x)$ (where $x$ is a variable) is represented by a term-graph $t$ of root $\alpha$ s.t. $\mathcal{N}(t) = \{\alpha, \beta, \delta\}$, $l_t(\alpha) = f$, $l_t(\beta) = a$, $1_t(\alpha) = \beta$, $2_t(\alpha) = \delta$. The feature $i$ where $i \in \mathbb{N}$ denotes the $i$-th argument of a function.

[1] The differences are mainly due to the fact that in contrast to [7] we make no distinction between actions affecting the *name* of a node, and actions only redirecting its *edges*. Consequently, the reduction strategies are less flexible, but on the other hand this yields a much simpler and more intuitive framework. Moreover we also slightly modify some of the definitions in order to properly handle non orthogonal rewrite systems.

We denote by $dom(t)$ the set of nodes $\alpha$ s.t. $a_t(\alpha)$ is defined for at least one feature $a$ (the other nodes can be seen as variables).

Let $t, s$ be two term-graphs. We write $t \subseteq s$ iff $t$ is included in $s$ i.e. iff $\mathcal{N}(t) \subseteq \mathcal{N}(s)$ and $a_t(\alpha) = a_s(\alpha)$, for every node $\alpha$ s.t. $a_t(\alpha)$ is defined.

## 2.2 $\mathcal{N}$-Mappings

An $\mathcal{N}$*-mapping* $\sigma$ is a *total* function from $\mathcal{N}$ to $\mathcal{N}$ s.t. for every $\alpha \in \mathcal{P}$ and for every $\beta \in \mathcal{N}$, $\sigma(\beta) = \alpha \Leftrightarrow \alpha = \beta$ (predefined nodes are left unchanged and no non predefined node can be mapped to a predefined node).

An $\mathcal{N}$-mapping $\sigma$ is said to be *compatible* with a term-graph $t$ if for all $\alpha, \beta \in \mathcal{N}(t)$ s.t. $\sigma(\alpha) = \sigma(\beta)$ and for all $a \in \mathcal{F}$, if $a_t(\alpha)$ and $a_t(\beta)$ are defined, then $\sigma(a_t(\alpha)) = \sigma(a_t(\beta))$.

In this case, $\sigma(t)$ denotes the term-graph $s$ defined as follows: $\mathcal{N}(s) \stackrel{\text{def}}{=} \{\sigma(\alpha) \mid \alpha \in \mathcal{N}(t)\}$, and $a_s(\sigma(\alpha)) \stackrel{\text{def}}{=} \sigma(a_t(\alpha))$.

An $\mathcal{N}$-mapping $\sigma$ is said to be a *renaming* if it is injective and $\sigma(\alpha) \succ \sigma(\beta)$ for any $\alpha \succ \beta$. Note that by definition a renaming is compatible with any term-graph $t$.

An $\mathcal{N}$*-relation* $\Delta$ is a relation on the nodes of a term-graph which is independent from the *names* of the nodes. Formally, it is a function mapping every term-graph $t$ to a relation $\Delta_t$ on the nodes of $t$ s.t. for every renaming $\eta$ and for every pair of nodes $\alpha, \beta$ occurring in $t$ we have $\alpha \; \Delta_t \; \beta$ iff $\eta(\alpha) \; \Delta_{\eta(t)} \; \eta(\beta)$.

One of the simplest examples of an $\mathcal{N}$-relation is the relation $\geq_t$ defined as the smallest reflexive and transitive relation s.t. for all term-graphs $t$, for all $a \in \mathcal{F}$ and for all $\alpha \in dom(a_t)$, we have $\alpha \geq_t a_t(\alpha)$ ($\alpha \geq_t \beta$ iff there is a path from $\alpha$ to $\beta$ in $t$).

## 2.3 Actions

The definitions of actions and rewrite rules are close to the ones of [6,7], but slightly simpler. An *action* has one of the following forms:

- an **edge redirection/creation** $\alpha \gg_a \beta$ where $\alpha, \beta$ are nodes, $a$ is a feature and $\alpha \notin \mathcal{P}$. This means that the value of $a(\alpha)$ is changed to $\beta$. This can be viewed as an edge redirection: the edge starting from $\alpha$ and labeled by $a$ is redirected to point to $\beta$. The edge and nodes are created if they do not exist. If $a = l$ then we assume that $\beta \in \Sigma$.
- a **global redirection** $\alpha \gg \beta$ where $\alpha$ and $\beta$ are nodes and $\alpha \notin \mathcal{P}$. This means that all edges pointing to $\alpha$ are redirected to $\beta$.

Note that predefined nodes cannot be redirected. The result of applying an action $\epsilon$ to a term-graph $t$ is denoted by $\epsilon[t]$ and is defined as the term-graph $s$ s.t.:

- If $\epsilon = \alpha \gg_a \beta$ then $\mathcal{N}(s) \stackrel{\text{def}}{=} \mathcal{N}(t) \cup \{\alpha, \beta\}$, $a_s(\alpha) \stackrel{\text{def}}{=} \beta$ and for all features $b$ and all nodes $\gamma$ we have $b_s(\gamma) \stackrel{\text{def}}{=} b_t(\gamma)$ iff $a \neq b$ or $\gamma \neq \alpha$.
- If $\epsilon = \alpha \gg \beta$ then $\mathcal{N}(s) \stackrel{\text{def}}{=} \mathcal{N}(t) \cup \{\alpha, \beta\}$ and for all features $a$ and all nodes $\gamma$, $a_s(\gamma) \stackrel{\text{def}}{=} \beta$ if $a_t(\gamma) = \alpha$ and $a_s(\gamma) \stackrel{\text{def}}{=} a_t(\gamma)$ otherwise.

If $\varsigma$ is a finite sequence of actions, then $\varsigma[t]$ is defined inductively as follows: $\varsigma[t] \stackrel{\text{def}}{=} t$ if $\varsigma$ is empty and $(\epsilon;\varsigma)[t] \stackrel{\text{def}}{=} \varsigma[\epsilon[t]]$ (where $\epsilon$ is an action and ; denotes the concatenation operator). $\mathcal{N}(\varsigma)$ denotes the set of nodes occurring in $\varsigma$. $dom(\varsigma)$ denotes the set of nodes $\alpha$ s.t. $\varsigma$ contains an action of the form $\alpha \gg \beta$ or $\alpha \gg_a \beta$.

### 2.4 A Linear Notation for Actions and Term-Graphs

For the sake of conciseness and readability, we introduce another notation for denoting sequences of actions. They will be denoted as *terms with labels.*

The term $\alpha{:}f(a_1 \Rightarrow \beta_1{:}t_1,\ldots,a_n \Rightarrow \beta_n{:}t_n)$ will be used as an abbreviation for denoting the sequence of actions: $(\alpha \gg_l f);(\alpha \gg_{a_1} \beta_1);\ldots;(\alpha \gg_{a_n} \beta_n);\tau_1;\ldots;\tau_n$ where $\tau_1,\ldots,\tau_n$ are the sequences of actions corresponding to the terms $\beta_1{:}t_1,\ldots,\beta_n{:}t_n$, respectively. For instance, $\alpha{:}cons(car \Rightarrow \beta{:}0, cdr \Rightarrow \gamma{:}nil)$ denotes the sequence: $\alpha \gg_l cons; \alpha \gg_{car} \beta; \alpha \gg_{cdr} \gamma; \beta \gg_l 0; \gamma \gg_l nil$. The nodes $\alpha,\beta_1,\ldots,\beta_n$ can be left unspecified, and in this case they are simply replaced by arbitrary nodes not occurring elsewhere: for instance $cons(car \Rightarrow 0)$ denotes a sequence of the form $\alpha \gg_l cons; \alpha \gg_{car} \beta; \beta \gg_l 0$ where $\alpha,\beta$ are arbitrarily chosen nodes. $f(t_1,\ldots,t_n)$ is syntactic sugar for $f(1 \Rightarrow t_1,\ldots,n \Rightarrow t_n)$ (where $1,\ldots,n \in \mathcal{F}$).

The same notation can be used to denote term-graphs. Indeed, a term-graph $t$ can be described by giving a sequence of actions $\varsigma$ s.t. $\varsigma[\emptyset]=t$, where $\emptyset$ denotes the empty graph (no nodes and no edges). For instance, $\alpha{:}a$, which denotes the action $\alpha \gg_l a$, will also denote a term-graph $t$ s.t. $\mathcal{N}(t)=\{\alpha\}$ where $l_t(\alpha)=a$ and $b_t(\alpha)$ is undefined if $b \neq l$. The term $\alpha$ denotes the term-graph reduced to the unique node $\alpha$, with no edges.

The operator ";", used to denote composition of actions, is also used to denote a union of term-graphs, for instance $\alpha{:}f(\delta);\beta{:}g(\delta)$ denotes a term-graph $t$ of nodes $\alpha,\beta,\delta$, s.t. $l_t(\alpha)=f, l_t(\beta)=g$, $1_t(\alpha)=1_t(\beta)=\delta$. By convention, the root of the term-graph is the first mentioned node. For instance $root(t)=\alpha$.

$\beta{:}g(\delta);\alpha{:}f(\delta)$ and $\delta;\beta{:}g(\delta);\alpha{:}f(\delta)$ both denote the same term-graph, but with the roots $\beta$ and $\delta$ respectively (note that the term-graphs may contain nodes that are not reachable from the root).

## 3 Term-Graph Rewriting

**Definition 2.** *(Rewrite Rule) A* term-graph rewrite rule *is an expression of the form $L \rightarrow R \mid \phi$ where $L$ is a rooted term-graph, $R$ is a sequence of actions, and $\phi$ is a conjunction of disequations between nodes $\bigwedge_{i=1}^{n} \alpha_i \neq \beta_i$.*

*A* graph rewrite system *(GRS for short) is a set of rewrite rules.*

A rule is said to be *admissible* iff the following conditions hold:

- For every node $\alpha$ occurring in $L$, we have $root(L) \geq_L \alpha$. Any node occurring in the left-hand side must be reachable from the root[2].

[2] This is an important condition since otherwise the same rule could be applied in several different ways at the *same* node, which would make the system trivially non confluent in general.

- We have $l_L(root(L)) \in \mathcal{D}$ and for every node $\alpha \neq root(L)$ if $l_L(\alpha)$ is defined then $l_L(\alpha) \in \mathcal{C}$. The root is the only node that is labeled by a non constructor symbol. This condition is usual in constructor-based rewrite systems.
- For every action $\alpha \gg_l f$ in $R$ s.t. $f \in \mathcal{D}$, either $\alpha = root(L)$ or $\alpha$ does not occur in $L$. Moreover if $\alpha \gg \beta \in R$ then $\beta \notin \Sigma$. Only the created (new) nodes are allowed to be labeled by a defined symbol[3].

**In the following we always assume that the rules are admissible**
In practice, rather than writing the right-hand side as a sequence of actions, we often prefer to use the linear notation introduced in Section 2.4 which is clearer. For instance, we shall write $\alpha{:}a \rightarrow \alpha{:}f(a)$, instead of $\alpha{:}a \rightarrow \alpha \gg_l f; \alpha \gg_1 \beta; \beta \gg_l a$. If this notation is used, then we always implicitly assume that the root of the left-hand side is redirected to the root of the term-graph denoted by the right-hand side. For instance, $\alpha{:}f(a) \rightarrow \beta{:}g(b)$ denotes the rule: $\alpha{:}f(a) \rightarrow \alpha \gg \beta; \beta \gg_l g; \beta \gg_1 \gamma; \gamma \gg_l b$ (the action $\alpha \gg \beta$ is added at the beginning of the sequence). Using these conventions any term rewrite rule (in the usual sense) can be seen as a term-graph rewrite rule. These conventions (as the ones in Section 2.4) are only introduced as syntactic sugar to make the notations and examples clearer and easier to understand. They do not affect the semantics.

A substitution ($\mathcal{N}$-mapping) $\sigma$ is said to be a *solution* of a conjunction of disequations $\phi = \bigwedge_{i=1}^{n}(\alpha_i \neq \beta_i)$ iff for every $i \in [1..n]$, $\sigma(\alpha_i) \neq \sigma(\beta_i)$. The set of solutions of $\phi$ is denoted by $sol(\phi)$.

**Definition 3.** *Let $\rho : L \rightarrow R \mid \phi$ be a rule. A $\rho$-*matcher for *a term-graph $t$ (*at *a node $\alpha \in \mathcal{N}(t)$) is an $\mathcal{N}$-mapping $\sigma$ compatible with $L$ satisfying the following conditions.*

1. *$\sigma(L) \subseteq t$, i.e. $\sigma(L)$ must be a subgraph of $t$.*
2. *$\sigma \in sol(\phi)$ and $\alpha = \sigma(root(L))$.*
3. *Let $N$ be the set of nodes occurring in $R$ but not in $L$. $N$ corresponds to the nodes that are* created *by the rewrite rule. $\sigma$ maps the nodes in $N$ to pairwise distinct nodes not occurring in $t$ s.t.:*
   - *If $\beta, \gamma$ are two nodes in $N$ s.t. $\beta \succeq \gamma$ then $\sigma(\beta) \succeq \sigma(\gamma)$. This means that the newly created nodes should be ordered as specified in the rewrite rule.*
   - *For every node $\beta$ occurring in $t$, and for every node $\gamma$ in $N$, $\beta \prec \sigma(\gamma)$ iff $\beta \preceq \alpha$. This means that the newly created nodes inherit the priority of the parent node $\alpha$.*

$\rho$-matchers can be easily computed using standard matching algorithms. If $\sigma$ is a $\rho$-matcher for $t$, where $\rho = L \rightarrow R \mid \phi$, then we denote by $\rho^{\sigma}[t]$ the term-graph $\sigma(R)[t]$ obtained by applying the sequence of actions $\sigma(R)$ on $t$.

Let $\mathcal{R}$ be a GRS. Let $t$ be a term-graph. We define the following relations between nodes (implicitly depending on $\mathcal{R}$):

[3] This condition is not really restrictive since one can easily redirect an existing node $\alpha$ to a new node $\beta$ before relabeling it.

- $\alpha \rhd_t \beta$ iff there is a rule $\rho : L \to R \mid \phi \in \mathcal{R}$ and a $\rho$-matcher $\sigma$ at $\alpha$ s.t. $\beta \in \sigma(L)$. This means that a rule that **depends** on the node $\beta$ is applicable on $\alpha$
- $\alpha \gg_t \beta$ iff there is a rule $\rho : L \to R \mid \phi \in \mathcal{R}$ and a $\rho$-matcher $\sigma$ at $\alpha$ s.t. $\beta \in dom(\sigma(R))$. This means that a rule **affecting** $\beta$ can be applied on $\alpha$.

*Example 1.* Assume that $\mathcal{R} = \{f(\alpha{:}0, \beta{:}s(\delta)) \to \beta{:}2\}$. Let $t = \lambda_1{:}f(\lambda_2{:}0, \lambda_3{:}s(\lambda_4{:}s(\lambda_5{:}0)))$. Then we have $\lambda_1 \rhd_t \lambda_1, \lambda_2, \lambda_3, \lambda_4$ and $\lambda_2 \not\rhd_t \lambda_5$ (since $\lambda_5$ does not occur in the part of $t$ that is in the image of the left-hand side of the rule in $\mathcal{R}$). We have $\lambda_1 \gg_t \lambda_1, \lambda_3$ and $\lambda_1 \not\gg_t \lambda_2, \lambda_4, \lambda_5$.

**Definition 4.** *Let $\mathcal{R}$ be a set of rewrite rules. An $\mathcal{R}$-*dependency schema *is a pair $\xi = (\gg^*, \rhd^*)$ of $\mathcal{N}$-relations s.t. for every term-graph $t$, $\gg_t^*, \rhd_t^*$ contain $\gg_t$ and $\rhd_t$ respectively.*

*Let $\rho = L \to R \mid \phi$ be a rule and let $\sigma$ be a $\rho$-matcher for a term-graph $t$ at $\alpha$. We write $\sigma \bowtie_t^\xi \beta$ if $\alpha \prec \beta$ and there exists a node $\gamma$ s.t. either $\gamma \in dom(\sigma(R))$ and $\beta \rhd_t^* \gamma$ or $\gamma \in \mathcal{N}(\sigma(L))$ and $\beta \gg_t^* \gamma$.*

Informally, the intended meaning of $\alpha \gg_t^* \gamma$ is that the reduction of $\alpha$ **may affect** the node $\gamma$, possibly after some reduction steps. $\alpha \rhd_t^* \gamma$ means that the value of the node $\alpha$ **may depend** on the value of $\gamma$. As we shall see, for both relations, reduction should be taken into account (see Definition 5). For instance, if *length* is the function defined as usual on lists (see the rules $\#_1$ and $\#_2$ below), and $t$ is the term $\alpha{:}length(cons(0, cons(1, cons(2, \beta))))$, then we should have $\alpha \rhd_t^* \beta$, although $\alpha \not\rhd_t \beta$ (the value of $\alpha$ depends on $\beta$, but only after some reduction steps).

$$\#_1 : length(cons(\alpha, \beta)) \to s(length(\beta)) \quad \#_2 : length(nil) \to 0$$

$\sigma \bowtie_t^\xi \beta$ expresses the fact that there is a potential "conflict" between the rule $\rho$ corresponding to $\sigma$ and the node $\beta$ (according to the considered dependency schema): either $\rho$ redirects a node $\gamma$ on which $\beta$ (possibly) depends, or $\rho$ uses a node $\gamma$ which may be affected by $\beta$. From an intuitive point of view, $\gamma$ may be seen as a resource that is shared between $\rho$ and $\beta$.

Let $\mathcal{R}$ be a rewrite system and let $\xi$ be an $\mathcal{R}$-dependency schema. A matcher $\sigma$ for $t$ at a node $\alpha$ is said to be *eligible* if there is no node $\beta$ in $t$ s.t. $\sigma \bowtie_t^\xi \beta$. We define the three following rewriting relations.

- $t \to_{\mathcal{R}} s$ iff there exist a rule $\rho \in \mathcal{R}$ and a $\rho$-matcher $\sigma$ for $t$ s.t. $s = \rho^\sigma[t]$. This is the basic rewriting relation, close in spirit to the one used for terms. It is non confluent even if $\mathcal{R}$ is orthogonal [3] and even if all nodes only affect themselves.
- $t \xrightarrow{\succeq}_{\mathcal{R}} s$ iff there exist a rule $\rho \in \mathcal{R}$ and a $\rho$-matcher $\sigma$ for $t$ at a node $\alpha$ s.t. $s = \rho^\sigma[t]$ and if for every node $\beta$ in $t$, we have $\alpha \succeq \beta$ or $l_t(\beta) \in \mathcal{C}$. This means that the rules are applied only on maximal reducible nodes (according to the ordering $\succeq$). This relation is called *strict rewriting.* It is deterministic and one can view strict rewrite systems as purely imperative programs, in the sense that the order in which the actions are performed is entirely specified.

- $t \xrightarrow{\xi}_{\mathcal{R}} s$ iff there exist a rule $\rho \in \mathcal{R}$ and an *eligible* $\rho$-matcher $\sigma$ for $t$ at a node $\alpha$ s.t. $s = \rho^\sigma[t]$. $\xrightarrow{\xi}_{\mathcal{R}}$ is a rewriting relation that is more flexible than $\xrightarrow{\succ}_{\mathcal{R}}$ but as we shall see is also sufficiently strong to preserve confluence of orthogonal systems. The basic idea is that a rule $\rho$ may be applied on a non-maximal node $\alpha$ only if $\rho$ does not interfere with the reduction of the nodes $\beta \succ \alpha$.

Obviously, we have $\xrightarrow{\succ}_{\mathcal{R}} \subseteq \xrightarrow{\xi}_{\mathcal{R}} \subseteq \rightarrow_{\mathcal{R}}$. $\rightarrow_{\mathcal{R}}$ depends only on $\mathcal{R}$, $\xrightarrow{\succ}_{\mathcal{R}}$ depends on $\mathcal{R}$ and $\succ$. $\xrightarrow{\xi}_{\mathcal{R}}$ depends on $\mathcal{R}, \succ$ and $\xi$. If $\alpha \gg_t^* \beta$ and $\alpha \rhd_t^* \beta$ for every pair of nodes $(\alpha, \beta)$ then $\xrightarrow{\xi}_{\mathcal{R}}$ coincides with $\xrightarrow{\succ}_{\mathcal{R}}$.

The following definition states additional semantic conditions on $\xi$ that are needed to ensure confluence.

Let $t, s$ be two term-graphs s.t. $t \xrightarrow{\xi}_{\mathcal{R}} s$. Let $\alpha$ be a node in $s$ and let $\beta$ be the node on which the rule is applied. We denote by $\alpha^-_{t \to s}$ the node defined as follows: $\alpha^-_{t \to s} \stackrel{\text{def}}{=} \alpha$ if $\alpha \in \mathcal{N}(t)$ and $\alpha^-_{t \to s} \stackrel{\text{def}}{=} \beta$ otherwise. $\alpha^-_{t \to s}$ denotes the *ancestor* of the node $\alpha$.

**Definition 5.** *An $\mathcal{N}$-relation $\Delta$ is said to be* invariant *for an $\mathcal{R}$-dependency schema $\xi$ if the following holds: if $\sigma$ is an eligible $\rho$-matcher for $t$, $s = \rho^\sigma[t]$ and $\beta \, \Delta_s \, \gamma$ then $\beta^-_{t \to s} \, \Delta_t \, \gamma^-_{t \to s}$. An $\mathcal{R}$-dependency schema $\xi = (\gg^*, \rhd^*)$ is* invariant *if $\gg^*, \rhd^*$ are invariant for $\xi$.*

Of course, the above property cannot be checked automatically in general (see Section 4 for syntactic criteria). The intuitive idea is that one cannot discover new interferences during the derivation: if $\alpha, \beta$ are related at some point in the derivation, then either $\alpha, \beta$ or their ancestors were already in relation before.

**Examples**

We provide some simple examples of rewriting rules (defined using the above linear relation for the sake of conciseness and readability):

**In situ append:**

$a_1 : \mathit{append}(\alpha{:}\mathit{nil}, \beta) \rightarrow \beta; \alpha \gg \beta$ % $\alpha$ is redirected to $\beta$

$a_2 : \mathit{append}(\alpha{:}\mathit{cons}(\beta, \delta), \gamma) \rightarrow \alpha; \mathit{append}(\delta, \gamma)$ % Apply *append* on the tail

**In situ increment of all the elements of a list:**

$i_1 : \mathit{inc}(\alpha{:}\mathit{nil}) \rightarrow \alpha$ $\quad i_2 : \mathit{inc}(\alpha{:}\mathit{cons}(\beta, \delta)) \rightarrow \alpha; \alpha \gg_1 s(\beta); \mathit{inc}(\delta)$

**In situ list reversal:**

$r_1 : \mathit{rev}(\alpha) \rightarrow \beta{:}\mathit{rev}'(\alpha, \mathit{nil}); \alpha \gg \beta$ $\quad r_2 : \mathit{rev}'(\alpha{:}\mathit{nil}, \beta) \rightarrow \beta$

$r_3 : \mathit{rev}'(\alpha{:}\mathit{cons}(\beta, \delta), \gamma) \rightarrow \mathit{rev}'(\delta, \alpha); \alpha \gg_2 \gamma$

**Check whether an element occurs in a (possibly circular) list**

$f_1 : \mathit{find}(\alpha, \beta) \rightarrow \lambda_1{:}\mathit{find}'(\alpha, \beta); \lambda_2{:}\mathit{clean}(\alpha)$

$f_2 : \mathit{find}'(\alpha, \beta{:}\mathit{cons}(\alpha', \beta')) \rightarrow \mathit{find}'(\alpha, \beta'); \beta \gg_l \mathit{mark} \mid \alpha \neq \alpha'$

$f_3 : \mathit{find}'(\alpha, \mathit{nil}) \rightarrow \mathit{false}$ $\quad f_4 : \mathit{find}'(\alpha, \mathit{cons}(\alpha, \beta)) \rightarrow \mathit{true}$

$f_5 : \mathit{find}'(\alpha, \mathit{mark}(\alpha', \beta)) \rightarrow \mathit{false}$ $\quad f_6 : \mathit{clean}(\alpha{:}\mathit{nil}) \rightarrow \alpha$

$f_7 : \mathit{clean}(\alpha{:}\mathit{cons}(\beta, \delta)) \rightarrow \alpha$ $\quad f_8 : \mathit{clean}(\alpha{:}\mathit{mark}(\beta, \delta)) \rightarrow \mathit{clean}(\delta); \alpha \gg_l \mathit{cons}$

The function $\mathit{find}'(\alpha, \beta)$ explores the list until $\alpha$ is found. The nodes are marked in order to avoid looping. The function *clean* removes the marks. In this example, the ordering of the nodes in the right-hand side is crucial. We assume that $\lambda_1 \succ \lambda_2$. Here are some examples of reductions ($\lambda_1, \lambda_2, \ldots$ are new nodes created during the derivation):

$$append(\alpha{:}cons(\beta, \delta{:}nil), \alpha) \to_{a_1} \alpha{:}cons(\beta, \delta{:}nil); \lambda_1{:}append(\delta, \alpha) \to_{a_2} \alpha{:}cons(\beta, \alpha)$$

$$\begin{aligned} rev(\alpha{:}cons(\beta_1, \delta{:}cons(\beta_2, nil))) &\to_{r_1} rev'(\alpha{:}cons(\beta_1, \delta{:}cons(\beta_2, nil)), nil) \\ &\to_{r_3} rev'(\delta{:}cons(\beta_2, nil), \alpha{:}cons(\beta_1, nil)) \\ &\to_{r_3} rev'(nil, \delta{:}cons(\beta_2, \alpha{:}cons(\beta_1, nil))) \\ &\to_{r_2} \delta{:}cons(\beta_2, \alpha{:}cons(\beta_1, nil)) \end{aligned}$$

**Confluence** is an important property from a programming point of view, because it ensures that any object has a unique normal form (thus the defined symbols encode functions).

We write $t \equiv s$ iff there exists a renaming $\eta$ for $t$ s.t. $\eta(t) = s$.

**Definition 6.** *A rewrite system is said to be* weak orthogonal *if for every pair of distinct rules* $\rho : L \to R \mid \phi$ *and* $\pi : L' \to R' \mid \psi$ *and for every* $\mathcal{N}$*-mapping* $\sigma \in sol(\phi) \cap sol(\psi)$ *compatible with* $L, L'$ *s.t.* $\sigma(root(L)) = \sigma(root(L'))$*, we have* $\sigma(R) = \eta(\sigma(R'))$ *for some renaming* $\eta$ *s.t.* $\eta(\alpha) = \alpha$ *for every node* $\alpha$ *occurring in* $\sigma(L)$ *or* $\sigma(L')$.

**Theorem 1.** *(Confluence of Weak Orthogonal Systems Modulo Renaming) Let* $\mathcal{R}$ *be a weak orthogonal rewrite system. Let* $\xi$ *be an invariant* $\mathcal{R}$*-dependency schema.* $\overset{\xi}{\to}_{\mathcal{R}}{}^{*} \cup \equiv$ *is confluent. Thus, if* $t \overset{\xi}{\to}{}^{*}_{\mathcal{R}} s$*,* $t \overset{\succ}{\to}{}^{*}_{\mathcal{R}} s'$ *and if* $s, s'$ *are irreducible (w.r.t.* $\overset{\xi}{\to}_{\mathcal{R}}$*), then* $s \equiv s'$.

## 4 An Example of an Invariant Dependency Schema

From a practical point of view, we want the relation $\overset{\xi}{\to}_{\mathcal{R}}$ to be as weak as possible, thus the relations $\gg^*$ and $\rhd^*$ must be as strong as possible. From a purely theoretical point of view one could take the smallest invariant relations, but of course these "ideal" relations are not computable. For instance, let $\mathcal{R}$ be a GRS containing the rules $f(\alpha{:}0, 0) \to \alpha{:}1$ and $f(\alpha{:}0, 1) \to \alpha$, and let $t = \beta{:}f(\alpha{:}0, s)$ where $s$ is an arbitrary term-graph. If $\gg^*$ denotes the smallest invariant dependency schema, then $\beta \gg^*_t \alpha$ iff $s \overset{\succ}{\to}_{\mathcal{R}} 0$ (since the first rule is the only one that affects $\alpha$). Since $s$ is arbitrary, $\gg^*$ is non computable.

In this section, we give an example of a tractable invariant dependency schema. The provided relations are significantly stronger than the ones we proposed in [3] or in [7]. For this purpose, we need to introduce additional definitions and notations.

Let $f$ be a function symbol. An $f$*-rule* is a rule $L \to R \mid \phi$ s.t. $l_L(root(L)) = f$.

Let $\rho = L \to R \mid \phi$ be a rule. $\geq_\rho$ denotes the smallest transitive and reflexive relation containing $\geq_L$ s.t. if $\alpha \gg_a \beta \in R$ or $\alpha \gg \beta \in R$ then $\alpha \geq_\rho \beta$. Intuitively, $\alpha \geq_\rho \beta$ states that an application of the rule $\rho$ can create a path from $\alpha$ to $\beta$.

$\geq_{\mathcal{R}}$ denotes the smallest reflexive and transitive relation s.t. $(f, a) \geq_{\mathcal{R}} (g, b)$ if $\rho = L \to R \mid \phi$ is an $f$-rule and if there exist three nodes $\alpha, \beta, \delta$ s.t. $\alpha \gg_l g$

and $\alpha \gg_b \delta$ are actions in $R$, $\delta \geq_\rho \beta$ and $a_L(root(L)) \geq_L \beta$. Intuitively, this means that the function $f$ calls another function $g$ and that, moreover, a node $\beta$ initially reachable from feature $a$ of the node labeled by $f$ may become reachable from the feature $b$ of $g$.

A rule $L \to R \mid \phi$ is said to *produce a side-effect on a feature* $a$ if $dom(R)$ contains a node $\alpha$ occurring in $L$ but distinct from $root(L)$, s.t. $a_L(root(L)) \geq_L \alpha$.

An $\mathcal{F}$-family of sets of function symbols $(E_a)_{a\in\mathcal{F}}$ is said to be $\geq_\mathcal{R}$*-closed* iff for all $g \in E_b$ and for all $f$ s.t. $(f,a) \geq_\mathcal{R} (g,b)$, we have $f \in E_a$.

$\mathcal{SE}(\mathcal{R})$ is the smallest $\geq_\mathcal{R}$-closed $\mathcal{F}$-family of function symbols s.t. if there exists an $f$-rule producing a side-effect on $a$ then $f \in \mathcal{SE}(\mathcal{R})_a$.

**Definition 7.** *We denote by $\chi$ the dependency schema defined as follows.*
$\chi = (\gg^*, \rhd^*)$, *where $\gg_t^*, \rhd_t^*$ and $\geq_t^a$ are the smallest reflexive relations s.t.:*

- *If $a_t(\alpha) \geq_t \beta$ and $l_t(\alpha) \in \mathcal{D}$ then $\alpha \geq_t^a \beta$.*
- *If $\alpha \geq_t^a \beta$, $\gamma \succ \alpha$, $\gamma \gg_t^* \beta$ and $\gamma \rhd_t^* \delta$ then $\alpha \geq_t^a \delta$.*
- *If $\alpha \geq_t^a \beta$ for some feature $a$, then $\alpha \rhd_t^* \beta$.*
- *If $l_t(\alpha) \in \mathcal{SE}(\mathcal{R})_a$ and $\alpha \geq_t^a \beta$ then $\alpha \gg_t^* \beta$.*

Intuitively $\alpha \geq_t^a \beta$ expresses the fact that $\alpha$ is labeled by a defined symbol and that $\beta$ is reachable (or may become reachable after some rewriting steps) from the feature $a$ of the node $\alpha$. The first item corresponds straightforwardly to this definition. The second item is slightly more complicated and states that $\beta$ may become reachable (at some point) if there exists a node $\gamma$ that both affects a node reachable from $\alpha$ (from feature $a$) and depends on $\beta$. Indeed, reducing the node $\gamma$ may create a path from $\alpha$ to $\beta$.

Using this relation, $\gg^*$ and $\rhd^*$ are easy to define. The third item states that $\alpha$ depends on a node $\beta$ only if a path exists (or may be created) from a feature of $\alpha$ to $\beta$. The fourth item states that $\alpha$ affects $\beta$ only if $\alpha$ is labeled by a defined symbol performing a side effect on a feature $a$ and if $\beta$ is reachable (or may become reachable) from the feature $a$ of $\alpha$.

*Example 2.* Let $\mathcal{R} = \{f(\alpha,\beta) \to g(\beta,\alpha), g(s(\alpha),s(\beta)) \to g(\alpha,\beta), g(\alpha{:}0,\beta) \to \beta \gg_1 0\}$. We have $(f,1) \geq_\mathcal{R} (g,2)$ and $(f,2) \geq_\mathcal{R} (g,1)$. $g$ performs a side effect on feature 2 (third rule) thus $g \in \mathcal{SE}(\mathcal{R})_2$ and $f \in \mathcal{SE}(\mathcal{R})_1$.

Let $t = \alpha{:}f(s(s(\beta)), s(s(\beta')))$. We have $\alpha \gg_t^* \beta$ and $\alpha \not\gg_t^* \beta'$ (but $\alpha \rhd_t^* \beta'$).

Given a term-graph $t$, it is easy to compute the relations $\gg_t^*$ and $\rhd_t^*$ in $\chi$ (in polynomial time w.r.t. the size of $t$).

**Theorem 2.** *$\chi$ is an invariant dependency schema.*

## 5 Needed Rewriting

This section is the most important part of the paper. We provide a reduction strategy which is optimal (for a particular class of GRS) in the sense that the only nodes that are reduced are *needed*, i.e. are reduced in every derivation to normal form[4]. Moreover, the length of the derivation is also minimal. This last point makes an interesting difference with term rewrite systems in which this property does not hold[5]. In order to formally define these notions, we need to introduce additional definitions and notations.

Let $\mathcal{R}$ be a GRS, let $\xi$ be an $\mathcal{R}$-dependency schema. If $t$ is a term-graph and $\alpha \in \mathcal{N}(t)$, we denote by $[t]^{\xi}_{\alpha}$ the maximal term-graph included in $t$ s.t. for every node $\beta \in dom([t]^{\xi}_{\alpha})$ and for every $\gamma \in \mathcal{N}(t)$ s.t. $\gamma \succ \alpha$ or $l_t(\alpha) \notin \mathcal{D}$, we have $\gamma \not\gg^*_t \beta$.

Intuitively, $[t]^{\xi}_{\alpha}$ denotes the part of the term-graph $t$ that will surely not change until $\alpha$ is reduced. Thus if there exists a node $\gamma$ in $t$ s.t. $\gamma \gg^*_t \beta$ (i.e. "$\gamma$ possibly affects $\beta$") for some node $\beta \in dom(s)$ then $\alpha$ must be reduced before $\gamma$, which is possible only if $\alpha$ is labeled by a defined symbol and $\alpha \succeq \gamma$. The definition is slightly different for constructor nodes (second subcondition) since in this case $\alpha$ is never reduced.

*Example 3.* Let $f, g$ be defined symbols and let $s, s', 0$ be constructors. Assume that $f$ performs side effects on feature 1 and that $g$ does not perform any side effect.

Let $t = \alpha{:}s(\alpha'{:}g(s'(0)), \beta{:}s'(\delta{:}s'(0))); \gamma{:}g(\beta); \lambda{:}f(\delta)$. Then $[t]^{\chi}_{\alpha} = \alpha{:}s(\alpha', s'(\delta)); s'(0)$. Notice that the argument of $\alpha'$ remains in the term-graph (as a disconnected subgraph $s'(0)$) but that the edges starting from $\alpha'$ are removed. Indeed, we have $\alpha' \gg^*_t \alpha'$ and $\lambda \gg^*_t \delta$ (but $\gamma \not\gg^*_t \beta$).

If $t$ is a term-graph, we denote by $\hat{t}$ the term-graph obtained from $t$ by removing all the nodes that are not reachable from the root. A term-graph $t$ is said to be a *value* if $\hat{t}$ contains no defined symbols **and** if $\hat{t} \subseteq [t]^{\xi}_{root(t)}$. This means that the part of the term-graph that is reachable from the root cannot be affected by a derivation (this implies in particular that $\hat{t}$ contains no defined symbol, *but this condition is not sufficient*, see Example 4).

**Definition 8.** *Let $\mathcal{R}$ be a GRS and let $\xi$ be an invariant $\mathcal{R}$-dependency schema. A node $\alpha$ is said to be $\xi$-*needed *for a term-graph $t_0$ if for every sequence $t_0, \ldots, t_n$ s.t. $t_n$ is a value and s.t. for every $i \in [1..n]$ $t_i = \rho_i^{\sigma_i}[t_{i-1}]$ for some rule $\rho_i \in \mathcal{R}$ and some eligible $\rho_i$-matcher $\sigma_i$ at a node $\beta_i$, there exists $i \in [1..n]$ s.t. $\beta_i = \alpha$.*

It should be emphasized that neededness is relative to a given dependency schema. Defining a needed rewriting strategy for the *smallest* possible dependency schema $\xi$ is obviously impossible, since $\xi$ is non computable as soon as side effects are possible as explained in Section 4 (even if we strongly restrict the class of GRS).

---

[4] Note that this notion of neededness implicitly assumes that rooted term-graphs are considered and that one is only interested by the part of term-graph that is reachable from the root. If the term-graphs are not rooted then obviously *any node is needed* (since at least it defines its own value).

[5] This is due to the fact that term-graph rewrite rules do not duplicate the terms. For instance, if we apply the rule $f(x) \to g(x, x)$ on the term $f(t)$, we obtain a term like $g(\alpha{:}t, \alpha)$. $t$ is not duplicated hence it will not have to be reduced twice. It should be noticed that collapsing (i.e. the merging of identical subgraphs) is *incorrect* in our context.

The definition of the reduction strategy is more complicated than in the usual case, because one has to handle the dependencies between the nodes. In the standard case, the only node that can affect a given node $\alpha$ is $\alpha$ itself (no side-effect). This is not the case here. Thus, a node may be needed even if it is non reachable from the root.

*Example 4.* Let $\mathcal{R} \stackrel{\text{def}}{=} \{\rho : g(\alpha{:}0) \rightarrow \alpha{:}1\}$ and $t = \delta{:}c(\beta{:}0); \gamma{:}g(\beta)$, where $\delta$ is the root ($c$ is a constructor). $\gamma$ is non reachable from $\delta$, but it is obviously needed for computing the normal form. The only applicable rule is $\rho$, yielding: $\delta{:}c(\beta{:}1)$.

Even if $\xi$ is computable, defining a needed strategy is not easy as illustrated by the following:

*Example 5.* $\mathcal{R} \stackrel{\text{def}}{=} \{\rho_1 : f(s(\alpha), \beta) \rightarrow f(\alpha, \beta), \rho_2 : f(\alpha{:}0, 0) \rightarrow \alpha{:}s(0), \rho_3 : f(\alpha{:}0, 1) \rightarrow \alpha, \pi_1 : g(\beta{:}s(\alpha), 0) \rightarrow \beta{:}0, \pi_2 : g(\beta, 1) \rightarrow \beta\}$.

Let $t = \alpha; \gamma{:}f(s(s(\beta{:}s(\alpha{:}0))), s_1); \delta{:}g(\beta, s_2)$ (the root is $\alpha$). Assume that we know that $\delta \not\gg_t^* \alpha$. In order to decide whether $\gamma$ is $\xi$-needed or not, we have to evaluate one of the terms $s_1$ or $s_2$. If $s_1 \rightarrow 1$ then obviously $\gamma \not\gg_t^* \alpha$ (since $\rho_2$ will never be applicable). Similarly, if $s_2 \rightarrow 0$ then the link between $\gamma$ and $\alpha$ is cut by the rule $\pi_1$, thus $\gamma \not\gg_t^* \alpha$. If $s_1 \rightarrow 0$ and $s_2 \rightarrow 1$ then we have $\gamma \gg_t^* \alpha$. But $s_1, s_2$ are not *both* needed if $\gamma \not\gg_t^* \alpha$.

We could overcome this problem by imposing very strong restrictions on the considered GRS. However, we prefer to impose additional conditions on the dependency schemata, which are not too restrictive, because they are satisfied by the relations proposed in Section 4:

**Definition 9.** *Let $\mathcal{R}$ be a GRS. An invariant $\mathcal{R}$-dependency schema $\xi$ is said to be* strongly invariant, *if for every rule $\rho = L \rightarrow R \mid \phi \in \mathcal{R}$ and for every pair of term-graphs $(t, s)$ s.t. $s = \rho^\sigma[t]$ for some $\rho$-matcher $\sigma$ at $\alpha$, then for every $\beta, \delta \in \mathcal{N}(t)$ s.t. $\beta \neq \alpha$, $\beta \gg_t^* \delta$ (resp. $\beta \rhd_t^* \delta$) and $\beta \not\gg_s^* \delta$ (resp. $\beta \ntriangleright_s^* \delta$) we have $\alpha \gg_t^* \delta$. Moreover there exists a node $\gamma \in dom(\sigma(R))$ s.t. $\beta \rhd_t^* \gamma$.*

It is easy to see that the dependency schema $\chi$ introduced in Section 4 is strongly invariant.

## 5.1 Relevancy

The definition of the strategy requires great care. For instance, one cannot assume that any node affecting the root is needed. Consider the term: $\alpha{:}0; \beta{:}g(\alpha); \beta'{:}g(\alpha)$. Then both $\beta$ and $\beta'$ affect $\alpha$ (according to the previous definitions), but only the maximal node (according to the priority ordering) is actually needed for sure (the other one may actually be useless).

A rooted term-graph $t$ is said to be a *line* if $\hat{t} = t$ and if for every node $\alpha$ in $t$ there exists at most one feature $a \neq l$ s.t. $a_t(\alpha)$ is defined. $s$ is said to be a *line of* $t$ if $s$ is a line, $s \subseteq t$ and $root(t) = root(s)$.

*Example 6.* Let $t = \alpha{:}f(\beta{:}s(\gamma), \lambda{:}s(0))$. The term-graphs $s_1 = \alpha{:}f(1 \Rightarrow \beta{:}s(\gamma))$ and $s_2 = \alpha{:}f(2 \Rightarrow \lambda)$ are two lines of $t$. $s_1$ is maximal, $s_2$ is not.

If $t$ is a term-graph, $\mathcal{R}$ is a GRS, and $\xi$ is an $\mathcal{R}$-dependency schema, we denote by $\langle t\rangle_\alpha^\xi$ the maximal subgraph $s$ of $[t]_\alpha^\xi$ s.t. for every node $\beta$ of $s$:

- Either $\alpha = root(t)$, $l_t(\alpha) \in \mathcal{C}$ and $\alpha \geq_t \beta$.
- Or there exists a rule $L \to R \mid \phi \in \mathcal{R}$, a line $l$ of $L$ and a $\mathcal{N}$-mapping $\sigma \in sol(\phi)$ s.t. $\sigma(root(L)) = \alpha$, $\sigma(l) \subseteq t$ and $\beta \in \sigma(l)$.

Intuitively, $\langle t\rangle_\alpha^\xi$ denotes the part of $[t]_\alpha^\xi$ that is currently "useful" w.r.t. the GRS $\mathcal{R}$ (a node is useful either because it is reachable from the root or because it occurs in the left hand side of a rewrite rule).

*Example 7.* Let $t = \beta{:}f(\alpha{:}s(\alpha'{:}0), s(\alpha''{:}0))$. Let $\mathcal{R} = \{f(\alpha{:}s, s(\beta{:}0)) \to \beta\}$. Then $\langle t\rangle_\beta^\xi = f(\alpha, s(\alpha''))$. The node $\alpha'$ is useless.

Let $\alpha$ be a node. We denote by $w^\xi(t,\alpha)$ the set of nodes $\beta$ of $t$ s.t. $\beta \gg_t^* \alpha$. $w^\xi(t,\alpha)$ denotes the set of nodes that possibly affect $\alpha$ ($w$ stands for "write").

**Definition 10.** *Let $\mathcal{R}$ be a GRS and let $\xi$ be a strongly invariant dependency schema. The set of nodes that are* $\xi$-relevant *in $t$ is the smallest set of nodes $\alpha$ s.t. $l_t(\alpha) \in \mathcal{D}$ and one of the following conditions holds:*

1. *Either there exist $\beta, \delta$ s.t. $\beta \prec \alpha$, $\beta$ is either $root(t)$ or $\xi$-relevant in $t$, $\delta$ is a node in $\langle t\rangle_\beta^\xi$ and $\alpha$ is the maximal node in $w^\xi(t,\delta)$. Intuitively, $\alpha$ is relevant, because it affects a node $\delta$ which is useful for reducing $\beta$.*
2. *Or $\beta$ is $\xi$-relevant, $w^\xi(t,\delta) = \emptyset$ for every node $\delta$ in $\langle t\rangle_\beta^\xi$ and there exists a node $\lambda$ in $\langle t\rangle_\beta^\xi$ s.t. $\beta \gg_t \lambda$, $\alpha \succ \beta$ and $\alpha \rhd_t^* \lambda$. $\alpha$ is relevant because since it is of greater priority than $\beta$ and since it depends on a node $\lambda$ which is affected by $\beta$, it prevents the reduction on $\beta$ (by definition of $\xrightarrow{\xi}_{\mathcal{R}}$).*

**Definition 11.** *(Needed Reduction Strategy) Let $\mathcal{R}$ be a GRS and let $\xi$ be a strongly invariant $\mathcal{R}$-dependency schema. We write $t \overset{\xi}{\hookrightarrow}_{\mathcal{R}} s$ iff there exist a rule $\rho \in \mathcal{R}$ and an eligible $\rho$-matcher $\sigma$ for $t$ at a $\xi$-relevant node $\alpha$ s.t. $s = \rho^\sigma[t]$.*

The next theorem shows that $\overset{\xi}{\hookrightarrow}_{\mathcal{R}}$ is normalizing, and that the length of the obtained derivations are minimal.

**Theorem 3.** *Let $\mathcal{R}$ be a GRS. Let $t$ be a term-graph and let $\xi$ be a strongly invariant $\mathcal{R}$-dependency schema. Assume that there exists a sequence $t_1, \ldots, t_n$ of term-graphs s.t. $t_1 = t$, $t_n$ is a value and $t_i \xrightarrow{\xi}_{\mathcal{R}} t_{i+1}$, for every $i \in [1..n-1]$.*
*There exists a sequence $s_1, \ldots, s_m$ s.t. $s_1 = t$, $s_m$ is a value, $\hat{s_m} = \hat{t_n}$, $m \leq n$ and $s_i \overset{\xi}{\hookrightarrow}_{\mathcal{R}} s_{i+1}$ for every $i \in [1..m-1]$.*

## 5.2 Neededness

Now, we introduce a class of GRS for which the above strategy only reduces needed nodes.

We denote by $\mathcal{L}(t)$ the set of maximal (w.r.t. $\subseteq$) lines of $t$. If $\rho = L \to R \mid \phi, \pi = G \to D \mid \psi$ are two rules, we write $\rho \sqsubset \pi$ iff for every pair of lines $(l, l') \in (\mathcal{L}(L), \mathcal{L}(G))$ and for every $\mathcal{N}$-mapping $\sigma$ s.t. $\sigma(l) \subseteq \sigma(l')$, we have $\sigma(l) = \sigma(l')$.

A set of rules $\mathcal{R}$ is said to be *elementary* if for every pair of rules $\rho, \pi \in \mathcal{R}$ we have $\rho \sqsubset \pi$ (and $\pi \sqsubset \rho$). This condition is easy to check. Intuitively, it expresses the fact that the rules $\rho$ and $\pi$ potentially depend on the same set of nodes. For instance, the system $\{f(\alpha{:}s(\alpha)) \to 0, f(\alpha'{:}s(\beta')) \to 1 \mid \alpha' \neq \beta'\}$ is elementary, but $\{f(\alpha{:}s(\beta), 0) \to 0, f(\alpha'{:}s(s(\beta')), 1) \to 1\}$ is not (since the sequence $\alpha{:}s(\beta)$ is *strictly* contained into $\alpha'{:}s(s(\beta'))$, up to an adequate $\mathcal{N}$-mapping).

*Remark 1.* The above condition may seem very restrictive. For instance, a GRS as simple as $\{f(a, s(\alpha)) \to 0, f(b, \alpha) \to 1\}$ is not elementary. However it can easily be transformed into an elementary system: $\{f(a, \alpha) \to f'(\alpha), f(b, \alpha) \to 1, f'(s(\alpha)) \to 0\}$. This process can be generalized: if we restrict ourselves to term rewrite systems, it is obvious that any inductively sequential system (i.e. any system having a definitional tree [2]) can be automatically transformed into an elementary system. Thus our results apply to any strongly/inductively sequential system [9,10], since these notions coincide for constructor-based rules [8]. Actually, the notion of elementary GRS can be seen as a generalization of the notion of definitional tree (note that elementary GRS are not necessarily orthogonal).

**Theorem 4.** *Let $\mathcal{R}$ be a GRS. Let $\xi$ be a strongly invariant $\mathcal{R}$-dependency schema. If $\mathcal{R}$ is elementary, then every $\xi$-relevant node $\beta$ for $t$ is $\xi$-needed for $t$.*

## 6 Examples

We apply the above needed rewriting strategy on the following examples, using the dependency schema defined in Section 4. By convention, $\beta_1 \succ \beta_2 \succ \ldots$. Relevant nodes are marked with a box. The root is always the first mentioned node in the term-graph.

We use the rules defined in Section 3 and the following ones[6]:

$$\begin{array}{llll}
cadr: & cadr(cons(\alpha, cons(\alpha', \beta))) \to \alpha' & car: & car(cons(\alpha, \beta)) \to \alpha \\
nf_2: & nf(s(\delta), \alpha{:}cons(\alpha', \beta)) \to \alpha; nf(\delta, \beta) & nf_1: & nf(0, \alpha{:}cons(\beta, \delta)) \to \alpha{:}nil \\
+_1: & plus(s(\alpha), \beta) \to plus(\alpha, s(\beta)) & +_2: & plus(0, \beta) \to \beta \\
l_0: & l_0(\alpha{:}nil) \to \alpha{:}cons(0, \lambda{:}nil); l_0(\lambda) & &
\end{array}$$

$$\begin{array}{ll}
 & \boxed{\beta_3}{:}cadr(\alpha{:}nil); \beta_2{:}inc(\alpha); \boxed{\beta_1}{:}l_0(\alpha) \\
\to_{l_0} & \boxed{\beta_3}{:}cadr(\alpha{:}cons(0, \lambda_1{:}nil)); \boxed{\beta_2}{:}inc(\alpha); \boxed{\lambda_2}{:}l_0(\lambda_1) \\
\to_{i_2} & \boxed{\beta_3}{:}cadr(\alpha{:}cons(s(0), \lambda_1{:}nil)); \lambda_3{:}inc(\lambda_1); \boxed{\lambda_2}{:}l_0(\lambda_1) \\
\to_{l_0} & \boxed{\beta_3}{:}cadr(\alpha{:}cons(s(0), \lambda_1{:}cons(0, \lambda_4{:}nil))); \boxed{\lambda_3}{:}inc(\lambda_1); \lambda_5{:}l_0(\lambda_4) \\
\to_{i_2} & \boxed{\beta_3}{:}cadr(\alpha{:}cons(s(0), \lambda_1{:}cons(s(0), \lambda_4{:}nil))); \lambda_6{:}inc(\lambda_4); \lambda_5{:}l_0(\lambda_4) \\
\to_{cadr} & s(0); \alpha{:}cons(s(0), \lambda_1{:}cons(s(0), \lambda_4{:}nil)); \lambda_6{:}inc(\lambda_4); \lambda_5{:}l_0(\lambda_4)
\end{array}$$

6 *cadr*, *car*, *plus* and *length* are standard. If $\alpha$ is a list containing at least $m$ elements, then $nf(m, \alpha)$ is the list of the $m$ first elements of $\alpha$. $l_0$ returns an infinite list of 0's.

At this point the normal form is known: $s(0)$. The remaining nodes are irrelevant. Note that the strict strategy diverges since the computation of $l_0(\mathit{nil})$ does not terminate (infinite list of 0's). We provide a second example:

$$
\begin{array}{ll}
 & \boxed{\beta_4}{:}\mathit{plus}(\boxed{\beta_1}{:}\mathit{car}(\delta{:}\mathit{cons}(s(0), \mathit{cons}(0, \mathit{nil}))), \beta_3{:}\mathit{length}(\delta)); \beta_2{:}\mathit{nf}(\beta_1, \delta) \\
\to_{\mathit{car}} & \boxed{\beta_4}{:}\mathit{plus}(\gamma{:}s(0), \beta_3{:}\mathit{length}(\delta{:}\mathit{cons}(\gamma, \mathit{cons}(0, \mathit{nil})))); \beta_2{:}\mathit{nf}(\gamma, \delta) \\
\to_{+_1} & \boxed{\beta_4}{:}\mathit{plus}(\gamma'{:}0, s(\beta_3{:}\mathit{length}(\delta{:}\mathit{cons}(\gamma{:}s(\gamma'), \mathit{cons}(0, \mathit{nil}))))); \beta_2{:}\mathit{nf}(\gamma, \delta) \\
\to_{+_2} & \beta_4{:}s(\boxed{\beta_3}{:}\mathit{length}(\delta{:}\mathit{cons}(\gamma{:}s(0), \mathit{cons}(0, \mathit{nil})))); \boxed{\beta_2}{:}\mathit{nf}(\gamma, \delta) \\
\to_{\mathit{nf}_2} & \beta_4{:}s(\boxed{\beta_3}{:}\mathit{length}(\delta{:}\mathit{cons}(\gamma{:}s(\gamma'{:}0), \delta'{:}\mathit{cons}(0, \mathit{nil})))); \lambda_1{:}\mathit{nf}(\gamma', \delta') \\
\to_{\#_1} & \beta_4{:}s(s(\boxed{\lambda_2}{:}\mathit{length}(\delta'{:}\mathit{cons}(0, \mathit{nil})))); \boxed{\lambda_1}{:}\mathit{nf}(\gamma', \delta'); \delta{:}\mathit{cons}(\gamma{:}s(\gamma'{:}0), \delta') \\
\to_{\mathit{nf}_1} & \beta_4{:}s(s(\boxed{\lambda_2}{:}\mathit{length}(\delta'{:}\mathit{nil}))); \delta{:}\mathit{cons}(\gamma{:}s(\gamma'{:}0), \delta'); 0; \mathit{nil} \\
\to_{\#_2} & \beta_4{:}s(s(0)); \delta'{:}\mathit{nil}; \delta{:}\mathit{cons}(\gamma{:}s(\gamma'{:}0), \delta'); 0; \mathit{nil}
\end{array}
$$

We obtain the normal form $s(s(0))$. Notice that the first $+$ rule is applicable although the first argument $\gamma$ occurs behind the scope of the function *nf* because *nf* does not perform side effects on its *first* argument. Thus $\beta_2 \not\gg^* \gamma$. This would not be the case with the dependency schema presented in [7].

## 7 Conclusion

We have presented a general framework for handling rewrite rules operating on term-graphs (encoding *pointer-based data-structures*), which is simpler (but similar to) the one presented in [7]. We have presented a new tractable way of detecting *interferences* between the defined nodes in the considered term-graph rewrite systems. Then we have provided a *reduction strategy* which is *optimal* for a class of term-graph rewrite rules, w.r.t. both the length of the derivations *and* the set of reduced nodes (only needed nodes are reduced). Our results extend the scope of declarative languages by allowing the programmer to define algorithms that physically affect pointer-based data-structures (as in imperative programming). Lazy reduction strategies can now be applied, similar to the ones that were already known for terms (and used by rewrite-based languages such as Haskell). From a practical side, we are now implementing a first prototype of our approach. From a more theoretical point of view, it would be interesting to extend these results to non constructor-based systems. This would require to extend the notion of *strong sequentiality* to GRS.

## References


1. Antoy, S.: Definitional trees. In: Kirchner, H., Levi, G. (eds.) Algebraic and Logic Programming; Third International Conference; Proceedings, Berlin, Germany, pp. 143–157. Springer, Berlin, Germany (1992)
2. Antoy, S.: Definitional trees. In: Kirchner, H., Levi, G. (eds.) ALP 1992. LNCS, vol. 632, pp. 143–157. Springer, Heidelberg (1992)

3. Caferra, R., Echahed, R., Peltier, N.: Rewriting term-graphs with priority. In: Proceedings of the Eighth ACM SIGPLAN Symposium on Principles and Practice of Declarative Programming, pp. 109–120. ACM Press, New York (2006)
4. Echahed, R., Janodet, J.-C.: Admissible graph rewriting and narrowing. In: Proceedings of 15th International Conference and Symposium on Logic Programming, Manchester, pp. 325–340. MIT Press, Cambridge (1998)
5. Echahed, R., Janodet, J.C.: Parallel admissible graph rewriting. In: Fiadeiro, J.L. (ed.) WADT 1998. LNCS, vol. 1589, pp. 122–138. Springer, Heidelberg (1999)
6. Echahed, R., Peltier, N.: Narrowing data-structures with pointers. In: Corradini, A., Ehrig, H., Montanari, U., Ribeiro, L., Rozenberg, G. (eds.) ICGT 2006. LNCS, vol. 4178, pp. 92–106. Springer, Heidelberg (2006)
7. Echahed, R., Peltier, N.: Non Strict Confluent Rewrite Systems for Data-Structures with Pointers. In: Baader, F. (ed.) RTA 2007. LNCS, vol. 4533, pp. 137–152. Springer, Heidelberg (2007)
8. Hanus, M., Lucas, S., Middeldorp, A.: Strongly sequential and inductively sequential term rewriting systems. Information Processing Letters 67(1), 1–8 (1998)
9. Huet, G., Levy, J.-J.: Computations in orthogonal rewriting systems. In: Lassez, J.-L., Plotkin, G. (eds.) Computational Logic: Essays in Honor of Alan Robinson, pp. 395–443. MIT Press, Cambridge (1991)
10. Klop, J.W., Middeldorp, A.: Sequentiality in orthogonal term rewriting systems. J. Symb. Comput. 12(2), 161–195 (1991)
11. Schorr, H., Waite, W.M.: An Efficient Machine Independent Procedure for Garbage Collection in Various List Structures. Communication of the ACM 10, 501–506 (1967)
12. Thompson, S.: Haskell: The Craft of Functional Programming, 2nd edn. Addison-Wesley, Reading (1999)

# Effectively Checking the Finite Variant Property*

Santiago Escobar[1], José Meseguer[2], and Ralf Sasse[2]

[1] Universidad Politécnica de Valencia, Spain
sescobar@dsic.upv.es
[2] University of Illinois at Urbana-Champaign, USA
{meseguer,rsasse}@cs.uiuc.edu

**Abstract.** An equational theory decomposed into a set $B$ of equational axioms and a set $\Delta$ of rewrite rules has the *finite variant* (FV) *property* in the sense of Comon-Lundh and Delaune iff for each term $t$ there is a finite set $\{t_1, \ldots, t_n\}$ of $\rightarrow_{\Delta,B}$-normalized instances of $t$ so that any instance of $t$ normalizes to an instance of some $t_i$ modulo $B$. This is a very useful property for cryptographic protocol analysis, and for solving both unification and disunification problems. Yet, at present the property has to be established by hand, giving a separate mathematical proof for each given theory: no checking algorithms seem to be known. In this paper we give both a necessary and a sufficient condition for FV from which we derive an algorithm ensuring the sufficient condition, and thus FV. This algorithm can check automatically a number of examples of FV known in the literature.

## 1 Introduction

The *finite variant* (FV) *property* is a useful property of a rewrite theory $\mathcal{R} = (\Sigma, B, \Delta)$ with signature $\Sigma$, rewrite rules $\Delta$, and equational axioms $B$ introduced by Comon-Lundh and Delaune in [2]. Very simply, it states the existence of a finite set of pairs $(t_i, \theta_i)$ for a given term $t$ such that: (i) $t_i$ is the $\rightarrow_{\Delta,B}$-normal form of $t\theta_i$, and (ii) for any normalized substitution $\rho$, the $\rightarrow_{\Delta,B}$-normal form of $t\rho$ is, up to $B$-equivalence, a substitution instance of some $t_i$. Comon-Lundh and Delaune list several important applications in [2], including formal reasoning about cryptographic protocol security using constraints [3], and reducing disunification problems modulo $\Delta \uplus B$ (when rules in $\Delta$ are viewed as equations) to disunification problems modulo $B$.

We have studied in detail how, if a rewrite theory $\mathcal{R} = (\Sigma, B, \Delta)$ is confluent, terminating, and coherent modulo the axioms $B$, and has the FV property, one can define an efficient narrowing strategy, which we call *variant narrowing*, to obtain a finitary unification algorithm modulo $\Delta \uplus B$ if a finitary $B$-unification

* S. Escobar has been partially supported by the EU (FEDER) and the Spanish MEC under grant TIN2007-68093-C02-02, and Integrated Action HA 2006-0007. J. Meseguer and R. Sasse have been partially supported by the ONR Grant N00014-02-1-0715, and by the NSF Grants IIS 07-20482 and CNS 07-16638.

A. Voronkov (Ed.): RTA 2008, LNCS 5117, pp. 79–93, 2008.

algorithm exists [6]. We agree with Comon-Lundh and Delaune [2] that if an efficient, dedicated $\Delta \uplus B$-unification algorithm is known, using the FV property to generate unifiers is usually much less efficient. But such an efficient, dedicated algorithm may not be known at all. Furthermore, for common equational axioms such as $AC$, it is well-known that narrowing modulo $AC$ almost never terminates [2]. Typically it does not terminate even when $\mathcal{R} = (\Sigma, B, \Delta)$ has the FV property; yet, existence of a *finite*, complete set of narrowing-generated unifiers is guaranteed by a *bound* on the depth of the narrowing tree that has to be explored [6]. Therefore, we view the FV property as the basis of an attractive method for obtaining finitary unification algorithms in many cases where no dedicated algorithm is known, and narrowing itself would almost certainly be nonterminating and therefore would yield an infinitary algorithm.

For all the above reasons: for reasoning about cryptographic protocols, to solve disunification problems, and, in our view, to solve also unification problems, it would be very useful to be able to *check* in an effective way whether a given rewrite theory $\mathcal{R} = (\Sigma, B, \Delta)$ has the FV property. This is the main question that we ask and we provide an answer for in this paper: is there an effective *algorithm* that can ensure that $\mathcal{R} = (\Sigma, B, \Delta)$ has the FV property?

We approach this main goal by stages. In Section 4, we give a necessary and a sufficient condition for FV. The necessary condition, which we abbreviate to FVNS is the absence of infinite *variant-preserving narrowing sequences.* The sufficient condition is the conjunction of FVNS with a second condition which we call *variant-preservingness* (VP). So we have a chain of implications

$$(\mathit{FVNS} \wedge \mathit{VP}) \Rightarrow \mathit{FV} \Rightarrow \mathit{FVNS}$$

This chain of implications then provides a useful division of labor for arriving in Section 5 at the desired checking algorithms. Since checking FVNS and VP ensures FV, we need algorithms checking both of these properties. It turns out that, under mild conditions on $B$, VP is a *decidable* property, so we have an algorithm for it. Instead, for FVNS we have a situation strongly analogous to what happens with the use of the dependency pairs (DP) method [1] for termination proofs: the DP method is sound and complete for termination, yet termination is undecidable. The point, of course, is that one usually cannot compute the *exact* dependency graph, but can nevertheless compute an *estimated* dependency graph and use it in termination proofs. This analogy is not far-fetched at all, since in fact we were inspired by the DP-method (in its "modulo" version as developed by Giesl and Kapur in [7]) to develop a DP-like analysis of the theory $\mathcal{R} = (\Sigma, B, \Delta)$ from which we derive our desired algorithm for checking FVNS.

We discuss several examples of theories that have the FV property. In particular, we show that for all the examples presented in [2] that were there proved to have the FV property by mathematical arguments given for each specific theory, our checking method can *automatically* prove the FV property. In [5], we also provide a method for disproving the FV property and show that all the examples presented in [2] that were there disproved to have the FV property are automatically disproved by our method. At the end of the paper we summarize our contributions, and discuss future work and applications, including applications to the formal analysis of cryptographic protocols modulo equational properties. All proofs can be found in [5].

## 2 Preliminaries

We follow the classical notation and terminology from [13] for term rewriting and from [10,11] for rewriting logic and order-sorted notions. We assume an S-sorted family $\mathcal{X} = \{\mathcal{X}_s\}_{s \in S}$ of disjoint variable sets with each $\mathcal{X}_s$ countably infinite. $\mathcal{T}_{\Sigma}(\mathcal{X})_s$ is the set of terms of sort s, and $\mathcal{T}_{\Sigma,s}$ is the set of ground terms of sort s. We write $\mathcal{T}_{\Sigma}(\mathcal{X})$ and $\mathcal{T}_{\Sigma}$ for the corresponding term algebras. For a term $t$ we write $Var(t)$ for the set of all variables in $t$. The set of positions of a term $t$ is written $Pos(t)$, and the set of non-variable positions $Pos_{\Sigma}(t)$. The root position of a term is $\Lambda$. The subterm of $t$ at position $p$ is $t|_p$ and $t[u]_p$ is the term $t$ where $t|_p$ is replaced by $u$. A *substitution* $\sigma$ is a sorted mapping from a finite subset of $\mathcal{X}$, written $Dom(\sigma)$, to $\mathcal{T}_{\Sigma}(\mathcal{X})$. The set of variables introduced by $\sigma$ is $Ran(\sigma)$. The identity substitution is *id*. Substitutions are homomorphically extended to $\mathcal{T}_{\Sigma}(\mathcal{X})$. The application of a substitution $\sigma$ to a term $t$ is denoted by $t\sigma$. The restriction of $\sigma$ to a set of variables $V$ is $\sigma|_V$. Composition of two substitutions is denoted by $\sigma\sigma'$. We call a substitution $\sigma$ a *renaming* if there is another substitution $\sigma^{-1}$ such that $\sigma\sigma^{-1}|_{Dom(\sigma)} = id$.

A $\Sigma$*-equation* is an unoriented pair $t = t'$, where $t, t' \in \mathcal{T}_{\Sigma}(\mathcal{X})_s$ for some sort $s \in S$. Given $\Sigma$ and a set $E$ of $\Sigma$-equations such that $\mathcal{T}_{\Sigma,s} \neq \emptyset$ for every sort s, order-sorted equational logic induces a congruence relation $=_E$ on terms $t, t' \in \mathcal{T}_{\Sigma}(\mathcal{X})$ (see [11]). Throughout this paper we assume that $\mathcal{T}_{\Sigma,s} \neq \emptyset$ for every sort s. An *equational theory* $(\Sigma, E)$ is a set of $\Sigma$-equations.

The $E$*-subsumption* preorder $\leq_E$ (or $\leq$ if $E$ is understood) holds between $t, t' \in \mathcal{T}_{\Sigma}(\mathcal{X})$, denoted $t \leq_E t'$ (meaning that $t$ is more general than $t'$ modulo $E$), if there is a substitution $\sigma$ such that $t\sigma =_E t'$; such a substitution $\sigma$ is said to be an $E$*-match* from $t$ to $t'$. For substitutions $\sigma, \rho$ and a set of variables $V$ we define $\sigma|_V =_E \rho|_V$ if $x\sigma =_E x\rho$ for all $x \in V$; $\sigma|_V \leq_E \rho|_V$ if there is a substitution $\eta$ such that $(\sigma\eta)|_V =_E \rho|_V$.

An $E$*-unifier* for a $\Sigma$-equation $t = t'$ is a substitution $\sigma$ such that $t\sigma =_E t'\sigma$. For $Var(t) \cup Var(t') \subseteq W$, a set of substitutions $CSU_E(t = t')$ is said to be a *complete* set of unifiers of the equation $t =_E t'$ away from $W$ if: (i) each $\sigma \in CSU_E(t = t')$ is an $E$-unifier of $t =_E t'$; (ii) for any $E$-unifier $\rho$ of $t =_E t'$ there is a $\sigma \in CSU_E(t = t')$ such that $\sigma|_W \leq_E \rho|_W$; (iii) for all $\sigma \in CSU_E(t = t')$, $Dom(\sigma) \subseteq (Var(t) \cup Var(t'))$ and $Ran(\sigma) \cap W = \emptyset$. An $E$-unification algorithm is *complete* if for any equation $t = t'$ it generates a complete set of $E$-unifiers. Note that this set needs not be finite. A unification algorithm is said to be *finitary* and complete if it always terminates after generating a finite and complete set of solutions.

A *rewrite rule* is an oriented pair $l \to r$, where $l \notin \mathcal{X}$, and $l, r \in \mathcal{T}_{\Sigma}(\mathcal{X})_s$ for some sort $s \in S$. An *(unconditional) order-sorted rewrite theory* is a triple $\mathcal{R} = (\Sigma, E, R)$ with $\Sigma$ an order-sorted signature, $E$ a set of $\Sigma$-equations, and $R$ a set of rewrite rules. The rewriting relation on $\mathcal{T}_{\Sigma}(\mathcal{X})$, written $t \to_R t'$ or

$t \xrightarrow{p}_R t'$ holds between $t$ and $t'$ iff there exist $p \in Pos_\Sigma(t)$, $l \to r \in R$ and a substitution $\sigma$, such that $t|_p = l\sigma$, and $t' = t[r\sigma]_p$. The relation $\to_{R/E}$ on $\mathcal{T}_\Sigma(\mathcal{X})$ is $=_E; \to_R; =_E$. Note that $\to_{R/E}$ on $\mathcal{T}_\Sigma(\mathcal{X})$ induces a relation $\to_{R/E}$ on $\mathcal{T}_{\Sigma/E}(\mathcal{X})$ by $[t]_E \to_{R/E} [t']_E$ iff $t \to_{R/E} t'$. The transitive closure of $\to_{R/E}$ is denoted by $\to^+_{R/E}$ and the transitive and reflexive closure of $\to_{R/E}$ is denoted by $\to^*_{R/E}$. We say that a term $t$ is $\to_{R/E}$-irreducible (or just $R/E$-irreducible) if there is no term $t'$ such that $t \to_{R/E} t'$.

For substitutions $\sigma, \rho$ and a set of variables $V$ we define $\sigma|_V \to_{R/E} \rho|_V$ if there is $x \in V$ such that $x\sigma \to_{R/E} x\rho$ and for all other $y \in V$ we have $y\sigma =_E y\rho$. A substitution $\sigma$ is called *$R/E$-normalized* (or normalized) if $x\sigma$ is $R/E$-irreducible for all $x \in V$. We say a rewrite step $t \xrightarrow{p}_{R/E} s$ is *normalized* if the substitution $\sigma$, s.t. $t =_E t'$ and $t'|_p = l\sigma$, is $R/E$-normalized.

We say that the relation $\to_{R/E}$ is *terminating* if there is no infinite sequence $t_1 \to_{R/E} t_2 \to_{R/E} \cdots \to_{R/E} \cdots$. We say that the relation $\to_{R/E}$ is *confluent* if whenever $t \to^*_{R/E} t'$ and $t \to^*_{R/E} t''$, there exists a term $t'''$ such that $t' \to^*_{R/E} t'''$ and $t'' \to^*_{R/E} t'''$. An order-sorted rewrite theory $\mathcal{R} = (\Sigma, E, R)$ is confluent (resp. terminating) if the relation $\to_{R/E}$ is confluent (resp. terminating). In a confluent, terminating, order-sorted rewrite theory, for each term $t \in \mathcal{T}_\Sigma(\mathcal{X})$, there is a unique (up to $E$-equivalence) $R/E$-irreducible term $t'$ obtained from $t$ by rewriting to canonical form, which is denoted by $t \to^!_{R/E} t'$ or $t{\downarrow}_{R/E}$ (when $t'$ is not relevant).

## 3 Narrowing and Variants

Since $E$-congruence classes can be infinite, $\to_{R/E}$-reducibility is undecidable in general. Therefore, $R/E$-rewriting is usually implemented [9] by $R, E$-rewriting. We assume the following properties on $R$ and $E$:

1. $E$ is *regular*, i.e., for each $t = t'$ in $E$, we have $Var(t) = Var(t')$, and *sort-preserving*, i.e., for each substitution $\sigma$, we have $t\sigma \in \mathcal{T}_\Sigma(\mathcal{X})_{\mathsf{s}}$ if and only if $t'\sigma \in \mathcal{T}_\Sigma(\mathcal{X})_{\mathsf{s}}$, and all variables in $Var(t)$ have a top sort.
2. $E$ has a finitary and complete unification algorithm.
3. For each $t \to t'$ in $R$ we have $Var(t') \subseteq Var(t)$.
4. $R$ is *sort-decreasing*, i.e., for each $t \to t'$ in $R$, each $\mathsf{s} \in \mathsf{S}$, and each substitution $\sigma$, $t'\sigma \in \mathcal{T}_\Sigma(\mathcal{X})_{\mathsf{s}}$ implies $t\sigma \in \mathcal{T}_\Sigma(\mathcal{X})_{\mathsf{s}}$.
5. The rewrite rules $R$ are *confluent and terminating modulo* $E$, i.e., the relation $\to_{R/E}$ is confluent and terminating.

**Definition 1 (Rewriting modulo).** [14] *Let $\mathcal{R} = (\Sigma, E, R)$ be an order-sorted rewrite theory satisfying properties* (1)–(5). *We define the relation $\to_{R,E}$ on $\mathcal{T}_\Sigma(\mathcal{X})$ by $t \to_{R,E} t'$ iff there is a $p \in Pos_\Sigma(t)$, $l \to r$ in $R$ and substitution $\sigma$ such that $t|_p =_E l\sigma$ and $t' = t[r\sigma]_p$.*

Note that, since $E$-matching is decidable, $\to_{R,E}$ is decidable. Notions such as confluence, termination, irreducible terms, normalized substitution, and normalized rewrite steps are defined in a straightforward manner for $\to_{R,E}$. Note that since $R$ is confluent and terminating (modulo $E$), the relation $\to^!_{R,E}$ is decidable, i.e., it terminates and produces a unique term (up to $E$-equivalence) for each initial term $t$, denoted by $t\downarrow_{R,E}$. Of course $t \to_{R,E} t'$ implies $t \to_{R/E} t'$, but the converse need not hold. To prove completeness of $\to_{R,E}$ w.r.t. $\to_{R/E}$ we need the following additional *coherence* assumption; we refer the reader to [7] for coherence completion algorithms.

6. $\to_{R,E}$ is *E-coherent* [9], i.e., $\forall t_1, t_2, t_3$ we have $t_1 \to_{R,E} t_2$ and $t_1 =_E t_3$ implies $\exists t_4, t_5$ such that $t_2 \to^*_{R,E} t_4$, $t_3 \to^+_{R,E} t_5$, and $t_4 =_E t_5$.

Narrowing generalizes rewriting by performing unification at non-variable positions instead of the usual matching. The essential idea behind narrowing is to *symbolically* represent the rewriting relation between terms as a narrowing relation between more general terms.

**Definition 2 (Narrowing modulo).** *(see, e.g.,* [9,12]*) Let $\mathcal{R} = (\Sigma, E, R)$ be an order-sorted rewrite theory satisfying properties* (1)–(6). *Let $CSU_E(u = u')$ provide a finitary, and complete set of unifiers for any pair of terms $u, u'$. The $R, E$-narrowing relation on $\mathcal{T}_\Sigma(\mathcal{X})$ is defined as $t \overset{p,\sigma}{\leadsto}_{R,E} t'$ (or $\overset{\sigma}{\leadsto}$ or $\leadsto_\sigma$ if $p, R, E$ are understood) if there is $p \in Pos_\Sigma(t)$, a (possibly renamed) rule $l \to r$ in $R$ s.t. $Var(l) \cap Var(t) = \emptyset$, and $\sigma \in CSU_E(t|_p = l)$ such that $t' = (t[r]_p)\sigma$.*

In the following, we introduce the notion of variant and finite variant property.

**Definition 3 (Decomposition).** [6] *Let $(\Sigma, E)$ be an order-sorted equational theory. We call $(\Delta, B)$ a* decomposition *of $E$ if $E = B \uplus \Delta$ and $(\Sigma, B, \overrightarrow{\Delta})$ is an order-sorted rewrite theory satisfying properties* (1)–(6), *where rules $\overrightarrow{\Delta}$ are an oriented version of $\Delta$.*

*Example 1 (Exclusive Or).* The following equational theory, denoted $\mathcal{R}_\oplus$, is a presentation of the exclusive or operator together with the cancellation equations for public key encryption/decryption.

$$X \oplus 0 = X \ (1) \quad pk(K, sk(K, M)) = M \ (4) \quad X \oplus (Y \oplus Z) = (X \oplus Y) \oplus Z \ (6)$$
$$X \oplus X = 0 \ (2) \quad sk(K, pk(K, M)) = M \ (5) \quad X \oplus Y = Y \oplus X \ (7)$$
$$X \oplus X \oplus Y = Y \ (3)$$

This equational theory $(\Sigma, E)$ has a decomposition into $\Delta$ containing the oriented version of equations (1)–(5) and $B$ containing the last two associativity and commutativity equations (6)–(7) for $\oplus$. Note that equations (1)–(2) are not $AC$-coherent, but adding equation (3) is sufficient to recover that property.

We recall the notions of *variant, finite variants*, and the *finite variant property* proposed by Comon and Delaune in [2].

**Definition 4 (Variants).** [2] *Given a term $t$ and an order-sorted equational theory $E$, we say that $(t', \theta)$ is an $E$-*variant *of $t$ if $t\theta =_E t'$, where $Dom(\theta) \subseteq Var(t)$ and $Ran(\theta) \cap Var(t) = \emptyset$.*

**Definition 5 (Complete set of variants).** [2] *Let* $(\Delta, B)$ *be a decomposition of an order-sorted equational theory* $(\Sigma, E)$. *A* complete set of $E$-variants *(up to renaming) of a term* $t$, *denoted* $V_{\Delta,B}(t)$, *is a set* $S$ *of* $E$*-variants of* $t$ *such that, for each substitution* $\sigma$, *there is a variant* $(t', \rho) \in S$ *and a substitution* $\theta$ *such that: (i)* $t'$ *is* $\Delta, B$*-irreducible, (ii)* $(t\sigma){\downarrow}_{\Delta,B} =_B t'\theta$, *and (iii)* $(\sigma{\downarrow}_{\Delta,B})|_{Var(t)} =_B (\rho\theta)|_{Var(t)}$.

**Definition 6 (Finite variant property).** [2] *Let* $(\Delta, B)$ *be a decomposition of an order-sorted equational theory* $(\Sigma, E)$. *Then* $E$, *and thus* $(\Delta, B)$, *has the* finite variant (FV) property *if for each term* $t$, *there exists a finite and complete set of* $E$*-variants, denoted* $FV_{\Delta,B}(t)$. *We will call* $(\Delta, B)$ *a* finite variant decomposition *if* $(\Delta, B)$ *has the finite variant property.*

Comon and Delaune characterize the finite variant property in terms of the following boundedness property, which is *equivalent* to FV.

**Definition 7 (Boundedness property).** [2] *Let* $(\Delta, B)$ *be a decomposition of an order-sorted equational theory* $(\Sigma, E)$. $(\Delta, B)$ *satisfies the* boundedness property *(BP) if for every term* $t$ *there exists an integer* $n$, *denoted by* $\#_{\Delta,B}(t)$, *such that for every* $\Delta, B$*-normalized substitution* $\sigma$ *the normal form of* $t\sigma$ *is reachable by a* $\Delta, B$*-rewriting derivation whose length can be bounded by* $n$ *(thus independently of* $\sigma$*), i.e.,* $\forall t, \exists n, \forall \sigma$ *s.t.* $t(\sigma{\downarrow}_{\Delta,B}) \xrightarrow{\leq n}_{\Delta,B} (t\sigma){\downarrow}_{\Delta,B}$.

**Theorem 1.** [2] *Let* $(\Delta, B)$ *be a decomposition of an order-sorted equational theory* $(\Sigma, E)$. *Then,* $(\Delta, B)$ *satisfies the boundedness property if and only if* $(\Delta, B)$ *is a finite variant decomposition of* $(\Sigma, E)$.

Obviously, if for a term $t$, the minimal length of a rewrite sequence to the canonical form of an instance $t\sigma$, with $\sigma$ normalized, cannot be bounded, the theory does not have the finite variant property. It is easy to see that for the addition equations $0 + Y = Y$, and $s(X) + Y = s(X + Y)$, the term $t = X + Y$, and the substitution $\sigma_n = \{X \mapsto s^n(0), Y \mapsto Y\}$, $n \in \mathbb{N}$, this is the case, and therefore, since $FV \Leftrightarrow BP$, the addition theory lacks the finite variant property.

We can effectively compute a complete set of variants in the following form.

**Proposition 1 (Computing the Finite Variants).** [6] *Let* $(\Delta, B)$ *be a finite variant decomposition of an order-sorted equational theory* $(\Sigma, E)$. *Let* $t \in \mathcal{T}_\Sigma(\mathcal{X})$ *and* $\#_{\Delta,B}(t) = n$. *Then,* $(s, \sigma) \in FV_{\Delta,B}(t)$ *if and only if there is a narrowing derivation* $t \overset{\sigma}{\rightsquigarrow}{}^{\leq n}_{\Delta,B} s$ *such that* $s$ *is* $\to_{\Delta,B}$*-irreducible and* $\sigma$ *is* $\to_{\Delta,B}$*-normalized.*

*Example 2.* The equational theory from Example 1 has the boundedness property. Thus, we use Proposition 1 to get the $E$-variants of $t{=}M{\oplus}sk(K,pk(K,M))$. As $t \to^!_{\Delta,B} 0$ we have $t \overset{id}{\rightsquigarrow}{}^!_{\Delta,B} 0$. Therefore, $(0, id) \in FV_{\Delta,B}(t)$ and it is the only element of the complete set of $E$-variants as no more general narrowing sequences are possible. For $s = X \oplus sk(K,pk(K,Y))$ we get (i) $s \overset{id}{\rightsquigarrow}{}^*_{\Delta,B} X \oplus Y$, (ii) $s \rightsquigarrow^*_{\{X \mapsto Z \oplus U, Y \mapsto U\},\Delta,B} Z$, (iii) $s \rightsquigarrow^*_{\{X \mapsto U, Y \mapsto Z \oplus U\},\Delta,B} Z$,

(iv) $s \leadsto^*_{\{X \mapsto U \oplus Z_1, Y \mapsto U \oplus Z_2\},\Delta,B} Z_1 \oplus Z_2$, and (v) $s \leadsto^*_{\{X \mapsto U, Y \mapsto U\},\Delta,B} 0$, so $(X \oplus Y, id)$, $(Z, \{X \mapsto Z \oplus U, Y \mapsto U\})$, $(Z, \{X \mapsto U, Y \mapsto Z \oplus U\})$, $(Z_1 \oplus Z_2, \{X \mapsto U \oplus Z_1,\ Y \mapsto U \oplus Z_2\})$, and $(0, \{X \mapsto U, Y \mapsto U\})$, are the $E$-variants. As no more general narrowing sequences are possible, these make up a complete set of $E$-variants. Note that (iv) is an instance of (i) and it is not necessary for a minimal and complete set of variants.

*Example 3.* Consider again Example 1. For this theory, narrowing clearly does not terminate because $Z_1 \oplus Z_2 \leadsto_{\{Z_1 \mapsto X_1 \oplus Z_1',\ Z_2 \mapsto X_1 \oplus Z_2'\},\Delta,B} Z_1' \oplus Z_2'$ and this can be repeated infinitely often. However, if we always assume that we are interested only in a normalized substitution, which is the case, for any narrowing sequence obtained in the previous form, there is a one-step rewriting sequence that provides the same result. That is, given the narrowing sequence

$$Z_1 \oplus Z_2 \leadsto_{\{Z_1 \mapsto X_1 \oplus Z_1', Z_2 \mapsto X_1 \oplus Z_2'\},\Delta,B} Z_1' \oplus Z_2' \leadsto_{\{Z_1' \mapsto X_1' \oplus Z_1'', Z_2' \mapsto X_1' \oplus Z_2''\},\Delta,B} Z_1'' \oplus Z_2''$$

and its corresponding rewrite sequence

$$X_1 \oplus X_1' \oplus Z_1'' \oplus X_1 \oplus X_1' \oplus Z_2'' \to_{\Delta,B} X_1' \oplus Z_1'' \oplus X_1' \oplus Z_2'' \to_{\Delta,B} Z_1'' \oplus Z_2''$$

we can also reduce it to the same normal form using only one application of (3) and the following normalized substitution $\rho = \{X \mapsto X_1 \oplus X_1', Y \mapsto Z_1'' \oplus Z_2''\}$. The trick is that rule (3) allows combining all pairs of canceling terms and thus gets rid of all of them at once.

## 4 Sufficient and Necessary Conditions for FV

Deciding whether an equational theory has the finite variant property is a non-trivial task, since we have to decide whether we can stop generating normalized substitution instances by narrowing for each term. Intuitively, since the theory is convergent, we only have to focus on normalized substitutions and, since it has the boundedness property, we can compute the variants in a bottom-up manner. Moreover, any rewrite sequence with a normalized substitution will be captured by a narrowing sequence leading to the same variant (i.e., irreducible term). Our algorithm for checking that an equational theory has the finite variant property is based on two notions: (i) a new notion called *variant-preservingness* (VP) that ensures that an intuitive bottom-up generation of variants is complete; and (ii) that there are no infinite sequences when we restrict ourselves to such intuitive bottom-up generation of variants (FVNS). In what follows, we show that $(VP \wedge FVNS) \Rightarrow FV \Rightarrow FVNS$.

Variant–preservingness (VP) ensures that we can perform an intuitive bottom-up[1] generation of variants. The following notion is useful.

[1] Note that this is not the same as innermost narrowing nor innermost narrowing up to some bound. Consider Example 5 where innermost narrowing does not terminate for term $c(f(X), X)$, since it looks for an innermost narrowing redex each time. A bottom-up generation of invariants does terminate (see Proposition 1) providing terms $c(f(X), X)$ and $c(X', f(X'))$. Even in the case of innermost narrowing with a bound, it will miss the term $c(f(X), X)$.

**Definition 8 (Variant–pattern).** *Let $\mathcal{R} = (\Sigma, E, R)$ be an order-sorted rewrite theory satisfying properties* (1)–(6). *We call a term $f(t_1, \ldots, t_n)$ a* variant–pattern *if all subterms $t_1, \ldots, t_n$ are $\rightarrow_{R,E}$-irreducible. We will say a term $t$* has a variant–pattern *if there is a variant–pattern $t'$ s.t. $t' =_E t$.*

It is worth pointing out that whether a term has a variant–pattern is decidable, assuming a finitary and complete $E$-unification procedure: given a term $t$, $t$ has a variant–pattern $t'$ iff there is a symbol $f \in \Sigma$ with arity $k$ and variables $X_1, \ldots, X_k$ of the appropriate top sorts and there is a substitution $\theta \in CSU_E(t = f(X_1, \ldots, X_k))$ such that $\theta$ is normalized, where $t' = f(X_1, \ldots, X_k)\theta$. In the case of a term $t$ rooted by a free symbol, $t$ has a variant–pattern if it is already a variant–pattern, i.e., every argument of the root symbol must be irreducible. And, in the case of a term $t$ rooted by an $AC$ symbol, we only have to consider in the previous algorithm the same $AC$ symbol at the root of $t$, instead of every symbol.

**Definition 9 (Variant–preserving).** *Let $\mathcal{R} = (\Sigma, E, R)$ be an order-sorted rewrite theory satisfying properties* (1)–(6). *We say that the theory $\mathcal{R}$ is* variant–preserving *(VP) if for any variant–pattern $t$, either $t$ is $\rightarrow_{R,E}$-irreducible or there is a normalized $\rightarrow_{R,E}$ step at the top position.*

Note that a theory can have the finite variant property even if it is not variant-preserving.

*Example 4.* Consider the following equational theory $f(a, b, X) = c$, where symbol $f$ is $AC$ and $X$ is a variable. The narrowing relation $\rightsquigarrow_{R,E}$ terminates for any term but the theory does not have the variant-preserving property, e.g., given the term $t = f(X, Y)$ and any normalized substitution $\theta \in \{X \mapsto f(a^n), Y \mapsto f(b^n, Z)\}$ for $n \geq 2$, there is no normalized reduction for $t\theta$. However, the theory does have the boundedness property, and therefore FV, since for any term rooted by $f$ (which is the only non-constant symbol), its normal form can be obtained in at most one step.

We characterize variant–preservingness in Section 5.1. A theory that already has the variant–preserving property, if there is no infinite $E$-narrowing sequence, clearly has the finite variant property. However, if infinite $E$-narrowing sequences exist, a theory may still have the finite variant property.

*Example 5.* Consider the equational theory $f(f(X)) = X$, which is well-known to be non-terminating for narrowing, i.e.,

$$c(f(X), X) \rightsquigarrow_{\{X \mapsto f(X')\},R,E} c(X', f(X')) \rightsquigarrow_{\{X' \mapsto f(X'')\},R,E} c(f(X''), X'') \cdots$$

When we consider all possible instances of term $c(f(X), X)$ for normalized substitutions, we obtain term $c(f(X), X)$ itself and the sequence $c(f(X), X) \rightsquigarrow_{\{X \mapsto f(X')\},R,E} c(X', f(X'))$. The theory does have the boundedness property, and therefore FV, since for any term and a normalized substitution, a bound is the number of $f$ symbols in the term.

Not all the narrowing sequences are relevant for the finite variant property, as shown in the previous example, and thus we must identify the relevant ones.

**Definition 10 (Variant–preserving sequences).** *Let $\mathcal{R} = (\Sigma, E, R)$ be an order-sorted rewrite theory satisfying properties (1)–(6). A rewrite sequence $t_0 \overset{p_1}{\to}_{R,E} t_1 \cdots \overset{p_n}{\to}_{R,E} t_n$ is called* variant–preserving *if $t_{i-1}|_{p_i}$ has a variant-pattern for $i \in \{1, \ldots, n\}$ and there is no sequence $t_0 \to^m_{R,E} t'_m$ such that $m < n$ and $t_n =_E t'_m$. A narrowing sequence $t_0 \overset{p_1,\sigma_1}{\leadsto}_{R,E} t_1 \cdots \overset{p_n,\sigma_n}{\leadsto}_{R,E} t_n$, $\sigma = \sigma_1 \cdots \sigma_n$, is called* variant–preserving *if $\sigma$ is $\to_{R,E}$-normalized and $t_0\sigma \overset{p_1}{\to}_{R,E} t_1\sigma \cdots \overset{p_n}{\to}_{R,E} t_n$ is variant–preserving.*

The set of variant–preserving sequences is not computable in general. However, we provide sufficient conditions in Section 5.

*Example 6.* The infinite narrowing sequence of Example 5 is not variant–preserving, since for any finite prefix of length greater than 1 the computed substitution is non-normalized. The only variant-preserving sequences for term $c(f(X), X)$ are the term itself and the one-step sequence with substitution $\{X \mapsto f(X')\}$.

*Example 7.* For Example 3, the narrowing sequence

$$Z_1 \oplus Z_2 \leadsto_{\{Z_1 \mapsto X_1 \oplus Z'_1, Z_2 \mapsto X_1 \oplus Z'_2\}, R,E} Z'_1 \oplus Z'_2 \leadsto_{\{Z'_1 \mapsto X'_1 \oplus Z''_1, Z'_2 \mapsto X'_1 \oplus Z''_2\}, R,E} Z''_1 \oplus Z''_2$$

is not a variant-preserving sequence, since the alternative rewrite sequence $X_1 \oplus X'_1 \oplus Z''_1 \oplus X_1 \oplus X'_1 \oplus Z''_2 \to_{R,E} Z''_1 \oplus Z''_2$ is shorter.

We prove that using variant–preserving sequences is sound and complete.

**Theorem 2 (Computing with variant–preserving sequences).** *Let $\mathcal{R} = (\Sigma, E, R)$ be an order-sorted rewrite theory satisfying properties (1)–(6) that also has the finite variant property. Let $t \in \mathcal{T}_\Sigma(\mathcal{X})$ and $\#_{R,E}(t) = n$. Then, $(s, \sigma) \in FV_{R,E}(t)$ if and only if there is a variant–preserving narrowing derivation $t \overset{\sigma}{\leadsto}{}^{\leq n}_{R,E} s$ such that $s$ is $\to_{R,E}$-irreducible.*

The following result provides sufficient conditions for the finite variant property.

**Theorem 3 (Sufficient conditions for FV).** *Let $\mathcal{R} = (\Sigma, E, R)$ be an order-sorted rewrite theory satisfying properties (1)–(6). If (i) $\mathcal{R}$ is variant–preserving (VP), and (ii) there is no infinite variant–preserving narrowing sequence (FVNS), then $\mathcal{R}$ satisfies the finite variant property.*

Note that variant-preservingness is not a *necessary* condition for FV, as shown in Example 4. However, the absence of infinite variant–preserving narrowing sequences is a *necessary* condition for FV.

**Theorem 4 (Necessary condition for FV).** *Let $\mathcal{R} = (\Sigma, E, R)$ be an order-sorted rewrite theory satisfying properties (1)–(6). If there is an infinite variant–preserving narrowing sequence, then $\mathcal{R}$ does not satisfy the finite variant property.*

## 5 Checking the Finite Variant Property

In the following, we show that the variant-preserving property is clearly checkable, in Section 5.1, but the absence of infinite variant-preserving narrowing sequences is not computable in general, and we approximate such property, in Section 5.2, by a checkable one using the dependency pairs technique of [7] for the modulo case.

### 5.1 Checking Variant–Preservingness

The following class of equational theories is relevant. The notion of $E$-descendants (given in [5]) is a straightforward extension of the standard notion of descendant for rules. Given $t =_E s$ and $p \in Pos(t)$, we write $p\backslash\!\!\backslash_s$ for the $E$-descendants of $p$ in $s$.

**Definition 11 (Upper-$E$-coherence).** *Let $\mathcal{R} = (\Sigma, E, R)$ be an order-sorted rewrite theory satisfying properties* (1)–(5). *We say $\mathcal{R}$ is* upper-$E$-coherent *if for all $t_1, t_2, t_3$ we have $t_1 \stackrel{p}{\to}_{R,E} t_2$, $t_1 =_E t_3$, $p > \Lambda$, and $p\backslash\!\!\backslash_{t_3} = \emptyset$ implies that for all $p' \leq p$ such that $p'\backslash\!\!\backslash_{t_3} = \emptyset$, there exist $t'_3, t_4, t_5$ such that $t_1 \stackrel{p'}{\to}_{R,E} t'_3$, $t_2 \to^*_{R,E} t_4$, $t'_3 \to^*_{R,E} t_5$, and $t_4 =_E t_5$.*

Assuming $E$-coherence, checking upper-$E$-coherence consists of taking term $t$ for each equation $t = t' \in E$ (or reverse), finding a position $p \in Pos(t)$ s.t. $p > \Lambda$ and a substitution $\sigma$ s.t. $t\sigma|_p$ is $\to_{R,E}$-reducible and then, let $p = p_1.\cdots.p_k$, for $i \in \{1, \ldots, k-1\}$, $t\sigma|_{p_i}$ must be $\to_{R,E}$-reducible. In general, upper-$E$-coherence implies $E$-coherence but not vice versa, as shown below.

*Example 8.* Let us consider the rewrite theory $R = \{g(f(X)) \to d,\ a \to c\}$ and $E = \{g(f(f(a))) = g(b)\}$. For the term $t = g(f(f(a)))$, subterm $a$ is reducible, $t =_E g(b)$, but subterms $f(f(a))$ and $f(a)$ are not reducible and thus the theory is not upper-$E$-coherent. However, the theory is trivially $E$-coherent because of the use of symbol $g$ at the top of both sides of the equation.

Now, we can provide an algorithm for checking variant–preservingness.

**Theorem 5 (Checking Variant–preservingness).** *Let $\mathcal{R} = (\Sigma, E, R)$ be an order-sorted rewrite theory satisfying properties* (1)–(6) *that is upper-$E$-coherent. $\mathcal{R}$ has the variant–preserving property iff for all $l \to r, l' \to r' \in R$ (possibly renamed s.t. $Var(l) \cap Var(l') = \emptyset$) and for all $X \in Var(l)$, the term $t = l\theta$, where $\theta = \{X \mapsto l'\}$ such that $\theta$ is an order-sorted substitution, satisfies that either (i) $t$ does not have a variant–pattern, or (ii) otherwise there is a normalized reduction on $t$.*

In [5], the variant-preservingness property for the exclusive or theory is proved. The upper-$E$-coherence condition is necessary, as shown below.

*Example 9.* The theory of Example 8 satisfies the conditions of Theorem 5 but it is not variant–preserving. That is, $g(f(a))$ does not have a variant–pattern. However, $g(b)$ is a variant–pattern, it is reducible, but it is not $\to_{R,E}$-reducible with a normalized substitution.

## 5.2 Checking Finiteness of Variant–Preserving Narrowing Sequences

First, we need to extend the notion of defined symbol. An equation $u = v$ is called *collapsing* if $v \in \mathcal{X}$ or $u \in \mathcal{X}$. We say a theory is *collapse-free*[2] if all its equations are non-collapsing.

**Definition 12 (Defined Symbols for Rewriting Modulo Equations).** [7] *Let $\mathcal{R} = (\Sigma, E, R)$ be an order-sorted rewrite theory with $E$ collapse-free. Then the set of defined symbols $D$ is the smallest set such that $D = \{root(l) \mid l \to r \in R\} \uplus \{root(v) \mid u = v \in E \text{ or } v = u \in E, root(u) \in D\}$.*

In order to correctly approximate the dependency relation between defined symbols in the theory, we need to extend the equational theory in the following way.

**Definition 13 (Adding Instantiations).** [7] *Given an order-sorted rewrite theory $\mathcal{R} = (\Sigma, E, R)$, let $Ins_E(R)$ be a set containing only rules of the form $l\sigma \to r\sigma$ (where $\sigma$ is a substitution and $l \to r \in R$). $Ins_E(R)$ is called an instantiation of $R$ for the equations $E$ iff $Ins_E(R)$ is the smallest set such that: (a) $R \subseteq Ins_E(R)$, (b) for all $l \to r \in R$, all $v$ such that $u = v \in E$ or $v = u \in E$, and all $\sigma \in CSU_E(v = l)$, there exists a rule $l' \to r' \in Ins_E(R)$ and a variable renaming $\nu$ such that $l\sigma =_E l'\nu$ and $r\sigma =_E r'\nu$.*

Note that when $E = \emptyset$ or $E$ contains only $AC$ or $C$ axioms, $Ins_E(R) = R$. Dependency pairs are obtained as follows. Since we are dealing with the modulo case, it will be notationally more convenient to use terms directly in dependency pairs, without the usual capital letters for the top symbols.

**Definition 14 (Dependency Pair).** [1] *Let $\mathcal{R} = (\Sigma, E, R)$ be an order-sorted rewrite theory. If $l \to C[g(t_1, \ldots, t_m)]$ is a rule of $Ins_E(\mathcal{R})$ with $C$ a context and $g$ a defined symbol in $Ins_E(\mathcal{R})$, then $\langle l, g(t_1, \ldots, t_m)\rangle$ is called a* dependency pair *of $\mathcal{R}$.*

*Example 10 (Abelian Group).* This presentation of Abelian group theory, called $\mathcal{R}_* = (\Sigma, E, R)$, has been shown to satisfy the finite variant property in [2]. The operators $\Sigma$ are $\_*\_$, $(\_)^{-1}$, and 1. The set of equations $E$ consists of associativity and commutativity for $*$. The rules $R$ are:

$$x * 1 \to x \quad (8)$$
$$1^{-1} \to 1 \quad (9)$$
$$x * x^{-1} \to 1 \quad (10)$$
$$x^{-1} * y^{-1} \to (x * y)^{-1} \quad (11)$$
$$(x * y)^{-1} * y \to x^{-1} \quad (12)$$
$$x^{-1^{-1}} \to x \quad (13)$$
$$(x^{-1} * y)^{-1} \to x * y^{-1} \quad (14)$$
$$x * (x^{-1} * y) \to y \quad (15)$$
$$x^{-1} * (y^{-1} * z) \to (x * y)^{-1} * z \quad (16)$$
$$(x * y)^{-1} * (y * z) \to x^{-1} * z \quad (17)$$

[2] Note that regularity does not imply collapse-free, e.g. equation 1 of Example 1 is regular but also collapsing.

The AC-dependency pairs for this rewrite theory are as follows. The other rules not mentioned here do not give rise to an AC-dependency pair[3].

| | | | |
|---|---|---|---|
| (11)$a$: | $\langle x^{-1} * y^{-1}\,, (x*y)^{-1}\rangle$ | (11)$b$: | $\langle x^{-1} * y^{-1}\,, x*y\rangle$ |
| (14)$a$: | $\langle (x^{-1} * y)^{-1}\,, x*y^{-1}\rangle$ | (14)$b$: | $\langle (x^{-1} * y)^{-1}\,, y^{-1}\rangle$ |
| (16)$a$: | $\langle x^{-1} * y^{-1} * z\,, (x*y)^{-1} * z\rangle$ | (16)$b$: | $\langle x^{-1} * y^{-1} * z\,, (x*y)^{-1}\rangle$ |
| (16)$c$: | $\langle x^{-1} * y^{-1} * z\,, x*y\rangle$ | (12)$a$: | $\langle (x*y)^{-1} * y\,, x^{-1}\rangle$ |
| (17)$a$: | $\langle (x*y)^{-1} * y * z\,, x^{-1} * z\rangle$ | (17)$b$: | $\langle (x*y)^{-1} * y * z\,, x^{-1}\rangle$ |

The relevant notions are chains of dependency pairs and the dependency graph.

**Definition 15 (Chain).** [1] *Let $\mathcal{R} = (\Sigma, E, R)$ be an order-sorted rewrite theory. A sequence of dependency pairs $\langle s_1, t_1\rangle\langle s_2, t_2\rangle \cdots \langle s_n, t_n\rangle$ of $\mathcal{R}$ is an $\mathcal{R}$-*chain *if there is a substitution $\sigma$ such that $t_j\sigma \to^*_{R,E} s_{j+1}\sigma$ holds for every two consecutive pairs $\langle s_j, t_j\rangle$ and $\langle s_{j+1}, t_{j+1}\rangle$ in the sequence.*

**Definition 16 (Dependency Graph).** [1] *Let $\mathcal{R} = (\Sigma, E, R)$ be an order-sorted rewrite theory. The dependency graph of $\mathcal{R}$ is the directed graph whose nodes (vertices) are the dependency pairs of $R$ and there is an arc (directed edge) from $\langle s, t\rangle$ to $\langle u, v\rangle$ if $\langle s, t\rangle\langle u, v\rangle$ is a chain.*

As in the dependency pair technique [1], the variant–preserving chains are not computable in general and an approximation must be performed. The notion of *connectable terms* as defined in [1] can be easily extended to the variant–preserving case, and the *estimated dependency graph* [1] can be computed using the CAP and REN procedures [1]. We omit this in the paper for lack of space but such an estimated dependency graph has been used in all examples.

*Example 11.* In [5], the dependency graph for Example 10 is shown. It was created with AProVE. We see that there are self-loops on (11)b, (14)b, (16)a, (16)c and (17)a. (11)a has a loop with (14)a, (14)a has a loop with (16)b, and so on. It is a very highly connected graph.

In order to correctly approximate the bound for the finite variant property, we include rules without defined symbols in their right-hand sides as extra dependency pairs, that we call *dummy*.

**Definition 17 (Dummy dependency pairs).** *Let $\mathcal{R} = (\Sigma, E, R)$ be an order-sorted rewrite theory. If for a rule $l \to r \in R$ the right-hand side $r$ does not contain a defined symbol then $\langle l, r\rangle$ is a* dummy *dependency pair of $\mathcal{R}$.*

*Example 12 (Abelian group variant–preserving dependency pairs).* Building upon the AC-dependency pairs computed in Example 10 we need to add these dummy dependency pairs, to the set of dependency pairs from the prior example:

| | | |
|---|---|---|
| (8)$a$ : $\langle x * 1\,, x\rangle$ | (9)$a$ : $\langle 1^{-1}\,, 1\rangle$ | (10)$a$ : $\langle x * x^{-1}\,, 1\rangle$ |
| (13)$a$ : $\langle {x^{-1}}^{-1}\,, x\rangle$ | (15)$a$ : $\langle x * x^{-1} * y\,, y\rangle$ | |

[3] We have used the AProVE tool [8] to generate the dependency pairs. AProVE first applies the coherence algorithm of [7] to this example which is unnecessary here and thus we drop the dependency pairs created that way.

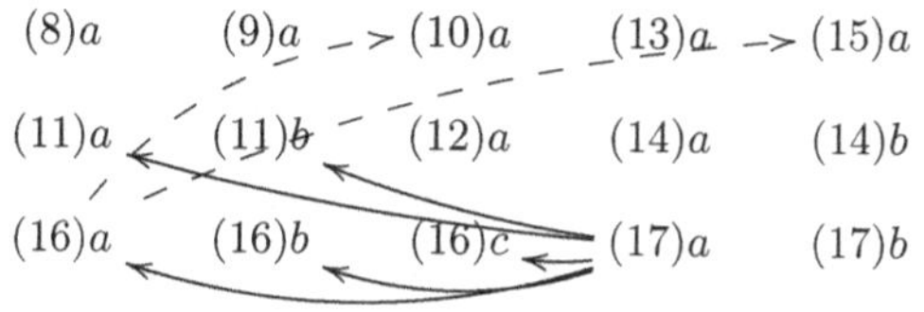


**Fig. 1.** Variant–preserving dependency graph

**Definition 18 (Cycle).** [1] *A nonempty set $\mathcal{P}$ of dependency pairs is called a* cycle *if, for any two dependency pairs $\langle s,t\rangle, \langle u,v\rangle \in \mathcal{P}$, there is a nonempty path from $\langle s,t\rangle$ to $\langle u,v\rangle$ and from $\langle u,v\rangle$ to $\langle s,t\rangle$ in the dependency graph that traverses dependency pairs from $\mathcal{P}$ only.*

As already demonstrated in the previous section, not all the rewriting (narrowing) sequences are relevant for the finite variant property.

**Definition 19 (Variant–preserving chain).** *Let $\mathcal{R} = (\Sigma, E, R)$ be an order-sorted rewrite theory. A chain of dependency pairs $\langle s_1,t_1\rangle\langle s_2,t_2\rangle \cdots \langle s_n,t_n\rangle$ of $\mathcal{R}$ is a* variant–preserving chain *if there is a substitution $\sigma$ such that $\sigma$ is $\to_{R,E}$-normalized and the following rewrite sequence obtainable from the chain $s_1\sigma \to_{R,E} C_1[t_1]\sigma \to^*_{R,E} C_1[s_2]\sigma \to_{R,E} C_1[C_2[t_2]]\sigma \to^*_{R,E} \cdots \to^*_{R,E} C_1[C_2[\cdots C_{n-1}[s_n]]]\sigma \to_{R,E} C_1[C_2[\cdots C_{n-1}[C_n[t_n]]]]\sigma$ is variant–preserving.*

The notions of a cycle, the dependency graph and the estimated dependency graph are easily extended to the variant–preserving case. The following straightforward result approximates the absence of infinite narrowing sequences.

**Proposition 2 (Checking Finiteness of the VP Narrowing sequences).** *Let $\mathcal{R} = (\Sigma, E, R)$ be a variant–preserving, order-sorted rewrite theory. Let $E$ contain only linear, non-collapsing equations. If the estimated dependency graph does not contain any variant–preserving cycle, then there are no infinite variant–preserving narrowing sequences.*

Note that the conditions that the axioms are non-collapsing and linear are necessary for completeness of the dependency graph, we refer the reader to [7] for explanations.

*Example 13 (Abelian group variant–preserving dependency pair graph).* We can show the variant–preserving dependency graph of Example 12 in Figure 1. As you can see in the picture, all the cycles have disappeared, because they involved non-normalized substitutions, or terms without a variant–pattern, or could be shortened.

Finally, we are able to provide an approximation result for the absence of infinite variant–preserving narrowing sequences. Also, we are able to compute a bound for each defined symbol thanks to a notion of *rank*.

**Definition 20 (Rank).** *The* rank *of a dependency pair p, denoted* $rank_{R,E}(p)$*, is the length of the longest variant–preserving chain starting from p. For a rule* $l \to r \in R$ *giving rise to dependency pairs* $dp_1, dp_2, \ldots, dp_n$*, its rank is* $rank_{R,E}(l \to r) = (rank_{R,E}(dp_1) - 1) + (rank_{R,E}(dp_2) - 1) + \ldots + (rank_{R,E}(dp_n) - 1) + 1$*. For a defined symbol f, its rank is* $rank_{R,E}(f) = max\{rank_{R,E}(l \to r) \mid l \to r \in R, root(l) = f\}$*. For a term t, its rank is* $rank_{R,E}(t) = \Sigma_{f \in \mathcal{D}}(rank_{R,E}(f) * \#_f(t))$ *where* $\mathcal{D}$ *is the set of defined symbols in* $\mathcal{R}$ *and* $\#_f(t)$ *is the number of appearances of f in t.*

Any cycle in the variant–preserving dependency graph of course gives the rank $\infty$ to all dependency pairs involved in the cycle. For any symbol $f$ it is obvious that $rank_{R,E}(f) \geq 1$ iff $f$ is a defined symbol.

Note that the dependency graph is not necessarily transitive for purposes of rank calculation.

*Example 14 (Abelian group variant–preserving dependency pair graph rank).* Consider again Example 13. The rank for the dependency pairs (17)a and (16)a is 2, the rank of all other dependency pairs is 1. Note that (17)a has rank 2 as according to Example 13 there is no variant–preserving chain of length 3 as in this case the graph is not transitive. Thus the rank of rule (17) is 2, which means that the rank of $*$ is 2 and the rank of $^{-1}$ is 1. Thus the rank for any term $t$ is $(\#_*(t) \times 2) + \#_{^{-1}}(t)$.

In [5], we show VP for Abelian group and Diffie-Hellman, and the finite variant property for Diffie-Hellman. The proof of our final result for this section is trivial by Theorem 4, since if the rank of all symbols in the signature is finite, there are no cycles in the estimated dependency graph and we know for sure that there is no infinite variant-preserving rewrite sequence.

**Theorem 6 (Approximation for the finite variant property).** *Let* $\mathcal{R} = (\Sigma, E, R)$ *be a variant–preserving, order-sorted rewrite theory. Let E contain only linear, non-collapsing equations. If for all defined symbols f we have that* $rank_{R,E}(f)$ *is finite, then* $\mathcal{R}$ *has the finite variant property.*

## 6 Conclusions

We have recalled Comon-Lundh and Delaune's finite variant property (FV) and summarized some of its applications. Our main two contributions have been: (i) giving new necessary conditions and new sufficient conditions for FV; and (ii) deriving from these conditions an algorithm for checking FV. To the best of our knowledge, no such algorithms were known before. The algorithms can certainly be improved. For example, more accurate ways of computing the effective dependency graph will help the checking of FV. Regarding implementations, we plan to implement these algorithms for frequently used equational axioms $B$ such as $\emptyset$, C, AC, and their combinations, so that they can be used in conjunction with the already-implemented variant narrowing algorithm described in [6] to derive finitary unification algorithms. This will provide a key component of the Maude-NPA [4], a tool for the analysis of cryptographic protocols modulo algebraic properties.

## References


1. Arts, T., Giesl, J.: Termination of term rewriting using dependency pairs. Theor. Comput. Sci. 236(1-2), 133–178 (2000)
2. Comon-Lundh, H., Delaune, S.: The finite variant property: How to get rid of some algebraic properties. In: Giesl, J. (ed.) RTA 2005. LNCS, vol. 3467, pp. 294–307. Springer, Heidelberg (2005)
3. Comon-Lundh, H., Shmatikov, V.: Intruder deductions, constraint solving and insecurity decision in presence of exclusive or. In: LICS, pp. 271–280. IEEE Computer Society, Los Alamitos (2003)
4. Escobar, S., Meadows, C., Meseguer, J.: A rewriting-based inference system for the NRL protocol analyzer and its meta-logical properties. Theor. Comput. Sci. 367(1-2), 162–202 (2006)
5. Escobar, S., Meseguer, J., Sasse, R.: Effectively checking or disproving the finite variant property. Technical Report UIUCDCS-R-2008-2960, Department of Computer Science - University of Illinois at Urbana-Champaign (April 2008)
6. Escobar, S., Meseguer, J., Sasse, R.: Variant narrowing and equational unification. In: 7th Int'l Workshop on Rewriting Logic and its Applications (to appear, 2008)
7. Giesl, J., Kapur, D.: Dependency pairs for equational rewriting. In: Middeldorp, A. (ed.) RTA 2001. LNCS, vol. 2051, pp. 93–108. Springer, Heidelberg (2001)
8. Giesl, J., Schneider-Kamp, P., Thiemann, R.: Automatic termination proofs in the dependency pair framework. In: Furbach, U., Shankar, N. (eds.) IJCAR 2006. LNCS (LNAI), vol. 4130, pp. 281–286. Springer, Heidelberg (2006)
9. Jouannaud, J.-P., Kirchner, C., Kirchner, H.: Incremental construction of unification algorithms in equational theories. In: Díaz, J. (ed.) ICALP 1983. LNCS, vol. 154, pp. 361–373. Springer, Heidelberg (1983)
10. Meseguer, J.: Conditioned rewriting logic as a united model of concurrency. Theor. Comput. Sci. 96(1), 73–155 (1992)
11. Meseguer, J.: Membership algebra as a logical framework for equational specification. In: Parisi-Presicce, F. (ed.) WADT 1997. LNCS, vol. 1376, pp. 18–61. Springer, Heidelberg (1998)
12. Meseguer, J., Thati, P.: Symbolic reachability analysis using narrowing and its application to verification of cryptographic protocols. Higher-Order and Symbolic Computation 20(1–2), 123–160 (2007)
13. TeReSe (ed.): Term Rewriting Systems. Cambridge University Press, Cambridge (2003)
14. Viry, P.: Equational rules for rewriting logic. Theor. Comput. Sci. 285(2), 487–517 (2002)

# Dependency Pairs for Rewriting with Built-In Numbers and Semantic Data Structures*

Stephan Falke and Deepak Kapur

Computer Science Department, University of New Mexico, Albuquerque, NM, USA
{spf,kapur}@cs.unm.edu

**Abstract.** This paper defines an expressive class of constrained equational rewrite systems that supports the use of semantic data structures (e.g., sets or multisets) and contains built-in numbers, thus extending our previous work presented at CADE 2007 [6]. These rewrite systems, which are based on normalized rewriting on constructor terms, allow the specification of algorithms in a natural and elegant way. Built-in numbers are helpful for this since numbers are a primitive data type in every programming language. We develop a dependency pair framework for these rewrite systems, resulting in a flexible and powerful method for showing termination that can be automated effectively. Various powerful techniques are developed within this framework, including a subterm criterion and reduction pairs that need to consider only subsets of the rules and equations. It is well-known from the dependency pair framework for ordinary rewriting that these techniques are often crucial for a successful automatic termination proof. Termination of a large collection of examples can be established using the presented techniques.

## 1 Introduction

Term rewriting provides a powerful framework for specifying algorithms in the form of rewrite systems that operate on data structures generated using constructors. Many algorithms operate on semantic data structures like finite sets, multisets, or sorted lists (e.g., using Java's collection classes or the OCaml extension Moca [3]). Constructors used to generate such data structures satisfy certain properties, i.e., they are not free. For example, finite sets can be generated using the empty set, singleton sets, and set union. Set union is associative (A), commutative (C), idempotent (I), and has the empty set as unit element (U). Such semantic data structures can be modeled using equational axioms. For sorted lists of numbers we also need to use arithmetic constraints on numbers for specifying relations on constructors.

Building upon our earlier work [6], this paper introduces constrained equational rewrite systems which have three components: (i) $\mathcal{R}$, a set of constrained rewrite rules operating on semantic data structures, (ii) $\mathcal{S}$, a set of constrained rewrite rules on constructors, and (iii) $\mathcal{E}$, a set of equations on constructors.

* Partially supported by NSF grant CCF-0541315.

A. Voronkov (Ed.): RTA 2008, LNCS 5117, pp. 94–109, 2008.

Here, (ii) and (iii) are used for modeling semantic data structures where normalization with $\mathcal{S}$ yields normal forms that are unique up to equivalence w.r.t. $\mathcal{E}$. The constraints for $\mathcal{R}$ and $\mathcal{S}$ are quantifier-free formulas from Presburger arithmetic. Rewriting in a constrained equational rewrite system is done using a combination of normalized rewriting [19] with validity checking of instantiated constraints. Notice that the OCaml extension Moca [3] uses the same strategy for semantic data structures. For a further generalization where the rules from $\mathcal{R}$ are allowed to contain conditions in addition to constraints we refer to [7].

*Example 1.* This example shows a mergesort algorithm that takes a set and returns a sorted list of the elements of the set. For this, sets are constructed using $\emptyset$, $\langle\cdot\rangle$ (a singleton set) and $\cup$, where we use the following sets $\mathcal{S}$ and $\mathcal{E}$.

$$\mathcal{E} = \{\ x \cup (y \cup z) \approx (x \cup y) \cup z,\quad x \cup y \approx y \cup x\ \}$$
$$\mathcal{S} = \{\ x \cup \emptyset \to x,\quad x \cup x \to x\ \}$$

The mergesort algorithm is given by the following constrained rewrite rules.

$$\begin{array}{rl}
\mathsf{merge}(\mathsf{nil}, y) \to y & \mathsf{merge}(x, \mathsf{nil}) \to x \\
\mathsf{merge}(\mathsf{cons}(x, xs), \mathsf{cons}(y, ys)) \to & \mathsf{cons}(y, \mathsf{merge}(\mathsf{cons}(x, xs), ys))\ [\![x > y]\!] \\
\mathsf{merge}(\mathsf{cons}(x, xs), \mathsf{cons}(y, ys)) \to & \mathsf{cons}(x, \mathsf{merge}(xs, \mathsf{cons}(y, ys)))\ [\![x \not> y]\!] \\
\mathsf{msort}(\emptyset) \to \mathsf{nil} & \mathsf{msort}(\langle x \rangle) \to \mathsf{cons}(x, \mathsf{nil}) \\
\mathsf{msort}(x \cup y) \to & \mathsf{merge}(\mathsf{msort}(x), \mathsf{msort}(y))
\end{array}$$

If rewriting modulo $\mathcal{E} \cup \mathcal{S}$ (or $\mathcal{E} \cup \mathcal{S}$-extended rewriting) is used with these constrained rewrite rules, then the resulting rewrite relation does not terminate since $\mathsf{msort}(\emptyset) \sim_{\mathcal{E}\cup\mathcal{S}} \mathsf{msort}(\emptyset \cup \emptyset) \to_{\mathcal{R}} \mathsf{merge}(\mathsf{msort}(\emptyset), \mathsf{msort}(\emptyset))$. ◊

An important property of constrained equational rewrite systems is termination. While automated termination methods work well for establishing termination of rewrite systems defined on free data structures, they do not easily extend to semantic data structures. Dependency pair methods for showing termination of AC-rewrite systems have been developed [17,20], and [9] generalized the dependency pair method to equational rewriting under the restriction that the equations are collapse-free (thus disallowing idempotency and unit elements), regular (i.e., the same variables occur on both sides), and linear.

In this paper, we extend the dependency pair framework [11] to constrained equational rewrite systems and present various termination techniques within this framework. This paper is a significant improvement over [6] in two directions:

1. The rewrite relation is a strict generalization of the rewrite relation considered in [6] since numbers are built-in. The resulting class of rewrite systems is highly expressive since domain-specific knowledge about numbers is available. This is helpful since most functional and imperative programming languages include numbers as a primitive data type. This paper only allows natural numbers, but an extension to integers is currently being developed.
2. Even if restricted to the rewrite relation considered in [6], the termination techniques presented in this paper strictly generalize the corresponding techniques presented in our earlier work since we present the following termination techniques that are not yet available in [6]:

- Section 4.2 presents a subterm criterion in the spirit of [15], which is a relatively simple yet surprisingly powerful termination technique.
- In Section 4.3 we show that a technique based on reduction pairs has to consider only subsets of $\mathcal{R}$, $\mathcal{S}$, and $\mathcal{E}$ as determined by the dependencies between function symbols. It is well-known from ordinary rewriting [13,15] that this is often crucial for automatic termination proofs.
- Section 4.4 shows that polynomial interpretations with negative coefficients are applicable in this setting. We can thus also argue about termination due to a bounded increase, which is so far only possible for innermost termination of ordinary rewriting [14].

This paper is organized as follows. In Section 2, the rewrite relation is defined. In Section 3, we present a characterization of termination of constrained equational rewrite systems that is based on dependency pairs, and we extend the dependency pair framework to constrained equational rewriting. Section 4 discusses various termination techniques within this framework, including the improvements discussed above. The proofs omitted from this paper can be found in the full version [5], which also contains a large collection of examples.

## 2 Normalized Equational Rewriting with Constraints

We assume familiarity with the concepts and notations of term rewriting [2]. We consider terms over two sorts, nat and univ, and we assume an initial signature $\mathcal{F}_{\mathcal{PA}} = \{0, 1, +\}$ with sorts $0, 1 : \texttt{nat}$ and $+ : \texttt{nat} \times \texttt{nat} \to \texttt{nat}$. Properties of natural numbers are modelled using the set $\mathcal{PA} = \{x+(y+z) \approx (x+y)+z,\ x+y \approx y+x,\ x+0 \approx x\}$ of equations. For each $k \in \mathbb{N} - \{0\}$, we denote the term $1+\ldots+1$ (with $k$ occurrences of 1) by $\mathsf{k}$.

We then extend $\mathcal{F}_{\mathcal{PA}}$ by a finite sorted signature $\mathcal{F}$. We omit stating the sorts explicitly in examples if they can be inferred. In the following we assume that all terms, contexts, context replacements, substitutions, rewrite rules, equations, etc. are sort correct. For any syntactic construct $c$ we let $\mathcal{V}(c)$ denote the set of variables occurring in $c$. Similarly, $\mathcal{F}(c)$ denotes the function symbols occurring in $c$. The root symbol of a term $s$ is denoted by $\text{root}(s)$. The root position of a term is denoted by $\lambda$. For an arbitrary set $\mathcal{E}$ of equations and terms $s, t$ we write $s \to_{\mathcal{E}} t$ iff there exist an equation $u \approx v \in \mathcal{E}$, a substitution $\sigma$, and a position $p \in \mathcal{P}os(s)$ such that $s|_p = u\sigma$ and $t = s[v\sigma]_p$. The symmetric closure of $\to_{\mathcal{E}}$ is denoted by $\vdash\dashv_{\mathcal{E}}$, and the reflexive transitive closure of $\vdash\dashv_{\mathcal{E}}$ is denoted by $\sim_{\mathcal{E}}$. For two terms $s, t$ we write $s \sim_{\mathcal{E}}^{>\lambda} t$ iff $s = f(s_1, \ldots, s_n)$ and $t = f(t_1, \ldots, t_n)$ such that $s_i \sim_{\mathcal{E}} t_i$ for all $1 \leq i \leq n$, i.e., if equations are only applied below the root.

An *atomic $\mathcal{PA}$-constraint* has the form $\top$ (truth), $s \simeq t$ (equality) or $s > t$ (greater) for terms $s, t \in \mathcal{T}(\mathcal{F}_{\mathcal{PA}}, \mathcal{V})$. The set of *$\mathcal{PA}$-constraints* is defined to be the closure of the set of atomic $\mathcal{PA}$-constraints under $\neg$ (negation) and $\wedge$ (conjunction). Validity (the constraint is true for all assignments) and satisfiability (the constraint is true for some assignment) of $\mathcal{PA}$-constraints are defined as usual, where we take the set of natural numbers as universe of concern. We also speak of $\mathcal{PA}$-validity and $\mathcal{PA}$-satisfiability. These properties are decidable [22].

We consider *constrained rewrite rules*, which are ordinary rewrite rules together with a $\mathcal{PA}$-constraint $C$, i.e., expressions of the form $l \to r[\![C]\!]$ for terms $l, r \in \mathcal{T}(\mathcal{F} \cup \mathcal{F}_{\mathcal{PA}}, \mathcal{V})$ and a $\mathcal{PA}$-constraint $C$ such that $\text{root}(l) \in \mathcal{F}$ and $\mathcal{V}(r) \subseteq \mathcal{V}(l)$. In a rule $l \to r[\![\top]\!]$ the constraint $\top$ is omitted. For a set $\mathcal{R}$ of constrained rewrite rules, the set of *defined symbols* is given by $\mathcal{D}(\mathcal{R}) = \{f \mid f = \text{root}(l) \text{ for some } l \to r[\![C]\!] \in \mathcal{R}\}$. The set of *constructors* is $\mathcal{C}(\mathcal{R}) = \mathcal{F} - \mathcal{D}(\mathcal{R})$. Notice that according to this definition, the symbols from $\mathcal{F}_{\mathcal{PA}}$ are considered to be neither defined symbols nor constructors.

Properties of non-free data structures are modeled using constructor equations and constrained constructor rules. A *constructor equation* has the form $u \approx v$ with $u, v \in \mathcal{T}(\mathcal{C}(\mathcal{R}), \mathcal{V})$ such that $u$ and $v$ are linear and $\mathcal{V}(u) = \mathcal{V}(v)$. A *constrained constructor rule* is a constrained rewrite rule $l \to r[\![C]\!]$ with $l, r \in \mathcal{T}(\mathcal{C}(\mathcal{R}), \mathcal{V})$.

Constructor equations and constrained constructor rules give rise to the following rewrite relation, which is based on extended rewriting [21] and requires that the $\mathcal{PA}$-constraint of the constrained constructor rule is $\mathcal{PA}$-valid after being instantiated by the matcher used for rewriting.

**Definition 2 (Constructor Rewrite Relation, $\mathcal{PA}$-based Substitutions).** *Let $\mathcal{E}$ be a finite set of constructor equations and let $\mathcal{S}$ be a finite set of constrained constructor rules. Then $s \to_{\mathcal{PA}\|\mathcal{E}\backslash\mathcal{S}} t$ iff there exist a rule $l \to r[\![C]\!] \in \mathcal{S}$, a position $p \in \mathcal{P}os(s)$, and a $\mathcal{PA}$-*based substitution *$\sigma$ (i.e., $\sigma(x) \in \mathcal{T}(\mathcal{F}_{\mathcal{PA}}, \mathcal{V})$ for all variables $x$ of sort* nat*) such that*

*(i) $s|_p \sim_{\mathcal{E}\cup\mathcal{PA}} l\sigma$, (ii) $C\sigma$ is $\mathcal{PA}$-valid, and (iii) $t = s[r\sigma]_p$.*

The reason for restricting substitutions to be $\mathcal{PA}$-based is that then $\mathcal{PA}$-validity of the instantiated $\mathcal{PA}$-constraint can be decided by a decision procedure for $\mathcal{PA}$-validity. We write $s \to^{>\lambda}_{\mathcal{PA}\|\mathcal{E}\backslash\mathcal{S}} t$ iff $s \to_{\mathcal{PA}\|\mathcal{E}\backslash\mathcal{S}} t$ at a position $p \neq \lambda$, and $s \xrightarrow{!}{}^{>\lambda}_{\mathcal{PA}\|\mathcal{E}\backslash\mathcal{S}} t$ iff $s$ reduces to $t$ in zero or more $\to^{>\lambda}_{\mathcal{PA}\|\mathcal{E}\backslash\mathcal{S}}$ steps such that $t$ is a normal form w.r.t. $\to^{>\lambda}_{\mathcal{PA}\|\mathcal{E}\backslash\mathcal{S}}$.

We combine constrained rewrite rules, constrained constructor rules, and constructor equations into a *constrained equational system*. These systems are a strict generalization of the equational systems considered in [6] since they allow the use of $\mathcal{PA}$-constraints. The system given in Example 1 is a constrained equational system.

**Definition 3 (Constrained Equational System (CESs)).** *A* constrained equational system *(CES) has the form $(\mathcal{R}, \mathcal{S}, \mathcal{E})$ for finite sets $\mathcal{R}$ of constrained rewrite rules, $\mathcal{S}$ of constrained constructor rules, and $\mathcal{E}$ of constructor equations such that*

1. *$\mathcal{S}$ is right-linear (i.e., $r$ is linear for all $l \to r[\![C]\!] \in \mathcal{S}$),*
2. *$\sim_{\mathcal{E}\cup\mathcal{PA}}$ commutes over $\to_{\mathcal{PA}\|\mathcal{E}\backslash\mathcal{S}}$ (i.e., the inclusion $\sim_{\mathcal{E}\cup\mathcal{PA}} \circ \to_{\mathcal{PA}\|\mathcal{E}\backslash\mathcal{S}} \subseteq \to_{\mathcal{PA}\|\mathcal{E}\backslash\mathcal{S}} \circ \sim_{\mathcal{E}\cup\mathcal{PA}}$ is satisfied), and*
3. *$\to_{\mathcal{PA}\|\mathcal{E}\backslash\mathcal{S}}$ is convergent modulo $\sim_{\mathcal{E}\cup\mathcal{PA}}$ (i.e., $\to_{\mathcal{PA}\|\mathcal{E}\backslash\mathcal{S}}$ is terminating and the inclusion $\leftarrow^*_{\mathcal{PA}\|\mathcal{E}\backslash\mathcal{S}} \circ \to^*_{\mathcal{PA}\|\mathcal{E}\backslash\mathcal{S}} \subseteq \to^*_{\mathcal{PA}\|\mathcal{E}\backslash\mathcal{S}} \circ \sim_{\mathcal{E}\cup\mathcal{PA}} \circ \leftarrow^*_{\mathcal{PA}\|\mathcal{E}\backslash\mathcal{S}}$ is satisfied).*

Here, the commutation property intuitively states that if $s \sim_{\mathcal{E}\cup\mathcal{PA}} s'$ and $s' \to_{\mathcal{PA}\|\mathcal{E}\backslash\mathcal{S}} t'$, then $s \to_{\mathcal{PA}\|\mathcal{E}\backslash\mathcal{S}} t$ for some $t \sim_{\mathcal{E}\cup\mathcal{PA}} t'$. If $\mathcal{S}$ does not already satisfy this property then it can be achieved by adding *extended rules* [21,9].

Some commonly used data structures and their specifications in our framework are listed below, where $\langle\cdot\rangle$ creates a singleton set or multiset, respectively. This list should not be considered exhaustive, i.e., there are further semantic data structures satisfying the conditions of Definition 3. The rule marked by "(∗)" is needed to make $\sim_{\mathcal{E}\cup\mathcal{PA}}$ commute over $\to_{\mathcal{PA}\|\mathcal{E}\backslash\mathcal{S}}$.

| | Constructors | $\mathcal{E}$ | $\mathcal{S}$ |
|---|---|---|---|
| Sorted lists | nil, cons | | $\mathsf{cons}(x, \mathsf{cons}(y, zs)) \to \mathsf{cons}(y, \mathsf{cons}(x, zs))[\![x > y]\!]$ |
| Multisets | $\emptyset$, ins | $\mathsf{ins}(x, \mathsf{ins}(y, zs)) \approx \mathsf{ins}(y, \mathsf{ins}(x, zs))$ | |
| Multisets | $\emptyset, \langle\cdot\rangle, \cup$ | $x \cup (y \cup z) \approx (x \cup y) \cup z$<br>$x \cup y \approx y \cup x$ | $x \cup \emptyset \to x$ |
| Sets | $\emptyset$, ins | $\mathsf{ins}(x, \mathsf{ins}(y, zs)) \approx \mathsf{ins}(y, \mathsf{ins}(x, zs))$ | $\mathsf{ins}(x, \mathsf{ins}(x, ys)) \to \mathsf{ins}(x, ys)$ |
| Sets | $\emptyset, \langle\cdot\rangle, \cup$ | $x \cup (y \cup z) \approx (x \cup y) \cup z$<br>$x \cup y \approx y \cup x$ | $x \cup \emptyset \to x \quad x \cup x \to x$<br>$(x \cup x) \cup y \to x \cup y \quad (*)$ |
| Sorted sets | $\emptyset$, ins | | $\mathsf{ins}(x, \mathsf{ins}(y, zs)) \to \mathsf{ins}(y, \mathsf{ins}(x, zs))[\![x > y]\!]$<br>$\mathsf{ins}(x, \mathsf{ins}(y, zs)) \to \mathsf{ins}(x, zs)[\![x \simeq y]\!]$ |

The rewrite relation corresponding to a CES is an extension of the normalized rewrite relation used in [6], which in turn is based on [19]. Notice that the redex is normalized by $\to^{>\lambda}_{\mathcal{PA}\|\mathcal{E}\backslash\mathcal{S}}$ before the matcher $\sigma$ is considered. Also notice that the restriction to $\mathcal{PA}$-based substitution enforces a kind of innermost rewriting for function symbols with resulting sort `nat`.

**Definition 4 (Rewrite Relation).** *Let $(\mathcal{R}, \mathcal{S}, \mathcal{E})$ be a CES and let $s, t$ be terms. Then $s \xrightarrow{\mathcal{S}}_{\mathcal{PA}\|\mathcal{E}\backslash\mathcal{R}} t$ iff there exist a constrained rewrite rule $l \to r[\![C]\!] \in \mathcal{R}$, a position $p \in \mathcal{P}os(s)$, and a $\mathcal{PA}$-based substitution $\sigma$ such that*

$$(i)\ s|_p \xrightarrow{!}{}^{>\lambda}_{\mathcal{PA}\|\mathcal{E}\backslash\mathcal{S}} \circ \sim^{>\lambda}_{\mathcal{E}\cup\mathcal{PA}} l\sigma, \quad (ii)\ C\sigma \text{ is } \mathcal{PA}\text{-valid, and} \quad (iii)\ t = s[r\sigma]_p.$$

*Example 5.* Continuing Example 1 we illustrate $\xrightarrow{\mathcal{S}}_{\mathcal{PA}\|\mathcal{E}\backslash\mathcal{R}}$. Notice that we add the rule $(x \cup x) \cup y \to x \cup y$ to $\mathcal{S}$ as indicated in the table above. Consider the term $t = \mathsf{msort}(\langle 1\rangle \cup (\langle 3\rangle \cup \langle 1\rangle))$ and the substitution $\sigma = \{x \mapsto \langle 3\rangle, y \mapsto \langle 1\rangle\}$. We get $t \xrightarrow{!}{}^{>\lambda}_{\mathcal{PA}\|\mathcal{E}\backslash\mathcal{S}} \mathsf{msort}(\langle 1\rangle \cup \langle 3\rangle) \sim^{>\lambda}_{\mathcal{E}\cup\mathcal{PA}} \mathsf{msort}(x \cup y)\sigma$ and thus $t \xrightarrow{\mathcal{S}}_{\mathcal{PA}\|\mathcal{E}\backslash\mathcal{R}} \mathsf{merge}(\mathsf{msort}(\langle 3\rangle), \mathsf{msort}(\langle 1\rangle))$. Continuing the reduction of this term for two more $\xrightarrow{\mathcal{S}}_{\mathcal{PA}\|\mathcal{E}\backslash\mathcal{R}}$ steps yields $\mathsf{merge}(\mathsf{cons}(3, \mathsf{nil}), \mathsf{cons}(1, \mathsf{nil}))$. Using $\sigma = \{x \mapsto 3, xs \mapsto \mathsf{nil}, y \mapsto 1, ys \mapsto \mathsf{nil}\}$ and the first rule for merge this term reduces to $\mathsf{cons}(1, \mathsf{merge}(\mathsf{nil}, \mathsf{cons}(3, \mathsf{nil})))$ because the instantiated constraint

$(x > y)\sigma = (3 > 1)$ is $\mathcal{PA}$-valid. With one further $\xrightarrow{\mathcal{S}}_{\mathcal{PA}\|\mathcal{E}\backslash\mathcal{R}}$ step we finally obtain the term $\mathsf{cons}(1, \mathsf{cons}(3, \mathsf{nil}))$. ◊

# 3 Dependency Pairs

In the following, we extend the dependency pair method in order to show termination of rewriting with CESs. The definition of a dependency pair is essentially the well-known one [1], with the only difference that the dependency pair inherits the constraint of the rule that it is created from. As usual, we introduce a signature $\mathcal{F}^\sharp$, containing for each function symbol $f \in \mathcal{D}(\mathcal{R})$ the function symbol $f^\sharp$ with the same arity and sorts as $f$. For a term $t = f(t_1, \ldots, t_n)$ we denote the term $f^\sharp(t_1, \ldots, t_n)$ by $t^\sharp$, and for a non-trivial context $D = f(t_1, \ldots, E, \ldots, t_n)$ (where $E$ is also a context) the context $f^\sharp(t_1, \ldots, E, \ldots, t_n)$ is denoted by $D^\sharp$.

**Definition 6 (Dependency Pairs).** *Let $(\mathcal{R}, \mathcal{S}, \mathcal{E})$ be a CES. The* dependency pairs *of $\mathcal{R}$ are* $\mathsf{DP}(\mathcal{R}) = \{l^\sharp \to t^\sharp[\![C]\!] \mid t$ *is a subterm of* $r$ *with* $\mathrm{root}(t) \in \mathcal{D}(\mathcal{R})$ *for some* $l \to r[\![C]\!] \in \mathcal{R}\}$.

In order to verify termination we rely on the notion of *chains*. Intuitively, a dependency pair corresponds to a recursive call, and a chain represents a possible sequence of calls in a reduction w.r.t. $\xrightarrow{\mathcal{S}}_{\mathcal{PA}\|\mathcal{E}\backslash\mathcal{R}}$. In the following we always assume that different (occurrences of) dependency pairs are variable disjoint, and we consider substitutions whose domain may be infinite. Additionally, we assume that all substitutions have $\mathcal{T}(\mathcal{F} \cup \mathcal{F}_{\mathcal{PA}}, \mathcal{V})$ as codomain.

**Definition 7 ((Minimal) $(\mathcal{P}, \mathcal{R}, \mathcal{S}, \mathcal{E})$-Chains).** *Let $(\mathcal{R}, \mathcal{S}, \mathcal{E})$ be a CES and let $\mathcal{P}$ be a set of dependency pairs. A (possibly infinite) sequence of dependency pairs $s_1 \to t_1[\![C_1]\!], s_2 \to t_2[\![C_2]\!], \ldots$ from $\mathcal{P}$ is a* $(\mathcal{P}, \mathcal{R}, \mathcal{S}, \mathcal{E})$-chain *iff there exists a $\mathcal{PA}$-based substitution $\sigma$ such that $t_i\sigma \xrightarrow{\mathcal{S}}{}^{*}_{\mathcal{PA}\|\mathcal{E}\backslash\mathcal{R}} \circ \xrightarrow{!}{}^{>\lambda}_{\mathcal{PA}\|\mathcal{E}\backslash\mathcal{S}} \circ \sim^{>\lambda}_{\mathcal{E}\cup\mathcal{PA}} s_{i+1}\sigma$ and the instantiated $\mathcal{PA}$-constraint $C_i\sigma$ is $\mathcal{PA}$-valid for all $i \geq 1$. The above $(\mathcal{P}, \mathcal{R}, \mathcal{S}, \mathcal{E})$-chain is* minimal *iff $t_i\sigma$ does not start an infinite $\xrightarrow{\mathcal{S}}_{\mathcal{PA}\|\mathcal{E}\backslash\mathcal{R}}$-reduction for any $i \geq 1$.*

Here, $\xrightarrow{\mathcal{S}}{}^{*}_{\mathcal{PA}\|\mathcal{E}\backslash\mathcal{R}}$ corresponds to reductions occurring strictly below the root of $t_i\sigma$ (notice that $\mathrm{root}(t_i) \in \mathcal{F}^\sharp$), and $\xrightarrow{!}{}^{>\lambda}_{\mathcal{PA}\|\mathcal{E}\backslash\mathcal{S}} \circ \sim^{>\lambda}_{\mathcal{E}\cup\mathcal{PA}}$ corresponds to normalization and matching before applying $s_{i+1} \to t_{i+1}[\![C_i]\!]$ at the root position.

*Example 8.* This example is a variation of an example in [23], modified to operate on sets and to use built-in natural numbers. Sets are modelled using the constructors $\emptyset$ and $\mathsf{ins}$ as in Section 2 and the function $\mathsf{nats}$ is defined so that $\mathsf{nats}(x, y)$ returns the set $\{z \mid x \leq z \leq y\}$.

$$
\begin{array}{rlrl}
\mathsf{inc}(\emptyset) & \to \emptyset & \mathsf{inc}(\mathsf{ins}(x, ys)) & \to \mathsf{ins}(x + 1, \mathsf{inc}(ys)) \\
\mathsf{nats}(0, 0) & \to \mathsf{ins}(0, \emptyset) & \mathsf{nats}(0, y + 1) & \to \mathsf{ins}(0, \mathsf{nats}(1, y + 1)) \\
\mathsf{nats}(x + 1, 0) & \to \emptyset & \mathsf{nats}(x + 1, y + 1) & \to \mathsf{inc}(\mathsf{nats}(x, y))
\end{array}
$$

We get four dependency pairs in $\mathsf{DP}(\mathcal{R})$.

$$\mathsf{inc}^\sharp(\mathsf{ins}(x, ys)) \to \mathsf{inc}^\sharp(ys) \quad (1)$$

$$\mathsf{nats}^\sharp(0, y+1) \to \mathsf{nats}^\sharp(1, y+1) \quad (2)$$

$$\mathsf{nats}^\sharp(x+1, y+1) \to \mathsf{inc}^\sharp(\mathsf{nats}(x, y)) \quad (3)$$

$$\mathsf{nats}^\sharp(x+1, y+1) \to \mathsf{nats}^\sharp(x, y) \quad (4)$$

Using the fourth dependency pair twice, we can construct the $(\mathsf{DP}(\mathcal{R}), \mathcal{R}, \mathcal{S}, \mathcal{E})$-chain $\mathsf{nats}^\sharp(x+1, y+1) \to \mathsf{nats}^\sharp(x, y)$, $\mathsf{nats}^\sharp(x'+1, y'+1) \to \mathsf{nats}^\sharp(x', y')$ by considering the $\mathcal{PA}$-based substitution $\sigma = \{x \to 1, y \to 1, x' \to 0, y' \to 0\}$. ◊

Using chains, we obtain the following characterization of termination. This is the key result of the dependency pair approach. The proof is similar to the case of ordinary rewriting and can be found in the full version of this paper [5].

**Theorem 9.** *Let $(\mathcal{R}, \mathcal{S}, \mathcal{E})$ be a CES. Then $\xrightarrow{\mathcal{S}}_{\mathcal{PA}\|\mathcal{E}\backslash\mathcal{R}}$ is terminating if and only if there are no infinite minimal $(\mathsf{DP}(\mathcal{R}), \mathcal{R}, \mathcal{S}, \mathcal{E})$-chains.*

In the next section we present a number of techniques for showing absence of infinite chains. In order to show soundness of these techniques independently, and in order to obtain flexibility on the order in which these techniques are applied, we follow the spirit of [11] and use a dependency pair framework for the termination analysis of CESs. This framework operates on *DP problems* $(\mathcal{P}, \mathcal{R}, \mathcal{S}, \mathcal{E})$, where $\mathcal{P}$ is a finite set of dependency pairs and $(\mathcal{R}, \mathcal{S}, \mathcal{E})$ is a CES. DP problems are transformed using *DP processors*. Here, a DP processor is a function that takes a DP problem as input and returns a finite set of DP problems as output. The DP processor $\mathsf{Proc}$ is *sound* iff for all DP problems $(\mathcal{P}, \mathcal{R}, \mathcal{S}, \mathcal{E})$ with an infinite minimal $(\mathcal{P}, \mathcal{R}, \mathcal{S}, \mathcal{E})$-chain there exists a DP problem $(\mathcal{P}', \mathcal{R}', \mathcal{S}', \mathcal{E}') \in \mathsf{Proc}(\mathcal{P}, \mathcal{R}, \mathcal{S}, \mathcal{E})$ with an infinite minimal $(\mathcal{P}', \mathcal{R}', \mathcal{S}', \mathcal{E}')$-chain.

For a termination proof of the CES $(\mathcal{R}, \mathcal{S}, \mathcal{E})$ we now start with the initial DP problem $(\mathsf{DP}(\mathcal{R}), \mathcal{R}, \mathcal{S}, \mathcal{E})$ and recursively apply sound DP processors. If all resulting DP problems have been transformed into the empty set, then termination has been shown.

## 4 DP Processors

This section introduces various sound DP processors. Section 4.1 introduces the estimated dependency graph, which determines which dependency pairs may potentially follow each other in a chain. Section 4.2 presents a subterm criterion in the spirit of [15]. The DP processor of Section 4.3 uses $\mathcal{PA}$-reduction pairs in order to remove dependency pairs. It makes use of the dependencies between function symbols and thus needs to consider only subsets of $\mathcal{R}$, $\mathcal{S}$, and $\mathcal{E}$. Finally, Section 4.4 shows how polynomial interpretations with negative coefficients can be used as $\mathcal{PA}$-reduction pairs. The DP processors presented here are strictly more powerful than the corresponding DP processors presented in our previous work [6]. It is well-known from ordinary rewriting that the refined techniques are often crucial for a successful automatic termination proof [13,15]. Additional DP processors are presented in [5].

### 4.1 Estimated Dependency Graphs

The DP processor introduced in this section decomposes a DP problem into several independent DP problems. The processor relies on the notion of *estimated dependency graphs* in order to determine which dependency pairs may potentially follow each other in a chain. Estimated dependency graphs are also used in the dependency pair method for ordinary rewriting [1].

To estimate whether the dependency pair $s_1 \to t_1[\![C_1]\!]$ may be followed by the dependency pair $s_2 \to t_2[\![C_2]\!]$ in a chain, subterms of $t_1$ which might be reduced by $\xrightarrow{\mathcal{S}}_{\mathcal{PA}\|\mathcal{E}\backslash\mathcal{R}}$ are abstracted by a fresh variable. Then it is checked whether the term obtained from $t_1$ in this way and $s_2$ are $\mathcal{E} \cup \mathcal{S} \cup \mathcal{PA}$-unifiable.

**Definition 10 (Estimated Dependency Graphs).** *Let $(\mathcal{P}, \mathcal{R}, \mathcal{S}, \mathcal{E})$ be a DP problem. The* estimated $(\mathcal{P}, \mathcal{R}, \mathcal{S}, \mathcal{E})$-dependency graph $\mathsf{EDG}(\mathcal{P}, \mathcal{R}, \mathcal{S}, \mathcal{E})$ *has the dependency pairs in $\mathcal{P}$ as nodes and there is an arc from $s_1 \to t_1[\![C_1]\!]$ to $s_2 \to t_2[\![C_2]\!]$ iff $\textsc{cap}(t_1)$ and $s_2$ are $\mathcal{E} \cup \mathcal{S} \cup \mathcal{PA}$-unifiable with a $\mathcal{PA}$-based unifier $\mu$ such that $s_1\mu$ and $s_2\mu$ are normal forms w.r.t. $\to^{>\lambda}_{\mathcal{PA}\|\mathcal{E}\backslash\mathcal{S}}$ and $C_1\mu$ and $C_2\mu$ are $\mathcal{PA}$-valid. Here, $\textsc{cap}$ is defined by*[1]

$$\textsc{cap}(x) = \begin{cases} x & \textit{if } x \textit{ is a variable of sort } \texttt{nat} \\ y & \textit{if } x \textit{ is a variable of sort } \texttt{univ} \end{cases}$$

$$\textsc{cap}(f(t_1, \ldots, t_n)) = \begin{cases} f(\textsc{cap}(t_1), \ldots, \textsc{cap}(t_n)) & \textit{if } f \notin \mathcal{D}(\mathcal{R}) \\ y & \textit{if } f \in \mathcal{D}(\mathcal{R}) \end{cases}$$

*where $y$ is the next variable in an infinite list $y_1, y_2, \ldots$ of fresh variables.*

It can be shown that this estimate is correct, i.e., if $s_1 \to t_1[\![C_1]\!]$ may potentially be followed by $s_2 \to t_2[\![C_2]\!]$ in a $(\mathcal{P}, \mathcal{R}, \mathcal{S}, \mathcal{E})$-chain, then there is a corresponding arc in $\mathsf{EDG}(\mathcal{P}, \mathcal{R}, \mathcal{S}, \mathcal{E})$. Computing $\mathsf{EDG}$ is still hard in general and an implementation could use weaker estimates.

Notice that every infinite $(\mathcal{P}, \mathcal{R}, \mathcal{S}, \mathcal{E})$-chain contains an infinite tail that stays within one *strongly connected component (SCC)* of $\mathsf{EDG}(\mathcal{P}, \mathcal{R}, \mathcal{S}, \mathcal{E})$, and it is thus sufficient to prove the absence of infinite chains for all SCCs separately.

*Example 11.* Recall Example 8 and the dependency pairs (1)–(4). We obtain the following estimated dependency graph $\mathsf{EDG}(\mathsf{DP}(\mathcal{R}), \mathcal{R}, \mathcal{S}, \mathcal{E})$.

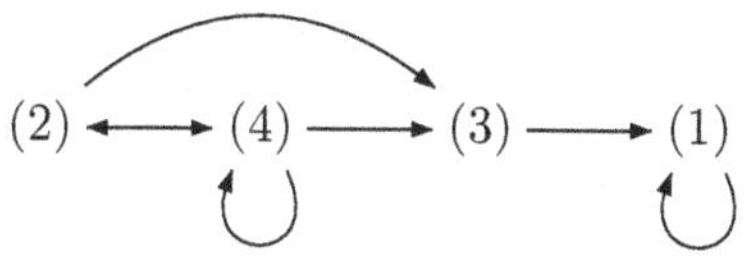


The graph contains two SCCs and it thus suffices to consider the DP problems $(\{(1)\}, \mathcal{R}, \mathcal{S}, \mathcal{E})$ and $(\{(2), (4)\}, \mathcal{R}, \mathcal{S}, \mathcal{E})$ separately. ◊

[1] The full version of this paper [5] considers a slightly refined estimation based on the function TCAP to abstract subterms that might be reduced, where TCAP is similar to the function of the same name used for ordinary rewriting in [12].

**Theorem 12 (DP Processor Based on** $\mathsf{EDG}$**).** *Let* $\mathsf{Proc}$ *be the DP processor with* $\mathsf{Proc}(\mathcal{P}, \mathcal{R}, \mathcal{S}, \mathcal{E}) = \{(\mathcal{P}_1, \mathcal{R}, \mathcal{S}, \mathcal{E}), \ldots, (\mathcal{P}_n, \mathcal{R}, \mathcal{S}, \mathcal{E})\}$*, where* $\mathcal{P}_1, \ldots, \mathcal{P}_n$ *are the SCCs of* $\mathsf{EDG}(\mathcal{P}, \mathcal{R}, \mathcal{S}, \mathcal{E})$*.*[2] *Then* $\mathsf{Proc}$ *is sound.*

### 4.2 Subterm Criterion

The subterm criterion [15] is a relatively simple technique which is none the less surprisingly powerful. In contrast to the DP processor introduced later in Section 4.3, it only needs to consider the dependency pairs in $\mathcal{P}$ and no rules from $\mathcal{R}$ and $\mathcal{S}$ or equations from $\mathcal{E}$ when operating on a DP problem $(\mathcal{P}, \mathcal{R}, \mathcal{S}, \mathcal{E})$. The subterm criterion can thus be used to easily handle many DP problems that do not require the more powerful DP processor introduced in Section 4.3.

For ordinary rewriting, the subterm criterion applies a *simple projection* [15] which collapses a term $f^\sharp(t_1, \ldots, t_n)$ to one of its direct subterms.

**Definition 13 (Simple Projections).** *A* simple projection *is a function* $\pi$ *that assigns to every* $f^\sharp \in \mathcal{F}^\sharp$ *with* $n$ *arguments a position* $i \in \{1, \ldots, n\}$*. The function that maps* $f^\sharp(t_1, \ldots, t_n)$ *to* $t_{\pi(f^\sharp)}$ *is also denoted by* $\pi$*.*

For ordinary rewriting the subterm criterion works as follows [15]. If, after applying a simple projection, the right-hand side is a subterm of the left-hand side for all dependency pairs, then all dependency pairs where this subterm relation is strict may be deleted from a DP problem. Crucial for this is the fact that the syntactic subterm relation $\rhd$ is well-founded.

Notice that the subterm relation modulo $\sim_{\mathcal{E} \cup \mathcal{PA}}$ is not well-founded since $\mathcal{PA}$ is collapsing. Thus, the subterm criterion in our framework uses more sophisticated subterm relations, depending on the sorts of the terms. For terms from $\mathcal{T}(\mathcal{F}_{\mathcal{PA}}, \mathcal{V})$ we use a semantic subterm relation, which also makes use of the constraints that are attached to the dependency pairs.

**Definition 14 (Subterms for Sort** $\mathtt{nat}$**).** *Let* $s, t \in \mathcal{T}(\mathcal{F}_{\mathcal{PA}}, \mathcal{V})$ *and let* $C$ *be a* $\mathcal{PA}$*-constraint. Then* $s[\![C]\!] \rhd_{\mathtt{nat}} t[\![C]\!]$ *iff* $C \Rightarrow s > t$ *is* $\mathcal{PA}$*-valid and* $s[\![C]\!] \unrhd_{\mathtt{nat}} t[\![C]\!]$ *iff* $C \Rightarrow s \geq t$ *is* $\mathcal{PA}$*-valid.*

For example, $x[\![x \geq y+1]\!] \rhd_{\mathtt{nat}} y[\![x \geq y+1]\!]$ since $x \geq y+1 \Rightarrow x > y$ is $\mathcal{PA}$-valid.

If a term $s$ has sort $\mathtt{univ}$, then we take the subterm relation modulo $\sim_{\mathcal{E} \cup \mathcal{PA}}$. Thus, if $s, t$ are terms such that $s$ has sort $\mathtt{univ}$ and $t$ has sort $\mathtt{univ}$ or $\mathtt{nat}$, then $t$ is a subterm of $s$ iff there exist terms $s', t'$ with $s \sim_{\mathcal{E} \cup \mathcal{PA}} s'$ and $t' \sim_{\mathcal{E} \cup \mathcal{PA}} t$ such that $t'$ is a syntactic subterm of $s'$.

**Definition 15 (Subterms for Sort** $\mathtt{univ}$**).** *Let* $(\mathcal{R}, \mathcal{S}, \mathcal{E})$ *be a CES and let* $s, t$ *be terms such that* $s$ *has sort* $\mathtt{univ}$ *and* $t$ *is arbitrary. Then* $t$ *is a* strict subterm *of* $s$*, written* $s \rhd_{\mathtt{univ}} t$*, iff* $s \sim_{\mathcal{E} \cup \mathcal{PA}} \circ \rhd \circ \sim_{\mathcal{E} \cup \mathcal{PA}} t$*. The term* $t$ *is a* subterm *of* $s$*, written* $s \unrhd_{\mathtt{univ}} t$*, iff* $s \rhd_{\mathtt{univ}} t$ *or* $s \sim_{\mathcal{E} \cup \mathcal{PA}} t$*.*

[2] Notice, in particular, that $\mathsf{Proc}(\emptyset, \mathcal{R}, \mathcal{S}, \mathcal{E}) = \emptyset$.

The relation $\rhd_{\mathtt{univ}}$ is not well-founded in general. Thus, in order to make this relation well-founded, we require $\mathcal{E}$ to be *size preserving*. All data structures given in Section 2 satisfy this property.

**Definition 16 (Size Preserving).** *Let $(\mathcal{R}, \mathcal{S}, \mathcal{E})$ be a CES. Then $\mathcal{E}$ is* size preserving *iff $|u| = |v|$ for all $u \approx v \in \mathcal{E}$, where $|t|$ denotes the number of function symbols in the term $t$.*

Before formally stating a DP processor based on the subterm criterion we illustrate it on an example.

*Example 17.* Continuing Example 11 we need to handle the two DP problems $(\{(1)\}, \mathcal{R}, \mathcal{S}, \mathcal{E})$ and $(\{(2), (4)\}, \mathcal{R}, \mathcal{S}, \mathcal{E})$. For the first DP problem consider the simple projection $\pi(\mathsf{inc}^\sharp) = 1$. For (1) we get $\pi(\mathsf{inc}^\sharp(\mathsf{ins}(x, ys))) = \mathsf{ins}(x, ys) \rhd_{\mathtt{univ}} ys = \pi(\mathsf{inc}^\sharp(ys))$ and the only dependency pair can be removed from the DP problem. For the second DP problem, apply the simple projection $\pi(\mathsf{nats}^\sharp) = 2$. Then we have $\pi(\mathsf{nats}^\sharp(0, y + 1)) = y + 1 \ \unrhd_{\mathtt{nat}} \ y + 1 = \pi(\mathsf{nats}^\sharp(1, y + 1))$ for (2) and $\pi(\mathsf{nats}^\sharp(x + 1, y + 1)) = y + 1 \ \rhd_{\mathtt{nat}} y \ = \pi(\mathsf{nats}^\sharp(x, y))$ for (4). Thus, (4) can be deleted and the resulting DP problem is handled by Theorem 12 since the estimated dependency graph $\mathsf{EDG}(\{(2)\}, \mathcal{R}, \mathcal{S}, \mathcal{E})$ does not contain any SCCs. ◊

**Theorem 18 (DP Processor Based on the Subterm Criterion).** *Let $\pi$ be a simple projection. Then* Proc *is sound, where* $\mathsf{Proc}(\mathcal{P}, \mathcal{R}, \mathcal{S}, \mathcal{E}) =$

- $\{(\mathcal{P} - \mathcal{P}', \mathcal{R}, \mathcal{S}, \mathcal{E})\}$, *if $\mathcal{E}$ is size preserving and $\mathcal{P}' \subseteq \mathcal{P}$ such that*
  - $\pi(s) \rhd_{\mathtt{univ}} \pi(t)$ *or* $\pi(s)[\![C]\!] \rhd_{\mathtt{nat}} \pi(t)[\![C]\!]$ *for all* $s \to t[\![C]\!] \in \mathcal{P}'$, *and*
  - $\pi(s) \unrhd_{\mathtt{univ}} \pi(t)$ *or* $\pi(s)[\![C]\!] \unrhd_{\mathtt{nat}} \pi(t)[\![C]\!]$ *for all* $s \to t[\![C]\!] \in \mathcal{P} - \mathcal{P}'$.
- $\{(\mathcal{P}, \mathcal{R}, \mathcal{S}, \mathcal{E})\}$, *otherwise.*

### 4.3 $\mathcal{PA}$-Reduction Pairs and Function Dependencies

**$\mathcal{PA}$-reduction pairs.** The dependency pair framework for ordinary rewriting makes heavy use of reduction pairs $(\succsim, \succ)$ [16]. If all dependency pairs are decreasing w.r.t. $\succsim \cup \succ$, then those decreasing w.r.t. $\succ$ may be deleted from a DP problem. Contrary to Section 4.2, the rewrite rules have to be considered for this as well since (a certain subset of) $\mathcal{R}$ needs to be decreasing w.r.t. $\succsim$.

In our setting we can relax the requirement that $\succsim$ needs to be monotonic[3]. We still require $\succsim$ to be *$\mathcal{F}$-monotonic*, i.e., $s \succsim t$ implies $D[s] \succsim D[t]$ for all contexts $D$ over $\mathcal{F} \cup \mathcal{F}_{\mathcal{PA}}$ and all terms $s, t \in \mathcal{T}(\mathcal{F} \cup \mathcal{F}_{\mathcal{PA}}, \mathcal{V})$. Monotonicity for contexts with a symbol $f^\sharp \in \mathcal{F}^\sharp$ at its root is only required for positions where a reduction with $\to_{\mathcal{PA}\|\mathcal{E}\backslash\mathcal{S}}$ or $\xrightarrow{\mathcal{S}}_{\mathcal{PA}\|\mathcal{E}\backslash\mathcal{R}}$ is possible. A similar observation was already used in [1] for proving innermost termination of ordinary rewriting.

**Definition 19 ($f^\sharp$-monotonic Relations).** *Let $\bowtie$ be a relation on terms and let $f^\sharp \in \mathcal{F}^\sharp$. Then $\bowtie$ is* $f^\sharp$-monotonic at position $i$ *iff $s \bowtie t$ implies $D^\sharp[s] \bowtie D^\sharp[t]$ for all contexts $D = f(s_1, \ldots, s_{i-1}, \Box, s_{i+1}, \ldots, s_n)$ over $\mathcal{F} \cup \mathcal{F}_{\mathcal{PA}}$ and all terms $s, t \in \mathcal{T}(\mathcal{F} \cup \mathcal{F}_{\mathcal{PA}}, \mathcal{V})$.*

[3] A relation $\bowtie$ on terms is *monotonic* iff $s \bowtie t$ implies $D[s] \bowtie D[t]$ for all contexts $D$.

When considering the DP problem $(\mathcal{P}, \mathcal{R}, \mathcal{S}, \mathcal{E})$, the reduction pair needs to be $f^\sharp$-monotonic at position $i$ only if $\mathcal{P}$ contains a dependency pair of the form $s \to f^\sharp(t_1, \ldots, t_i, \ldots, t_n)[\![C]\!]$ where $t_i \notin \mathcal{T}(\mathcal{F}_{\mathcal{PA}}, \mathcal{V})$. The reason for this is that no instance of a term from $\mathcal{T}(\mathcal{F}_{\mathcal{PA}}, \mathcal{V})$ can be reduced using $\xrightarrow{\mathcal{S}}_{\mathcal{PA}\|\mathcal{E}\backslash\mathcal{R}}$ or $\to_{\mathcal{PA}\|\mathcal{E}\backslash\mathcal{S}}$ in a $(\mathcal{P}, \mathcal{R}, \mathcal{S}, \mathcal{E})$-chain since the substitution used for the chain is $\mathcal{PA}$-based. It thus suffices to require $f^\sharp$-monotonicity for all *reducible positions*, which are determined by the DP problem under consideration.

**Definition 20 (Reducible Positions).** *Let $\mathcal{P}$ be a set of dependency pairs and let $f^\sharp \in \mathcal{F}^\sharp$. Then the set of* reducible positions *is defined by $\mathcal{R}ed\mathcal{P}os(f^\sharp, \mathcal{P}) = \{i \mid$ there exists $s \to f^\sharp(t_1, \ldots, t_i, \ldots, t_n)[\![C]\!] \in \mathcal{P}$ such that $t_i \notin \mathcal{T}(\mathcal{F}_{\mathcal{PA}}, \mathcal{V})\}$.*

Finally, reduction pairs need to satisfy a property that relates $\vdash\!\dashv_{\mathcal{PA}}$ and $\gtrsim \cap \gtrsim^{-1}$.

**Definition 21 ($\mathcal{PA}$-compatible Relations).** *Let $\bowtie$ be a relation on terms. Then $\bowtie$ is $\mathcal{PA}$-*compatible *iff, for all terms $s, t \in \mathcal{T}(\mathcal{F} \cup \mathcal{F}_{\mathcal{PA}}, \mathcal{V})$, $s \vdash\!\dashv_{\mathcal{PA}} t$ implies $D[s] \bowtie \cap \bowtie^{-1} D[t]$ for all contexts $D$ over $\mathcal{F} \cup \mathcal{F}_{\mathcal{PA}}$ and $D^\sharp[s] \bowtie \cap \bowtie^{-1} D^\sharp[t]$ for all contexts $D \neq \Box$ over $\mathcal{F} \cup \mathcal{F}_{\mathcal{PA}}$.*

Our notion of $\mathcal{PA}$-reduction pairs depends on the DP problem under consideration. It is similar to the notion of *generalized reduction pairs* [14] in the sense that full monotonicity is not required. However, the generalized reduction pairs of [14] are only applicable in the context of innermost termination of ordinary rewriting.

**Definition 22 ($\mathcal{PA}$-reduction Pairs).** *Let $(\mathcal{P}, \mathcal{R}, \mathcal{S}, \mathcal{E})$ be a DP problem and let $\gtrsim$ and $\succ$ be relations on terms such that $\succ$ is well-founded, $\gtrsim$ is $\mathcal{F}$-monotonic and $\mathcal{PA}$-compatible, and $\gtrsim$ is $f^\sharp$-monotonic at position $i$ for all $f^\sharp \in \mathcal{F}^\sharp$ and all $i \in \mathcal{R}ed\mathcal{P}os(f^\sharp, \mathcal{P})$. Then $(\gtrsim, \succ)$ is a $\mathcal{PA}$-*reduction pair *for $(\mathcal{P}, \mathcal{R}, \mathcal{S}, \mathcal{E})$ iff $\succ$ is compatible with $\gtrsim$, i.e., iff $\gtrsim \circ \succ \subseteq \succ$ or $\succ \circ \gtrsim \subseteq \succ$. The relation $\gtrsim \cap \gtrsim^{-1}$ is denoted by $\sim$.*

Section 4.4 shows how this definition allows the use of polynomial interpretations with negative coefficients

Notice that we do not require $\gtrsim$ or $\succ$ to be stable[4]. Indeed, stability for all substitutions is not needed since we only need this property for certain $\mathcal{PA}$-based substitutions that can be used in $(\mathcal{P}, \mathcal{R}, \mathcal{S}, \mathcal{E})$-chains. These substitutions are indirectly given by the constraints of the dependency pairs and rules that are to be oriented.

**Definition 23 ($\mathcal{PA}$-reduction Pairs on Constrained Terms).** *Let $(\gtrsim, \succ)$ be a $\mathcal{PA}$-reduction pair. Let $s, t$ be terms and let $C$ be a $\mathcal{PA}$-constraint. Then $s[\![C]\!] \gtrsim t[\![C]\!]$ iff $s\sigma \gtrsim t\sigma$ for all $\mathcal{PA}$-based substitutions $\sigma$ such that $C\sigma$ is $\mathcal{PA}$-valid. $s[\![C]\!] \succ t[\![C]\!]$ is defined analogously.*

---

[4] A relation $\bowtie$ is *stable* iff $s \bowtie t$ implies $s\sigma \bowtie t\sigma$ for all substitutions $\sigma$.

**Function Dependencies.** The DP processor based on $\mathcal{PA}$-reduction pairs makes use of the observation that only certain subsets of $\mathcal{R}$, $\mathcal{S}$, and $\mathcal{E}$ need to be considered. The corresponding result for ordinary rewriting is due to [13,15], where it is also shown that the power of the corresponding DP processor is strictly increased. The main idea is to show that each $(\mathcal{P}, \mathcal{R}, \mathcal{S}, \mathcal{E})$-chain can be transformed into a sequence that only uses subsets $\mathcal{R}' \subseteq \mathcal{R}$, $\mathcal{S}' \subseteq \mathcal{S}$, and $\mathcal{E}' \subseteq \mathcal{E}$. This sequence will not necessarily be a $(\mathcal{P}, \mathcal{R}, \mathcal{S}, \mathcal{E})$-chain in our setting, but this is not needed for soundness. A complete proof of the technical result in this section is given in the full version of this paper [5]. The subsets of $\mathcal{R}$, $\mathcal{S}$, and $\mathcal{E}$ are based on the dependencies between function symbols. Similar definitions are also used in [13,15].

**Definition 24 (Function Dependencies).** *Let $(\mathcal{P}, \mathcal{R}, \mathcal{S}, \mathcal{E})$ be a DP problem where $\mathcal{E}$ is size preserving. For two function symbols $f, g \in \mathcal{F}$ let $f \sqsupset_{(\mathcal{P},\mathcal{R},\mathcal{S},\mathcal{E})} g$ iff (i) there exists a rule $l \to r[\![C]\!] \in \mathcal{R} \cup \mathcal{S}$ such that $\text{root}(l) = f$ and $g \in \mathcal{F}(r)$, or (ii) there exists an equation $u \approx v$ or $v \approx u$ in $\mathcal{E}$ such that $\text{root}(u) = f$ and $g \in \mathcal{F}(u \approx v)$. Let $\Delta(\mathcal{P}, \mathcal{R}, \mathcal{S}, \mathcal{E}) = \mathcal{F}_{\mathcal{PA}} \cup \{g \mid f \sqsupset^*_{(\mathcal{P},\mathcal{R},\mathcal{S},\mathcal{E})} g$ for some $f \in \mathcal{F}$ with resulting sort $\texttt{nat}$ or some $f \in \mathcal{F}(\text{rhs}(\mathcal{P})) - \mathcal{F}^\sharp\}$, where $\text{rhs}(\mathcal{P})$ denotes the set of right-hand sides of the dependency pairs in $\mathcal{P}$.*

Subsets $\Delta \subseteq \mathcal{F} \cup \mathcal{F}_{\mathcal{PA}}$ give rise to subsets of $\mathcal{R}$, $\mathcal{S}$, and $\mathcal{E}$ in the obvious way.

**Definition 25 ($\mathcal{R}(\Delta)$, $\mathcal{S}(\Delta)$, and $\mathcal{E}(\Delta)$).** *Let $\Delta \subseteq \mathcal{F} \cup \mathcal{F}_{\mathcal{PA}}$. For $\mathcal{Q} \in \{\mathcal{R}, \mathcal{S}\}$ we define $\mathcal{Q}(\Delta) = \{l \to r[\![C]\!] \in \mathcal{Q} \mid \text{root}(l) \in \Delta\}$ and we let $\mathcal{E}(\Delta) = \{u \approx v \in \mathcal{E} \mid \text{root}(u) \in \Delta$ or $\text{root}(v) \in \Delta\}$.*

*Example 26.* Continuing Example 11, recall the DP problems $(\{(1)\}, \mathcal{R}, \mathcal{S}, \mathcal{E})$ and $(\{(2), (4)\}, \mathcal{R}, \mathcal{S}, \mathcal{E})$. These DP problems are hard to handle if all of $\mathcal{R}, \mathcal{S}$, and $\mathcal{E}$ need to be considered. Using function dependencies we get $\Delta = \mathcal{F}_{\mathcal{PA}}$ for both DP problems. Thus, $\mathcal{R}$, $\mathcal{S}$, and $\mathcal{E}$ do not need to be considered at all when handling these DP problems. Using the $\mathcal{PA}$-reduction pair $(\succsim_{\mathcal{P}ol}, \succ_{\mathcal{P}ol})$ based on a polynomial interpretation with $\mathcal{P}ol(\mathsf{inc}^\sharp) = x_1$, $\mathcal{P}ol(\mathsf{ins}) = x_2 + 1$, $\mathcal{P}ol(\mathsf{nats}^\sharp) = x_2$, $\mathcal{P}ol(\mathsf{0}) = 0$, $\mathcal{P}ol(\mathsf{1}) = 1$ and $\mathcal{P}ol(+) = x_1 + x_2$, the DP problem $(\{(1)\}, \mathcal{R}, \mathcal{S}, \mathcal{E})$ is transformed into the trivial DP problem $(\emptyset, \mathcal{R}, \mathcal{S}, \mathcal{E})$, while the DP problem $(\{(2), (4)\}, \mathcal{R}, \mathcal{S}, \mathcal{E})$ is transformed into $(\{(2)\}, \mathcal{R}, \mathcal{S}, \mathcal{E})$ whose estimated dependency graph does not contain any SCCs. ◊

The following DP processor is based on $\mathcal{PA}$-reduction pairs and function dependencies. Again, recall that $\mathcal{E}$ is size preserving for all data structures given in Section 2.

**Theorem 27 (DP Processor Based on Function Dependencies).** *The DP processor* Proc *is sound, where* $\mathsf{Proc}(\mathcal{P}, \mathcal{R}, \mathcal{S}, \mathcal{E}) =$

- $\{(\mathcal{P} - \mathcal{P}', \mathcal{R}, \mathcal{S}, \mathcal{E})\}$, *if $(\succsim, \succ)$ is a $\mathcal{PA}$-reduction pair for $(\mathcal{P}, \mathcal{R}, \mathcal{S}, \mathcal{E})$, $\mathcal{E}$ is size preserving, $\mathcal{P}' \subseteq \mathcal{P}$, and $\Delta = \Delta(\mathcal{P}, \mathcal{R}, \mathcal{S}, \mathcal{E})$ such that*
  - *$s[\![C]\!] \succ t[\![C]\!]$ for all $s \to t[\![C]\!] \in \mathcal{P}'$,*

  - $s[\![C]\!] \gtrsim t[\![C]\!]$ *for all* $s \to t[\![C]\!] \in (\mathcal{P} - \mathcal{P}') \cup \mathcal{R}(\Delta) \cup \mathcal{S}(\Delta) \cup \mathcal{R}_\Pi$, *and*[5]
  - $u[\![\top]\!] \sim v[\![\top]\!]$ *for all* $u \approx v \in \mathcal{E}(\Delta)$.[6]
- $\{(\mathcal{P}, \mathcal{R}, \mathcal{S}, \mathcal{E})\}$, *otherwise.*

If $\mathcal{E}$ is not size preserving a similar DP processor can be used which has to consider all of $\mathcal{R}$, $\mathcal{S}$, and $\mathcal{E}$. In this case, $\mathcal{R}_\Pi$ does not need to be oriented.

### 4.4 Generation of $\mathcal{PA}$-Reduction Pairs

To take advantage of the relaxed requirements on monotonicity and stability and in order to make use of the constraints attached to dependency pairs and rules, we propose to use relations based on polynomial interpretations [18] with polynomials containing coefficients from $\mathbb{Z}$. A similar kind of polynomial interpretations was used in [14] in the context of innermost termination of ordinary rewriting. The polynomial interpretations with coefficients in $\mathbb{Z}$ used in [15] are also similar but require the use of "max", which makes reasoning about them more complicated. Furthermore, the approach of [15] requires that rewrite rules are treated like equations, which is very restrictive. Using our approach, rewrite rules do not need to be oriented as equations.

A $\mathcal{PA}$*-polynomial interpretation* maps

1. $\mathcal{F}_{\mathcal{PA}}$ to polynomials over $\mathbb{N}$ in the natural way, i.e., $\mathcal{P}ol(0) = 0$, $\mathcal{P}ol(1) = 1$, and $\mathcal{P}ol(+) = x_1 + x_2$,
2. $\mathcal{F}$ to polynomials over $\mathbb{N}$ where $\mathcal{P}ol(f) \in \mathbb{N}[x_1, \ldots, x_n]$ if $f$ has arity $n$, and
3. $\mathcal{F}^\sharp$ to polynomials over $\mathbb{Z}$ where $\mathcal{P}ol(f^\sharp) \in \mathbb{Z}[x_1, \ldots, x_n]$ if $f^\sharp$ has arity $n$.

This mapping is extended to terms by letting $[x]_{\mathcal{P}ol} = x$ for all variables $x \in \mathcal{V}$ and $[f(t_1, \ldots, t_n)]_{\mathcal{P}ol} = \mathcal{P}ol(f)([t_1]_{\mathcal{P}ol}, \ldots, [t_n]_{\mathcal{P}ol})$ for all $f \in \mathcal{F} \cup \mathcal{F}^\sharp$.

$\mathcal{PA}$-polynomial interpretations generate relations on terms in the following way. Here, $\geq^{(\mathbb{Z},\mathbb{N})}$ and $>^{(\mathbb{Z},\mathbb{N})}$ mean that the respective relation holds in the integers for all instantiations of the variables by *natural numbers.*

**Definition 28 (Relations $\succ_{\mathcal{P}ol}$ and $\gtrsim_{\mathcal{P}ol}$).** *Let $\mathcal{P}ol$ be a $\mathcal{PA}$-polynomial interpretation. Then $\succ_{\mathcal{P}ol}$ is defined by $s \succ_{\mathcal{P}ol} t$ iff $[s]_{\mathcal{P}ol} \geq^{(\mathbb{Z},\mathbb{N})} 0$ and $[s]_{\mathcal{P}ol} >^{(\mathbb{Z},\mathbb{N})} [t]_{\mathcal{P}ol}$. Similarly, $\gtrsim_{\mathcal{P}ol}$ is defined by $s \gtrsim_{\mathcal{P}ol} t$ iff $[s]_{\mathcal{P}ol} \geq^{(\mathbb{Z},\mathbb{N})} [t]_{\mathcal{P}ol}$.*

The relations $\gtrsim_{\mathcal{P}ol}$ and $\succ_{\mathcal{P}ol}$ give rise to $\mathcal{PA}$-reduction pairs, where $f^\sharp$-monotonicity is translated into conditions on the polynomial $\mathcal{P}ol(f^\sharp)$.

**Theorem 29.** *Let $(\mathcal{P}, \mathcal{R}, \mathcal{S}, \mathcal{E})$ be a DP problem and let $\mathcal{P}ol$ be a $\mathcal{PA}$-polynomial interpretation. Then $(\gtrsim_{\mathcal{P}ol}, \succ_{\mathcal{P}ol})$ is a $\mathcal{PA}$-reduction pair if $\mathcal{P}ol(f^\sharp)$ is weakly increasing in all $x_i$ for which $i \in \mathcal{R}ed\mathcal{P}os(f^\sharp, \mathcal{P})$.*

In order to show $s[\![C]\!] \succ_{\mathcal{P}ol} t[\![C]\!]$ it suffices to show that $C \Rightarrow [s]_{\mathcal{P}ol} \geq 0$ and $C \Rightarrow [s]_{\mathcal{P}ol} > [t]_{\mathcal{P}ol}$ are $(\mathbb{Z}, \mathbb{N})$-valid, i.e., true in the integers for all instantiations of the variables by natural numbers. While this is undecidable in general

[5] Here, $\mathcal{R}_\Pi = \{\Pi(x, y) \to x,\ \Pi(x, y) \to y\}$ for a fresh function symbol $\Pi$.

[6] This condition ensures $u\sigma \sim v\sigma$ for all $\mathcal{PA}$-based substitutions $\sigma$.

it is decidable if all polynomials are linear. Automatic generation of suitable $\mathcal{PA}$-polynomial interpretations is possible by adapting techniques developed for ordinary polynomial interpretations based on solving non-linear Diophantine constraints [4,8].

*Example 30.* One of the leading examples from [14] (obtained from the imperative program `while (x > y) { y = y + 1; }`) can be given more elegantly using built-in numbers. In this example we have $\mathcal{E} = \mathcal{S} = \emptyset$ and there is only a single rewrite rule:

$$\mathsf{eval}(x, y) \to \mathsf{eval}(x, y+1)\ [\![x > y]\!]$$

The only dependency pair is identical to this rule, but with $\mathsf{eval}$ replaced by $\mathsf{eval}^\sharp$. Since $\mathcal{R}(\Delta) = \emptyset$ we do not need consider the rewrite rule when applying the DP processor of Theorem 27, i.e., it suffices to find a $\mathcal{PA}$-reduction pair $(\gtrsim, \succ)$ such that $\mathsf{eval}^\sharp(x, y)[\![x > y]\!] \succ \mathsf{eval}^\sharp(x, y+1)[\![x > y]\!]$. For this, consider a $\mathcal{PA}$-polynomial interpretation with $\mathcal{P}ol(\mathsf{eval}^\sharp) = x_1 - x_2$. We then have $\mathsf{eval}^\sharp(x, y)[\![x > y]\!] \succ_{\mathcal{P}ol} \mathsf{eval}^\sharp(x, y+1)[\![x > y]\!]$ since $x > y \Rightarrow x - y \geq 0$ and $x > y \Rightarrow x - y > x - (y+1)$ are $(\mathbb{Z}, \mathbb{N})$-valid. ◊

This example demonstrates that built-in numbers are useful for termination proofs of imperative programs operating on numbers since the termination proof is much simpler than the one given in [14]. A large collection of termination proofs for imperative programs is given in the full version of this paper [5].

## 5 Conclusions and Future Work

We have presented the notion of constrained equational systems for modeling algorithms. Constrained equational systems support semantic data structures and contain built-in numbers. These systems are a strict generalization of the equational systems presented in [6], which do no contain built-in numbers. Since virtually all programming languages have numbers as a primitive data type this extension is helpful for modeling algorithms. We have presented a dependency pair framework for proving termination of constrained equational systems and developed several DP processors within this framework. Most of these DP processors are not available in our previous work [6], while it is well-known from ordinary rewriting that the techniques employed in these DP processors are often crucial for a successful automatic termination proof [13,15]. The techniques of this paper have been successfully used on a large collection of examples [5].

Termination is only one among several important properties of constrained equational systems. We will study other properties as well, specifically confluence and sufficient completeness. Orthogonal to this we will investigate how the rewrite relation can be generalized by considering other built-in theories, in particular integers instead of natural numbers. The proposed method has not been implemented yet, but we believe that it can be easily implemented within a termination tool like AProVE [10].

## References


1. Arts, T., Giesl, J.: Termination of term rewriting using dependency pairs. TCS 236(1–2), 133–178 (2000)
2. Baader, F., Nipkow, T.: Term Rewriting and All That. Cambridge University Press, Cambridge (1998)
3. Blanqui, F., Hardin, T., Weis, P.: On the implementation of construction functions for non-free concrete data types. In: De Nicola, R. (ed.) ESOP 2007. LNCS, vol. 4421, pp. 95–109. Springer, Heidelberg (2007)
4. Contejean, E., Marché, C., Tomás, A.P., Urbain, X.: Mechanically proving termination using polynomial interpretations. JAR 34(4), 325–363 (2005)
5. Falke, S., Kapur, D.: Dependency pairs for rewriting with built-in numbers and semantic data structures. Technical Report TR-CS-2007-21, Department of Computer Science, University of New Mexico (2007), `http://www.cs.unm.edu/research/tech-reports/`
6. Falke, S., Kapur, D.: Dependency pairs for rewriting with non-free constructors. In: Pfenning, F. (ed.) CADE 2007. LNCS (LNAI), vol. 4603, pp. 426–442. Springer, Heidelberg (2007)
7. Falke, S., Kapur, D.: Operational termination of conditional rewriting with built-in numbers and semantic data structures. Technical Report TR-CS-2007-22, Department of Computer Science, University of New Mexico (2007), `http://www.cs.unm.edu/research/tech-reports/`
8. Fuhs, C., Giesl, J., Middeldorp, A., Schneider-Kamp, P., Thiemann, R., Zankl, H.: SAT solving for termination analysis with polynomial interpretations. In: Marques-Silva, J., Sakallah, K.A. (eds.) SAT 2007. LNCS, vol. 4501, pp. 340–354. Springer, Heidelberg (2007)
9. Giesl, J., Kapur, D.: Dependency pairs for equational rewriting. In: Middeldorp, A. (ed.) RTA 2001. LNCS, vol. 2051, pp. 93–108. Springer, Heidelberg (2001)
10. Giesl, J., Schneider-Kamp, P., Thiemann, R.: AProVE 1.2: Automatic termination proofs in the dependency pair framework. In: Furbach, U., Shankar, N. (eds.) IJCAR 2006. LNCS (LNAI), vol. 4130, pp. 281–286. Springer, Heidelberg (2006)
11. Giesl, J., Thiemann, R., Schneider-Kamp, P.: The dependency pair framework: Combining techniques for automated termination proofs. In: Baader, F., Voronkov, A. (eds.) LPAR 2004. LNCS (LNAI), vol. 3452, pp. 301–331. Springer, Heidelberg (2005)
12. Giesl, J., Thiemann, R., Schneider-Kamp, P.: Proving and disproving termination of higher-order functions. In: Gramlich, B. (ed.) FroCos 2005. LNCS (LNAI), vol. 3717, pp. 216–231. Springer, Heidelberg (2005)
13. Giesl, J., Thiemann, R., Schneider-Kamp, P., Falke, S.: Mechanizing and improving dependency pairs. JAR 37(3), 155–203 (2006)
14. Giesl, J., Thiemann, R., Swiderski, S., Schneider-Kamp, P.: Proving termination by bounded increase. In: Pfenning, F. (ed.) CADE 2007. LNCS (LNAI), vol. 4603, pp. 443–459. Springer, Heidelberg (2007)
15. Hirokawa, N., Middeldorp, A.: Tyrolean termination tool: Techniques and features. IC 205(4), 474–511 (2007)
16. Kusakari, K., Nakamura, M., Toyama, Y.: Argument filtering transformation. In: Nadathur, G. (ed.) PPDP 1999. LNCS, vol. 1702, pp. 47–61. Springer, Heidelberg (1999)
17. Kusakari, K., Toyama, Y.: On proving AC-termination by AC-dependency pairs. IEICE Transactions on Information and Systems E84-D(5), 604–612 (2001)

18. Lankford, D.S.: On proving term rewriting systems are Noetherian. Memo MTP-3, Mathematics Department, Louisiana Tech University, Ruston (1979)
19. Marché, C.: Normalized rewriting: An alternative to rewriting modulo a set of equations. JSC 21(3), 253–288 (1996)
20. Marché, C., Urbain, X.: Modular and incremental proofs of AC-termination. JSC 38(1), 873–897 (2004)
21. Peterson, G.E., Stickel, M.E.: Complete sets of reductions for some equational theories. J. ACM 28(2), 233–264 (1981)
22. Presburger, M.: Über die Vollständigkeit eines gewissen Systems der Arithmetik ganzer Zahlen, in welchem die Addition als einzige Operation hervortritt. In: Comptes Rendus du Premier Congrès de Mathématiciens des Pays Slaves, pp. 92–101 (1929)
23. Steinbach, J., Kühler, U.: Check your ordering—termination proofs and open problems. Technical Report SR-90-25, Universität Karlsruhe (1990)

# Maximal Termination*

Carsten Fuhs[1], Jürgen Giesl[1], Aart Middeldorp[2], Peter Schneider-Kamp[1], René Thiemann[2], and Harald Zankl[2]

[1] LuFG Informatik 2, RWTH Aachen University, Germany
[2] Institute of Computer Science, University of Innsbruck, Austria

**Abstract.** We present a new approach for termination proofs that uses polynomial interpretations (with possibly negative coefficients) together with the "maximum" function. To obtain a powerful automatic method, we solve two main challenges: (1) We show how to adapt the latest developments in the dependency pair framework to our setting. (2) We show how to automate the search for such interpretations by integrating "max" into recent SAT-based methods for polynomial interpretations. Experimental results support our approach.

## 1 Introduction

The use of *polynomial interpretations* [13] is standard in automated termination analysis of term rewrite systems (TRSs). This is especially true for termination proofs in the popular *dependency pair (DP) framework* [1,4,6,9] that is implemented in most automated termination tools for TRSs.

A *polynomial interpretation* $\mathcal{P}ol$ maps every $n$-ary function symbol $f$ to a polynomial $f_{\mathcal{P}ol}$ over $n$ variables $x_1, \ldots, x_n$. The mapping is extended to terms by defining $[x]_{\mathcal{P}ol} = x$ for variables $x$ and $[f(t_1, ..., t_n)]_{\mathcal{P}ol} = f_{\mathcal{P}ol}([t_1]_{\mathcal{P}ol}, ..., [t_n]_{\mathcal{P}ol})$. If $\mathcal{P}ol$ is clear from the context, we also write $[t]$ instead of $[t]_{\mathcal{P}ol}$. Traditionally, one uses polynomials with *natural* coefficients from $\mathbb{N} = \{0, 1, 2, \ldots\}$. Then $[t] \in \mathbb{N}$ for every ground term $t$. For example, consider the interpretation $\mathcal{P}ol$ with $\mathsf{0}_{\mathcal{P}ol} = 0$, $\mathsf{s}_{\mathcal{P}ol} = x_1 + 1$, and $\mathsf{minus}_{\mathcal{P}ol} = x_1$. Then $[\mathsf{minus}(\mathsf{s}(x), \mathsf{s}(y))]_{\mathcal{P}ol} = x + 1$.

An interpretation $\mathcal{P}ol$ induces an order $\succ_{\mathcal{P}ol}$ and quasi-order $\succsim_{\mathcal{P}ol}$ where $s \succ_{\mathcal{P}ol} t$ ($s \succsim_{\mathcal{P}ol} t$) iff $[s] > [t]$ ($[s] \geqslant [t]$) holds for all instantiations of variables with natural numbers. So with $\mathcal{P}ol$ above we have $\mathsf{minus}(\mathsf{s}(x), \mathsf{s}(y)) \succ_{\mathcal{P}ol} \mathsf{minus}(x, y)$. Recently, two extensions to *integer* polynomials were proposed:

(a) [7] used polynomial interpretations with integer coefficients where ground terms could also be mapped to arbitrary integers. However, this approach only works for analyzing *innermost* instead of *full* termination.
(b) [10] proposed interpretations of the form $\max(p, 0)$ where $p$ is a polynomial with integer coefficients. Thus, ground terms are still mapped to numbers from $\mathbb{N}$. So one could define $\mathsf{minus}_{\mathcal{P}ol} = \max(x_1 - x_2, 0)$ which would result in $\mathsf{minus}(\mathsf{s}(x), \mathsf{s}(y)) \approx_{\mathcal{P}ol} \mathsf{minus}(x, y)$. Here $\approx_{\mathcal{P}ol}$ denotes the equivalence relation associated with $\succsim_{\mathcal{P}ol}$, where for any quasi-order $\succsim$ we have $\approx \, = \, \succsim \cap \precsim$.

* Supported by the DFG (Deutsche Forschungsgemeinschaft) under grant GI 274/5-2 and the FWF (Austrian Science Fund) project P18763.

A. Voronkov (Ed.): RTA 2008, LNCS 5117, pp. 110–125, 2008.

The drawback is that the approach of [10] was not easy to automate and that it could only be combined with a weak version of the DP technique.

In this paper, we present a new approach which improves upon (a) and (b):

- It uses integer polynomials together with the function "max", where ground terms are only mapped to natural numbers, as in [10]. But in contrast to [10], we permit arbitrary combinations of polynomials and "max", e.g., "$p + \max(q, \max(r, s))$" where $p, q, r, s$ are integer polynomials. And in contrast to [7], integer polynomials may be used for interpreting *any* function symbol.
- It uses the newest and most powerful version of the DP technique as in [7].
- In contrast to [7], it can also prove *full* instead of *innermost* termination.
- In contrast to [10], we show how to search for arbitrary polynomial interpretations with "max" automatically in an efficient way using SAT solving.

After recapitulating the DP framework in Sect. 2, Sect. 3 extends it to handle *non-monotonic* quasi-orders like integer polynomial orders with "max". Sect. 4 shows how to search for such interpretations automatically using SAT solving. Sect. 5 discusses our implementation in the provers AProVE [5] and $\mathsf{T_TT_2}$ [17].

## 2 Dependency Pairs

For a TRS $\mathcal{R}$, the *defined* symbols $\mathcal{D}$ are the root symbols of left-hand sides of rules. All other function symbols are called *constructors*. For every defined symbol $f \in \mathcal{D}$, we introduce a fresh *tuple symbol* $f^\sharp$ with the same arity. To ease readability, we often write $F$ instead of $f^\sharp$, etc. If $t = f(t_1, \ldots, t_n)$ with $f \in \mathcal{D}$, we write $t^\sharp$ for $f^\sharp(t_1, \ldots, t_n)$. If $\ell \to r \in \mathcal{R}$ and $t$ is a subterm of $r$ with defined root symbol, then the rule $\ell^\sharp \to t^\sharp$ is a *dependency pair* of $\mathcal{R}$. We denote the set of all dependency pairs of $\mathcal{R}$ by $\mathit{DP}(\mathcal{R})$.

*Example 1.* Consider the TRS $\mathsf{SUBST}$ from [8] and [18, Ex. 6.5.42]:

$$\begin{array}{lll} \lambda(x) \circ y \to \lambda(x \circ (1 \star (y \circ \uparrow))) & \mathsf{id} \circ x \to x & 1 \circ (x \star y) \to x \\ (x \star y) \circ z \to (x \circ z) \star (y \circ z) & 1 \circ \mathsf{id} \to 1 & \uparrow \circ (x \star y) \to y \\ (x \circ y) \circ z \to x \circ (y \circ z) & \uparrow \circ \mathsf{id} \to \uparrow & \end{array}$$

The dependency pairs are

$$\begin{array}{llr} \lambda(x) \circ^\sharp y \to x \circ^\sharp (1 \star (y \circ \uparrow)) \quad (1) & & (x \star y) \circ^\sharp z \to y \circ^\sharp z \\ \lambda(x) \circ^\sharp y \to y \circ^\sharp \uparrow \quad (2) & & (x \circ y) \circ^\sharp z \to x \circ^\sharp (y \circ z) \\ (x \star y) \circ^\sharp z \to x \circ^\sharp z & & (x \circ y) \circ^\sharp z \to y \circ^\sharp z \end{array}$$

The main result of the DP framework states that a TRS $\mathcal{R}$ is terminating iff there is no infinite *minimal $\mathit{DP}(\mathcal{R})$-chain*. For any set of dependency pairs $\mathcal{P}$, a *minimal $\mathcal{P}$-chain* is a sequence of (variable renamed) pairs $s_1 \to t_1, s_2 \to t_2, \ldots$ from $\mathcal{P}$ such that there is a substitution $\sigma$ (with possibly infinite domain) where $t_i\sigma \to_{\mathcal{R}}^* s_{i+1}\sigma$ and where all $t_i\sigma$ are terminating w.r.t. $\mathcal{R}$.

The DP framework has several techniques (so-called *DP processors*) to prove absence of infinite chains. Thm. 2 recapitulates one of the most important processors, the so-called *reduction pair processor*. It uses *reduction pairs* $(\succsim, \succ)$ to compare terms. Here, $\succsim$ is a stable monotonic quasi-order and $\succ$ is a stable well-founded order, where $\succsim$ and $\succ$ are compatible (i.e., $\succ \circ \succsim \subseteq \succ$ or $\succsim \circ \succ \subseteq \succ$).

If $\mathcal{P}$ is the current set of dependency pairs,[1] then the reduction pair processor generates inequality constraints which should be satisfied by a reduction pair $(\succsim, \succ)$. The constraints require that all DPs in $\mathcal{P}$ are strictly or weakly decreasing and all *usable rules* $\mathcal{U}(\mathcal{P})$ are weakly decreasing. Then one can delete all strictly decreasing DPs from $\mathcal{P}$. Afterwards, the reduction pair processor can be applied again to the remaining set of DPs (possibly using a different reduction pair). This process is repeated until all DPs have been removed.

The *usable rules* include all rules that can reduce the terms in right-hand sides of $\mathcal{P}$ when their variables are instantiated with normal forms. To ensure that it suffices to regard only the *usable* rules instead of *all* rules in the reduction pair processor, one has to demand that $\succsim$ is $\mathcal{C}_\varepsilon$*-compatible*, i.e., that $\mathsf{c}(x, y) \succsim x$ and $\mathsf{c}(x, y) \succsim y$ holds for a fresh function symbol $\mathsf{c}$ [6,10]. This requirement is satisfied by virtually all quasi-orders used in practice.[2]

**Theorem 2 ([6,10]).** *Let $(\succsim, \succ)$ be a reduction pair where $\succsim$ is $\mathcal{C}_\varepsilon$-compatible. Then the following DP processor Proc is sound (i.e., if there is no infinite minimal Proc($\mathcal{P}$)-chain, then there is also no infinite minimal $\mathcal{P}$-chain):*

$$Proc(\mathcal{P}) = \begin{cases} \mathcal{P} \setminus \succ & \text{if } \mathcal{P} \subseteq \succ \cup \succsim \text{ and } \mathcal{U}(\mathcal{P}) \subseteq \succsim \\ \mathcal{P} & \text{otherwise} \end{cases}$$

*For any function symbol $f$, let $Rls(f) = \{\ell \to r \in \mathcal{R} \mid \mathrm{root}(\ell) = f\}$. For any term $t$, the* usable rules *$\mathcal{U}(t)$ are the smallest set such that*

$$\mathcal{U}(f(t_1, \ldots, t_n)) = Rls(f) \cup \bigcup\nolimits_{\ell \to r \in Rls(f)} \mathcal{U}(r) \cup \bigcup\nolimits_{i=1}^{n} \mathcal{U}(t_i)$$

*For a set of dependency pairs $\mathcal{P}$, its usable rules are $\mathcal{U}(\mathcal{P}) = \bigcup_{s \to t \in \mathcal{P}} \mathcal{U}(t)$.*

*Example 3.* For the TRS of Ex. 1, we use the reduction pair $(\succsim_{\mathcal{P}ol}, \succ_{\mathcal{P}ol})$ with

$$\lambda_{\mathcal{P}ol} = x_1 + 1 \qquad \star_{\mathcal{P}ol} = \max(x_1, x_2)$$
$$\circ_{\mathcal{P}ol} = \circ^\sharp_{\mathcal{P}ol} = x_1 + x_2 \qquad 1_{\mathcal{P}ol} = \mathsf{id}_{\mathcal{P}ol} = \uparrow_{\mathcal{P}ol} = 0$$

Then all (usable) rules and dependency pairs are weakly decreasing (w.r.t. $\succsim_{\mathcal{P}ol}$). Furthermore, the DPs (1) and (2) are strictly decreasing (w.r.t. $\succ_{\mathcal{P}ol}$) and can be removed by Thm. 2. Afterwards, we use the following interpretation where the remaining DPs are strictly decreasing and the rules are still weakly decreasing:

$$\circ^\sharp_{\mathcal{P}ol} = x_1 \qquad \star_{\mathcal{P}ol} = \max(x_1, x_2) + 1$$
$$\circ_{\mathcal{P}ol} = x_1 + x_2 + 1 \qquad \lambda_{\mathcal{P}ol} = 1_{\mathcal{P}ol} = \mathsf{id}_{\mathcal{P}ol} = \uparrow_{\mathcal{P}ol} = 0$$

Termination of SUBST cannot be proved with Thm. 2 using reduction pairs based on linear polynomial interpretations, cf. [3]. Thus, this example shows the

[1] For readability, we consider sets of DPs instead of *DP problems* [4]. This suffices to present our new results, since the DP processors of this paper only modify the DPs.

[2] An exception are *equivalences* like $\approx$, which are usually not $\mathcal{C}_\varepsilon$-compatible [10].

usefulness of polynomial interpretations with "max". Up to now, only restricted forms of such interpretations were available in termination tools. For example, already in 2004, T$_\mathsf{T}$T used interpretations like $\max(x_1 - x_2, 0)$, but no tool offered arbitrary interpretations with polynomials and "max" like $\max(x_1, x_2) + 1$.

While SUBST's original termination proof was very complicated [8], easier proofs were developed later, using the techniques of *distribution elimination* or *semantic labeling* [18]. Indeed, the only tool that could prove termination of SUBST automatically up to now (TPA [12]) used semantic labeling.[3] In contrast, Ex. 3 shows that there is an even simpler proof without semantic labeling.

## 3 Termination with Integer Polynomials and "max"

Our aim is to use polynomial interpretations with *integer* polynomials, together with the function "max". More precisely, we want to use interpretations that map $n$-ary function symbols to arbitrary functions from $\mathbb{N}^n \to \mathbb{N}$. But Ex. 4 demonstrates that such interpretations may not be used in Thm. 2, since then $\succsim_{\mathcal{P}ol}$ is not monotonic, and thus, $(\succsim_{\mathcal{P}ol}, \succ_{\mathcal{P}ol})$ is not a reduction pair.

*Example 4.* Consider this non-terminating TRS (inspired by [7, Ex. 4]):

$$\mathsf{f}(\mathsf{s}(x), x) \to \mathsf{f}(\mathsf{s}(x), \mathsf{round}(x))$$

$$\mathsf{round}(\mathsf{0}) \to \mathsf{0} \qquad \mathsf{round}(\mathsf{s}(\mathsf{0})) \to \mathsf{s}(\mathsf{0})$$

$$\mathsf{round}(\mathsf{0}) \to \mathsf{s}(\mathsf{0}) \qquad \mathsf{round}(\mathsf{s}(\mathsf{s}(x))) \to \mathsf{s}(\mathsf{s}(\mathsf{round}(x)))$$

Here, $\mathsf{round}(x)$ evaluates to $x$ if $x$ is odd and to $x$ or $\mathsf{s}(x)$ otherwise. We use the interpretation $\mathcal{P}ol$ with $\mathsf{F}_{\mathcal{P}ol} = x_1 + \max(x_1 - x_2, 0)$, $\mathsf{ROUND}_{\mathcal{P}ol} = x_1$, $\mathsf{0}_{\mathcal{P}ol} = 0$, and $\mathsf{s}_{\mathcal{P}ol} = \mathsf{round}_{\mathcal{P}ol} = x_1 + 1$, where F and ROUND are the tuple symbols for f and round, respectively. Then all DPs are strictly decreasing and the usable round-rules are weakly decreasing. So if we were allowed to use $\mathcal{P}ol$ in Thm. 2, then we could remove all DPs and falsely prove termination.

Ex. 4 shows the reason for unsoundness when dropping the requirement of monotonicity of $\succsim$. Thm. 2 requires $\ell \succsim r$ for all usable rules $\ell \to r$. This is meant to ensure that all reductions with usable rules will weakly decrease the reduced term (w.r.t. $\succsim$). However, this only holds if the quasi-order $\succsim$ is monotonic. For instance in Ex. 4, we have $\mathsf{round}(\mathsf{0}) \succsim_{\mathcal{P}ol} \mathsf{0}$, but $\mathsf{F}(\mathsf{s}(\mathsf{0}), \underline{\mathsf{round}(\mathsf{0})}) \not\succsim_{\mathcal{P}ol} \mathsf{F}(\mathsf{s}(\mathsf{0}), \underline{\mathsf{0}})$.

In [10], this problem was solved by requiring $\ell \approx r$ instead of $\ell \succsim r$. Then such rules are not just weakly decreasing but *equivalent* w.r.t. $\succsim$. This requirement is not satisfied in Ex. 4 as $\mathsf{round}(\mathsf{0}) \not\approx_{\mathcal{P}ol} \mathsf{0}$. In general, this equivalence even has to be required for *all* rules $\ell \to r$ (not just the usable ones), since the step from *all* rules to the *usable* rules in the proof of Thm. 2 also relies on the monotonicity of $\succsim$. Thus, up to now one had to apply the following reduction pair processor when using non-monotonic reduction pairs. The soundness of this processor immediately results from [4, Thm. 28] and [10, Thm. 23 and Cor. 31],

[3] For the semantic labeling, TPA uses only a (small) fixed set of functions, including certain fixed polynomials and the function "max". So in contrast to our automation in Sect. 4, TPA does not search for arbitrary combinations of polynomials and "max".

cf. [3].[4] Here, a *non-monotonic* reduction pair $(\succsim, \succ)$ consists of a stable quasi-order $\succsim$ and a compatible stable well-founded order $\succ$. But we do not require monotonicity of $\succsim$ (and $\succsim$ does not have to be $\mathcal{C}_\varepsilon$-compatible either). However, the equivalence relation $\approx$ associated with $\succsim$ must be monotonic.[5]

**Theorem 5.** *Let $(\succsim, \succ)$ be a non-monotonic reduction pair. Then Proc is sound:*

$$Proc(\mathcal{P}) = \begin{cases} \mathcal{P} \setminus \succ & \text{if } \mathcal{P} \subseteq \succ \cup \succsim \text{ and (a) or (b) holds:} \\ & \text{(a) } \mathcal{P} \cup \mathcal{U}(\mathcal{P}) \text{ is non-duplicating and } \mathcal{U}(\mathcal{P}) \subseteq \approx \\ & \text{(b) } \mathcal{R} \subseteq \approx \\ \mathcal{P} & \text{otherwise} \end{cases}$$

However, demanding $\ell \approx r$ for the usable rules as in Thm. 5(a) is a very strong requirement which makes the termination proof fail in many examples, cf. Ex. 11 and 12. Therefore, as already suggested in [7], one should take into account on which positions the quasi-order $\succsim$ is monotonically *increasing* resp. *decreasing*. If a defined function symbol $f$ occurs at a monotonically *increasing* position in the right-hand side of a dependency pair, then one should require $\ell \succsim r$ for all $f$-rules. If $f$ is at a *decreasing* position, one requires $r \succsim \ell$. Finally, if $f$ is at a position which is neither increasing nor decreasing, one requires $\ell \approx r$.

To modify our definition of usable rules accordingly, we need a *monotonicity specification* which specifies which arguments of a symbol have to be increasing ("$\Uparrow$") or decreasing ("$\Downarrow$"). Afterwards, we search for a (non-monotonic) reduction pair that is *compatible* with the monotonicity specification.

**Definition 6.** *A* monotonicity specification *is a mapping $\nu$ which assigns to every function symbol $f$ and every $i \in \{1, ..., \text{arity}(f)\}$ a subset of $\{\Uparrow, \Downarrow\}$. A reduction pair $(\succsim, \succ)$ is* $\nu$-compatible *iff*

- *if $\Uparrow \in \nu(f,i)$ then $\succsim$ is* monotonically increasing *on $f$'s $i$-th argument, i.e., $t_i \succsim s_i$ implies $f(t_1, ..., t_i, ..., t_n) \succsim f(t_1, ..., s_i, ..., t_n)$ for all terms $t_1, ..., t_n, s_i$*
- *if $\Downarrow \in \nu(f,i)$ then $\succsim$ is* monotonically decreasing *on $f$'s $i$-th argument, i.e., $t_i \succsim s_i$ implies $f(t_1, ..., t_i, ..., t_n) \precsim f(t_1, ..., s_i, ..., t_n)$ for all terms $t_1, ..., t_n, s_i$*
- *if $\nu(f,i) = \{\Uparrow, \Downarrow\}$ then[6] additionally $\succsim$ must be* independent *on $f$'s $i$-th argument, i.e., $f(t_1, ..., t_i, ..., t_n) \approx f(t_1, ..., s_i, ..., t_n)$ for all terms $t_1, ..., t_n, s_i$*

*We call $f$* $\nu$-dependent *on its $i$-th argument iff $\nu(f,i) \neq \{\Uparrow, \Downarrow\}$. The concept of* monotonicity *can be extended to positions in a term where $\nu(t, \varepsilon) = \{\Uparrow\}$ and*

---

[4] An alternative to Thm. 5(a) is presented in [10, Thm. 40] for reduction pairs $(\succsim_{\mathcal{P}ol}, \succ_{\mathcal{P}ol})$ based on polynomial interpretations. Here, "non-duplication of $\mathcal{P} \cup \mathcal{U}(\mathcal{P})$" is replaced by "$\mathcal{P}ol$-right-linearity of $\mathcal{P} \cup \mathcal{U}(\mathcal{P})$". So for every right-hand side $r$ there must be a linear term $r'$ with $r \approx_{\mathcal{P}ol} r'$ where $r'$ differs from $r$ only in the variables.

[5] Triples like $(\approx, \succsim, \succ)$ were called "reduction triples" in [10]. "Non-monotonic reduction pairs" are also related to the "general reduction pairs" in [7], but there $\succ$ did not have to be well founded. Consequently, the notion of stability was weakened too.

[6] Note that this condition is implied by the first two conditions whenever $\succsim$ is total on ground terms and whenever $s\sigma \succsim t\sigma$ for all ground substitutions $\sigma$ implies $s \succsim t$.

$$\nu(f(t_1,...,t_n),i\,p) = \begin{cases} \{\Uparrow,\Downarrow\} & \textit{if } \nu(f,i)=\{\Uparrow,\Downarrow\} \textit{ or } \nu(t_i,p)=\{\Uparrow,\Downarrow\} \\ \{\Uparrow\} & \textit{if } \nu(f,i)=\nu(t_i,p)=\{\Uparrow\} \textit{ or } \nu(f,i)=\nu(t_i,p)=\{\Downarrow\} \\ \{\Downarrow\} & \textit{if either } \nu(f,i)=\{\Uparrow\} \textit{ and } \nu(t_i,p)=\{\Downarrow\} \\ & \qquad \textit{or } \nu(f,i)=\{\Downarrow\} \textit{ and } \nu(t_i,p)=\{\Uparrow\} \\ \varnothing & \textit{otherwise} \end{cases}$$

*A position $p$ in a term $t$ is called* $\nu$-dependent *iff $\nu(t,p) \neq \{\Uparrow,\Downarrow\}$.*

**Definition 7 (General Usable Rules [7]).** *Let $\nu$ be a monotonicity specification. For any TRS $U$, we define $U^{\{\Uparrow,\Downarrow\}} = \varnothing$, $U^{\{\Uparrow\}} = U$, $U^{\{\Downarrow\}} = U^{-1} = \{r \to \ell \mid \ell \to r \in U\}$, and $U^{\varnothing} = U \cup U^{-1}$. For any term $t$, we define the* general usable rules *$\mathcal{GU}(t)$ as the smallest set such that*[7]

$$\mathcal{GU}(f(t_1,\ldots,t_n)) = \mathit{Rls}(f) \cup \bigcup_{\ell\to r\in \mathit{Rls}(f)} \mathcal{GU}(r) \cup \bigcup_{i=1}^{n} \mathcal{GU}^{\nu(f,i)}(t_i)$$

*For a set of DPs $\mathcal{P}$, we define $\mathcal{GU}(\mathcal{P}) = \bigcup_{s\to t\in\mathcal{P}} \mathcal{GU}(t)$. Moreover, we let $\mathcal{U}^{contr}(t)$ be those rules of $\mathcal{R}$ that contributed to $\mathcal{GU}(t)$, i.e., $\mathcal{U}^{contr}(t) = \{\ell\to r\in \mathcal{R} \mid \ell\to r\in\mathcal{GU}(t) \text{ or } r\to\ell\in\mathcal{GU}(t)\}$. Similarly, $\mathcal{U}^{contr}(\mathcal{P}) = \bigcup_{s\to t\in\mathcal{P}} \mathcal{U}^{contr}(t)$.*[8]

*Example 8.* In Ex. 4, as $\mathsf{F}_{\mathcal{P}ol} = x_1 + \max(x_1 - x_2, 0)$, $\succsim_{\mathcal{P}ol}$ is monotonically decreasing on F's second argument. So $(\succsim_{\mathcal{P}ol}, \succ_{\mathcal{P}ol})$ is $\nu$-compatible for the monotonicity specification $\nu$ with $\nu(\mathsf{F},2) = \{\Downarrow\}$ and $\nu(\mathsf{F},1) = \nu(\mathsf{ROUND},1) = \nu(\mathsf{s},1) = \nu(\mathsf{round},1) = \{\Uparrow\}$. Due to $\nu(\mathsf{F},2) = \{\Downarrow\}$, the general usable rules are the reversed round-rules. Thus, we cannot falsely prove termination with $\mathcal{P}ol$ anymore, since $\mathcal{P}ol$ does not make the *reversed* round-rules weakly decreasing; for example, we have $\mathsf{0} \prec_{\mathcal{P}ol} \mathsf{round}(\mathsf{0})$.

Our goal is to show that with the modified definition of usable rules above, Thm. 2 can also be used for non-monotonic reduction pairs. However, this is not true in general as shown by the following counterexample, cf. [10, Ex. 32].

*Example 9.* Consider the following famous TRS of Toyama [16]:

$$\mathsf{f}(\mathsf{0},\mathsf{1},x) \to \mathsf{f}(x,x,x) \qquad \mathsf{g}(x,y) \to x \qquad \mathsf{g}(x,y) \to y$$

We use a monotonicity specification $\nu$ with $\nu(\mathsf{F},1) = \{\Downarrow\}$, $\nu(\mathsf{F},2) = \{\Uparrow\}$, $\nu(\mathsf{F},3) = \{\Uparrow,\Downarrow\}$ and a $\nu$-compatible reduction pair $(\succsim_{\mathcal{P}ol}, \succ_{\mathcal{P}ol})$ where $\mathsf{F}_{\mathcal{P}ol} = \max(x_2 - x_1, 0)$, $\mathsf{0}_{\mathcal{P}ol} = 0$, and $\mathsf{1}_{\mathcal{P}ol} = 1$. The only DP is strictly decreasing and there is no (general) usable rule. Hence, one would falsely conclude termination.

To obtain a sound criterion, we therefore impose certain requirements on all rules $\ell \to r \in \mathcal{P} \cup \mathcal{U}^{contr}$. To this end, we need the following notions.

- A rule $\ell \to r$ is $\nu$-*more monotonic* ($\nu$-MM) if variables occur at *more monotonic* positions on the right-hand side than on the left-hand side. More precisely, for every $\nu$-dependent position $p$ of $r$ with $r|_p = x$ there is a position $q$ of $\ell$ such that $\ell|_q = x$ and $\nu(\ell,q) \subseteq \nu(r,p)$. However, each position of $\ell$ can only be used once, i.e., for different positions $p$ and $p'$ of $r$ we must choose different positions $q$ and $q'$ of $\ell$. To define this notion formally, let $\mathcal{P}os_x^{\nu}(t)$

[7] Note that $\mathcal{GU}(t)$ is no longer a subset of $\mathcal{R}$. We nevertheless refer to $\mathcal{GU}(t)$ as "usable" rules in order to keep the similarity to Thm. 2.

[8] $\mathcal{U}^{contr}$ are the "usable rules w.r.t. an argument filtering" from [6].

be the set of all $\nu$-dependent positions $p$ of $t$ with $t|_p = x$. Then a rule $\ell \to r$ is *$\nu$-MM* if for each variable $x$ there is an injective mapping $\alpha$ from $\mathcal{P}os^\nu_x(r)$ to $\mathcal{P}os^\nu_x(\ell)$ such that $\nu(\ell, \alpha(p)) \subseteq \nu(r, p)$ for all $p \in \mathcal{P}os^\nu_x(r)$.
So for the right-hand side of the DP in Ex. 9, we have $\mathcal{P}os^\nu_x(\mathsf{F}(x, x, x)) = \{1, 2\}$. Hence, $x$ would have to occur on at least two different $\nu$-dependent positions $q$ and $q'$ in the left-hand side $\mathsf{F}(0, 1, x)$. Moreover, we would need $\nu(\mathsf{F}(0, 1, x), q) \subseteq \nu(\mathsf{F}(x, x, x), 1) = \{\Downarrow\}$ and $\nu(\mathsf{F}(0, 1, x), q') \subseteq \nu(\mathsf{F}(x, x, x), 2) = \{\Uparrow\}$. However, this DP is not $\nu$-MM as $\mathcal{P}os^\nu_x(\mathsf{F}(0, 1, x)) = \varnothing$.
- $\ell \to r$ is *weakly $\nu$-MM* if for each $x$ with $\mathcal{P}os^\nu_x(\ell) \neq \varnothing$, there is an injective mapping $\alpha$ from $\mathcal{P}os^\nu_x(r)$ to $\mathcal{P}os^\nu_x(\ell)$ such that $\nu(\ell, \alpha(p)) \subseteq \nu(r, p)$ for all $p \in \mathcal{P}os^\nu_x(r)$. So in contrast to $\nu$-MM, now we also permit variables that occur at dependent positions of $r$, but not at any dependent position of $\ell$. Therefore, the DP of Ex. 9 is weakly $\nu$-MM.
- $\ell \to r$ is *$\nu$-right-linear* ($\nu$-RL) if all variables occur at most once at a $\nu$-dependent position in $r$. Formally, $\ell \to r$ is *$\nu$-RL* iff for all $x \in \mathcal{V}(r)$: $|\mathcal{P}os^\nu_x(r)| \leqslant 1$. So the DP in Ex. 9 is not $\nu$-RL since $x$ occurs twice at $\nu$-dependent positions in the right-hand side.

A TRS is *(weakly) $\nu$-MM* resp. *$\nu$-RL* iff all its rules satisfy that condition.

We now extend the processor from Thm. 2 to non-monotonic reduction pairs. Thm. 10 shows that to remove all strictly decreasing DPs, it is still sufficient if the (general) usable rules are weakly decreasing, provided that $\mathcal{P} \cup \mathcal{U}^{contr}(\mathcal{P})$ satisfies $\nu$-MM. Alternatively, one can also require weak $\nu$-MM and $\nu$-RL.

As shown in [7], if one only wants to prove *innermost* termination, then Thm. 10 can be used even without the conditions (weak) $\nu$-MM and $\nu$-RL. However, we now extend this result to *full* termination. Of course, if $\mathcal{P} \cup \mathcal{U}^{contr}(\mathcal{P})$ is not (weakly) $\nu$-MM resp. $\nu$-RL and one wants to prove full termination with a non-monotonic reduction pair, then one has to use Thm. 5 instead.

**Theorem 10.** *Let $\nu$ be a monotonicity specification and let $(\succsim, \succ)$ be a $\nu$-compatible non-monotonic reduction pair. Then Proc is sound:*[9]

$$Proc(\mathcal{P}) = \begin{cases} \mathcal{P} \setminus \succ & \text{if } \mathcal{P} \subseteq \succ \cup \succsim,\ \mathcal{GU}(\mathcal{P}) \subseteq \succsim, \text{ and one of (a) or (b) holds:} \\ & \quad \text{(a) } \mathcal{P} \cup \mathcal{U}^{contr}(\mathcal{P}) \text{ is } \nu\text{-MM} \\ & \quad \text{(b) } \mathcal{P} \cup \mathcal{U}^{contr}(\mathcal{P}) \text{ is weakly } \nu\text{-MM and } \nu\text{-RL} \\ \mathcal{P} & \text{otherwise} \end{cases}$$

*Example 11.* To modify Ex. 4 into a terminating TRS, we replace the $\mathsf{f}$-rule by

$$\mathsf{f}(\mathsf{s}(x), x) \to \mathsf{f}(\mathsf{s}(x), \mathsf{round}(\mathsf{s}(x)))$$

similar to [7, Ex. 9]. We use the monotonicity specification from Ex. 8. The interpretation $\mathcal{P}ol$ from Ex. 4 is modified by defining $\mathsf{round}_{\mathcal{P}ol} = x_1$. Then $(\succsim_{\mathcal{P}ol}, \succ_{\mathcal{P}ol})$ is $\nu$-compatible, all DPs are strictly decreasing, and the (general) usable rules (i.e., the *reversed* $\mathsf{round}$-rules) are weakly decreasing. Moreover, all rules in $\mathcal{P} \cup \mathcal{U}^{contr}(\mathcal{P})$ are $\nu$-MM. Thus, by Thm. 10(a) we can transform the initial DP problem $\mathcal{P} = DP(\mathcal{R})$ into $\mathcal{P} \setminus \succ = \varnothing$ and prove termination.

---

9 The proof can be found in [3].

In contrast, this was not possible by the method of [10] which requires $\ell \approx r$ for all usable rules. There is no (possibly non-monotonic) reduction pair that satisfies $\mathsf{round}(0) \approx 0 \approx \mathsf{s}(0)$ and $\mathsf{F}(\mathsf{s}(x), x) \succ \mathsf{F}(\mathsf{s}(x), \mathsf{round}(\mathsf{s}(x)))$. The method of [7] can only prove innermost termination of this example. However, this TRS does not belong to a known class of TRSs where innermost termination implies termination. So in fact, up to now all tools failed on this example.

*Example 12.* The following example illustrates Thm. 10(b):

$$\begin{array}{rlrl}
\mathsf{p}(0) & \to 0 & \mathsf{minus}(x, 0) & \to x \\
\mathsf{p}(\mathsf{s}(x)) & \to x & \mathsf{minus}(\mathsf{s}(x), \mathsf{s}(y)) & \to \mathsf{minus}(x, y) \\
\mathsf{div}(0, \mathsf{s}(y)) & \to 0 & \mathsf{minus}(x, \mathsf{s}(y)) & \to \mathsf{p}(\mathsf{minus}(x, y)) \\
\mathsf{div}(\mathsf{s}(x), \mathsf{s}(y)) & \multicolumn{3}{l}{\to \mathsf{s}(\mathsf{div}(\mathsf{minus}(\mathsf{s}(x), \mathsf{s}(y)), \mathsf{s}(y)))} \\
\mathsf{log}(\mathsf{s}(0), \mathsf{s}(\mathsf{s}(y))) & \multicolumn{3}{l}{\to 0} \\
\mathsf{log}(\mathsf{s}(\mathsf{s}(x)), \mathsf{s}(\mathsf{s}(y))) & \multicolumn{3}{l}{\to \mathsf{s}(\mathsf{log}(\mathsf{div}(\mathsf{minus}(x, y), \mathsf{s}(\mathsf{s}(y))), \mathsf{s}(\mathsf{s}(y))))}
\end{array}$$

We use a monotonicity specification $\nu$ with $\nu(\mathsf{s}, 1) = \nu(\mathsf{p}, 1) = \nu(\mathsf{minus}, 1) = \nu(\mathsf{MINUS}, 1) = \nu(\mathsf{div}, 1) = \nu(\mathsf{DIV}, 1) = \nu(\mathsf{LOG}, 1) = \{\Uparrow\}$, $\nu(\mathsf{minus}, 2) = \{\Downarrow\}$, $\nu(\mathsf{P}, 1) = \nu(\mathsf{MINUS}, 2) = \nu(\mathsf{div}, 2) = \nu(\mathsf{DIV}, 2) = \nu(\mathsf{LOG}, 2) = \{\Uparrow, \Downarrow\}$, and the interpretation $\mathsf{p}_{\mathcal{P}ol} = \max(x_1 - 1, 0)$, $\mathsf{minus}_{\mathcal{P}ol} = \max(x_1 - x_2, 0)$, $0_{\mathcal{P}ol} = \mathsf{P}_{\mathcal{P}ol} = 0$, $\mathsf{s}_{\mathcal{P}ol} = \mathsf{MINUS}_{\mathcal{P}ol} = \mathsf{div}_{\mathcal{P}ol} = \mathsf{LOG}_{\mathcal{P}ol} = x_1 + 1$, $\mathsf{DIV}_{\mathcal{P}ol} = x_1 + 2$. Now $(\succsim_{\mathcal{P}ol}, \succ_{\mathcal{P}ol})$ is $\nu$-compatible, all DPs except $\mathsf{MINUS}(x, \mathsf{s}(y)) \to \mathsf{MINUS}(x, y)$ are strictly decreasing, and the remaining DP and the usable p-, minus-, and div-rules are weakly decreasing. In addition, all DPs and usable rules are weakly $\nu$-MM and $\nu$-RL. Hence, by Thm. 10(b) we can remove all DPs except $\mathsf{MINUS}(x, \mathsf{s}(y)) \to \mathsf{MINUS}(x, y)$. Afterwards, we use $\mathsf{MINUS}_{\mathcal{P}ol'} = x_2$ and $\mathsf{s}_{\mathcal{P}ol'} = x_1 + 1$ to delete this remaining DP. (Now there are no usable rules.) Hence, termination is proved.

Note that here, Thm. 10(a) does not apply as the DP $\mathsf{DIV}(\mathsf{s}(x), \mathsf{s}(y)) \to \mathsf{DIV}(\mathsf{minus}(\mathsf{s}(x), \mathsf{s}(y)), \mathsf{s}(y))$ is not $\nu$-MM: the first occurrence of $y$ in the right-hand side is at a non-increasing position, whereas the only occurrence of $y$ in the left-hand side is at a $\nu$-independent, and thus increasing position.

The technique of [10] cannot handle the DP $\mathsf{LOG}(\ldots) \to \mathsf{LOG}(\mathsf{div}(\ldots), \ldots)$, because it would have to find an interpretation which makes the div-rules equivalent. In contrast, Thm. 10 only requires a weak decrease for the div-rules. Indeed, all existing termination tools failed on this example.

## 4 Automation

The most efficient implementations to search for polynomial interpretations are based on SAT solving [2]. However, [2] only handled the search for polynomial interpretations with natural coefficients as well as interpretations of the form $\max(p - n, 0)$ where $p$ is a polynomial with natural coefficients and $n \in \mathbb{N}$. So we permitted interpretations like $\max(x_1 - 1, 0)$, but not interpretations like $\max(x_1 - x_2, 0)$ (as needed in Ex. 11 and 12) or $\max(x_1, x_2)$ (as needed in Ex. 1).

We want to use SAT solvers to search for *arbitrary* interpretations using polynomials and "max". Compared to existing related approaches, there are two challenges: the additional use of "max" in polynomial interpretations (Sect. 4.1) and the handling of non-monotonic quasi-orders and general usable rules (Sect. 4.2).

### 4.1 Automating Polynomial Interpretations with "max"

We start with encoding the "classical" reduction pair processor of Thm. 2 as a SAT problem. This is simpler than encoding Thm. 10, because in Thm. 2 we use a monotonic reduction pair $(\succsim_{\mathcal{P}ol}, \succ_{\mathcal{P}ol})$ and thus, the applicability conditions and the usable rules $\mathcal{U}$ do not depend on a monotonicity specification. But in contrast to our earlier encoding from [2], now $\mathcal{P}ol$ can be an interpretation that combines polynomials and "max" arbitrarily.[10]

**Definition 13 (max-polynomial).** *Let $\mathcal{V}$ be the set of variables. The set of* max-polynomials $\mathbb{P}_M$ *over a set of numbers $M$ is the smallest set such that*

- $M \subseteq \mathbb{P}_M$ *and* $\mathcal{V} \subseteq \mathbb{P}_M$
- *if* $p, q \in \mathbb{P}_M$, *then* $p + q \in \mathbb{P}_M$, $p - q \in \mathbb{P}_M$, $p * q \in \mathbb{P}_M$, *and* $\max(p, q) \in \mathbb{P}_M$

At the moment, we only consider interpretations $\mathcal{P}ol$ that map every function symbol to a max-polynomial over $\mathbb{N}$ that does not contain any subtraction "$-$". Obviously, then $(\succsim_{\mathcal{P}ol}, \succ_{\mathcal{P}ol})$ is a $\mathcal{C}_\varepsilon$-compatible (monotonic) reduction pair.

To find such interpretations automatically, one starts with an *abstract* polynomial interpretation. It maps each function symbol to a max-polynomial over a set $\mathcal{A}$ of *abstract* coefficients. In other words, one has to determine the degree and the shape of the max-polynomial, but the actual coefficients are left open. For example, for the TRS of Ex. 1 we could use an abstract polynomial interpretation $\mathcal{P}ol$ where $\star_{\mathcal{P}ol} = \max(a_1\, x_1 + a_2\, x_2,\ a'_1\, x_1 + a'_2\, x_2)$, $\uparrow_{\mathcal{P}ol} = b$, $\circ_{\mathcal{P}ol} = x_1 + x_2$, etc.[11] Here, $a_1, a_2, a'_1, a'_2, b$ are abstract coefficients.

Now to apply the reduction pair processor of Thm. 2, we have to find an instantiation of the abstract coefficients satisfying the following condition. Then all dependency pairs that are strictly decreasing (i.e., $[s] \geqslant [t] + 1$) can be removed.

$$\bigwedge_{s \to t\, \in\, \mathcal{P}} [s]_{\mathcal{P}ol} \geqslant [t]_{\mathcal{P}ol} \;\wedge \bigvee_{s \to t\, \in\, \mathcal{P}} [s]_{\mathcal{P}ol} \geqslant [t]_{\mathcal{P}ol} + 1 \;\wedge \bigwedge_{\ell \to r\, \in\, \mathcal{U}(\mathcal{P})} [\ell]_{\mathcal{P}ol} \geqslant [r]_{\mathcal{P}ol} \quad (3)$$

Here, all rules in $\mathcal{P} \cup \mathcal{U}(\mathcal{P})$ are variable-renamed to have pairwise different variables. The polynomials $[s]_{\mathcal{P}ol}$, $[t]_{\mathcal{P}ol}$, etc. are again max-polynomials over $\mathcal{A}$. So with the interpretation $\mathcal{P}ol$ above, to make the last rule of Ex. 1 weakly decreasing (i.e., $\uparrow \circ (x \star y) \succsim_{\mathcal{P}ol} y$) we obtain the inequality $[\uparrow \circ (x \star y)]_{\mathcal{P}ol} \geqslant [y]_{\mathcal{P}ol}$:

$$b + \max(a_1\, x + a_2\, y,\ a'_1\, x + a'_2\, y) \geqslant y \quad (4)$$

We have to find an instantiation of the abstract coefficients $a_1, a_2, \ldots$ such that (4) holds for *all* instantiations of the variables $x$ and $y$. In other words, the variables from $\mathcal{V}$ occurring in such inequalities are universally quantified.

[10] Of course, in an analogous way, one can also integrate the "minimum" function and indeed, we did this in our implementations.

[11] Here we already fixed $\circ$'s interpretation to simplify the presentation. Our implementations use heuristics to determine when to use an interpretation with "max".

Several techniques have been proposed to transform such inequalities further in order to remove such universally quantified variables [11]. However, the existing techniques only operate on inequalities without "max". Therefore, we now present new inference rules to eliminate "max" from such inequalities.

Our inference rules operate on *conditional* constraints of the form

$$p_1 \geqslant q_1 \wedge \ldots \wedge p_n \geqslant q_n \;\Rightarrow\; p \geqslant q \tag{5}$$

Here, $n \geqslant 0$ and $p_1, ..., p_n, q_1, ..., q_n$ are polynomials with abstract coefficients without "max". In contrast, $p$, $q$ are max-polynomials with abstract coefficients.

The first inference rule eliminates an inner occurrence of "max" from the inequality $p \geqslant q$. If $p$ or $q$ have a sub-expression $\max(p', q')$ where $p'$ and $q'$ do not contain "max", then we can replace this sub-expression by $p'$ or $q'$ when adding the appropriate condition $p' \geqslant q'$ or $q' \geqslant p' + 1$, respectively.

**I. Eliminating "max"**

$$\frac{p_1 \geqslant q_1 \wedge \ldots \wedge p_n \geqslant q_n \;\Rightarrow\; \ldots \max(p', q') \ldots}{\begin{array}{ll} p_1 \geqslant q_1 \wedge \ldots \wedge p_n \geqslant q_n \wedge p' \geqslant q' & \Rightarrow\; \ldots \; p' \; \ldots \quad \wedge \\ p_1 \geqslant q_1 \wedge \ldots \wedge p_n \geqslant q_n \wedge q' \geqslant p' + 1 & \Rightarrow\; \ldots \; q' \; \ldots \end{array}}$$

if $p'$ and $q'$ do not contain "max"

Obviously, by repeated application of inference rule (I), all occurrences of "max" can be removed. In our example, the constraint (4) is transformed into the following new constraint that does not contain "max" anymore.

$$a_1\,x + a_2\,y \geqslant a_1'\,x + a_2'\,y \;\Rightarrow\; b + a_1\,x + a_2\,y \geqslant y \qquad \wedge \tag{6}$$

$$a_1'\,x + a_2'\,y \geqslant a_1\,x + a_2\,y + 1 \;\Rightarrow\; b + a_1'\,x + a_2'\,y \geqslant y \tag{7}$$

Since the existing methods for eliminating universally quantified variables only work for *unconditional* inequalities, the next inference rule eliminates the conditions $p_i \geqslant q_i$ from a constraint of the form (5).[12] To this end, we introduce two new abstract polynomials $\overline{p}$ and $\overline{q}$ (that do not contain "max"). The polynomial $\overline{q}$ over the variables $x_1, ..., x_n$ is used to "measure" the polynomials $p_1, ..., p_n$ resp. $q_1, ..., q_n$ in the premise of (5) and the unary polynomial $\overline{p}$ measures the polynomials $p$ and $q$ in the conclusion of (5). We write $\overline{q}[p_1, ..., p_n]$ to denote the result of instantiating the variables $x_1, ..., x_n$ in $\overline{q}$ by $p_1, ..., p_n$, etc.

**II. Eliminating Conditions**

$$\frac{p_1 \geqslant q_1 \wedge \ldots \wedge p_n \geqslant q_n \;\Rightarrow\; p \geqslant q}{\overline{p}[p] - \overline{p}[q] \geqslant \overline{q}[p_1, \ldots, p_n] - \overline{q}[q_1, \ldots, q_n]}$$

if $\overline{q}$ and $\overline{p}$ do not contain "max", $\overline{p}$ is strictly monotonic, and $\overline{q}$ is weakly monotonic

Here, the monotonicity conditions mean that $x > y \Rightarrow \overline{p}[x] > \overline{p}[y]$ must hold and similarly that $x_1 \geqslant y_1 \wedge \ldots \wedge x_n \geqslant y_n \Rightarrow \overline{q}[x_1, \ldots, x_n] \geqslant \overline{q}[y_1, \ldots, y_n]$.

[12] Such conditional polynomial constraints also occur in other applications, e.g., in the termination analysis of logic programs. Indeed, we used a rule similar to inference rule (II) in the tool Polytool for termination analysis of logic programs [15]. However, Polytool only applies classical polynomial interpretations without "max".

To see why Rule (II) is sound, let $\overline{p}[p]-\overline{p}[q] \geqslant \overline{q}[p_1,\dots,p_n]-\overline{q}[q_1,\dots,q_n]$ hold and assume that there is an instantiation $\sigma$ of all variables in the polynomials with numbers that refutes $p_1 \geqslant q_1 \wedge \dots \wedge p_n \geqslant q_n \;\Rightarrow\; p \geqslant q$. Now $p_1\sigma \geqslant q_1\sigma \wedge \dots \wedge p_n\sigma \geqslant q_n\sigma$ implies $\overline{q}[p_1,\dots,p_n]\sigma \geqslant \overline{q}[q_1,\dots,q_n]\sigma$ by weak monotonicity of $\overline{q}$. Hence, $\overline{p}[p]\sigma - \overline{p}[q]\sigma \geqslant 0$. Since the instantiation $\sigma$ is a counterexample to our original constraint, we have $p\sigma \not\geqslant q\sigma$ and thus $p\sigma < q\sigma$. But then strict monotonicity of $\overline{p}$ would imply $\overline{p}[p]\sigma - \overline{p}[q]\sigma < 0$ which gives a contradiction.

If we choose[13] the abstract polynomials $\overline{p} = c\,x_1$ and $\overline{q} = d\,x_1$ for (6) and $\overline{p} = c'\,x_1$ and $\overline{q} = d'\,x_1$ for (7), then (6) and (7) are transformed into the following unconditional inequalities. (Note that we also have to add the inequalities $c \geqslant 1$ and $c' \geqslant 1$ to ensure that $\overline{p}$ is strictly monotonic.)

$$c\cdot(b + a_1\,x + a_2\,y) - c\cdot y \geqslant d\cdot(a_1\,x + a_2\,y) - d\cdot(a_1'\,x + a_2'\,y) \quad \wedge \qquad (8)$$
$$c'\cdot(b + a_1'\,x + a_2'\,y) - c'\cdot y \geqslant d'\cdot(a_1'\,x + a_2'\,y) - d'\cdot(a_1\,x + a_2\,y + 1) \qquad (9)$$

Of course, such inequalities can be transformed into inequalities with 0 on their right-hand side. For example, (8) is transformed to

$$(c\,a_1 - d\,a_1 + d\,a_1')\,x \;+\; (c\,a_2 - c - d\,a_2 + d\,a_2')\,y \;+\; c\,b \;\geqslant\; 0 \qquad (10)$$

Thus, we now have to ensure non-negativeness of "polynomials" over variables like $x, y$, where the "coefficients" are polynomials over the abstract variables like $c\,a_1 - d\,a_1 + d\,a_1'$. To this end, it suffices to require that all these "coefficients" are $\geqslant 0$ [11]. In other words, now one can eliminate all universally quantified variables like $x, y$ and (10) is transformed into the *Diophantine constraint*

$$c\,a_1 - d\,a_1 + d\,a_1' \geqslant 0 \quad \wedge \quad c\,a_2 - c - d\,a_2 + d\,a_2' \geqslant 0 \quad \wedge \quad c\,b \geqslant 0$$

**III. Eliminating Universally Quantified Variables**

$$\frac{p_0 + p_1\,x_1^{e_{11}}\dots x_n^{e_{n1}} + \dots + p_k\,x_1^{e_{1k}}\dots x_n^{e_{nk}} \geqslant 0}{p_0 \geqslant 0 \;\wedge\; p_1 \geqslant 0 \;\wedge \dots \wedge\; p_k \geqslant 0}$$ if the $p_i$ neither contain "max" nor any variable from $\mathcal{V}$

To search for suitable values for the abstract coefficients that satisfy the resulting Diophantine constraints, one fixes an upper bound for these values. Then we showed in [2] how to translate such Diophantine constraints into a satisfiability problem for propositional logic which can be handled by SAT solvers efficiently. In our example, the constraints resulting from the initial inequality (4) are for example satisfied by $a_1 = 1$, $a_2 = 0$, $a_1' = 0$, $a_2' = 1$, $b = 0$, $c = 1$, $d = 1$, $c' = 1$, $d' = 0$. With these values, the abstract interpretation $\max(a_1\,x_1 + a_2\,x_2,\; a_1'\,x_1 + a_2'\,x_2)$ for $\star$ is turned into the concrete interpretation $\max(x_1, x_2)$.

[13] A good heuristic is to choose $\overline{q} = b_1x_1 + \dots + b_nx_n$ where all $b_i$ are from $\{0,1\}$ and $\overline{p} = a\cdot x_1$ where $1 \leqslant a \leqslant \max(\Sigma_{i=1}^{n} b_i,\; 1)$.

### 4.2 Automating Thm. 10

Now we show how to automate the improved reduction pair processor of Thm. 10. As before, our aim is to translate the resulting constraints into Diophantine constraints and further into propositional satisfiability problems.

Again, we start with an *abstract* polynomial interpretation $\mathcal{P}ol$. But since the values for the abstract coefficients can now be from $\mathbb{Z}$, we add the constraint

$$[f] \geqslant 0 \qquad \text{for all function symbols } f \tag{11}$$

to ensure the well-foundedness of the resulting order. In the TRS of Ex. 12, we could start with an abstract interpretation where $\mathsf{minus}_{\mathcal{P}ol} = \max(m_1 x_1 + m_2 x_2, m_0)$. Here, $m_0, m_1, m_2$ are abstract coefficients which can later be instantiated by integers. Thus, we obtain the constraint $\max(m_1 x_1 + m_2 x_2, m_0) \geqslant 0$.

The challenge when automating Thm. 10 is that the general usable rules $\mathcal{GU}$ and the conditions (weakly) $\nu$-MM and $\nu$-RL depend on the (yet unknown) monotonicity specification $\nu$, which itself enforces constraints on the quasi-order $\succsim_{\mathcal{P}ol}$ that one searches for. Nevertheless, if one uses max-polynomial interpretations, then the search for reduction pairs can still be mechanized efficiently. More precisely, we show how to encode all conditions of Thm. 10 as a formula which is independent of $\nu$. In other words, this formula only contains Diophantine and Boolean variables. The latter are used to encode $\nu$. The formula has the form

$$\mathit{Orient} \wedge \mathit{Usable} \wedge \big(\mathit{More} \;\vee\; (\mathit{Wmore} \wedge \mathit{Rlinear})\big) \wedge \mathit{Compat} \wedge \mathit{Depend} \tag{12}$$

where *Orient* requires that the DPs and general usable rules are weakly decreasing and at least one DP is strictly decreasing. Here, we use Boolean variables that state which rules are usable and *Usable* ensures that these variables have the correct values. *More*, *Wmore*, and *Rlinear* correspond to $\nu$-MM, weak $\nu$-MM, and $\nu$-RL, respectively. *Compat* requires that $\succsim_{\mathcal{P}ol}$ is $\nu$-compatible. Finally, the formula *Depend* computes the sets $\nu(t,p)$ from the monotonicity specification $\nu$.

We start with defining *Depend*. To represent a monotonicity specification $\nu$, for every function symbol $f$ of arity $n$ and every $1 \leqslant i \leqslant n$ we introduce two Boolean variables $\Uparrow_{f,i}$ and $\Downarrow_{f,i}$ which encode the set $\nu(f,i)$. So $\Uparrow_{f,i}$ is *true* iff $\Uparrow \in \nu(f,i)$ and likewise for $\Downarrow_{f,i}$. *Depend* is the conjunction of the following formulas for every term $t$ in $\mathcal{P} \cup \mathcal{U}(\mathcal{P})$ and every position $p$ of $t$. They introduce two Boolean variables $\Uparrow_{t,p}$ and $\Downarrow_{t,p}$ to encode the sets $\nu(t,p)$ according to Def. 6.

$$\begin{array}{ll}
\Uparrow_{t,\varepsilon} & \Leftrightarrow \mathit{true} \\
\Uparrow_{f(t_1,\ldots,t_n),ip} & \Leftrightarrow (\Uparrow_{f,i} \wedge \Uparrow_{t_i,p}) \quad \vee \quad (\Downarrow_{f,i} \wedge \Downarrow_{t_i,p}) \quad \vee \\
 & \phantom{\Leftrightarrow} (\Uparrow_{f,i} \wedge \Downarrow_{f,i}) \quad \vee \quad (\Uparrow_{t_i,p} \wedge \Downarrow_{t_i,p}) \\
\Downarrow_{t,\varepsilon} & \Leftrightarrow \mathit{false} \\
\Downarrow_{f(t_1,\ldots,t_n),ip} & \Leftrightarrow (\Uparrow_{f,i} \wedge \Downarrow_{t_i,p}) \quad \vee \quad (\Downarrow_{f,i} \wedge \Uparrow_{t_i,p}) \quad \vee \\
 & \phantom{\Leftrightarrow} (\Uparrow_{f,i} \wedge \Downarrow_{f,i}) \quad \vee \quad (\Uparrow_{t_i,p} \wedge \Downarrow_{t_i,p})
\end{array}$$

Next we define *Usable*. We use two Boolean variables $\mathsf{us}_f$ and $\overline{\mathsf{us}}_f$ for every defined symbol $f$. Here, $\mathsf{us}_f$ (resp. $\overline{\mathsf{us}}_f$) is *true* if the $f$-rules (resp. reversed $f$-rules) are usable according to Def. 7. So whenever an $f$ occurs at a non-decreasing position of a right-hand side of $\mathcal{P}$ then the $f$-rules are usable. Similarly, if $f$ occurs at a non-increasing position, then the reversed $f$-rules are usable. Moreover, if (possibly reversed) $f$-rules are already usable then this may yield new usable rules due to right-hand sides of $f$-rules. Here, one has to keep the direction of the rules for non-decreasing positions and reverse the direction for non-increasing positions. This gives rise to the following formula *Usable*.

$$\bigwedge_{s\to t\in\mathcal{P},\, t|_p=f(\ldots),\, f \text{ defined}} (\neg\Downarrow_{t,p} \Rightarrow \mathsf{us}_f) \;\wedge\; (\neg\Uparrow_{t,p} \Rightarrow \overline{\mathsf{us}}_f) \quad\wedge$$

$$\bigwedge_{\ell\to r\in Rls(f),\, r|_p=g(\ldots),\, g \text{ defined}} (\mathsf{us}_f \Rightarrow (\neg\Downarrow_{r,p} \Rightarrow \mathsf{us}_g) \wedge (\neg\Uparrow_{r,p} \Rightarrow \overline{\mathsf{us}}_g)) \wedge (\overline{\mathsf{us}}_f \Rightarrow (\neg\Downarrow_{r,p} \Rightarrow \overline{\mathsf{us}}_g) \wedge (\neg\Uparrow_{r,p} \Rightarrow \mathsf{us}_g))$$

With the Boolean variables $\mathsf{us}_f$ and $\overline{\mathsf{us}}_f$ we can easily formalize that the rules in $\mathcal{P} \cup \mathcal{GU}(\mathcal{P})$ are weakly decreasing and that at least one pair is strictly decreasing. We obtain the following constraint *Orient* which is analogous to (3).

$$\bigwedge_{s\to t\in\mathcal{P}} [s]_{Pol} \geqslant [t]_{Pol} \quad\wedge\quad \bigvee_{s\to t\in\mathcal{P}} [s]_{Pol} \geqslant [t]_{Pol} + 1 \quad\wedge$$

$$\bigwedge_{\ell\to r\in\mathcal{R},\, f=\mathrm{root}(\ell)} (\mathsf{us}_f \Rightarrow [\ell]_{Pol} \geqslant [r]_{Pol}) \quad\wedge\quad (\overline{\mathsf{us}}_f \Rightarrow [r]_{Pol} \geqslant [\ell]_{Pol})$$

To ensure that $\mathcal{P} \cup \mathcal{U}^{contr}(\mathcal{P})$ is $\nu$-RL, we interpret the Boolean values *true* and *false* as 1 and 0. Then we express $\nu$-RL as a Diophantine constraint which we solve in the same way as the ones obtained from *Orient* later on. For any variable $x$, any term $t$, and any set $M \subseteq \{\Uparrow, \Downarrow\}$, let $\#_x^M(t)$ be a polynomial that describes the number of occurrences of $x$ in $t$ at positions $p$ where $\nu(t,p) = M$. Thus, $\#_x^{\varnothing}(t) = \sum_{t|_p=x}(\neg\Uparrow_{t,p} \wedge \neg\Downarrow_{t,p})$ and $\#_x^{\{\Uparrow\}}(t)$, $\#_x^{\{\Downarrow\}}(t)$, $\#_x^{\{\Uparrow,\Downarrow\}}(t)$ are defined accordingly. Moreover, $\#_x(t) = \sum_{t|_p=x}(\neg\Uparrow_{t,p} \vee \neg\Downarrow_{t,p})$ encodes the number of occurrences of $x$ at dependent positions of $t$. Then the constraint *Rlinear* is:

$$\bigwedge_{s\to t\in\mathcal{P},\, x\in\mathcal{V}(s)} \#_x(t) \leqslant 1 \quad\wedge \bigwedge_{\ell\to r\in\mathcal{R},\, x\in\mathcal{V}(\ell),\, f=\mathrm{root}(\ell)} (\mathsf{us}_f \vee \overline{\mathsf{us}}_f \Rightarrow \#_x(r) \leqslant 1)$$

*More* and *Wmore* ensure that $\mathcal{P} \cup \mathcal{U}^{contr}(\mathcal{P})$ is (weakly) $\nu$-MM. For every rule $\ell \to r$ and every variable $x$ at a $\nu$-dependent position $p$ of $r$, this variable must also occur at a unique less monotonic "partner" position $q$ of $\ell$. Thus, we could require $\#_x^{\varnothing}(r) \leqslant \#_x^{\varnothing}(\ell)$, $\#_x^{\{\Uparrow\}}(r) \leqslant \#_x^{\{\Uparrow\}}(\ell)$, and $\#_x^{\{\Downarrow\}}(r) \leqslant \#_x^{\{\Downarrow\}}(\ell)$. However, these requirements would be too strong, because they ignore the possibility that the "partner" position in $\ell$ may also be *strictly less* monotonic than the one in $r$. Therefore, for every rule $\ell \to r$ we introduce two new Diophantine variables $pt_x^{\Uparrow}$ and $pt_x^{\Downarrow}$ which stand for the number of those positions $p \in \mathcal{P}os_x^{\nu}(r)$ with $\nu(r,p) = \{\Uparrow\}$ (resp. $\nu(r,p) = \{\Downarrow\}$) where the "partner" position $q \in \mathcal{P}os_x^{\nu}(\ell)$ is non-monotonic (i.e., $\nu(\ell,q) = \varnothing$). Then *Wmore* is the following formula:

$$\bigwedge_{s\to t\in\mathcal{P},\, x\in\mathcal{V}(t)} (\#_x(s) \geqslant 1 \Rightarrow mm(s \to t, x)) \wedge \bigwedge_{\ell\to r\in\mathcal{R},\, x\in\mathcal{V}(r),\, f=\mathrm{root}(\ell)} ((\mathsf{us}_f \vee \overline{\mathsf{us}}_f) \wedge \#_x(\ell) \geqslant 1 \Rightarrow mm(\ell \to r, x))$$

where $mm(\ell \to r, x)$ is the following formula to encode $\nu$-MM. Its first part ensures that $\ell$ contains enough non-monotonic occurrences of $x$ to "cover" all occurrences of $x$ in $r$ that have a non-monotonic "partner" position in $\ell$.

$$\#_x^{\varnothing}(r)+pt_x^{\Uparrow}+pt_x^{\Downarrow} \leqslant \#_x^{\varnothing}(\ell) \wedge \#_x^{\{\Uparrow\}}(r) \leqslant pt_x^{\Uparrow}+\#_x^{\{\Uparrow\}}(\ell) \wedge \#_x^{\{\Downarrow\}}(r) \leqslant pt_x^{\Downarrow}+\#_x^{\{\Downarrow\}}(\ell)$$

Now *More* results from *Wmore* by removing the premises "$\#_x(\cdot) \geqslant 1$".

*Compat* ensures that whenever the Boolean variable $\Uparrow_{f,i}$ is *true*, then $f_{\mathcal{P}ol}$ is a max-polynomial that is (weakly) monotonically increasing on its $i$-th argument (similarly for $\Downarrow_{f,i}$). We express such monotonicity conditions by the *partial derivatives* of $f_{\mathcal{P}ol}$. If $f_{\mathcal{P}ol}$ is differentiable (i.e., $f_{\mathcal{P}ol}$ contains no "max"), then $\succsim_{\mathcal{P}ol}$ is monotonically increasing on $f$'s $i$-th argument iff $\frac{\partial f_{\mathcal{P}ol}}{\partial x_i} \geqslant 0$ (similarly for monotonic decrease). If $f_{\mathcal{P}ol}$ is a max-polynomial, then it is in general not differentiable, but *piecewise differentiable* and *continuous*. Then

> $\succsim_{\mathcal{P}ol}$ is monotonically increasing (resp. decreasing) on $f$'s $i$-th argument iff $\frac{\partial f_{\mathcal{P}ol}}{\partial x_i} \geqslant 0$ (resp. $\frac{\partial f_{\mathcal{P}ol}}{\partial x_i} \leqslant 0$) holds for all values where $\frac{\partial f_{\mathcal{P}ol}}{\partial x_i}$ is defined.

For instance, $\max(x_1-1,2)$ is not differentiable at $x_1 = 3$. We have $\frac{\partial \max(x_1-1,2)}{\partial x_1} = 0$ for $x_1 < 3$ and $\frac{\partial \max(x_1-1,2)}{\partial x_1} = 1$ for $x_1 > 3$. But as $\frac{\partial \max(x_1-1,2)}{\partial x_1} \geqslant 0$ whenever it is defined, the function $\max(x_1-1,2)$ is indeed monotonically increasing.

Therefore we introduce a new function symbol $\mathrm{der}_x$ for partial derivatives. Here, $\mathrm{der}_x(p)$ stands for $\frac{\partial p}{\partial x}$ whenever $p$ is a function depending on $x$. However, at the moment the expressions $\mathrm{der}_x(p)$ are not "evaluated". Thus, we can also write $\mathrm{der}_x(p)$ if $p$ is not differentiable. Then, *Compat* is the conjunction of the following constraints for all function symbols $f$ and all $1 \leqslant i \leqslant \mathrm{arity}(f)$:

$$(\Uparrow_{f,i} \ \Rightarrow\ \mathrm{der}_{x_i}(f_{\mathcal{P}ol}) \geqslant 0) \quad \wedge \quad (\Downarrow_{f,i} \ \Rightarrow\ 0 \geqslant \mathrm{der}_{x_i}(f_{\mathcal{P}ol}))$$

This is indeed sufficient to guarantee that $(\succsim_{\mathcal{P}ol}, \succ_{\mathcal{P}ol})$ is $\nu$-compatible. In particular, $\Uparrow_{f,i} \wedge \Downarrow_{f,i}$ now implies $\mathrm{der}_{x_i}(f_{\mathcal{P}ol}) = 0$, which ensures that $\succsim_{\mathcal{P}ol}$ is independent on $f$'s $i$-th argument. Thus, the third condition of Def. 6 is always satisfied for quasi-orders like $\succsim_{\mathcal{P}ol}$, cf. Footnote 6.

So to automate Thm. 10,[14] we start with the constraint (12) instead of (3). In addition, we need the constraints of the form (11). Then we again apply the inference rules (I) - (III) in order to obtain Diophantine constraints.

However, now inequalities also contain "$\mathrm{der}_x(p)$" for max-polynomials $p$. Here, we apply Rule (I) repeatedly in order to eliminate "max". So by Rule (I), the constraint $\mathrm{der}_{x_1}(\max(m_1x_1 + m_2x_2, m_0)) \geqslant 0$ would be transformed into

$$\begin{array}{lll} (m_1x_1 + m_2x_2 \geqslant m_0 & \Rightarrow\ \mathrm{der}_{x_1}(m_1x_1 + m_2x_2) \geqslant 0) & \wedge \\ (m_0 \geqslant m_1x_1 + m_2x_2 + 1 & \Rightarrow\ \mathrm{der}_{x_1}(m_0) \geqslant 0) & \end{array} \tag{13}$$

[14] The automation of Thm. 5 works as for Thm. 2. To automate the combination of Thm. 5 and Thm. 10, one first generates the constraints for Thm. 10 and tries to solve them. If one does not find a solution, one checks whether $\mathcal{P} \cup \mathcal{U}(\mathcal{P})$ is non-duplicating. In this case, one uses Thm. 5(a) and otherwise, one uses Thm. 5(b).

To eliminate "$\text{der}_x$" afterwards, we need the following rule for partial derivation:

| **IV. Eliminating "der"** | |
|---|---|
| $\dfrac{\ldots \text{der}_{x_i}(p_0 + p_1\, x_1^{e_{11}} \ldots x_n^{e_{n1}} + \cdots + p_k\, x_1^{e_{1k}} \ldots x_n^{e_{nk}}) \ldots}{\ldots p_1\, e_{i1}\, x_1^{e_{11}} \ldots x_i^{e_{i1}-1} \ldots x_n^{e_{n1}} + \cdots + p_k\, e_{ik}\, x_1^{e_{1k}} \ldots x_i^{e_{ik}-1} \ldots x_n^{e_{nk}}}$ | if the $p_i$ neither contain "max" nor any variable from $\mathcal{V}$ |

So in (13), one could replace $\text{der}_{x_1}(m_1x_1 + m_2x_2)$ by $m_1$ and $\text{der}_{x_1}(m_0)$ by 0.

## 5 Experiments and Conclusion

We showed how to use integer polynomial interpretations with "max" in termination proofs with DPs and developed a method to encode the resulting search problems into SAT. All our results are implemented in the systems AProVE and $\mathsf{T_TT_2}$. While AProVE and $\mathsf{T_TT_2}$ were already the two most powerful termination provers for TRSs at the *International Competition of Termination Tools* 2007 [14], our contributions increase the power of both tools considerably without affecting their efficiency. More precisely, when using a time limit of 1 minute per example, AProVE and $\mathsf{T_TT_2}$ can now automatically prove termination of 15 additional examples from the *Termination Problem Data Base* that is used for the competitions. Several of these examples had not been proven terminating by any tool at the competitions before. Moreover, AProVE and $\mathsf{T_TT_2}$ now also succeed on all examples from this paper (i.e., Ex. 1, 11, and 12), whereas all previous tools from the competitions failed (with the exception of TPA that could already solve Ex. 1). Our experiments also show the advantages over the earlier related contributions of [7,10] which were already implemented in AProVE and $\mathsf{T_TT_2}$, respectively. To run the AProVE implementation via a web-interface and for further details, we refer to `http://aprove.informatik.rwth-aachen.de/eval/maxpolo`.

## References

1. Arts, T., Giesl, J.: Termination of term rewriting using dependency pairs. Theoretical Computer Science 236, 133–178 (2000)
2. Fuhs, C., Giesl, J., Middeldorp, A., Schneider-Kamp, P., Thiemann, R., Zankl, H.: SAT solving for termination analysis with polynomial interpretations. In: Marques-Silva, J., Sakallah, K.A. (eds.) SAT 2007. LNCS, vol. 4501, pp. 340–354. Springer, Heidelberg (2007)
3. Fuhs, C., Giesl, J., Middeldorp, A., Schneider-Kamp, P., Thiemann, R., Zankl, H.: Maximal termination. Technical Report AIB-2008-03, RWTH Aachen, Germany (2008), `http://aib.informatik.rwth-aachen.de`
4. Giesl, J., Thiemann, R., Schneider-Kamp, P.: The dependency pair framework: Combining techniques for automated termination proofs. In: Baader, F., Voronkov, A. (eds.) LPAR 2004. LNCS (LNAI), vol. 3452, pp. 301–331. Springer, Heidelberg (2005)
5. Giesl, J., Thiemann, R., Schneider-Kamp, P.: AProVE 1.2: Automatic termination proofs in the dependency pair framework. In: Furbach, U., Shankar, N. (eds.) IJCAR 2006. LNCS (LNAI), vol. 4130, pp. 281–286. Springer, Heidelberg (2006)

6. Giesl, J., Thiemann, R., Schneider-Kamp, P., Falke, S.: Mechanizing and improving dependency pairs. Journal of Automated Reasoning 37(3), 155–203 (2006)
7. Giesl, J., Thiemann, R., Swiderski, S., Schneider-Kamp, P.: Proving termination by bounded increase. In: Pfenning, F. (ed.) CADE 2007. LNCS (LNAI), vol. 4603, pp. 443–459. Springer, Heidelberg (2007)
8. Hardin, T., Laville, A.: Proof of termination of the rewriting system SUBST on CCL. Theoretical Computer Science 46(2,3), 305–312 (1986)
9. Hirokawa, N., Middeldorp, A.: Automating the dependency pair method. Information and Computation 199(1,2), 172–199 (2005)
10. Hirokawa, N., Middeldorp, A.: Tyrolean Termination Tool: Techniques and features. Information and Computation 205(4), 474–511 (2007)
11. Hong, H., Jakuš, D.: Testing positiveness of polynomials. Journal of Automated Reasoning 21(1), 23–38 (1998)
12. Koprowski, A.: TPA: Termination proved automatically. In: Pfenning, F. (ed.) RTA 2006. LNCS, vol. 4098, pp. 257–266. Springer, Heidelberg (2006)
13. Lankford, D.: On proving term rewriting systems are Noetherian. Technical Report MTP-3, Louisiana Technical University, Ruston, LA, USA (1979)
14. Marché, C., Zantema, H.: The termination competition. In: Baader, F. (ed.) RTA 2007. LNCS, vol. 4533, pp. 303–313. Springer, Heidelberg (2007)
15. Nguyen, M., De Schreye, D., Giesl, J., Schneider-Kamp, P.: Polytool: Polynomial interpretations as a basis for termination analysis of logic programs. KU Leuven (2008)
16. Toyama, Y.: Counterexamples to the termination for the direct sum of term rewriting systems. Information Processing Letters 25, 141–143 (1987)
17. $\mathrm{T_TT_2}$, `http://colo6-c703.uibk.ac.at/ttt2`
18. Zantema, H.: Termination. In: Terese (ed.) Term Rewriting Systems, ch. 6, pp. 181–259. Cambridge University Press, Cambridge (2003)

# Usable Rules for Context-Sensitive Rewrite Systems*

Raúl Gutiérrez[1], Salvador Lucas[1], and Xavier Urbain[2]

[1] DSIC, Universidad Politécnica de Valencia, Spain
{rgutierrez , slucas}@dsic.upv.es
[2] Cédric-CNAM, ENSIIE, France
urbain@ensiie.fr

**Abstract.** Recently, the dependency pairs (DP) approach has been generalized to context-sensitive rewriting (CSR). Although the *context-sensitive dependency pairs (CS-DP) approach* provides a very good basis for proving termination of CSR, the current developments basically correspond to a ten-years-old DP approach. Thus, the task of adapting all recently introduced dependency pairs techniques to get a more powerful approach becomes an important issue. In this direction, *usable rules* are one of the most interesting and powerful notions. Actually usable rule have been investigated in connection with proofs of *innermost termination* of CSR. However, the existing results apply to a quite restricted class of systems. In this paper, we introduce a notion of usable rules that can be used in proofs of termination of CSR with arbitrary systems. Our benchmarks show that the performance of the CS-DP approach is much better when such usable rules are considered in proofs of termination of CSR.

**Keywords:** Dependency pairs, term rewriting, termination.

## 1 Introduction

During the last decade, the impressive advances in techniques for proving termination of rewriting (remarkably the dependency pairs approach [6,10,13,14]) have succeeded in solving termination problems that stood out of reach for a long time. Roughly speaking, given a Term Rewriting System (TRS) $\mathcal{R}$, the dependency pairs associated to $\mathcal{R}$ give rise to a new TRS $\mathsf{DP}(\mathcal{R})$ which (together with $\mathcal{R}$) determines the so-called *dependency chains* whose finiteness characterizes termination of $\mathcal{R}$. The dependency pairs can be presented as a *dependency graph*, where the absence of infinite chains can be analyzed by considering the *cycles* in the graph. Basically, given a *cycle* $\mathfrak{C} \subseteq \mathsf{DP}(\mathcal{R})$ in the dependency graph, we require $l \succsim r$ for *all* rules in the TRS $\mathcal{R}$, $u \succsim v$ or $u \sqsupset v$ for all dependency pairs $u \to v \in \mathfrak{C}$ and $u \sqsupset v$ for *at least one* $u \to v \in \mathfrak{C}$. Here, $\succsim$ is a stable

* Work partially supported by the EU (FEDER) and the Spanish MEC, under grants TIN 2004-7943-C04-02, TIN 2007-68118-C02 and HA 2006-0007.

A. Voronkov (Ed.): RTA 2008, LNCS 5117, pp. 126–141, 2008.

and monotonic quasi-ordering on terms and $\sqsupset$ is a *well-founded* ordering; both of them can be different for the different cycles in the dependency graph.

Termination problems with many rules require more time for getting an answer. Even worse: since termination proofs are usually constrained to succeed within a given (often short) time-out, the proof could get lost due to a lack of time. For those reasons, techniques leading to increase the efficiency (and also the power) of the dependency pairs method, like *usable rules*, appear like a key issue. Usable rules $\mathcal{U}(\mathcal{R},\mathfrak{C}) \subseteq \mathcal{R}$ are associated to a given *cycle* $\mathfrak{C}$ of the dependency graph for $\mathcal{R}$. For particular (but widely used) classes of quasi-orderings $\succeq$, we can restrict the comparisons $l \succeq r$ to rules $l \to r$ in $\mathcal{U}(\mathcal{R},\mathfrak{C})$ instead of using $\mathcal{R}$. Since $\mathcal{U}(\mathcal{R},\mathfrak{C})$ is (usually) smaller than $\mathcal{R}$, proofs of termination often become easier in this way. Usable rules were introduced ten years ago by Arts and Giesl for proving termination of innermost rewriting [5]. The adaptation of the idea to (unrestricted) rewriting [14,17] took some years. A possible reason for that is that the proof of soundness for the innermost and for the unrestricted cases are totally different. The proof of soundness in [14,17] relies on a transformation in which all infinite (minimal) rewrite sequences can be simulated by using a *restricted* set of rules. This transformation was devised by Gramlich for a completely different purpose [15]. Later, Urbain [24] used it (with some modifications) to prove termination of rewriting modules. Finally, Hirokawa and Middeldorp [17] and (independently) Thiemann *et al.* [14] combined this idea with the idea of *usable rules* leading to an improved framework for proving termination of rewriting.

In this paper, we extend the notion of usable rule to the recently introduced dependency pairs approach for context-sensitive rewriting (CS-DPs [2,3]). Proving termination of *context-sensitive rewriting* (*CSR* [18,20]) is an interesting problem with many applications in the fields of term rewriting and programming languages (see [8,12,19,20,22] for further motivations). In *CSR*, a *replacement map* (i.e., a mapping $\mu : \mathcal{F} \to \wp(\mathbb{N})$ satisfying $\mu(f) \subseteq \{1,\ldots,k\}$, for each $k$-ary symbol $f$ of a signature $\mathcal{F}$) is used to discriminate the argument positions on which the rewriting steps are allowed; rewriting at the topmost position is always possible. The following example gives a first intuition of CSR and CS-DPs; full details are given below.

*Example 1.* Consider the following TRS $\mathcal{R}$ borrowed from [7, Example 4.7.37]. The program zips two lists of integers into a single one but instead of pairing the components it rather computes their quotients:

$$\begin{array}{rlrl}
\mathtt{sel}(\mathtt{0},\mathtt{cons}(x,xs)) \to x & (1) & \mathtt{sel}(\mathtt{s}(n),\mathtt{cons}(x,xs)) \to \mathtt{sel}(n,xs) & (7)\\
\mathtt{minus}(x,\mathtt{0}) \to x & (2) & \mathtt{minus}(\mathtt{s}(x),\mathtt{s}(y)) \to \mathtt{minus}(x,y) & (8)\\
\mathtt{quot}(\mathtt{0},\mathtt{s}(y)) \to \mathtt{0} & (3) & \mathtt{quot}(\mathtt{s}(x),\mathtt{s}(y)) \to \mathtt{s}(\mathtt{quot}(\mathtt{minus}(x,y),\mathtt{s}(y))) & (9)\\
\mathtt{zWquot}(\mathtt{nil},x) \to \mathtt{nil} & (4) & \mathtt{from}(x) \to \mathtt{cons}(x,\mathtt{from}(\mathtt{s}(x))) & (10)\\
\mathtt{zWquot}(x,\mathtt{nil}) \to \mathtt{nil} & (5) & \mathtt{tail}(\mathtt{cons}(x,xs)) \to xs & (11)\\
\mathtt{head}(\mathtt{cons}(x,xs)) \to x & (6) & &
\end{array}$$

$$\mathtt{zWquot}(\mathtt{cons}(x,xs),\mathtt{cons}(y,ys)) \to \mathtt{cons}(\mathtt{quot}(x,y),\mathtt{zWquot}(xs,ys)) \quad (12)$$

with $\mu(\mathtt{cons}) = \{1\}$ and $\mu(f) = \{1,\ldots,ar(f)\}$ for all other symbols $f \in \mathcal{F}$. The set of CS-DPs of $\mathcal{R}$ is:

$$\begin{array}{rlrl}
\mathtt{MINUS}(\mathtt{s}(x),\mathtt{s}(y)) & \to \mathtt{MINUS}(x,y) & \mathtt{SEL}(\mathtt{s}(n),\mathtt{cons}(x,xs)) & \to \mathtt{SEL}(n,xs)\\
\mathtt{QUOT}(\mathtt{s}(x),\mathtt{s}(y)) & \to \mathtt{MINUS}(x,y) & \mathtt{ZWQUOT}(\mathtt{cons}(x,xs),\mathtt{cons}(y,ys)) & \to \mathtt{QUOT}(x,y)\\
\mathtt{QUOT}(\mathtt{s}(x),\mathtt{s}(y)) & \to \mathtt{QUOT}(\mathtt{minus}(x,y),\mathtt{s}(y)) & \mathtt{SEL}(\mathtt{s}(n),\mathtt{cons}(x,xs)) & \to xs\\
\mathtt{TAIL}(\mathtt{cons}(x,xs)) & \to xs & &
\end{array}$$

Note that non-$\mu$-replacing subterms in right-hand sides (e.g., $\mathtt{from}(\mathtt{s}(x))$ in rule (10)) are *not* considered to build the CS-DPs. Also, in sharp contrast with the unrestricted case, *collapsing* dependency pairs like $\mathtt{TAIL}(\mathtt{cons}(x,xs)) \to xs$ (where the right-hand side is a variable) are introduced.

Regarding proofs of termination of *innermost CSR*, the straightforward adaptation of usable rules to the context-sensitive setting only works for the so-called *conservative systems* (see [4]) where collapsing dependency pairs do not occur. In Section 3, we show that the standard adaptation does *not* work when proofs of termination of *CSR* are attempted. In Section 4, we provide a general notion of usable rules for proving termination of *CSR*. Although we follow the same proof style, our proof of soundness differs from those in [14,15,17,24] in several aspects that we clarify below. In Section 5, we prove that it is possible to use the *standard* (simpler) notion of usable rules [14,17] in proofs of termination of *CSR* for a restricted class of CS-TRSs: the *strongly* conservative systems. Section 6 provides experimental evaluations and Section 7 concludes. Complete proofs are given in [16].

## 2 Preliminaries

We assume knowledge about standard definitions and notations for term rewriting (including dependency pairs) as given in, e.g., [23]. In the following, we provide some definitions and notation on CSR [18,20] and CS-DPs [2,3].

*Context-Sensitive Rewriting.* Given a TRS $\mathcal{R} = (\mathcal{F}, R)$, we consider the signature $\mathcal{F}$ as the disjoint union $\mathcal{F} = \mathcal{C} \uplus \mathcal{D}$ of *constructors* symbols $c \in \mathcal{C}$ and *defined* symbols $f \in \mathcal{D}$ where $\mathcal{D} = \{root(l) \mid l \to r \in R\}$ and $\mathcal{C} = \mathcal{F} - \mathcal{D}$. A mapping $\mu : \mathcal{F} \to \wp(\mathbb{N})$ is a *replacement map* (or $\mathcal{F}$-map) if $\forall f \in \mathcal{F},\ \mu(f) \subseteq \{1,\ldots,ar(f)\}$ [18]. Let $M_{\mathcal{F}}$ be the set of all $\mathcal{F}$-maps ($M_{\mathcal{R}}$ for the $\mathcal{F}$-maps of a TRS $\mathcal{R} = (\mathcal{F}, R)$). A binary relation $R$ on terms in $\mathcal{T}(\mathcal{F},\mathcal{X})$ is $\mu$-monotonic if $tRs$ implies $f(t_1,\ldots,t_{i-1},t,\ldots,t_n)\,R\,f(t_1,\ldots,t_{i-1},s,\ldots,t_n)$ for all $f \in \mathcal{F}, i \in \mu(f)$, and $t,s,t_1,\ldots,t_n \in \mathcal{T}(\mathcal{F},\mathcal{X})$. The set of *$\mu$-replacing positions* $\mathcal{P}os^{\mu}(t)$ of $t \in \mathcal{T}(\mathcal{F},\mathcal{X})$ is: $\mathcal{P}os^{\mu}(t) = \{\epsilon\}$, if $t \in \mathcal{X}$ and $\mathcal{P}os^{\mu}(t) = \{\epsilon\} \cup \bigcup_{i\in\mu(root(t))} i.\mathcal{P}os^{\mu}(t|_i)$, if $t \notin \mathcal{X}$. The set of *$\mu$-replacing* variables of $t$ is $\mathcal{V}ar^{\mu}(t) = \{x \in \mathcal{V}ar(t) \mid \exists p \in \mathcal{P}os^{\mu}(t), t|_p = x\}$. The $\mu$-replacing subterm relation $\trianglerighteq_{\mu}$ is defined by $t \trianglerighteq_{\mu} s$ if there is $p \in \mathcal{P}os^{\mu}(t)$ such that $s = t|_p$. We write $t \rhd_{\mu} s$ if $t \trianglerighteq_{\mu} s$ and $t \neq s$. We write

$t \rhd_{\not\mu} s$ to denote that $s$ is a non-$\mu$-replacing strict subterm of $t$: $t \rhd_{\not\mu} s$ if there is $p \in \mathcal{P}os(t) - \mathcal{P}os^\mu(t)$ such that $s = t|_p$. We say that $f \in \mathcal{F}$ is a *hidden symbol* in $l \to r \in R$ if there exists a term $t \in \mathcal{T}(\mathcal{F}, \mathcal{X})$ s.t. $r \rhd_{\not\mu} t$ and $root(t) = f$. We say that a variable $x$ is *migrating* in $l \to r \in R$ if $x \in \mathcal{V}ar^\mu(r) - \mathcal{V}ar^\mu(l)$. In *context-sensitive rewriting* (*CSR* [18]), we (only) rewrite terms at $\mu$-*replacing* positions: $t$ $\mu$-rewrites to $s$, written $t \hookrightarrow_\mu s$ (or $t \hookrightarrow_{\mathcal{R},\mu} s$), if $t \xrightarrow{p}_{\mathcal{R}} s$ and $p \in \mathcal{P}os^\mu(t)$. A TRS $\mathcal{R}$ is $\mu$-terminating if $\hookrightarrow_\mu$ is terminating. A term $t$ is $\mu$-terminating if there is no infinite $\mu$-rewrite sequence $t = t_1 \hookrightarrow_\mu t_2 \hookrightarrow_\mu \cdots$. A pair $(\mathcal{R}, \mu)$ (or triple $(\mathcal{F}, \mu, R)$) where $\mathcal{R} = (\mathcal{F}, R)$ is a TRS and $\mu \in M_{\mathcal{R}}$ is often called a CS-TRS. We denote $\mathcal{H}(\mathcal{R}, \mu)$ (or just $\mathcal{H}$, if there is no ambiguity) the set of all hidden symbols in $(\mathcal{R}, \mu)$.

*Context-Sensitive Dependency Pairs.* Given a TRS $\mathcal{R} = (\mathcal{F}, R) = (\mathcal{C} \uplus \mathcal{D}, R)$ and $\mu \in M_{\mathcal{R}}$, the set of context-sensitive dependency pairs (CS-DPs) is $\mathsf{DP}(\mathcal{R}, \mu) = \mathsf{DP}_{\mathcal{F}}(\mathcal{R}, \mu) \cup \mathsf{DP}_{\mathcal{X}}(\mathcal{R}, \mu)$, where $\mathsf{DP}_{\mathcal{F}}(\mathcal{R}, \mu)$ and $\mathsf{DP}_{\mathcal{X}}(\mathcal{R}, \mu)$ are obtained as follows: let $f(t_1, \ldots, t_m) \to r \in R$ and $s \in \mathcal{T}(\mathcal{F}, \mathcal{X})$ be such that $r \trianglerighteq_\mu s$. Then (1) if $s = g(s_1, \ldots, s_n)$, for some $g \in \mathcal{D}$, $s_1, \ldots, s_n \in \mathcal{T}(\mathcal{F}, \mathcal{X})$ and $l \ntriangleright_\mu s$, then $f^\sharp(t_1, \ldots, t_m) \to g^\sharp(s_1, \ldots, s_n) \in \mathsf{DP}_{\mathcal{F}}(R, \mu)$; (2) if $s = x \in \mathcal{V}ar^\mu(r) - \mathcal{V}ar^\mu(l)$, then $f^\sharp(t_1, \ldots, t_m) \to x \in \mathsf{DP}_{\mathcal{X}}(R, \mu)$. Here, $f^\sharp$ and $g^\sharp$ are new fresh symbols (called *tuple* symbols) associated to the symbols $f$ and $g$ respectively. The CS-DPs in $\mathsf{DP}_{\mathcal{X}}(\mathcal{R}, \mu)$ are called the *collapsing* CS-DPs. Let $\mathcal{F}^\sharp = \mathcal{F} \cup \{f^\sharp \mid f \in \mathcal{F}\}$. We extend $\mu \in M_{\mathcal{F}}$ into $\mu^\sharp \in M_{\mathcal{F}^\sharp}$ by $\mu^\sharp(f) = \mu^\sharp(f^\sharp) = \mu(f)$ for each $f \in \mathcal{F}$. As usual, for $t = f(t_1, \ldots, t_n) \in \mathcal{T}(\mathcal{F}, \mathcal{X})$, we write $t^\sharp$ to denote the *marked* term $f^\sharp(t_1, \ldots, t_n)$. Let $\mathcal{T}^\sharp(\mathcal{F}, \mathcal{X}) = \{t^\sharp \mid t \in \mathcal{T}(\mathcal{F}, \mathcal{X}) - \mathcal{X}\}$ be the set of marked terms. We will also use the set $\mathcal{P}^\sharp(\mathcal{F}, \mathcal{X}) = \mathcal{T}^\sharp(\mathcal{F}, \mathcal{X}) \times (\mathcal{T}^\sharp(\mathcal{F}, \mathcal{X}) \cup \mathcal{X})$. Given $t = f^\sharp(t_1, \ldots, t_k) \in \mathcal{T}^\sharp(\mathcal{F}, \mathcal{X})$, we write $t^\natural$ to denote the *unmarked* term $f(t_1, \ldots, t_k) \in \mathcal{T}(\mathcal{F}, \mathcal{X})$. As usual, capital letters denote marked symbols in examples. A set of pairs $\mathcal{P} \subseteq \mathcal{P}^\sharp(\mathcal{F}, \mathcal{X})$ is decomposed into collapsing and non-collapsing pairs ($\mathcal{P}_{\mathcal{X}}$ and $\mathcal{P}_{\mathcal{F}}$, respectively): $\mathcal{P}_{\mathcal{X}} = \{u \to v \in \mathcal{P} \mid v \in \mathcal{X}\}$ and $\mathcal{P}_{\mathcal{F}} = \mathcal{P} - \mathcal{P}_{\mathcal{X}}$.

Let $\mathcal{R} = (\mathcal{F}, R)$ be a TRS, $\mathcal{P} \subseteq \mathcal{P}^\sharp(\mathcal{F}, \mathcal{X})$ and $\mu \in M_{\mathcal{F}}$. An $(\mathcal{R}, \mathcal{P}, \mu^\sharp)$-chain is a finite or infinite sequence of pairs $u_i \to v_i \in \mathcal{P}$, for $i \geq 1$ such that there is a substitution $\sigma$ satisfying both:

1. $\sigma(v_i) \hookrightarrow^*_{\mathcal{R},\mu^\sharp} \sigma(u_{i+1})$, if $u_i \to v_i \in \mathcal{P}_{\mathcal{F}}$, and
2. if $u_i \to v_i = u_i \to x_i \in \mathcal{P}_{\mathcal{X}}$, then there is $s_i \in \mathcal{T}(\mathcal{F}, \mathcal{X})$ such that $\sigma(x_i) \trianglerighteq_\mu s_i$ and $s_i^\sharp \hookrightarrow^*_{\mathcal{R},\mu^\sharp} \sigma(u_{i+1})$.

where $\mathcal{V}ar(v_i) \cap \mathcal{V}ar(u_j) = \varnothing$ for all $i \neq j$ (renaming if necessary). Let $\mathcal{M}_{\infty,\mu}$ be the set of minimal non-$\mu$-terminating terms. Then, $t \in \mathcal{M}_{\infty,\mu}$ if $t$ is non-$\mu$-terminating and every strict $\mu$-replacing subterm of $t$ is terminating. We say that an $(\mathcal{R}, \mathcal{P}, \mu^\sharp)$-chain is *minimal* if for all $i \geq 1$ $\sigma(v_i)$ (whenever $u_i \to v_i \in \mathcal{P}_{\mathcal{F}}$), $s_i^\sharp$ (whenever $u_i \to v_i \in \mathcal{P}_{\mathcal{X}}$) are $\mu$-terminating w.r.t. $\mathcal{R}$. A CS-TRS $\mathcal{R} = (\mathcal{F}, \mu, R)$ is $\mu$-terminating if and only if there is no infinite minimal $(\mathcal{R}, \mathsf{DP}(\mathcal{R}, \mu), \mu^\sharp)$-chain. For finite CS-TRSs, the CS-DPs can be presented as a *context-sensitive dependency graph* (CS-DG); there is an arc from $u \to v \in \mathsf{DP}_{\mathcal{F}}(\mathcal{R}, \mu)$ to $u' \to$

$v' \in \mathsf{DP}(\mathcal{R},\mu)$ if there is a substitution $\sigma$ such that $\sigma(v) \hookrightarrow^*_{\mathcal{R},\mu} \sigma(u')$; and, there is an arc from $u \to v \in \mathsf{DP}_{\mathcal{X}}(\mathcal{R},\mu)$ to $u' \to v' \in \mathsf{DP}(\mathcal{R},\mu)$ if $root(u')^\natural \in \mathcal{H}$. We consider the *strongly connected components* in this graph. A $\mu$-reduction pair $(\succsim, \sqsupset)$ consists of a stable and weakly $\mu$-monotonic quasi-ordering $\succsim$, and a stable and well-founded ordering $\sqsupset$ satisfying $\succsim \circ \sqsupset \subseteq \sqsupset$ or $\sqsupset \circ \succsim \subseteq \sqsupset$. From now on, we assume that all CS-TRSs are finite.

## 3 Basic Usable Rules

Consider a set of pairs $\mathcal{P}$ and a CS-TRS $(\mathcal{R},\mu)$. Then, the set of usable rules is the smallest set of rules from $\mathcal{R}$ which are needed to capture all the infinite minimal $(\mathcal{R},\mathcal{P},\mu^\sharp)$-chains. The rules that are responsible for generating the chains between pairs are those rules rooted by symbols that appear in the right-hand side of the pairs below the root symbol. This concept is captured by the definition of direct dependency [14,17,24]:

**Definition 1 (Direct Dependency [14,17]).** *Given a TRS $\mathcal{R} = (\mathcal{F}, R)$, we say that $f \in \mathcal{F}$* directly depends *on $g \in \mathcal{F}$, written $f \rhd_d g$, if there is a rule $l \to r \in R$ with $f = root(l)$ and $g$ occurs in $r$.*

The set of defined symbols in a term $t$ is $\mathcal{D}Fun(t) = \{f \mid \exists p \in \mathcal{P}os(t), f = root(t|_p) \in \mathcal{D}\}$. Let $\rhd^*_d$ be the transitive and reflexive closure of $\rhd_d$. Then, we have:

**Definition 2 (Usable Rules [14,17]).** *For a set $\mathcal{G}$ of symbols we denote by $\mathcal{R} \mid \mathcal{G}$ the set of rewriting rules $l \to r \in \mathcal{R}$ with $root(l) \in \mathcal{G}$. The set $\mathcal{U}(\mathcal{R},t)$ of* usable rules *of a term $t$ is defined as $\mathcal{R} \mid \{g \mid f \rhd^*_d g$ for some $f \in \mathcal{D}Fun(t)\}$. If $\mathcal{P}$ is a set of dependency pairs then $\mathcal{U}(\mathcal{R},\mathcal{P}) = \bigcup_{l \to r \in \mathcal{P}} \mathcal{U}(\mathcal{R}, r)$.*

The set $\mathcal{U}(\mathcal{R},\mathcal{P})$ can be used instead of $\mathcal{R}$ when looking for a reduction pair that proves termination of $\mathcal{R}$ [14,17]. Let us now focus on CS-TRSs.

A first attempt to give a notion of usable rules for CSR is given in [4] (basic usable rules) for proofs of *innermost* termination. The results in [4] show that the straightforward generalization of Definition 2 to *CSR* (see Definition 4 below) only applies to *conservative* CS-TRSs and cycles (of CS-DPs), that is, systems having only conservative rules [22]: a rule $l \to r \in R$ is *conservative* if $\mathcal{V}ar^\mu(r) \subseteq \mathcal{V}ar^\mu(l)$. First, we adapt Definition 1 to the CSR setting as follows:

**Definition 3 (Basic $\mu$-Dependency).** *Given a CS-TRS $(\mathcal{F},\mu,R)$, we say that $f \in \mathcal{F}$ has a* basic $\mu$-dependency *on $g \in \mathcal{F}$, written $f \blacktriangleright_{d,\mu} g$, if there is $l \to r \in R$ with $f = root(l)$ and $g$ occurs in $r$ at a $\mu$-replacing position.*

This leads to a straightforward extension of Definition 2. The set of *$\mu$-replacing* defined symbols in a term $t$ is $\mathcal{D}Fun^\mu(t) = \{f \mid \exists p \in \mathcal{P}os^\mu(t), f = root(t|_p) \in \mathcal{D}\}$. Then, we have[1]:

[1] Note that, due to the focus on innermost CSR, [4, Def. 5] slightly differs from ours.

**Definition 4 (Basic Context-Sensitive Usable Rules).** *Let* $\mathcal{R} = (\mathcal{F}, R)$ *be a TRS and* $\mu \in M_{\mathcal{R}}$*. The set* $\mathcal{U}_B(\mathcal{R}, \mu, t)$ *of* basic context-sensitive usable rules *of a term* $t$ *is defined as* $\mathcal{R} \mid \{g \mid f \blacktriangleright^*_{d,\mu} g \text{ for some } f \in \mathcal{D}Fun^{\mu}(t)\}$*, where* $\blacktriangleright^*_{d,\mu}$ *is the transitive and reflexive closure of* $\blacktriangleright_{d,\mu}$*. If* $\mathcal{P} \subseteq \mathcal{P}^\sharp(\mathcal{F}, \mathcal{X})$*, then* $\mathcal{U}_B(\mathcal{R}, \mu^\sharp, \mathcal{P}) = \bigcup_{l \to r \in \mathcal{P}} \mathcal{U}_B(\mathcal{R}, \mu^\sharp, r)$.

*Example 2.* (Continuing Example 1) The cycles in the CS-DG are:

$$\{\mathtt{SEL}(\mathtt{s}(n), \mathtt{cons}(x, xs)) \to \mathtt{SEL}(n, xs)\} \qquad (C_1)$$
$$\{\mathtt{MINUS}(\mathtt{s}(x), \mathtt{s}(y)) \to \mathtt{MINUS}(x, y)\} \qquad (C_2)$$
$$\{\mathtt{QUOT}(\mathtt{s}(x), \mathtt{s}(y)) \to \mathtt{QUOT}(\mathtt{minus}(x, y), \mathtt{s}(y))\} \qquad (C_3)$$

Consider the cycle $C_3$; then, $\mathcal{U}_B(\mathcal{R}, \mu^\sharp, C_3)$ contains the following rules:

$$\mathtt{minus}(x, \mathtt{0}) \to x \qquad \mathtt{minus}(\mathtt{s}(x), \mathtt{s}(y)) \to \mathtt{minus}(x, y)$$

However, as we are going to see, and in sharp contrast with [4], Definition 4 does not lead to a correct approach for proving termination of *CSR*, *even for conservative TRSs.*

*Example 3.* Consider the TRS $\mathcal{R} = \{\mathtt{f}(\mathtt{c}(x), x) \to \mathtt{f}(x, x), \mathtt{b} \to \mathtt{c}(\mathtt{b})\}$ [4] together with $\mu(\mathtt{f}) = \{1, 2\}$ and $\mu(\mathtt{c}) = \varnothing$. Note that $(\mathcal{R}, \mu)$ is conservative (and innermost $\mu$-terminating, see [4]).

We have a single cycle $C = \{\mathtt{F}(\mathtt{c}(x), x) \to \mathtt{F}(x, x)\}$. According to Definition 4, we have no usable rules because $\mathtt{F}(x, x)$ contains no symbol in $\mathcal{F}$. We could wrongly conclude $\mu$-termination of $(\mathcal{R}, \mu)$, but we have the infinite minimal $(\mathcal{R}, C, \mu^\sharp)$-chain $\underline{\mathtt{F}(\mathtt{c}(\mathtt{b}), \mathtt{b})} \to \mathtt{F}(\underline{\mathtt{b}}, \mathtt{b}) \hookrightarrow \underline{\mathtt{F}(\mathtt{c}(\mathtt{b}), \mathtt{b})} \to \cdots$.

In the following, we develop a correct definition of usable rules that can be applied to *arbitrary* CS-TRSs.

## 4 Termination of CS-TRSs with Usable Rules

As shown in [14,17], considering the set of *usable rules* instead of all the rules suffices for proving termination of $(\mathcal{R}, \mathcal{P})$-chains (or $\mathcal{P}$-minimal sequences in [17]). In [14,17], an *interpretation* of terms as sequences of their possible reducts is used[2]. The definition of the transformation requires adding new fresh (list constructor) symbols $\bot, \mathtt{g} \notin \mathcal{F}$ and the (projection) rules $\mathtt{g}(x, y) \to x$, $\mathtt{g}(x, y) \to y$ (the $\pi$-rules). In this way, infinite minimal $(\mathcal{R}, \mathcal{P})$-chains can be represented as infinite $(\mathcal{U}(\mathcal{R}, \mathcal{P}) \cup \pi, \mathcal{P})$-chains. We recall here the interpretation definition.

**Definition 5 (Interpretation [14,17]).** *Let* $\mathcal{R} = (\mathcal{F}, R)$ *be a TRS and* $\mathcal{G} \subseteq \mathcal{F}$*. Let* $>$ *be an arbitrary total ordering over* $\mathcal{T}(\mathcal{F}^\sharp \cup \{\bot, \mathtt{g}\}, \mathcal{X})$ *where* $\bot$ *is a new constant symbol and* $\mathtt{g}$ *is a new binary symbol. The interpretation* $I_{\mathcal{G}}$ *is a mapping*

[2] This method goes back to [15].

*from terminating terms in $\mathcal{T}(\mathcal{F}^\sharp, \mathcal{X})$ to terms in $\mathcal{T}(\mathcal{F}^\sharp \cup \{\perp, \mathsf{g}\}, \mathcal{X})$ defined as follows:*

$$I_\mathcal{G}(t) = \begin{cases} t & \text{if } t \in \mathcal{X} \\ f(I_\mathcal{G}(t_1), \ldots, I_\mathcal{G}(t_n)) & \text{if } t = f(t_1 \ldots t_n) \text{ and } f \notin \mathcal{G} \\ \mathsf{g}(f(I_\mathcal{G}(t_1), \ldots, I_\mathcal{G}(t_n)), t') & \text{if } t = f(t_1 \ldots t_n) \text{ and } f \in \mathcal{G} \end{cases}$$

*where* $\quad t' = \mathit{order}\,(\{I_\mathcal{G}(u) \mid t \to_\mathcal{R} u\})$

$$\mathit{order}(T) = \begin{cases} \perp, & \text{if } T = \varnothing \\ \mathsf{g}(t, \mathit{order}(T - \{t\})) & \text{if } t \text{ is minimal in } T \text{ w.r.t. } > \end{cases}$$

The set of symbols $\mathcal{G} \subseteq \mathcal{F}$ in Definition 5 is intended to represent the set of 'non-usable symbols', i.e., symbols which do not occur in the usable rules of the considered set of pairs $\mathcal{P}$. In rewriting, when considering infinite minimal $(\mathcal{R}, \mathcal{P})$-chains, we only deal with terminating terms over $\mathcal{R}$. The interpretation in Definition 5 is defined only for terminating terms because non-terminating terms would yield an infinite term which, actually, does *not* belong to $\mathcal{T}(\mathcal{F}^\sharp \cup \{\perp, \mathsf{g}\}, \mathcal{X})$.

Similarly, we aim at defining a $\mu$-interpretation $I_{\mathcal{G},\mu}$ that allows us to associate an infinite $(\mathcal{U}(\mathcal{R}, \mu^\sharp, \mathcal{P}) \cup \pi, \mathcal{P}, \mu^\sharp)$-chain to each infinite minimal $(\mathcal{R}, \mathcal{P}, \mu^\sharp)$-chain. Actually, the main problem is that $(\mathcal{R}, \mathcal{P}, \mu^\sharp)$-chains contain non-$\mu$-terminating terms in non-$\mu$-replacing positions which are potentially able to reach $\mu$-replacing positions: subterms at a $\mu$-replacing position are $\mu$-terminating, but we do not know anything about subterms at non-$\mu$-replacing positions. Hence, we have to define our $\mu$-interpretation $I_{\mathcal{G},\mu}$ both on $\mu$-terminating and non-$\mu$-terminating terms. In [3], we have investigated the structure of infinite $\mu$-rewriting sequences issued from minimal non-$\mu$-terminating terms. Intuitively, one of the main results in [3] states that terms at non-$\mu$-replacing positions in the right-hand side of the rules are essential to track infinite minimal $(\mathcal{R}, \mathcal{P}, \mu^\sharp)$-chains involving collapsing CS-DPs (see [3, Proposition 3.6]). These terms, by definition, are formed by *hidden symbols*. This observation gives us the key to generalize Definition 5 properly. Following Definition 5, a $\mu$-terminating but non-terminating term generates an infinite list. For this reason, $I_\mathcal{G}$ (as a mapping from finite into finite terms) is *not* defined for non-terminating terms.

Regarding our $\mu$-interpretation, if we consider the rules headed by hidden symbols as *usable*, then we are avoiding such infinite $\mu$-interpretations of $\mu$-terminating terms. A non-$\mu$-terminating term $t$ (below a non-$\mu$-replacing position) is treated as if its root symbol does not belong to $\mathcal{G}$, because if it occurs in the $(\mathcal{R}, \mathcal{P}, \mu^\sharp)$-chain at a $\mu$-replacing position, then $t \trianglerighteq_\mu s$ and $s^\sharp$ becomes the next term in the chain. To simulate all possible derivations of the terms over $(\mathcal{R}, \mu)$ we also need to add to the system the $\pi$-rules. Our new $\mu$-interpretation is:

**Definition 6 ($\mu$-Interpretation).** *Let $\mathcal{R} = (\mathcal{F}, \mu, R)$ be a CS-TRS, $\mathcal{G} \subseteq \mathcal{F}$ be such that $\mathcal{G} \cap \mathcal{H} = \varnothing$. Let $>$ be an arbitrary total ordering over $\mathcal{T}(\mathcal{F}^\sharp \cup \{\perp, \mathsf{g}\}, \mathcal{X})$ where $\perp$ is a new constant symbol and $\mathsf{g}$ is a new binary symbol (with $\mu(\mathsf{g}) = \{1, 2\}$). The $\mu$-interpretation $I_{\mathcal{G},\mu}$ is a mapping from arbitrary terms in $\mathcal{T}(\mathcal{F}^\sharp, \mathcal{X})$*

*to terms in* $\mathcal{T}(\mathcal{F}^\sharp \cup \{\bot, \mathsf{g}\}, \mathcal{X})$ *defined as follows:*

$$I_{\mathcal{G},\mu}(t) = \begin{cases} t & \text{if } t \in \mathcal{X} \\ f(I_{\mathcal{G},\mu}(t_1), \ldots, I_{\mathcal{G},\mu}(t_n)) & \text{if } t = f(t_1 \ldots t_n) \text{ and } f \notin \mathcal{G} \\ & \text{or } t \text{ is non-}\mu\text{-terminating} \\ \mathsf{g}(f(I_{\mathcal{G},\mu}(t_1), \ldots, I_{\mathcal{G},\mu}(t_n)), t') & \text{if } t = f(t_1 \ldots t_n) \text{ and } f \in \mathcal{G} \\ & \text{and } t \text{ is } \mu\text{-terminating} \end{cases}$$

*where* $t' = order\left(\{I_{\mathcal{G},\mu}(u) \mid t \hookrightarrow_{(\mathcal{R},\mu)} u\}\right)$

$$order(T) = \begin{cases} \bot, & \text{if } T = \varnothing \\ \mathsf{g}(t, order(T - \{t\})) & \text{if } t \text{ is minimal in } T \text{ w.r.t. } > \end{cases}$$

The set $\mathcal{G} \subseteq \mathcal{F}$ in Definition 6 corresponds to the set of non-usable symbols as discussed below. Now, we prove that $I_{\mathcal{G},\mu}$ is well-defined. The most important difference (and essential in our proof) among our $\mu$-interpretation and all previous ones [14,15,17,24] is that $I_{\mathcal{G},\mu}$ is well-defined both for $\mu$-terminating or non-$\mu$-terminating terms.

**Lemma 1.** *Let* $\mathcal{R} = (\mathcal{F}, R)$ *be a TRS,* $\mu \in M_{\mathcal{F}}$ *and let* $\mathcal{G} \subseteq \mathcal{F} - \mathcal{H}$*. Then,* $I_{\mathcal{G},\mu}$ *is well-defined.*

Now, we define an appropriate notion of direct $\mu$-dependency. This is not straightforward as shown in the next example.

*Example 4.* Consider the following conservative non-$\mu$-terminating CS-TRS $\mathcal{R} = \{\mathtt{a}(x,y) \to \mathtt{b}(x,x), \mathtt{d}(x,\mathtt{e}) \to \mathtt{a}(x,x), \mathtt{b}(x,\mathtt{c}) \to \mathtt{d}(x,x), \mathtt{c} \to \mathtt{e}\}$ with $\mu(\mathtt{a}) = \mu(\mathtt{d}) = \{1,2\}$, $\mu(\mathtt{b}) = \{1\}$ and $\mu(\mathtt{c}) = \mu(\mathtt{e}) = \varnothing$. The only cycle consists of the dependency pairs $C = \{\mathtt{A}(x,y) \to \mathtt{B}(x,x), \mathtt{D}(x,\mathtt{e}) \to \mathtt{A}(x,x), \mathtt{B}(x,\mathtt{c}) \to \mathtt{D}(x,x)\}$.

According to Definition 4, we have no basic usable rules because the right-hand sides of the dependency pairs have no defined symbols. Since we do not consider the rule $\mathtt{c} \to \mathtt{e}$ as usable, we would assume $\mathcal{G} = \{\mathtt{a},\mathtt{b},\mathtt{c},\mathtt{d},\mathtt{e}\}$. Then, we *cannot* simulate the infinite minimal $(\mathcal{R}, \mathcal{P}, \mu^\sharp)$-chain $\underline{\mathtt{A}(\mathtt{c},\mathtt{c})} \hookrightarrow \underline{\mathtt{B}(\mathtt{c},\mathtt{c})} \hookrightarrow \mathtt{D}(\mathtt{c},\underline{\mathtt{c}}) \hookrightarrow \underline{\mathtt{D}(\mathtt{c},\mathtt{e})} \hookrightarrow \underline{\mathtt{A}(\mathtt{c},\mathtt{c})} \hookrightarrow \cdots$ because we have:

$$s = I_{\mathcal{G},\mu}(\mathtt{A}(\mathtt{c},\mathtt{c})) = \underline{\mathtt{A}(g(\mathtt{c},g(\mathtt{e},\bot)),g(\mathtt{c},g(\mathtt{e},\bot)))} \hookrightarrow \mathtt{B}(g(\mathtt{c},g(\mathtt{e},\bot)),g(\mathtt{c},g(\mathtt{e},\bot))) = t$$

The interpreted term $g(\mathtt{c},g(\mathtt{e},\bot))$ at the $\mu$-replacing position 1 of $s$ is 'moved' to a non-$\mu$-replacing position 2 of $t$. Hence, we cannot reduce $t$ on the second argument of $\mathtt{B}$ to obtain the term $\mathtt{B}(g(\mathtt{c},g(\mathtt{e},\bot)),\mathtt{c})$ required for applying the next CS-DP $(\mathtt{B}(x,\mathtt{c}) \to \mathtt{D}(x,x))$ which continues the previous $(\mathcal{R}, \mathcal{P}, \mu)$-chain.

In order to avoid this problem, we modify Definition 3 to take into account symbols occurring at non-$\mu$-replacing positions in the *left-hand sides* of the rules.

**Definition 7 ($\mu$-Dependency).** *Given a CS-TRS* $\mathcal{R} = (\mathcal{F}, \mu, R)$*, we say that* $f \in \mathcal{F}$ directly $\mu$-depends *on* $g \in \mathcal{F}$*, written* $f \rhd_{d,\mu} g$*, if there is a rule* $l \to r \in R$ *with* $f = root(l)$ *and (1)* $g$ *occurs in* $r$ *at a* $\mu$*-replacing position or (2)* $g$ *occurs in* $l$ *at a non-*$\mu$*-replacing position.*

Remarkably, condition (2) in Definition 7 is not very problematic in practice because most programs are *constructor systems*, which means that no defined symbols occur below the root in the left-hand side of the rules.

Now we are ready to define our notion of usable rules. The set of *non-$\mu$-replacing* defined symbols in a term $t$ is $\mathcal{NDF}un^{\mu}(t) = \{f \mid \exists p \in \mathcal{P}os(t) \text{ and } p \notin \mathcal{P}os^{\mu}(t), f = root(t|_p) \in \mathcal{D}\}$.

**Definition 8 (Context-Sensitive Usable Rules).** *Let $\mathcal{R} = (\mathcal{F}, R)$ be a TRS, $\mu \in M_{\mathcal{R}}$, and $\mathcal{P} \subseteq \mathcal{P}^{\sharp}(\mathcal{F}, \mathcal{X})$. The set $\mathcal{U}(\mathcal{R}, \mu^{\sharp}, \mathcal{P})$ of* context-sensitive usable rules *for $\mathcal{P}$ is given by* $\mathcal{U}(\mathcal{R}, \mu^{\sharp}, \mathcal{P}) = \mathcal{U}_{\mathcal{H}}(\mathcal{R}, \mu) \cup \bigcup_{l \to r \in \mathcal{P}} \mathcal{U}_E(\mathcal{R}, \mu^{\sharp}, l \to r)$.

$$\begin{array}{lll} \textit{where } \mathcal{U}_E(\mathcal{R}, \mu, l \to r) & = \mathcal{R} \mid \{g \mid f \rhd^{*}_{d,\mu} g \textit{ for some } f \in \mathcal{DF}un^{\mu}(r) \cup \mathcal{NDF}un^{\mu}(l)\} \\ \mathcal{U}_{\mathcal{H}}(\mathcal{R}, \mu) & = \mathcal{R} \mid \{g \mid f \rhd^{*}_{d,\mu} g \textit{ for some } f \in \mathcal{H}\} \end{array}$$

Note that $\mathcal{U}_E$ *extends* the notion of usable rules in Definition 2, by taking into account not only dependencies with symbols on the right-hand sides of the rules, but *also with some symbols in proper subterms of the left-hand sides.* We call $\mathcal{U}_E(\mathcal{R}, \mu)$ the set of *extended* usable rules. On the other hand, $\mathcal{U}_{\mathcal{H}}$ is the set of usable rules corresponding to the hidden symbols. Now, we are ready to formulate and prove our main result in this section.

**Theorem 1.** *Let $\mathcal{R} = (\mathcal{F}, R)$ be a TRS, $\mathcal{P} \subseteq \mathcal{P}^{\sharp}(\mathcal{F}, \mathcal{X})$, and $\mu \in M_{\mathcal{F}}$. If there exists a $\mu$-reduction pair $(\gtrsim, \sqsupset)$ such that $\mathcal{U}(\mathcal{R}, \mu^{\sharp}, \mathcal{P}) \cup \pi \subseteq \gtrsim$, $\mathcal{P} \subseteq \gtrsim \cup \sqsupset$, and*

1. *If $\mathcal{P}_{\mathcal{X}} = \varnothing$, then $\mathcal{P} \cap \sqsupset \neq \varnothing$*
2. *If $\mathcal{P}_{\mathcal{X}} \neq \varnothing$, then $\trianglerighteq_{\mu} \subseteq \gtrsim$, and*
   *(a) $\mathcal{P} \cap \sqsupset \neq \varnothing$ and $f(x_1, \ldots, x_k) \gtrsim f^{\sharp}(x_1, \ldots, x_k)$ for all $f^{\sharp}$ in $\mathcal{P}$, or*
   *(b) $f(x_1, \ldots, x_k) \sqsupset f^{\sharp}(x_1, \ldots, x_k)$ for all $f^{\sharp}$ in $\mathcal{P}$.*

*Let $P_{\sqsupset} = \{u \to v \in \mathcal{P} \mid u \sqsupset v\}$. Then there are no infinite minimal $(\mathcal{R}, \mathcal{P}, \mu^{\sharp})$-chains whenever:*

1. *there are no infinite minimal $(\mathcal{R}, \mathcal{P} \setminus \mathcal{P}_{\sqsupset}, \mu^{\sharp})$-chains in in case (1) and in case (2a).*
2. *there are no infinite minimal $(\mathcal{R}, (\mathcal{P} \setminus \mathcal{P}_{\mathcal{X}}) \setminus \mathcal{P}_{\sqsupset}, \mu^{\sharp})$-chains in case (2b).*

*Proof (Sketch).* By contradiction. Assume that there exists an infinite minimal $(\mathcal{R}, \mathcal{P}, \mu^{\sharp})$-chain $\mathcal{A}$ but there is no infinite minimal $(\mathcal{R}, \mathcal{P} \setminus \mathcal{P}_{\sqsupset}, \mu^{\sharp})$-chains in case (1) and (2a), or there is no infinite minimal $(\mathcal{R}, (\mathcal{P} \setminus \mathcal{P}_{\mathcal{X}}) \setminus \mathcal{P}_{\sqsupset}, \mu^{\sharp})$-chains in case (2b). We can assume that there is a $\mathcal{P}' \subseteq \mathcal{P}$ such that $\mathcal{A}$ has a tail $\mathcal{B}$ where all pairs are used infinitely often:

$$t_1 \hookrightarrow^{*}_{\mathcal{R},\mu} u_1 \to_{\mathcal{P}'} \circ \trianglerighteq^{\sharp}_{\mu} t_2 \hookrightarrow^{*}_{\mathcal{R},\mu} u_2 \to_{\mathcal{P}'} \circ \trianglerighteq^{\sharp}_{\mu} \cdots$$

where $s \trianglerighteq^{\sharp}_{\mu} t$ for $s \in \mathcal{T}(\mathcal{F}, \mathcal{X})$ and $t \in \mathcal{T}^{\sharp}(\mathcal{F}, \mathcal{X})$ means that $s \trianglerighteq_{\mu} t^{\natural}$.

Let $\sigma$ be a substitution, we denote by $\sigma_{I_{\mathcal{G},\mu}}$ the substitution that assigns to each variable $x$ the term $I_{\mathcal{G},\mu}(\sigma(x))$ and let $\mathcal{G}$ be the set of defined symbols of $\mathcal{R} \backslash \mathcal{U}(\mathcal{R}, \mu^{\sharp}, \mathcal{P})$. We show that after applying $I_{\mathcal{G},\mu}$ we get an infinite $(\mathcal{U}(\mathcal{R}, \mu^{\sharp}, \mathcal{P}) \cup \pi, \mathcal{P}', \mu^{\sharp})$-chain. All terms in the infinite chain are $\mu$-terminating w.r.t. $(\mathcal{R}, \mu)$. We proceed by induction. Let $i \geq 1$.

– If we consider the step $u_i \to_{\mathcal{P}'} \circ \trianglerighteq^{\sharp}_{\mu} t_{i+1}$, we have two possibilities:

1. There is $l \to r \in \mathcal{P}'_{\mathcal{F}}$, then we get:

$$I_{\mathcal{G},\mu}(u_i) \hookrightarrow^*_{\pi} \sigma_{\mathcal{I}_{\mathcal{G},\mu}}(l) \to_{\mathcal{P}'_{\mathcal{F}}} \sigma_{\mathcal{I}_{\mathcal{G},\mu}}(r) = I_{\mathcal{G},\mu}(r) = I_{\mathcal{G},\mu}(t_{i+1})$$

2. There is an $l \to x \in \mathcal{P}'_{\mathcal{X}}$, then we get:

$$I_{\mathcal{G},\mu}(u_i) \hookrightarrow^*_{\pi} \sigma_{\mathcal{I}_{\mathcal{G},\mu}}(l) \to_{\mathcal{P}'_{\mathcal{X}}} \sigma_{\mathcal{I}_{\mathcal{G},\mu}}(x) = I_{\mathcal{G},\mu}(\sigma(x))$$

and $I_{\mathcal{G},\mu}(\sigma(x)) \trianglerighteq_{\mu} I_{\mathcal{G},\mu}(t^{\natural}_{i+1})$

– If we consider $t_i \hookrightarrow^*_{\mathcal{R},\mu} u_i$. We get $I_{\mathcal{G},\mu}(t_i) \hookrightarrow^*_{\mathcal{U}(\mathcal{R},\mu^{\sharp},\mathcal{P})\cup\pi} I_{\mathcal{G},\mu}(u_i)$.

Therefore we get the infinite $(\mathcal{U}(\mathcal{R},\mu^{\sharp},\mathcal{P}),\mathcal{P}',\mu^{\sharp})$-chain:

$$I_{\mathcal{G},\mu}(t_1) \hookrightarrow^*_{\mathcal{U}(\mathcal{R},\mu^{\sharp},\mathcal{P})\cup\pi} I_{\mathcal{G},\mu}(u_1) \to_{\mathcal{P}'} \circ \trianglerighteq^{\sharp}_{\mu} I_{\mathcal{G},\mu}(t_2) \hookrightarrow^*_{\mathcal{U}(\mathcal{R},\mu^{\sharp},\mathcal{P})\cup\pi} I_{\mathcal{G},\mu}(u_2) \to_{\mathcal{P}'} \cdots$$

Using the premises of the theorem, by monotonicity and stability of $\gtrsim$, we would have that $I_{\mathcal{G},\mu}(t_i) \gtrsim I_{\mathcal{G},\mu}(u_i)$ for all $i \geq 1$. By stability of $\sqsupset$ (and of $\gtrsim$), we have that $I_{\mathcal{G},\mu}(u_i)(\gtrsim \cup \sqsupset)I_{\mathcal{G},\mu}(t_{i+1})$ for all $i \geq 1$ and $I_{\mathcal{G},\mu}(u_i) \sqsupset I_{\mathcal{G},\mu}(t_{i+1})$ for all $j \in J$ for an infinite set $J = \{j_1, \ldots, j_n, \ldots\}$ of natural numbers $j_1 < j_2 < \ldots < j_n < \ldots$. Now, since $\gtrsim \circ \sqsupset \subseteq \sqsupset$ or $\sqsupset \circ \gtrsim \subseteq \sqsupset$, we would obtain an infinite sequence consisting of infinitely many $\sqsupset$-steps. We obtain a contradiction to the well-foundedness of $\sqsupset$. □

*Remark 1.* Notice that (as expected) $\mathcal{U}(\mathcal{R},\mathcal{P},\mu_{\top}) = \mathcal{U}(\mathcal{R},\mathcal{P})$, i.e., our usable rules for CS-TRSs $(\mathcal{R},\mu)$ coincide with the standard definition (see Definition 2) when $\mu = \mu_{\top}$ is considered (here, $\mu_{\top}(f) = \{1,\ldots,ar(f)\}$ for all symbols $f \in \mathcal{F}$, i.e., no replacement restriction is associated to any symbol).

Thanks to Theorem 1, we do not need to make all rules in $\mathcal{R}$ compatible with the weak component $\gtrsim_{\mathcal{P}}$ of a reduction pair $(\gtrsim_{\mathcal{P}}, \sqsupset_{\mathcal{P}})$ associated to a given set of pairs $\mathcal{P}$. We just need to consider $\mathcal{U}(\mathcal{R},\mu^{\sharp},\mathcal{P})$ (together with the $\pi$-rules).

*Example 5.* (Continuing Examples 1 and 2) Since $\mathcal{H} \cap \mathcal{D} = \{\texttt{from}, \texttt{zWquot}\}$, we have that $\mathcal{U}(\mathcal{R},\mu^{\sharp},C_1)$ is:

$$\begin{array}{rr}
\texttt{minus}(x,\texttt{0}) \to x & \texttt{minus}(\texttt{s}(x),\texttt{s}(y)) \to \texttt{minus}(x,y) \\
\texttt{quot}(\texttt{0},\texttt{s}(y)) \to \texttt{0} & \texttt{quot}(\texttt{s}(x),\texttt{s}(y)) \to \texttt{s}(\texttt{quot}(\texttt{minus}(x,y),\texttt{s}(y))) \\
\texttt{zWquot}(\texttt{nil},x) \to \texttt{nil} & \texttt{from}(x) \to \texttt{cons}(x,\texttt{from}(\texttt{s}(x))) \\
\texttt{zWquot}(x,\texttt{nil}) \to \texttt{nil} &
\end{array}$$

$$\texttt{zWquot}(\texttt{cons}(x,xs),\texttt{cons}(y,ys)) \to \texttt{cons}(\texttt{quot}(x,y),\texttt{zWquot}(xs,ys))$$

According to Theorem 1, the following polynomial interpretation (computed by MU-TERM [1,21]) shows the absence of infinite $(\mathcal{R},C_1,\mu^{\sharp})$-chains.

$$\begin{array}{rrr}
[\texttt{s}](x) = x+1 & [\texttt{quot}](x,y) = x+y & [\texttt{minus}](x,y) = 0 \\
[\texttt{from}](x) = 0 & [\texttt{sel}](x,y) = 0 & [\texttt{zWquot}](x,y) = x+y \\
[\texttt{cons}](x,y) = 0 & [\texttt{0}](x,y) = 0 & [\texttt{nil}](x,y) = 1 \\
[\texttt{SEL}](x,y) = x & &
\end{array}$$

Note that, if the rules for `sel` were present, we could not find a linear polynomial interpretation for solving this cycle.

*Remark 2.* When considering Definition 8 (usable rules for *CSR*) and Definition 2 (standard usable rules), one can observe that, despite the fact that *CSR* is a *restriction* of rewriting, we can obtain *more* usable rules in the context-sensitive case. Examples 3 and 4 show that this is because rules associated to hidden symbols that do *not* occur in the right-hand sides of the dependency pairs in the considered cycle can play an essential role in capturing infinite $\mu$-rewrite sequences. Thus, for terminating TRSs $\mathcal{R}$, it could be sometimes easier to find a proof of $\mu$-termination of the CS-TRS $(\mathcal{R}, \mu)$ if we ignore the replacement map $\mu$.

# 5 Improving Usable Rules

According to the discussion in Section 3, the notion of basic usable rules is not correct even for conservative systems. Still, since $\mathcal{U}_B(\mathcal{R}, \mu, \mathcal{P})$ is contained in (and is usually smaller than) $\mathcal{U}(\mathcal{R}, \mu, \mathcal{P})$, it is interesting to identify a class of CS-TRSs where basic usable rules can be safely used. Then, we consider a more restrictive kind of conservative CS-TRSs: the *strongly conservative* CS-TRSs.

**Definition 9.** *Let $\mathcal{F}$ be a signature, $\mu \in M_{\mathcal{F}}$ and $t \in \mathcal{T}(\mathcal{F}, \mathcal{X})$. We denote $\mathcal{V}ar^{\not\mu}(t)$ the set of variables in $t$ occurring at non-$\mu$-replacing positions, i.e., $\mathcal{V}ar^{\not\mu}(t) = \{x \in \mathcal{V}ar(t) \mid t \rhd_{\not\mu} x\}$.*

**Definition 10 (Strongly Conservative).** *Let $\mathcal{R}$ be a TRS and $\mu \in M_{\mathcal{R}}$. A rule $l \to r$ is strongly conservative if it is conservative and $\mathcal{V}ar^{\mu}(l) \cap \mathcal{V}ar^{\not\mu}(l) = \mathcal{V}ar^{\mu}(r) \cap \mathcal{V}ar^{\not\mu}(r) = \varnothing$; and $\mathcal{R}$ is strongly conservative if all rules in $\mathcal{R}$ are strongly conservative.*

Linear CS-TRSs trivially satisfy $\mathcal{V}ar^{\mu}(l) \cap \mathcal{V}ar^{\not\mu}(l) = \mathcal{V}ar^{\mu}(r) \cap \mathcal{V}ar^{\not\mu}(r) = \varnothing$. Hence, linear conservative CS-TRSs are strongly conservative. Note that the CS-TRSs in Examples 1 and 3 are not strongly conservative.

Theorem 2 below is the other main result of this paper. It shows that basic usable rules in Definition 4 can be used to improve proofs of termination of *CSR* for strongly conservative CS-TRSs. As discussed in Section 4, if we consider minimal $(\mathcal{R}, \mathcal{P}, \mu^{\sharp})$-chains, then we deal with $\mu$-terminating terms w.r.t. $(\mathcal{R}, \mu)$. We know that any $\mu$-replacing subterm is $\mu$-terminating, but we do not know anything about non-$\mu$-replacing subterms. However, dealing with strongly conservative CS-TRSs, we ensure that non-$\mu$-replacing subterms cannot become $\mu$-replacing after $\mu$-rewriting(s) above them. Hence, we develop a new basic $\mu$-interpretation $I'_{\mathcal{G},\mu}$ where non-$\mu$-replacing positions are not interpreted. In contrast to $I'_{\mathcal{G},\mu}$ (but closer to $I_{\mathcal{G}}$) our new basic $\mu$-interpretation is defined now for $\mu$-terminating terms only.

**Definition 11 (Basic $\mu$-Interpretation).** *Let $(\mathcal{F}, \mu, R)$ be a CS-TRS and $\mathcal{G} \subseteq \mathcal{F}$. Let $>$ be an arbitrary total ordering over $\mathcal{T}(\mathcal{F}^{\sharp} \cup \{\bot, \mathsf{g}\}, \mathcal{X})$ where $\bot$ is a new constant symbol and $\mathsf{g}$ is a new binary symbol. The basic $\mu$-interpretation $I'_{\mathcal{G},\mu}$ is* *a mapping from $\mu$-terminating terms in $\mathcal{T}(\mathcal{F}^\sharp,\mathcal{X})$ to terms in $\mathcal{T}(\mathcal{F}^\sharp\cup\{\bot,\mathsf{g}\},\mathcal{X})$ defined as follows:*

$$I'_{\mathcal{G},\mu}(t)=\begin{cases} t & \text{if } t\in\mathcal{X}\\ f(I'_{\mathcal{G},\mu,f,1}(t_1),\ldots,I'_{\mathcal{G},\mu,f,n}(t_n)) & \text{if } t=f(t_1\ldots t_n) \text{ and } f\notin\mathcal{G}\\ \mathsf{g}(f(I'_{\mathcal{G},\mu,f,1}(t_1),\ldots,I'_{\mathcal{G},\mu,f,n}(t_n)),t') & \text{if } t=f(t_1\ldots t_n) \text{ and } f\in\mathcal{G}\end{cases}$$

$$\text{where}\quad I'_{\mathcal{G},\mu,f,i}(t)=\begin{cases} I'_{\mathcal{G},\mu}(t) & \text{if } i\in\mu(f)\\ t & \text{if } i\notin\mu(f)\end{cases}$$

$$t'=order\left(\{I'_{\mathcal{G},\mu}(u)\mid t\hookrightarrow_{\mathcal{R},\mu}u\}\right)$$

$$order(T)=\begin{cases}\bot, & \text{if } T=\varnothing\\ \mathsf{g}(t,order(T-\{t\})) & \text{if } t \text{ is minimal in } T \text{ w.r.t. } >\end{cases}$$

It is easy to prove that the basic $\mu$-interpretation is well-defined (finite) for all $\mu$-terminating terms.

**Lemma 2.** *For each $\mu$-terminating term $t$, the term $I'_{\mathcal{G},\mu}(t)$ is finite.*

For the proof of our next theorem, we need some auxiliary definitions and results.

**Definition 12.** *Let $(\mathcal{R},\mu)$ be a CS-TRS and $\sigma$ be a substitution and let $\mathcal{G}\subseteq\mathcal{F}$. We denote by $\sigma_{I'_{\mathcal{G},\mu}}:\mathcal{T}(\mathcal{F},\mathcal{X})\to\mathcal{T}(\mathcal{F},\mathcal{X})$ a function that, given a term $t$ replaces occurrences of $x\in\mathcal{V}ar(t)$ at position $p$ in $t$ by either $I'_{\mathcal{G},\mu}(\sigma(x))$ if $p\in\mathcal{P}os^\mu(t)$, or $\sigma(x)$ if $p\notin\mathcal{P}os^\mu(t)$.*

**Proposition 1.** *Let $(\mathcal{R},\mu)$ be a CS-TRS and $\sigma$ be a substitution and let $\mathcal{G}\subseteq\mathcal{F}$. Let $t$ be a term such that $\mathcal{V}ar^\mu(t)\cap\mathcal{V}ar^{\not\mu}(t)=\varnothing$. Let $\overline{\sigma}_{I'_{\mathcal{G},\mu},t}$ be a substitution given by*

$$\overline{\sigma}_{I'_{\mathcal{G},\mu},t}(x)=\begin{cases} I'_{\mathcal{G},\mu}(\sigma(x)) & \text{if } x\in\mathcal{V}ar^\mu(t)\\ \sigma(x) & \text{otherwise}\end{cases}$$

*Then, $\overline{\sigma}_{I'_{\mathcal{G},\mu},t}(t)=\sigma_{I'_{\mathcal{G},\mu}}(t)$.*

The following theorem shows that we can safely consider the basic usable rules (with the $\pi$-rules) for proving termination of strongly conservative CS-TRSs.

**Theorem 2.** *Let $\mathcal{R}=(\mathcal{F},R)$ be a TRS, $\mathcal{P}\subseteq\mathcal{P}^\sharp(\mathcal{F},\mathcal{X})$, and $\mu\in M_\mathcal{F}$. If $\mathcal{P}\cup\mathcal{U}_B(\mathcal{R},\mu^\sharp,\mathcal{P})$ is strongly conservative and there exists a $\mu$-reduction pair $(\gtrsim,\sqsupset)$ such that $\mathcal{U}_B(\mathcal{R},\mu^\sharp,\mathcal{P})\cup\pi\subseteq\gtrsim$, $\mathcal{P}\subseteq\gtrsim$, and $\mathcal{P}\cap\sqsupset\neq\varnothing$. Let $\mathcal{P}_\sqsupset=\{u\to v\in\mathcal{P}\mid u\sqsupset v\}$. Then there are no infinite minimal $(\mathcal{R},\mathcal{P},\mu^\sharp)$-chains whenever there are no infinite minimal $(\mathcal{R},\mathcal{P}\setminus\mathcal{P}_\sqsupset,\mu^\sharp)$-chains.*

*Proof (Sketch).* By contradiction. Assume that there exists an infinite minimal $(\mathcal{R},\mathcal{P},\mu^\sharp)$-chain $\mathcal{A}$ but there is no infinite minimal $(\mathcal{R},\mathcal{P}\setminus\mathcal{P}_\sqsupset,\mu^\sharp)$-chains. We can assume that there is a $\mathcal{P}'\subseteq\mathcal{P}$ such that $\mathcal{A}$ has a tail $\mathcal{B}$ where all pairs are used infinitely often:

$$t_1\hookrightarrow^*_{\mathcal{R},\mu}u_1\to_{\mathcal{P}'}t_2\hookrightarrow^*_{\mathcal{R},\mu}u_2\to_{\mathcal{P}'}\cdots$$

After applying the basic $\mu$-interpretation $I'_{\mathcal{G},\mu}$ we obtain an infinite $(\mathcal{U}_B(\mathcal{R},\mu^\sharp,\mathcal{P}) \cup \pi, \mathcal{P}', \mu^\sharp)$-chain. Since all terms in the infinite $(\mathcal{R},\mathcal{P}',\mu^\sharp)$-chain are $\mu$-terminating w.r.t. $(\mathcal{R},\mu)$, we can indeed apply the basic $\mu$-interpretation $I'_{\mathcal{G},\mu}$. Let $i \geq 1$.

- If we consider the pair step $u_i \to_{\mathcal{P}'} t_{i+1}$ we can obtain the following sequence:

$$I'_{\mathcal{G},\mu}(u_i) \hookrightarrow^*_\pi \sigma_{I'_{\mathcal{G},\mu}}(l) \hookrightarrow^*_\pi \overline{\sigma}_{I'_{\mathcal{G},\mu},r}(l) \to_{\mathcal{P}'} \overline{\sigma}_{I'_{\mathcal{G},\mu},r}(r) = \sigma_{I'_{\mathcal{G},\mu}}(r) = I'_{\mathcal{G},\mu}(t_{i+1})$$

- If we consider the rewrite sequence $t_i \hookrightarrow^*_{\mathcal{R},\mu} u_i$. All terms in it are $\mu$-terminating, then we get $I'_{\mathcal{G},\mu}(t_i) \hookrightarrow^*_{\mathcal{U}_B(\mathcal{R},\mu^\sharp,\mathcal{P})\cup\pi} I'_{\mathcal{G},\mu}(u_i)$.

So we obtain the infinite $\mu$-rewrite sequence:

$$I'_{\mathcal{G},\mu}(t_1) \hookrightarrow^*_{\mathcal{U}_B(\mathcal{R},\mu^\sharp,\mathcal{P})\cup\pi} I'_{\mathcal{G},\mu}(u_1) \hookrightarrow^*_\pi \circ \to_{\mathcal{P}'} I'_{\mathcal{G},\mu}(t_2) \hookrightarrow^*_{\mathcal{U}_B(\mathcal{R},\mu^\sharp,\mathcal{P})\cup\pi} \cdots$$

Using the premise of the theorem, it is transformed into an infinite sequence consisting of $\gtrsim$ and infinitely many $\sqsupset$ steps. Using the stability condition, this contradicts the well-foundedness of $\sqsupset$. □

*Example 6.* (Continuing Examples 1, 2 and 5) Cycle $C_1$ is not strongly conservative, but cycles $C_2$ and $C_3$ are strongly conservative. Thus, we can use their basic usable rules. Cycle $C_2$ has no usable rules and we can easily find a polynomial interpretation to show the absence of infinite minimal $(\mathcal{R}, C_2, \mu^\sharp)$-chains:

$$[\mathtt{s}](x) = x + 1 \qquad [\mathtt{MINUS}](x, y) = y$$

The basic usable rules $\mathcal{U}_B(\mathcal{R}, \mu^\sharp, C_3)$ for $C_3$ are strongly conservative (see Example 2). The following polynomial interpretation proves the absence of infinite $(\mathcal{R}, C_3, \mu^\sharp)$-chains:

$$[\mathtt{0}] = 0 \qquad [\mathtt{s}](x) = x + 1 \qquad [\mathtt{minus}](x, y) = x \qquad [\mathtt{QUOT}](x, y) = x$$

Since we dealt with cycle $C_1$ in Example 5, $\mu$-termination of $\mathcal{R}$ is proved. Until now, no tool for proving termination of *CSR* could find a proof for this $\mathcal{R}$ in Example 1. Thanks to the results in this paper, which have been implemented in MU-TERM, we can easily prove $\mu$-termination of $\mathcal{R}$ now.

## 6 Experiments

The techniques described in the previous sections have been implemented as part of the tool MU-TERM [1,21]. In order to make clear the real contribution of the new technique to the performance of the tool, we have implemented three different versions of MU-TERM: (1) a basic version without any kind of usable rules, (2) a second version implementing the results about usable rules described in [4], and (3) a final version that implements the usable rules described in this paper (we do not use the notion in [4] even if the TRS is conservative and innermost equivalent). Version (2) of MU-TERM proves termination of *CSR* as termination

**Table 1.** Comparative among the three MU-TERM versions

| Tool Version | Proved | Total Time | Average Time |
|---|---|---|---|
| **No Usable Rules** | 44/90 | 6.11s | 0.14s |
| **Innermost Usable Rules** | 52/90 | 11.75s | 0.23s |
| **Usable Rules** | 64/90 | 8.91s | 0.14s |

**Table 2.** Comparative over the 44 examples

| Tool Version | Proved | Total Time | Average Time |
|---|---|---|---|
| **No Usable Rules** | 44/90 | 6.11s | 0.14s |
| **Innermost Usable Rules** | 44/90 | 5.03s | 0.11s |
| **Usable Rules** | 44/90 | 3.57s | 0.08s |

of innermost *CSR* when the TRS is orthogonal (see [4,11]), 37 systems, and as termination of *CSR without usable rules* in the rest of cases. In order to keep the set of experiments simple (but still meaningful), we only use linear interpretations with coefficients in $\{0, 1\}$. The usual practice shows that this is already quite powerful (see [9] for recent benchmarks in this sense). The benchmarks have been executed in a completely automatic way with a timeout of 1 minute on each of the 90 examples in the Context-Sensitive Rewriting subcategory of the 2007 Termination Competition[3]. A complete report of our experiments can be found in:

`http://www.dsic.upv.es/~rgutierrez/muterm/rta08/benchmarks.html`

Table 1 summarizes our results. Our notion of usable rules works pretty well: we are able to prove 20 more examples than without any usable rules, and 12 more than with the restricted notion in [4]. Furthermore, a comparison over the 44 examples solved by all the three versions of MU-TERM, we see that version (3) of MU-TERM is 43% faster than (1) and 27% faster than (2) (see Table 2).

## 7 Conclusions

We have investigated how *usable rules* can be used to improve termination proofs of CSR when the (context-sensitive) dependency pairs approach is used to achieve the proof. In contrast to [4], the straightforward extension of the standard notion of usable rules (called here *basic* usable rules, see Definition 4) does not work for *CSR* even for the quite restrictive class of conservative (cycles of) CS-TRSs. We have shown how to adapt the notion of usable rules for their use with arbitrary CS-TRSs (Definition 8). Theorem 1 shows that the new notion of usable rules can be used in proofs of termination of CS-TRSs. Here, although the proof uses a transformation in the very same style than [14,17], the definition of the transformation is quite different from the usual one in that it applies to

[3] See `http://www.lri.fr/~marche/termination-competition/2007`

arbitrary terms, not only terminating ones. To our knowledge, this is the first time that Gramlich's transformation [15] is adapted and used in that way. We have also introduced the notion of strongly conservative rule and CS-TRS (Definition 10). Theorem 2 shows that basic usable rules can be used in proofs of termination involving strongly conservative cycles and rules. Although we follow the proof scheme in [14,17], a number of subtleties have to be carefully addressed before getting a correct adaptation of the proof.

We have implemented our techniques as part of the tool MU-TERM [1,21]. Our experiments show that usable rules are helpful to improve proofs of termination of *CSR*. Regarding the previous work on usable rules for innermost *CSR* [4], this paper provides a fully general definition which is not restricted to conservative systems. Actually, as we show in our experiments, our framework is more powerful in practice than trying to prove termination of *CSR* as innermost termination of *CSR* with the restricted notion of usable rules in [4]. Actually, our results provide a basis for refining the notion of usable rules in the *innermost* setting, thus hopefully allowing a generalization of the results in [4].

Finally, usable rules were an essential ingredient for MU-TERM in winning the context-sensitive subcategory of the 2007 competition of termination tools.

## References

1. Alarcón, B., Gutiérrez, R., Iborra, J., Lucas, S.: Proving Termination of Context-Sensitive Rewriting with MU-TERM. ENTCS 188, 105–115 (2007)
2. Alarcón, B., Gutiérrez, R., Lucas, S.: Context-Sensitive Dependency Pairs. In: Arun-Kumar, S., Garg, N. (eds.) FSTTCS 2006. LNCS, vol. 4337, pp. 297–308. Springer, Heidelberg (2006)
3. Alarcón, B., Gutiérrez, R., Lucas, S.: Improving the Context-Sensitive Dependency Graph. ENTCS 188, 91–103 (2007)
4. Alarcón, B., Lucas, S.: Termination of Innermost Context-Sensitive Rewriting Using Dependency Pairs. In: Konev, B., Wolter, F. (eds.) FroCos 2007. LNCS (LNAI), vol. 4720, pp. 73–87. Springer, Heidelberg (2007)
5. Arts, T., Giesl, J.: Proving Innermost Normalisation Automatically. In: Comon, H. (ed.) RTA 1997. LNCS, vol. 1232, pp. 157–171. Springer, Heidelberg (1997)
6. Arts, T., Giesl, J.: Termination of Term Rewriting Using Dependency Pairs. Theoretical Computer Science 236(1–2), 133–178 (2000)
7. Borralleras, C.: Ordering-Based Methods for Proving Termination Automatically. PhD thesis, Departament de Llenguatges i Sistemes Informàtics, UPC (2003)
8. Durán, F., Lucas, S., Meseguer, J., Marché, C., Urbain, X.: Proving Operational Termination of Membership Equational Programs. Higher-Order and Symbolic Computation (to appear, 2008)
9. Fuhs, C., Giesl, J., Middeldorp, A., Schneider-Kamp, P., Thiemann, R., Zankl, H.: SAT Solving for Termination Analysis with Polynomial Interpretations. In: Marques-Silva, J., Sakallah, K.A. (eds.) SAT 2007. LNCS, vol. 4501, pp. 340–354. Springer, Heidelberg (2007)
10. Giesl, J., Arts, T., Ohlebusch, E.: Modular Termination Proofs for Rewriting Using Dependency Pairs. Journal of Symbolic Computation 34(1), 21–58 (2002)
11. Giesl, J., Middeldorp, A.: Innermost Termination of Context-Sensitive Rewriting. In: Ito, M., Toyama, M. (eds.) DLT 2002. LNCS, vol. 2450, pp. 231–244. Springer, Heidelberg (2003)

12. Giesl, J., Middeldorp, A.: Transformation Techniques for Context-Sensitive Rewrite Systems. Journal of Functional Programming 14(4), 379–427 (2004)
13. Giesl, J., Thiemann, R., Schneider-Kamp, P.: The Dependency Pair Framework: Combining Techniques for Automated Termination Proofs. In: Baader, F., Voronkov, A. (eds.) LPAR 2004. LNCS (LNAI), vol. 3452, pp. 301–331. Springer, Heidelberg (2005)
14. Giesl, J., Thiemann, R., Schneider-Kamp, P., Falke, S.: Mechanizing and Improving Dependency Pairs. Journal of Automatic Reasoning 37(3), 155–203 (2006)
15. Gramlich, B.: Generalized Sufficient Conditions for Modular Termination of Rewriting. Applicable Algebra in Engineering, Communication and Computing 5, 131–151 (1994)
16. Gutiérrez, R., Lucas, S., Urbain, X.: Usable Rules for Context-Sensitive Rewrite System, DSIC-II/03/08. Technical report, UPV (2008)
17. Hirokawa, N., Middeldorp, A.: Tyrolean Termination Tool: Techniques and Features. Information and Computation 205(4), 474–511 (2007)
18. Lucas, S.: Context-Sensitive Computations in Functional and Functional Logic Programs. Journal of Functional and Logic Programming 1998(1), 1–61 (1998)
19. Lucas, S.: Termination of on-demand rewriting and termination of OBJ programs. In: Proc. of PPDP 2001, pp. 82–93. ACM Press, New York (2001)
20. Lucas, S.: Context-Sensitive Rewriting Strategies. Information and Computation 178(1), 293–343 (2002)
21. Lucas, S.: MU-TERM: A Tool for Proving Termination of Context-Sensitive Rewriting. In: van Oostrom, V. (ed.) RTA 2004. LNCS, vol. 3091, pp. 200–209. Springer, Heidelberg (2004), `http://zenon.dsic.upv.es/muterm/`
22. Lucas, S.: Proving Termination of Context-Sensitive Rewriting by Transformation. Information and Computation 204(12), 1782–1846 (2006)
23. Ohlebusch, E.: Advanced Topics in Term Rewriting. Springer, Heidelberg (2002)
24. Urbain, X.: Modular & Incremental Automated Termination Proofs. Journal of Automated Reasoning 32(4), 315–355 (2004)

# Combining Equational Tree Automata over AC and ACI Theories*

Joe Hendrix[1] and Hitoshi Ohsaki[2]

[1] University of Illinois at Urbana-Champaign
jhendrix@uiuc.edu
[2] National Institute of Advanced Industrial Science and Technology
ohsaki@ni.aist.go.jp

**Abstract.** In this paper, we study combining equational tree automata in two different senses: (1) whether decidability results about equational tree automata over disjoint theories $\mathcal{E}_1$ and $\mathcal{E}_2$ imply similar decidability results in the *combined theory* $\mathcal{E}_1 \cup \mathcal{E}_2$; (2) checking emptiness of a language obtained from the *Boolean combination* of regular equational tree languages. We present a negative result for the first problem. Specifically, we show that the intersection-emptiness problem for tree automata over a theory containing at least one AC symbol, one ACI symbol, and 4 constants is undecidable despite being decidable if either the AC or ACI symbol is removed. Our result shows that decidability of intersection-emptiness is a *non-modular* property even for the union of disjoint theories. Our second contribution is to show a decidability result which implies the decidability of two open problems: (1) If idempotence is treated as a rule $f(x,x) \to x$ rather than an equation $f(x,x) = x$, is it decidable whether an AC tree automata accepts an idempotent normal form? (2) If $\mathcal{E}$ contains a single ACI symbol and arbitrary free symbols, is emptiness decidable for a Boolean combination of regular $\mathcal{E}$-tree languages?

## 1 Introduction

Tree automata are a theoretical tool with applications in many areas, including sufficient completeness of algebraic specifications [2, 7], protocol verification [4, 5], type inference [3], and theorem proving [13]. Many different frameworks have been proposed for addressing these applications as each framework must balance the often competing goals of expressive power and tractability of different operations. In our own applications [6, 7], the most important properties are a decidable emptiness problem, and closure under Boolean operations and equational congruences. Regular tree automata satisfy two of these properties, however they are not closed under arbitrary equational congruences. For example, the set of terms equivalent modulo associativity to a term in a regular tree language may not be a regular tree language [16].

Many extensions to tree automata have been proposed to remedy this problem, including multitree automata [14], equational tree automata [16], and two-way alternating equational tree automata [25]. These extensions allow one to

* Research supported by ONR Grant N00014-02-1-0715.

A. Voronkov (Ed.): RTA 2008, LNCS 5117, pp. 142–156, 2008.

recognize terms equivalent modulo an equational theory, however multitree automata are only defined for AC theories and the other frameworks lack closure under Boolean operations. Due to this problem, *propositional tree automata* were proposed in [9]. They are closed under both an equational theory and Boolean operations — but have an undecidable emptiness problem.

A separate issue in equational tree automata is that few properties are decidable for arbitrary theories. Consequently, most work on equational tree automata focuses on particular equational theories where one or more symbols satisfies combinations of specific equations such as associativity (A), commutativity (C), and idempotence (I). This restriction is unavoidable due to decidability issues, but leaves open the question as to whether these results can be *combined.* For example, tree automata over a theory $\mathcal{E}_{\mathsf{AC}}$ with an AC symbol and free symbols are effectively closed under intersection [18], and tree automata over a theory $\mathcal{E}_{\mathsf{ACI}}$ with an ACI symbol and free symbols are also effectively closed under intersection [24]. Does this imply that tree automata over the combined theory $\mathcal{E}_{\mathsf{AC}} \cup \mathcal{E}_{\mathsf{ACI}}$ are effectively closed under intersection as well?

Our first contribution is to show that tree automata over $\mathcal{E}_{\mathsf{AC}} \cup \mathcal{E}_{\mathsf{ACI}}$ are *not* effectively closed under intersection. Moreover, the intersection-emptiness problem, which is decidable for tree automata over $\mathcal{E}_{\mathsf{AC}}$ and $\mathcal{E}_{\mathsf{ACI}}$ separately, is undecidable for tree automata over the combined theory $\mathcal{E}_{\mathsf{AC}} \cup \mathcal{E}_{\mathsf{ACI}}$. We obtain this result by showing that every *alternating tree language* [25] over a theory $\mathcal{E}$ can be effectively expressed as the intersection of two regular tree languages over a theory $\mathcal{E}'$ containing $\mathcal{E}$ and an additional ACI symbol. Since the emptiness problem for alternating AC-tree automata is undecidable [25], it follows that so is intersection-emptiness for regular tree automata over $\mathcal{E}_{\mathsf{AC}} \cup \mathcal{E}_{\mathsf{ACI}}$. Since emptiness is always decidable for regular equational tree automata, it follows that regular tree automata over $\mathcal{E}_{\mathsf{AC}} \cup \mathcal{E}_{\mathsf{ACI}}$ are not effectively closed under intersection.

Our result implies that both the decidability of intersection-emptiness and effective closure under intersection are *non-modular* properties, even for disjoint theories. Modularity is an important property to have, because it aids in the process of decomposing complex problems into simpler parts which can be reasoned about separately. For example, the Shostak [21] and Nelson-Oppen [15] combination methods have been fundamental to the development of general-purpose theorem provers that combine the capabilities of many different decision procedures. Given the importance of modularity, we decided to further analyze how the interaction between the AC symbol and ACI symbol led to undecidability.

Our second contribution is to define a restricted class of tree automata over a theory $\mathcal{E}$ with AC and ACI symbols which are closed under equational congruences. We further show that the emptiness problem is decidable for the Boolean closure of tree languages in that class — a problem which we call the *propositional emptiness problem* as it closely relates to the emptiness problem for propositional tree automata. The tree automata in the restricted class we consider are called *AC-intersection free* and subjects each ACI symbol $+$ in $\mathcal{E}$ to one of two constraints: (1) either the clauses in the automaton where $+$ appears must satisfy certain syntactic restrictions to avoid simulating the intersection clauses of alternating tree automata; or (2) the idempotence equation $x + x = x$ in $\mathcal{E}$ must be treated as a rewrite rule $x+x \to x$ as in the *tree automata with normalization* framework of [17]. In that framework, some of the equations in $\mathcal{E}$ may be treated as rewrite rules in a confluent and terminating rewrite theory $\mathcal{R}$. Rather than computing the congruence closure of the tree language modulo $\mathcal{E}$, terms are first normalized by rewriting with $\mathcal{R}$ modulo the remaining equations $\mathcal{E}' \subseteq \mathcal{E}$, and then checked for membership in the underlying equational tree languages $\mathcal{L}(\mathcal{A}/\mathcal{E}')$. Their framework has different semantics than standard equational tree automata, but is often able to obtain better closure and decidability properties.

An important consequence of our second contribution is that it solves two open problems: (1) We show that the emptiness problem is decidable for tree automata with normalization over idempotence rules and AC equations. This problem was mentioned in [17] and left unsolved. (2) We show that the propositional emptiness problem is decidable for equation tree automata over the theory $\mathcal{E}_{\mathsf{ACI}}$ containing a single ACI symbol and arbitrary free symbols. This problem is interesting, because equational tree automata over $\mathcal{E}_{\mathsf{ACI}}$ are not closed under complementation [23]. Its decidability also has a further implication: propositional emptiness is a non-modular property. Our earlier undecidability result implies that propositional emptiness is undecidable for equational tree automata over $\mathcal{E}_{\mathsf{AC}} \cup \mathcal{E}_{\mathsf{ACI}}$, while propositional emptiness is decidable for $\mathcal{E}_{\mathsf{AC}}$ [18].

One underlying goal in this work is to develop better tree automata techniques for *non-linear theories*. This is important in applications such as sufficient completeness checking where existing techniques either do not support rewriting modulo axioms [2] or are restricted to left-linear rewrite rules [7]. Although sufficient completeness checking is undecidable in general for specifications with non-linear rules and rewriting modulo AC [12], our decidability results show that sufficient completeness is decidable modulo AC when every non-linear rule in the specification has the form $f(x,x) \to r$. It would be interesting to see if the techniques presented here can be extended to other forms of non-linear rules.

This paper is organized as follows. In Section 2, we review basic concepts from rewriting and tree automata. In Section 3, we show how alternating tree languages can be expressed as the intersection of two regular tree languages. In Section 4, we define a subclass of equational tree automata, which we call *AC-intersection free*, and state a decidability result which solves the two open problems discussed previously, and in Section 5, we present our algorithm for showing the previous decidability result. Finally, we discuss related work and suggest avenues for future research in Section 6. Some proofs have been omitted due to space limitations and are available in [8].

## 2 Preliminary Definitions

We assume the reader is familiar with term rewriting as well as tree automata [1].

**Equational and Rewrite Theories.** An *equational theory* $\mathcal{E} = (F, E)$ consists of a signature $F$ together with a set of equations $l = r$ with $l, r \in T_F(X)$. For each term $t \in T_F(X)$, we let $\mathsf{root}(t) \in F$ denote the top-most symbol, and let $[t]_{\mathcal{E}}$

denote the equivalence class of terms equal to $t$ with respect to the equivalence relation $=_{\mathcal{E}}$ induced by $\mathcal{E}$. We just write $[t]$ for $[t]_{\mathcal{E}}$ when the theory can be inferred from the context, and we let $T_{\mathcal{E}}$ denote the $F$-algebra whose universe $T_{\mathcal{E}}$ consists of the equivalence classes of $T_F$ formed by $=_{\mathcal{E}}$.

A *rewrite theory* $\mathcal{R}$ is a set of *rewrite rules* of the form $l \to r$ with $l, r \in T_F(X)$. A term $t \in T_F(X)$ rewrites to $u \in T_F(X)$ modulo $\mathcal{E}$, denoted $t \to_{\mathcal{R}/\mathcal{E}} u$ if there is rule $l \to r \in R$, context $C$, and substitution $\theta$ such that $t =_{\mathcal{E}} C[l\theta]$ and $u =_{\mathcal{E}} C[r\theta]$. A term $t$ is *$\mathcal{R}/\mathcal{E}$-irreducible* if it cannot be further rewritten. We write $t{\downarrow_{\mathcal{R}/\mathcal{E}}}u$ if there is a term $v \in T_F(X)$ such that $t \to^*_{\mathcal{R}/\mathcal{E}} v$ and $u \to^*_{\mathcal{R}/\mathcal{E}} v$. A rewrite theory $\mathcal{R}$ is *terminating* modulo $\mathcal{E}$ if $\to^+_{\mathcal{R}/\mathcal{E}}$ is well-founded. $\mathcal{R}$ is *confluent* if $t \to^*_{\mathcal{R}/\mathcal{E}} u$ and $t \to^*_{\mathcal{R}/\mathcal{E}} v$ implies $u{\downarrow_{\mathcal{R}/\mathcal{E}}}v$. If $\mathcal{R}$ is terminating and confluent modulo $\mathcal{E}$, then for all $t \in T_F$, there is an $\mathcal{R}/\mathcal{E}$-irreducible term $t{\downarrow_{\mathcal{R}/\mathcal{E}}} \in T_F$ that is unique up to $=_{\mathcal{E}}$. We let $\mathsf{Can}_{\mathcal{R}/\mathcal{E}} \subseteq T_{\mathcal{E}}$ denote the canonical term algebra whose universe is the set of $\mathcal{E}$-equivalence classes of $\mathcal{R}/\mathcal{E}$-irreducible terms.

In this paper, we restrict our attention to equational theories $\mathcal{E}$ only containing axioms with the following forms:

| | | |
|---|---|---|
| $(x+y)+z = x+(y+z)$ | $x+y = y+x$ | $x+x = x$ |
| associativity | commutativity | idempotence |

Relative to an equational theory $\mathcal{E}$, if a symbol $f \in F$ does not appear in any of the equations, we say it is a *free symbol.* If $f \in F$ appears in associativity and commutativity equations but no other equations, we say that it is an *AC symbol.* Finally, if $f \in F$ appears in associativity, commutativity, and idempotence equations, we say that it is an *ACI symbol.* We shall restrict our attention to equational theories where each symbol is a free, AC, or ACI symbol.

**Tree Automata.** We treat tree automata as collections of Horn clauses of particular forms as in [25]. A *regular $\mathcal{E}$-tree automaton* $\mathcal{A}$ is a finite set of Horn clauses each with the form:

$$p(f(x_1, \ldots, x_n)) \Leftarrow p_1(x_1), \ldots, p_n(x_n) \qquad \text{regular clause}$$

where $f \in F$ has arity $n$ and $p, p_1, \ldots, p_n$ are elements of a finite set of unary predicate symbols called the *states* of the automaton. In some definitions, tree automata may also contain *$\epsilon$-clauses* of the form $p(x) \Leftarrow q(x)$, but these can be eliminated without loss of expressive power. We write $\mathcal{A}/\mathcal{E} \vdash p(t)$ if $p(t)$ is entailed by the axioms in $\mathcal{A} \cup \mathcal{E}$. For an equational theory $\mathcal{E} = (F, \varnothing)$ with no equations, we write $\mathcal{A} \vdash p(t)$ for $\mathcal{A}/\mathcal{E} \vdash p(t)$.

We keep the acceptance condition separate from the automaton itself, and since the automaton only recognizes languages that are closed modulo $\mathcal{E}$, we define languages as subsets of $T_{\mathcal{E}}$ rather than $T_F$. For each state $p$ belonging to $\mathcal{A}$, the *language recognized by p in $\mathcal{A}$*, denoted $\mathcal{L}_p(\mathcal{A}/\mathcal{E}) \subseteq T_{\mathcal{E}}$, is defined by

$$\mathcal{L}_p(\mathcal{A}/\mathcal{E}) = \{\, [t] \in T_{\mathcal{E}} \mid \mathcal{A}/\mathcal{E} \vdash p(t) \,\}. \tag{1}$$

One fundamental result from [25] about regular $\mathcal{E}$-tree automata is:

**Theorem 1.** *For each theory $\mathcal{E}$ and regular $\mathcal{E}$-tree automaton $\mathcal{A}$,*

$$\mathcal{A}/\mathcal{E} \vdash p(t) \iff (\exists u \in [t]_{\mathcal{E}})\, \mathcal{A} \vdash p(u).$$

For an arbitrary theory $\mathcal{E}$, the class of languages recognized by regular $\mathcal{E}$-tree automata is closed under union, but not under intersection or complementation. Motivated by this fact, an equational tree automata framework called *propositional tree automata* is introduced in [9] that is effectively closed under Boolean operations in all theories. The key idea is to use a propositional formula rather than a set of final states as the acceptance condition for defining the language recognized by the automaton. In this paper, we present an alternative formalization that preserves the basic idea. Given a tree automaton $\mathcal{A}$ with states $Q$, we extend (1) from languages $\mathcal{L}_p(\mathcal{A}/\mathcal{E})$ recognized by a state $p$ to languages $\mathcal{L}_\phi(\mathcal{A}/\mathcal{E})$ recognized by a propositional formula $\phi$ constructed from atomic predicates $Q$ and Boolean connectives $\wedge$ and $\neg$:

$$\mathcal{L}_{\phi_1 \wedge \phi_2}(\mathcal{A}/\mathcal{E}) = \mathcal{L}_{\phi_1}(\mathcal{A}/\mathcal{E}) \cap \mathcal{L}_{\phi_2}(\mathcal{A}/\mathcal{E}) \qquad \mathcal{L}_{\neg \phi_1}(\mathcal{A}/\mathcal{E}) = T_{\mathcal{E}} \setminus \mathcal{L}_{\phi_1}(\mathcal{A}/\mathcal{E}).$$

There are many decision problems that have been studied in the context of tree automata. The *membership problem* for $\mathcal{E}$ is the problem of deciding for an equivalence class $[t] \in T_{\mathcal{E}}$, $\mathcal{E}$-tree automaton $\mathcal{A}$ and state $p$ in $\mathcal{A}$ whether $[t] \in \mathcal{L}_p(\mathcal{A}/\mathcal{E})$. The *emptiness problem* for $\mathcal{E}$ is the problem of deciding for an $\mathcal{E}$-tree automaton $\mathcal{A}$ and state $p$ whether $\mathcal{L}_p(\mathcal{A}/\mathcal{E}) = \varnothing$. This problem is decidable in linear time for an arbitrary theory $\mathcal{E}$ using Theorem 1 and standard tree automata techniques [1]. The *intersection-emptiness problem* for $\mathcal{E}$ is the problem of deciding for an $\mathcal{E}$-tree automaton $\mathcal{A}$ and states $p_1, \ldots, p_n$ of $\mathcal{A}$ whether $\mathcal{L}_{p_1}(\mathcal{A}/\mathcal{E}) \cap \cdots \cap \mathcal{L}_{p_n}(\mathcal{A}/\mathcal{E}) = \varnothing$. Finally, the *propositional emptiness problem* for $\mathcal{E}$ is the problem of deciding for an $\mathcal{E}$-tree automaton $\mathcal{A}$ with states $Q$ and propositional formula $\phi$ over atomic predicates $Q$ whether $\mathcal{L}_\phi(\mathcal{A}/\mathcal{E}) = \varnothing$.

It is known that both the intersection-emptiness and propositional emptiness problem is decidable for regular equational tree automata over a theory $\mathcal{E}_{\mathsf{AC}}$ with AC and free symbols [16]. In contrast, both intersection-emptiness and propositional emptiness are undecidable for regular equational tree automata over a theory $\mathcal{E}_{\mathsf{A}}$ with associative and free symbols [18]. As an example of a tree automata framework where intersection-emptiness is decidable and propositional emptiness is undecidable, we refer the reader to the *monotone AC tree automata* framework of [19].

## 3 Alternating Tree Automata

One extension to tree automata is the alternating tree automata framework of [22] which was extended to the equational case in [25]. In a Horn-clause representation, an *alternating tree automaton* is a tree automaton which in addition to regular clauses, may also contain *intersection clauses* of the form:

$$p(x) \Leftarrow p_1(x), p_2(x) \qquad \text{intersection clause.}$$

Alternating $\mathcal{E}$-tree automata are closed under both intersection and union, but are not always closed under complementation. If $\mathcal{E}$ is the free theory, i.e., $\mathcal{E} = (F, \varnothing)$, then the class of languages recognized by alternating and regular automata coincide. However, this is often not the case for other theories. For example, alternating AC-tree automata are strictly more powerful than regular AC-automata. In particular, the emptiness problem is undecidable for alternating AC-tree automata [25].

Our first new result in this paper is to show that every alternating $\mathcal{E}$-tree language is isomorphic to the intersection of two regular $\mathcal{E}'$-tree languages where $\mathcal{E}'$ is the theory obtained by adding a fresh ACI symbol $\circ$ to $\mathcal{E}$.

**Theorem 2.** *Let $\mathcal{E} = (F, E)$ and $\mathcal{E}' = (F', E')$ be equational theories such that $\mathcal{E}'$ contains the symbols and equations in $\mathcal{E}$ and adds a fresh ACI operator $\circ$.*

*Given an alternating $\mathcal{E}$-tree automaton $\mathcal{A}$ with states $Q$, one can effectively construct a regular $\mathcal{E}'$-tree automaton $\mathcal{B}$ containing the states $Q$ and an additional fresh state $k$ such that*

- *For all $p \in Q$ and $t \in T_F$, $\mathcal{A}/\mathcal{E} \vdash p(t) \iff \mathcal{B}/\mathcal{E}' \vdash p(t)$.*
- *For all $t \in T_{F'}$, $\mathcal{B}/\mathcal{E}' \vdash k(t) \iff T_F \cap [t]_{\mathcal{E}'} \neq \varnothing$.*

*Proof.* Let $\mathcal{B}$ be the automaton containing the following clauses:

- $\mathcal{B}$ contains all of the clauses in $\mathcal{A}$ that are not intersection clauses;
- for each intersection clause $p(x) \Leftarrow p_1(x), p_2(x)$ in $\mathcal{A}$, $\mathcal{B}$ contains the clause $p(x_1 \circ x_2) \Leftarrow p_1(x_1), p_2(x_2)$; and
- for each symbol $f \in F$ with arity $n$, $\mathcal{B}$ contains the clause $k(f(x_1, \ldots, x_n)) \Leftarrow k(x_1), \ldots, k(x_n)$.

We first show that $\mathcal{A}/\mathcal{E} \vdash p(t)$ implies $\mathcal{B}/\mathcal{E}' \vdash p(t)$ for all $p \in Q$. Since $\mathcal{B}$ contains all the clauses in $\mathcal{A}$ other than the intersection clauses, all we need to show is that $\mathcal{B} \cup \mathcal{E}'$ entails each intersection clause $q(x) \Leftarrow q_1(x), q_2(x)$ in $\mathcal{A}$. This is immediate, because $\mathcal{B}$ must contain the clause $q(x_1 \circ x_2) \Leftarrow q_1(x_1), q_2(x_2)$, and so $\mathcal{B}$ entails $q(x \circ x) \Leftarrow q_1(x), q_2(x)$. The theory $\mathcal{E}'$ contain the axiom $x \circ x = x$, and thus $\mathcal{B} \cup \mathcal{E}'$ entails $q(x) \Leftarrow q_1(x), q_2(x)$.

We now show that $\mathcal{B}/\mathcal{E}' \vdash p(t)$ implies $\mathcal{A}/\mathcal{E} \vdash p(t)$ for all $p \in Q$. If $\mathcal{B}/\mathcal{E}' \vdash p(t)$ then by Theorem 1, there is a term $u \in T_{F'}$ such that $t =_{\mathcal{E}'} u$ such that $\mathcal{B} \vdash p(u)$. We construct a term $v \in T_F$ such that $u =_{\mathcal{E}'} v$ and $\mathcal{A}/\mathcal{E} \vdash p(v)$. Since $t =_{\mathcal{E}'} u =_{\mathcal{E}'} v$ and neither $t$ nor $v$ contain the added symbol $\circ$, it is not difficult to show that $t =_{\mathcal{E}} v$, and thus $\mathcal{A}/\mathcal{E} \vdash p(t)$.

We construct the term $v \in T_F$ from the proof that $\mathcal{B} \vdash p(u)$ by analyzing the proof bottom-up starting from the leaves. Each inference step that does not use a clause containing the idempotence symbol $\circ$ has a direct corresponding inference step using the clauses in $\mathcal{A}$ and can be handled easily. On the other hand, given an inference step of the form

$$\frac{\mathcal{B} \vdash q_1(u_1) \qquad \mathcal{B} \vdash q_2(u_2)}{\mathcal{B} \vdash q(u_1 \circ u_2)}$$

with $q(x_1 \circ x_2) \Leftarrow q_1(x_1), q_2(x_2)$ in $\mathcal{B}$, we first observe that $u_1 =_{\mathcal{E}'} u_2 =_{\mathcal{E}'} u_1 \circ u_2$, because $u_1 \circ u_2$ is a subterm of $u$, and $u$ is equivalent to $t \in T_F$ which does not contain the symbol $\circ$. By induction, we know that for $i \in [1, 2]$, there is a term $v_i \in T_F$ such that $u_i =_{\mathcal{E}'} v_i$ and $\mathcal{A}/\mathcal{E} \vdash q_i(v_i)$. As $v_1 =_{\mathcal{E}'} u_1 =_{\mathcal{E}'} u_2 =_{\mathcal{E}'} v_2$ and both $v_1$ and $v_2$ are in $T_F$, it follows that $v_1 =_{\mathcal{E}} v_2$, and thus $\mathcal{A}/\mathcal{E} \vdash p_2(v_1)$. By using the intersection clause $p(x) \Leftarrow p_1(x), p_2(x)$ in $\mathcal{A}$, it follows that $\mathcal{A}/\mathcal{E} \vdash p(v_1)$ and thus we are done as $v_1 =_{\mathcal{E}} u_1 =_{\mathcal{E}} u_1 \circ u_2$.

Finally, we show that $\mathcal{B}/\mathcal{E}' \vdash k(t)$ if and only if $T_F \cap [t]_{\mathcal{E}'} \neq \varnothing$ for all $t \in T_{F'}$, by observing that $\mathcal{B} \vdash k(u)$ iff $u$ is in $T_F$, and so by Theorem 1,

$$\mathcal{B}/\mathcal{E}' \vdash k(t) \iff (\exists u \in [t]_{\mathcal{E}'})\, \mathcal{B} \vdash k(u) \iff T_F \cap [t]_{\mathcal{E}'} \neq \varnothing.$$

□

From this theorem, it follows that for each $p \in Q$, the languages $\mathcal{L}_p(\mathcal{A}/\mathcal{E})$ and $\mathcal{L}_p(\mathcal{B}/\mathcal{E}') \cap \mathcal{L}_k(\mathcal{B}/\mathcal{E}')$ are isomorphic with the bijective mapping

$$h_p : [t]_{\mathcal{E}} \in \mathcal{L}_p(\mathcal{A}/\mathcal{E}) \mapsto [t]_{\mathcal{E}'} \in \mathcal{L}_p(\mathcal{B}/\mathcal{E}') \cap \mathcal{L}_k(\mathcal{B}/\mathcal{E}').$$

Although this connection between alternating and regular languages seems worth further study, our main interest in this result is that allows us to use the result in [25] about the undecidability of emptiness for alternating AC-tree automata to show that intersection-emptiness is undecidable for regular tree automata over a theory $\mathcal{E}$ with both AC and ACI symbols.

**Corollary 1.** *If $\mathcal{E}$ is an equational theory with at least 4 constants, an AC symbol, and an ACI symbol, then the intersection-emptiness problem for regular tree automata over $\mathcal{E}$ is undecidable.*

*Proof.* Let $\mathcal{E}_{\mathsf{AC}}$ denote the equational theory obtained by removing the ACI symbol from $\mathcal{E}$. The theory $\mathcal{E}_{\mathsf{AC}}$ is *torsion-free* according to the definition in [25] with regard to the 4 constants, and consequently the emptiness problem is undecidable for alternating $\mathcal{E}_{\mathsf{AC}}$-tree automata by Prop. 11 in [25]. By Theorem 2, for each alternating automaton $\mathcal{A}$, we can construct a regular $\mathcal{E}$-tree automaton $\mathcal{B}$ such that $\mathcal{L}_p(\mathcal{A}/\mathcal{E}_{\mathsf{AC}}) = \varnothing$ iff $\mathcal{L}_p(\mathcal{B}/\mathcal{E}) \cap \mathcal{L}_k(\mathcal{B}/\mathcal{E}) = \varnothing$. □

The theory $\mathcal{E}$ in the previous statement can be partitioned into disjoint theories $\mathcal{E}_{\mathsf{AC}}$ and $\mathcal{E}_{\mathsf{ACI}}$ where $\mathcal{E}_{\mathsf{AC}}$ contains the AC symbol and $\mathcal{E}_{\mathsf{ACI}}$ contains the ACI symbol and the constants are split freely between them. Intersection-emptiness is decidable for both $\mathcal{E}_{\mathsf{AC}}$ [18] and $\mathcal{E}_{\mathsf{ACI}}$ [24], but as the previous statement shows it is undecidable for $\mathcal{E} = \mathcal{E}_{\mathsf{AC}} \cup \mathcal{E}_{\mathsf{ACI}}$. It follows that intersection-emptiness is a non-modular property for equational tree automata even for combinations of disjoint theories.

## 4 AC-Intersection Free Tree Automata

Having shown that intersection-emptiness is undecidable in general for equational tree automata over a theory $\mathcal{E}$ with AC and ACI symbols, we have decided to search for a restricted subclass of equational tree automata over $\mathcal{E}$ for which not only is intersection-emptiness decidable, but so is the propositional emptiness problem. Our search for this class began by trying to eliminate the main culprit that led to the undecidability result in Cor. 1 — the ability of clauses with ACI symbols to simulate the intersection clauses of an alternating AC-tree automata.

The solution we have found is to subject each ACI symbol $\circ$ in $\mathcal{E}$ to one of two constraints: (1) either the clauses in the automaton where $\circ$ appears must satisfy certain syntactic restrictions explained below; or (2) the idempotence equation $x \circ x = x$ in $\mathcal{E}$ must be treated as a rewrite rule $x \circ x \to x$ as in the *tree automata with normalization* framework of [17]. We first define the syntactic restrictions:

**Definition 1.** *Let $\mathcal{E}$ be an equational theory $\mathcal{E}$ in which each symbol is AC, ACI, or free. A regular $\mathcal{E}$-tree automaton $\mathcal{A}$ is* AC-intersection free *iff for each clause in $\mathcal{A}$ with the form $p(x_1 \circ x_2) \Leftarrow p_1(x_1), p_2(x_2)$ where $\circ \in F$ is an ACI symbol, it is the case that for all $q_1, q_2 \in Q$, and AC or ACI symbols $+ \neq \circ$,*

$$p_1(x_1 + x_2) \Leftarrow q_1(x_1), q_2(x_2) \in \mathcal{A} \implies p(x_1 + x_2) \Leftarrow q_1(x_1), q_2(x_2) \in \mathcal{A}.$$

The intuition behind this definition is that if an intersection clause $p(x) \Leftarrow p_1(x), p_2(x)$ is entailed by a clause $p(x_1 \circ x_2) \Leftarrow p_1(x_1), p_2(x_2)$ with an ACI symbol $\circ$, then we can disregard it in considering terms whose root symbol is an AC or ACI symbol $+ \neq \circ$. See our technical report [8] for further details.

One important observation is that AC-intersection free automata are closed under disjoint unions — that is given two AC-intersection free $\mathcal{E}$-tree automata $\mathcal{A}$ and $\mathcal{B}$ such that the states have been renamed so that the states in $\mathcal{A}$ and $\mathcal{B}$ are disjoint, the union $\mathcal{E}$-tree automaton $\mathcal{C} = \mathcal{A} \cup \mathcal{B}$ is also AC-intersection free. Moreover, $\mathcal{L}_p(\mathcal{A}/\mathcal{E}) = \mathcal{L}_p(\mathcal{C}/\mathcal{E})$ for each state $p$ in $\mathcal{A}$, and $\mathcal{L}_q(\mathcal{B}/\mathcal{E}) = \mathcal{L}_q(\mathcal{C}/\mathcal{E})$ for each state $q$ in $\mathcal{A}$. Since we will soon show that the propositional emptiness problem is decidable for AC-intersection free automata, it follows that the emptiness of an arbitrary Boolean combination of AC-intersection free tree languages is decidable even if the languages are defined in different automata.

This syntactic restriction may be too strong in some applications, and so we also study a different approach to handling idempotence equations that is suggested by the tree automata with normalization framework of [17]. A *tree automaton with normalization* (TAN) $\mathcal{A}$ is equipped with a rewrite system $\mathcal{R}$ that is confluent and terminating modulo an equational theory $\mathcal{E}$. A term $t$ is accepted by TAN $\mathcal{A}$ if its normal form $[t{\downarrow}_{\mathcal{R}/\mathcal{E}}]$ is in the underlying equational tree language $\mathcal{L}(\mathcal{A}/\mathcal{E})$. This framework borrows the fundamental idea in term rewriting, namely that some of the equations in a theory $\mathcal{E}'$ are best handled by orienting them as rewrite rules in a rewrite system $\mathcal{R}$ in a way so that $\mathcal{R}$ is confluent and terminating modulo the remaining equations $\mathcal{E} \subseteq \mathcal{E}'$. As $\mathcal{R}$ is terminating and confluent modulo $\mathcal{E}$, the language is closed with respect to both the equations in $\mathcal{E}$ and the equations obtained from the rules in $\mathcal{R}$.

Our interest in the TAN framework stems from the fact that if $\mathcal{R}_\mathsf{I}$ is a rewrite system containing idempotence rules $f(x, x) \to x$ for some of the AC symbols in a theory $\mathcal{E}$ with free, AC, and ACI symbols, then $\mathcal{R}_\mathsf{I}$ is confluent and terminating modulo $\mathcal{E}$. This suggests that as an alternative to the restrictions in Def. 1, we can treat some of the idempotence equations as rules, and still have a class of tree automata closed modulo both the equations in $\mathcal{E}$ and the underlying equations in $\mathcal{R}_{\mathsf{I}}$. By handling the idempotence equations as rules, we avoid the problem of simulating intersection clauses, because that simulation relies on applying idempotence in the direction $x \to x + x$.

By requiring that each ACI symbol either satisfies the syntactic constraints in the definition of AC-intersection free automata, or treats the idempotence equation as a rule as in the tree automata with normalization approach, we describe an algorithm in the next section whose correctness implies the following:

**Theorem 3.** *Let $\mathcal{E}$ be a theory with free, AC, and ACI symbols, and let $\mathcal{R}_{\mathsf{I}}$ be a set of rewrite rules which may contain an idempotence rule for any of the AC symbols in $\mathcal{E}$.*

*For each AC-intersection free $\mathcal{E}$-tree automaton $\mathcal{A}$, and propositional formula $\phi$ over the states in $\mathcal{A}$, the following problem is decidable:*

$$\mathcal{L}_\phi(\mathcal{A}/\mathcal{E}) \cap \mathsf{Can}_{\mathcal{R}_{\mathsf{I}}/\mathcal{E}} = \varnothing.$$

In other words, we can decide whether the language $\mathcal{L}_\phi(\mathcal{A}/\mathcal{E})$ contains an $\mathcal{R}_{\mathsf{I}}/\mathcal{E}$-irreducible equivalence class $[t] \in \mathsf{Can}_{\mathcal{R}_{\mathsf{I}}/\mathcal{E}}$. This theorem simultaneously settles two open questions:

The first open question is the emptiness problem for tree automata with normalization over an equational theory $\mathcal{E}_{\mathsf{AC}}$ with free and AC symbols and a rewrite system $\mathcal{R}_{\mathsf{I}}$ containing idempotence equations for the AC symbols in $\mathcal{E}_{\mathsf{AC}}$. Specifically, we want to decide whether $\mathsf{Can}_{\mathcal{R}_{\mathsf{I}}/\mathcal{E}_{\mathsf{AC}}} \cap L_p(\mathcal{A}/\mathcal{E}_{\mathsf{AC}}) = \varnothing$ for each $\mathcal{E}_{\mathsf{AC}}$-tree automaton $\mathcal{A}$ and state $p$ in $\mathcal{A}$. The problem was mentioned in [17], but left unsolved. Theorem 3 solves this problem, because $\mathcal{E}_{\mathsf{AC}}$ contains no ACI symbols and thus every $\mathcal{E}_{\mathsf{AC}}$-tree automaton is AC-intersection free. One observation made in [17] is that the decidability of the emptiness problem for tree automata with normalization only depends on the left hand sides of the rules in $\mathcal{R}$. It follows that if the emptiness problem is decidable when $\mathcal{R}$ contains idempotence rules $x + x \to x$, it is also decidable when $\mathcal{R}$ contains nilpotence rules $x + x \to 0$.

The second open question settled by Theorem 3 is the problem of deciding the propositional emptiness of equational tree automata over a theory $\mathcal{E}_{\mathsf{ACI}}$ with a single ACI symbol and free symbols. This problem is interesting, because equational tree automata over $\mathcal{E}_{\mathsf{ACI}}$ are not closed under complementation [23], and so the propositional emptiness problem is not reducible to the regular emptiness problem in this theory. Theorem 3 solves this problem, because $\mathcal{E}_{\mathsf{ACI}}$ contains only a single ACI symbol, and thus every $\mathcal{E}_{\mathsf{ACI}}$-tree automaton is AC-intersection free. Solving the propositional emptiness problem also shows that both subsumption ($\mathcal{L}_p(\mathcal{A}/\mathcal{E}_{\mathsf{ACI}}) \subseteq \mathcal{L}_q(\mathcal{B}/\mathcal{E}_{\mathsf{ACI}})$) and universality ($\mathcal{L}_p(\mathcal{A}/\mathcal{E}_{\mathsf{ACI}}) = T_{\mathcal{E}_{\mathsf{ACI}}}$) are decidable. Both problems appear to be open. Additionally, since intersection-emptiness is undecidable for equational tree automata over $\mathcal{E}_{\mathsf{AC}} \cup \mathcal{E}_{\mathsf{ACI}}$ due to Cor. 1, it follows that propositional emptiness over $\mathcal{E}_{\mathsf{AC}} \cup \mathcal{E}_{\mathsf{ACI}}$ is undecidable as well. However, propositional emptiness is decidable for $\mathcal{E}_{\mathsf{AC}}$ [18] and implied to be decidable for

$\mathcal{E}_{\mathsf{ACI}}$ by Theorem 3. It follows that propositional emptiness is also a *non-modular property* for the combination of disjoint theories.

## 5 Decision Procedure

In this section, we define an algorithm that solves the decision problem posed in Theorem 3. Due to space limitations, we only present the algorithm here, and not the complete correctness proof which is available in our technical report [8].

For this section, $\mathcal{E} = (F, E)$ denotes a theory in which each symbol is AC, ACI, or free, $\mathcal{R}_{\mathsf{I}}$ denotes a rewrite system where the only axioms are idempotence rules of the form $x + x \to x$ for an AC symbol $+ \in F$, and $\mathcal{A}$ denotes a regular AC-intersection free $\mathcal{E}$-tree automaton with states $Q$. It is sometimes useful to treat all of the idempotence equations as rules. We let $\mathcal{E}_{\mathsf{AC}} \subseteq \mathcal{E}$ denote the theory containing only the AC equations in $\mathcal{E}$, and we let $\hat{\mathcal{R}}_{\mathsf{I}}$ denote the rewrite system containing the rules in $\mathcal{R}_{\mathsf{I}}$ as well as a rule $x+x \to x$ for each equation $x+x = x$ in $\mathcal{E}$. $\hat{\mathcal{R}}_{\mathsf{I}}$ is terminating and confluent modulo $\mathcal{E}_{\mathsf{AC}}$, so for all $\mathcal{R}_{\mathsf{I}}/\mathcal{E}$-irreducible terms $t, u \in T_F$, $t =_{\mathcal{E}} u$ iff $t{\downarrow}_{\hat{\mathcal{R}}_{\mathsf{I}}/\mathcal{E}_{\mathsf{AC}}} =_{\mathcal{E}_{\mathsf{AC}}} u{\downarrow}_{\hat{\mathcal{R}}_{\mathsf{I}}/\mathcal{E}_{\mathsf{AC}}}$. For all $[t], [u] \in \mathsf{Can}_{\mathcal{R}_{\mathsf{I}}/\mathcal{E}}$, we say that $[t]$ is a *flattened subterm* of $[u]$, denoted $[t] \trianglelefteq_{\mathsf{flat}} [u]$, if either:

- $u{\downarrow}_{\hat{\mathcal{R}}_{\mathsf{I}}/\mathcal{E}_{\mathsf{AC}}} =_{\mathcal{E}_{\mathsf{AC}}} f(u_1, \ldots, u_n)$ with $f$ a free symbol and $t{\downarrow}_{\hat{\mathcal{R}}_{\mathsf{I}}/\mathcal{E}_{\mathsf{AC}}} =_{\mathcal{E}_{\mathsf{AC}}} u_i$ for some $i \in [1, n]$, or
- $u{\downarrow}_{\hat{\mathcal{R}}_{\mathsf{I}}/\mathcal{E}_{\mathsf{AC}}} =_{\mathcal{E}_{\mathsf{AC}}} u_1 + \cdots + u_n$ with $+$ an AC or ACI symbol, $n \geq 2$, $\mathsf{root}(u_i) \neq +$ for all $i \in [1, n]$, and $t{\downarrow}_{\hat{\mathcal{R}}_{\mathsf{I}}/\mathcal{E}_{\mathsf{AC}}} =_{\mathcal{E}_{\mathsf{AC}}} u_j$ for some $j \in [1, n]$.

Our algorithm is similar to the subset construction algorithm in [9] for checking the propositional emptiness of equational tree automata over A and AC symbols. For each $[t] \in T_{\mathcal{E}}$, the *profile* of $[t]$, denoted $\mathsf{profile}([t])$, is a pair that contains all the information about $[t]$ relevant to the algorithm.

**Definition 2.** *Let* $\mathsf{profile} : T_{\mathcal{E}} \to F \times \mathcal{P}(Q)$ *be the function such that:*

$$\mathsf{profile}([t]) = (\mathsf{root}(t{\downarrow}_{\hat{\mathcal{R}}_{\mathsf{I}}/\mathcal{E}_{\mathsf{AC}}}), \mathsf{states}_{\mathcal{A}/\mathcal{E}}([t])).$$

*where* $\mathsf{states}_{\mathcal{A}/\mathcal{E}}([t]) = \{\, p \in Q \mid \mathcal{A}/\mathcal{E} \vdash p(t) \,\}$.

Note that $\mathsf{root}(t{\downarrow}_{\hat{\mathcal{R}}_{\mathsf{I}}/\mathcal{E}_{\mathsf{AC}}})$ is uniquely determined as $\mathcal{E}_{\mathsf{AC}}$ only contains associativity and commutativity axioms which do not change the root symbol of a term. We have shown in [9] how to compute $\mathsf{states}_{\mathcal{A}/\mathcal{E}}([t])$ when $\mathcal{E}$ contains A and AC symbols.

For an automaton $\mathcal{B}$ with states $Q'$ over a theory $\mathcal{E}' = (F', E')$ with free, A, and AC symbols, we presented a semi-algorithm in [9] for constructing the set

$$\mathsf{det}(\mathcal{B}) = \{\, (f, P) \in F' \times \mathcal{P}(Q') \mid (\exists [t] \in T_{\mathcal{E}'})\, \mathsf{root}([t]) = f \wedge \mathsf{states}_{\mathcal{B}/\mathcal{E}'}([t]) = P \,\}.$$

By computing this set, we can decide if $\mathcal{L}_{\phi}(\mathcal{B}/\mathcal{E}') \neq \varnothing$ by checking for a profile $(f, P) \in \mathsf{det}(\mathcal{B})$ such that $P \models \phi$ where $P \models \phi$ is defined inductively:

$$P \models \phi_1 \wedge \phi_2 \text{ iff } P \models \phi_1 \text{ and } P \models \phi_2 \qquad P \models \neg\phi \text{ iff } P \not\models \phi \qquad P \models p \text{ iff } p \in P$$

For solving the problem in Theorem 3, this approach is inadequate for two reasons: (1) We want to decide whether $\mathsf{Can}_{\mathcal{R}_{\mathsf{l}}/\mathcal{E}} \cap \mathcal{L}_{\phi}(\mathcal{A}/\mathcal{E}) = \varnothing$ rather than deciding whether $\mathcal{L}_{\phi}(\mathcal{A}/\mathcal{E}) = \varnothing$. (2) Both $\mathcal{E}$ and $\mathcal{R}$ may contain idempotence axioms, and idempotence appears to require constructing a structure which in addition to enable checking if there *exists* a term with a particular profile, also enables checking *how many* distinct terms have that profile. We illustrate this with an example. Let $\mathcal{E}_{\mathsf{ACI}}$ be the theory containing an ACI symbol $\circ$ and constants $a$, $b$, and $c$, and let $\mathcal{B}$ be the $\mathcal{E}_{\mathsf{ACI}}$-tree automaton with the rules:

$$p_1(a) \quad p_1(b) \quad p_2(x_1 \circ x_2) \Leftarrow p_1(x_1), p_1(x_2) \quad p_3(x_1 \circ x_2) \Leftarrow p_1(x_1), p_2(x_2).$$

In this automaton, one can observe that

$$\mathcal{L}_{p_2}(\mathcal{B}/\mathcal{E}_{\mathsf{ACI}}) = \mathcal{L}_{p_3}(\mathcal{B}/\mathcal{E}_{\mathsf{ACI}}) = \{\, [a], [b], [a \circ b] \,\},$$

and consequently $\mathcal{L}_{p_3 \wedge \neg p_2}(\mathcal{B}/\mathcal{E}_{\mathsf{ACI}}) = \varnothing$. Now consider the automaton $\mathcal{B}'$ containing the clauses in $\mathcal{B}$ and the additional clause $p_1(c)$. One can observe that $\mathcal{L}_{p_3 \wedge \neg p_2}(\mathcal{B}'/\mathcal{E}_{\mathsf{ACI}}) = \{\, [a \circ b \circ c] \,\}$. The language $\mathcal{L}_{p_3 \wedge \neg p_2}(\mathcal{B}'/\mathcal{E}_{\mathsf{ACI}})$ is not empty, because there are 3 distinct elements in $\mathcal{L}_{p_1}(\mathcal{B}'/\mathcal{E}_{\mathsf{ACI}})$, whereas $\mathcal{L}_{p_1}(\mathcal{B}/\mathcal{E}_{\mathsf{ACI}})$ only contains 2 elements. If we generalize this idea, it is not difficult to show that for any positive integer $n \in \mathbb{N}$ and tree automaton $\mathcal{B}$ over $\mathcal{E}$ with a state $p$, we can construct a tree automaton $\mathcal{B}'_n$ over the theory $\mathcal{E}'$ containing $\mathcal{E}$ as well as a fresh ACI symbol $\circ$ and a formula $\phi_n$ over the states in $\mathcal{B}'_n$ such that

$$\mathcal{L}_{\phi_n}(\mathcal{B}'_n/\mathcal{E}') \neq \varnothing \iff |\mathcal{L}_p(\mathcal{B}/\mathcal{E})| \geq n.$$

In this work, we construct the directed graph $(D_{\mathcal{A}}, \trianglelefteq_{\mathcal{A}})$ where

$$D_{\mathcal{A}} = \{\, d \in F \times \mathcal{P}(Q) \mid (\exists [t] \in \mathsf{Can}_{\mathcal{R}_{\mathsf{l}}/\mathcal{E}})\ \mathsf{profile}([t]) = d \,\},$$

and $\trianglelefteq_{\mathcal{A}}$ contains an edge $d_1 \trianglelefteq_{\mathcal{A}} d_2$ iff there are $[t], [u] \in \mathsf{Can}_{\mathcal{R}_{\mathsf{l}}/\mathcal{E}}$ such that $\mathsf{profile}([t]) = d_1$, $\mathsf{profile}([u]) = d_2$, and $[t] \trianglelefteq_{\mathsf{flat}} [u]$. The edge relation $\trianglelefteq_{\mathcal{A}}$ is used in counting the number of equivalence classes with a given profile. To given an example of the directed graph, in the automaton $\mathcal{B}'$ described above:

$$D_{\mathcal{B}'} = \{\, (a, \{\, p_1, p_2, p_3 \,\}), (b, \{\, p_1, p_2, p_3 \,\}), (c, \{\, p_1, p_2, p_3 \,\}), (\circ, \{\, p_2, p_3 \,\}), (\circ, \{\, p_3 \,\}) \,\},$$

and $\trianglelefteq_{\mathcal{B}'}$ contains the following edges:

$$\begin{array}{ll}
(a, \{\, p_1, p_2, p_3 \,\}) \trianglelefteq_{\mathcal{B}'} (\circ, \{\, p_2, p_3 \,\}) & (a, \{\, p_1, p_2, p_3 \,\}) \trianglelefteq_{\mathcal{B}'} (\circ, \{\, p_3 \,\}) \\
(b, \{\, p_1, p_2, p_3 \,\}) \trianglelefteq_{\mathcal{B}'} (\circ, \{\, p_2, p_3 \,\}) & (b, \{\, p_1, p_2, p_3 \,\}) \trianglelefteq_{\mathcal{B}'} (\circ, \{\, p_3 \,\}) \\
(c, \{\, p_1, p_2, p_3 \,\}) \trianglelefteq_{\mathcal{B}'} (\circ, \{\, p_2, p_3 \,\}) & (c, \{\, p_1, p_2, p_3 \,\}) \trianglelefteq_{\mathcal{B}'} (\circ, \{\, p_3 \,\})
\end{array}$$

Our approach is to incrementally construct $(D_{\mathcal{A}}, \trianglelefteq_{\mathcal{A}})$. We start with the empty graph $(D_0, \trianglelefteq_0) = (\varnothing, \varnothing)$ and apply inference rules to form increasing larger subgraphs $(D_1, \trianglelefteq_1) \subseteq (D_2, \trianglelefteq_2) \subseteq \cdots \subseteq (D_{\mathcal{A}}, \trianglelefteq_{\mathcal{A}})$ until saturation. This process terminates with a unique final graph as the size of $D_{\mathcal{A}}$ is at most $|F| \times 2^{|Q|}$, and the construction process is monotonic. Each profile graph $(D, \trianglelefteq) \subseteq (D_{\mathcal{A}}, \trianglelefteq_{\mathcal{A}})$ can be viewed as representing the possibly infinite subset of $\mathsf{Can}_{\mathcal{R}_{\mathsf{l}}/\mathcal{E}}$ that is already explored:

**Definition 3.** *For each graph* $(D, \trianglelefteq) \subseteq (D_{\mathcal{A}}, \trianglelefteq_{\mathcal{A}})$, *let* $\mathsf{Can}_{D,\trianglelefteq}$ *denote the smallest set containing each* $[t] \in \mathsf{Can}_{\mathcal{R}_\mathsf{I}/\mathcal{E}}$ *if* $\mathsf{profile}([t]_\mathcal{E}) \in D$ *and for all* $[u] \in \mathsf{Can}_{\mathcal{R}_\mathsf{I}/\mathcal{E}}$,

$$[u] \trianglelefteq_{\mathsf{flat}} [t] \implies [u] \in \mathsf{Can}_{D,\trianglelefteq} \wedge \mathsf{profile}([u]) \trianglelefteq \mathsf{profile}([t]).$$

*Furthermore, for each* $d \in D$, *we let* $\mathsf{profile}^{-1}_{D,\trianglelefteq}(d)$ *denote the elements in* $\mathsf{Can}_{D,\trianglelefteq}$ *with profile* $d$, *i.e.,* $\mathsf{profile}^{-1}_{D,\trianglelefteq}(d) = \{\, [t] \in \mathsf{Can}_{D,\trianglelefteq} \mid \mathsf{profile}([t]) = d \,\}$.

In [8], we show that $\mathsf{Can}_{D_\mathcal{A},\trianglelefteq_\mathcal{A}} = \mathsf{Can}_{\mathcal{R}_\mathsf{I}/\mathcal{E}}$, and consequently the graph $(D_\mathcal{A}, \trianglelefteq_\mathcal{A})$ can be viewed as the graph where every $\mathcal{R}_\mathsf{I}/\mathcal{E}$-irreducible term has been explored.

For each free symbol $f \in F$, we define a function $\mathsf{states}_f$ which computes the states of a term $f(t_1, \ldots, t_n)$ when the states for each term $t_i$ are already known:

**Definition 4.** *Given a free symbol* $f \in F$ *with arity* $n$, *we define the function* $\mathsf{states}_f : \mathcal{P}(Q)^n \to \mathcal{P}(Q)$ *such that for* $P_1, \ldots, P_n \subseteq Q$, $\mathsf{states}_f(P_1, \ldots, P_n) \subseteq Q$ *is the smallest set containing a state* $p \in Q$ *if either:*

- $\mathcal{A}$ *contains* $p(f(x_1, \ldots, x_n)) \Leftarrow p_1(x_1), \ldots, p_n(x_n)$ *with* $p_i \in P_i$ *for* $i \in [1, n]$,
- *or* $\mathcal{A}$ *contains* $p(x_1 \circ x_2) \Leftarrow p_1(x_1), p_2(x_2)$ *with* $\circ$ *an ACI-symbol in* $\mathcal{E}$ *and* $p_1, p_2 \in \mathsf{states}_f(P_1, \ldots, P_n)$.

Similar to [9], we define a context free grammar $G(+)$ for each AC or ACI symbol $+ \in F$. Intuitively, the grammar captures inferences in the automaton $\mathcal{A}$ over *flattened* terms with the form $t_1 + \cdots + t_n$ where $\mathsf{root}(t_i) \neq +$ for $i \in [1, n]$.

**Definition 5.** *For an AC or ACI symbol* $+ \in F$, $G(+)$ *is the CFG grammar with terminals* $\Sigma(+) = (F \backslash \{+\}) \times \mathcal{P}(Q)$, *non-terminals* $Q$, *and production rules*

$$\begin{aligned} G(+) = \{\, & p := p_1 p_2 \mid p(x_1 + x_2) \Leftarrow p_1(x_1), p_2(x_2) \in \mathcal{A} \,\} \\ \cup \{\, & p := (f, P) \mid (f, P) \in \Sigma(+) \wedge p \in P \,\}. \end{aligned}$$

For each state $p \in Q$, we let $\mathcal{L}_p(G(+))$ denote the language generated by $p$ using the rules in $G(+)$. For each non-terminal $p \in Q$, there effectively exists a Presburger formula $\psi_{G(+),p}(\vec{x})$ with free variables $\vec{x} = \{\, x_d \,\}_{d \in \Sigma(+)}$ whose models $M(\psi_{G(+),p}) \subseteq \mathbb{N}^{\Sigma(+)}$ equal the commutative image of $\mathcal{L}_p(G(+))$ [20], i.e.,

$$M(\psi_{G(+),p}) = \{\, \#(w) \mid w \in \mathcal{L}_p(G(+)) \,\}.$$

where $\# : \Sigma(+)^* \to \mathbb{N}^{\Sigma(+)}$ maps each string to the vector counting the number of occurrences of each letter in the string. We use $\psi_{G(+),p}$ to define the formula $\psi_{G(+),P}$ which, as proven in [8], identifies terms whose profile is $(+, P)$. For each AC symbol $+ \in F$ and each symbol $\circ \in F$ idempotent in $\mathcal{E}$ or $\mathcal{R}$,

$$\begin{aligned} \psi_{+,P}(\vec{x}) &= \bigwedge_{p \in P} \psi_{G(+),p}(\vec{x}) \wedge \bigwedge_{p \in Q \backslash P} \neg \psi_{G(+),p}(\vec{x}) \wedge \sum_{x_d \in \vec{x}} x_d \geq 2 \\ \psi_{\circ,P}(\vec{x}) &= \bigwedge_{p \in P} (\exists \vec{y})\, \vec{x} \sqsubseteq \vec{y} \wedge \psi_{G(\circ),p}(\vec{y}) \wedge \bigwedge_{p \in Q \backslash P} \neg (\exists \vec{y})\, \vec{x} \sqsubseteq \vec{y} \wedge \psi_{G(\circ),p}(\vec{y}) \wedge \sum_{x_d \in \vec{x}} x_d \geq 2. \end{aligned}$$

where $\vec{x} \sqsubseteq \vec{y}$ is the formula $\displaystyle\bigwedge_{d \in \Sigma(+)} x_d \leq y_d \wedge ((y_d > 0) \Rightarrow (x_d > 0))$.

Starting with the empty graph $(D_0, \trianglelefteq_0) = (\varnothing, \varnothing)$, we freely apply either of the rules below to construct $(D_{i+1}, \trianglelefteq_{i+1})$ from $(D_i, \trianglelefteq_i)$ subject to the condition that a rule may only be applied if the resulting graph $(D_{i+1}, \trianglelefteq_{i+1})$ is distinct from $(D_i, \trianglelefteq_i)$. The rules are applied until completion to obtain the graph $(D_*, \trianglelefteq_*)$.

$$\frac{\text{choose free symbol } f \in F \text{ and } (f_1, P_1), \ldots, (f_n, P_n) \in D_i}{\begin{array}{l} D_{i+1} := D_i \cup \{ (f, \mathsf{states}_f(P_1, \ldots, P_n)) \} \\ \trianglelefteq_{i+1} := \trianglelefteq_i \cup \{ ((f_j, P_j), (f, \mathsf{states}_f(P_1, \ldots, P_n))) \mid j \in [1, n] \} \end{array}}$$

$$\frac{\text{choose AC or ACI symbol } + \in F \text{ and } P \subseteq Q \text{ s.t. } (\exists \vec{x})\ \psi_{+,P,D_i,\trianglelefteq_i}(\vec{x})}{\begin{array}{l} D_{i+1} := D_i \cup \{ (f, P) \} \\ \trianglelefteq_{i+1} := \trianglelefteq_i \cup \{ (d, (f, P)) \mid d \in D_i \wedge (\exists \vec{x})\ \psi_{+,P,D_i,\trianglelefteq_i}(\vec{x}) \wedge x_d > 0 \} \end{array}}$$

where for each symbol $\circ \in F$ that is idempotent in $\mathcal{E}$ or $\mathcal{R}$ and each AC symbol $+ \in F$ that is not idempotent in $\mathcal{E}$ or $\mathcal{R}$, we let

$$\psi_{\circ,P,D_i,\trianglelefteq_i}(\vec{x}) = \psi_{\circ,P}(\vec{x}) \wedge \bigwedge_{d \in \Sigma(\circ) \setminus D_i} x_d = 0 \wedge \bigwedge_{d \in \Sigma(\circ) \cap D_i} x_d \leq \mathsf{cnt}_{D_i,\trianglelefteq_i}(d)$$

$$\psi_{+,P,D_i,\trianglelefteq_i}(\vec{x}) = \psi_{+,P}(\vec{x}) \wedge \bigwedge_{d \in \Sigma(+) \setminus D_i} x_d = 0 \wedge \bigwedge_{\substack{d \in \Sigma(+) \cap D_i \\ \mathsf{cnt}_{D_i,\trianglelefteq_i}(d) = 0}} x_d = 0.$$

**Fig. 1.** Inference System for Constructing $(D_*, \trianglelefteq_*)$

We next introduce a function $\mathsf{cnt}_{D,\trianglelefteq} : D \to \mathbb{N} \cup \{ \omega \}$ which for each graph $(D, \trianglelefteq) \subseteq (D_\mathcal{A}, \trianglelefteq_\mathcal{A})$ and profile $d \in D$, returns an estimate of the number of elements in $\mathsf{Can}_{\mathcal{R}_1/\mathcal{E}}$ with the profile $d$. Due to space constraints, we refer the reader to [8] for the precise definition. The basic idea is to first check if $d \trianglelefteq^+ d$. If so, $\mathsf{profile}^{-1}_{D,\trianglelefteq}$ may be finite. However, as shown in [8], $\left|\mathsf{profile}^{-1}_{D_\mathcal{A},\trianglelefteq_\mathcal{A}}(d)\right| = \omega$, and we let $\mathsf{cnt}_{D,\trianglelefteq}(d) = \omega$. Otherwise $d \not\trianglelefteq^+ d$, and $\mathsf{cnt}_{D,\trianglelefteq}(d)$ is obtained by evaluating $\mathsf{cnt}_{D,\trianglelefteq}$ on smaller elements $d' \trianglelefteq d$ and using fundamental combinatoric properties of sets and multisets. The fundamental requirement is that for each $d \in D$,

$$\left|\mathsf{profile}^{-1}_{D,\trianglelefteq}(d)\right| \leq \mathsf{cnt}_{D,\trianglelefteq}(d) \leq \left|\mathsf{profile}^{-1}_{D_\mathcal{A},\trianglelefteq_\mathcal{A}}(d)\right|.$$

For correctness purposes, any value in the range is sufficient. The proof that our procedure always shows *emptiness* only requires that $\mathsf{cnt}_{D,\trianglelefteq}(d)$ is at most the total number of elements in $\mathsf{Can}_{D_\mathcal{A},\trianglelefteq_\mathcal{A}}$ with a profile $d$, while the proof that our procedure always shows *non-emptiness* only requires that $\mathsf{cnt}_{D,\trianglelefteq}(d)$ is at least the number of explored elements in $\mathsf{Can}_{D,\trianglelefteq}$ with a profile $d$.

The algorithm for constructing the profile graph $(D_*, \trianglelefteq_*)$ is given Fig. 1. In our technical report [8], we prove that $(D_*, \trianglelefteq_*) = (D_\mathcal{A}, \trianglelefteq_\mathcal{A})$, and thus

**Theorem 4.** *The graph $(D_\mathcal{A}, \trianglelefteq_\mathcal{A})$ is effectively constructable.* □

Theorem 3 can be as a corollary of this theorem as $\mathsf{Can}_{\mathcal{R}_1/\mathcal{E}} \cap \mathcal{L}_\phi(\mathcal{A}/\mathcal{E}) \neq \varnothing$ iff $\exists (f, P) \in D_\mathcal{A}$ such that $P \models \phi$.

## 6 Related Work and Conclusions

Our main contributions in this paper are: (1) We have shown that an alternating equational tree language can be expressed as the intersection of two regular equational tree languages by adding a fresh ACI symbol to the theory. This implies that intersection-emptiness is undecidable for regular equational tree automata over a theory with both AC and ACI symbols. (2) We studied modularity in equational tree automata and have shown that both intersection-emptiness and propositional emptiness are non-modular properties even for disjoint theories. (3) We presented a subclass of regular equational tree automata over theories with AC and ACI symbols and have shown propositional emptiness is decidable for that subclass. This result further implies that propositional emptiness is decidable for equational tree automata with one ACI symbol and tree automata with normalization over a rewrite theory with idempotence rules and AC symbols.

One of our goals was to obtain decidability results over non-linear theories. In this direction there are numerous papers on extending tree automata techniques to better handle non-linearity in adding constraints to the automata rules [1, Chapter 4] as well as extending that idea to handle some equational theories [11]. The problem of deciding whether a non-equational tree language accepts an irreducible term for any set of linear or non-linear rules was shown in [2], however the approach used here is quite different. The technique of counting the number of distinct terms was influenced by similar issues in deciding the emptiness of multitree automata [14], and our realization that Presburger arithmetic is useful in the ACI case was inspired by the generalization of Parikh's theorem to arbitrary Kleene algebras in [10].

Although we have solved two open problems, our work suggests additional questions that are worth exploring, including: (1) Can we obtain positive modularity results by imposing stronger conditions on the theories such as linearity or collapse-freeness? (2) Can the semi-decision procedure for the associative case in [9] be extended to handle AC-intersection free automata over theories with any combination of associativity, commutativity, and idempotence? (3) Although ground reducibility modulo AC is undecidable in general for non-linear rules [12], for what other non-linear rules is ground reducibility modulo AC decidable?

**Acknowledgments.** The authors would like to thank the referees for comments which helped to improve the paper.

## References

[1] Comon, H., Dauchet, M., Gilleron, R., Löding, C., Jacquemard, F., Lugiez, D., Tison, S., Tommasi, M.: Tree automata techniques and applications (2007), `http://www.grappa.univ-lille3.fr/tata`

[2] Comon, H., Jacquemard, F.: Ground reducibility is EXPTIME-complete. Information and Computation 187(1), 123–153 (2003)

[3] Devienne, P., Talbot, J.-M., Tison, S.: Solving classes of set constraints with tree automata. In: Smolka, G. (ed.) CP 1997. LNCS, vol. 1330, pp. 62–76. Springer, Heidelberg (1997)

[4] Feuillade, G., Genet, T., Viet Triem Tong, V.: Reachability analysis over term rewriting systems. J. Autom. Reasoning 33(3–4), 341–383 (2004)

[5] Genet, T., Klay, F.: Rewriting for cryptographic protocol verification. In: McAllester, D. (ed.) CADE 2000. LNCS, vol. 1831, pp. 271–290. Springer, Heidelberg (2000)
[6] Hendrix, J., Meseguer, J.: On the completeness of context-sensitive order-sorted specifications. In: Baader, F. (ed.) RTA 2007. LNCS, vol. 4533, pp. 229–245. Springer, Heidelberg (2007)
[7] Hendrix, J., Meseguer, J., Ohsaki, H.: A sufficient completeness checker for linear order-sorted specifications modulo axioms. In: Furbach, U., Shankar, N. (eds.) IJCAR 2006. LNCS (LNAI), vol. 4130, pp. 151–155. Springer, Heidelberg (2006)
[8] Hendrix, J., Ohsaki, H.: Combining equational tree automata over AC and ACI theories. Technical Report UIUCDCS-R-2008-2940, University of Illinois (2008)
[9] Hendrix, J., Ohsaki, H., Viswanathan, M.: Propositional tree automata. In: Pfenning, F. (ed.) RTA 2006. LNCS, vol. 4098, pp. 50–65. Springer, Heidelberg (2006)
[10] Hopkins, M.W., Kozen, D.: Parikh's theorem in commutative kleene algebra. In: Proc. of LICS 1999, pp. 394–401. IEEE Computer Society, Los Alamitos (1999)
[11] Jacquemard, F., Rusinowitch, M., Vigneron, L.: Tree automata with equality constraints modulo equational theories. In: Furbach, U., Shankar, N. (eds.) IJCAR 2006. LNCS (LNAI), vol. 4130, pp. 557–571. Springer, Heidelberg (2006)
[12] Kapur, D., Narendran, P., Rosenkrantz, D., Zhang, H.: Sufficient-completeness, ground-reducibility and their complexity. Acta Inf. 28(4), 311–350 (1991)
[13] Klarlund, N., Møller, A.: MONA Version 1.4 User Manual. BRICS, Department of Computer Science, University of Aarhus, Notes Series NS-01-1. Revision of BRICS NS-98-3 (January 2001), `http://www.brics.dk/mona/`
[14] Lugiez, D.: Multitree automata that count. Theoretical Comput. Sci. 333(1–2), 225–263 (2005)
[15] Nelson, G., Oppen, D.C.: Simplification by cooperating decision procedures. ACM Trans. Program. Lang. Syst. 1(2), 245–257 (1979)
[16] Ohsaki, H.: Beyond regularity: Equational tree automata for associative and commutative theories. In: Fribourg, L. (ed.) CSL 2001 and EACSL 2001. LNCS, vol. 2142, pp. 539–553. Springer, Heidelberg (2001)
[17] Ohsaki, H., Seki, H.: Languages modulo normalization. In: Konev, B., Wolter, F. (eds.) FroCos 2007. LNCS (LNAI), vol. 4720, pp. 221–236. Springer, Heidelberg (2007)
[18] Ohsaki, H., Takai, T.: Decidability and closure properties of equational tree languages. In: Tison, S. (ed.) RTA 2002. LNCS, vol. 2378, pp. 114–128. Springer, Heidelberg (2002)
[19] Ohsaki, H., Talbot, J.-M., Tison, S., Roos, Y.: Monotone AC-tree automata. In: Sutcliffe, G., Voronkov, A. (eds.) LPAR 2005. LNCS (LNAI), vol. 3835, pp. 337–351. Springer, Heidelberg (2005)
[20] Parikh, R.J.: On context-free languages. J. ACM 13(4), 570–581 (1966)
[21] Shostak, R.E.: Deciding combinations of theories. J. ACM 31(1), 1–12 (1984)
[22] Slutzki, G.: Alternating tree automata. Theoretical Comput. Sci. 41, 305–318 (1985)
[23] Verma, K.N.: On closure under complementation of equational tree automata for theories extending AC. In: Vardi, M.Y., Voronkov, A. (eds.) LPAR 2003. LNCS, vol. 2850, pp. 183–197. Springer, Heidelberg (2003)
[24] Verma, K.N.: Two-way equational tree automata for AC-like theories: Decidability and closure properties. In: Nieuwenhuis, R. (ed.) RTA 2003. LNCS, vol. 2706, pp. 180–196. Springer, Heidelberg (2003)
[25] Verma, K.N., Goubault-Larrecq, J.: Alternating two-way AC-tree automata. Information and Computation 205(6), 817–869 (2007)

# Closure of Hedge-Automata Languages by Hedge Rewriting

Florent Jacquemard[1] and Michael Rusinowitch[2]

[1] INRIA Futurs and LSV, UMR CNRS, ENS Cachan, France
florent.jacquemard@lsv.ens-cachan.fr
[2] LORIA & INRIA Lorraine, UMR 7503
rusi@loria.fr

**Abstract.** We consider rewriting systems for unranked ordered terms, i.e. trees where the number of successors of a node is not determined by its label, and is not a priori bounded. The rewriting systems are defined such that variables in the rewrite rules can be substituted by hedges (sequences of terms) instead of just terms. Consequently, this notion of rewriting subsumes both standard term rewriting and word rewriting.

We investigate some preservation properties for two classes of languages of unranked ordered terms under this generalization of term rewriting. The considered classes include languages of hedge automata (HA) and some extension (called CF-HA) with context-free languages in transitions, instead of regular languages.

In particular, we show that the set of unranked terms reachable from a given HA language, using a so called inverse context-free rewrite system, is a HA language. The proof, based on a HA completion procedure, reuses and combines known techniques with non-trivial adaptations. Moreover, we prove, with different techniques, that the closure of CF-HA languages with respect to restricted context-free rewrite systems, the symmetric case of the above rewrite systems, is a CF-HA language. As a consequence, the problems of ground reachability and regular hedge model checking are decidable in both cases. We give several counter examples showing that we cannot relax the restrictions.

## 1 Introduction

In many applications the system states can be modeled by words or trees, sets of configurations by word or tree languages and the transitions of the system can be represented by rewrite rules. In this setting verifying whether a system can enter a set of unsafe states can be expressed as a reachability problem. This approach to the analysis of infinite-state systems requires the computation of the closure of languages under rewrite rules or at least an over-approximation of this closure. Since the usually considered languages are regular the approach is called *regular model checking* [2,1]. Regular model checking has been quite successful in protocol and hardware verification. For increasing the scope of regular model checking it is therefore important to be able to derive new classes of languages and rewrite systems such that the rewrite closure is computable.

A. Voronkov (Ed.): RTA 2008, LNCS 5117, pp. 157–171, 2008.

Unranked trees as well as ordered sequences of unranked trees called *hedges* [13,14,5] are flexible structures that are quite appealing to represent XML documents where the number of nodes can be modified, for instance when these nodes correspond to database records. Unranked trees have also been employed to model multithreaded recursive program configurations where the number of parallel processes is unbounded [3,18]. Hedge-automata (HA) are considered now as the natural model of automata for unranked trees. A hedge automaton is a variation of tree automata for hedges. Given a hedge, a hedge automaton assigns some state to a node whenever the *sequence* of states of the siblings belong to some specified word language (sometimes called horizontal language).

Although regular model checking with languages for words and *ranked* trees (where function symbols have fixed arity) has been widely investigated, very few results are available for *unranked* trees and almost none exists on the *computation of exact reachability sets* for HA languages.

In this paper we tackle the problem above by proving (Theorem 1) that we can compute a HA for recognizing the rewrite closure of a language defined by a given HA, for the class of rewrite systems with inverse context-free rules, which are rules whose right-hand side is of type $f(x)$ where $x$ is a variable. Hence in that case we can compute the exact reachability set from the initial one. The rewriting notion that we consider here for unranked terms generalizes ranked term rewriting and is close to the one that has been introduced by [22]. The idea is that the variables in the rewrite rules can be substituted by hedges (sequences of terms) instead of just terms. Moreover our results cannot be derived from related ones on ranked terms (*e.g.* [15]) using encodings of unranked terms into ranked ones (such as the *First-Child-Next-Sibling* encoding or the encoding used in stepwise automata [4]). Relaxing the condition in the definition in the above class of rewrite systems leads to counterexamples (Propositions 3–6).

We have also considered a more general class of automata for unranked ordered trees, called CF-HA, where word context-free languages are used instead of regular ones at the horizontal level. We show (Theorem 2) that CF-HA are preserved by rewrite closure using context-free rewrite rules. Context-free rewrite rules are the symmetric case of inverse context-free rules, *i.e.* rules with left-hand-side of the form $f(x)$. Some additional restrictions are assumed for this result, they cannot be relaxed as shown by the counter examples in Proposition 7–10.

*Related works.* Whether the rewrite closure of regular ranked trees languages is regular too is a problem that has been addressed in [19,7,9,15,21,20,6]. An important breakthrough of the proof in [15] (against former results) is that it works for TRS which are not left-linear. H. Ohsaki introduces equational tree automata for associative and commutative theories in [16] and study their closure properties for Boolean operations. T. Touili has studied the regular model checking problem for HA [22]. She shows how to compute the image of a HA language in one step of rewriting by a right-linear rewrite system. She also gives a procedure to compute an over-approximation of the rewrite closure of a HA. We rather compute exactly this closure for a class of non-linear rewrite systems.

Our first main result (Theorem 1) can be viewed as a non trivial generalization of both [15] and [22], with proof techniques extending both former constructions.

C. Löding and A. Spelten [11] compute exact rewrite closure of HA for extensions of ground term rewriting and prefix word rewriting. These results cannot be compared to ours since in our case variables (that can be substituted by arbitrarily large hedges) allow non local hedge transformations.

There exists other rewriting notions like the top-down XML transformations [12] or the relabeling transducers of [18] but they do not cover our notion since either they use specific hedge traversal strategies or they are structure-preserving.

*Layout of the paper.* In Section 2 we introduce terms, hedges and the related rewriting concepts. In particular we define hedge rewriting systems (HRS) and context-free rewrite rules. In Section 3 we recall the hedge-automata classes HA and CF-HA that we shall investigate. In Section 4 we show that the class of HA languages, (*i.e.* recognized by HA) is preserved by rewrite closure for rewriting systems containing rules that are inverse context-free. In Section 5 we show that a class of context-free hedge rewrite systems preserves CF-HA languages. In both Sections 4 and 5, we also exhibit some counter-examples obtained when trying to relax the conditions on rules.

## 2 Hedge Rewriting

We consider a finite alphabet $\Sigma$ and an infinite set of variables $\mathcal{X}$. The set of *terms* over $\Sigma$ and $\mathcal{X}$ is $\mathcal{T}(\Sigma,\mathcal{X}) := \mathcal{X} \cup \{f(h) \mid f \in \Sigma, h \in \mathcal{H}(\Sigma,\mathcal{X})\}$ and the set $\mathcal{H}(\Sigma,\mathcal{X})$ of *hedges* over $\Sigma$ and $\mathcal{X}$ is the set of finite (possibly empty) sequences of terms of $\mathcal{T}(\Sigma,\mathcal{X})$. When $h$ is empty, $f()$ will be simply written $f$. We will sometimes consider a term as a hedge of length one, *i.e.* consider that $\mathcal{T}(\Sigma,\mathcal{X}) \subset \mathcal{H}(\Sigma,\mathcal{X})$. The sets of ground terms (terms without variables) and ground hedges are respectively denoted $\mathcal{T}(\Sigma)$ and $\mathcal{H}(\Sigma)$. A hedge $h \in \mathcal{H}(\Sigma,\mathcal{X})$ is called *linear* if every variable of $\mathcal{X}$ occurs at most once in $h$.

The set of variables occurring in a term $t \in \mathcal{T}(\Sigma,\mathcal{X})$ is denoted $var(t)$. A *substitution* $\sigma$ is a mapping from $\mathcal{X}$ to $\mathcal{H}(\Sigma,\mathcal{X})$ of finite domain. The application of a substitution $\sigma$ to a hedge $h \in \mathcal{H}(\Sigma,\mathcal{X})$, denoted $h\sigma$, is the homomorphic extension of $\sigma$ to $\mathcal{H}(\Sigma,\mathcal{X})$, defined, for $t_1,\ldots,t_n \in \mathcal{T}(\Sigma,\mathcal{X})$, with $n \geq 0$, by $(t_1 \ldots t_n)\sigma := t_1\sigma \ldots t_n\sigma$ and $f(h)\sigma := f(h\sigma)$.

The set of *positions* $\mathcal{P}os(t)$ of a term $t \in \mathcal{T}(\Sigma,\mathcal{X})$ is a set of sequences of positive integers. The empty sequence, denoted $\varepsilon$, is the root position of a term. The subterm of $t$ at position $p$, denoted $t|_p$, is defined by $f(t_1 \ldots t_n)|_{ip} := t_i|_p$ if $i \leq n$ and, $f(h)|_\varepsilon := f(h)$. The replacement in $t \in \mathcal{T}(\Sigma,\mathcal{X})$ of the subterm at position $p$ by $t' \in \mathcal{T}(\Sigma,\mathcal{X})$ is denoted $t[t']_p$. The *depth* of a term is the maximal length of one of its positions.

A *context* is a linear hedge of $\mathcal{H}(\Sigma,\{x\})$, denoted $C[x]$. The application of a context $C[x]$ to a hedge $h$ is defined by $C[h] := C\{x \mapsto h\}$.

A hedge rewriting system (HRS) is a set of rewrite rules of the form $\ell \to r$ where $\ell \in \mathcal{T}(\Sigma,\mathcal{X}) \setminus \mathcal{X}$ and $r \in \mathcal{T}(\Sigma,\mathcal{X})$ ($\ell$ and $r$ are respectively called *lhs* and *rhs* of the rule). The rewrite relation $\xrightarrow[\mathcal{R}]{}$ of an HRS $\mathcal{R}$ is the binary relation on $\mathcal{H}(\Sigma, \mathcal{X})$ defined by $h \xrightarrow[\mathcal{R}]{} h'$ iff $h = (t_1 \ldots t_n)$, there exists $i \leq n$, a position $p \in \mathcal{P}os(t_i)$, a rule $\ell \to r \in \mathcal{R}$ and a substitution $\sigma$ such that $t_i|_p = \ell\sigma$ and $h' = t_1 \ldots t_{i-1} t_i[r\sigma] t_{i+1} \ldots t_n$. The reflexive and transitive closure of $\xrightarrow[\mathcal{R}]{}$ is denoted $\xrightarrow[\mathcal{R}]{*}$.

*Example 1.* With $\mathcal{R} = \{g(x) \to x\}$, $\xrightarrow[\mathcal{R}]{}$ associates to a term $g(h)$ the hedge $h$ of its arguments. With $\mathcal{R} = \{g(x) \to g(axb)\}$, $g(c) \xrightarrow[\mathcal{R}]{*} g(a^n c b^n)$ for every $n \geq 0$.

Given a set of terms $L \subseteq \mathcal{T}(\Sigma)$ and an HRS $\mathcal{R}$, we note $\mathcal{R}^*(L)$ the set $\{t \in \mathcal{T}(\Sigma) \mid \exists s \in L, s \xrightarrow[\mathcal{R}]{*} t\}$. We restrict to terms (instead of hedges) because we are mainly interested in term languages below.

A rewrite rule $\ell \to r$ is called *left-linear* (resp. *right-linear*, *linear*) if $\ell$ (resp. $r$, both) is linear, *left-ground* (resp. *right-ground*) if $\ell \in \mathcal{T}(\Sigma)$ (resp. $r \in \mathcal{T}(\Sigma)$), *collapsing* if $r \in var(\ell)$, it is called *context-free* if $\ell = f(x)$ with $x \in \mathcal{X}$ (it is not required that $x \in var(r)$ however) and *inverse context-free* if $r \to \ell$ is context-free, *prefix* (resp. *postfix*) if $r = g(t_0 \ldots t_n\, x)$ (resp. $r = g(x\, t_0 \ldots t_n)$) with $x \in var(\ell)$ and no variable of $\ell$ occurs in the terms $t_0, \ldots, t_n$. A rewrite system is said to have one of the above properties if all its rules have this property.

*Example 2.* We give a few applications of our rewrite rules in the vein of [22]. A context-free rule $\mathsf{doc}(x) \to \mathsf{doc}(\mathsf{a} x \bar{\mathsf{a}})$ can be employed to introduce tags in an XML document. An inverse context-free rule can be used to eliminate comments $\mathsf{doc}(x\ \mathsf{comment}\ y\ \overline{\mathsf{comment}}) \to \mathsf{doc}(x)$. Non left-linear inverse context-free rules are quite useful for processing list of items as in: $\mathsf{doc}(\mathsf{todo}\ x\ \overline{\mathsf{todo}}\ y\ \mathsf{done}\ x\ \overline{\mathsf{done}}) \to \mathsf{doc}(y)$.

Note that hedge rewriting cannot be reduced to term rewriting through encoding of unranked trees into ranked trees like the First-Child/Next-Sibling encoding, or the encoding used in stepwise automata (see details in the companion report [10]).

## 3 Hedge-Automata, Context-Free Hedge-Automata

We recall now the definition of hedge-automata [13] (denoted HA) and the less known class of context-free hedge automata (denoted CF-HA) introduced in [17] and where they are shown to recognize the closure of regular (ranked) tree languages modulo associativity.

A *hedge automaton* (resp. *context-free hedge automaton*) is a tuple $\mathcal{A} = (Q, \Sigma, Q^{\mathsf{f}}, \Delta)$ where $Q$ is a finite set of states, $\Sigma$ is an unranked alphabet, $Q^{\mathsf{f}} \subseteq Q$ is a set of final states, and $\Delta$ is a set of transitions of the form $f(L) \to q$ where $f \in \Sigma, q \in Q$ and $L \subseteq Q^*$ is a regular word language (resp. a context-free word language). When $\Sigma$ is clear from the context it is omitted in the tuple specifying $\mathcal{A}$.

We define the move relation between ground hedges in $\mathcal{T}(\Sigma \cup Q)$ as follows: for every terms $t, t'$ we have $t \xrightarrow[\mathcal{A}]{} t'$ if there exists a context $C[x]$ and a transition $f(L) \to q$ in $\Delta$ such that $t = C[f(q_1 \ldots q_n)]$, $q_1 \ldots q_n \in L$ and $t' = C[q]$. The relation $\xrightarrow[\mathcal{A}]{*}$ is the transitive closure of $\xrightarrow[\mathcal{A}]{}$. Following [22], we extend $\xrightarrow[\mathcal{A}]{}$ to terms of $\mathcal{T}(\Sigma \cup 2^{Q^*})$ as follows: $C\big[f(L_1 \ldots L_n)\big] \xrightarrow[\mathcal{A}]{} C[q]$ if there exists a rule $f(L) \to q$ in $\mathcal{A}$ such that $L_1 \ldots L_n \subseteq L$ (in this definition, a lone state $q$ is considered as a singleton set $\{q\}$).

The language denoted by $L(\mathcal{A}, q)$ is the set of ground terms $t \in \mathcal{T}(\Sigma)$ such that $t \xrightarrow[\mathcal{A}]{*} q$. A term is accepted by $\mathcal{A}$ if there is $q \in Q^{\mathsf{f}}$ such that $t \in L(\mathcal{A}, q)$. The language denoted by $L(\mathcal{A})$ is the set of terms accepted by $\mathcal{A}$.

It is know that for both classes of automata [13,17] membership and emptiness problems are decidable. Moreover HA are closed under Boolean operations.

We call a HA or CF-HA $\mathcal{A} = (Q, Q^{\mathsf{f}}, \Delta)$ *normalized* if for every $f \in \Sigma$ and every $q \in Q$, there is at most one transition rule $f(L_{f,q}) \to q$ in $\Delta$. Every HA (resp. CF-HA) can be transformed into a normalized HA (resp. CF-HA) in polynomial time by replacing every two rules $f(L_1) \to q$ and $f(L_2) \to q$ by $f(L_1 \cup L_2) \to q$.

A HA $\mathcal{A} = (Q, Q^{\mathsf{f}}, \Delta)$ is called *deterministic* iff for all two transitions rules $f(L_1) \to q_1$ and $f(L_2) \to q_2$ in $\Delta$, either $L_1 \cap L_2 = \emptyset$ or $q_1 = q_2$. It is called *complete* if for all $f \in \Sigma$ and and $w \in Q^*$, there exists at least one rule $f(L) \to q \in \Delta$ such that $w \in L$. When $\mathcal{A}$ is deterministic (resp. complete), for all $t \in \mathcal{T}(\Sigma)$, there exists at most (resp. at least) one state $q \in Q$ such that $t \in L(\mathcal{A}, q)$.

Every HA can be completed by adding a sink state (and using the closure properties of regular languages). A determinization procedure (with a subset construction) which preserves completeness is described in Section 4.1 (see also [4]).

### 3.1 Epsilon- and Collapsing Transitions

We can extend HA and CF-HA with *ε-transitions* of the form $q \to q'$, where $q$ and $q'$ are states, without augmenting the respective expressiveness of these classes. We also consider the extensions of HA (resp. CF-HA), with *collapsing transitions* of the form $L \to q$ where $L$ is a regular (resp. CF) language and $q$ is a state. The move relation for the extended set of transitions is defined as for HA and CF-HA for standard transition and by $C[q_1 \ldots q_k] \xrightarrow[\mathcal{A}]{} C[q]$ if $L \to q$ is a collapsing transition of $\mathcal{A}$ and $q_1 \ldots q_k \in L$. Note that the collapsing transition $L \to q$ is never applied at the root position (i.e. the above context $C$ cannot be a variable) because HA and CF-HA are limited to the recognition of terms only (and not hedges).

Unlike ε-transitions, collapsing transitions strictly extend HA in expressiveness. However, we show that they can be eliminated for CF-HA.

**Proposition 1.** *For every extended HA or CF-HA with collapsing transitions $\mathcal{A}$, there exists a CF-HA $\mathcal{A}'$ (without collapsing transitions) such that $L(\mathcal{A}') = L(\mathcal{A})$.*

*Proof.* Assume that $L \to q$ is a collapsing transition of $\mathcal{A}$. Then we get a CF-HA $\mathcal{A}'$ such that $L(\mathcal{A}') = L(\mathcal{A})$ by replacing every transition $f(L_1) \to q_1$ by the transition $f(L_2) \to q_1$ where $L_2$ is the context-free word language generated by the grammar $G_2$ as follows. We consider a context-free grammar $G$ for $L$ (resp.

$G_1$ for $L_1$) with axiom $X$ (resp. $X_1$). The axiom of $G_2$ is $X_1$ and the set of productions in $G_2$ contains *i*) $G[q \leftarrow X_q] \cup G_1[q \leftarrow X_q]$ *i.e.* the terminal $q$ is replaced by a non terminal $X_q$ and *ii*) we add to these rules the production: $X_q := q \mid X$. We can iterate this construction to eliminate all collapsing transitions. □

**Proposition 2.** *There exists an extended HA with collapsing transitions whose language is not a HA language.*

*Proof.* Consider the extended HA $\mathcal{A} = (\{q, q_a, q_b, q_f\}, \{g, a, b, c\}, \{q_f\}, \Delta)$ where

$$\Delta = \{c \to q,\ a \to q_a,\ b \to q_b,\ g(q) \to q_f,\ q_a q q_b \to q\}$$

Its recognized language is $\{g(a^n c b^n) \mid n \geq 0\}$ and this is not a HA language. □

### 3.2 Decision Problems

The problem of *ground reachability* and *ground joinability* are to decide that, given two ground terms $s, t \in \mathcal{T}(\Sigma)$ and a HRS $\mathcal{R}$, whether, $s \xrightarrow[\mathcal{R}]{*} t$, respectively, $s \xrightarrow[\mathcal{R}]{*} \circ \xleftarrow[\mathcal{R}]{*} t$.

*Regular hedge model checking* is the problem to decide, given two HA languages $L_{\mathsf{init}}$ and $L_{\mathsf{err}}$ and a HRS $\mathcal{R}$ whether $\mathcal{R}^*(L_{\mathsf{init}})$ contains a term of $L_{\mathsf{err}}$. Ground reachability is reducible to regular hedge model-checking. Indeed, given $s$, $t$ and $\mathcal{R}$, $s \xrightarrow[\mathcal{R}]{*} t$ iff $\mathcal{R}^*(\{s\}) \cap \{t\} \neq \emptyset$. Note also that if ground-reachability (hence regular hedge model-checking) is undecidable for a class of HRS, then $\mathcal{R}^*(L)$ is not recursive in general when $\mathcal{R}$ is in this class and $L$ is a HA or CF-HA. Indeed, by definition $s \xrightarrow[\mathcal{R}]{*} t$ iff $t \in \mathcal{R}^*(\{s\})$.

## 4 Closure of Regular Hedge Automata Languages

In this section, we prove one result of preservation of HA language for a class of HRS, and give several counter example showing that the restrictions defining this class of HRS are necessary.

### 4.1 Inverse Context-Free Rewrite Rules

**Theorem 1.** *The closure $\mathcal{R}^*(L)$ of a HA language $L \subseteq \mathcal{T}(\Sigma)$ under rewriting by an inverse context-free HRS $\mathcal{R}$ is a HA language.*

*Proof.* Let $\mathcal{A} = (Q, Q^{\mathsf{f}}, \Delta)$ be a complete and normalized HA recognizing $L$. We shall construct below a finite sequence of HA $(\mathcal{A}_i)_{0 \leq i \leq h}$ whose last element recognizes $\mathcal{R}^*(L)$. Our construction uses elements of [15] and [22], but it is not a simple combination of both. Indeed, on one side we generalize [22] to an unbounded number of rewriting steps, and on the other side we generalize [15] to unranked tree languages. Both generalizations are non-trivial and require new constructions and new conditions.

For each $f \in \Sigma$, $q \in Q$, we note $L_{f,q}$ the language in the transition (assumed unique) $f(L_{f,q}) \to q \in \Delta$. We construct first from $\mathcal{A}$ a deterministic, complete and normalized HA $\mathcal{A}_\mathsf{d} = (Q_\mathsf{d}, Q^\mathsf{f}_\mathsf{d}, \Delta_\mathsf{d})$ recognizing $L$. The HA $\mathcal{A}_\mathsf{d}$ is obtained by a subset construction, see *e.g.* [4], with $Q_\mathsf{d} := 2^Q$, $Q^\mathsf{f}_\mathsf{d} := \{s \in Q_\mathsf{d} \mid s \cap Q^\mathsf{f} \neq \emptyset\}$ and $\Delta_\mathsf{d} := \big\{f(L_{f,s}) \to s \bigm| f \in \Sigma, s \subseteq Q\big\}$ where $L_{f,s} := \big(\bigcap_{q \in s} S_{f,q}\big) \setminus \big(\bigcup_{q \notin s} S_{f,q}\big)$ and $S_{f,q} = \big\{s_1 \ldots s_n \in Q^*_\mathsf{d} \bigm| \exists q_1 \in s_1, \ldots, q_n \in s_n, q_1 \ldots q_n \in L_{f,q}\big\}$[1].

Next, following the approach of [22], we define first the set of languages of $Q^*_\mathsf{d}$ that will be used in the transitions of the $\mathcal{A}_i$'s constructed below. However, we must consider here a bigger set than [22] in order to deal with non linear variables in *lhs* of rules. Let $\mathcal{L}$ be the smallest set of subsets of $Q^*_\mathsf{d}$ such that

*i*. all $L_{f,s}$ (for $f \in \Sigma$ and $s \in Q_\mathsf{d}$) and $Q^*_\mathsf{d}$ are in $\mathcal{L}$,
*ii*. if $L \in \mathcal{L}$ and $u, v \in Q^*_\mathsf{d}$, then $u^{-1}\, L\, v^{-1} \in \mathcal{L}$, where

$$u^{-1}\, L\, v^{-1} := \{w \in Q^*_\mathsf{d} \mid uwv \in L\},$$

*iii*. if $L_1, L_2 \in \mathcal{L}$ then $L_1 \cap L_2 \in \mathcal{L}$,
*iv*. if $L_1, L_2 \in \mathcal{L}$ then $L_1 \setminus L_2 \in \mathcal{L}$.

Note that the condition $Q^*_\mathsf{d} \in \mathcal{L}$ in *i* together with *iii* and *iv* imply that $\mathcal{L}$ is also closed under union (if $L_1, L_2 \in \mathcal{L}$ then $L_1 \cup L_2 \in \mathcal{L}$), by De Morgan's Law.

Let us show that $\mathcal{L}$ is finite and that all its members are regular languages. First, let us note that $\mathcal{L}_1$, the smallest set satisfying *i* and *ii* above, is a finite set of regular languages of $Q^*_\mathsf{d}$, since every $L_{f,q}$ is regular by hypothesis. The closure $\mathcal{L}_2$ of $\mathcal{L}_1$ under *iii* and then *iv* is also a finite set of regular languages. The following lemma shows that $\mathcal{L}_2$ fulfills *ii*, *i.e.* that $\mathcal{L}_2 = \mathcal{L}$.

**Lemma 1.** *For all* $L_1, L_2 \subseteq Q^*_\mathsf{d}$, $u_1, u_2, v_1, v_2, u, v \in Q^*$, *we have*
$u^{-1}(u_1^{-1}\, L_1\, v_1^{-1} \cap u_2^{-1}\, L_2\, v_2^{-1})\, v^{-1} = (u_1 u)^{-1}\, L_1\, (v v_1)^{-1} \cap (u_2 u)^{-1}\, L_2\, (v v_2)^{-1}$,
$u^{-1}(u_1^{-1}\, L_1\, v_1^{-1} \setminus u_2^{-1}\, L_2\, v_2^{-1})\, v^{-1} = (u_1 u)^{-1}\, L_1\, (v v_1)^{-1} \setminus (u_2 u)^{-1}\, L_2\, (v v_2)^{-1}$.

*Proof.* The set in the left-hand-side of the first identity in Lemma 1 is $A = \big\{\ell \mid u\ell v \in \{\ell' \mid u_1 \ell' v_1 \in L_1 \text{ and } u_2 \ell' v_2 \in L_2\}\big\}$, and the set in its right hand side is $B = \big\{\ell \mid u_1 u \ell v v_1 \in L_1 \text{ and } u_2 u \ell v v_2 \in L_2\big\}$. If $\ell \in A$, then $u_1 u \ell v v_1 \in L_1$ and $u_2 u \ell v v_2 \in L_2$, hence $\ell \in B$. Conversely, if $\ell \in B$, then $u\ell v \in u_1^{-1}\, L_1\, v_1^{-1} \cap u_2^{-1}\, L_2\, v_2^{-1}$, hence $\ell \in A$. The proof is very similar for the identity with the complementation. □

Let us now construct the HA $\mathcal{A}_0, \ldots, \mathcal{A}_h$ as announced. The set of states and final states of each of these HA are respectively $Q_\mathsf{d}$ and $Q^\mathsf{f}_\mathsf{d}$. We give below an iterative construction of the respective transition sets $\Delta_i$, $0 \le i \le h$.

Let $\Delta_0 = \Delta_\mathsf{d}$. Assume that $\Delta_i$ has been constructed and contains one transition $f(L^i_{f,s}) \to s$ for every $f \in \Sigma$ and $s \in Q_\mathsf{d}$; $\Delta_{i+1}$ is obtained from $\Delta_i$ as follows: choose (non deterministically) an inverse context-free rewrite rule $\ell \to g(x) \in \mathcal{R}$, and a substitution $\tau : var(\ell) \cup \{x\} \to \{L' \in \mathcal{L} \mid \forall s_1 \ldots s_k \in L', \forall j \le k, L(\mathcal{A}_i, s_j) \neq \emptyset\}$, such that $\ell\tau \xrightarrow[\mathcal{A}_i]{*} s' \in Q_\mathsf{d}$. Let $L' = x\tau$ (note that if the variable $x$ does not occur in $\ell$, then $L'$ is an arbitrary language of $\mathcal{L}$ of sequences of states reachable by $\mathcal{A}_i$); $\Delta_{i+1}$ is obtained as follows: for each $s \in Q_\mathsf{d}$,

[1] Note that $S_{f,q}$ and $L_{f,s}$ are indeed regular languages, see [4].

1. replace the rule $g(L^i_{g,s}) \to s$ by $g(L^i_{g,s} \cap L') \to s \cup s'$ and $g(L^i_{g,s} \setminus L') \to s$
2. after this operation, normalize the set of transition rules with the operation described in Section 3 (page 161). (Note that if $s' \subseteq s$ then the normalization merges the 2 rules and regenerate $g(L^i_{g,s}) \to s$.)

The idea behind this construction is that if $s'$ is reachable from a lhs $\ell\tau$ of rewrite rule, then the states in $s'$ must also be reachable from the corresponding rhs $g(x\tau)$. Note that for all transitions $g(L) \to s$ produced by the algorithm, we have $L \in \mathcal{L}$ (even after normalization), according to the closure properties of this set. Since $\mathcal{L}$ and the set of states $s$ is finite (no new state is added) this shows that the construction terminates say with a HA $\mathcal{A}_h$ that will be denoted $\mathcal{A}^*$.

We can also show the following invariant: *every $\mathcal{A}_i$ constructed in the algorithm is deterministic, complete and normalized.* Indeed, assume that $\mathcal{A}_i$ has these properties. If $s' \subseteq s$ no transition is added and the invariant is trivially preserved; hence we can assume now $s' \nsubseteq s$. If another rule $g(L^i_{g,s\cup s'}) \to s \cup s'$ was in $\Delta_i$ it is merged with $g(L^i_{g,s} \cap L') \to s \cup s'$ by normalization producing the rule $g((L^i_{g,s} \cap L') \cup L^i_{g,s\cup s'}) \to s \cup s'$. Hence there is at most one $L^{i+1}_{g,s\cup s'} = (L^i_{g,s} \cap L') \cup L^i_{g,s\cup s'}$ such that $g(L^{i+1}_{g,s\cup s'}) \to s \cup s' \in \Delta_{i+1}$. Note also that there is at most one $L^{i+1}_{g,s} = L^i_{g,s} \setminus L'$ such that $g(L^{i+1}_{g,s}) \to s \in \Delta_{i+1}$. It is easy to see (from the fact that $\mathcal{A}_i$ is deterministic and normalized) that $L^{i+1}_{g,s\cup s'}$, $L^{i+1}_{g,s}$, and $L^{i+1}_{g,s''}$, for all $s'' \notin \{s, s'\}$, are pairwise disjoint, hence $\mathcal{A}_{i+1}$ is deterministic. From the facts that $\mathcal{A}_i$ is complete and that $L^i_{g,s} \cap L'$ and $L^i_{g,s} \setminus L'$ form a partition of $L^i_{g,s}$, we deduce that $\mathcal{A}_{i+1}$ is also complete.

We show in [10] that $L(\mathcal{A}^*) = \mathcal{R}^*(L)$. Let us simply sketch the proof here for space reasons.

The proof of the direction $L(\mathcal{A}^*) \subseteq \mathcal{R}^*(L)$ relies on the following lifting lemma.

**Lemma 2.** *For all $i \geq 0$, $t \in \mathcal{T}(\Sigma, \mathcal{X})$, $\sigma : var(t) \to \mathcal{H}(\Sigma)$ and $\theta : var(t) \to Q^*_{\mathsf{d}}$ such that for all $x \in var(t)$, $x\sigma$ and $x\theta$ have the same length, if $t\theta \xrightarrow[\mathcal{A}_i]{*} s_0 \in Q_{\mathsf{d}}$, and for all $x \in var(t)$, all components $(x\theta)|_j$ of $x\theta$ (state of $Q_{\mathsf{d}}$) and $q \in (x\theta)|_j$, there exists $u \in L(\mathcal{A}_i, q)$ such that $u \xrightarrow[\mathcal{R}]{*} (x\sigma)|_j$, then for all $q' \in s_0$, there exists $v \in L(\mathcal{A}_i, q')$ s. t. $v \xrightarrow[\mathcal{R}]{*} t\sigma$.*

Lemma 2 is proved (see [10]) by induction on $i$, and, for the induction step, by a second induction on the number of applications of a rule of $\Delta_{i+1} \setminus \Delta_i$ in the reduction $t\theta \xrightarrow[\mathcal{A}_{i+1}]{*} s_0$. Intuitively, every such application corresponds to a rewrite step in $v \xrightarrow[\mathcal{R}]{*} t\sigma$. Now, for the particular case of Lemma 2 where $t \in \mathcal{T}(\Sigma)$, we have that if $t \xrightarrow[\mathcal{A}_i]{*} s_0$, for some $i$ and $s_0 \in Q^{\mathsf{f}}_{\mathsf{d}}$, for all $q^{\mathsf{f}} \in s_0$, where $q^{\mathsf{f}}$ is a final state of $\mathcal{A}$, there exists $u \in L(\mathcal{A}, q^{\mathsf{f}}) \subseteq L(\mathcal{A})$ such that $u \xrightarrow[\mathcal{R}]{*} t$. This terminates the proof of the direction $L(\mathcal{A}^*) \subseteq \mathcal{R}^*(L)$.

For the direction $L(\mathcal{A}^*) \supseteq \mathcal{R}^*(L)$, assume that $t \in L(\mathcal{A})$ and that $t \xrightarrow[\mathcal{R}]{*} t'$. We show in [10] that $t' \in L(\mathcal{A}_i)$ for some $i$ by induction on the length of the rewrite sequence. □

**Corollary 1.** *Ground reachability, ground joinability and regular hedge model-checking are decidable for inverse context-free HRS.*

We present in the next subsections (4.2–4.4) some counter examples showing that relaxing the assumption on $\mathcal{R}$ in Theorem 1 invalidate the result.

### 4.2 Collapsing Rewrite Rules

Collapsing rules preserve regularity of term languages [15] when the function symbols are ranked. Indeed, in this case, if $\mathcal{R}$ is left-linear and collapsing, a tree automaton (TA) recognizing $L$ can be completed into a TA recognizing $\mathcal{R}^*(L)$ just by the iterated addition of $\varepsilon$-transitions of the form $x\tau \to q$ when there is $\ell \to x \in \mathcal{R}$ and a substitution $\tau : var(\ell) \to Q$ such that $\ell\tau \xrightarrow[\mathcal{A}]{*} q$. When $\mathcal{R}$ is just collapsing (not left-linear), the construction requires determinism and hence is more complicated but the idea is the same [15].

In the case of unranked terms and HA, if we want to follow the principles of the construction of Section 4.1, we need to add *collapsing transitions* and not just $\varepsilon$-transitions. But the addition of collapsing transitions does not preserve HA languages (Proposition 1). The following proposition shows that the above construction is actually not possible for collapsing rewrite rules.

**Proposition 3.** *$\mathcal{R}^*(L)$ is not a HA language in general when $L$ is a HA language and $\mathcal{R}$ is a linear collapsing HRS.*

*Proof.* We use the principle of the construction in the proof of Proposition 1. Let $\Sigma = \{f, g, a, b, c\}$, let $L$ be the language of the HA

$$\mathcal{A} = \big(\{q, q_a, q_b, q_f\}, \{q_f\}, \{c \to q, a \to q_a, b \to q_b, g(q_a q q_b) \to q, f(q) \to q_f\}\big)$$

and let $\mathcal{R} = \{g(x) \to x\}$. Assume that $\mathcal{R}^*(L)$ is a HA language. Its intersection with the HA language $\{f(a^*cb^*)\}$ is $\{f(a^n cb^n) \mid n \geq 0\}$. It is not a HA language. This contradicts the fact that HA languages are closed under intersection. □

Note that the completion of the above $\mathcal{A}$, following the procedure in the proof of Theorem 1, would add the collapsing transition $q_a q q_b \to q$.

### 4.3 Flat Linear Rewrite Rules

In the case of ranked terms, it is known [15] that regularity of tree languages is preserved under rewriting with systems with right-linear rules of the form $\ell \to f(u_1, \ldots, u_n)$ where $f$ has arity $n$ and each $u_i$ $(i \leq n)$ is either a ground term or a variable of $var(\ell)$. We call such a rule *flat* if its *lhs* and *rhs* both have depth one. Note that this class of TRS is not captured by the HRS of Theorem 1 (when restricted to ranked terms). The above regularity preservation result is no longer true for unranked terms.

**Proposition 4.** *$\mathcal{R}^*(L)$ is not a HA language in general when $L$ is a a HA language and $\mathcal{R}$ is a context-free, linear and flat HRS. Moreover, it can be assumed that all the rules of $\mathcal{R}$ are prefix or postfix.*

*Proof.* Let us consider the context-free HRS $\mathcal{R} = \{g(x) \to g(axb)\}$ of Example 1, and the HA language $L = \{g(c)\}$. The language $\mathcal{R}^*(L) = \{g(a^n cb^n) \mid n \geq 0\}$ is not HA. We can transform the above $\mathcal{R}$ into $\mathcal{R}' = \{g(x) \to g'(ax), g'(y) \to g(yb)\}$ whose rules are prefix or postfix (and linear) and which is such that $\mathcal{R}'^*(L) \cap \mathcal{T}(\{g, a, b\}) = \mathcal{R}^*(L)$. □

Note that the language in the above proof is recognized by a CF-HA. We shall show below (Theorem 2 in Section 5) that context-free HRS like the $\mathcal{R}$ above preserve CF-HA languages.

We show now the stronger result that the closure of a HA language under rewriting with a flat HRS, even linear, is neither HA, nor CF-HA and actually not even recursive.

**Proposition 5.** *$\mathcal{R}^*(L)$ is not recursive in general when $L$ is a HA language and $\mathcal{R}$ is a linear and flat HRS whose rules contain at most two variables.*

*Proof.* We reduce the blank accepting problem for TM to ground reachability for an HRS. Let $\mathcal{M}$ be a TM with a tape alphabet $\Gamma$ and a state set $S$ and let $\Sigma = \Gamma \cup S \cup \{g\}$. A configuration of $\mathcal{M}$ is represented by a term $g(w)$ where $w$ is a word of $\Gamma^* S \Gamma^*$ (the position of the state symbol indicates the position of the head of $\mathcal{M}$ and the rest represents the contents of the tape). We assume, *wlog* unique blank initial and final configurations, respectively $c_\mathsf{i}$ and $c_\mathsf{f}$. We consider a HRS $\mathcal{R}$ containing one rule for each transition of $\mathcal{M}$. For instance, $\mathcal{R}$ contains a rule $f(xasy) \to f(xs'a'y)$ corresponding to a transition $s, a \to L, s', a'$ (with $s, s' \in S$ and $a, a' \in \Gamma$) and $f(xasby) \to f(xa'bs'y)$ to the transition $s, a \to R, s'$. The blank tape is accepted by $\mathcal{M}$ iff $c_\mathsf{i} \xrightarrow[\mathcal{R}]{*} c_\mathsf{f}$. □

As a consequence, regular hedge model checking is undecidable for the HRS of Proposition 5, according to the remarks in Section 3.2.

### 4.4 Rewrite Rules with Flat and One-Variable or Ground Right-Hand-Sides

If we relax the inverse context-free condition, with only one variable allowed in the *rhs* of rules, but possibly with two occurrences, both at depth 1, then the result of Theorem 1, again, is not valid anymore.

**Proposition 6.** *$\mathcal{R}^*(L)$ is not recursive in general when $L$ is a HA language and $\mathcal{R}$ is a HRS whose rhs of rules are ground or of the form $d(xx)$.*

We reduce in [10], the blank accepting problem for a TM to ground reachability for a HRS with right-ground (but not left-linear) rules and a rule $d(xx) \to d'(xx)$.

## 5 Closure of Context-Free Hedge Automata Languages

It has been observed [8] that in several cases, one class of word rewrite system preserves regularity and its symmetric class preserves context-free languages. In this section, we prove a similar result by showing that a restricted case of context-free HRS, *i.e.* of the symmetric version of the systems considered in Section 4, preserve CF-HA languages. We give next some counterexamples showing that the restrictions are necessary for this result.

### 5.1 Linear Restricted Context-Free Rewrite Rules

We call a HRS $\mathcal{R}$ *restricted context-free* if it is context-free, and moreover, for all rule $f(x) \to r \in \mathcal{R}$, $x$ can occurs in $r$ only at depth at most 1. Note that this definition includes the case of collapsing rules $f(x) \to x$.

**Theorem 2.** *The closure $\mathcal{R}^*(L)$ of a CF-HA language $L$ under rewriting by a linear restricted context-free HRS $\mathcal{R}$ is a CF-HA language.*

*Proof.* Let $\mathcal{A}_L = (Q_L, Q_L^{\mathsf{f}}, \Delta_L)$ be a normalized CF-HA recognizing $L$. We shall construct an extended CF-HA $\mathcal{A}'$ with collapsing transitions (see Section 3.1 for the definition) recognizing $\mathcal{R}^*(L)$. The result follows then from Proposition 1.

First, let us construct for each rule $f(x) \to g(r_1 \dots r_n) \in \mathcal{R}$ and every subterm $r \neq x$ amongst $r_1, \dots, r_n$ (let us denote $rhs(\mathcal{R})$ the set of such subterms) a CF-HA (with collapsing transitions) $\mathcal{A}_r = (Q_r, \emptyset, \Delta_r)$ characterizing the set of ground instances of $r$. We have in $\mathcal{A}_r$ one state $q_u \in Q_r$ for each non-variable subterm $u$ of $r$, and a universal state $q_\forall \in Q_r$. Below, for every subterm $u$ of $r$, we shall write $q_u$ to denote either the state $q_u$ if $u$ is not a variable or $q_\forall$ otherwise. The set of final states of $\mathcal{A}_r$ is left unspecified. It is indeed not relevant to our purpose since $\mathcal{A}_r$ is only used as a part of the CF-HA $\mathcal{A}'$ constructed below. The transition set $\Delta_r$ contains one rule $f(q_{u_1} \dots q_{u_n}) \to q_{f(u_1 \dots u_n)}$ for each subterm $f(u_1 \dots u_n)$ of $r$ (as specified above, $q_i$ is $q_\forall$ if $u_i$ is a variable and $q_i$ is a state $q_{u_i}$ otherwise). It contains moreover one collapsing transition $q_\forall^* \to q_\forall$ and one transition rule $f(q_\forall^*) \to q_\forall$ for each $f \in \Sigma$. The states sets $Q_r$ and $Q_L$ are assumed pairwise disjoint. Let $\mathcal{A} := (Q, Q_L^{\mathsf{f}}, \Delta)$ with

$$Q := Q_L \uplus \biguplus_{r \in rhs(\mathcal{R})} Q_r \text{ and } \Delta := \Delta_L \uplus \biguplus_{r \in rhs(\mathcal{R})} \Delta_r.$$

For each $f \in \Sigma$, $q \in Q$, let $L_{f,q}$ be the context-free language in the transition (assumed unique) $f(L_{f,q}) \to q \in \Delta$, and let $\mathcal{G}_{f,q} = (Q, N_{f,q}, I_{f,q}, P_{f,q})$ be a CF grammar generating $L_{f,q}$, with alphabet (set of terminal symbols) $Q$, set of non terminal symbols $N_{f,q}$, axiom $I_{f,q} \in N_{f,q}$, and set of production rules $P_{f,q}$. The sets of non-terminals $N_{f,q}$ are assumed pairwise disjoint.

We complete the grammars $\mathcal{G}_{f,q}$ with new non-terminals $I'_{f,q}$ and some sets $P'_{f,q}$ of new production rules containing:

*i.* $I'_{f,q} := I_{f,q}$ for all $f \in \Sigma$, $q \in Q$,
*ii.* $I'_{g,q} := q_{r_1} \dots q_{r_n} I'_{f,q}\, q_{s_1} \dots q_{s_m}$ for each rule $f(x) \to g(r_1 \dots r_n x s_1 \dots s_m) \in \mathcal{R}$, with $n, m \geq 0$, and $x \notin var(r_1, \dots, r_n, s_1, \dots, s_m)$, and
*iii.* $I'_{g,q} := q_{r_1} \dots q_{r_n}$ (with $n > 0$), or $I'_{g,q} := \varepsilon$ (with $n = 0$), for each rule $f(x) \to g(r_1 \dots r_n) \in \mathcal{R}$ with $x \notin var(r_1, \dots, r_n)$, if $L(\mathcal{A}, q) \cap f\big(\mathcal{H}(\Sigma)\big) \neq \emptyset$.

Note that in the cases *ii* and *iii* cover all the cases of linear restricted context-free rewrite rules, except the collapsing rules.

Let $N = \bigcup_{f\in\Sigma,q\in Q} (N_{f,q} \cup \{I'_{f,q}\})$ and $P = \bigcup_{f\in\Sigma,q\in Q} (P_{f,q} \cup P'_{f,q})$.

Let us clean up these sets: if the language generated by a CF grammar $(Q, N, I'_{f,q}, P)$ is empty then we remove $I'_{f,q}$ from $N$ and all the productions of $P$ which contain $I'_{f,q}$. We iterate this operation, until there is no remaining non-terminals generating an empty language in $N$ (note that the construction stops since we only remove non-terminals and productions). Let us note $N'$ and $P'$ the sets of non-terminals and productions obtained. For each $f \in \Sigma$, $q \in Q$, let $\mathcal{G}'_{f,q} = (Q, N', I'_{f,q}, P')$, and let $L'_{f,q}$ be its language.

Finally, $\mathcal{A}' = (Q, Q^{\mathsf{f}}_{L}, \Delta')$ is obtained by the addition of collapsing transitions corresponding to the collapsing rewrite rules in $\mathcal{R}$

$$\Delta' = \{f(L'_{f,q}) \to q \mid f \in \Sigma, q \in Q, L'_{f,q} \neq \emptyset\} \cup \{L'_{f,q} \to q \mid f(x) \to x \in \mathcal{R}\}$$

We show in [10] that $L(\mathcal{A}') = \mathcal{R}^*(L(\mathcal{A}))$.

The proof of the direction $\subseteq$ is by induction on the number of application of collapsing transitions other than $q^*_{\forall} \to q_{\forall}$ in a reduction by $\mathcal{A}'$. For the base case, we need to consider the occurrences of non-terminals $I'_{g,q}$ in the derivations with the grammars $\mathcal{G}'_{f,q}$. Intuitively every occurrence of such $I'_{g,q}$ corresponds to a rewrite step with a context-free rule of $\mathcal{R}$.

The proof of the direction $\supseteq$ is by induction on the length of a rewrite sequence $u \xrightarrow[\mathcal{R}]{*} t$ for $u \in L(\mathcal{A})$. □

**Corollary 2.** *Reachability and regular hedge model-checking are decidable for linear restricted context-free HRS.*

*Proof.* The intersection of an CF-HA language and a HA languages is a CF-HA language, and emptiness of CF-HA is decidable. □

It is shown in [17] that the languages of CF-HA are closures of regular tree languages modulo associativity of one or several binary function symbols. Therefore, the above results are also valid for these languages.

### 5.2 Linear Context-Free Rewrite Rules

Context-free HRS are named after context-free tree grammars, whose production rules have the form $N(x_1, \ldots, x_n) \to r$ where $N$ is a non-terminal of arity $n$ (from a finite set $\mathcal{N}$), $x_1, \ldots, x_n \in \mathcal{X}$ and $r \in \mathcal{T}(\Sigma \cup \mathcal{N}, \{x_1, \ldots, x_n\})$. Note that our definition of context-free HRS is restricted to unary non-terminals. However, even for this case of unary non-terminals and right-linear rewrite rules, the result of Theorem 2 cannot be generalized to context-free HRS.

**Proposition 7.** *$\mathcal{R}^*(L)$ is not a CF-HA language in general when $L$ is a CF-HA language and $\mathcal{R}$ is a linear context-free HRS.*

*Proof.* Let us consider the context-free HRS: $\mathcal{R} = \{f(x) \to g(f(ax))\}$ and let $L = \{f(c)\}$. The set $\mathcal{R}^*(L)$ is $\{\underbrace{g(g(\ldots g(}_{n} f(a^n c)))) \mid n \in \mathbb{N}\}$.
Using a pumping argument, we can show that it is not a CF-HA language. □

The above counter-example shows the importance for Theorem 2 of the condition, in the definition of restricted context-free HRS, that the variable $x$ in a *lhs* of rule occurs at a *shallow* position in the corresponding *rhs*.

### 5.3 Restricted Context-Free Rewrite Rules

If we keep the restricted context-free condition (the variable $x$ in the *lhs* of a rule occurs at a shallow position in the corresponding *rhs*) but we drop the linearity condition, we also lose the CF-HA preservation result of Theorem 2.

**Proposition 8.** *$\mathcal{R}^*(L)$ is not a CF-HA language in general when L is a CF-HA language and $\mathcal{R}$ is a restricted context-free HRS.*

*Proof.* Let $\mathcal{R} = \{f(x) \to f(xx)\}$ and $L = \{f(a)\}$. We have that $\mathcal{R}^*(L) = \{f(a^n) \mid n = 2^k, k \geq 0\}$ which is not a CF-HA language. Assume indeed that this language is recognized by a CF-HA $(Q, Q^{\mathsf{f}}, \Delta)$. It means that $\Delta$ contains a transition $f(L) \to q$ where $L$ is a context-free language of words of $Q^*$ of length $2^k$, $k \geq 0$. The image of $L$ under the strictly alphabetic homomorphism which translates every state $q \in Q$ into $a$ is context-free. As it is a one letter language, it is also regular. But it is well known that this language $\{a^n \mid n = 2^k, k \geq 0\}$ is actually not regular. □

### 5.4 Mixing Inverse CF and Restricted CF Rewrite Rules

We show now that the results of Theorems 1 and 2 cannot be combined. In other terms, for some HRS containing both linear inverse context-free and restricted context-free rules, the set of descendants of a HA language is not a HA language, neither a CF-HA language and even not recursive.

**Proposition 9.** *$\mathcal{R}^*(L)$ is not recursive in general when L is a HA language and $\mathcal{R}$ is a HRS whose rules are either inverse context-free or restricted context-free and contain only one variable.*

*Proof.* We reduce the Post Correspondence Problem (PCP). Let us consider an instance $\mathcal{P} = \{\langle u_i, v_i\rangle \mid i \leq n, u_i, v_i \in \Gamma^*\}$ of PCP on an finite alphabet $\Gamma$. The problem is to find a sequence $i_1, \ldots, i_k \leq n$ such that $u_{i_1} \ldots u_{i_k} = v_{i_1} \ldots v_{i_k}$

Let $\mathcal{R}$ be an HRS containing a rule $f_0(x) \to f_0(\widetilde{u_i} x v_i)$ for each pair $\langle u_i, v_i\rangle \in \mathcal{P}$ ($\widetilde{u_i}$ is the mirror image of $u_i$), and two rules $f_0(axa) \to f_1(x)$ and $f_1(axa) \to f_1(x)$ for each $a \in \Gamma$. We assume that $f_0$, $f_1$, and $c$ are symbols not in $\Gamma$. We have that $f_0(c) \xrightarrow[\mathcal{R}]{*} f_1(c)$ iff $\mathcal{P}$ has a solution. □

Moreover, as we have shown that context-free HRS do not preserve HA languages (Proposition 4), the symmetric also holds for inverse-context-free HRS and CF-HA languages.

**Proposition 10.** *$\mathcal{R}^*(L)$ is not recursive in general when $L$ is a CF-HA language and $\mathcal{R}$ is an inverse context-free HRS.*

*Proof.* Let $\mathcal{R}_1$ be the subset of the context-free rewrite rules of the HRS of the above proof of Proposition 9, and $\mathcal{R}_2$ be the subset of the other rules. Note that $\mathcal{R}_2$ is an inverse context-free HRS.

By Theorem 2, $L = \mathcal{R}_1^*(\{f_0(c)\})$ is a CF-HA language. Like in the proof of Proposition 9, we have that $f_1(c) \in \mathcal{R}_2^*(L)$ iff the PCP has a solution. Hence, because of the decidability of the membership problem for CF-HA, $\mathcal{R}_2^*(L)$ cannot be a CF-HA language. □

## 6 Conclusion

We have shown that HA and CF-HA languages are preserved by rewrite closure for interesting classes of non ground hedge rewriting rules. These rules allow us for instance to modify the structure of XML documents when processing them. We plan to extend our results to non ordered unranked trees by considering sheaves automata as in [5] or commutative hedge automata (see [3] for application to process rewrite systems).

Regularity preservation has been studied in the case of ranked terms for transducing term rewriting system, *i.e.* rewrite rules corresponding to transducers rules [20]. A generalization of such classes of TRS to hedge rewriting seems conceptually close to XML transformations [12] and we plan to study the preservation of HA or CF-HA languages w.r.t. to such HRS.

## References

1. Abdulla, P.A., Jonsson, B., Nilsson, M., Saksena, M.: A survey of regular model checking. In: Gardner, P., Yoshida, N. (eds.) CONCUR 2004. LNCS, vol. 3170, pp. 35–48. Springer, Heidelberg (2004)
2. Bouajjani, A., Jonsson, B., Nilsson, M., Touili, T.: Regular model checking. In: Emerson, E.A., Sistla, A.P. (eds.) CAV 2000. LNCS, vol. 1855, pp. 403–418. Springer, Heidelberg (2000)
3. Bouajjani, A., Touili, T.: On computing reachability sets of process rewrite systems. In: Giesl, J. (ed.) RTA 2005. LNCS, vol. 3467, pp. 484–499. Springer, Heidelberg (2005)
4. Comon, H., Dauchet, M., Gilleron, R., Jacquemard, F., Lugiez, D., Tison, S., Tommasi, M.: Tree automata techniques and applications (Last release October 12, 2007), `http://www.grappa.univ-lille3.fr/tata`
5. Dal-Zilio, S., Lugiez, D.: XML schema, tree logic and sheaves automata. In: Nieuwenhuis, R. (ed.) RTA 2003. LNCS, vol. 2706, pp. 246–263. Springer, Heidelberg (2003)
6. Durand, I., Sénizergues, G.: Bottom-up rewriting is inverse recognizability preserving. In: Baader, F. (ed.) RTA 2007. LNCS, vol. 4533, pp. 107–121. Springer, Heidelberg (2007)
7. Gilleron, R., Tison, S.: Regular tree languages and rewrite systems. Fundamenta Informaticae 24(1/2), 157–176 (1995)

8. Hofbauer, D., Waldmann, J.: Deleting string rewriting systems preserve regularity. Theor. Comput. Sci. 327(3), 301–317 (2004)
9. Jacquemard, F.: Decidable approximations of term rewriting systems. In: Ganzinger, H. (ed.) RTA 1996. LNCS, vol. 1103, pp. 362–376. Springer, Heidelberg (1996)
10. Jacquemard, F., Rusinowitch, M.: Rewrite closure of hedge-automata languages. Research Report LSV-08-05, Laboratoire Spécification et Vérification, ENS Cachan, France (2007), `http://www.lsv.ens-cachan.fr/Publis`
11. Löding, C., Spelten, A.: Transition graphs of rewriting systems over unranked trees. In: Kučera, L., Kučera, A. (eds.) MFCS 2007. LNCS, vol. 4708, pp. 67–77. Springer, Heidelberg (2007)
12. Martens, W., Neven, F.: On the complexity of typechecking top-down XML transformations. Theor. Comput. Sci. 336(1), 153–180 (2005)
13. Murata, M.: Hedge Automata: a Formal Model for XML Schemata (2000), `http://www.horobi.com/Projects/RELAX/Archive/hedge_nice.html`
14. Murata, M., Lee, D., Mani, M.: Taxonomy of xml schema languages using formal language theory. In: Extreme Markup Languages (2001)
15. Nagaya, T., Toyama, Y.: Decidability for left-linear growing term rewriting systems. In: Narendran, P., Rusinowitch, M. (eds.) RTA 1999. LNCS, vol. 1631, pp. 256–270. Springer, Heidelberg (1999)
16. Ohsaki, H.: Beyond the regularity: Equational tree automata for associative and commutative theories. In: Fribourg, L. (ed.) CSL 2001 and EACSL 2001. LNCS, vol. 2142. Springer, Heidelberg (2001)
17. Ohsaki, H., Seki, H., Takai, T.: Recognizing boolean closed A-tree languages with membership conditional rewriting mechanism. In: Nieuwenhuis, R. (ed.) RTA 2003. LNCS, vol. 2706, pp. 483–498. Springer, Heidelberg (2003)
18. d'Orso, J., Touili, T.: Regular hedge model checking. In: Proc. of the 4th IFIP Int. Conf. on Theoretical Computer Science (TCS 2006), IFIP (2006)
19. Salomaa, K.: Deterministic Tree Pushdown Automata and Monadic Tree Rewriting Systems. J. of Comp. and System Sci. 37, 367–394 (1988)
20. Seki, H., Takai, T., Fujinaka, Y., Kaji, Y.: Layered Transducing Term Rewriting System and Its Recognizability Preserving Property. In: Tison, S. (ed.) RTA 2002. LNCS, vol. 2378, pp. 98–113. Springer, Heidelberg (2002)
21. Takai, T., Kaji, Y., Seki, H.: Right-linear finite path overlapping term rewriting systems effectively preserve recognizability. In: Bachmair, L. (ed.) RTA 2000. LNCS, vol. 1833, pp. 246–260. Springer, Heidelberg (2000)
22. Touili, T.: Computing transitive closures of hedge transformations. In: Proc. 1st Int. Workshop on Verification and Evaluation of Computer and Communication Systems (VECOS 2007). eWIC Series. British Computer Society (2007)

# On Normalisation of Infinitary Combinatory Reduction Systems

Jeroen Ketema

Research Institute of Electrical Communication, Tohoku University
2-1-1 Katahira, Aoba-ku, Sendai 980-8577, Japan
jketema@nue.riec.tohoku.ac.jp

**Abstract.** For fully-extended, orthogonal infinitary Combinatory Reduction Systems, we prove that terms with perpetual reductions starting from them do not have (head) normal forms. Using this, we show that

1. needed reduction strategies are normalising for fully-extended, orthogonal infinitary Combinatory Reduction Systems, and that
2. weak and strong normalisation coincide for such systems as a whole and, in case reductions are non-erasing, also for terms.

## 1 Introduction

Infinitary higher-order rewrite systems extend infinitary TRSs (iTRSs) [1,2] with bound variables and nestings. Their introduction invalidates the Strip Lemma. Hence, new proof techniques are required to obtain confluence and normalisation results. The latter of these are the subject of this paper.

Failure of the Strip Lemma was first observed by Kennaway et al. [3] in infinitary $\lambda$-calculus. To prove confluence modulo certain subterms for this system, while avoiding the Strip Lemma, Kennaway et al. [3] and Kennaway and De Vries [2] use resp. a non-collapsing variant of the $\beta$-rule and standard reductions. To prove a similar confluence result for the infinite extension of Combinatory Reduction Systems (CRSs) [4], i.e. for infinitary CRSs (iCRSs) [5,6,7], Van Oostrom's technique of essential rewrite steps [8] was adapted.

Below we give an abstract formulation of Van Oostrom's technique under the name *projection pairs*. With the help of these pairs we show for fully-extended, orthogonal iCRSs that terms with perpetual reductions starting from them, i.e. reductions with an infinite number of root-steps, do not have (head) normal forms — the known proofs for iTRSs by Kennaway et al. [1] and by Klop and De Vrijer [9] do not carry over due to dependence on the Strip Lemma.

Using the above fact, we prove our main results: Needed reductions normalise for fully-extended, orthogonal iCRSs and weak and strong normalisation coincide for such systems as a whole and, in case of non-erasing reductions, also for terms.

*Needed Reductions.* Normalisation of needed reductions implies that any reduction strategy contracting only needed redexes, i.e. redexes a residual of which is contracted in every reduction to normal form, yields a normal form. Hence, these strategies are useful to obtain normal forms. We extend the classical result by Huet and Lévy [10] who show the same for orthogonal TRSs. This also extends

A. Voronkov (Ed.): RTA 2008, LNCS 5117, pp. 172–186, 2008.

identical results for orthogonal iTRSs by Kennaway et al. [1] and for orthogonal higher-order systems by Glauert and Khasidashvili [11].

*Uniform Normalisation.* Uniform normalisation [12], i.e. the coincidence of weak and strong normalisation, is special in the case of orthogonal iTRSs. As shown by Klop and De Vrijer [9], the property holds without any restrictions. As such, iTRSs behave different from TRSs, which need to be non-erasing [13].

However, the result for TRSs concerns terms and not systems. This result does not carry over to iTRSs, as noted by Kennaway et al. [14] and by Klop and De Vrijer [9]. Partial recovery is possible by considering non-erasing reductions instead of non-erasing rules, as indicated by Kennaway et al. [14]. We extend both this recovery and the result concerning systems to fully-extended, orthogonal iCRSs.

*Overview.* We give some preliminaries in Sect. 2. In Sect. 3, normalisation is introduced. Projection pairs are defined in Sect. 4 and used in Sect. 5 to obtain the result regarding perpetual reductions. In Sects. 6 and 7 we prove normalisation of needed reductions and uniform normalisation. We conclude in Sect. 8.

## 2 Preliminaries

We outline some basic facts concerning iCRSs; see [2,1,5,6,7] for more detailed accounts. Throughout, we denote the first infinite ordinal by $\omega$, and arbitrary ordinals by $\alpha$, $\beta$, $\gamma$, .... By $\mathbb{N}$ we denote the natural numbers including zero.

**Terms and Substitutions.** Let $\Sigma$ be a signature with each element of finite arity. Moreover, assume a countably infinite set of variables and, for each finite arity, a countably infinite set of meta-variables — countably infinite sets suffice given 'Hilbert hotel'-style renaming.

Infinite terms are usually defined by metric completion [15,1,5]. Here, we give the shorter, but equivalent, definition from [6]:

**Definition 2.1.** *The set of* meta-terms *is defined by interpreting the following rules coinductively, where $s$ and $s_1$, ..., $s_n$ are again meta-terms:*

1. *each variable $x$ is a meta-term,*
2. *if $x$ is a variable, then $[x]s$ is a meta-term,*
3. *if $Z$ is an $n$-ary meta-variable, then $Z(s_1, \ldots, s_n)$ is a meta-term, and*
4. *if $f \in \Sigma$ is $n$-ary, then $f(s_1, \ldots, s_n)$ is a meta-term.*

*The set of* finite *meta-terms, a subset of the set of meta-terms, is the set inductively defined by the above rules. A* term *is a meta-term without meta-variables and a* context *is a meta-term over $\Sigma \cup \{\Box\}$.*

We consider (meta-)terms modulo $\alpha$-equivalence. A meta-term of the form $[x]s$ is called an *abstraction*; a variable $x$ in $s$ is called *bound* in $[x]s$. Meta-terms with meta-variables only occur in rewrite rules; rewriting itself is defined over terms. We have that $Z(Z(\ldots))$, and $Z([x]Z'([y]Z(\ldots)))$ are meta-terms. Moreover, $[x]f(Z(x))$ is a finite meta-term and $[x]x$ is a finite term.

The set of *positions* [5] of a meta-term $s$, denoted $\mathcal{P}os(s)$, is a set of *finite* strings over $\mathbb{N}$, with each string denoting the 'location' of a subterm in $s$. If $p$ is a position of $s$, then $s|_p$ is the *subterm of s at* position $p$. The length of $p$ is denoted $|p|$. There exists a well-founded order $<$ on positions: $p < q$ iff $p$ is a proper prefix of $q$. The concatenation of positions $p$ and $q$ is denoted $p \cdot q$.

A *valuation* [4], denoted $\bar{\sigma}$, substitutes terms for meta-variables in meta-terms and is defined by coinductively interpreting the rules of valuations for CRSs [5]. In CRSs, applying a valuation to a meta-term yields a unique term. This is not the case for iCRSs [5]. To alleviate this problem, the set of meta-terms satisfying the so-called 'finite chains property' is defined in [5]:

**Definition 2.2.** *Let $s$ be a meta-term. A* chain *in $s$ is a sequence of (context, position)-pairs $(C_i[\Box], p_i)_{i<\alpha}$, with $\alpha \leq \omega$, such that for each $(C_i[\Box], p_i)$ there exists a term $t_i$ with $C_i[t_i] = s|_{p_i}$ and $p_{i+1} = p_i \cdot q$ where $q$ is the position of the hole in $C_i[\Box]$. A* chain of meta-variables *in $s$ is such that for each $i < \alpha$ it holds that $C_i[\Box] = Z(t_1, \ldots, t_n)$ with $t_j = \Box$ for exactly one $1 \leq j \leq n$.*

*The meta-term $s$ is said to satisfy the* finite chains property *if no infinite chain of meta-variables occurs in $s$.*

Remark that $\Box$ only occurs in $C_i[\Box]$ if $i + 1 < \alpha$, otherwise $C_i[\Box] = s|_{p_i}$. The meta-term $[x_1]Z_1([x_2]Z_2(\ldots[x_n]Z_n(\ldots)))$ e.g. satisfies the finite chains property, while $Z(Z(\ldots Z(\ldots)))$ does not. Finite meta-terms always satisfy the finite chains property. The following is shown in [5]:

**Proposition 2.3.** *Let $s$ be a meta-term satisfying the finite chains property and let $\bar{\sigma}$ be a valuation. There is a unique term that is the result of applying $\bar{\sigma}$ to $s$.*

**Rewriting.** To define rewriting, recall that a *pattern* is a finite meta-term each meta-variable of which has distinct bound variables as arguments and that a meta-term is *closed* if all variables occur bound [4].

**Definition 2.4.** *A* rewrite rule *is a pair of closed meta-terms $(l, r)$, denoted $l \to r$, with $l$ a finite pattern of the form $f(s_1, \ldots, s_n)$ and $r$ satisfying the finite chains property such that all meta-variables occurring in $r$ also occur in $l$.*

*An* infinitary Combinatory Reduction System (iCRS) *is a pair $\mathcal{C} = (\Sigma, R)$ with $\Sigma$ a signature and $R$ a set of rewrite rules.*

Left-linearity and orthogonality are defined as for CRSs [4], by virtue of left-hand sides of rewrite rules being finite. A rewrite rule is *collapsing* if the root of its right-hand side is a meta-variable. Moreover, a pattern is *fully-extended*, if, for each meta-variable $Z$ and abstraction $[x]s$ with an occurrence of $Z$ in its scope, $x$ is an argument of that occurrence of $Z$; a rewrite rule is *fully-extended* if its left-hand side is and an iCRS is *fully-extended* if all its rewrite rules are.

**Definition 2.5.** *A* rewrite step *is a pair of terms $(s, t)$ denoted $s \to t$ and adorned with a context $C[\Box]$, a rewrite rule $l \to r$, and a valuation $\bar{\sigma}$ such that $s = C[\bar{\sigma}(l)]$ and $t = C[\bar{\sigma}(r)]$. The term $\bar{\sigma}(l)$ is called an $l \to r$-*redex*. It* occurs *at position $p$ and depth $|p|$ in $s$, where $p$ is the position of the hole in $C[\Box]$.*

*A position $q$ of $s$ occurs in the* redex pattern *of the redex at position $p$ if $q \geq p$ and if there does not exist a position $q'$ with $q \geq p \cdot q'$ such that $q'$ is the position of a meta-variable in $l$.*

Both $\bar{\sigma}(l)$ and $\bar{\sigma}(r)$ are well-defined, as left- and right-hand sides of rewrite rules satisfy the finite chains property (left-hand sides because they are finite).

In addition to collapsing rewrite rules, a redex and a rewrite step are *collapsing* if the employed rewrite rule is. Using rewrite steps, we define reductions:

**Definition 2.6.** *A* transfinite reduction *with domain $\alpha > 0$ is a sequence of terms $(s_\beta)_{\beta<\alpha}$ such that $s_\beta \to s_{\beta+1}$ for all $\beta + 1 < \alpha$. In case $\alpha = \alpha' + 1$, the reduction is* closed *and of length $\alpha'$. In case $\alpha$ is a limit ordinal, the reduction is* open *and of length $\alpha$.*

*The reduction is* weakly *or* Cauchy continuous *if for every limit ordinal $\gamma < \alpha$ it holds that $s_\beta$ converges to $s_\gamma$ as $\beta$ approaches $\gamma$ from below. The reduction is* weakly *or* Cauchy convergent *if it is weakly continuous and closed.*

*For each rewrite step $s_\beta \to s_{\beta+1}$, let $d_\beta$ denote the depth of the contracted redex. The reduction is* strongly continuous *if it is weakly continuous and if, for every limit ordinal $\gamma < \alpha$, the depth $d_\beta$ tends to infinity as $\beta$ approaches $\gamma$ from below. The reduction is* strongly convergent *if strongly continuous and closed.*

Consider the rules $a \to a$ and $f(Z) \to g(f(Z))$ and the term $f(a)$. The following reduction of length $\omega$ is both weakly and strongly continuous:

$$f(a) \to f(a) \to \cdots \to f(a) \to \cdots .$$

Extending the reduction with $f(a)$ yields a weakly convergent reduction but not a strongly convergent one. The reduction

$$f(a) \to g(f(a)) \to \cdots \to g^n(f(a)) \to \cdots g^\omega ,$$

also of length $\omega$ and where $g^\omega$ denotes $g(g(\ldots g(\ldots)))$, is strongly convergent.

Reductions are ranged over by $D$, $S$, and $T$. We mostly consider *strongly convergent* reductions: By $s \twoheadrightarrow^\alpha t$, resp. $s \twoheadrightarrow^{\leq\alpha} t$, we denote a strongly convergent reduction of length $\alpha$, resp. of length at most $\alpha$. By $s \twoheadrightarrow t$, resp. $s \to^* t$, we denote a strongly convergent reduction of arbitrary length, resp. of finite length.

Across strongly convergent reductions we assume that a position that occurs in the redex pattern of a contracted redex does not have any descendants; likewise for residuals [5]. We write $P/(s \twoheadrightarrow t)$ for the descendants of a set of positions $P \subseteq \mathcal{P}os(s)$ across a strongly convergent reduction $s \twoheadrightarrow t$ and $\mathcal{U}/(s \twoheadrightarrow t)$ for the residuals of a set $\mathcal{U}$ of subterms of $s$ across $s \twoheadrightarrow t$.

In the remainder we appeal to a number of properties of iCRSs. The first is immediate by the proof of the compression property in [5].

**Theorem 2.7 (Compression).** *For every fully-extended, left-linear iCRS, if $s \twoheadrightarrow^\alpha t$, then $s \twoheadrightarrow^{\leq\omega} t$. Moreover, if $s \twoheadrightarrow^\alpha t$ has a root-step, then does $s \twoheadrightarrow^{\leq\omega} t$.*

Assuming orthogonality, let $\mathcal{U}$ be a set of redexes of a term $s$. A *development* of $\mathcal{U}$ is a reduction $s \twoheadrightarrow t$ each step of which contracts a residual of a redex in $\mathcal{U}$. A development $s \twoheadrightarrow t$ is *complete* if $\mathcal{U}/(s \twoheadrightarrow t) = \emptyset$; in this case we also write $s \Rightarrow t$, where the arrow is adorned with $\mathcal{U}$ as needed. We have the following:

**Proposition 2.8 (See [6]).** *Let $s$ be a term in an orthogonal iCRS. If $\mathcal{U}$ is a set of redexes of $s$ with a complete development $s \Rightarrow t$ and if $v$ is a redex of $s$, then the following diagram exists:*

$$\begin{array}{ccc} s & \xrightarrow{\;v\;} & t' \\ \Downarrow \mathcal{U} & & \Downarrow \mathcal{U}/(s \to^{v} t') \\ t & \underset{v/(s \Rightarrow t)}{\Longrightarrow} & s' \end{array}$$

A term $s$ is *hypercollapsing* if for all $s \twoheadrightarrow t$ there exists a $t \twoheadrightarrow t'$ such that $t'$ is a collapsing redex. We write $s \sim_{hc} t$ if $t$ can be obtained from $s$ by replacing hypercollapsing subterms in $s$ by other hypercollapsing subterms. We have:

**Theorem 2.9.** *Fully-extended, orthogonal iCRSs are confluent modulo $\sim_{hc}$, i.e. if $s \twoheadrightarrow s'$ and $t \twoheadrightarrow t'$ with $s \sim_{hc} t$, then $s' \twoheadrightarrow s''$ and $t' \twoheadrightarrow t''$ with $s'' \sim_{hc} t''$.*

The above is shown in [6] under assumption that rewrite rules have finite right-hand sides; in [7] the result is extended to allow for infinite right-hand sides.

## 3 Weak and Strong Normalisation

We define (head) normal forms together with weak and strong normalisation. Ample motivation for the definitions is given by Klop and De Vrijer [9].

**Definition 3.1.** *A term $s$ is a* normal form *if no redexes occur in $s$ and a* head normal form *if it is not reducible to a redex by a strongly convergent reduction. In addition, $s$ is* weakly normalising *if a strongly convergent reduction exists from $s$ to a normal form and $s$ is* strongly normalising *if for all open strongly continuous reductions starting in $s$ there exists a term that extends the reduction such that it becomes strongly convergent.*

*An iCRS is* weakly normalising, *resp.* strongly normalising, *if all terms are.*

Consider again the rules $a \to a$ and $f(Z) \to g(f(Z))$, introduced below Definition 2.6. The term $f(a)$ is weakly normalising by the second reduction below the definition; $g^\omega$ is a normal form. The term is not strongly normalising, as the first reduction below the definition cannot be extended such that it becomes strongly convergent. On the other hand, $f(x)$ is strongly normalising, as the only open strongly continuous reduction starting from it is

$$f(x) \to g(f(x)) \to \cdots \to g^n(f(x)) \to \cdots,$$

which extends to a strongly convergent reduction by adding $g^\omega$.

The definition of weak normalisation is taken from finitary rewriting. To understand strong normalisation, consider the following proposition, which is immediate by the fact that strongly convergent reductions have a finite number of reduction steps at each depth [1,5]:

**Proposition 3.2.** *An open strongly continuous reduction extends to a strongly convergent one iff the number of reduction steps is finite at every depth.*

Hence, the definition of strong normalisation from finitary rewriting is relaxed: A finite number of steps in total implies a finite number of steps at each depth. Given that a term does not need to have a maximum depth in the current setting, this seems a reasonable way to relax the definition.

As in the finite case, strong normalisation implies weak normalisation: To start, remark that any strongly normalising term reduces to a head normal form, otherwise it has an open strongly continuous reduction starting from it with an infinite number of root-steps. Next, as the same holds for each subterm of the head normal form, again by strong normalisation, iteration gives a term each subterm of which is a head normal form, i.e. it gives a normal form.

## 4 Projection Pairs

We give an abstract formulation of Van Oostrom's technique of essential rewrite steps [8] and its adaptation to iCRSs [6,7]. This requires an auxiliary definition:

**Definition 4.1.** *Let $s$ and $t$ be terms and $P \subseteq \mathcal{P}os(s)$. The set $P$ is a* prefix set *of $s$ if $P$ is* finite *and if all prefixes of positions in $P$ are also in $P$. Moreover, $t$* mirrors *$s$ in $P$, if for all $p \in P$ it holds that $p \in \mathcal{P}os(t)$ and $root(t|_p) = root(s|_p)$.*

Van Oostrom's technique is a termination argument focusing on prefix sets $P$ and finite sequences of complete developments $D$, i.e. reductions $D$ consisting of a finite number of such developments. Given a prefix set $P$ of the final term of $D$, the defined measure assigns to $D$ a tuple of natural numbers of the same length as the sequence. In addition, a map is defined which, intuitively, given $P$ yields a prefix set of its initial term such that the function symbols that occur at the positions in obtained prefix set are those 'responsible', across $D$, for what occurs at the positions of $P$ in the final term of $D$.

At the core of the technique lies a projection. Given a reduction step from the initial term of $D$, which is called *essential* in case it occurs in the obtained prefix of the initial term of $D$ and *inessential* otherwise, the projection yields a finite sequence of complete developments $D'$, starting in the term created by the reduction step, such that the final term of $D'$ mirrors the final one of $D$ in $P$. The projection is such that the measure decreases in case of an essential reduction step and stays equal otherwise, facilitating the termination argument.

Moving away from tuples, the measure and the map on prefixes can abstractly be defined as follows:

**Definition 4.2.** *Given a well-founded order $\prec$, a* projection pair *is a pair $(\mu, \varepsilon)$ of maps over finite sequences of complete developments $D$ and prefix sets $P$ of the final term of the chosen $D$ such that:*

- *$\mu_P(D)$ maps to an element of the well-founded order $\prec$, and*
- *$\varepsilon_P(D)$ maps to a prefix of the initial term of $D$,*

*and such that if $D'$ is a sequence of complete developments strictly shorter than $D$ with $P'$ a prefix set of the final term of $D'$, then $\mu_{P'}(D') \prec \mu_P(D)$.*

The map $\mu$ is the measure and $\varepsilon$ is the map for prefix sets. The measure requires a sequence that is strictly shorter than $D$ to map to a smaller element in the well-founded order. Although of a technical nature, this property is easily obtained in case tuples are used to define the well-founded order, as described above, and the tuples are first compared length-wise and next lexicographically.

The existence of the projection mentioned above can now be formulated as the soundness of a projection pair:

**Definition 4.3.** *Let $\prec$ be a well-founded order. A projection pair $(\mu, \varepsilon)$ is* sound *iff for each finite sequence of complete development $D$, prefix set $P$ of the final term of $D$, and $s \twoheadrightarrow t$, with $s$ the initial term of $D$, it holds that:*

- *if $s \twoheadrightarrow t$ consists of a* single *step contracting a redex $u$ at a position in $\varepsilon_P(D)$, with no residual from $u/D$ occurring at a position in $P$, then there exists a $D'$ such that $\mu_P(D') \prec \mu_P(D)$, and*
- *if $s \twoheadrightarrow t$* only *contracts redexes at positions* outside *$\varepsilon_P(D)$, then there exists a $D'$ such that $\mu_P(D') = \mu_P(D)$ and $\varepsilon_P(D') = \varepsilon_P(D)$,*

*where in both cases $D'$ is a finite sequence of complete developments with initial term $t$ such that the final term of $D'$ mirrors the final one of $D$ in $P$.*

In the first clause the redex is essential and in the second clause all are inessential. Intuitively, the restriction in the first clause stating that no residual from $u/D$ occurs in $P$ ensures that the projection preserves $P$. Together the clauses formalise the intuition behind $\varepsilon$, i.e. that $P$ only depends on positions in $\varepsilon_P(D)$. The map is constant for reductions contracting only redexes outside $\varepsilon_P(D)$ and, obviously, any term in such a reduction mirrors all the other terms in $\varepsilon_P(D)$.

*Remark 4.4.* The first clause of Definition 4.3 deals neither with reductions where residuals from $u/D$ occur in $P$ nor with infinite reductions. In the next section, we deal with the first through the restriction on strictly shorter sequences of complete developments and with the second through strong convergence.

This leaves to show that sound projection pairs actually exist. For *fully-extended, orthogonal* iCRSs this is done in [6] in case all rewrite rules have finite right-hand sides. In [7] the result is extended to iCRSs that allow for infinite right-hand sides. The lengthy definitions from [6] and [7] are omitted here; the abstract definitions suffice.

## 5 Perpetual Reductions

To show our main results, we prove that terms with perpetual reductions starting from them do not have (head) normal forms. Except for the final lemma of this section, the proofs in this section differ from the proofs for head normal forms [1] and normal forms [9] of iTRSs, which depend on the Strip Lemma.

We assume *fully-extended, orthogonal* iCRSs. Perpetual reductions, not to be mistaken for perpetual reduction strategies [16], are defined as in [1]:

**Definition 5.1.** *A* perpetual reduction *is an open strongly continuous reduction with an infinite number of root-steps.*

Any perpetual reduction can be 'compressed' to one of length $\omega$:

**Lemma 5.2.** *Let $s$ be a term. If there is a perpetual reduction starting from $s$, then there also is a perpetual reduction of length $\omega$ starting from it.*

*Proof.* By definition, we may write a perpetual reduction starting from $s$ as:

$$s = s_0 \twoheadrightarrow s'_0 \rightarrow s_1 \twoheadrightarrow s'_1 \rightarrow s_2 \twoheadrightarrow \cdots ,$$

with $s'_i \rightarrow s_{i+1}$ a root-step and no root-steps occurring in $s_i \twoheadrightarrow s'_i$ for each $i \in \mathbb{N}$.

We inductively define a perpetual reduction of length $\omega$:

$$s = t_0 \rightarrow^* t'_0 \rightarrow t_1 \rightarrow^* t'_1 \rightarrow t_2 \rightarrow^* \cdots ,$$

where for all $i \in \mathbb{N}$ we have that $t'_i \rightarrow t_{i+1}$ is a root-step and $t_i \rightarrow^* t'_i$ is finite and without root-steps. First, define $t_0 = s_0 = s$. Next, assume we have defined a term $t_i$ with $t_i \twoheadrightarrow s_i$. Compression of $t_i \twoheadrightarrow s_i \twoheadrightarrow s'_i \rightarrow s_{i+1}$ yields a reduction $t_i \rightarrow^* t'_i \rightarrow t_{i+1} \twoheadrightarrow^{\leq\omega} s_{i+1}$ with $t'_i \rightarrow t_{i+1}$ a root-step and $t_i \rightarrow^* t'_i$ finite and without root-steps. We thus obtain a perpetual reduction with the required properties. □

The following lemma, which projects perpetual reductions over single steps, is the iCRS analogue of Proposition 17 in [9]. Its proof is the only in the current paper explicitly dealing with nestings; in all other cases these are either 'hidden' by the current result or the use of projection pairs.

**Lemma 5.3.** *Let $s$ and $t$ be terms with $s \rightarrow t$. If there is a perpetual reduction starting from $s$, then there is a perpetual reduction starting from $t$.*

*Proof.* Define $s_0 = s$, $t_0 = t$, and suppose $u$ is the redex contracted in $s \rightarrow t$. By Lemma 5.2, we may write the perpetual reduction starting from $s_0$ as:

$$s_0 \rightarrow^* s'_0 \rightarrow s_1 \rightarrow^* s'_1 \rightarrow s_2 \rightarrow^* \cdots ,$$

where for all $i \in \mathbb{N}$, we have that $s'_i \rightarrow s_{i+1}$ is a root-step and $s_i \rightarrow^* s'_i$ is finite and without root-steps. By repeated application of Proposition 2.8, we obtain:

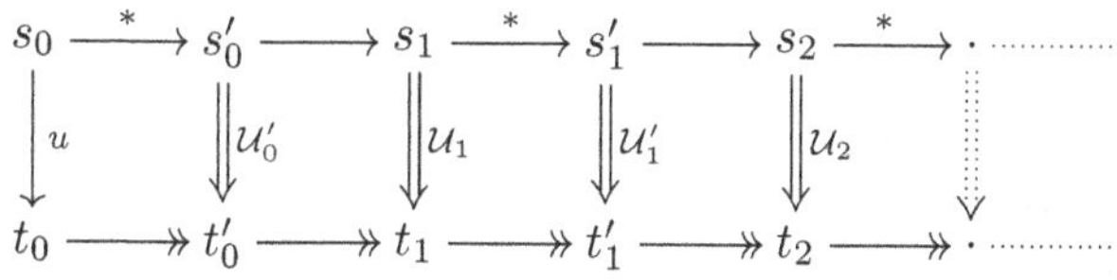

Write $S_i$ for $s_i \rightarrow^* s'_i \rightarrow s_{i+1} \rightarrow^* \cdots$ and $T_i$ for $t_i \twoheadrightarrow t'_i \twoheadrightarrow t_{i+1} \twoheadrightarrow \cdots$. If we can show for each $i \in \mathbb{N}$ that a root-step occurs in $T_i$, then an infinite number of root-steps occur in $T_0$, implying that the reduction is perpetual.

To show that a root-step occurs in $T_i$ we distinguish two cases: (1) a root-step occurs in $S_i$ not contracting a residual of $u$, and (2) all root-steps in $S_i$ contract a residual of $u$. We deal with each of these cases in turn:

1. In this case there exists a root-step $s'_j \rightarrow s_{j+1}$ with $j \geq i$ such that the contracted redex, say $v$, is not a residual of $u$. Since $\mathcal{U}'_j$ contracts only residuals of $u$, we have by orthogonality that a residual of $v$ occurs at the root of $t'_j$ and that no other residuals of $v$ occur in $t'_j$. By construction, $t'_j \twoheadrightarrow t_{j+1}$ contracts precisely all residuals of $v$. Hence, $t'_j \twoheadrightarrow t_{j+1}$ is a root-step.
2. In this case, the infinite number of root-steps of $S_i$ each contract a residual of $u$. Hence, all terms in $S_i$ have a chain of residuals of $u$ at the root and $u$ is collapsing. All the chains are finite, as only a finitely many steps occur

before each term and as right-hand sides of rewrite rules only allow for finite chains of meta-variables.

Residuals of $u$ cannot create further nestings of other residuals of $u$: This requires a residual of $u$ to occur on the path between the redex pattern and a bound variable of another residual of $u$. Such a situation cannot occur by definition of rewrite rules and valuations. Thus, for each step following $s_i$, we have that each residual in the chain at the root of $s_i$ has at most one residual. Eventually, no residuals are left, as an infinite number of root-steps occur in $S_i$. Since the residuals always occur in a chain starting at the root, the last residual is contracted by means of a root-step, say $s'_j \to s_{j+1}$.

Suppose now that no redex contracted in $s_i \to^* s_{j+1}$ has a residual occurring at the root of one of the terms in $t_i \twoheadrightarrow t_{j+1}$. As $u$ is collapsing and as each development of $\mathcal{U}_k$ and $\mathcal{U}'_k$ contracts only residuals of $u$, which occur in finite chains, it follows that a fixed function symbol occurs at the root of each the terms in $t_i \twoheadrightarrow t_{j+1}$. Moreover, as residuals of $u$ cannot create further nestings of other residuals of $u$, the fixed function symbol also occurs at the root of $s_{j+1}$, i.e. no residual of $u$ occurs at the root of $s_{j+1}$, contradiction. Hence, a root-step occurs in $t_i \twoheadrightarrow t_{j+1}$.

As required, we have that a root-step occurs in each $T_i$. Hence, $T_0$ is a perpetual reduction starting from $t_0 = t$. □

We next show that reduction to a redex is preserved if no root-steps occur. In the proof we assume the existence of a sound projection pair $(\mu, \varepsilon)$, which is possible in case of fully-extended, orthogonal iCRSs, as remarked in Sect. 4.

**Lemma 5.4.** *If no root-steps occur in $s \twoheadrightarrow t$ and $s$ reduces to a redex, then $t$ reduces to a redex.*

*Proof.* Using ordinal induction, we show that every term $s_\alpha$ in $s \twoheadrightarrow t$ reduces to a redex by a finite sequence of complete developments $D_\alpha$. Denote by $P_\alpha$ the set of positions of the redex pattern at the root of the final term of $D_\alpha$; to facilitate the induction we also show for $\beta \leq \alpha$ that either $\mu_{P_\alpha}(D_\alpha) \prec \mu_{P_\beta}(D_\beta)$ or $\mu_{P_\alpha}(D_\alpha) = \mu_{P_\beta}(D_\beta)$ and $\varepsilon_{P_\alpha}(D_\alpha) = \varepsilon_{P_\beta}(D_\beta)$.

For $s_0 = s$, it follows by assumption that $s_0$ reduces to a redex. In fact, by strong convergence and compression, $s_0$ reduces to a redex by a finite reduction $D_0$. As any finite reduction is a finite sequence of complete developments, where each set of redexes is a singleton set, the result follows.

For $s_{\alpha+1}$ there are two cases, depending on the occurrence of a residual of $u$, the redex contracted in $s_\alpha \to s_{\alpha+1}$, at the root of the final term of $D_\alpha$:

- If no residual of $u$ occurs at the root of the final term of $D_\alpha$, the result is immediate by soundness of the pair $(\mu, \varepsilon)$ and the induction hypothesis.
- If a residual of $u$ does occur at the root of the final term of $D_\alpha$, a root-step not contracting a residual of $u$ occurs in $D_\alpha$. Otherwise, no residual of $u$ occurs at the root of the final term of $D_\alpha$, because $s_\alpha \to s_{\alpha+1}$ is not a root-step. Hence, there is a finite sequence $D'_\alpha$ of complete developments, strictly shorter than $D_\alpha$, that has a redex at the root of its final term which is not a

residual of $u$. By definition of projection pairs, $\mu_{P'_\alpha}(D'_\alpha) \prec \mu_{P_\alpha}(D_\alpha)$, where $P'_\alpha$ is the set of positions of the redex pattern at the root of the final term of $D'_\alpha$. The case in which no residual of $u$ occurs at the root of the final term of the complete development now applies and the result follows.

For $s_\alpha$, with $\alpha$ a limit ordinal, it follows by the induction hypothesis, strong convergence, and the well-foundedness of $\prec$ that there exists a $\beta < \alpha$ such that all steps in $s_\beta \twoheadrightarrow s_\alpha$ occur at positions outside $\varepsilon_{P_\beta}(D_\beta)$. Hence, the result follows by the second clause of Definition 4.3. □

Using the above we can prove the result we are after, which generalises Proposition 8.9 in [1] for head normal forms and Corollary 20 in [9] for normal forms:

**Lemma 5.5.** *Let $s$ be a term. If $s$ has a perpetual reduction starting from it, then $s$ does not have a (head) normal form.*

*Proof.* Assume a perpetual reduction starting from $s$ and let $s \twoheadrightarrow t$ be arbitrary. By compression and strong convergence, we may write $s \rightarrow^* t' \twoheadrightarrow^{\leq\omega} t$, where all root-steps occur in $s \rightarrow^* t'$. By repeated application of Lemma 5.3, there exists a perpetual reduction starting from $t'$. Thus, $t'$ reduces to a redex. Since $t' \twoheadrightarrow t$ contains no root-steps, we have by Lemma 5.4 that $t$ also reduces to a redex. As $s \twoheadrightarrow t$ is arbitrary, it follows that $s$ does not have a (head) normal form. □

The reverse of the above lemma only holds for head normal forms: Suppose the term $s$ does not have a head normal form. Hence, each reduct of $s$ reduces to a redex. Repeatedly contracting the redexes obtained yields a perpetual reduction.

In case of normal forms consider the rule $a \rightarrow a$. The term $f(a)$ does not have a normal form, as the term reduces to itself, but no perpetual reduction starts from the term either, as $f(a)$ is a head normal form.

## 6 Needed Reductions

Assuming again *fully-extended*, *orthogonal* iCRSs, we show that needed reductions are normalising. We define needed redexes and reductions as in [1]:

**Definition 6.1.** *A redex $u$ in a term $s$ is* needed *if in every strongly convergent reduction from $s$ to normal form some residual of $u$ is contracted. A* needed reduction *is a weakly continuous reduction contracting only needed redexes.*

Non-neededness is due to the erasure of residuals. As in the finite case, this can be the result of the absence of residuals after a certain rewrite step, while residuals did occur earlier. In addition, a redex can also be 'pushed out' of a term by an infinite reduction. To see this, consider the rules $a \rightarrow a$ and $f(Z) \rightarrow g(f(Z))$ from Sect. 2. The $a \rightarrow a$-redex in the term $f(a)$ can be 'pushed out' of the term in the reduction to the normal form $g^\omega$ without contracting it.

We next proceed in two steps: First, we show that a term with a normal form has a needed redex. Thereafter, we prove the actual result.

**Existence of Needed Redexes.** To prove that a term with a normal form has a needed redex, we adapt a proof by Middeldorp [17], who shows for TRSs that a non-root-stable term has a root-needed redex. The proof deviates from the one by Huet and Lévy [10] and its analogue for iTRSs by Kennaway et al. [1]; it does not require the introduction of external redexes, although the redex eventually identified in Lemma 6.4 has the property of being external.

We start by proving the iCRS analogues of Lemmas 3.3 and 4.2 in [17], where we write $s \mathrel{\check{\twoheadrightarrow}} t$ in case all contracted redexes in $s \twoheadrightarrow t$ occur below the root.

**Lemma 6.2.** *Let $s \mathrel{\check{\twoheadrightarrow}} s'$ and $t \mathrel{\check{\twoheadrightarrow}} t'$. If $s \sim_{hc} t$, where it suffices to replace hypercollapsing subterms below the root, then $s' \mathrel{\check{\twoheadrightarrow}} s''$ and $t' \mathrel{\check{\twoheadrightarrow}} t''$ with $s'' \sim_{hc} t''$, where it also suffices to replace hypercollapsing subterms below the root.*

*Proof.* Let $s \sim_{hc} t$, where it suffices to replace hypercollapsing subterms below the root. By assumption, $s = f(s_1, \ldots, s_n) \mathrel{\check{\twoheadrightarrow}} f(s'_1, \ldots, s'_n) = s'$ and $t = f(t_1, \ldots, t_n) \mathrel{\check{\twoheadrightarrow}} f(t'_1, \ldots, t'_n) = t'$. Moreover, $s_i \sim_{hc} t_i$ for all $1 \leq i \leq n$. Hence, by Theorem 2.9 it holds for all $1 \leq i \leq n$ that $s'_i \twoheadrightarrow s''_i$ and $t'_i \twoheadrightarrow t''_i$ with $s''_i \sim_{hc} t''_i$. The result follows by defining $s'' = f(s''_1, \ldots, s''_n)$ and $t'' = f(t''_1, \ldots, t''_n)$. □

**Lemma 6.3.** *Let $s$ be a term. If $s$ reduces to a redex, then the rule used to contract the first such redex is independent of the reduction.*

*Proof.* Suppose $s$ reduces to a redex. We may assume that all rewrite steps occur below the root, otherwise $s$ reduces to a redex by a shorter reduction. Let $s \mathrel{\check{\twoheadrightarrow}} \bar{\sigma}_1(l_1)$ and $s \mathrel{\check{\twoheadrightarrow}} \bar{\sigma}_2(l_2)$, where $l_1$ and $l_2$ are left-hand sides of rewrite rules. By Lemma 6.2 and since $s \sim_{hc} s$, there exist $\bar{\sigma}_1(l_1) \mathrel{\check{\twoheadrightarrow}} t_1$ and $\bar{\sigma}_2(l_2) \mathrel{\check{\twoheadrightarrow}} t_2$ with $t_1 \sim_{hc} t_2$, where it is suffices to replace hypercollapsing subterms below the root. Since a redex at the root cannot be destroyed by either replacing hypercollapsing subterms below the root or contracting of redexes below the root, by orthogonality and fully-extendedness, we have $l_1 = l_2$. □

We now show the presence of needed redexes in terms with normal forms. The proof is based on the one of Theorem 4.3 in [17], although the induction employed there no longer applies as terms may be infinite:

**Lemma 6.4.** *Let $s$ be a term which is not a normal form. If $s$ has a normal form, then $s$ has a needed redex.*

*Proof.* Suppose $s$ has a normal form. As $s$ is not a normal form, there exists a minimal position $p$ in $s$ such that $s|_p$ is not a head normal form. There are two possibilities: either $s|_p$ is a redex or not.

If $s|_p$ is a redex, it is needed: By minimality of $p$, $s|_q$ is a head normal form for each $q < p$. Hence, by orthogonality and fully-extendedness, residuals of $s|_p$ cannot be erased or occur at increasingly greater depths in the reducts of $s$.

If $s|_p$ is not a redex, it reduces to one, otherwise $s|_p$ is a head normal form. By Lemma 6.3, the rule used in the first redex to which $s|_p$ reduces is independent of the reduction. Assume $l$ is the left-hand side of this rule. Since $s|_p$ is not a redex, there exists a non-root position $q$ in the intersection of $\mathcal{P}os(s|_p)$ and the set of positions in the redex pattern of $l$ such that $root(s|_{p \cdot q}) \neq root(l|_q)$ — if $q$ would be the root position, then $s|_p$ reduces to a redex by a shorter reduction. Consider $s|_{p \cdot q}$. If $s|_{p \cdot q}$ is a redex, then it is needed, otherwise we can reduce $s$ to a normal form without reducing $s|_p$ to a redex, which is impossible by minimality of $p$. If

$s|_{p \cdot q}$ is not a redex, then the argument for $s|_p$ can be repeated with $p$ replaced by $p \cdot q$. Repeating the argument, a needed redex must eventually be encountered. If not, then $s|_p$ is a head normal form, contradicting assumptions. □

**Normalisation.** To prove that needed reductions are normalising, we need to show that these reductions are strongly convergent for terms with normal forms. To this end, we first prove the iCRS analogues of Theorem 8.10 and Corollary 8.11 in [1]: Reductions outside subterms without a head normal form are strongly convergent and redexes in that do occur in such subterms are never needed.

**Lemma 6.5.** *Reductions in which all contracted redexes occur outside subterms without a head normal form are strongly convergent.*

*Proof (Sketch).* Identical to the proof of Theorem 8.10 in [1]: A non-strongly convergent reduction yields a subterm with a perpetual reduction starting from it. By Lemma 5.5 this implies the subterm is without a head normal form. □

**Lemma 6.6.** *Let $s$ be a term with a normal form. A redex in $s$ which occurs in a subterm without a head normal form is never needed.*

*Proof (Sketch).* Identical to the proof of Corollary 8.11 in [1], employing the previous lemma instead of Theorem 8.10 in [1]. □

Our intermediate result is now easily obtained and is the iCRS analogue of Corollary 8.12 in [1]:

**Lemma 6.7.** *Let $s$ be a term with a normal form. Every needed reduction starting from $s$ is strongly convergent.*

*Proof.* By Lemma 6.6 no needed redexes occur in subterms without a head normal form. Hence, the result follows by Lemma 6.5. □

By the previous lemma and Lemma 6.4, we now immediately obtain:

**Theorem 6.8.** *In fully-extended, orthogonal iCRSs, needed reductions of terms with normal forms are strongly convergent and normalising.*

Although needed reductions are countable, by definition of strong convergence, no bound exists on the maximum length of such reductions. To see this, consider the rule $f(Z) \to g(Z)$ and the term $f^\omega$, i.e. $f(f(\ldots f(\ldots)))$. Obviously, all redexes in $f^\omega$ are needed with respect to the unique normal form $g^\omega$. Assume that $\delta$ is a bijection between any countable, infinite ordinal $\alpha$ and $\mathbb{N}$ and note that for each depth there is precisely one position in $f^\omega$. Define $(s_\beta)_{\beta<\alpha+1}$ with $s_0 = f^\omega$ and $s_\alpha = g^\omega$ such that $s_\beta \to s_{\beta+1}$ contracts the redex at depth $\delta(\beta)$. As $\delta$ is a bijection, all rewrite steps in the reduction $(s_\beta)_{\beta<\alpha+1}$ of length $\alpha$ exist and by Proposition 3.2 the reduction is strongly convergent.

*Remark 6.9.* Needed reductions are not hypernormalising, i.e. if a *finite* number of arbitrary steps occur between each step contracting a needed redex, then the obtained reduction need not to be strongly convergent. This is contrary to the finite higher-order case [8].

To see this, consider the rules $a \to f(a)$, $b \to b$, and $g(Z, Z') \to Z$. Moreover, consider for each $n \in \mathbb{N}$ the term $g(f^n(a), b)$, where the root-redex and the

$a \to f(a)$-redex are needed, but where the $b \to b$-redex is not. We have a reduction contracting a needed redex in every other step:

$$g(a, \underline{b}) \to g(\underline{a}, b) \to g(f(a), \underline{b}) \to g(f(\underline{a}), b) \to \cdots \\ \to g(f^n(a), \underline{b}) \to g(f^n(\underline{a}), b) \to g(f^{n+1}(a), \underline{b}) \to \cdots ,$$

where the contracted redexes are underlined. The reduction is not strongly convergent, as an infinite number of $b$ redexes are contracted at a single depth.

Although hypernormalisation does not hold, not all is lost: Needed-fair reductions, i.e. reductions in which each needed redex that is a residual of another needed redex is contracted within a finite number of steps, are normalising [7].

## 7 Uniform Normalisation

We next consider uniform normalisation of iCRSs, i.e. the coincidence of weak and strong normalisation. Both the global and local variant are considered, i.e. we consider both iCRSs as a whole and individual terms. As before, we assume *fully-extended, orthogonal* iCRSs.

**Global Uniform Normalisation.** Like orthogonal iTRSs [9], fully-extended, orthogonal iCRSs are uniformly normalising. To show this, we need the following lemma, which is the iCRS analogue of Proposition 21 in [9] and whose proof is identical to the proof of that proposition.

**Lemma 7.1.** *If there exists an open strongly continuous reduction with an infinite number of steps at a certain depth, then there exists a perpetual reduction.*

We can now prove the iCRS analogue of Theorem 22 in [9]:

**Theorem 7.2.** *A fully-extended, orthogonal iCRS is weakly normalising iff it is strongly normalising.*

*Proof (Sketch).* Identical to the proof of Theorem 22 in [9]: That strong normalisation implies weak normalisation is explained on p. 177. For the reverse, reason by contradiction, employing in turn Lemmas 7.1 and 5.5. □

**Local Uniform Normalisation.** Uniform normalisation does not hold for terms, even under assumption of non-erasure [14,9], i.e. assuming that all variables occurring on the left-hand sides of rules also occur on their right-hand sides. This is contrary to TRSs [13]. That weak normalisation does not imply strong normalisation is the result of iCRSs being both infinite and higher-order.

From the perspective of infinitary rewriting, failure is due to subterms being 'pushed out' of terms (see also Sect. 6). Given the non-erasing rules $a \to a$ and $f(Z) \to g(f(Z))$, it follows that $f(a)$ reduces to the normal form $g^\omega$, but repeatedly contracting the $a \to a$-redex in $f(a)$ yields an open strongly continuous reduction of length $\omega$ with an infinite number of reductions at a single depth.

From the perspective of higher-order rewriting, failure is due to erasure by certain variables not occurring bound. Consider $a \to a$ and $f([x]Z(x), Z') \to Z(Z')$. The term $f([x]y, a)$ is weakly normalising, for we have $f([x]y, a) \to y$, where $a$ is erased as $x$ does not occur bound in $[x]y$. The term is not strongly normalising; to see this, repeatedly contract the $a \to a$-redex in $f([x]y, a)$.

As observed by Kennaway et al. [14], albeit without a proof, uniform normalisation holds for terms in iTRSs if all possible reductions are non-erasing. The same holds for iCRSs; to see this we first define non-erasing reductions:

**Definition 7.3.** *A reduction* $s \twoheadrightarrow t$ *is* non-erasing *if for every every subterm* $s'|_p$ *of term* $s'$ *in* $s \twoheadrightarrow t$ *either (1) a residual of* $s'|_p$ *occurs in* $t$ *or (2) a descendant of* $p$ *occurs in, or is a variable bound by, the redex pattern of a redex contracted in the suffix* $s' \twoheadrightarrow t$ *of* $s \twoheadrightarrow t$.

Remark that the second condition applies to a specific residual of a subterm. Any other residual must still satisfy either the first or second condition.

Strengthening the observation by Kennaway et al. [14] slightly, we obtain:

**Theorem 7.4.** *In fully-extended, orthogonal iCRSs, weak and strong normalisation coincide for terms with only non-erasing reductions starting from them.*

*Proof (Sketch).* That strong normalisation implies weak normalisation is explained on p. 177. For the reverse, reason by contradiction, employing in turn Lemma 5.5 and Theorem 2.9. □

It is in general undecidable if a term has only non-erasing reductions starting from it. Hence, sufficient, decidable criteria are called for. In the case of iTRSs an obvious criterion is the non-erasure of rules in combination with non-depth increasingness, i.e. each variable occurring on the left-hand side of a rule also occurs on its right-hand side and does so at depth lesser or equal depth.

The criterion no longer suffices for iCRSs. From above, consider the rules $a \to a$ and $f([x]Z(x), Z') \to Z(Z')$ and the term $f([x]y, a)$. Both rules are non-erasing and non-depth increasing, while $f([x]y, a)$ is not uniformly normalising.

## 8 Conclusion

Using Van Oostrom's technique of essential redexes [8], we showed that terms with perpetual reductions starting from them do not have (head) normal forms. As such, we avoided the use of the Strip Lemma, which is traditionally employed [1,9], but which no longer holds in the higher-order case.

With the help of the above, we showed that needed reductions are normalising for fully-extended, orthogonal iCRSs, extending the classical result by Huet and Lévy [10] and similar ones for iTRSs by Kennaway et al. [1] and for higher-order systems by Glauert and Khasidashvili [11]. We also proved that uniform normalisation holds for these iCRSs and, in case of non-erasing reductions, also for terms, extending results by Klop and De Vrijer [9] and Kennaway et al. [14].

A number of questions remain. For example, what is the relation between strong normalisation in infinite systems — both iTRSs and iCRSs — and root-stabilisation in finite systems [17]? What about weak orthogonality in the case of needed reductions? And, in the case of uniform normalisation can fully-extendedness be dropped or orthogonality be replaced by weak orthogonality?

The dissimilar definitions of finite and infinite reductions pose a problem in the case of root-stabilisation. Fully-extendedness cannot be dropped in case of needed reductions, as Van Raamsdonk [18] already shows for finite systems. This also implies that making the current theory more abstract might be difficult.

*Acknowledgements.* The author wishes to express his gratitude to Jakob Grue Simonsen for his useful comments and his earlier collaboration. Moreover, the author wants to thank Yoshihito Toyama for his support and the referees (of earlier versions) of the paper for comments helping to improve the presentation.

## References

1. Kennaway, R., Klop, J.W., Sleep, R., de Vries, F.J.: Transfinite reductions in orthogonal term rewriting systems. I&C 119(1), 18–38 (1995)
2. Kennaway, R., de Vries, F.J.: Infinitary rewriting. In: [19], ch.12
3. Kennaway, J.R., Klop, J.W., Sleep, M.R., de Vries, F.J.: Infinitary lambda calculus. TCS 175(1), 93–125 (1997)
4. Klop, J.W., van Oostrom, V., van Raamsdonk, F.: Combinatory reduction systems: introduction and survey. TCS 121(1&2), 279–308 (1993)
5. Ketema, J., Simonsen, J.G.: Infinitary combinatory reduction systems. In: Giesl, J. (ed.) RTA 2005. LNCS, vol. 3467, pp. 438–452. Springer, Heidelberg (2005)
6. Ketema, J., Simonsen, J.G.: On confluence of infinitary combinatory reduction systems. In: Sutcliffe, G., Voronkov, A. (eds.) LPAR 2005. LNCS (LNAI), vol. 3835, pp. 199–214. Springer, Heidelberg (2005)
7. Ketema, J., Simonsen, J.G.: Infinitary combinatory reduction systems. Technical Report D-558, Department of Computer Science, University of Copenhagen (2006)
8. van Oostrom, V.: Normalisation in weakly orthogonal rewriting. In: Narendran, P., Rusinowitch, M. (eds.) RTA 1999. LNCS, vol. 1631, pp. 60–74. Springer, Heidelberg (1999)
9. Klop, J.W., de Vrijer, R.: Infinitary normalization. In: We Will Show Them: Essays in Honour of Dov Gabbay, vol. 2, pp. 169–192. College Publications (2005)
10. Huet, G., Lévy, J.J.: Computations in orthogonal rewriting systems. In: Computational Logic: Essays in honor of Alan Robinson, pp. 395–443. MIT Press, Cambridge (1991)
11. Glauert, J., Khasidashvili, Z.: Relative normalization in orthogonal expression reduction systems. In: Lindenstrauss, N., Dershowitz, N. (eds.) CTRS 1994. LNCS, vol. 968, pp. 144–165. Springer, Heidelberg (1995)
12. Khasidashvili, Z., Ogawa, M.: Perpetuality and uniform normalization. In: Hanus, M., Heering, J., Meinke, K. (eds.) ALP 1997 and HOA 1997. LNCS, vol. 1298, pp. 240–255. Springer, Heidelberg (1997)
13. Klop, J.W.: Term rewriting systems. In: Handbook of Logic in Computer Science, vol. 2, pp. 1–116. Oxford University Press, Oxford (1992)
14. Kennaway, R., van Oostrom, V., de Vries, F.J.: Meaningless terms in rewriting. The Journal of Functional and Logic Programming 1 (1999)
15. Arnold, A., Nivat, M.: The metric space of infinite trees. Algebraic and topological properties. Fundamenta Informaticae 3(4), 445–476 (1980)
16. Klop, J.W., van Oostrom, V., de Vrijer, R.: Orthogonality. In: [19], ch. 4
17. Middeldorp, A.: Call by need computations to root-stable form. In: POPL 1997, pp. 94–105 (1997)
18. van Raamsdonk, F.: Confluence and Normalisation for Higher-Order Rewriting. PhD thesis, Vrije Universiteit, Amsterdam (1996)
19. Terese (ed.): Term Rewriting Systems. Cambridge Tracts in Theorectical Computer Science, vol. 55. Cambridge University Press, Cambridge (2003)

# Innermost Reachability and Context Sensitive Reachability Properties Are Decidable for Linear Right-Shallow Term Rewriting Systems

Yoshiharu Kojima and Masahiko Sakai

Graduate School of Information Science, Nagoya University
Furo-cho, Chikusa-ku, Nagoya, 464-8603 Japan
{kojima@trs.cm.,sakai@}is.nagoya-u.ac.jp

**Abstract.** A reachability problem is a problem used to decide whether $s$ is reachable to $t$ by $R$ or not for a given two terms $s$, $t$ and a term rewriting system $R$. Since it is known that this problem is undecidable, effort has been devoted to finding subclasses of term rewriting systems in which the reachability is decidable. However few works on decidability exist for innermost reduction strategy or context-sensitive rewriting.

In this paper, we show that innermost reachability and context-sensitive reachability are decidable for linear right-shallow term rewriting systems. Our approach is based on the tree automata technique that is commonly used for analysis of reachability and its related properties.

## 1 Introduction

The reachability problem is a problem used to decide whether $s$ is reachable to $t$ by $R$ or not for a given two terms: $s$, $t$, and a term rewriting systems (TRS) $R$. Since it is known that this problem is undecidable even if restricted to linear TRS or to shallow TRS [7], effort has been made to find subclasses of TRSs in which the reachability is decidable. Reachability properties for several subclasses of TRSs have been proved to be decidable [2,6,10,11,3]. These results are based on the more powerful property of effective preservation of regularity. We say a rewrite relation effectively preserves regularity if it is possible to construct a tree automaton (TA) which recognizes a set of terms reachable from some term in the regular set defined by a given TA. It is easy to see that the reachability property for TRSs is decidable if the TRSs effectively preserve regularity.

Innermost reduction, a strategy that rewrites innermost redexes, is used for call-by-value computation. Context-sensitive reduction [8] is a strategy in which rewritable positions are indicated by specifying arguments of function symbols. For innermost reduction strategy, recently Godoy and Huntingford showed that reachability and joinability with respect to innermost reduction for (possibly non-linear) shallow TRSs are decidable [5]. In this case the proof method is not based on tree automata techniques. This paper shows that innermost reduction and context-sensitive reduction effectively preserve regularity for linear right-shallow term rewriting systems: hence, innermost reachability, innermost

A. Voronkov (Ed.): RTA 2008, LNCS 5117, pp. 187–201, 2008.

joinability, context-sensitive joinability and context-sensitive reachability are decidable for this class.

## 2 Preliminary

Let $F$ be a set of function symbols with fixed arity and $X$ be an enumerable set of variables. The arity of function symbol $f$ is denoted by $\mathrm{ar}(f)$. Function symbols with $\mathrm{ar}(f) = 0$ are *constants*. The set of *terms*, defined in the usual way, is denoted by $\mathcal{T}(F, X)$. A term is *linear* if no variable occurs more than once in the term. The set of variables occurring in $t$ is denoted by $\mathrm{Var}(t)$. A term $t$ is *ground* if $\mathrm{Var}(t) = \emptyset$. The set of ground terms is denoted by $\mathcal{T}(F)$.

A *position* in a term $t$ is defined, as usual, as a sequence of positive integers, and the set of all positions in a term $t$ is denoted by $\mathrm{Pos}(t)$, where the empty sequence $\varepsilon$ is used to denote root position. The depth of a position $p$ is denoted by $|p|$. A term $t$ is *shallow* if every variable occurs at depth 0 or 1 in $t$. The *subterm* of $t$ at position $p$ is denoted by $t|_p$, and $t[t']_p$ represents the term obtained from $t$ by replacing the subterm $t|_p$ by $t'$.

A *substitution* $\sigma$ is a mapping from $X$ to $\mathcal{T}(F, X)$ whose domain $\mathrm{Dom}(\sigma) = \{x \in X \mid x \neq \sigma(x)\}$ is finite. We sometimes represent $\sigma$ as $\{x_1 \mapsto t_1, \ldots, x_n \mapsto t_n\}$ where $x_i \in \mathrm{Dom}(\sigma)$ and $t_i = \sigma(x_i)$. The term obtained by applying a substitution $\sigma$ to a term $t$ is written as $t\sigma$.

A *rewrite rule* is an ordered pair of terms in $\mathcal{T}(F, X)$, written as $l \to r$, where $l \notin X$ and $\mathrm{Var}(l) \supseteq \mathrm{Var}(r)$. We say that variables $x \in \mathrm{Var}(l) \setminus \mathrm{Var}(r)$ are *erasing*. A *term rewriting system (over $F$)* (TRS) is a finite set of rewrite rules. Rewrite relation $\xrightarrow[R]{}$ induced by a TRS $R$ is as follows: $s \xrightarrow[R]{} t$ if and only if $s = s[l\sigma]_p$, and $t = s[r\sigma]_p$ for some rule $l \to r \in R$, with substitution $\sigma$ and position $p \in \mathrm{Pos}(s)$. We call $l\sigma$ *redex*. We sometimes write $\xrightarrow[R]{}^p$ by presenting the position $p$ explicitly.

A rewrite rule $l \to r$ is *left-linear* (resp. *right-linear*, *linear*, *right-shallow*) if $l$ is linear (resp. $r$ is linear, $l$ and $r$ are linear, $r$ is shallow). A TRS $R$ is *left-linear* (resp. *right-linear*, *linear*, *right-shallow*) if every rule in R is left-linear (resp. right-linear, linear, right-shallow).

A *tree automaton* (TA) is a 4-tuple $\mathcal{A} = (F, Q, Q^f, \Delta)$ where $Q$ is a finite set of states, $Q^f\,(\subseteq Q)$ is a set of final states, and $\Delta$ is a finite set of transition rules of the forms $f(q_1, \ldots, q_n) \to q$ or $q_1 \to q$ where $f \in F$ with $\mathrm{ar}(f) = n$, and $q_1, \ldots, q_n, q \in Q$. We can regard $\Delta$ as a TRS over $F \cup Q$. The rewrite relation induced by $\Delta$ is called a *transition relation* denoted by $\xrightarrow[\Delta]{}$ or $\xrightarrow[\mathcal{A}]{}$. We say that a term $s$ $(\in \mathcal{T}(F))$ is *accepted* by $\mathcal{A}$ if $s \xrightarrow[\mathcal{A}]{*} q \in Q^f$. The set of all terms accepted by $\mathcal{A}$ is denoted by $\mathcal{L}(\mathcal{A})$. We say $\mathcal{A}$ *recognizes* $\mathcal{L}(\mathcal{A})$. We use a notation $\mathcal{L}(\mathcal{A}, q)$ or $\mathcal{L}(\Delta, q)$ to represent the set $\{s \mid s \xrightarrow[\mathcal{A}]{*} q\}$. A TA $\mathcal{A}$ is *deterministic* if $s \xrightarrow[\mathcal{A}]{*} q$ and $s \xrightarrow[\mathcal{A}]{*} q'$ implies $q = q'$ for any $s \in \mathcal{T}(F)$. A TA $\mathcal{A}$ is *complete* if there exists $q \in Q$ such that $s \xrightarrow[\mathcal{A}]{*} q$ for any $s \in \mathcal{T}(F)$. A set $T$ of terms is *regular* if there exists a TA $\mathcal{A}$ such that $T = \mathcal{L}(\mathcal{A})$.

Let $\to$ be a binary relation on a set $\mathcal{T}(F)$. We say $s \in \mathcal{T}(F)$ is a *normal form* (with respect to $\to$) if there exists no term $t \in \mathcal{T}(F)$ such that $s \to t$. We use $\circ$ to denote the composition of two relations. We write $\xrightarrow{*}$ for the reflexive and transitive closure of $\rightarrow$. We also write $\xrightarrow{n}$ for the relation $\rightarrow \circ \cdots \circ \rightarrow$ that is composed of $n$ $\rightarrow$'s. The set of *reachable terms* from a term in $T$ is defined by $\rightarrow[T] = \{t \mid s \in T,\ s \xrightarrow{*} t\}$. We say that a reduction $\rightarrow$ *effectively preserves regularity* if a tree automata $\mathcal{A}_*$ that satisfies $\mathcal{L}(\mathcal{A}_*) = \rightarrow[\mathcal{L}(\mathcal{A})]$ can be effectively constructed from an automata $\mathcal{A}$. The *reachability problem* (resp. *joinability problem*) with respect to $\rightarrow$ is a problem that decides whether $s \xrightarrow{*} s'$ (resp. $s \xrightarrow{*} \circ \xleftarrow{*} s'$) or not, for given terms $s$ and $s'$.

**Theorem 1 ([4]).** *Let $\rightarrow$ be a relation on terms that effectively preserves recognizability. Then both reachability and joinability properties with respect to $\rightarrow$ are decidable.*

## 3 Regularity Preservation for Innermost Reduction

We say a step rewrite $s \xrightarrow[R]{}^{p} t$ is *innermost* if all proper subterms of $s|_p$ are normal forms. We write $\xrightarrow[R]{}_{\text{in}}$ for the innermost rewrite relation induced by $R$.

This section shows that innermost reduction $\xrightarrow[R]{}_{\text{in}}$ effectively preserves regularity if $R$ is a linear right-shallow TRS. In order to show the property, we prepare a procedure $\mathrm{P_{in}}$ that inputs a TA $\mathcal{A}$ and a TRS $R$ and outputs a TA $\mathcal{A}_*$, and show that $\mathcal{A}_*$ recognizes a set $\xrightarrow[R]{}_{\text{in}}[\mathcal{L}(\mathcal{A})]$. The procedure almost follows the procedure in [6]. The main difference is the construction of states. Each state in the resulting automata consists of a pair of states. The first state originates in the input automata and remembers a reachable set. The second state remembers whether the corresponding terms are a normal form or not, which is necessary because every proper subterm of the innermost redex must be a normal form. First we show an example.

*Example 2.* Let $R = \{a \rightarrow b,\ f(x) \rightarrow g(x)\}$ and $\mathcal{A}$ be a TA such that $\mathcal{L}(\mathcal{A}) = \{f(a)\}$ defined by a finite state $\{q_{fa}\}$, and transition rules $\{a \rightarrow q_a,\ f(q_a) \rightarrow q_{fa}\}$. The procedure produces the following TA defined by final states: $Q_*^f = \{\langle q_{fa}, u_a\rangle,\ \langle q_{fa}, u_b\rangle\}$ and transition rules: $\Delta_* = \{a \rightarrow \langle q_a, u_a\rangle, b \rightarrow \langle q_a, u_b\rangle, f(\langle q_a, u_a\rangle) \rightarrow \langle q_{fa}, u_a\rangle, f(\langle q_a, u_b\rangle) \rightarrow \langle q_{fa}, u_a\rangle, g(\langle q_a, u_b\rangle) \rightarrow \langle q_{fa}, u_b\rangle\}$. Here $u_b$ is a state for normal forms.

The TA $\mathcal{A}_*$ accepts terms $f(a)$, $f(b)$ and $g(b)$ in $\xrightarrow[R]{}_{\text{in}}[\{f(a)\}]$, and does not accept $g(a)$. □

We show the procedure $\mathrm{P_{in}}$ in Figure 1, where we use a notation $\mathrm{RS}(R)$ for the set of all *non-variable direct subterms of the right-hand sides* of rules in TRS $R$: that is, $\mathrm{RS}(R) = \{r_i \notin X \mid l \rightarrow f(r_1, \ldots, r_n) \in R\}$. Note that $\mathrm{RS}(R)$ is a set of ground terms if $R$ is right shallow.

*Example 3.* Let us follow how procedure $\mathrm{P_{in}}$ works. Consider $R$ and $\mathcal{A}$ in Example 2.

In the initialization step, we have $\Delta_{\mathrm{RS}} = \emptyset$, $Q_{\mathrm{NF}}^f = \{u_b\}, \Delta_{\mathrm{NF}} = \{a \rightarrow u_a, b \rightarrow u_b, f(u_a) \rightarrow u_a, f(u_b) \rightarrow u_a, g(u_a) \rightarrow u_a, g(u_b) \rightarrow u_b\}$, $Q_* =$

**Input** TA $\mathcal{A} = \langle F, Q, Q^f, \Delta \rangle$ and left-linear right-shallow TRS $R$ over $F$.
**Output** TA $\mathcal{A}_* = \langle F, Q_*, Q_*^f, \Delta_* \rangle$ such that $\mathcal{L}(\mathcal{A}_*) = \xrightarrow[R]{}_{\text{in}}[\mathcal{L}(\mathcal{A})]$ if $R$ is right-linear.
**Step 1 (initialize)** 1. Prepare a TA $\mathcal{A}_{\text{RS}} = \langle F, Q_{\text{RS}}, Q_{\text{RS}}^f, \Delta_{\text{RS}} \rangle$ such that $Q_{\text{RS}}^f = \{q^t \mid t \in \text{RS}(R)\}$ and $\mathcal{L}(\mathcal{A}_{\text{RS}}, q^t) = \{t\}$.
2. Prepare a deterministic complete TA $\mathcal{A}_{\text{NF}} = \langle F, Q_{\text{NF}}, Q_{\text{NF}}^f, \Delta_{\text{NF}} \rangle$ such that
   - $\mathcal{L}(\mathcal{A}_{\text{NF}})$ is the set $\text{NF}_R$ ($\subseteq \mathcal{T}(F)$) of all ground normal forms
   - $\mathcal{L}(\mathcal{A}_{\text{NF}}, q) \neq \emptyset$ for any $q \in Q_{\text{NF}}$.
3. Let
   - $k := 0$
   - $Q_* = (Q \uplus Q_{\text{RS}}) \times Q_{\text{NF}}$
   - $Q_*^f = Q^f \times Q_{\text{NF}}$
   - $\Delta_0 = \{f(\langle q_1, u_1 \rangle, \ldots, \langle q_n, u_n \rangle) \to \langle q, u \rangle \mid f(q_1, \ldots, q_n) \to q \in \Delta \uplus \Delta_{\text{RS}},\ f(u_1, \ldots, u_n) \to u \in \Delta_{\text{NF}}\}$

**Step 2** Let $\Delta_{k+1}$ be transition rules produced by augmenting transition rules of $\Delta_k$ by the following inference rules:

$$\frac{f(l_1, ..., l_n) \to g(r_1, ..., r_m) \in R,\ f(\langle q_1, u_1 \rangle, \ldots, \langle q_n, u_n \rangle) \to \langle q, u \rangle \in \Delta_k}{g(\langle q_1', u_1' \rangle, \ldots, \langle q_m', u_m' \rangle) \to \langle q, u' \rangle \in \Delta_{k+1}}$$

if there exists $\theta : X \to (Q \uplus Q_{\text{RS}}) \times Q_{\text{NF}}^f$ such that

- $l_i\theta \xrightarrow[\Delta_k]{*} \langle q_i, u_i \rangle$ and $u_i \in Q_{\text{NF}}^f$ for all $1 \leq i \leq n$,
- $\langle q_j', u_j' \rangle = \begin{cases} r_j\theta & \cdots\ r_j \in X \\ \langle q^{r_j}, u'' \rangle & \cdots\ r_j \notin X,\ u'' \in Q_{\text{NF}} \end{cases}$ for all $1 \leq j \leq m$, and
- $g(u_1', \ldots, u_m') \xrightarrow[\Delta_{\text{NF}}]{} u'$.

and

$$\frac{f(l_1, ..., l_n) \to x \in R,\ f(\langle q_1, u_1 \rangle, \ldots, \langle q_n, u_n \rangle) \to \langle q, u \rangle \in \Delta_k}{\langle q', u' \rangle \to \langle q, u' \rangle \in \Delta_{k+1}}$$

if there exists $\theta : X \to (Q \uplus Q_{\text{RS}}) \times Q_{\text{NF}}^f$ such that

- $l_i\theta \xrightarrow[\Delta_k]{*} \langle q_i, u_i \rangle$ for all $1 \leq i \leq n$, and
- $\langle q', u' \rangle = x\theta$.

**Step 3** If $\Delta_{k+1} = \Delta_k$ then stop and set $\Delta_* = \Delta_k$. Otherwise, $k := k + 1$, and go to step 2.

**Fig. 1.** Procedure $\text{P}_{\text{in}}$

$\{q_a, q_{fa}\} \times \{u_a, u_b\}$, $Q_*^f = \{q_{fa}\} \times \{u_a, u_b\}$, $\Delta_0 = \{a \to \langle q_a, u_a \rangle, f(\langle q_a, u_a \rangle) \to \langle q_{fa}, u_a \rangle, f(\langle q_a, u_b \rangle) \to \langle q_{fa}, u_a \rangle\}$.

The saturation steps stop at $k = 1$ and we have $\Delta_1 = \Delta_0 \cup \{b \to \langle q_a, u_b \rangle,\ g(\langle q_a, u_b \rangle) \to \langle q_{fa}, u_b \rangle\}$, $\Delta_2 = \Delta_1$ □

The procedure $\mathrm{P_{in}}$ eventually terminates at some $k$, because rewrite rules in $R$ and states $Q_*$ are finite, and hence, possible transitions rules are finite. Apparently $\Delta_0 \subset \cdots \subset \Delta_k = \Delta_{k+1} = \cdots$. A measurement of transitions of $\Delta_*$ is defined as $||s \xrightarrow[\Delta_0]{} t|| = 0$ and $||s \xrightarrow[\Delta_{i+1}\setminus\Delta_i]{} t|| = i+1$ for $i \geq 0$. This is extended on transition sequences as a multiset:

$$||s_0 \xrightarrow[\Delta_*]{} s_1 \xrightarrow[\Delta_*]{} \cdots \xrightarrow[\Delta_*]{} s_n|| = \{||s_i \xrightarrow[\Delta_*]{} s_{i+1}|| \mid 0 \leq i < n\}.$$

Now we can define an order $\sqsupset$ on transition sequences by $\Delta_*$, which is necessary in proofs.

$$\alpha \sqsupset \beta \overset{\text{def}}{\Leftrightarrow} ||\alpha|| >_{\text{mul}} ||\beta||$$

where $>_{\text{mul}}$ is the multiset extension of $>$ on $\mathbb{N}$.

**Proposition 4.** *(a) $s \xrightarrow[\Delta]{*} q$ if and only if $s \xrightarrow[\Delta_0]{*} \langle q, u\rangle$ for some $u \in Q_{NF}$.*
*(b) $s \xrightarrow[\Delta_*]{*} \langle q, u\rangle$ implies $s \xrightarrow[\Delta_{\mathrm{NF}}]{*} u$*

*Proof.* Direct consequence of the construction of $\Delta_0$ and the completeness of $\mathcal{A}_{\mathrm{NF}}$. □

**Lemma 5.** *Let $\alpha : s[\langle q, u\rangle]_p \xrightarrow[\Delta_k]{*} \langle q', u'\rangle$. If $k = 0$ or $u \in Q_{NF} \setminus Q^f_{NF}$ then there exists $v' \in Q_{NF}$ such that $\beta : s[\langle q, v\rangle]_p \xrightarrow[\Delta_k]{*} \langle q', v'\rangle$ and $\alpha \sqsupseteq \beta$ for any $v \in Q_{NF}$.*

*Proof.* If $k = 0$, it trivially holds from the construction of $\Delta_0$. We prove in the case $u \in Q_{\mathrm{NF}} \setminus Q^f_{\mathrm{NF}}$ by induction on steps $n$ of transition $s[\langle q, v\rangle]_p \xrightarrow[\Delta_k]{n} \langle q', v'\rangle$. Since the lemma is trivial in the case $n = 0$, let $n > 0$. Then we have two cases according to the form of the transition rule applied in the last step.

1. In the case that

$$s[\langle q, u\rangle]_p \xrightarrow[\Delta_k]{n-1} f(\langle q_1, u_1\rangle, \ldots, \langle q_n, u_n\rangle) \xrightarrow[\Delta_k]{} \langle q', u'\rangle, \tag{1}$$

   the position $p$ can be represented as $ip'$ for $1 \leq i \leq n$. Here $u_i \notin Q^f_{\mathrm{NF}}$ follows from $u \notin Q^f_{\mathrm{NF}}$, Proposition 4 (b), and the construction of $\Delta_{\mathrm{NF}}$. Since $\alpha_i : (s|_i)[\langle q, u\rangle]_{p'} \xrightarrow[\Delta_k]{*} \langle q_i, u_i\rangle$, we have $\beta_i : (s|_i)[\langle q, v\rangle]_{p'} \xrightarrow[\Delta_k]{*} \langle q_i, v''\rangle$ and $\alpha_i \sqsupseteq \beta_i$ for some $v'' \in Q_{\mathrm{NF}}$ by the induction hypothesis. Thus we have $s[\langle q, v\rangle]_p \xrightarrow[\Delta_k]{*} f(\ldots, \langle q_{i-1}, u_{i-1}\rangle, \langle q_i, v''\rangle, \langle q_{i+1}, u_{i+1}\rangle, \ldots)$
   (a) If the transition rule in the last step of (1) is in $\Delta_0$, we also have $f(\ldots, \langle q_{i-1}, u_{i-1}\rangle, \langle q_i, v''\rangle, \langle q_{i+1}, u_{i+1}\rangle, \ldots) \to \langle q', v'\rangle \in \Delta_0$ from the construction, where $v'$ is determined by $f(\ldots, u_{i-1}, v'', u_{i+1}, \ldots) \to v' \in \Delta_{\mathrm{NF}}$.
   (b) Otherwise we assume that the transition rule in the last step of (1) is in $\Delta_k \setminus \Delta_{k-1}$ without loss of generality. It is known that the rule is produced by the first inference rule in Step 2 and hence $r_i \notin X$; otherwise $u_i \in Q^f_{\mathrm{NF}}$ follows from $\langle q_i, u_i\rangle = r_i\theta$, which contradicts $u_i \notin Q^f_{\mathrm{NF}}$. Thus $f(\ldots, \langle q_{i-1}, u_{i-1}\rangle, \langle q_i, v''\rangle, \langle q_{i+1}, u_{i+1}\rangle, \ldots) \to \langle q', v'\rangle \in \Delta_k \setminus \Delta_{k-1}$ and $\alpha \sqsupseteq \beta$.

2. In the case that $s[\langle q,u\rangle]_p \xrightarrow[\Delta_k]{n-1} \langle q_1,u_1\rangle \xrightarrow[\Delta_k]{} \langle q',u'\rangle$, we can show the lemma similarly to the previous case. □

**Lemma 6.** *Let $R$ be left-linear and right-shallow. Then $s \xrightarrow[\Delta_*]{*} \langle q,u\rangle$ and $s \xrightarrow[R]{n}_{\text{in}} t$ imply $t \xrightarrow[\Delta_*]{*} \langle q,u'\rangle$ for some $u' \in Q_{NF}$.*

*Proof.* We present the proof in the case where $n = 1$. Let $s \xrightarrow[\Delta_k]{*} \langle q,u\rangle$ and $s = s[l\sigma]_p \xrightarrow[R]{}_{\text{in}} s[r\sigma]_p = t$ for some rewrite rule $l \to r \in R$.

1. Consider the case where the rewrite rule is in the form $f(l_1,\ldots,l_n) \to g(r_1,\ldots,r_m)$. Since this rewrite rule is left-linear, $s \xrightarrow[\Delta_k]{*} \langle q,u\rangle$ is represented as $s = s[f(l_1\sigma,\ldots,l_n\sigma)]_p \xrightarrow[\Delta_k]{*} s[f(l_1\theta,\ldots,l_n\theta)]_p \xrightarrow[\Delta_k]{*} s[f(\langle q_1,u_1\rangle,\ldots,\langle q_n,u_n\rangle)]_p \xrightarrow[\Delta_k]{} s[\langle q'',u''\rangle]_p \xrightarrow[\Delta_k]{*} \langle q,u\rangle$ for some $\theta : X \to Q_*$. Note that $u'' \notin Q^f_{\text{NF}}$ since $s|_p$ is not a normal form.
We have $l_i\theta \xrightarrow[\Delta_k]{*} \langle q_i,u_i\rangle$, where $u_i \in Q^f_{\text{NF}}$ since each $l_i\sigma$ is a normal form. We also have $f(u_1,\ldots,u_n) \xrightarrow[\Delta_{\text{NF}}]{*} u''$ by Proposition 4 (b). From the construction of $\mathcal{A}_*$, there exists a transition rule $g(\langle q'_1,u'_1\rangle,\ldots,\langle q'_n,u'_m\rangle) \to \langle q'',v'\rangle \in \Delta_{k+1}$ such that
$$\langle q'_j,u'_j\rangle = \begin{cases} r_j\theta & \cdots\ r_j \in X \\ \langle q^{r_j},v''\rangle & \cdots\ r_j \notin X \text{ where } r_j \xrightarrow[\Delta_0]{*} \langle q^{r_j},v''\rangle \end{cases}$$
   (a) For $j$ such that $r_j \in X$, we have $l_i|_{p'} = r_j$ for some $i$ and $p'$. Hence $r_j\sigma = l_i|_{p'}\sigma \xrightarrow[\Delta_k]{*} l_i|_{p'}\theta = r_j\theta = \langle q'_j,u'_j\rangle$.
   (b) For $j$ such that $r_j \notin X$, we have $r_j\sigma = r_j$ since $R$ is right-shallow, and $r_j \xrightarrow[\Delta_0]{*} \langle q^{r_j},v''\rangle = \langle q'_j,u'_j\rangle$.

   Therefore we have $t = s[g(r_1\sigma,\ldots,r_m\sigma)]_p \xrightarrow[\Delta_k]{*} s[g(\langle q'_1,u'_1\rangle,\ldots,\langle q'_m,u'_m\rangle)]_p \xrightarrow[\Delta_{k+1}]{} s[\langle q'',v'\rangle]_p$, and $s[\langle q'',v'\rangle]_p \xrightarrow[\Delta_k]{*} \langle q,u'\rangle$ for some $u' \in Q_{\text{NF}}$ by Lemma 5.
2. In the case that the rewrite rule is in the form $f(l_1,\ldots,l_n) \to x$, we can show the lemma similarly to the previous case. □

Now we obtain the following lemma.

**Lemma 7.** *If $R$ be left-linear and right-shallow, then $\mathcal{L}(\mathcal{A}_*) \supseteq \xrightarrow[R]{}_{\text{in}}[\mathcal{L}(\mathcal{A})]$.*

*Proof.* Let $s \xrightarrow[R]{*}_{\text{in}} t$ and $s \xrightarrow[\Delta]{*} q \in Q^f$. Then we have $s \xrightarrow[\Delta_0]{*} \langle q,u\rangle \in Q^f_*$ by Proposition 4 (a). Hence $t \xrightarrow[\Delta_*]{*} \langle q,u'\rangle \in Q^f_*$ by Lemma 6. □

**Lemma 8.** *Let $\Delta_*$ be generated from a right-linear right-shallow TRS. Then $\alpha : t\ (\xrightarrow[\Delta_0]{*} \circ \xrightarrow[\Delta_{k+1}]{})\ \langle q,u'\rangle$ implies $s \xrightarrow[R]{}_{\text{in}} t$, $\beta : s \xrightarrow[\Delta_k]{*} \langle q,u\rangle$ and $\alpha \sqsupset \beta$ for some term $s$ and $u \in Q_{NF}$.*

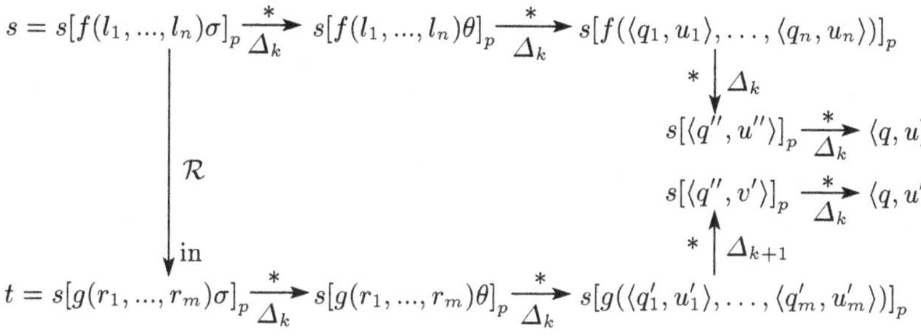

**Fig. 2.** The diagram of proof of lemma 6

*Proof.* Consider the case where the last transition rule applied in $\alpha$ is (in the form of) $g(\langle q'_1, u'_1\rangle, \ldots, \langle q'_m, u'_m\rangle) \to \langle q, u'\rangle$ and we assume that it is in $\Delta_{k+1} \setminus \Delta_k$ without loss of generality. Then $\alpha$ can be represented as $t = g(t_1, \ldots, t_m) \xrightarrow[\Delta_0]{*} g(\langle q'_1, u'_1\rangle, \ldots, \langle q'_m, u'_m\rangle) \xrightarrow[\Delta_{k+1}\setminus\Delta_k]{} \langle q, u'\rangle$, and the last transition rule applied is added by the first inference rule in the procedure. Hence there exist $f(l_1, \ldots, l_n) \to g(r_1, \ldots, r_m) \in R$, $f(\langle q_1, u_1\rangle, \ldots, \langle q_n, u_n\rangle) \to \langle q, u\rangle \in \Delta_k$, and $\theta : X \to (Q \uplus Q_{\mathrm{RS}}) \times Q^f_{\mathrm{NF}}$ such that

- $l_i\theta \xrightarrow[\Delta_k]{*} \langle q_i, u_i\rangle$ and $u_i \in Q^f_{\mathrm{NF}}$,
- $\langle q'_j, u'_j\rangle = r_j\theta$ if $r_j \in X$,
- $q'_j = q^{r_j}$ if $r_j \notin X$, and
- $\mathcal{L}(\Delta_0, x\theta) \neq \emptyset$ for each erasing variable.

Hence, we have the following:

1. For $j$ such that $r_j \in X$, we have $t_j \xrightarrow[\Delta_0]{*} \langle q'_j, u'_j\rangle = r_j\theta$.
2. For $j$ such that $r_j \notin X$, we have $t_j \xrightarrow[\Delta_{\mathrm{RS}}]{*} q^{r_j}$ hence $t_j = r_j$ from the construction of $\Delta_0$. Thus we have $t_j = r_j\theta$ since R is right-shallow.

Thus we have $g(t_1, \ldots, t_m) \xrightarrow[\Delta_0]{*} g(r_1\theta, \ldots, r_m\theta) \xrightarrow[\Delta_0]{*} g(\langle q'_1, u'_1\rangle, \ldots, \langle q'_m, u'_m\rangle)$. We define a substitution $\sigma : \mathrm{Var}(f(l_1, \ldots, l_n)) \to \mathcal{T}(F)$ as follows:

$$x\sigma = \begin{cases} t_j & \cdots \text{ if there exists } j \text{ such that } r_j = x \\ t' & \cdots \text{ otherwise, choose an arbitral } t' \text{ such that } t' \xrightarrow[\Delta_0]{*} x\theta, \end{cases}$$

where $\sigma$ is well-defined from the right-linearity of rewrite rules. We can construct $\beta : f(l_1, \ldots, l_n)\sigma \xrightarrow[\Delta_0]{*} f(l_1, \ldots, l_n)\theta \xrightarrow[\Delta_k]{*} f(\langle q_1, u_1\rangle, \ldots, \langle q_n, u_n\rangle) \xrightarrow[\Delta_k]{} \langle q, u\rangle$, where $\alpha \sqsupset \beta$. Since $u_i \in Q^f_{\mathrm{NF}}$, each $l_i\sigma$ is a normal form. Hence we have $f(l_1, \ldots, l_n)\sigma \xrightarrow[R]{}_{\mathrm{in}} g(r_1, \ldots, r_m)\sigma = g(t_1, \ldots, t_m) = t$. Therefore the lemma of the case follows by taking $s = f(l_1, \ldots, l_n)\sigma$.

For the case where the transition rule applied last in $\alpha$ is (in the form of) $\langle q', u''\rangle \to \langle q, u'\rangle$, the lemma can be shown as similar to the previous case. □

**Lemma 9.** *If $R$ be right-linear and right-shallow, then* $\mathcal{L}(\mathcal{A}_*) \subseteq \xrightarrow[R]{}_{\mathrm{in}}[\mathcal{L}(\mathcal{A})]$.

*Proof.* From Proposition 4(a), it is enough to show the claim that $\alpha : t \xrightarrow[\Delta_*]{*} \langle q, u' \rangle$ implies $s \xrightarrow[R]{*}_{\text{in}} t$, and $s \xrightarrow[\Delta_0]{*} \langle q, u \rangle$ for some $s \in \mathcal{T}(F)$ and $u \in Q_{\text{NF}}$. We prove it by induction on $\alpha$ with respect to $\sqsupset$.

1. Consider the case where the last transition rule applied in $\alpha$ is (in the form of) $g(\langle q'_1, u'_1 \rangle, \ldots, \langle q'_m, u'_m \rangle) \to \langle q, u \rangle \in \Delta_k$. Then $\alpha$ can be represented as $t = g(t_1, \ldots, t_m) \xrightarrow[\Delta_*]{*} g(\langle q'_1, u'_1 \rangle, \ldots, \langle q'_m, u'_m \rangle) \xrightarrow[\Delta_k]{} \langle q, u \rangle$. Since $\alpha \sqsupset (t_j \xrightarrow[\Delta_*]{*} \langle q'_j, u'_j \rangle)$, there exists $s_j$ for every $j$ such that $s_j \xrightarrow[R]{*}_{\text{in}} t_j$ and $s_j \xrightarrow[\Delta_0]{*} \langle q'_j, v'_j \rangle$ from the induction hypothesis. Here we have $v'_j \in Q_{\text{NF}} \setminus Q^f_{\text{NF}}$ or $s_j = t_j$ for each $j$ since $v'_i \in Q^f_{\text{NF}}$ implies that $s_i$ is a normal form. Hence $g(s_1, \ldots, s_m) \xrightarrow[R]{*}_{\text{in}} t$ and we have $\alpha' : g(s_1, \ldots, s_m) \xrightarrow[\Delta_0]{*} g(\langle q'_1, v'_1 \rangle, \ldots, \langle q'_m, v'_m \rangle) \xrightarrow[\Delta_k]{} \langle q, v \rangle$ and $\alpha \sqsupseteq \alpha'$ by applying Lemma 5 repeatedly to $g(\langle q'_1, u'_1 \rangle, \ldots, \langle q'_m, u'_m \rangle) \xrightarrow[\Delta_k]{} \langle q, u \rangle$.
   If $k = 0$ then the claim trivially holds by letting $s = g(s_1, \ldots, s_m)$. Hence let $k > 0$. Then we have $\beta : s' \xrightarrow[\Delta_{k-1}]{*} \langle q, v' \rangle$ with $\alpha' \sqsupset \beta$ and $s' \xrightarrow[R]{}_{\text{in}} g(s_1, \ldots, s_m)$ for some $s'$ and $v'$ by Lemma 8. Since $\alpha \sqsupset \beta$, the claim of this case follows from the induction hypothesis.
2. In the case where the last transition rule applied in $\alpha$ is (in the form of) $\langle q', u' \rangle \to \langle q, u \rangle \in \Delta_k$, we can show the lemma similarly to the previous case. □

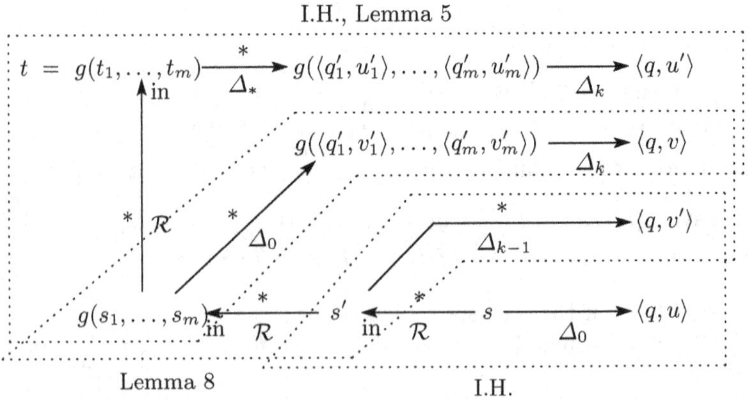


**Fig. 3.** The diagram of proof of lemma 9

We obtain the following theorems from Lemma 7, Lemma 9, and Theorem 1.

**Theorem 10.** *Innermost reduction for linear right-shallow TRSs effectively preserves regularity. Thus innermost reachability is a decidable property for linear right-shallow TRSs.*

# 4 Regularity Preservation for Context-Sensitive Reduction

A context-sensitive rewrite relation is a subrelation of a rewrite relation in which rewritable positions are indicated by specifying arguments of function symbols. A mapping $\mu : F \to \mathcal{P}(\mathbb{N})$ is said to be a *replacement map* (or $F$-map) if $\mu(f) \subseteq \{1, \ldots, \mathrm{ar}(f)\}$ for all $f \in F$. A *context-sensitive term rewriting system* (CS-TRS) is a pair $\mathcal{R} = (R, \mu)$ of a TRS and a replacement map. We say $R$ is an underlined TRS of $\mathcal{R}$. The set of *$\mu$-replacing positions* $\mathrm{Pos}^{\mu}(t)$ ($\subseteq \mathrm{Pos}(t)$) is recursively defined: $\mathrm{Pos}^{\mu}(t) = \{\varepsilon\}$ if $t$ is a constant or a variable, otherwise $\mathrm{Pos}^{\mu}(f(t_1, \ldots, t_n)) = \{\varepsilon\} \cup \{ip \mid i \in \mu(f), p \in \mathrm{Pos}^{\mu}(t_i)\}$. The rewrite relation induced by a CS-TRS $\mathcal{R}$ is defined: $s \underset{\mathcal{R}}{\hookrightarrow} t$ if and only if $s \underset{R}{\to}^{p} t$ and $p \in \mathrm{Pos}^{\mu}(t)$.

Similarly to the previous section, this section shows that a context-sensitive reduction $\underset{\mathcal{R}}{\hookrightarrow}$ effectively preserves recognizability if the underlined TRS $R$ of $\mathcal{R}$ is linear and right-shallow. In order to show the property, we prepare a procedure $\mathrm{P_{cs}}$ that inputs a TA $\mathcal{A}$ and a CS-TRS $\mathcal{R}$ and outputs a TA $\mathcal{A}_*$, and show that $\mathcal{L}(\mathcal{A}_*) = \underset{\mathcal{R}}{\hookrightarrow}[\mathcal{L}(\mathcal{A})]$.

The main idea is the introduction of an extra state $\tilde{q}$ for each state $q$. The former state $\tilde{q}$ is used for accepting terms in $\underset{\mathcal{R}}{\hookrightarrow}[\mathcal{L}(\mathcal{A}, q)]$, while the latter state $q$ keeps accepting terms only in $\mathcal{L}(\mathcal{A}, q)$.

*Example 11.* Let $\mathcal{R} = (R, \mu)$ where $R = \{a \to b, f(x) \to g(x)\}$ and $\mu(f) = \emptyset, \mu(g) = \{1\}$. Let $\mathcal{A}$ be a TA that recognizes $\{f(a)\}$ defined by a final state $\{q_{fa}\}$ and transition rules $\{a \to q_a, f(q_a) \to q_{fa}\}$. The procedure produces the following TA defined by final states: $Q_*^f = \{\tilde{q}_{fa}\}$, and transition rules: $\Delta_* = \{a \to q_a, a \to \tilde{q}_a, f(q_a) \to q_{fa}, f(q_a) \to \tilde{q}_{fa}, b \to \tilde{q}_a, g(\tilde{q}_a) \to \tilde{q}_{fa}\}$.

The TA $\mathcal{A}_*$ accepts terms $f(a)$, $g(a)$ and $g(b)$ in $\underset{\mathcal{R}}{\hookrightarrow}[\{f(a)\}]$, and does not accept $f(b)$. □

We show the procedure $\mathrm{P_{cs}}$ in Figure 4, where we use a notation $\widetilde{Q}$ to represent $\{\tilde{q} \mid q \in Q\}$, and $q^*$ to represent either $q$ or $\tilde{q}$.

*Example 12.* Let us follow how procedure $\mathrm{P_{cs}}$ works. Consider $R$ and $\mathcal{A}$ in Example 11.

In the initializing step, we have $\Delta_{\mathrm{RS}} = \emptyset$ , $Q_* = \{q_a, q_{fa}, \tilde{q}_a, \tilde{q}_{fa}\}$, $Q_*^f = \{\tilde{q}_{fa}\}$, $\Delta_0 = \Delta \cup \{a \to \tilde{q}_a, f(q_a) \to \tilde{q}_{fa}\}$.

The saturation steps stop at $k = 1$, we have $\Delta_1 = \Delta_0 \cup \{b \to \tilde{q}_a, g(\tilde{q}_a) \to \tilde{q}_{fa}\}$, $\Delta_2 = \Delta_1$. □

The procedure $\mathrm{P_{cs}}$ eventually terminates at some $k$, because rewrite rules in $R$ and states $Q_*$ are finite and hence possible transitions rules are finite. Apparently $\Delta_0 \subset \cdots \subset \Delta_k = \Delta_{k+1} = \cdots$.

We show several technical lemmas.

**Proposition 13.** *If* $t \xrightarrow[\Delta_k]{*} q \in Q \uplus Q_{RS}$, *then* $t \xrightarrow[\Delta_0]{*} q$.

**Input** TA $\mathcal{A} = \langle F, Q, Q^f, \Delta \rangle$ and right-shallow CS-TRS $\mathcal{R} = (R, \mu)$ over $F$.
**Output** TA $\mathcal{A}_* = \langle F, Q_*, Q^f_*, \Delta_* \rangle$ such that $\mathcal{L}(\mathcal{A}_*) = \underset{\mathcal{R}}{\hookrightarrow}[\mathcal{L}(\mathcal{A})]$, if $R$ is linear.
**Step 1 (initialize)** 1. Prepare a TA $\mathcal{A}_{\rm RS} = \langle F, Q_{\rm RS}, Q^f_{\rm RS}, \Delta_{\rm RS} \rangle$ that recognizes $\mathrm{RS}(R)$ (the same as the procedure $\mathrm{P_{in}}$). Here we assume $Q^f_{\rm RS} = \{q^t \mid t \in \mathrm{RS}(R)\}$ and $\mathcal{L}(\mathcal{A}_{\rm RS}, q^t) = \{t\}$.
2. Let
   - $k := 0$
   - $Q_* = (Q \uplus Q_{\rm RS}) \cup (\widetilde{Q \uplus Q_{\rm RS}})$
   - $Q^f_* = \widetilde{Q^f}$
   - $\Delta_0 = \Delta \cup \{\tilde{q}' \to \tilde{q} \mid q' \to q \in \Delta\}$

$$\cup \left\{ f(p_i, \ldots, p_n) \to \tilde{q} \;\middle|\; \begin{array}{l} f(q_1, \ldots, q_n) \to q \in \Delta, \\ p_i = \begin{cases} \tilde{q}_i & \cdots \text{ if } i \in \mu(f), \\ q_i & \cdots \text{ otherwise} \end{cases} \end{array} \right\}$$

**Step 2** Let $\Delta_{k+1}$ be transition rules produced by augmenting transition rules of $\Delta_k$ by following inference rules:

$$\frac{f(l_1, \ldots, l_n) \to g(r_1, \ldots, r_m) \in R \quad f(q^*_1, \ldots, q^*_n) \to \tilde{q} \in \Delta_k}{g(q'^*_1, \ldots, q'^*_m) \to \tilde{q} \in \Delta_{k+1}}$$

if there exists $\theta : X \to Q_*$ such that

- $l_i\theta \xrightarrow[\Delta_k]{*} q^*_i$ for all $1 \leq i \leq n$,
- $q'^*_j = \begin{cases} \tilde{p}_j & \cdots j \in \mu(g), r_j \in X, \; p^*_j = r_j\theta \\ p^*_j & \cdots j \notin \mu(g), r_j \in X, \; p^*_j = r_j\theta \\ \tilde{q}^{r_j} & \cdots j \in \mu(g), r_j \notin X \\ q^{r_j} & \cdots j \notin \mu(g), r_j \notin X \end{cases}$ for all $1 \leq j \leq m$

and

$$\frac{f(l_1, \ldots, l_n) \to x \in R \quad f(q^*_1, \ldots, q^*_n) \to \tilde{q} \in \Delta_k}{\tilde{q}' \to \tilde{q} \in \Delta_{k+1}}$$

if there exists $\theta : X \to Q_*$ such that

- $l_i\theta \xrightarrow[\Delta_k]{*} q^*_i$ for all $1 \leq i \leq n$, and
- $\tilde{q}' = \tilde{p}$ where $p^* = x\theta$.

**Step 3** If $\Delta_{k+1} = \Delta_k$ then stop and set $\Delta_* = \Delta_k$; Otherwise $k := k+1$ and goto step 2.

**Fig. 4.** Procedure $\mathrm{P_{cs}}$

*Proof.* The proposition follows from the fact that transition rules having $q \in Q \uplus Q_{\rm RS}$ on right-hand sides are in $\Delta$ or $\Delta_{\rm RS}$. □

**Proposition 14.** *$t \xrightarrow[\Delta_0]{*} \tilde{q} \in \widetilde{Q \uplus Q_{RS}}$ if and only if $t \xrightarrow[\Delta_0]{*} q \in Q \uplus Q_{RS}$.*

*Proof.* From construction of $\Delta_0$. □

**Proposition 15.** *If* $t \xrightarrow[\Delta_k]{*} q \in Q \uplus Q_{RS}$, *then* $t \xrightarrow[\Delta_k]{*} \tilde{q}$.

*Proof.* Let $t \xrightarrow[\Delta_k]{*} q$, then $t \xrightarrow[\Delta_0]{*} q$ by Proposition 13. The lemma follows from Proposition 14 and $\Delta_0 \subseteq \Delta_k$. □

**Lemma 16.** *If* $t[t']_p \xrightarrow[\Delta_k]{*} \tilde{q}$ *and* $p \in \mathrm{Pos}^{\mu}(t)$, *then there exists* $\tilde{q}'$ *such that* $t' \xrightarrow[\Delta_k]{*} \tilde{q}'$ *and* $t[\tilde{q}']_p \xrightarrow[\Delta_k]{*} \tilde{q}$.

*Proof.* We show the lemma by induction on the length $n$ of transition sequence $\alpha : t[t']_p \xrightarrow[\Delta_k]{n} \tilde{q}$. Let $t[t']_p \xrightarrow[\Delta_k]{n} \tilde{q}$ and $p \in \mathrm{Pos}^{\mu}(t)$.

1. If $p = \varepsilon$, then $t = t'$, and hence $t' \xrightarrow[\Delta_k]{*} \tilde{q}$ follows.
2. Consider the case $p = ip'$ for some $i \in \mathbb{N}$. Then $\alpha$ can be represented as $t[t']_p = f(\ldots, t_{i-1}, t_i[t']_{p'}, t_{i+1}, \ldots) \xrightarrow[\Delta_k]{n-1} f(\ldots, q^*_{i-1}, q^*_i, q^*_{i+1}, \ldots) \xrightarrow[\Delta_k]{} \tilde{q}$. Since $ip' = p \in \mathrm{Pos}^{\mu}(t)$, we have $i \in \mu(f)$. Hence $q^*_i = \tilde{q}_i$ follows from the construction of $\Delta_k$.
   By the induction hypothesis, there exists $\tilde{q}'$ such that $t' \xrightarrow[\Delta_k]{*} \tilde{q}'$ and $t_i[\tilde{q}']_{p'} \xrightarrow[\Delta_k]{*} \tilde{q}_i$. Here we have $t[\tilde{q}']_p = f(\ldots, t_{i-1}, t_i[\tilde{q}']_{p'}, t_{i+1}, \ldots) \xrightarrow[\Delta_k]{*} f(\ldots, q^*_{i-1}, \tilde{q}_i, q^*_{i+1}, \ldots) \xrightarrow[\Delta_k]{} \tilde{q}$. □

The following lemma is obtained from the above propositions and lemmas.

**Lemma 17.** *Let* $\mathcal{R}$ *be left-linear and right-shallow. Then* $s \xrightarrow[\Delta_*]{*} \tilde{q}$ *and* $s \xhookrightarrow[\mathcal{R}]{n} t$ *imply* $t \xrightarrow[\Delta_*]{*} \tilde{q}$.

*Proof.* We present the proof in the case $n = 1$. Let $s \xrightarrow[\Delta_k]{*} \tilde{q}$ and $s = s[l\sigma]_p \xhookrightarrow[\mathcal{R}]{} s[r\sigma]_p = t$ for some rewrite rule $l \to r \in R$, where $p \in \mathrm{Pos}^{\mu}(s)$. We have a transition sequence $s \xrightarrow[\Delta_k]{*} s[\tilde{q}']_p \xrightarrow[\Delta_k]{*} \tilde{q}$ by Lemma 16.

1. Consider the case where the rewrite rule is in the form $f(l_1, \ldots, l_n) \to g(r_1, \ldots, r_m)$. Since this rewrite rule is left-linear, $s \xrightarrow[\Delta_k]{*} \tilde{q}$ is represented as $s = s[f(l_1\sigma, \ldots, l_n\sigma)]_p \xrightarrow[\Delta_k]{*} s[f(l_1\theta, \ldots, l_n\theta)]_p \xrightarrow[\Delta_k]{*} s[f(q^*_1, \ldots, q^*_n)]_p \xrightarrow[\Delta_k]{} s[\tilde{q}']_p \xrightarrow[\Delta_k]{*} \tilde{q}$ for some $\theta : X \to Q_*$. Here we have $l_i\theta \xrightarrow[\Delta_k]{*} q^*_i$. From the construction of $\mathcal{A}_*$, there exists a transition rule $g(q'^*_1, \ldots, q'^*_n) \to \tilde{q}' \in \Delta_{k+1}$ such that
$$q'^*_j = \begin{cases} \tilde{p}_j & \cdots\ j \in \mu(g), r_j \in X \\ p^*_j & \cdots\ j \notin \mu(g), r_j \in X \\ \tilde{q}^{r_j} & \cdots\ j \in \mu(g), r_j \notin X \\ q^{r_j} & \cdots\ j \notin \mu(g), r_j \notin X \end{cases}$$
   where $p^*_j = r_j\theta$.
   (a) For $j$ such that $r_j \in X$, we have $l_i|_{p'} = r_j$ for some $i$ and $p'$. Hence $r_j\sigma = l_i|_{p'}\sigma \xrightarrow[\Delta_k]{*} l_i|_{p'}\theta = r_j\theta = p^*_j$. Since we have $r_j\sigma \xrightarrow[\Delta_k]{*} \tilde{p}_j$ by Proposition 15, we obtain $r_j\sigma \xrightarrow[\Delta_k]{*} q'^*_j$ in either case of $q'^*_j = \tilde{p}_j$ or $q'^*_j = p_j$.

(b) For $j$ such that $r_j \notin X$, we have $r_j\sigma = r_j$ from the shallowness of $r_j$ and also have $r_j \xrightarrow[\Delta_0]{*} q^{r_j}$. Thus $r_j\sigma \xrightarrow[\Delta_0]{*} q^{r_j}$. Since we have $r_j\sigma \xrightarrow[\Delta_0]{*} \tilde{q}^{r_j}$ by Proposition 15, we obtain $r_j\sigma \xrightarrow[\Delta_k]{*} q_j'^*$ in either case of $q_j'^* = \tilde{q}^{r_j}$ or $q_j'^* = q^{r_j}$ .

Therefore we have $t = s[g(r_1, \ldots, r_m)\sigma]_p \xrightarrow[\Delta_k]{*} s[g(q_1'^*, \ldots, q_m'^*)]_p \xrightarrow[\Delta_{k+1}]{} s[\tilde{q}']_p \xrightarrow[\Delta_k]{*} \tilde{q}$.

2. In the case where the rewrite rule is in the form $f(l_1, \ldots, l_n) \to x$, we can show the lemma similarly to the previous case. □

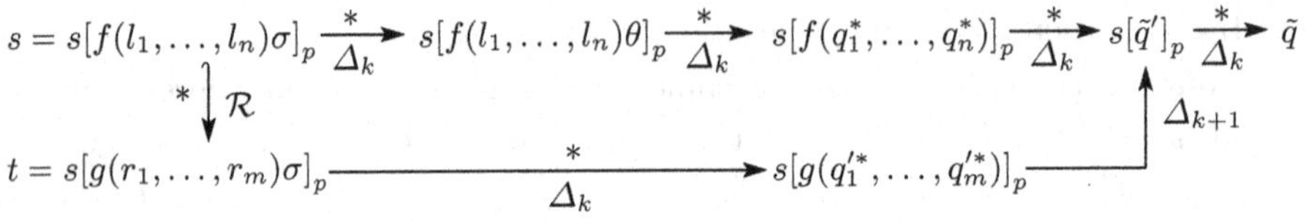


**Fig. 5.** The diagram of proof of lemma 17

**Lemma 18.** *If $\mathcal{R}$ is left-linear and right-shallow then $\mathcal{L}(\mathcal{A}_*) \supseteq \xhookrightarrow[\mathcal{R}]{}[\mathcal{L}(\mathcal{A})]$.*

*Proof.* Let $s \xhookrightarrow[\mathcal{R}]{*} t$ and $s \xrightarrow[\Delta]{*} q \in Q^f$. Since $s \xrightarrow[\Delta_0]{*} q$ from construction of $\Delta_0$, we have $s \xrightarrow[\Delta_0]{*} \tilde{q}$ by Proposition 15. Hence $t \xrightarrow[\Delta_*]{*} \tilde{q} \in Q_*^f$ by lemma17. □

**Lemma 19.** *Let $\Delta_*$ be generated from linear right-shallow CS-TRS. Then*

*1. $\alpha: \ t = g(t_1, \ldots, t_m) \xrightarrow[\Delta_*]{*} g(q_1'^*, \ldots, q_m'^*) \xrightarrow[\Delta_{k+1} \setminus \Delta_k]{} \tilde{q}$ where $t_j \xrightarrow[\Delta_0]{*} q_j'^*$ for all $j \in \mu(g)$, or*
*2. $\alpha: \ t \xrightarrow[\Delta_0]{*} \tilde{q}' \xrightarrow[\Delta_{k+1} \setminus \Delta_k]{} \tilde{q}$,*

*implies $s \xhookrightarrow[\mathcal{R}]{} t$, $\beta: \ s \xrightarrow[\Delta_*]{*} \tilde{q}$ and $\alpha \sqsupset \beta$ for some term $s$.*

*Proof.* Consider the first case. Since the last transition rule applied in $\alpha$ is introduced by the first inference rule in the procedure, there exist $f(l_1, \ldots, l_n) \to g(r_1, \ldots, r_m) \in R$, $f(q_1^*, \ldots, q_n^*) \to \tilde{q} \in \Delta_k$, and $\theta: \ X \to Q_*$ such that

- $l_i\theta \xrightarrow[\Delta_k]{*} q_i^*$ for all $1 \le i \le n$,
- $q_j'^* = \begin{cases} \tilde{p}_j & \cdots j \in \mu(g), r_j \in X, \ p_j^* = r_j\theta \\ p_j^* & \cdots j \notin \mu(g), r_j \in X, \ p_j^* = r_j\theta \\ \tilde{q}^{r_j} & \cdots j \in \mu(g), r_j \notin X \\ q^{r_j} & \cdots j \notin \mu(g), r_j \notin X \end{cases}$ for all $1 \le j \le m$, and
- $\mathcal{L}(\Delta_0, x\theta) \neq \emptyset$ for each erasing variable $x$.

We have the following:

1. For $j \in \mu(g)$ such that $r_j \in X$, we have $q_j'^* = \tilde{p}_j$ and $t_j \xrightarrow[\Delta_0]{*} q_j'^*$. Hence we have $t_j \xrightarrow[\Delta_0]{*} r_j\theta$ by Proposition 14.

2. For $j \notin \mu(g)$ such that $r_j \in X$, we have $t_j \xrightarrow[\Delta_*]{*} q_j'^* = r_j\theta$.
3. For $j \in \mu(g)$ such that $r_j \notin X$, we have $t_j \xrightarrow[\Delta_0]{*} q_j'^* = \tilde{q}^{r_j}$. Since $t_j \xrightarrow[\Delta_0]{*} q^{r_j}$ by Proposition 14, we have $t_j = r_j$ from the construction of $\Delta_0$. Therefore $t_j = r_j\theta$ follows from right-shallowness.
4. For $j \notin \mu(g)$ such that $r_j \notin X$, we have $t_j \xrightarrow[\Delta_*]{*} q_j'^* = q^{r_j}$. Since $t_j \xrightarrow[\Delta_0]{*} q^{r_j}$ by Proposition 13, we have $t_j = r_j$ from the construction of $\Delta_0$. Therefore $t_j = r_j\theta$ follows from right-shallowness.

Thus we have $g(t_1, \ldots, t_m) \xrightarrow[\Delta_*]{*} g(r_1\theta, \ldots, r_m\theta) \xrightarrow[\Delta_0]{*} g(q_1'^*, \ldots, q_m'^*)$.

We define a substitution $\sigma :\ \mathrm{Var}(f(l_1, \ldots, l_n)) \to \mathcal{T}(F)$ as follows:

$$x\sigma = \begin{cases} t_j & \cdots \text{ if there exists } j \text{ such that } r_j = x \\ t' & \cdots \text{ otherwise, choose an arbitral } t' \text{ such that } t' \xrightarrow[\Delta_0]{*} x\theta, \end{cases}$$

where $\sigma$ is well-defined from the right-linearity of rewrite rules. We can construct $\beta : f(l_1, \ldots, l_n)\sigma \xrightarrow[\Delta_*]{*} f(l_1, \ldots, l_n)\theta \xrightarrow[\Delta_k]{*} f(q_1^*, \ldots, q_n^*) \xrightarrow[\Delta_k]{} \tilde{q}$.

On the other hand, we have $f(l_1, \ldots, l_n)\sigma \underset{\mathcal{R}}{\hookrightarrow} g(r_1, \ldots, r_m)\sigma = g(t_1, \ldots, t_m) = t$. Therefore the lemma of this case follows by taking $f(l_1, \ldots, l_n)\sigma$ as $s$. Here $\alpha \sqsupset \beta$ follows from the left-linearity of rewrite rules.

For the case where the transition rule applied last in $\alpha$ is (in the form of) $\tilde{q}' \to \tilde{q} \in \Delta_{k+1} \setminus \Delta_k$, the lemma can be shown as similar to the previous case. □

**Lemma 20.** *If $\mathcal{R}$ be linear and right-shallow, then $\mathcal{L}(\mathcal{A}_*) \subseteq \underset{\mathcal{R}}{\hookrightarrow}[\mathcal{L}(\mathcal{A})]$.*

*Proof.* Let $t \xrightarrow[\Delta_*]{*} \tilde{q} \in Q_*^f$ and the following claim holds:

$$\alpha :\ t \xrightarrow[\Delta_*]{*} \tilde{q} \text{ implies } s \xrightarrow[\Delta_0]{*} \tilde{q} \text{ and } s \overset{*}{\underset{\mathcal{R}}{\hookrightarrow}} t.$$

Then, we have $s \xrightarrow[\Delta_0]{*} \tilde{q}$ and $s \overset{*}{\underset{\mathcal{R}}{\hookrightarrow}} t$ for some $s$. Hence $s \in \mathcal{L}(\mathcal{A})$ follows from Proposition 14.

In the sequel, we prove the claim by induction on $\alpha$ with respect to $\sqsupset$.

1. Consider the case that the last transition rule applied in $\alpha$ is (in the form of) $g(q_1'^*, \ldots, q_m'^*) \to \tilde{q} \in \Delta_k$. Then $\alpha$ can be represented as $t = g(t_1, \ldots, t_m) \xrightarrow[\Delta_*]{*} g(q_1'^*, \ldots, q_m'^*) \xrightarrow[\Delta_k]{} \tilde{q}$.
   (a) For $j \in \mu(g)$ such that $q_j'^* = \tilde{q}_j'$, there exists $s_j$ such that $s_j \overset{*}{\underset{\mathcal{R}}{\hookrightarrow}} t_j$ and $s_j \xrightarrow[\Delta_0]{*} \tilde{q}_j' = q_j'^*$ from the induction hypothesis, since $\alpha \sqsupset (t_j \xrightarrow[\Delta_*]{*} \tilde{q}_j')$.
   (b) For $j \notin \mu(g)$ such that $q_j'^* = \tilde{q}_j'$, we take $s_j$ as $t_j$.
   (c) For $j$ such that $q_j'^* = q_j'$, we have $t_j \xrightarrow[\Delta_0]{*} q_j' = q_j'^*$ by Proposition 14. We take $s_j$ as $t_j$.

   Now we have $g(s_1, \ldots, s_m) \overset{*}{\underset{\mathcal{R}}{\hookrightarrow}} g(t_1, \ldots, t_m) = t$ and $\alpha' :\ g(s_1, \ldots, s_m) \xrightarrow[\Delta_*]{*} g(q_1'^*, \ldots, q_{m'}'^*,) \xrightarrow[\Delta_k]{} \tilde{q}$, where $\alpha \sqsupseteq \alpha'$ and $s_j \xrightarrow[\Delta_0]{*} q_j'^*$ for all $j \in \mu(g)$.

In the subcase $k = 0$ we have no $j$ that satisfies (b), since $j \in \mu(g)$ if and only if $q_j'^* = \tilde{q}_j'$ from the construction of $\Delta_0$. Thus every transition rule used in $\alpha'$ is in $\Delta_0$. Therefore the claim trivially holds by letting $s = g(s_1, \ldots, s_m)$. In the subcase $k > 0$, we have $s' \hookrightarrow_{\mathcal{R}} g(s_1, \ldots, s_m)$ and $\beta : s' \xrightarrow[\Delta_*]{*} \tilde{q}$ for some $s'$ such that $\alpha' \sqsupset \beta$ by Lemma 19. Therefore the claim holds by the induction hypothesis since $\alpha \sqsupset \beta$.

2. In the case where the last transition rule applied in $\alpha$ is (in the form of) $q'^* \to q^* \in \Delta_k$, we can show it similarly to the previous case. □

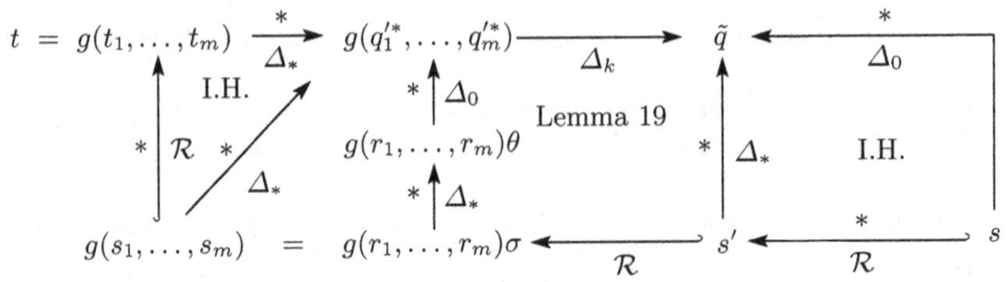


**Fig. 6.** The diagram of proof of lemma 20

The following theorem is proved by lemma 18, lemma 20 and theorem 1.

**Theorem 21.** *Context-sensitive reduction for linear right-shallow TRSs effectively preserves recognizability. Thus context-sensitive reachability is decidable for linear right-shallow TRSs.*

## 5 Discussion

The authors think that the left-linear restriction for the context-sensitive case will be removed by modifying the procedure $\mathrm{P_{cs}}$ similar to [11] if all variables that occur in $\mathrm{Pos}(r) \setminus \mathrm{Pos}^{\mu}(r)$ are left-linear. For the innermost case, similar modification may be possible. However, constructing $\mathcal{A}_{\mathrm{NF}}$ would be a barrier. Using automata with an equality test between brothers [1] is a possible direction

On the other hand, removing the right-linearity restriction is impossible for both cases, because there exists a counter example such as the following.

*Example 22.* For TRS $R = \{g(x) \to f(x,x)\}$ and a regular set $G = \{g(t) \mid t \in T(\{a,h\})\}$, $\to_R[G] = \to_{R\,\mathrm{in}}[G] = \{g(t), f(t,t) \mid t \in T(\{a,h\})\}$ is not a regular. □

## References

1. Caron, A.-C., Coquide, J.-L., Dauchet, M.: Encompassment Properties and Automata with Constraints. In: Kirchner, C. (ed.) RTA 1993. LNCS, vol. 690, pp. 328–342. Springer, Heidelberg (1993)
2. Comon, H.: Sequentiality, second-order monadic logic and tree automata. In: 10th annual IEEE symposium on logic in computer science (LICS 1995), pp. 508–517 (1995)

3. Durand, I., Sénizergues, G.: Bottom-Up Rewriting Is Inverse Recognizability Preserving. In: Baader, F. (ed.) RTA 2007. LNCS, vol. 4533, pp. 107–121. Springer, Heidelberg (2007)
4. Gilleron, R., Tison, S.: Regular tree languages and rewrite systems. Fundamenta Informaticae 24, 157–176 (1995)
5. Godoy, G., Huntingford, E.: Innermost-reachability and innermost-joinability are decidable for shallow term rewrite systems. In: Baader, F. (ed.) RTA 2007. LNCS, vol. 4533, pp. 184–199. Springer, Heidelberg (2007)
6. Jacquemard, F.: Decidable approximations of term rewriting systems. In: Ganzinger, H. (ed.) RTA 1996. LNCS, vol. 1103, pp. 362–376. Springer, Heidelberg (1996)
7. Jacquemard, F.: Reachability and confluence are undecidable for flat term rewriting systems. Information processing letters 87(5), 265–270 (2003)
8. Lucas, S.: Context-sensitive computations in functional and functional logic programs. Journal of Functional and Logic Programming 1998(1), 1–61 (1998)
9. Mitsuhashi, I., Oyamaguchi, M., Jacquemard, F.: The Confluence Problem for Flat TRSs. In: Calmet, J., Ida, T., Wang, D. (eds.) AISC 2006. LNCS (LNAI), vol. 4120, pp. 68–81. Springer, Heidelberg (2006)
10. Nagaya, T., Toyama, Y.: Decidability for left-linear growing term rewriting systems. In: Narendran, P., Rusinowitch, M. (eds.) RTA 1999. LNCS, vol. 1631, pp. 256–270. Springer, Heidelberg (1999)
11. Takai, T., Kaji, Y., Seki, H.: Right-linear finite path overlapping term rewriting systems effectively preserve recognizability. In: Bachmair, L. (ed.) RTA 2000. LNCS, vol. 1833, pp. 246–260. Springer, Heidelberg (2000)

# Arctic Termination ... Below Zero

Adam Koprowski[1] and Johannes Waldmann[2]

[1] Department of Computer Science
Eindhoven University of Technology
P.O. Box 513, 5600 MB Eindhoven, The Netherlands
[2] Hochschule für Technik, Wirtschaft und Kultur (FH) Leipzig
Fb IMN, PF 30 11 66, D-04251 Leipzig, Germany

**Abstract.** We introduce the arctic matrix method for automatically proving termination of term rewriting. We use vectors and matrices over the arctic semi-ring: natural numbers extended with $-\infty$, with the operations "max" and "plus". This extends the matrix method for term rewriting and the arctic matrix method for string rewriting. In combination with the Dependency Pairs transformation, this allows for some conceptually simple termination proofs in cases where only much more involved proofs were known before. We further generalize to arctic numbers "below zero": integers extended with $-\infty$. This allows to treat some termination problems with symbols that require a predecessor semantics. The contents of the paper has been formally verified in the Coq proof assistant and the formalization has been contributed to the CoLoR library of certified termination techniques. This allows formal verification of termination proofs using the arctic matrix method. We also report on experiments with an implementation of this method which, compared to results from 2007, outperforms TPA (winner of the certified termination competition for term rewriting), and in the string rewriting category is as powerful as Matchbox was but now all of the proofs are certified.

## 1 Introduction

One method of proving termination is interpretation into a well-founded algebra. Polynomial interpretations (over the naturals) are a well-known example of this approach. Another example is the recent development of the matrix method [17,7] that uses linear interpretations over vectors of naturals, or equivalently, $\mathbb{N}$-weighted automata. In [23,22] one of the authors extended this method (for string rewriting) to arctic automata, i.e. on the max/plus semi-ring on $\{-\infty\} \cup \mathbb{N}$. Its implementation in the termination prover Matchbox [21] contributed to this prover winning the string rewriting division of the 2007 termination competition [26].

The first contribution of the present work is a *generalization of arctic termination to term rewriting.* We use interpretations given by functions of the form $(x_1, \dots, x_n) \mapsto M_0 + M_1 \cdot x_1 + \dots + M_n \cdot x_n$. Here, $x_i$ are (column) vector variables, $M_0$ is a vector and $M_1, \dots, M_n$ are square matrices, where all entries are arctic numbers, and operations are understood in the arctic semi-ring.

A. Voronkov (Ed.): RTA 2008, LNCS 5117, pp. 202–216, 2008.

Since the max operation is not strictly monotone in single arguments, we obtain monotone interpretations only for the case when all function symbols are at most unary, i.e. string rewriting. For symbols of higher arity, arctic interpretations are weakly monotone. These cannot prove termination, but only top termination, where rewriting steps are only applied at the root of terms. This is a restriction but it fits with the framework of the dependency pairs method [2] that transforms a termination problem to a top termination problem.

The second contribution is a *generalization from arctic naturals to arctic integers*, i.e. $\{-\infty\} \cup \mathbb{Z}$. Arctic integers allow e.g. to interpret function symbols by the predecessor function and this matches the "intrinsic" semantics of some termination problems. There is previous work on polynomial interpretations with negative coefficients [14], where the interpretation for predecessor is also expressible using ad-hoc max operations. Using arctic integers, we obtain verified termination proofs for 10 of the 24 rewrite systems Beerendonk/* from TPDB, simulating imperative computations. Previously, they could only be handled by the method of Bounded Increase [12].

The third contribution is that definitions, theorems and proofs (excluding Section 5 with results on full termination) have been *formalized with the proof assistant Coq* [25]. This extends previous work [19] and will become part of the CoLoR project [4] that gathers formalizations of termination techniques and employs them to certify termination proofs found automatically. In 2007, the certified category of the termination competition was won by the termination prover TPA [18] that uses CoLoR.

A method to search for arctic interpretations is implemented for the termination prover Matchbox. It works by transformation to a boolean satisfiability problem and application of a state-of-the-art SAT solver (in this case, Minisat). For several termination problems that could not be solved in last year's certified termination competition it finds proofs via arctic interpretations and the new CoLoR version certifies them.

The paper is organized as follows. We present notation and basic facts on rewriting and the arctic semi-ring in Section 2. Then in Section 3 we describe what kind of functions we use for interpretation and in Section 4 we discuss the appropriate ordering relations. We present arctic interpretations for termination in Section 5, for top termination in Section 6 and the generalization to arctic integers in Section 7. We report on the formal verification in Section 8 and on performance of our implementation in Section 9. We present some discussion of the method, its limitations and related work in Section 10 and we conclude in Section 11.

## 2 Notation and Preliminaries

We follow the notation of [3] for term rewriting. The top one-step derivation relation of a rewriting system $\mathcal{R}$ is denoted by $\stackrel{\text{top}}{\to}_{\mathcal{R}}$ and the full one-step derivation relation is $\to_{\mathcal{R}}$. We often abbreviate these by $\mathcal{R}_{\text{top}}$ and $\mathcal{R}$, respectively. A relation $\to$ is terminating if it does not admit infinite descending chains $t_0 \to t_1 \to \ldots$, denoted as $\mathrm{SN}(\rightarrow)$. For relations $\rightarrow_1, \rightarrow_2$, we define $\rightarrow_1 / \rightarrow_2$ by $(\rightarrow_1) \circ (\rightarrow_2)^*$. If $\mathrm{SN}(\mathcal{R}/\mathcal{S})$, we say that $\mathcal{R}$ is terminating relative to $\mathcal{S}$.

We cite notation for monotone algebras [7]. A $k$-ary operation $[f]$ is *monotone* with respect to a relation $\rightarrow$, if it is monotone in each argument individually: $x_i \rightarrow x_i'$ implies $[f](x_1, \ldots, x_i, \ldots, x_k) \rightarrow [f](x_1, \ldots, x_i', \ldots, x_k)$. A *weakly monotone algebra* for a signature $\Sigma$ is a $\Sigma$-algebra $(A, [\cdot])$ with two relations $>, \gtrsim$ such that $>$ is well-founded, $> \cdot \gtrsim \subseteq >$ and for every $f \in \Sigma$, the operation $[f]$ is monotone with respect to $\gtrsim$. Such an algebra is called *extended monotone* if additionally each $[f]$ is monotone with respect to $>$. For terms $\ell, r$ with variables from a set $\mathcal{X}$, we write $[\ell] >_\alpha [r]$ to abbreviate $[\ell, \alpha] > [r, \alpha]$ for every $\alpha : \mathcal{X} \rightarrow A$. Now we present a slight variant of the main theorem from [7], for proving relative (top)-termination with monotone algebras:

**Theorem 1.** *Let $\mathcal{R}, \mathcal{R}', \mathcal{S}, \mathcal{S}'$ be TRSs over a signature $\Sigma$.*

1. *Let $(A, [\cdot], >, \gtrsim)$ be an extended monotone algebra such that: $[\ell] \gtrsim_\alpha [r]$ for every rule $\ell \rightarrow r \in \mathcal{R} \cup \mathcal{S}$ and $[\ell] >_\alpha [r]$ for every rule $\ell \rightarrow r \in \mathcal{R}' \cup \mathcal{S}'$. Then $\mathrm{SN}(\mathcal{R}/\mathcal{S})$ implies $\mathrm{SN}(\mathcal{R} \cup \mathcal{R}'/\mathcal{S} \cup \mathcal{S}')$.*
2. *Let $(A, [\cdot], >, \gtrsim)$ be a weakly monotone algebra such that: $[\ell] \gtrsim_\alpha [r]$ for every rule $\ell \rightarrow r \in \mathcal{R} \cup \mathcal{S}$ and $[\ell] >_\alpha [r]$ for every rule $\ell \rightarrow r \in \mathcal{R}'$. Then $\mathrm{SN}(\mathcal{R}_{\mathrm{top}}/\mathcal{S})$ implies $\mathrm{SN}(\mathcal{R}_{\mathrm{top}} \cup \mathcal{R}'_{\mathrm{top}}/\mathcal{S})$.* □

A *commutative semi-ring* [13] consists of a carrier $D$, two designated elements $d_0, d_1 \in D$ and two binary operations $\oplus, \otimes$ on $D$, such that both $(D, d_0, \oplus)$ and $(D, d_1, \otimes)$ are commutative monoids and multiplication distributes over addition: $\forall x, y, z \in D : x \otimes (y \oplus z) = (x \otimes y) \oplus (x \otimes z)$.

One example of semi-rings are the natural numbers with the standard operations. We will need the *arctic semi-ring* (also called the *max/plus algebra*) [9] with carrier $\mathbb{A}_\mathbb{N} \equiv \{-\infty\} \cup \mathbb{N}$, where semi-ring addition is the max operation with neutral element $-\infty$ and semi-ring multiplication is the standard plus operation with neutral element 0 ($x \otimes y = -\infty$ if either $x = -\infty$ or $y = -\infty$). We also consider these operations for arctic numbers *below zero* (ie. *arctic integers*), that is, on the carrier $\mathbb{A}_\mathbb{Z} \equiv \{-\infty\} \cup \mathbb{Z}$.

For any semi-ring $D$, we can consider the space of linear functions (square matrices) on $n$-dimensional vectors over $D$. These functions (matrices) again form a semi-ring (though a non-commutative one), and indeed we write $\oplus$ and $\otimes$ for its operations as well.

A semi-ring is *ordered* [8] by $\geq$ if $\geq$ is a partial order compatible with the operations: $\forall x \geq y, z : x \oplus z \geq y \oplus z$ and $\forall x \geq y, z : x \otimes z \geq y \otimes z$.

The standard semi-ring of natural numbers is ordered by the standard $\geq$ relation. The semi-ring of arctic naturals and arctic integers is ordered by $\geq$, being the reflexive closure of $>$ defined as $\ldots > 1 > 0 > -1 > \ldots > -\infty$. Note that standard integers with standard operations form a semi-ring but it is not ordered in this sense, as we have for instance $1 \geq 0$ but $1 * (-1) = -1 \not\geq 0 = 0 * (-1)$.

## 3 Max/Plus Linear Algebra

We consider vectors of arctic numbers. They form a monoid under component-wise arctic addition. For arctic matrices we define arctic addition and multiplication as usual. Square matrices form a non-commutative semi-ring with these operations. E.g. the $3 \times 3$ identity matrix is

$$\begin{pmatrix} 0 & -\infty & -\infty \\ -\infty & 0 & -\infty \\ -\infty & -\infty & 0 \end{pmatrix}$$

A square matrix $M$ then maps a (column) vector $x$ to a (column) vector $M \otimes x$ and this mapping is linear: $M \otimes (x \oplus y) = M \otimes x \oplus M \otimes y$. We use the following shape of vector-valued functions of several vector arguments:

$$f(x_1, \ldots, x_n) = f_0 \oplus f_1 \otimes x_1 \oplus \ldots \oplus f_n \otimes x_n.$$

Here, $x_i$ are column vectors, $f_0$ is a column vector and $f_1, \ldots, f_n$ are square matrices. We call this an *arctic linear function* (with linear factors $f_1, \ldots, f_n$ and absolute part $f_0$).

Note that for brevity in all the examples we use the following notation for such linear functions:

$$f(x_1, \ldots, x_n) = f_0 \oplus f_1 x_1 \oplus \ldots \oplus f_n x_n.$$

**Definition 2.** – *A number $a \in \mathbb{A}$ is called* finite *if $a > -\infty$.*
- *A number $a \in \mathbb{A}$ is called* positive *if $a \geq 0$.*
- *A vector $x = (x_1, \ldots, x_n) \in \mathbb{A}^n$ is called finite if $x_1$ is finite and it is called positive if $x_1$ is positive.*
- *A matrix $M \in \mathbb{A}^{m \times n}$ is called finite if $M_{1,1}$ is finite.*
- *A linear function $f$ is called* somewhere finite *if $\exists 0 \leq i \leq n : \text{finite}(f_i)$.*
- *A linear function $f$ is called* absolute positive *if* positive$(f_0)$. ◇

*Example 3.* Consider a linear function:

$$f(x, y) = \begin{pmatrix} 1 & -\infty \\ 0 & -\infty \end{pmatrix} x \oplus \begin{pmatrix} -\infty & -\infty \\ 0 & 1 \end{pmatrix} y \oplus \begin{pmatrix} -\infty \\ 0 \end{pmatrix}$$

which is somewhere finite, as the upper-leftmost element of the matrix coefficient of $x$ is 1, which is finite. It is not absolute positive, as the constant vector has $-\infty$ on its first position.

Evaluation of this function on some exemplary arguments yields:

$$f(\begin{pmatrix} -\infty \\ 0 \end{pmatrix}, \begin{pmatrix} 1 \\ -\infty \end{pmatrix}) = \begin{pmatrix} 1 & -\infty \\ 0 & -\infty \end{pmatrix} \begin{pmatrix} -\infty \\ 0 \end{pmatrix} \oplus \begin{pmatrix} -\infty & -\infty \\ 0 & 1 \end{pmatrix} \begin{pmatrix} 1 \\ -\infty \end{pmatrix} \oplus \begin{pmatrix} -\infty \\ 0 \end{pmatrix} = \begin{pmatrix} -\infty \\ 1 \end{pmatrix}.$$

◁

**Lemma 4.** *For numbers, vectors, matrices:*

- *if $a$ is finite and $b$ arbitrary, then $a \oplus b$ is finite.*
- *if $a$ is positive and $b$ arbitrary, then $a \oplus b$ is positive.*
- *if $a$ and $b$ are finite, then $a \otimes b$ is finite.* □

**Lemma 5.** *For a linear function $f$:*

1. *if $f$ is somewhere finite, then* $\forall x_1, \ldots, x_n : (\forall i : \text{finite}(x_i)) \Rightarrow \text{finite}(f(x_1, \ldots, x_n))$.
2. *if $f$ is absolute positive, then* $\forall x_1, \ldots, x_n : \text{positive}(f(x_1, \ldots, x_n))$. □

## 4 Orders on Max/Plus

Arctic addition (i.e., the max operation) is not strictly monotone in single arguments: we have e.g. $5 > 3$ but $5 \oplus 6 = 6 \not> 6 = 3 \oplus 6$. It is, however, "half strict" in the following sense: a strict increase in both arguments simultaneously gives a strict increase in the result, e.g. $5 > 3$ and $6 > 4$ implies $5 \oplus 6 > 3 \oplus 4$. Compared to the standard matrix method, this special property of arctic addition requires a somewhat different treatment of monotonicity. In several places where the standard matrix method needs just one strict inequality (among several non-strict ones), the arctic matrix method needs all inequalities to be strict. There is one exception: arctic addition is obviously strict if one argument is arctic zero, i.e., $-\infty$. This explains the definition of $\gg$ below. In this section, we consider arctic integers.

**Definition 6.**
- *We write $\geq$ for reflexive closure of the standard ordering $\ldots > 1 > 0 > -1 > \ldots > -\infty$ and extend this notation component-wise to vectors, matrices and linear functions.*
- *We write $a \gg b$ if $(a > b) \vee (a = b = -\infty)$, and we extend this notation component-wise to vectors, matrices and linear functions.* ◇

Note that $\gg \cdot \geq \subseteq \gg$, which is required to apply the monotone algebra theorem.

**Lemma 7.** *For arctic integers $a, a_1, a_2, b_1, b_2$,*

- *if $a_1 \geq a_2 \wedge b_1 \geq b_2$, then $a_1 \oplus b_1 \geq a_2 \oplus b_2$ and $a_1 \otimes b_1 \geq a_2 \otimes b_2$.*
- *if $a_1 \gg a_2 \wedge b_1 \gg b_2$, then $a_1 \oplus b_1 \gg a_2 \oplus b_2$.*
- *if $b_1 \gg b_2$, then $a \otimes b_1 \gg a \otimes b_2$.* □

The following lemma allows to establish order on results of two functions by comparison of their coefficients. It is the arctic counter-part of the absolute-positiveness criterion used for polynomial interpretations.

**Lemma 8.** *For linear functions $f, g$ with $f \geq g$ (resp. $f \gg g$), and for each tuple of vectors $x_1, \ldots, x_n$: $f(x_1, \ldots, x_n) \geq g(x_1, \ldots, x_n)$ (resp. $f(x_1, \ldots, x_n) \gg g(x_1, \ldots, x_n)$).* □

**Lemma 9.** *Every linear function $f$ is monotone with respect to $\geq$.*

*Proof.* For $x_i \geq x_i'$ we have:

$$f_0 \oplus f_1 \otimes x_1 \oplus \ldots f_i \otimes x_i \ldots \oplus f_n \otimes x_n \geq f_0 \oplus f_1 \otimes x_1 \oplus \ldots f_i \otimes x_i' \ldots \oplus f_n \otimes x_n$$

using Lemma 7 lifted to vectors. □

# 5 Full Arctic Termination

In this section we present a method of using arctic matrices to prove full termination (as opposed to top termination, see Section 6). For some fixed dimension $d$ we choose the algebra over the domain, $\mathbb{N} \times \mathbb{A}_{\mathbb{N}}^{d-1}$, that is over vectors of arctic naturals where the first position of the vector is finite. The algebra is ordered with $\gg$ and the ordering is well-founded due to restriction to finite elements on first vector positions. Function symbols are interpreted by linear arctic functions.

The following theorem provides a termination criterion with such monotone interpretations. A linear $\Sigma$-interpretation is an interpretation that associates an arctic linear function $[f]$ with every $f \in \Sigma$. As noted at the beginning of Section 4, "$\oplus$" is not strictly monotone. Therefore, a function of the shape $f_0 \oplus f_1 \otimes x_1 \oplus \ldots \oplus f_n \otimes x_n$ is monotone only if the $\oplus$ operation is essentially redundant. This happens in the following cases.

**Theorem 10.** *Let $\mathcal{R}, \mathcal{R}', \mathcal{S}, \mathcal{S}'$ be TRSs over a signature $\Sigma$ and $[\cdot]$ be a linear $\Sigma$-interpretation with coefficients in $\mathbb{A}_{\mathbb{N}}$. If:*

- *every function symbol has arity at most 1,*
- *for every constant $f \in \Sigma$, $[f]_0$ is finite,*
- *for every unary symbol $f \in \Sigma$, $[f]_0$ is the arctic zero vector and $[f]_1$ is finite,*
- *$[\ell] \geq [r]$ for every rule $\ell \to r \in \mathcal{R} \cup \mathcal{S}$,*
- *$[\ell] \gg [r]$ for every rule $\ell \to r \in \mathcal{R}' \cup \mathcal{S}'$ and*
- $\mathrm{SN}(\mathcal{R}/\mathcal{S})$.

*Then* $\mathrm{SN}(\mathcal{R} \cup \mathcal{R}'/\mathcal{S} \cup \mathcal{S}')$.

*Proof.* By Theorem 1.1. Note that, by Lemma 8, $[\ell] \geq [r]$ (resp. $[\ell] \gg [r]$) implies $[\ell] \geq_\alpha [r]$ (resp. $[\ell] \gg_\alpha [r]$). So we only need to show that $(\mathbb{N} \times \mathbb{A}_{\mathbb{N}}^{d-1}, [\cdot], \gg, \geq)$ is an extended monotone algebra. The order $\gg$ is well-founded on this domain as with every decrease we get a decrease in the first component of the vector, which differs from $-\infty$. It is an easy observation that, due to the first three premises of this theorem, such interpretations are monotone. Finally evaluation of interpretations stays within the domain by Lemma 5.1 as every $[f]$ is somewhere finite by assumption. □

For symbols of arity $n > 1$ there is no arctic linear function that is monotone, hence the arctic matrix method for full termination is only applicable for string rewriting (plus constants). As such, it had been described in [22] and had been applied by Matchbox in the 2007 termination competition. The following example illustrates the method.

*Example 11.* The relative termination problem `SRS/Waldmann/r2` is

$$\{\mathrm{c\,a\,c} \to \epsilon,\ \mathrm{a\,c\,a} \to \mathrm{a}^4\ /\ \epsilon \to \mathrm{c}^4\}.$$

In the 2007 termination competition, it had been solved by Jambox [6] via "self labeling" and by Matchbox via essentially the following arctic proof.

We use the following arctic interpretation

$$[\mathsf{a}](x) = \begin{pmatrix} 0 & 0 & -\infty \\ 0 & 0 & -\infty \\ 1 & 1 & 0 \end{pmatrix} x \oplus \begin{pmatrix} -\infty \\ -\infty \\ -\infty \end{pmatrix} \qquad [\mathsf{c}](x) = \begin{pmatrix} 0 & -\infty & -\infty \\ -\infty & -\infty & 0 \\ -\infty & 0 & -\infty \end{pmatrix} x \oplus \begin{pmatrix} -\infty \\ -\infty \\ -\infty \end{pmatrix}$$

It is immediate that $[\mathsf{c}]$ is a permutation (it swaps the second and third component of its argument vector), so $[\mathsf{c}]^2 = [\mathsf{c}]^4$ is the identity and we have $[\epsilon] = [\mathsf{c}]^4$. A short calculation shows that $[\mathsf{a}]$ is idempotent, so $[\mathsf{a}] = [\mathsf{a}^4]$. We compute

$$[\mathsf{c\,a\,c}](x) = \begin{pmatrix} 0 & -\infty & 0 \\ 1 & 0 & 1 \\ 0 & -\infty & 0 \end{pmatrix} x \quad [\mathsf{a\,c\,a}](x) = \begin{pmatrix} 1 & 1 & 0 \\ 1 & 1 & 0 \\ 2 & 2 & 1 \end{pmatrix} x \quad [\mathsf{a}^4](x) = \begin{pmatrix} 0 & 0 & -\infty \\ 0 & 0 & -\infty \\ 1 & 1 & 0 \end{pmatrix} x$$

and therefore $[\mathsf{c\,a\,c}] \geq [\epsilon]$ and $[\mathsf{a\,c\,a}] \gg [\mathsf{a}^4]$. Note that indeed we have point-wise $\gg$ and the top left entries of matrices are finite. This allows to remove one strict rule. The remaining strict rule can be removed by counting letters a. ◁

## 6 Arctic Top Termination

As explained earlier, there are no monotone linear arctic functions of more than one argument. We therefore change our attention from proving full termination to proving top termination. This fits with the Dependency Pairs method that replaces a full termination problem with an equivalent top termination problem.

The domain, as in Section 5, is $\mathbb{N} \times \mathbb{A}_{\mathbb{N}}^{d-1}$ for some fixed dimension $d$ and we use the ordering relations $\gg$ (strict) and $\geq$ (weak). The following theorem allows us to prove top termination in this setting:

**Theorem 12.** *Let $\mathcal{R}, \mathcal{R}', \mathcal{S}$ be TRSs over a signature $\Sigma$ and $[\cdot]$ be a linear $\Sigma$-interpretation with coefficients in $\mathbb{A}_{\mathbb{N}}$. If:*

- *for each $f \in \Sigma$, $[f]$ is somewhere finite,*
- *$[\ell] \geq [r]$ for every rule $\ell \to r \in \mathcal{R} \cup \mathcal{S}$,*
- *$[\ell] \gg [r]$ for every rule $\ell \to r \in \mathcal{R}'$ and*
- *$\mathrm{SN}(\mathcal{R}_{\mathrm{top}}/\mathcal{S})$.*

*Then* $\mathrm{SN}(\mathcal{R}_{\mathrm{top}} \cup \mathcal{R}'_{\mathrm{top}}/\mathcal{S})$.

*Proof.* By Theorem 1.2; we need to show that $(\mathbb{N} \times \mathbb{A}_{\mathbb{N}}^{d-1}, [\cdot], \gg, \geq)$ is a weakly monotone algebra. The proof is essentially the same as the proof of Theorem 10. Note that now we only need a weakly monotone algebra and indeed by allowing function symbols of arity $> 1$, we lose the strict monotonicity property. □

*Example 13.* Consider the rewriting system `secret05/tpa2`:

$$\begin{aligned} \mathsf{f}(\mathsf{s}(x), y) &\to \mathsf{f}(\mathsf{p}(\mathsf{s}(x) - y), \mathsf{p}(y - \mathsf{s}(x))), & \mathsf{p}(\mathsf{s}(x)) &\to x, \\ \mathsf{f}(x, \mathsf{s}(y)) &\to \mathsf{f}(\mathsf{p}(x - \mathsf{s}(y)), \mathsf{p}(\mathsf{s}(y) - x)), & x - 0 &\to x, \\ & & \mathsf{s}(x) - \mathsf{s}(y) &\to x - y. \end{aligned}$$

It was solved in the 2007 competition by AProVE [11] using narrowing followed by polynomial interpretations and by T$_\mathsf{T}$T2 [15] using polynomial interpretations with negative constants.

After the DP transformation 9 dependency pairs can be removed using polynomial interpretations leaving the essential two dependency pairs:

$$\begin{aligned}
\mathsf{f}^\sharp(\mathsf{s}(x), y) &\to \mathsf{f}^\sharp(\mathsf{p}(\mathsf{s}(x) - y), \mathsf{p}(y - \mathsf{s}(x))) \\
\mathsf{f}^\sharp(x, \mathsf{s}(y)) &\to \mathsf{f}^\sharp(\mathsf{p}(x - \mathsf{s}(y)), \mathsf{p}(\mathsf{s}(y) - x))
\end{aligned}$$

Now the arctic interpretation

$$\begin{aligned}
[\mathsf{f}^\sharp(x, y)] &= \begin{pmatrix} -\infty & -\infty \\ -\infty & -\infty \end{pmatrix} x \oplus \begin{pmatrix} 0 & 0 \\ -\infty & -\infty \end{pmatrix} y \oplus \begin{pmatrix} 0 \\ -\infty \end{pmatrix} & [0] &= \begin{pmatrix} 3 \\ 3 \end{pmatrix} \\
[x - y] &= \begin{pmatrix} 0 & -\infty \\ 0 & 0 \end{pmatrix} x \oplus \begin{pmatrix} -\infty & -\infty \\ 0 & 0 \end{pmatrix} y \oplus \begin{pmatrix} 0 \\ 0 \end{pmatrix} & [\mathsf{p}(x)] &= \begin{pmatrix} 0 & -\infty \\ 0 & -\infty \end{pmatrix} x \oplus \begin{pmatrix} -\infty \\ -\infty \end{pmatrix} \\
[\mathsf{f}(x, y)] &= \begin{pmatrix} 0 & 0 \\ 0 & -\infty \end{pmatrix} x \oplus \begin{pmatrix} 2 & 0 \\ 0 & -\infty \end{pmatrix} y \oplus \begin{pmatrix} 0 \\ -\infty \end{pmatrix} & [\mathsf{s}(x)] &= \begin{pmatrix} 0 & 0 \\ 2 & 1 \end{pmatrix} x \oplus \begin{pmatrix} 0 \\ 2 \end{pmatrix}
\end{aligned}$$

removes the second dependency pair as we have:

$$\begin{aligned}
[\mathsf{f}^\sharp(x, \mathsf{s}(y))] &= \begin{pmatrix} -\infty & -\infty \\ -\infty & -\infty \end{pmatrix} x \oplus \begin{pmatrix} 2 & 1 \\ -\infty & -\infty \end{pmatrix} y \oplus \begin{pmatrix} 2 \\ -\infty \end{pmatrix} \\
[\mathsf{f}^\sharp(\mathsf{p}(x - \mathsf{s}(y)), \mathsf{p}(\mathsf{s}(y) - x))] &= \begin{pmatrix} -\infty & -\infty \\ -\infty & -\infty \end{pmatrix} x \oplus \begin{pmatrix} 0 & 0 \\ -\infty & -\infty \end{pmatrix} y \oplus \begin{pmatrix} 0 \\ -\infty \end{pmatrix}
\end{aligned}$$

and it is weakly compatible with all the rules. The remaining dependency pair can be removed by a standard matrix interpretation of dimension two. ◁

## 7 ... Below Zero

We extend the domain of matrix and vector coefficients from $\mathbb{A}_\mathbb{N}$ (arctic naturals) to $\mathbb{A}_\mathbb{Z}$ (arctic integers). This allows to interpret some function symbols by the "predecessor" function $x \mapsto x - 1$, and so represents their "intrinsic" semantics. This is the same motivation as the one for allowing polynomial interpretations with negative coefficients [14].

We need to be careful though, as the relation $\gg$ on vectors of arctic integers is not well-founded.

**Theorem 14.** *Let $\mathcal{R}, \mathcal{R}', \mathcal{S}$ be TRSs over a signature $\Sigma$ and $[\cdot]$ be a linear $\Sigma$-interpretation with coefficients in $\mathbb{A}_\mathbb{Z}$. If:*

- *for each $f \in \Sigma$, $[f]$ is absolute positive,*
- *$[\ell] \geq [r]$ for every rule $\ell \to r \in \mathcal{R} \cup \mathcal{S}$,*
- *$[\ell] \gg [r]$ for every rule $\ell \to r \in \mathcal{R}'$ and*
- *$\mathrm{SN}(\mathcal{R}_{\mathrm{top}}/\mathcal{S})$.*

*Then $\mathrm{SN}(\mathcal{R}_{\mathrm{top}} \cup \mathcal{R}'_{\mathrm{top}}/\mathcal{S})$.*

*Proof.* The proof goes along the same lines as the proof of Theorem 12. Note however that as we are working with integers now, to ensure that we stay within the domain, we need a stronger assumption on interpretations; we get that property by Lemma 5.2. □

*Example 15.* Let us consider the Beerendonk/2.trs TRS from the TPDB [27], consisting of the following six rules:

$$\begin{aligned} \mathrm{cond}(\mathrm{true}, x, y) &\to \mathrm{cond}(\mathrm{gr}(x, y), \mathrm{p}(x), \mathrm{s}(y)), & \mathrm{gr}(\mathrm{s}(x), \mathrm{s}(y)) &\to \mathrm{gr}(x, y), \\ \mathrm{gr}(0, x) &\to \mathrm{false}, & \mathrm{gr}(\mathrm{s}(x), 0) &\to \mathrm{true}, \\ \mathrm{p}(0) &\to 0, & \mathrm{p}(\mathrm{s}(x)) &\to x \end{aligned}$$

This is a straightforward encoding of the following imperative program

```
while x > y do (x, y) := (x-1, y+1);
```

which is obviously terminating. However this TRS posed a serious challenge for the tools in the termination competition. Only AProVE could deal with this system (as well as a number of others coming from such transformations from imperative programs) using a specialized bounded increase method [12]. We will now show a termination proof for this system using the arctic below zero interpretations.

We begin by applying the dependency pair method and obtaining four dependency pairs, three of which can be easily removed (for instance using standard matrix or polynomial interpretations) leaving the following single dependency pair:

$$\mathrm{cond}^\sharp(\mathrm{true}, x, y) \to \mathrm{cond}^\sharp(\mathrm{gr}(x, y), \mathrm{p}(x), \mathrm{s}(y))$$

Now, consider the following arctic matrix interpretation:

$$\begin{aligned} [\mathrm{cond}^\sharp(x, y, z)] &= (0)x \oplus (0)y \oplus (-\infty)z \oplus (0), & [0] &= (0), \\ [\mathrm{cond}(x, y, z)] &= (0)x \oplus (2)y \oplus (-\infty)z \oplus (0), & [\mathrm{false}] &= (0), \\ [\mathrm{gr}(x, y)] &= (-1)x \oplus (-\infty)y \oplus (0), & [\mathrm{true}] &= (2), \\ [\mathrm{p}(x)] &= (-1)x \oplus (0), & [\mathrm{s}(x)] &= (2)x \oplus (3). \end{aligned}$$

With this interpretation we get a decrease for the dependency pair:

$$\begin{aligned} [\mathrm{cond}^\sharp(\mathrm{true}, x, y)] &= (\ \ 0)x \oplus (-\infty)y \oplus (2) \\ [\mathrm{cond}^\sharp(\mathrm{gr}(x, y), \mathrm{p}(x), \mathrm{s}(y))] &= (-1)x \oplus (-\infty)y \oplus (0) \end{aligned}$$

and all the original rules are oriented weakly. ◁

*Remark 16.* We discuss a variant that looks more liberal, but turns out to be equivalent to the one given here. We cannot allow $\mathbb{Z} \times \mathbb{A}_{\mathbb{Z}}^{d-1}$ for the domain, because it is not well-founded for $\gg$. So we can restrict the admissible range of negative values by some bound $c > -\infty$, and use the domain $\mathbb{A}_{\mathbb{Z}\geq c} \times \mathbb{A}_{\mathbb{Z}}^{d-1}$

where $\mathbb{A}_{\mathbb{Z} \geq c} := \{b \in \mathbb{A}_{\mathbb{Z}} \mid b \geq c\}$. Now to ensure that we stay within this domain we would demand that the first position of the constant vector of every interpretation is greater or equal than $c$.

Note however that this $c$ can be fixed to 0 without any loss of generality as every interpretation using lower values in those positions can be "shifted" upwards. For any interpretation $[\cdot]$ and arctic number $d$ construct an interpretation $[\cdot]'$ by $[t]' := [t] \otimes d$. This is obtained by going from $[f] = f_0 \oplus f_1 x_1 \oplus \dots f_k x_k$ to $[f]' = f_0 \otimes d \oplus f_1 x_1 \oplus \dots f_k x_k$. (A linear function with absolute part can be scaled by scaling the absolute part.) □

## 8 Certification

The certification has been carried out within the CoLoR library [4]: a library of termination techniques formalized in Coq. This library is then used by a tool Rainbow to transform termination proofs in the common termination proof format, designed within the CoLoR project, to actual Coq proofs certifying termination.

The basis of this work was the certification of the matrix interpretations method [19], which consists of formalizations of:

- a semi-ring structure,
- vectors and matrices over arbitrary semi-rings of coefficients,
- the monotone algebras framework and
- the matrix interpretation method.

The framework of monotone algebras was used without any changes at all. Vectors and matrices were formalized for arbitrary semi-rings, however all the results involving orders were developed for the usual orders on natural numbers, as used in the matrix interpretations method. So the first step in the certification process was to generalize the semi-ring structure to a semi-ring equipped with two orders $(>, \geq)$ and to adequately generalize results on vectors and matrices. Then the arctic semi-ring was developed in this setting.

As for the technique itself it has a lot in common with the technique of matrix interpretations. Therefore the common parts were extracted to a module `MatrixBasedInt` which was then specialized to the matrix interpretation method (`MatrixInt`) and to a basis for arctic based methods (`ArcticBasedInt`), which was narrowed down to the methods of arctic interpretations (`ArcticInt`) and arctic below-zero interpretations (`ArcticBZInt`). This hierarchy is depicted in Figure 1.

Considering the extension of the proof format in Rainbow it was minimal. The format for the matrix interpretation proofs was already developed in [19] and it essentially requires to provide matrix interpretations for all the function symbols in the signature. The format for arctic interpretations is the same except that:

- it indicates which matrix-based method is to be used, indicated by different XML tags (as the common proof format of CoLoR is specified using XML syntax),
- the entries of vectors and matrices are from a different domain.

The experimental data concerning certification results is presented in the following section.

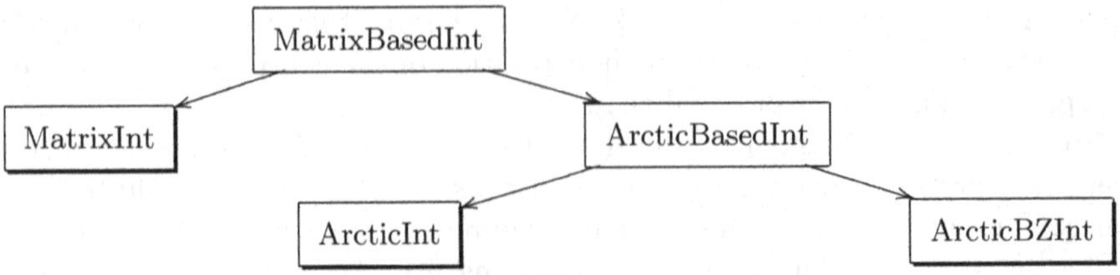


**Fig. 1.** Hierarchy of different matrix-based methods in CoLoR

# 9 Implementation

The implementation in Matchbox follows the scheme described in [7]. The constraint problem for the arctic interpretation is translated to a constraint problem for matrices, for arctic numbers and, finally, for Boolean variables. This is then solved by Minisat [5].

An arctic number is represented by a pair $a = (b; v_0, v_1, \ldots, v_n)$ where $b$ is a Boolean value and $v_0, \ldots, v_n$ is sequence of Booleans (all numbers have fixed bit-width). If $b$ is 1, then $a$ represents $-\infty$, if $b$ is 0, then $a$ represents the binary value of $v_0, \ldots, v_n$.

To represent integers, we use two's complement representation, i.e., the most significant bit is the "sign bit".

Note that implementation of max/plus operation is less expensive than standard plus/times: with a binary representation both max and plus can be computed (encoded) with a linear size formula (whereas a naive implementation of the standard multiplication requires quadratic size and asymptotically better schemes do not pay off for small bit widths).

It is useful to require the following, for each arctic number $a = (b, v)$: if the infinity bit $b$ is set, then $v = 0$. Then $(b, v) \oplus (b', v') = (b \wedge b', \max(v, v'))$. For $(b, v) \otimes (b', v')$ we compute $c = b \vee b'$, $u = (u_0, \ldots, u_n) = v + v'$ and the result is $(c; \neg c \wedge u_0, \ldots, \neg c \wedge u_n)$.

To represent arctic integers, we use a similar convention: if the infinity flag $b$ is set, we require that the number $v$ represents the lowest value of its range.

The following table lists the numbers of certified proofs that we obtain with DP transformation (without SCC decomposition, see below) and these matrix methods: (s)tandard, (a)rctic, below (z)ero. For comparison, we give the corresponding numbers for last year's winner of the (certified, where applicable) termination competition.

| problem set | time | s | sa | sz | saz | 2007 winner |
|---|---|---|---|---|---|---|
| 975 TRS | 1 min | 361 | 376 | 388 | 389 | TPA: 354 |
| | 10 min | 365 | 381 | 393 | 394 | |
| 517 SRS | 1 min | 178 | 312 | 298 | 320 | Matchbox: 337 |
| | 10 min | 185 | 349 | 323 | 354 | |

Runs were executed on a single core of an Intel X5365 processor running at 3GHz. All proofs will be made available for inspection at the Matchbox web page [21]. In all cases we used standard matrices of dimension 1 and 2 to remove rules before the DP transformation, and then matrix dimensions $d$ from 1 up; with numbers of bit width $\max(1, 4 - \lfloor d/2 \rfloor)$, and a timeout of $5 + 2^d$ seconds for each individual attempt.

It should be noted that TPA 2007 additionally used (non-linear) polynomial interpretations, and that Matchbox 2007 also used additional methods (e.g. RFC match-bounds) and was running uncertified.

Here, we count only verified proofs, so we are missing about 3 to 5 proofs where Coq does not finish in reasonable time. (This happened—for exactly the same problems—also in 2007.)

To certify termination of string rewriting, we use the standard transformation to a term rewriting system with all symbols unary. We do this for the original system $\mathcal{R}$ as well as for the system $\text{reverse}(\mathcal{R}) = \{\text{reverse}(l) \to \text{reverse}(r) \mid (l \to r) \in \mathcal{R}$. It is obvious (though presently not included in CoLoR) that this transformation preserves termination both ways. Half of the allotted time is spent for each of $\mathcal{R}$ and $\text{reverse}(\mathcal{R})$. This increases the score considerably (by about one third).

The dependency pairs transformation is often combined with a decomposition of the resulting top termination problem into independent subproblems; analyzing strongly connected components of the estimated dependency graph [10]. Currently, CoLoR provides only a simple graph approximation by top symbols of dependency pairs, but at the moment it is not efficient. Our current implementation therefore does not do decomposition. However, with only this simple graph approximation, this does not decrease power: note that an interpretation that removes rules from a maximal component in the DP graph (with no incoming arrows) can be extended to the complete graph by assigning constant zero to all top symbols not occurring in this component.

## 10 Discussion

Arctic naturals form a sub-semi-ring of arctic integers. So the question comes up whether Theorem 14 subsumes Theorem 12. Note that the prerequisites for both theorems are incomparable. Still there might be a method to construct from a somewhere-finite interpretation (above zero) an equivalent absolute-positive interpretation (below zero). We are not aware of any. Experience with implementation shows that it is useful to have both methods, especially for string rewriting. Naturals are easier to handle than integers because they do not require signed arithmetics. So typically we can increase the bit width or the matrix dimension for naturals. Our implementation finds several proofs according to Theorem 12 where it fails to find a proof according to Theorem 14 and vice versa.

It is interesting to ask whether the preconditions of Theorems 10,12,14 can be weakened. We discussed one variant in Remark 16. In general, a linear interpretation $[\cdot]$ with coefficients in $\mathbb{A}_\mathbb{N}$ ($\mathbb{A}_\mathbb{Z}$ respectively) is admissible for a termination proof if for each ground term $t$, the value $[t]$ is finite (positive, respectively). This is in fact a reachability problem for weighted (tree) automata. It is decidable for interpretations on arctic naturals, but it is undecidable for arctic integers (follows from a result of Krob [20] on tropical word automata). In our setting, we do not guess an interpretation and then decide whether it is admissible. Rather, we have to formulate the decision algorithm as part of the constraint system for the interpretation. Therefore we chose sharper conditions on interpretations that imply finiteness (positiveness, respectively) and have an easy constraint encoding.

Another question is the relation of the standard matrix method with the arctic matrix method(s). Performance of our implementation suggests that neither method subsumes the other, but this may well be a problem of computing resources, as we hardly reach matrix dimension 5 and bit width 3.

As for the relation to other termination methods (e.g. path orderings), the only information we have is that arctic (and other) matrix methods can do non-simple termination, while path orders and polynomial interpretations cannot; and on the other hand, the arctic matrix method implies a linear bound on derivational complexity (see below), which is easily surpassed by path orders and other interpretations.

The full arctic termination method bounds lengths of derivations:

**Lemma 17.** *For a rewriting system $\mathcal{R}$ that fulfils the requirements of Theorem 10 for $\mathcal{S} = \emptyset$, the derivational complexity of $\mathcal{R}$ is linear.*

*Proof.* For a finite arctic vector $x = (x_1, \ldots, x_k)$, define $|x| = \max(x_1, \ldots, x_k)$. Then $|x \oplus y| \leq \max(|x|, |y|)$ and $|x \otimes y^T| \leq |x| + |y|$.

For a finite arctic matrix $A$ of dimension $k \times k$, define $|A| = \max\{A_{i,j} \mid 1 \leq i, j \leq k\}$. Then $|A \otimes x| \leq |A| + |x|$ and $|A \otimes B| \leq |A| + |B|$.

For an interpretation $[\cdot]$ of some signature $\Sigma$, and any word $w \in \Sigma^*$, this implies that $|[w]| \leq c \cdot |w|$ where $c = \max\{|[f]| : f \in \Sigma\}$.

Now we remark that $u \to_{\mathcal{R}} v$ implies $[u] \gg [v]$, and $x \gg y$ implies $|x| > |y|$. Thus the derivational complexity of $\mathcal{R}$ is linear: any derivation starting from $u$ has at most $c \cdot |u|$ steps. □

This means that rewriting systems with higher derivational complexity (e.g. quadratic: $\{ab \to ba\}$, or exponential $\{ab \to b^2a\}$) do not admit an arctic termination proof. Note that both these systems admit a standard matrix proof.

It seems very difficult to combine this argument with the dependency pairs transformation, as it can drastically alter (i.e., reduce) derivational complexity.

*Example 18.* The following rewriting system [16] has a derivational complexity that is not primitive recursive:

$$\{\mathrm{s}(x) + (y + z) \to x + (\mathrm{s}(\mathrm{s}(y)) + z), \mathrm{s}(x) + (y + (z + w)) \to x + (z + (y + w))\}$$

and still it has, after DP transformation, an easy termination proof by "counting symbols" [7]. Note however that arctic interpretations cannot count globally: to compute the interpretation $[f(t_1, t_2)]$, it is impossible to add values from subtrees $[t_1], [t_2]$, as we can only take the maximum of $[t_1], [t_2]$. Yet we find an arctic proof, as follows. The given system is in fact an encoding of a length-preserving string rewriting system on the infinite alphabet $\mathbb{N}$. Both rules keep the right spine of terms (corresponding to the length of the simulated string) intact, so we can remove dependency pairs that shrink it, using the interpretation $[+](x, y) = y \otimes 1$. We are left with two dependency pairs (that directly correspond to the original rules). They can be handled by $[+](x, y) = x$ and $[s](x) = x \otimes 1$. So instead of numbers of symbols, we were just using path lengths. $\triangleleft$

Arctic interpretations subsume quasi-periodic interpretations [24]. This has been remarked in [22] for string rewriting and it easily extends to term rewriting.

Max/Plus polynomials have been used by Amadio [1] as quasi-interpretations (i.e. functions are weakly monotone), to bound the space complexity of derivations. Proving termination directly was not intended.

## 11 Conclusions

We presented the arctic interpretations method for proving termination of term rewriting. It is based on the matrix interpretation method [7] where the usual plus/times operations on $\mathbb{N}$ are generalized to an arbitrary semi-ring, in this case instantiated by the arctic semi-ring (max/plus algebra) on $\{-\infty\} \cup \mathbb{N}$.

We also generalized this to arctic integers. This generalization allowed us to solve 10 of Beerendonk/* examples that are difficult to prove terminating and thus far could only be solved by AProVE with the Bounded Increase [12] technique, dedicated to such class of problems coming from transformations from imperative programs.

Our presentation of the theory is accompanied by a formalization in the Coq proof assistant. By becoming part of the CoLoR project this formalization allows us to formally verify termination proofs involving the arctic matrix method. With this contribution CoLoR can now certify more than half of the systems that could be proven terminating in the 2007 competition in term rewriting and essentially all (and some more) systems in the string rewriting category.

We want to remark here that all performance data and all examples presented in this paper were collected from problems of TPDB 2007, and we did not "cook up" any special examples to show off the arctic method. The emphasis of these examples (in fact, of the whole paper) is not to provide termination proofs where none were known before, but rather to provide certified (and often conceptually simpler) termination proofs where only uncertified proofs were available up to now.

## References

1. Amadio, R.M.: Synthesis of max-plus quasi-interpretations. Fundamenta Informaticae 65(1-2), 29–60 (2005)
2. Arts, T., Giesl, J.: Termination of term rewriting using dependency pairs. TCS 236(1-2), 133–178 (2000)
3. Baader, F., Nipkow, T.: Term Rewriting and All That. Cambridge University Press, Cambridge (1998)
4. Blanqui, F., Delobel, W., Coupet-Grimal, S., Hinderer, S., Koprowski, A.: CoLoR, a Coq library on rewriting and termination. In: WST (2006), `http://color.loria.fr`

5. Eén, N., Sörensson, N.: An extensible sat-solver. In: Giunchiglia, E., Tacchella, A. (eds.) SAT 2003. LNCS, vol. 2919, pp. 502–518. Springer, Heidelberg (2004)
6. Endrullis, J.: Jambox, `http://joerg.endrullis.de`
7. Endrullis, J., Waldmann, J., Zantema, H.: Matrix interpretations for proving termination of term rewriting. Journal of Automated Reasoning (to appear, 2007)
8. Fuchs, L.: Partially Ordered Algebraic Systems. Addison-Wesley, Reading (1962)
9. Gaubert, S., Plus, M.: Methods and applications of (max, +) linear algebra. In: Reischuk, R., Morvan, M. (eds.) STACS 1997. LNCS, vol. 1200, pp. 261–282. Springer, Heidelberg (1997)
10. Giesl, J., Arts, T., Ohlebusch, E.: Modular termination proofs for rewriting using dependency pairs. Journal of Symbolic Computation 34(1), 21–58 (2002)
11. Giesl, J., Thiemann, R., Schneider-Kamp, P., Falke, S.: Automated termination proofs with AProVE. In: van Oostrom, V. (ed.) RTA 2004. LNCS, vol. 3091, pp. 210–220. Springer, Heidelberg (2004)
12. Giesl, J., Thiemann, R., Swiderski, S., Schneider-Kamp, P.: Proving termination by bounded increase. In: Pfenning, F. (ed.) CADE 2007. LNCS (LNAI), vol. 4603, pp. 443–459. Springer, Heidelberg (2007)
13. Golan, J.S.: Semirings and their Applications. Kluwer, Dordrecht (1999)
14. Hirokawa, N., Middeldorp, A.: Polynomial interpretations with negative coefficients. In: Buchberger, B., Campbell, J.A. (eds.) AISC 2004. LNCS (LNAI), vol. 3249, pp. 185–198. Springer, Heidelberg (2004)
15. Hirokawa, N., Middeldorp, A.: Tyrolean termination tool: Techniques and features. Information and Computation 205(4), 474–511 (2007)
16. Hofbauer, D., Lautemann, C.: Termination proofs and the length of derivations. In: Dershowitz, N. (ed.) RTA 1989. LNCS, vol. 355, pp. 167–177. Springer, Heidelberg (1989)
17. Hofbauer, D., Waldmann, J.: Termination of string rewriting with matrix interpretations. In: Pfenning, F. (ed.) RTA 2006. LNCS, vol. 4098, pp. 328–342. Springer, Heidelberg (2006)
18. Koprowski, A.: TPA: Termination proved automatically. In: Pfenning, F. (ed.) RTA 2006. LNCS, vol. 4098, pp. 257–266. Springer, Heidelberg (2006), `http://www.win.tue.nl/tpa`
19. Koprowski, A., Zantema, H.: Certification of proving termination of term rewriting by matrix interpretations. In: SOFSEM. LNCS, vol. 4910, pp. 328–339 (2008)
20. Krob, D.: The equality problem for rational series with multiplicities in the tropical semiring is undecidable. In: Kuich, W. (ed.) ICALP 1992. LNCS, vol. 623, pp. 101–112. Springer, Heidelberg (1992)
21. Waldmann, J.: Matchbox: A tool for match-bounded string rewriting. In: van Oostrom, V. (ed.) RTA 2004. LNCS, vol. 3091, pp. 85–94. Springer, Heidelberg (2004), `http://dfa.imn.htwk-leipzig.de/matchbox`
22. Waldmann, J.: Arctic termination. In: WST (2007)
23. Waldmann, J.: Weighted automata for proving termination of string rewriting. Journal of Automata, Languages and Combinatorics (to appear, 2007)
24. Zantema, H., Waldmann, J.: Termination by quasi-periodic interpretations. In: Baader, F. (ed.) RTA 2007. LNCS, vol. 4533, pp. 404–418. Springer, Heidelberg (2007)
25. The Coq proof assistant, `http://coq.inria.fr`
26. Termination competition, `http://www.lri.fr/~marche/termination-competition`
27. Termination problems data base, `http://www.lri.fr/~marche/tpdb`

# Logics and Automata for Totally Ordered Trees

Marco Kuhlmann[1] and Joachim Niehren[2]

[1] Uppsala University, Sweden
[2] INRIA, Lille, France

**Abstract.** A totally ordered tree is a tree equipped with an additional total order on its nodes. It provides a formal model for data that comes with both a hierarchical and a sequential structure; one example for such data are natural language sentences, where a sequential structure is given by word order, and a hierarchical structure is given by grammatical relations between words. In this paper, we study monadic second-order logic (MSO) for *totally ordered terms*. We show that the MSO satisfiability problem of unrestricted structures is undecidable, but give a decision procedure for practically relevant sub-classes, based on tree automata.

## 1 Introduction

A totally ordered tree is a tree equipped with an additional total order on its nodes. It provides a formal model for data that comes with both a hierarchical and a sequential structure. Depending on the application, the two structural aspects may be more or less dependent on each other: the total order may be obtained by a traversal of the tree, defined by a logic formula from tree relations, or completely independent.

The research reported in this paper is motivated by an application of totally ordered trees in computational linguistics, where they are used as formal models for *dependency structures*. A dependency structure is a representation of the syntactic structure of a natural-language sentence in terms of word-to-word dependencies, such as the dependency between a verb and its direct object. Such dependencies impose a tree-shaped hierarchical structure onto the words, while word order imposes a sequential structure. An example for a dependency structure is given in Fig. 1. Dependency-based representations have a long history in descriptive linguistics. Recently, they have received a lot of interest for many computational tasks, such as information extraction [1], machine translation [2], and, most prominently, data-driven parsing [3].

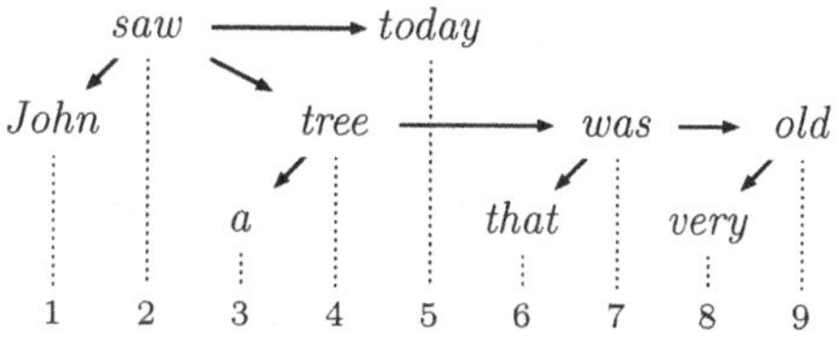


**Fig. 1.** A dependency structure

In this paper, we study monadic second-order logic (MSO) as a description language for *totally ordered terms (tots)*. Dependency structures can be understood as special cases of tots, where the sibling order is induced by the total order. Monadic second-order logic is generally useful in order to express properties of

A. Voronkov (Ed.): RTA 2008, LNCS 5117, pp. 217–231, 2008.

graphs or trees [4,5,6,7]. Here we are particularly interested in lifting Doner, Thatcher, and Wright's theorem on the equivalent expressiveness of MSO and tree automata for finite trees to tots [4,8]. This theorem has first been proved for ground terms, and got extended to various kinds of trees and graphs of bounded tree width [9,10,7].

The new problem that we are faced with, is to deal with the addition of a total order to a finite term structure. The easy cases of this problem are those where the total order is MSO-definable from the term structure: MSO for ground terms is decidable, and thus MSO for ground terms with MSO-defined total orders is decidable as well. In the first part of the paper, we prove that MSO for general tots is *not* decidable. We show this by a reduction of the MSO satisfiability problem for *grids*; this problem is well-known to be undecidable (while first-order logic of grids can encoded into Presburger arithmetics). In the second part of the paper, we restrict the classes of models of MSO formulas to sets of tots with *bounded gap-degree* [11]. This means that the descendant sets of nodes are segmented into a bounded number of intervals in the total order. Our main contribution is the result that MSO satisfiablity of sets of tots with bounded gap-degree is decidable. To establish this, we introduce an algebraization for these sets. This leads us to a notion of tree automaton for gap-bounded sets of tots, which we show to have the same expressiveness as MSO.

*Related work.* Our algebraic perspective on automata goes back to the work of Mezei and Wright from the 1960s [16]. It was generalized by Courcelle [17], and applied to many kinds of trees, including unranked sibling-ordered trees as they appear in the context of XML [18]. Courcelle has proposed two different algebraizations for graphs [7] for which MSO can be reduced to finite automata. Graphs of bounded tree or clique width belong to these algebras, so that MSO satisfiability is decidable for them.

Dependency structures can be used to quantify the generative capacity of many grammar formalisms for natural language [12]: they are more informative than strings, but less formalism-specific than parse trees. The formal properties of dependency structures and sets of such structures have been studied only recently [12]. Automata for these structures define a notion of regularity. For regular sets of dependency structures, there is a direct relation between the gap-degree measure and string-generative capacity. In particular, regular sets of dependency structures with gap-degree 0 correspond exactly to the context-free languages, and regular sets of structures with gap-degree at most 1 and an additional property called *well-nestedness* give the string languages generated by Lexicalized Tree Adjoining Grammars (TAGs). More generally, every string language obtained from a regular set of dependency structures with bounded gap-degree is semilinear and can be recognized in polynomial time. Without gap restrictions, parsing quickly becomes NP-hard [13]. Recent work has shown that most dependency structures required for the analysis of natural language have a small gap-degree [15]. However, not all natural languages can be described by sets of dependency structures with bounded gap-degree; counterexamples have been given for Czech [14].

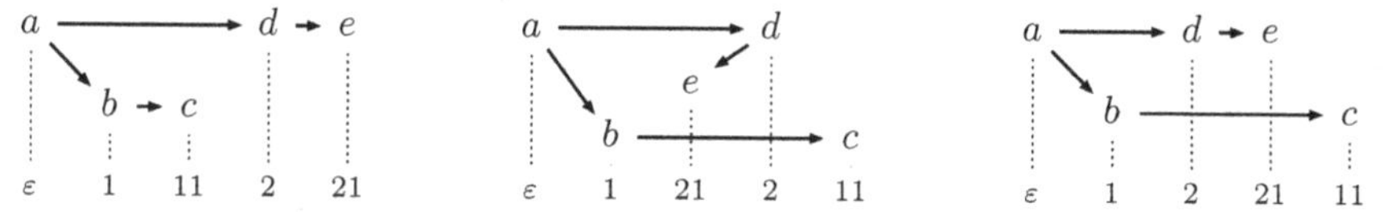

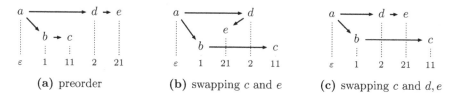

**(a)** preorder **(b)** swapping $c$ and $e$ **(c)** swapping $c$ and $d, e$

**Fig. 2.** Three total orders for the tree $a(b(c), d(e))$

## 2 Totally Ordered Terms

We start by introducing totally ordered terms, and MSO for their description. We make use of the following auxiliary notions and notations: The set of positive natural numbers is denoted by $\mathbb{N}$. Given some $n \in \mathbb{N}$, we write $[n]$ for the set $\{1, \ldots, n\}$. A *signature* $\Sigma$ is a non-empty, finite set of function symbols $\sigma$, each equipped with a fixed, non-negative arity, denoted by $\mathrm{ar}(\sigma)$. The *set of (ground) terms over* $\Sigma$ is the smallest set $T_\Sigma$ such that if $\sigma$ is a symbol of arity $m$ and for each $i \in [m]$, $t_i \in T_\Sigma$, then $\sigma(t_1, \ldots t_m) \in T_\Sigma$. The set of *nodes* of a term $t \in T_\Sigma$ is a set of addresses in $\mathbb{N}^*$: the root node of $t$ is addressed by the empty word $\varepsilon$, and the $i$th child of the node $\pi$ is addressed by the extended address $\pi i$:

$$nod(\sigma(t_1, \ldots, t_m)) = \{\varepsilon\} \cup \{\, i\pi \in \mathbb{N}^* \mid i \in [m],\ \pi \in nod(t_i)\,\}\,.$$

A *linearization* of a finite set $A$ is a word in $A^*$ in which each element of $A$ occurs exactly once. We note that the set of linearizations of $A$ is isomorphic to the set of total orders on $A$ in an obvious way.

**Definition 1.** *A* totally ordered term (tot) *over* $\Sigma$ *is a pair* $\tau = \langle t, w\rangle$ *where* $t \in T_\Sigma$ *is a term, and* $w$ *is a linearization of* $nod(t)$.

The set of all tots over $\Sigma$ is denoted by $TOT_\Sigma$. Three examples for tots are visualized in Fig. 2; each of them provides a different linearization for the same term $a(b(c), d(e))$. Solid edges depict the term structure, dotted lines project nodes to their position in the linearization. In Fig. 2a, the nodes of $a(b(c), d(e))$ are ordered by the preorder traversal of the underlying term structure. The two examples in Fig. 2b and 2c are derived by swapping $c$ with $e$ and $d, e$, respectively.

In the following, we often identify a term $t \in T_\Sigma$ with a relational structure $\langle nod(t)\,; (:\sigma)_{\sigma\in\Sigma}\rangle$: for each $m$-ary symbol $\sigma \in \Sigma$, this structure provides a labelling relation $:\sigma$ with arity $m+1$. Given a node $\pi \in nod(t)$, labelling $\pi : \sigma(\pi 1, \ldots, \pi n)$ holds if and only if $\pi$ is labelled by $\sigma$ in $t$. A tot $\tau = (t, w)$ can be viewed as a relational structure that extends the structure corresponding to $t$ by the total order $\preceq$ that is represented by the linearization $w$.

Monadic second-order logic (MSO) for tots is defined as usual. Let *Vars* be a set that contains infinitely many node variables $x, y, z \in Vars$, and infinitely many set variables $X, Y, Z \in Vars$. The formulas of MSO for tots are the following, where $\sigma \in \Sigma$ is a function symbol of arity $m$:

$$\phi, \phi' ::= x : \sigma(x_1, \ldots, x_m) \mid x \preceq y \mid x \in X \mid \phi \wedge \phi' \mid \neg\phi \mid \exists x.\, \phi \mid \exists X.\, \phi$$

These formulas are interpreted in the usual Tarskian style in the relational structures corresponding to tots. Given a tot $\tau = (t, w)$ and a variable assignment $\alpha\colon \mathit{Vars} \to \mathit{nod}(t)$, we write $\tau, \alpha \models \phi$ if and only if the formula $\phi$ evaluates to true in $\tau$ under the assignment $\alpha$, and $\tau \models \phi$ if $\tau, \alpha \models \phi$ holds for any assignment $\alpha$.

MSO formulas with $n$ free node variables over the signature $\Sigma$ define $n$-ary relations for all tots over $\Sigma$. Most basically, we can define node equality ($=$), equality-or-precedence ($\preceq$), and the child relation $\triangleleft$:

$$x \triangleleft y =_{\mathrm{def}} \bigvee_{\sigma \in \Sigma} \bigvee_{1 \leq i \leq m = \mathrm{ar}(\sigma)} \exists x_1, \dots, x_m.\, y = x_i \wedge x : \sigma(x_1, \dots, x_m)) \,.$$

The dominance relation $\triangleleft^*$ of a term is the reflexive, transitive closure of $\triangleleft$, and thus definable in MSO.[1] This is in contrast to first-order logic, where the dominance relation is usually added to the relational structure. Let $C \subseteq \mathit{TOT}_\Sigma$ be a set of tots over $\Sigma$. The *MSO-satisfiability problem of $C$* is the problem to decide whether $\exists \tau \in C.\, \tau \models \phi$, where $\phi$ is some closed MSO formula.

## 3 Undecidability of MSO Satisfiability

We now present our first technical result, which is valid for all signatures $\Sigma$ with at least one binary function symbol and one constant.

**Theorem 1.** *MSO satisfiability for the class of all tots over $\Sigma$ is undecidable.*

For the proof, we make use of a simple tool to obtain undecidability results for graph-like structures. Let $m, n \in \mathbb{N}$. The *grid of dimensions $m \times n$* is the structure

$$\begin{aligned} G_{m,n} &= \langle [m] \times [n]\,; \mathit{nextEast}, \mathit{nextSouth} \rangle\,, \quad \text{where} \\ \mathit{nextEast} &= \{\, \langle \langle i, j \rangle, \langle i', j' \rangle \rangle \mid i' = i + 1,\, j' = j \,\} \quad \text{and} \\ \mathit{nextSouth} &= \{\, \langle \langle i, j \rangle, \langle i', j' \rangle \rangle \mid i' = i,\, j' = j + 1 \,\}\,. \end{aligned}$$

We can view a grid as an $(m \times n)$-matrix in which we can navigate along columns (using the relation *nextEast*) and rows (using *nextSouth*). The *square grid of size $m$* is the grid of dimensions $m \times m$. The following result is standard:

**Fact [7].** *The MSO satisfiability problem of every class of graphs that contains infinitely many square grids is undecidable.*

To make use of this result to prove Theorem 1, we show how to encode grids as tots (Proposition 1), and that every closed MSO formula interpreted on grids can be translated into a closed MSO formula interpreted on tots that has the same models modulo encoding (Proposition 2).

Without loss of generality, we use the signature $\Sigma = \{\mathsf{cons}, \circ, \mathsf{nil}\}$ with $\mathrm{ar}(\mathsf{cons}) = 2$, $\mathrm{ar}(\circ) = 1$, and $\mathrm{ar}(\mathsf{nil}) = 0$, and employ the following naming scheme for nodes: for $0 \leq i$ and $1 \leq j$,

$$\pi_{i,j} = \mathbf{if}\ i = 0\ \mathbf{then}\ 1^{j-1}\ \mathbf{else}\ 1^{j-1} 2 1^{i-1}\,.$$

[1] To see this, note that $x$ dominates all elements of the set $Y$, where $Y$ is the least set satisfying the equation $Y = \{x\} \cup \{\, z \mid \exists y.\, y \in Y \to y \triangleleft z \,\}$.

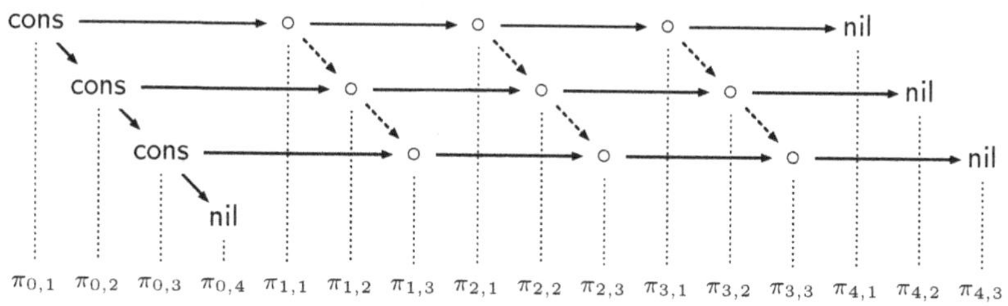


**Fig. 3.** The encoding of the square grid $G_{3,3}$. The solid lines represent the term structure; the dashed lines visualize the relation *nextSouth*.

The encoding of the general grid $G_{m,n}$ is the tot $[\![G_{m,n}]\!] = \langle t_{m,n}, w_{m,n}\rangle$, where

$$\begin{aligned} t_{m,n} &= \textbf{if } n = 1 \textbf{ then } \mathsf{cons}(\mathsf{nil}, \circ^m(\mathsf{nil})) \textbf{ else } \mathsf{cons}(t_{m,n-1}, \circ^m(\mathsf{nil})) \\ w_{m,n} &= \pi_{0,1} \cdots \pi_{0,n} \cdot \pi_{0,n+1} \cdot \pi_{1,1} \cdots \pi_{1,n} \cdots \pi_{m+1,1} \cdots \pi_{m+1,n} \end{aligned}$$

For illustration, Fig. 3 shows the tot that encodes the square grid $G_{3,3}$. We will use the nodes that are labelled with the constructor $\circ$ (i.e., the nodes of the form $\pi_{i,j}$, $i \in [m]$, $j \in [n]$) as representatives for the entries of the grid proper.

The following MSO-defined sets are of general interest for the axiomatization of grid encodings. We use definitions by formulas with free set variables that may be parametrized by a free node variable $x$, so that the corresponding sets are parametrized by the node value of $x$.

$$\begin{aligned} \mathit{FirstChild}(x) &= \{\, y \mid x : \circ(y) \,\} \cup \{\, y \mid \exists\, z.\, x : \mathsf{cons}(y, z) \,\} \\ \mathit{SecondChild}(x) &= \{\, y \mid \exists\, z.\, x : \mathsf{cons}(z, y) \,\}, \quad \mathit{Root} = \{\, x \mid \mathit{Father}(x) = \emptyset \,\} \\ \mathit{Father}(x) &= \{\, y \mid x \in \mathit{FirstChild}(y) \,\} \cup \{\, y \mid x \in \mathit{SecondChild}(y) \,\} \\ \mathit{Succ}(x) &= \{\, y \mid x \prec y,\ \neg\exists\, z.\, x \preceq z \preceq y \,\}, \quad \mathit{Cons} = \{\, x \mid x{:}\mathsf{cons}(\_,\_) \,\}, \\ \mathit{Grid} &= \{\, x \mid x : \circ(\_) \,\}, \quad \mathit{Nil} = \{\, x \mid x : \mathsf{nil} \,\}. \end{aligned}$$

We note that, since $\preceq$ is a total order, *Succ* is an injective partial function. The cons-labelled nodes provide the backbone of the tot; they 'glue' the lines of the grid together. The north of the grid is its first line; the south is its last line.

$$\begin{aligned} \mathit{Backbone} &= (\mathit{Root} \cup \mathit{FirstChild}(\mathit{Backbone})) - \mathit{Nil} \\ \mathit{LastBackbone} &= \{\, x \in \mathit{Backbone} \mid \mathit{FirstChild}(x) \subseteq \mathit{Nil} \,\} \\ \mathit{North} &= (\mathit{SecondChild}(\mathit{Root}) \cup \mathit{FirstChild}(\mathit{North})) \cap \mathit{Grid} \\ \mathit{South} &= (\mathit{SecondChild}(\mathit{LastBackbone}) \cup \mathit{FirstChild}(\mathit{South})) \cap \mathit{Grid} \end{aligned}$$

Recursive definitions are interpreted with respect to least fixed points, which is expressible in MSO, as is the reflexive-transitive closure of a relational image

$S(x)$, denoted by $S(x)^*$. We next define sets of nodes by MSO formulas that permit us to navigate through the grid:

$$\begin{aligned} \mathit{NextEast}(x) &= \{\, y \in \mathit{Grid} \cap \mathit{FirstChild}(x) \mid x \in \mathit{Grid} \,\} \\ \mathit{NextSouth}(x) &= \{\, y \in \mathit{Grid} \cap \mathit{Succ}(x) \mid x \in \mathit{Grid} - \mathit{South} \,\} \end{aligned}$$

The *anchor* of a node $x$ is the (uniquely determined) closest node on the backbone that dominates $x$. The anchors allow us to navigate along a grid's rows.

$$\begin{aligned} \mathit{Anchor}(x) &= \{\, y \in \mathit{Backbone} \mid x \in \mathit{FirstChild}^*(\mathit{SecondChild}(y)) \,\} \\ \mathit{NextRow}(x) &= \{\, y \in \mathit{Grid} \mid \mathit{FirstChild}(\mathit{Anchor}(x)) = \mathit{Anchor}(y) \,\} \\ \mathit{PreviousRow}(x) &= \{\, y \in \mathit{Grid} \mid \mathit{FirstChild}(\mathit{Anchor}(y)) = \mathit{Anchor}(x) \,\} \end{aligned}$$

We can now axiomatize the class of grid encodings in MSO:

(A1) The backbone consists of the nodes labelled with cons.

$\mathit{Backbone} = \mathit{Cons}$

(A2) For every node in the grid that does not lie in the south, taking the next step in the total order takes us to the next row.

$\forall x \in \mathit{Grid} - \mathit{South}.\, \exists y \in \mathit{NextRow}(x).\, \mathit{Succ}(x) = \{y\}$

(A3) Dually, every node in the grid that does not lie in the north can be reached from a node in the previous row.

$\forall x \in \mathit{Grid} - \mathit{North}.\, \exists y \in \mathit{PreviousRow}(x).\, \mathit{Succ}(y) = \{x\}$

(A4) The root node occupies the first position in the total order.

$\neg\exists x.\, \mathit{Succ}(x) = \mathit{Root}$

(A5) Every step to the first child of a cons node is a step in the total order.

$\forall x \in \mathit{Cons}.\, \mathit{Succ}(x) = \mathit{FirstChild}(x)$

(A6) The nil node terminating the backbone immediately precedes the north-western corner of the grid.

$\mathit{Succ}(\mathit{FirstChild}(\mathit{LastBackbone})) = \mathit{SecondChild}(\mathit{Root})$

(A7) The total order propagates along the rows of the grid.

$\forall x, y.\, \mathit{Succ}(\mathit{Father}(x)) = \mathit{Father}(y) \implies \mathit{Succ}(x) = \{y\}$

(A8) When being in the south, the next step with the total order leads to the first element of the next column eastwards.

$\forall x \in \mathit{North}.\, \mathit{Succ}(\mathit{NextSouth}^*(x) \cap \mathit{South}) = \mathit{FirstChild}(x)$

**Proposition 1.** *Every encoded grid satisfies the axioms A1–A8. Conversely, every tot that satisfies these axioms is a grid encoding.*

*Proof.* To verify that encoded grids satisfy the axioms A1–A8 is straightforward; here we show the converse. Let $\tau = (t, w)$ be a tot such that $\tau \models \mathrm{A1} \wedge \cdots \wedge \mathrm{A8}$. We show that there exists a grid $G_{m,n}$ such that $\tau = [\![G_{m,n}]\!]$. Axiom $A_1$ asserts that $\mathit{Backbone} = \mathit{Cons}$, which implies that $t$ has the following form, where $t_i = \mathsf{o}^{m_i}(\mathsf{nil})$ for some $m_i \geq 0$: $t = \mathsf{cons}(t_1, \mathsf{cons}(t_2, \ldots \mathsf{cons}(t_n, \mathsf{nil}) \ldots))$.

We have to show that all terms $t_i$ have the same length in order to prove that $t = t_{m,n}$. This property is less obvious since it depends on the total order among the nodes. For $n = 0$, it follows that $t = \mathsf{nil}$, and there is nothing to show, so we can assume $n \geq 1$. The form of $t$ implies that $\mathit{Backbone} = \{\, \pi_{0,j} \mid j \in [n] \,\}$,

$$\mathit{North} = \{\, \pi_{i,1} \mid i \in [m_1] \,\}, \qquad \text{and} \qquad \mathit{South} = \{\, \pi_{i,n} \mid i \in [m_n] \,\}.$$

**Lemma 1.** $m_1 = \cdots = m_n$

We first show that $|t_1| \leq |t_j|$ for all $j \in [n]$ by induction on $n$. The case $n = 1$ is trivial. For the case $n > 1$, it suffices to show that $|t_{n-1}| \leq |t_n|$. This follows from Axiom A2, which asserts that for each node $\pi$ in $t_{n-1}$, there exists a node $\pi'$ in $t_n$ such that $\mathit{Succ}(\pi) = \{\pi'\}$, and from the observation that $\mathit{Succ}$ is total for all nodes of $t_1$ by Axiom A4, functional and injective. Dually, using Axiom A3, we can show that $|t_j| \leq |t_1|$ for all $j \in [n]$. Consequently, $|t_1| = \cdots = |t_n|$.

It remains to show that $w$ is indeed the total order that is imposed on the encoded grid. The following series of Lemmas together with Axiom A4 establishes that the total order on $t$ is uniquely determined by the axioms. Since the order in the grid translation satisfies these axioms, it must be equal to this order.

**Lemma 2.** *$\forall j \in [n]. \mathit{Succ}(\pi_{0,j}) = \{\pi_{0,j+1}\}$ and $\mathit{Succ}(\pi_{0,n+1}) = \{\pi_{1,1}\}$*

Since for each $j \in [n]$, the node $\pi_{0,j}$ is a $\mathsf{cons}$ node, the first half of the claim is stated in Axiom A5. The successor of $\pi_{0,n+1}$ follows from Axiom A6 in combination with the observation that $\mathit{SecondChild}(\mathit{Root}) \neq \{\pi_{1,1}\}$; this is so since $\mathit{Root} \subseteq \mathit{Cons}$ by Axiom A6, and $n \geq 1$.

**Lemma 3.** $\forall i \in [m]. \forall j \in [n-1]. \mathit{Succ}(\pi_{i,j}) = \{\pi_{i,j+1}\}$

We show that $\mathit{Succ}(\mathit{Father}(\pi_{i,j})) = \mathit{Father}(\pi_{i,j+1})$, and from this deduce the claim by Axiom A7. The proof proceeds by induction on $i$. In the case that $i = 1$, using the definition of $\mathit{Father}$ and Lemma 2, we see that

$$\mathit{Succ}(\mathit{Father}(\pi_{1,j})) = \mathit{Succ}(\pi_{0,j}) = \{\pi_{0,j+1}\} = \mathit{Father}(\pi_{1,j+1}).$$

For $i > 1$, we may assume that $\mathit{Succ}(\pi_{i-1,j}) = \{\pi_{i-1,j+1}\}$. Thus,

$$\mathit{Succ}(\mathit{Father}(\pi_{i,j})) = \mathit{Succ}(\pi_{i-1,j}) = \{\pi_{i-1,j+1}\} = \mathit{Father}(\pi_{i,j+1}).$$

**Lemma 4.** $\forall i \in [m]. \mathit{Succ}(\pi_{i,n}) = \{\pi_{i+1,1}\}$

We start by proving the following auxiliary claim by induction on $j$:

$$\forall j \in [n]. \forall i \in [m]. \mathit{NextSouth}^*(\pi_{i,j}) \cap \mathit{South} = \{\pi_{i,n}\}$$

For $j = n$, this is obvious, given that $\pi_{i,n} \in \mathit{South}$. For $j < n$, we may assume that $\mathit{NextSouth}^*(\pi_{i,j+1}) \cap \mathit{South} = \{\pi_{i,n}\}$. Moreover, $\pi_{i,j} \in \mathit{Grid} - \mathit{South}$, which, by definition of $\mathit{NextSouth}$, implies that $\mathit{NextSouth}(\pi_{i,j}) = \mathit{Succ}(\pi_{i,j})$. Thus,

$$\begin{aligned} &\mathit{NextSouth}^*(\pi_{i,j}) \cap \mathit{South} = \mathit{NextSouth}^*(\mathit{NextSouth}(\pi_{i,j})) \cap \mathit{South} \\ &= \mathit{NextSouth}^*(\mathit{Succ}(\pi_{i,j})) \cap \mathit{South} \overset{\text{Lemma 3}}{=} \mathit{NextSouth}^*(\pi_{i+1,j}) \cap \mathit{South} = \{\pi_{i,n}\}. \end{aligned}$$

$$\begin{aligned} [\![ \mathit{nextEast}(x,y) ]\!]_0 &= y \in \mathit{NextEast}(x) & [\![ \forall x.\psi ]\!]_0 &= \forall x \in \mathit{Grid}.\, [\![\psi]\!]_0 \\ [\![ \mathit{nextSouth}(x,y) ]\!]_0 &= y \in \mathit{NextSouth}(x) & [\![ \forall X.\psi ]\!]_0 &= \forall X \subseteq \mathit{Grid}.\, [\![\psi]\!]_0 \\ [\![ x \in X ]\!]_0 &= x \in X \cap \mathit{Grid} & [\![ \psi \wedge \psi' ]\!]_0 &= [\![\psi]\!]_0 \wedge [\![\psi']\!]_0 \\ [\![ \neg\psi ]\!]_0 &= \neg [\![\psi]\!]_0 \end{aligned}$$

**Fig. 4.** Encoding MSO of grids into MSO of tots: the compositional part

Instantiating Axiom A8 with $\pi_{i,1} \in \mathit{North}$, we see that

$$\mathit{Succ}(\mathit{NextSouth}^*(\pi_{i,1}) \cap \mathit{South}) = \mathit{FirstChild}(\pi_{i,1}) = \{\pi_{i+1,1}\}\,.$$

Since $\mathit{NextSouth}^*(\pi_{i,1}) \cap \mathit{South} = \{\pi_{i,n}\}$ by the auxiliary claim above, this implies that $\mathit{Succ}(\pi_{i,n}) = \{\pi_{i+1,1}\}$.

Together with Axiom A4, the preceding Lemmas establish that $\mathit{Succ}(\pi)$ is uniquely determined for all $\pi \neq \pi_{m,n}$. This ends the proof of Proposition 1. □

Proposition 1 states that we can define the encodings of grids using MSO for tots. We now show how to express MSO formulas for grids using MSO formulas for tots. The translation of a closed MSO formula $\psi$ for grids is as follows:

$$[\![\psi]\!] =_{\mathrm{def}} \exists \mathit{Grid} \exists \mathit{NextSouth} \exists \mathit{NextEast} \exists \ldots.\ (\mathrm{Defs} \wedge \mathrm{A1} \wedge \cdots \wedge \mathrm{A8} \wedge [\![\psi]\!]_0)$$

where $[\![\psi]\!]_0$ is given in Fig. 4, and Defs contains the above definitions for all occuring set variables, all of which are existentially quantified in the outermost quantifier prefix.

**Proposition 2.** *For every closed formula $\psi$ from the set of all MSO formulas over grids and every grid $G$, $\psi \models G$ implies that $[\![\psi]\!] \models [\![G]\!]$.*

Proving this proposition from Proposition 1 is routine, since it is largely independent of the particularities of the encoding. Note however that the proposition does not hold for formulas with free variables. For these, quantification would need to be restricted to nodes in *Grid*, and exclude nodes in $\mathit{Cons} \cup \mathit{Nil}$.

## 4 Bounded Gap-Degree

Given the undecidability of the MSO satisfiability problem for the class of unrestricted tots, we are interested in restricted classes for which decidability can be obtained. A family of such classes that is relevant for applications in computational linguistics is obtained from the *gap-degree measure* [11,12].

Let $\tau$ be a tot, and let $\pi, \pi_1, \pi_2$ be nodes of $\tau$. The set of *descendants of* $\pi$, denoted by $\mathit{desc}(\pi)$, is the set of all nodes $\pi' \in \mathit{nod}(t)$ such that $\pi \lhd^* \pi'$. The *interval with endpoints* $\pi_1$ *and* $\pi_2$, denoted by $[\pi_1, \pi_2]$, is the set of all nodes $\pi' \in \mathit{nod}(\tau)$ such that $\pi_1 \preceq \pi' \preceq \pi_2$. Note that for terms ordered by a pre-order traversal, each descendant set forms an interval. In the general case, though, a descendant set $desc(\pi)$ may be partitioned into a sequence of (maximal) intervals, which we call the *segments* of $\pi$. We define MSO formulas $y_1 \equiv_x y_2$ stating that $y_1$ and $y_2$ belong to the same segment of $x$:

$$y_1 \equiv_x y_2 =_{\text{def}} \forall z. (y_1 \preceq z \preceq y_2 \lor y_2 \preceq z \preceq y_1) \to x \lhd^* z$$

This formula defines an equivalence relation $\equiv_\pi$ on the descendant set $desc(\pi)$ with the property that each equivalence class forms a (maximal) interval. We call such a relation a *segmentation.*

**Definition 2.** *Let $\tau$ be a tot, and let $\pi$ be a node of $\tau$. The* gap-degree of $\pi$, *$deg(\pi)$, is defined as the index of the relation $\equiv_\pi$, minus one. The* gap-degree *of $\tau$, $deg(\tau)$, is the maximum among the gap-degrees of its nodes. A set $L$ of tots is* gap-bounded, *if there is a constant $g_L$ such that $deg(\tau) \leq g_L$, for every $\tau \in L$.*

The tot in Fig. 2a has gap-degree 0, while those in Figs. 2b and 2c have gap-degree 1. In Figs. 2b and Figs. 2c, the node $b$ has two segments ($\{b\}$ and $\{c\}$); in Fig. 2a, it has only one ($\{b, c\}$).

**Lemma 5.** *The gap-degree of the tot $[\![G_{n,n}]\!]$ is $n + 1$.*

This Lemma shows that our undecidability proof fails when we restrict the models of our MSO formulas to classes of tots that are gap-bounded. Even better, this restriction implies decidability:

**Theorem 2.** *The MSO satisfiability problem of every gap-bounded class of tots is decidable.*

The remainder of this paper is concerned with the proof of this result. To do so, we establish a link between MSO for gap-bounded classes of tots on the one hand, and an algebraic notion of automata on the other.

## 5 Segmented Tots and Subtots

We develop a notion of 'substructure' for tots that will serve as intermediate results when constructing tots algebraically. They may be little more general than tots, in that their linearisations may be segmented, so we call them segmented tots.

For illustration, let us reconsider the tot in Fig. 2c. Its first subterm is $b(c)$. The linearization of the nodes of this subterm is segmented into two parts, while leaving a gap between the $b$-node and the $c$ node, into which external nodes may be plugged, such as the nodes of $d(e)$ when reconstructing the original tot.

A *$k$-segmented linearization* over a finite set $A$ is a $k$-tuple $\langle w_1, \ldots, w_k\rangle$ of non-empty words over $A$ such that $w_1 \cdots w_k$ forms a linearization. Every $k$-segmented linearization $w$ defines a total ordering $\preceq$ on $A$, which satisfies $a_1 \preceq a_2$ if $a_1$ occurs left of $a_2$ in $w$. In addition, it defines an equivalence relation $\equiv$ on $A$, such that $a_1 \equiv a_2$ if they occur in the same segment of $w$. The segments of $w$ are the equivalence classes of $\equiv$, so that the index of $\equiv$ is $k$. The equivalence classes form intervals with respect to the total ordering, i.e.:

$$a_1 \equiv a_2 \Rightarrow \forall a.\, (a_1 \preceq a \preceq a_2 \vee a_2 \preceq a \preceq a_1) \to a_1 \equiv a \equiv a_2$$

Conversely, every pair of total ordering $\preceq$ on $A$ and an equivalence relation $\equiv$ on $A$ of index $k$ that satisfy the above condition define an $k$-segmented linearization on $A$.

**Definition 3.** *A $k$-segmented* tot *over a signature $\Sigma$ is a pair $\langle t, w\rangle$ where $t$ is a term over $\Sigma$ and $w$ a $k$-segmented linearization of nod$(t)$.*

The relational structure corresponding to a segmented tot $(t, w)$ is the structure $\langle nod(t)\,;\preceq,\equiv,(:\sigma)_{\sigma\in\Sigma}\rangle$; it provides the total ordering $\preceq$ and the equivalence relation $\equiv$ defined by $w$ and the labeling predicates. The MSO of segmented tots is the MSO with symbols for all these relations and interpreted over these relational structures.

Substructures of segmented tots can now be defined by extending the notion of subterms. Let $\tau = \langle t, w\rangle$ be a segmented tot and node $\pi \in nod(t)$. The set $desc(\pi)$ is segmented by the relation $\equiv^{\pi}$, which is defined by an MSO formula with three free variables:

$$y_1 \equiv^{x} y_2 =_{\text{def}} y_1 \equiv y_2 \wedge \forall z.\, (y_1 \preceq z \preceq y_2 \vee y_2 \preceq z \preceq y_1) \to x \lhd^{*} z\,.$$

This means that segments of $desc(\pi)$ are either separated by nodes of $\tau$ external to $desc(\pi)$ or by the segmentation of $\tau$.

Let $t|_{\pi}$ the subterm of $t$ at node $\pi$, i.e., $t|_{\varepsilon} = t$ and $\sigma(t_1,\dots,t_m)|_{i\pi} = t_i|_{\pi}$, where $i \in [m] = \text{ar}(\sigma)$. Recall that $nod(t|_{\pi}) = \{\,\pi' \mid \pi\pi' \in nod(t)\,\}$. The subtot of $\tau$ at $\pi$ is given by the subterm $t_{|\pi}$ and the segmented linearization of its nodes induced by $\equiv^{\pi}$ and $\preceq$. The of gap-degree of a segmented tot $\tau$ is one less then the maximal number of segments of its subtots $\tau_{|\pi}$.

## 6 Segmented Words

We introduce segmented words and define operators for them that will be useful for the algebraization of tots. Let $A$ be a set and $k$ a natural numbers.

A *$k$-segmented word* over $A$ is a $k$-tuple of non-empty words over $A$. The *positions* of a $k$-segmented word $\psi = \langle a_1^1 \cdots a_1^{n_1}, \dots, a_k^1 \cdots a_k^{n_k}\rangle$ are pairs of natural numbers: $pos(\psi) = \{\,\langle i,j\rangle \mid i \in [k],\ j \in [n_i]\,\}$. The set of positions is totally ordered by the lexicographic order, i.e. $\langle i',j'\rangle < \langle i,j\rangle$ iff $i' < i$ or $i' = i \wedge j' < j$. It is partitioned into $k$ classes by the equivalence $\langle i',j'\rangle \equiv \langle i,j\rangle$ iff $i' = i$. For every $a \in A$, the relation $Q_a \subseteq pos(\psi)$ contains all positions $\langle i,j\rangle$ where $a$ occurs, i.e., $a_i^j = a$.

We now define operators for $k$-segmented words, where we assume $k \in [l]$ for a fixed natural number $l$. Let $W(k)$ be the set of $k$-segmented words over $A$. We define a multi-sorted algebra $\mathscr{S}$ with domains $W(1),\dots,W(l)$; the functions of this algebra are defined by segmented words. Each function first shuffles the segments of its arguments into a new segmented word, and then fuses adjacent segments to arrive at a segmented word with at most $l$ segments. To make this formal, we need three auxiliary notions: The *multiplicity* of a letter $a \in A$ in a segmented word $\psi$ over $A$ is the number of positions $\langle i, j\rangle \in pos(\psi)$ such that $\langle i, j\rangle \in Q_a$. A *multiset* over a finite set $B$ is a function $\mu \colon B \to \mathbb{N}$. A *k-segmented linearization of of a multiset* $\mu$ is a $k$-segmented word $\psi$ over $B$ such that for all $b \in B$, the multiplicity of $b$ in $\psi$ is $\mu(b)$.

Let $k \in [l]$ and $m \in \mathbb{N}$, and $k_1, \dots, k_m \in [l]$. Every $k$-segmented linearization $\psi \in ([m]^+)^k$ of the multiset $\{\, i \mapsto k_i \mid i \in [m]\,\}$ defines an operator $\psi^{\mathscr{S}}$ of type:

$$\psi^{\mathscr{S}} : W(k_1) \times \cdots \times W(k_m) \to W(k)\,.$$

To give a definition of this function, let the *occurrence number* $\mathrm{occ}(i, j) \in \mathbb{N}$ of a position $\langle i, j\rangle \in pos(\psi)$ with $\langle i, j\rangle \in Q_a$ be the number of positions $\langle i', j'\rangle \in pos(\psi)$ such that $\langle i', j'\rangle \leq \langle i, j\rangle$ and $\langle i', j'\rangle \in Q_a$. If $\psi {=} \langle a_1^1 \cdots a_1^{n_1}, \dots, a_k^1 \cdots a_k^{n_k}\rangle$, then the application of the function $\psi^{\mathscr{S}}$ to segmented words $S_i \in W(k_i)$, $i \in [m]$, is defined as

$$\psi^{\mathscr{S}}(\langle S_1^1, \dots, S_1^{k_1}\rangle, \dots, \langle S_m^1, \dots, S_m^{k_n}\rangle) \;=\; \langle T_1^1 \cdots T_1^{n_1}, \dots, T_k^1 \cdots T_k^{n_k}\rangle\,,$$

where $T_i^j = S_{a_i^j}^{\mathrm{occ}(i,j)}$. Note that, for each $i \in [k]$, $T_1^1 \cdots T_1^{n_i} \in A^+$ is a concatenation of $n_i$ words, while $\langle S_1^1, \dots, S_1^{k_i}\rangle \in W(k_i)$ is a $k_i$-tuple. The number of functions of $\mathscr{S}$ is infinite as long as we do not bound the maximal arity $m \in \mathbb{N}$. Furthermore, $\mathscr{S}$ does not contain any constant, so it is useless standalone.

## 7 Algebra of Segmented Tots

For the algebraization of tots, we fix a signature $\Sigma$ of function symbols and a gap bound $l \in \mathbb{N}$. For all $k \in \mathbb{N}$ let $T(k) \subseteq TOT_\Sigma$ be the set of tots over $\Sigma$ with $k$ segments. We define a multi-sorted algebra $\mathscr{T}$ with domains $T(1), \dots, T(l)$.

For every tot $\tau = \langle t, w\rangle$ in $T(k)$, let $term(\tau) = t$ and $seg(\tau) = w$. The algebra $\mathscr{T}$ provides operations of type $\langle \sigma, \psi\rangle^{\mathscr{T}} : T(k_1) \times \cdots \times T(k_m) \to T(k)$ for all $\sigma \in \Sigma$ of arity $m$, $k$-segmentation $\psi$ of the multiset $\{0 \mapsto 1\} \cup \{\, i \mapsto k_i \mid i \in [m]\,\}$, and $k, k_1, \dots, k_m \in [l]$. Specifically, we define the value $\langle \sigma, \psi\rangle^{\mathscr{T}}(\tau_1, \dots, \tau_n) = \tau$ as follows:

$$\begin{aligned} term(\tau) &= \sigma(term(\tau_1), \dots term(\tau_n)) \\ seg(\tau) &= \psi^{\mathscr{S}}(\langle \varepsilon\rangle, pfx_1(seg(\tau_1)), \dots, pfx_n(seg(\tau_n)))\,, \end{aligned}$$

where $pfx_i$ is the function that takes a sequence of nodes and prefixes every node in this sequence by the address $i$. This algebra has a finite number of functions, whose maximal arity is bounded by the maximal arity in $\Sigma$. The constants of this algebra are of the form $\langle \sigma, \langle 1\rangle\rangle$, where $\sigma$ is a constant in $\Sigma$.

**Proposition 3.** *Let $k \in [l]$. For every tot $\tau \in T_k$, there exist a symbol $\sigma \in \Sigma$ $ar(\sigma) = m$, natural numbers $k_1, \ldots, k_m \in [l]$, a $k$-segmentation $\psi$ of the multiset $\{0 \mapsto 1\} \cup \{\, i \mapsto k_i \mid i \in [m]\,\}$, and tots $\tau_1 \in T_{k_1}, \ldots, \tau_m \in T_{k_m}$ such that $\tau = \langle \sigma, \psi \rangle^{\mathscr{T}}(\tau_1, \ldots, \tau_m)$.*

This means that every $k$-segmented tot can be constructed from the operators of the algebra $\mathscr{T}$ if $k \in [l]$. For instance, the tot in Fig. 2b is equal to:

$$\langle a, \langle 0121 \rangle \rangle^{\mathscr{T}}(\langle b, \langle 0, 1 \rangle \rangle^{\mathscr{T}}(\langle c, \langle 0 \rangle \rangle^{\mathscr{T}}), \langle d, \langle 10 \rangle \rangle^{\mathscr{T}}(\langle e, \langle 0 \rangle \rangle^{\mathscr{T}}))\,.$$

The substructure at $b$, i.e., the first child of the root, has gap-degree 1. This is reflected by the segmented word $\langle 0, 1 \rangle$ whose comma matches the gap.

*Proof.* Let $\tau = (t, w)$, where $t = \sigma(t_1, \ldots, t_n)$ for some $\sigma \in \Sigma$, and $t_1, \ldots, t_n \in T_\Sigma$. For all $i \in [k]$, let $k_i$ be the number of segments of $\tau|_i$, so that

$$pfx_i(seg(\tau|_i)) \;=\; \langle S_i^1, \ldots, S_i^{k_i} \rangle$$

for some $k_i \in [l]$ and $S_i^1, \ldots, S_i^{k_i} \in nod(t)^+$. Furthermore, let $k_0 = 1$ and $S_0^1 = \varepsilon$. The set $nod(t)$ is partitioned into segments $S_i^j$, where $0 \leq i \leq n$ and $j \in [k_i]$. The segmented linearization $w$ of $\tau$ must be of the following form, where $T_i^j = S_{a_i^j}^{\mathrm{occ}(i,j)}$:

$$seg(\tau) \;=\; \langle T_1^1 \cdots T_1^{n_1}, \ldots, T_k^1 \cdots T_k^{n_k} \rangle\,.$$

This holds, since each segment of $seg(\tau)$ must be a concatenation of words of $S_i^j$s due to convexity, and since the segments of the same substructure must appear in their original order. Let

$$\psi \;=\; \langle a_1^1 \cdots a_1^{n_1}, \ldots, a_k^1 \cdots a_k^{n_k} \rangle\,.$$

It then follows that $\tau = \langle \sigma, \psi \rangle^{\mathscr{T}}(\tau|_1, \ldots, \tau|_n)$. □

The decomposition of segmented tots does not need to be unique; for example,

$$\langle a, \langle 011 \rangle \rangle^{\mathscr{T}}(\langle a, \langle 0, 1 \rangle \rangle^{\mathscr{T}}(\langle b, \langle 0 \rangle \rangle^{\mathscr{T}})) \;=\; \langle a, \langle 01 \rangle \rangle^{\mathscr{T}}(\langle a, \langle 01 \rangle \rangle^{\mathscr{T}}(\langle b, \langle 0 \rangle \rangle^{\mathscr{T}}))\,.$$

Both expressions describe the same tot with term $a(a(b))$ and the node order $\varepsilon \preceq 1 \preceq 11$. In the first expression, the subtot of the first child is artificially split into two subsequent segments, which are then fused without insertion of nodes into the gap. On the right, this artificial split is avoided. It is not difficult to see that every segmented tot can be constructed in a unique manner, when disallowing immediate repetitions in constructors $\psi$, i.e. segments in $\mathbb{N}^* ii \mathbb{N}^*$, for every $i \in \mathbb{N}$. Let $\mathscr{T}'$ be the restriction of $\mathscr{T}$ to operators $(\sigma, \psi)$ without immediate repetitions in $\psi$.

**Theorem 3.** *The algebra of $k$-segmented tots $\mathscr{T}'$ is isomorphic to a $[k]$-sorted term algebra $T_\Delta$.*

*Proof (Sketch).* The multi-sorted signature $\Delta$ contains all symbols $(\sigma, \psi)$ of type $k_1 \times \cdots \times k_m \to k$, where $\sigma \in \Sigma$ has arity $m$, and $\psi$ is a $k$-segmentation of the multiset $\{0 \mapsto 1\} \cup \{\, i \mapsto k_i \mid i \in [m]\,\}$ without immediate repetitions. The interpretation function $[\![\cdot]\!]^{\mathscr{T}'} : T_\Delta \to \mathscr{T}'$ is a homomorphism, which is onto by Proposition 3, and one-to-one since immediate repetitions are forbidden.

# 8 Automata for Segmented Tots

We define automata for the algebra of segmented tots $\mathscr{T}'$ as automata for the multi-sorted term algebra $T_\Delta$. Since well-sorted terms over $\Delta$ are recognizable, we can use standard tree automata that recognized languages of well-sorted terms.

We call a set of segmented tots $L \subseteq \mathscr{T}'$ with bounded gap-degree *recognizable*, if and only if the corresponding set of term encodings $\{ t \in T_\Delta \mid t^{\mathscr{T}'} \in L \}$ is recognizable by a tree automaton. Standard results on tree automata [8] show that recognizable set of segmented tots are closed under union, intersection, complementation and projection. Note also, that the set of all tots (i.e., segmented tots without gaps) is recognizable

**Theorem 4.** *A gap-bounded set of segmented tots is MSO-definable if and only if it is recognizable. The transformations from formulas to automata and back are effective.*

*Proof.* The transformation of automata for $T_\Delta$ into MSO for segmented tots needs to express the rules of the automata. This works as usual, except that one has to express the operators $\langle \sigma, \psi \rangle^{\mathscr{T}'}$ in the MSO of segmented tots. This is straightforward.

The transformation of MSO formulas to automata works as for Thatcher and Wright's theorem [4]: Boolean connectives and monadic second-order quantifiers are mapped to the closure operators for union, complementation and projection. What remains to show is that we can construct automata that check the atomic predicates $x \preceq y$ and $x \equiv y$. The fundamental insight in this construction is that it suffices to remember, by means of the state information that the tree automaton provides, for each node $\pi$, in which segments of the subtot $\tau|_\pi$ the nodes $\alpha(x), \alpha(y)$ occur (if any). Based on this information, the question whether $\alpha(x) \prec \alpha(y)$ can be decided by looking at the label $\langle \sigma, \psi \rangle$ at $\pi$: if $\alpha(x)$ occurs in the $j_x$th segment of the $i_x$th substructure of $\tau|_\pi$, and $\alpha(y)$ occurs in the $j_y$th segment of the $i_y$th substructure, then $\alpha(x) \prec \alpha(y)$ iff the $j_x$th occurrence of the symbol $i_x$ precedes the $j_y$th occurrence of the symbol $i_y$; if $\alpha(x)$ or $\alpha(y)$ does not occur in some substructure, but equals $\pi$, then the relevant occurrence is the (single) occurrence of the symbol 0 in $\psi$. A similar argument holds for the case $\alpha(x) \equiv \alpha(y)$. As long as the number of gaps (and hence, the number of segments) in a tot is bounded, the required state set is of bounded size.

One perhaps surprising consequence of this Theorem is that total orders of MSO-defined sets of tots with bounded gap-degree are always definable by MSO over terms without the order. Given a term $t \in T_\Sigma$, variables $x_1, x_2$ and nodes $\pi, \pi' \in \mathit{nod}(t)$, let $t * [x \mapsto \pi, y \mapsto \pi'] \in T_{\Sigma \times 2^{\{x,y\}}}$ be the tree obtained from $t$ by annotating all its nodes by the set of variables that are mapped to it. Similarly, we define $\tau * [x \mapsto \pi, y \mapsto \pi'] \in \mathit{TOT}_{\Sigma \times 2^{\{x,y\}}}$ to be the tot $\langle \mathit{term}(\tau) * [x \mapsto \pi, y \mapsto \pi'], \mathit{seg}(\tau) \rangle$ in which the variable assignment is annotated.

**Corollary 1.** *Let $\phi$ be a closed MSO formula for tots in $TOT_\Sigma$ with bounded gap degree. Then the following term language $\{ term(\tau) \in T_\Sigma \mid \tau \models \phi \}$ is regular as well as $\{ term(\tau) * [x \mapsto \alpha(x), y \mapsto \alpha(y)] \in T_{\Sigma \times 2^{\{x,y\}}} \mid \tau \models \phi \}$.*

*Proof.* The set $L_1 = \{ \tau \in TOT_\Sigma \mid \tau \models \phi \}$ is an MSO defined set of tots over $\Sigma$. Let $A_1$ be an tree automaton over $\Delta$ that recognizes $\{ s \in T_\Delta \mid s^{\mathcal{T}} \in L_1 \}$ according to Theorem 4. The projection $A_1$ to $\Sigma$ recognizes $\{ term(\tau) \in T_\Sigma \mid \tau \in L_1 \}$ as required, where the rules of the projected automaton $B_1$ are as follows:

$$\frac{\langle \sigma, \psi \rangle (p_1, \ldots, p_m) \to p \in Rules(A_1)}{\sigma(\langle p_1, \psi_1 \rangle, \ldots, \langle p_m, \psi_m \rangle) \to \langle p, \psi \rangle \in Rules(B_1)}$$

For the second statement, let $L_2 = \{ \tau * [x \mapsto \alpha(x), y \mapsto \alpha(y)] \in TOT_{\Sigma \times 2^{\{x,y\}}} \mid \tau, \alpha \models \phi \wedge x \preceq y \}$ and automaton $A_2$ recognize $\{ s \in T_{\Delta \times 2^{\{x,y\}}} \mid s^{\mathcal{T}} \in L_2 \}$ according to Theorem 4. The projection $A_2$ to $\Sigma \times 2^{\{x,y\}}$ recognizes the second term language of the corollary $\{ term(\tau) * [x \mapsto \alpha(x), y \mapsto \alpha(y)] \in T_{\Sigma \times 2^{\{x,y\}}} \mid \tau, \alpha \models \phi \wedge x \preceq y) \}$, where the projection automaton $B_2$ has the following rules:

$$\frac{\langle \langle \sigma, \psi \rangle, V \rangle (p_1, \ldots, p_m) \to p \in Rules(A_2)}{\langle \sigma, V \rangle (\langle p_1, \psi_1 \rangle, \ldots, \langle p_n, \psi_m \rangle) \to \langle p, \psi \rangle \in Rules(B_2)}$$

*Final remark.* We have shown that the MSO satisfiability problem for tots with bounded gap-degree is decidable, while the general case is undecidable. A question that we need to leave unanswered for now is whether the first-order satisfiability problem of general tots is decidable in contrast to the case of MSO.

## References


1. Culotta, A., Sorensen, J.: Dependency tree kernels for relation extraction. In: 42nd Annual Meeting of the ACL, pp. 423–429 (2004)
2. Quirk, C., Menezes, A., Cherry, C.: Dependency treelet translation: Syntactically informed phrasal SMT. In: 43rd Annual Meeting of the ACL, pp. 271–279 (2005)
3. Nivre, J., Hall, J., Kübler, S., McDonald, R., Nilsson, J., Riedel, S., Yuret, D.: The CoNLL 2007 shared task on dependency parsing. In: Joint Conference on Empirical Methods in NLP and Computational Natural Language Learning, pp. 915–932 (2007)
4. Thatcher, J.W., Wright, J.B.: Generalized finite automata with an application to a decision problem of second-order logic. Math. System Theory 2, 57–82 (1968)
5. Courcelle, B.: Handbook of graph grammars and computing by graph transformations. Foundations. Handbook of Graph Grammars, vol. 1 (1997)
6. Gottlob, G., Koch, C.: Monadic queries over tree-structured data. In: 17th Annual IEEE Symposium on Logic in Computer Science, pp. 189–202 (2002)
7. Courcelle, B.: Graph Grammars and Logic. Book in preparation (2008)
8. Comon, H., Dauchet, M., Gilleron, R., Löding, C., Jacquemard, F., Lugiez, D., Tison, S., Tommasi, M.: Tree automata techniques and applications (1997/2007)
9. Rabin, M.: Decidability of Second-Order Theories and Automata on Infinite Trees. Transactions of the American Mathematical Society 141, 1–35 (1969)

10. Gottlob, G., Koch, C.: Monadic datalog and the expressive power of languages for web information extraction. In: 21rd ACM PODS, pp. 17–28 (2002)
11. Plátek, M., Holan, T., Kuboň, V.: On relaxability of word order by D-Grammars. In: 3rd Int. Conf. on Combinatorics, Computability and Logic. DMTCS, pp. 159–174 (2001)
12. Kuhlmann, M.: Dependency Structures and Lexicalized Grammars. Doctoral dissertation, Saarland University, Saarbrücken, Germany (2007)
13. Koller, A., Striegnitz, K.: Generation as dependency parsing. In: 40th Annual Meeting of the ACL, pp. 17–24 (2002)
14. Holan, T., Kuboň, V., Oliva, K., Plátek, M.: Two useful measures of word order complexity. Work. on Processing of Dependency-Based Grammars, 21–29 (1998)
15. Kuhlmann, M., Nivre, J.: Mildly non-projective dependency structures. In: 21st COLING-ACL, Main Conference Poster Sessions, pp. 507–514 (2006)
16. Mezei, J., Wright, J.B.: Algebraic automata and context-free sets. Information and Control 11, 3–29 (1967)
17. Courcelle, B.: Recognizable sets of unrooted trees. In: Nivat, M., Podelski, A. (eds.) Tree Automata and Languages. Elsevier Science, Amsterdam (1992)
18. Carme, J., Niehren, J., Tommasi, M.: Querying unranked trees with stepwise tree automata. In: van Oostrom, V. (ed.) RTA 2004. LNCS, vol. 3091, pp. 105–118. Springer, Heidelberg (2004)

# Diagram Rewriting for Orthogonal Matrices: A Study of Critical Peaks*

Yves Lafont and Pierre Rannou

Institut de Mathématiques de Luminy, UMR 6206 du CNRS
Université de la Méditerranée (Aix-Marseille 2)

**Abstract.** *Orthogonal diagrams* represent decompositions of isometries of $\mathbb{R}^n$ into symmetries and rotations. Some convergent (that is noetherian and confluent) rewrite system for this structure was introduced by the first author. One of the rules is similar to Yang-Baxter equation. It involves a map $h : ]0,\pi[^3 \to ]0,\pi[^3$.

In order to obtain the algebraic properties of $h$, we study the confluence of critical peaks (or critical pairs) for our rewrite system. For that purpose, we introduce *parametric diagrams* describing the calculation of angles of rotations generated by rewriting. In particular, one of those properties is related to the *tetrahedron equation* (also called Zamolodchikov equation).

**Keywords:** critical pair, diagram rewriting, orthogonal matrix, Zamolodchikov.

## 1 Introduction

Diagrams are widely used for computation in various fields of mathematics and physics, such as category theory, knot theory, proof theory, quantum electrodynamics, relativity. Formally, a diagram is an element of the free 2-*monoid* (or strict monoidal category) generated by some 2-*computad* (or 2-*polygraph*). See for instance [Str76] or [Bur93]. See also [Gui06b] for an application of 3-polygraphs to proof theory.

Typical examples are *boolean circuits*, which are interpreted in the category of sets with cartesian product, and *quantum circuits*, which are interpreted in the category of (complex) vector spaces with tensor product. Here, we consider *orthogonal diagrams*, which are interpreted in the category of (real) vector spaces with direct sum.

The starting point of our study is *Euler decomposition* of 3-dimensional rotations, which can be generalized to a canonical decomposition of $n$-dimensional isometries. This decomposition is similar to the well-known decomposition of permutations into transpositions of type ($i$ $i$+1), and it is also related to the decomposition of invertible matrices into elementary ones, which is given by Gauss algorithm. See [Laf03].

It happens that our canonical decompositions are the normal forms for a convergent rewrite system. In that case, confluence follows immediately from the uniqueness of our decomposition. Nevertheless, it makes sense to study the confluence of critical peaks.

* This work has been partially supported by ANR project *Invariants algébriques des systèmes informatiques* (INVAL, ANR-05-BLAN-0267).

A. Voronkov (Ed.): RTA 2008, LNCS 5117, pp. 232–245, 2008.

Indeed, one of our rules is not given by explicit formulas. In fact, such formulas exist, but they are not very nice. Moreover, they define some operation $h : ]0, \pi[^3 \to ]0, \pi[^3$, for which diagrams are more suitable than formulas.

The confluence of our critical peaks yields identities between *parametric diagrams*. Such a parametric diagram is interpreted by some partial map $\phi : ]0, \pi[^p \to ]0, \pi[^q$. Therefore, we get an axiomatic description of $h$. We do not know whether $h$ is uniquely determined by those identities.

It is important to notice that diagram rewriting is more complicated than word or term rewriting. Because of the *interchange law*, which expresses an essential property of parallel composition, a finite rewrite system may generate infinitely many conflicts. It happens that, in certain cases, this infinity of conflicts can be reduced to a finite list. See [Laf03] for such examples. See also [Gui06a] for a study of termination in the framework of diagram rewriting.

Originally, this study was motivated by the following question: What is the algebraic theory of quantum circuits? Those circuits are built with unary and binary gates. There are two kinds of unary gates, which are interpreted by the following unitary matrices:

$$\begin{pmatrix} \cos\alpha & -\sin\alpha \\ \sin\alpha & \cos\alpha \end{pmatrix}, \qquad \begin{pmatrix} e^{i\alpha} & 0 \\ 0 & e^{-i\alpha} \end{pmatrix}.$$

These matrices satisfy identities which are similar to those for 3-dimensional rotations. See [Ran07]. This corresponds to a well-known relation between groups $\mathbf{SU}_2$ and $\mathbf{O}_3$.

Of course, the two theories are of different natures: In the case of quantum circuits, parallel composition corresponds to tensor product $\otimes$, whereas in the case of orthogonal diagrams, it corresponds to direct sum $\oplus$. Therefore, we cannot ensure that our study of orthogonal diagrams will help us to understand the algebraic theory of quantum circuits.

Here are some other motivations:

- There are potential applications in physics or in robotics.
- Orthogonal matrices play a central role in mathematics, and they provide a nice example of diagram rewriting system.
- There are interesting connections between rewriting and homology, where critical peaks play a crucial role. See [Laf07]. This should extend to diagram rewriting.

## 2 Rotations of $^3$

The matrix of an isometry is an orthogonal matrix. Such a matrix has determinant $\pm 1$. If it is 1, the isometry is a rotation. In particular, the following matrices correspond to rotations of respective axes $Ox$ and $Oz$ in $\mathbb{R}^3$:

$$R^x_\alpha = \begin{pmatrix} 1 & 0 & 0 \\ 0 & \cos\alpha & -\sin\alpha \\ 0 & \sin\alpha & \cos\alpha \end{pmatrix}, \qquad R^z_\alpha = \begin{pmatrix} \cos\alpha & -\sin\alpha & 0 \\ \sin\alpha & \cos\alpha & 0 \\ 0 & 0 & 1 \end{pmatrix}.$$

**Theorem 1.** *(Euler angles)*
*Any 3-dimensional rotation matrix can be decomposed as follows:* $R = R^x_\gamma R^z_\beta R^x_\alpha S$ *where* $\alpha, \beta, \gamma, \in [0, \pi[$ *and* $S$ *is the identity or an axial symmetry. Moreover, this decomposition is unique if the second angle is nonzero. Otherwise, the decomposition reduces to* $R^x_\alpha S$.

*Proof.* Consider a rotation matrix $R = \begin{pmatrix} a & * & * \\ b & * & * \\ c & * & * \end{pmatrix}$.

- Let $\gamma \in [0, \pi[$ be the angle between vectors $\begin{pmatrix} \pm 1 \\ 0 \end{pmatrix}$ and $\begin{pmatrix} b \\ c \end{pmatrix}$. Then $R = R^x_\gamma \begin{pmatrix} a & * & * \\ b' & * & * \\ 0 & * & * \end{pmatrix}$.
- Similarly, we get $\beta \in [0, \pi[$ such that $R = R^x_\gamma R^z_\beta \begin{pmatrix} u & 0 & 0 \\ 0 & * & * \\ 0 & * & * \end{pmatrix}$ with $u = \pm 1$.
- Finally, we get $\alpha \in [0, \pi[$ such that $R = R^x_\gamma R^z_\beta R^x_\alpha \begin{pmatrix} u & 0 & 0 \\ 0 & v & 0 \\ 0 & 0 & w \end{pmatrix}$ with $u, v, w = \pm 1$.

Moreover $uvw = \det R = 1$. Hence the last matrix $S$ is an axial symmetry. If $b = c = 0$, the first two steps are skipped. Q.E.D.

This decomposition is called the *(left) canonical decomposition* of $R$. It is *standard* if $\alpha, \beta, \gamma \neq 0$ and $S = \mathrm{Id}$. In that case, $\alpha, \beta, \gamma$ are called the *Euler angles* of the rotation.

By exchanging $Ox$ and $Oz$, we get the notions of *right canonical decomposition* and of *right standard decomposition*.

**Lemma 1.** *The canonical decomposition of a rotation matrix is standard if and only if its lower left coefficient and its upper right coefficient are positive.*

*Proof.* The decomposition is standard if and only if $\alpha, \beta, \gamma \neq 0$ and $u = v = w = 1$. Moreover, the lower left coefficient $a$ and the upper right coefficient $b$ are:

$$a = u \sin\gamma \sin\beta,$$
$$b = w \sin\beta \sin\alpha.$$

Since $\alpha, \beta, \gamma \in [0, \pi[$, we have $a, b > 0$ if and only if $\alpha, \beta, \gamma \neq 0$, and $u = w = 1$. Since $uvw = \det R = 1$, this means that the decomposition is standard. Q.E.D.

There is a similar lemma for the right canonical decomposition. Hence, we get:

**Corollary 1.** *The left canonical decomposition of a rotation matrix is standard if and only if its right canonical decomposition is standard.*

**Notation:** We write $\alpha, \beta, \gamma \rhd \alpha', \beta', \gamma'$ whenever the standard right decomposition with angles $\alpha, \beta, \gamma$ corresponds to the left decomposition with angles $\alpha', \beta', \gamma'$.

## 3 Orthogonal Diagrams

We introduce orthogonal diagrams. A diagram on $n$ wires is interpreted as an isometry of $\mathbb{R}^n = \mathbb{R} \oplus \cdots \oplus \mathbb{R}$ or, equivalently, as an orthogonal $n \times n$ matrix. Diagrams are oriented in the following way: inputs at the top and outputs at the bottom.

There are two kinds of gates, which represent isometries in low dimension:

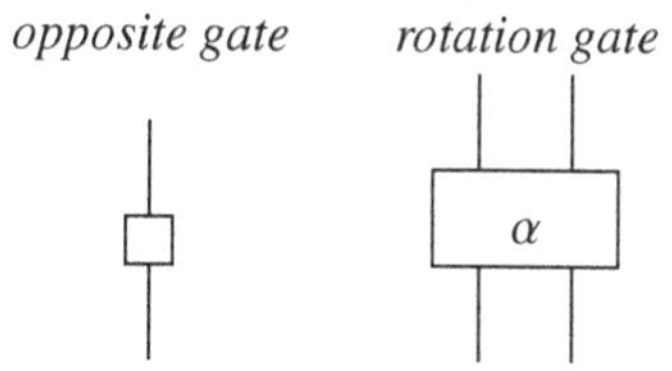


for $\alpha \in ]0, \pi[$

The first one is interpreted by scalar $-1$ and the second one by the following matrix:

$$R_\alpha = \begin{pmatrix} \cos\alpha & -\sin\alpha \\ \sin\alpha & \cos\alpha \end{pmatrix}.$$

Compositions of diagrams are interpreted as follows:

- Let $A$ and $B$ be diagrams respectively with $n$ and $m$ wires, interpreted by orthogonal matrices $M_A$ and $M_B$. Their parallel composition is the following diagram:

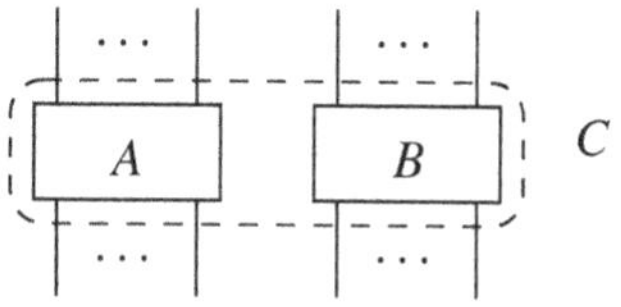


  It is interpreted by the matrix $M_C = M_A \oplus M_B = \left(\begin{array}{c|c} M_A & 0 \\ \hline 0 & M_B \end{array}\right)$.

- If $n = m$, the sequential composition of $A$ and $B$ is the following diagram:

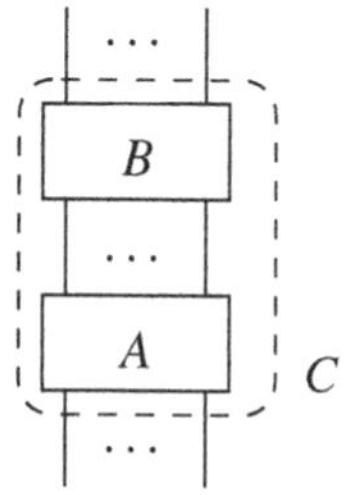


  It is interpreted by the matrix $M_C = M_A M_B$ (corresponding to $B$ followed by $A$).

**Remark:** The identity on $\mathbb{R}$ is represented by a wire. Hence, the matrix $Id_i \oplus M_A \oplus Id_j$ is represented by the following diagram:

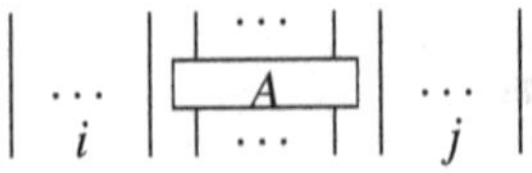


**Definition 1.** *Canonical diagrams are defined by induction on the number of wires:*

- *A canonical diagram on 1 wire is:*

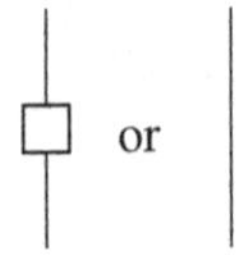


- *If $n > 0$, the general form of a canonical diagram on n wires is given in Figure 1, where $C_1$ and $C_{n-1}$ are canonical diagrams, and $0 \leq k \leq n-1$.*

**Theorem 2.** *Any isometry of $\mathbb{R}^n$ can be represented by a unique canonical diagram.*

*Proof.* By induction on $n$, using the same method as for Theorem 1. Q.E.D.

Note that, even for $n = 3$, this result is more general than Theorem 1, since it applies to isometries, not only to rotations.

**Remark:** A canonical diagram contains no sub-diagram of the following form:

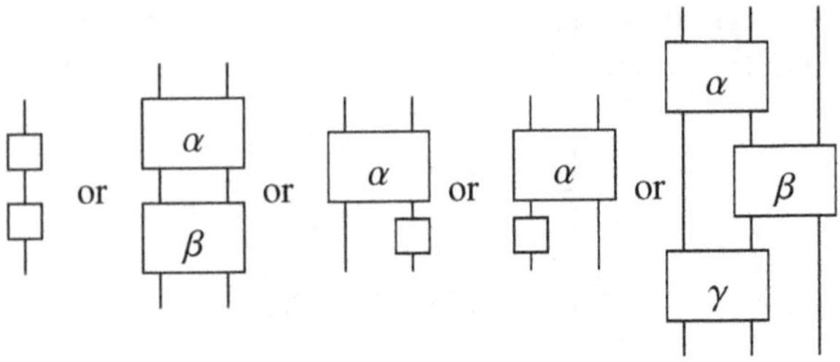


In fact, the converse holds: it is a consequence of Lemma 4, which is proved later.

We consider the rewrite system $\mathcal{R}$ given in Figure 2. Note that the left members of those rewrite rules are the above diagrams.

**Lemma 2.** *If a diagram D reduces to D′, then D and D′ have the same interpretation.*

*Proof.* It is obvious for the first rule. The next cases are checked by easy computation. For instance:

$$\begin{pmatrix}1 & 0\\ 0 & -1\end{pmatrix}\begin{pmatrix}\cos\alpha & -\sin\alpha\\ \sin\alpha & \cos\alpha\end{pmatrix} = \begin{pmatrix}\cos(\pi-\alpha) & -\sin(\pi-\alpha)\\ \sin(\pi-\alpha) & \cos(\pi-\alpha)\end{pmatrix}\begin{pmatrix}-1 & 0\\ 0 & 1\end{pmatrix},$$

$$\begin{pmatrix}\cos(\pi-\alpha) & -\sin(\pi-\alpha)\\ \sin(\pi-\alpha) & \cos(\pi-\alpha)\end{pmatrix}\begin{pmatrix}\cos\alpha & -\sin\alpha\\ \sin\alpha & \cos\alpha\end{pmatrix} = \begin{pmatrix}-1 & 0\\ 0 & -1\end{pmatrix},$$

$$\begin{pmatrix}\cos\beta & -\sin\beta\\ \sin\beta & \cos\beta\end{pmatrix}\begin{pmatrix}\cos\alpha & -\sin\alpha\\ \sin\alpha & \cos\alpha\end{pmatrix} = \begin{pmatrix}\cos(\alpha+\beta) & -\sin(\alpha+\beta)\\ \sin(\alpha+\beta) & \cos(\alpha+\beta)\end{pmatrix}.$$

For the last rule, this holds by definition of $\rhd$. Q.E.D.

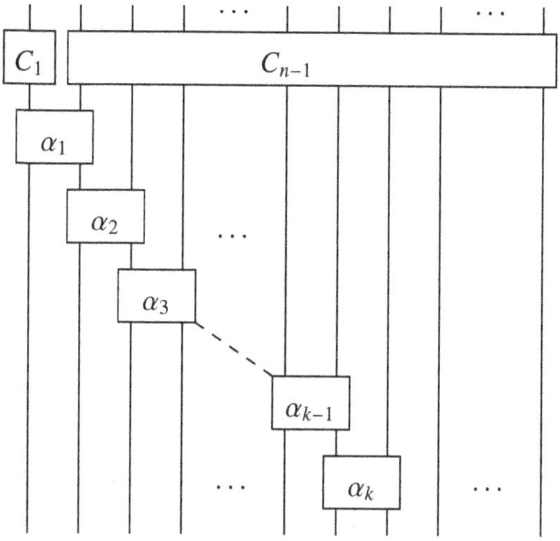


**Fig. 1.** General form of a canonical diagram

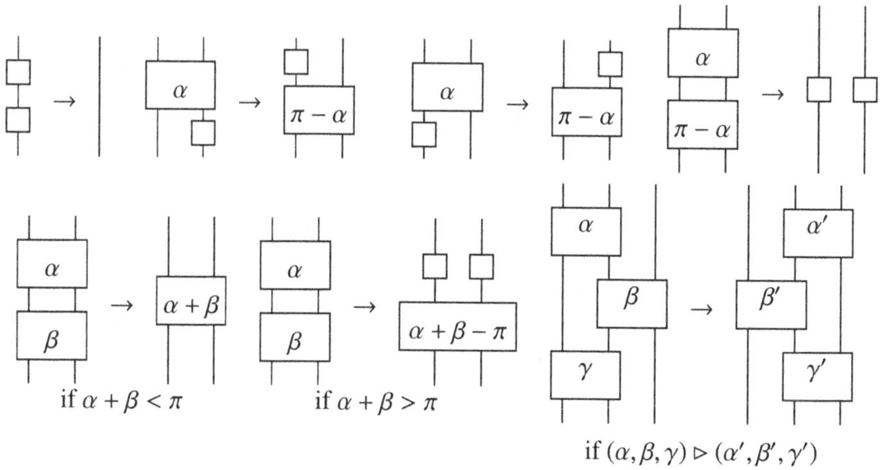


**Fig. 2.** Rules for orthogonal diagrams

We say that a gate $B$ is *immediately above* a gate $A$ if an output of $B$ is connected to an input of $A$. By transitive closure, we get the notion of gate *above* $A$.

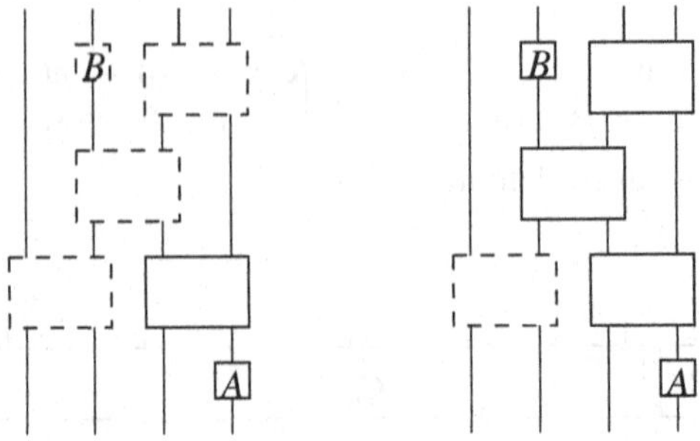


gate immediately above $A$ gates above $A$

For each diagram on $n$ wires we define a vector $(p_1, \ldots, p_n, q)$ as follows:

- $p_i$ is the number of occurrences of binary gates having an input on wire $i$;
- $q = \Sigma_{A\in\Xi}(f(A) + 1)$, where $\Xi$ is the set of occurrences of unary gates and $f(A)$ is the number of occurrences of binary gates above $A$.

For instance, $f(A) = 3$ and $f(B) = 0$ in the above diagram: The vector is $(1, 2, 3, 2, 5)$.

**Lemma 3.** *$\mathcal{R}$ is noetherian.*

*Proof.* The left lexicographic order on vectors defines a termination ordering. Q.E.D.

**Lemma 4.** *Every orthogonal diagram $D$ reduces to a unique canonical diagram $\widehat{D}$.*

*Proof.* By double induction on the number $n$ of wires and the number $m$ of gates.

If $m = 0$, then $D$ is canonical. Otherwise, it consists of some diagram $D'$ with $n$ wires and $m - 1$ gates, followed by some gate $A$. By induction hypothesis, $D'$ reduces to a canonical diagram $\widehat{D}'$. Hence, $D$ reduces to $\widehat{D}'$ followed by $A$.

Now there are several configurations, depending on the type and the position of $A$:

- if $A$ is an unary gate, there are three cases: see Figure 3;
- if $A$ is a binary gate, there are four cases: see Figure 4.

After reduction, we obtain a new diagram, where some unary gate may appear just below $C_1$ and some (unary or binary) gate may appear just below $C_{n-1}$. The first one can always be eliminated using the first rule, and the second one can be eliminated by applying the induction hypothesis for $n - 1$ wires.

Uniqueness follows from Theorem 2 and Lemma 2. Q.E.D.

To sum up, we have proved the following result:

**Theorem 3.** *$\mathcal{R}$ is convergent. In other words, it is noetherian and confluent.*

## 4 Critical Peaks

We consider an abstract version $\mathcal{A}$ of $\mathcal{R}$, where $]0, \pi[$ is replaced by an abstract set $P$. For that purpose, we need some partition $P^2 = \Delta^0 \cup \Delta^- \cup \Delta^+$ and four maps:

$$f : P \to P, \qquad g_- : \Delta^- \to P, \qquad g_+ : \Delta^+ \to P, \qquad h : P^3 \to P^3.$$

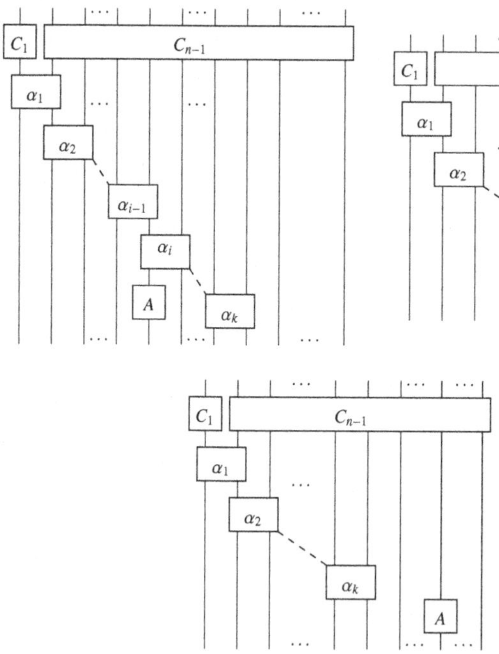


**Fig. 3.** Configurations for a unary gate

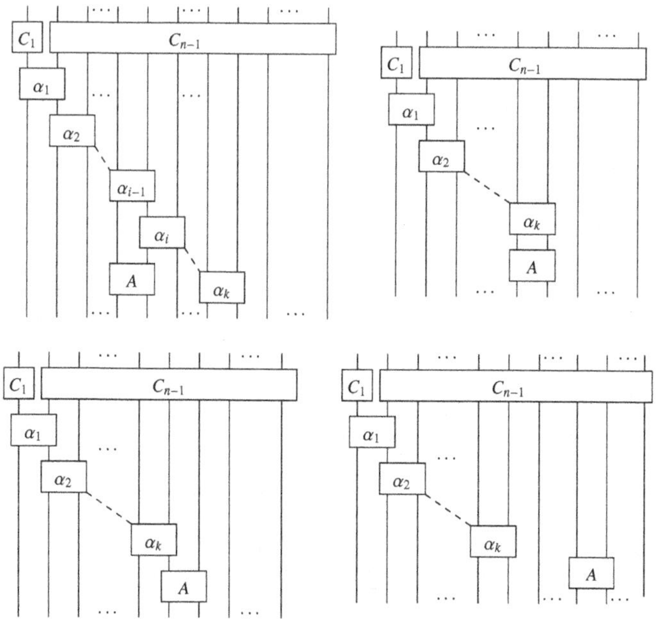


**Fig. 4.** Configurations for a rotation

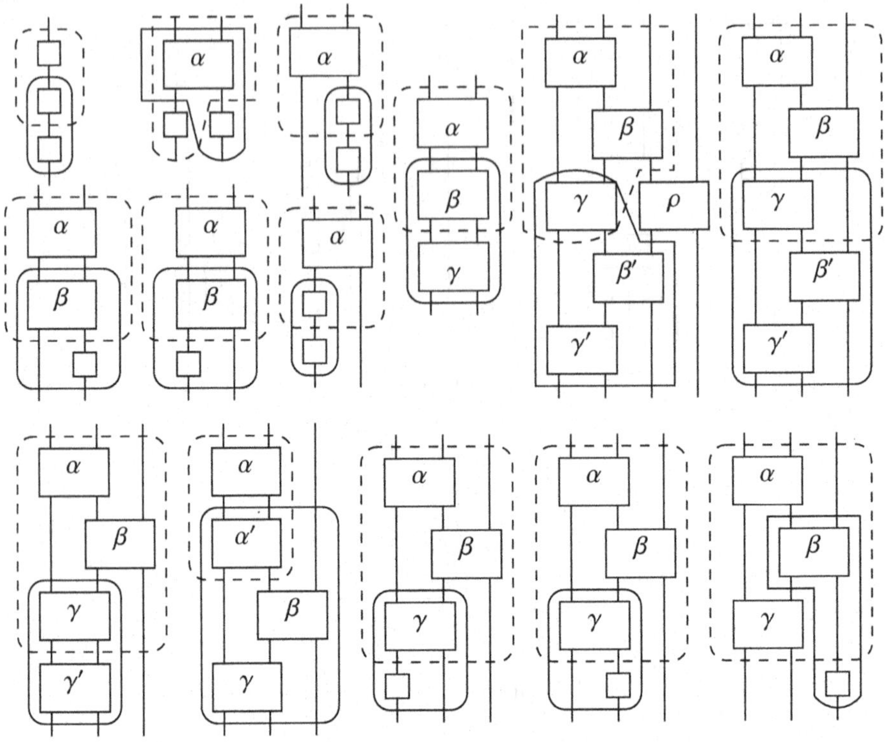


**Fig. 5.** Critical peaks

**Notation:** We write $\alpha \mid \beta$ if $(\alpha,\beta) \in \Delta^0$, $\alpha \smile \beta$ if $(\alpha,\beta) \in \Delta^-$, and $\alpha \frown \beta$ if $(\alpha,\beta) \in \Delta^+$.

The gates are the same as for $\mathcal{R}$, except that parameters belong to $P$. We have one rule in dimension 1, five in dimension 2, and one in dimension 3:

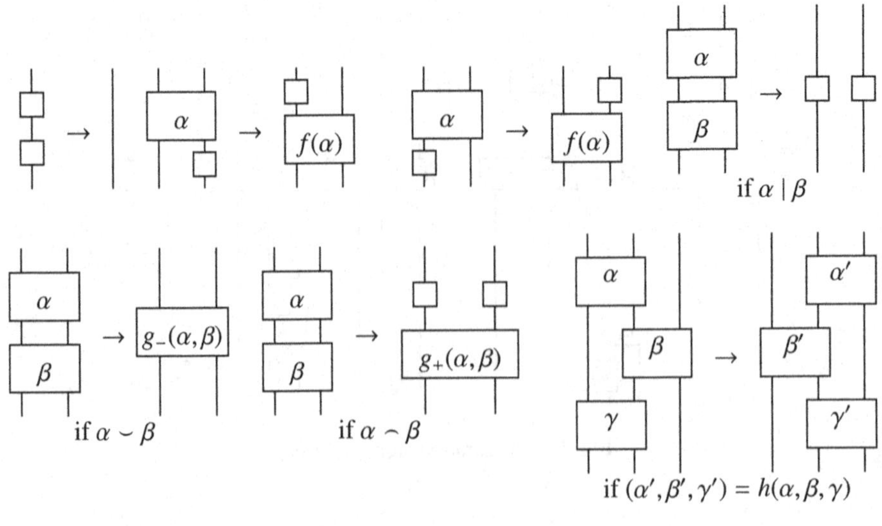

Note that the system $\mathcal{R}$ corresponds to the particular case where $P = ]0,\pi[$ and:

$$\alpha \mid \beta \text{ if } \alpha+\beta=\pi, \qquad \alpha \smile \beta \text{ if } \alpha+\beta<\pi, \qquad \alpha \frown \beta \text{ if } \alpha+\beta>\pi,$$

$$f(\alpha)=\pi-\alpha, \qquad g_-(\alpha,\beta)=\alpha+\beta, \qquad g_+(\alpha,\beta)=\alpha+\beta-\pi,$$

$$h(\alpha,\beta,\gamma)=(\alpha',\beta',\gamma') \text{ if } (\alpha,\beta,\gamma) \rhd (\alpha',\beta',\gamma').$$

The question is: what are the abstract properties of those maps that make $\mathcal{A}$ confluent? Those abstract properties will be expressed by means of *parametric diagrams*, which are interpreted as partial maps corresponding to calculations on parameters. Those new diagrams are built with the following gates:

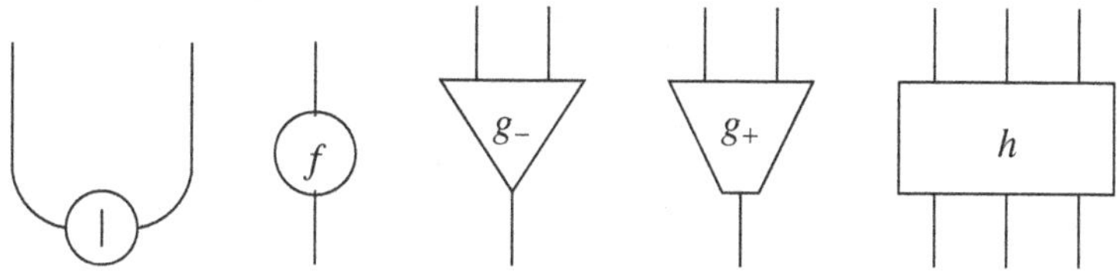


The first one is interpreted by the predicate $\alpha \mid \beta$. The other ones are interpreted by the corresponding (partial) maps. We shall omit labels in parametric diagrams.

Note that $\mathcal{A}$ is noetherian by the same argument as for Lemma 3. Moreover, a diagram is irreducible when it is canonical. See Figure 1.

**Definition 2.** *A* peak *is given by two distinct (one-step) reductions of the same diagram.*

For instance, the peaks of Figure 5 are symbolized by diagrams with circled redexes.

**Theorem 4.** *The following statements are equivalent:*

1. *$\mathcal{A}$ is confluent;*
2. *the peaks of Figure 5 are confluent;*
3. *the maps $f$, $g_-$, $g_+$, and $h$ satisfy the identities of Figures 7 and 8.*

*Proof.* The peaks of Figure 5 are called *critical peaks*.

Clearly, the confluence of $\mathcal{A}$ implies the confluence of all critical peaks. Conversely, assume $D$ reduces in one step to distinct diagrams. If the two rules apply to disjoint sub-diagrams of $D$, then the peak is trivially confluent. Otherwise, we have an overlap of redexes and there are two cases:

– if at least one of the rules is in dimension 1 or 2, the conflict appears in Figure 5;
– if both rules are in dimension 3, we get a *global conflict* of the following form:

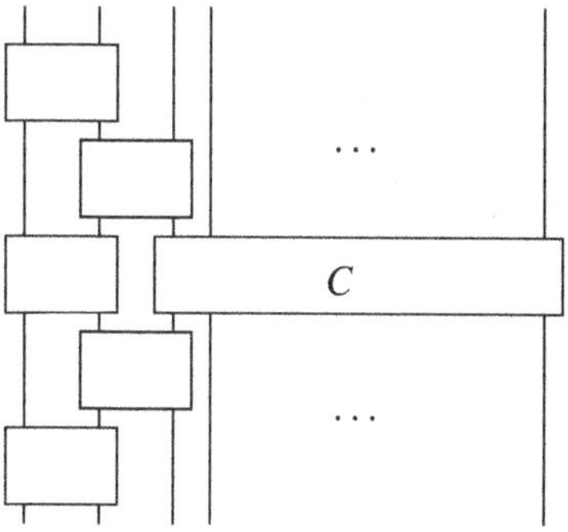

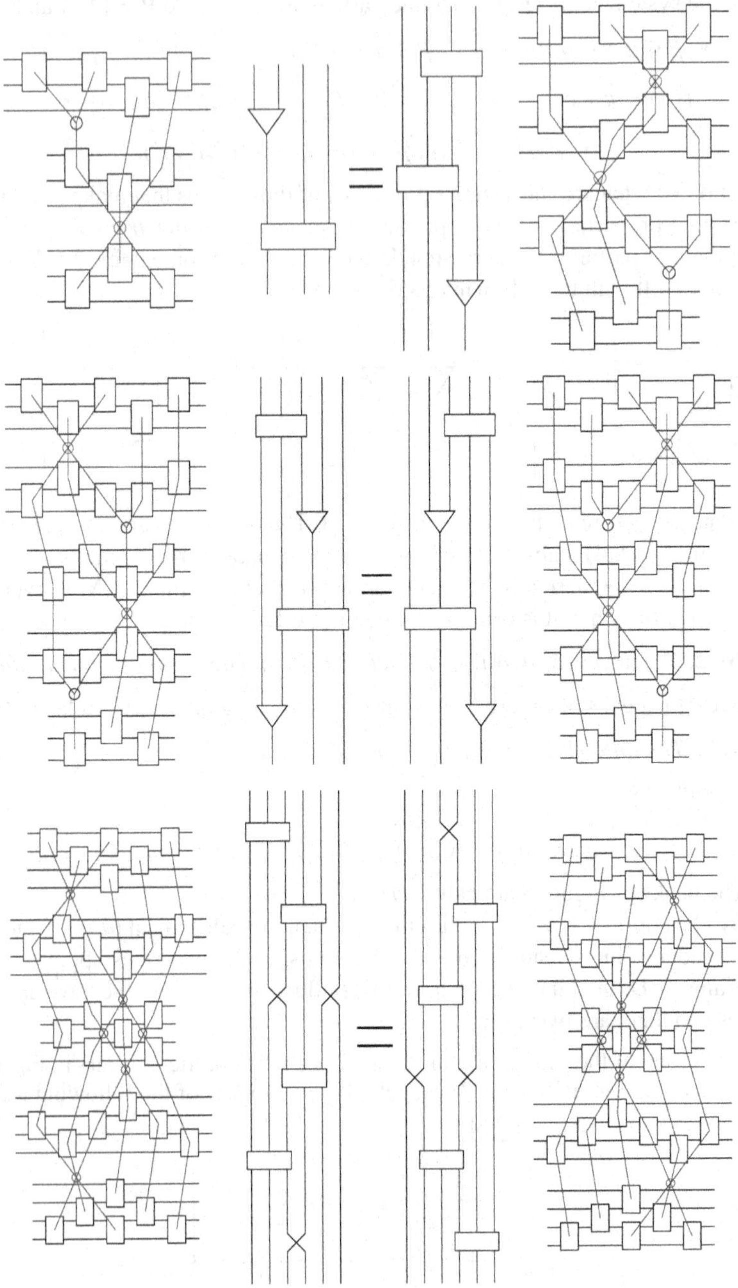


**Fig. 6.** Construction of parametric diagrams (three examples)

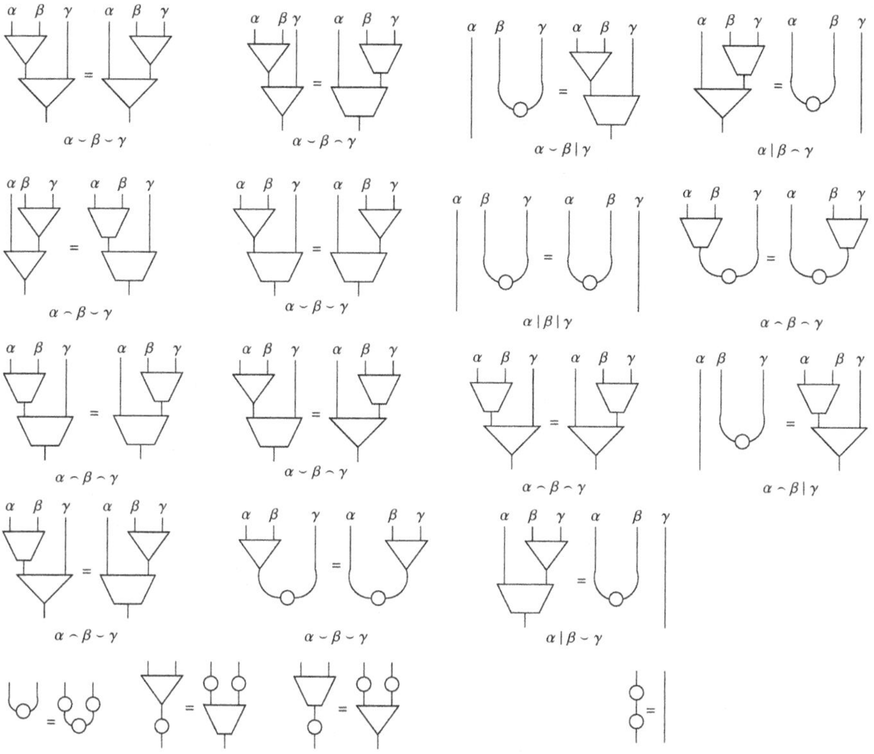


**Fig. 7.** Parametric identities 1

Here, $C$ stands for an arbitrary diagram. Although $C$ is not involved in the reductions, it cannot be eliminated. In fact, it suffices to consider the cases where $C$ is canonical. Then, we use interchange to reduce the study to the following peaks:

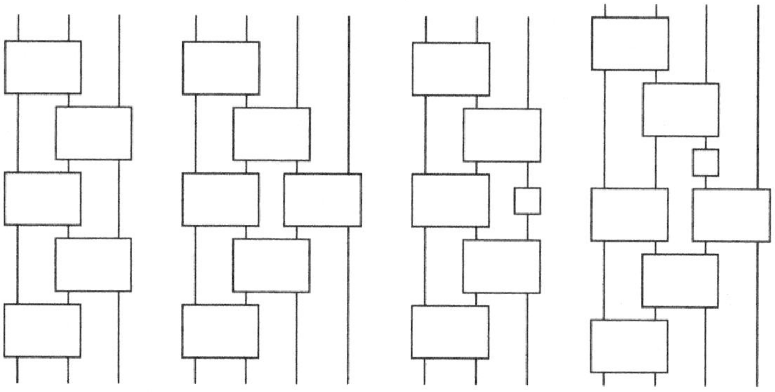

The first two peaks appear in Figure 5. Finally, the third case contains a smaller peak, and similarly for the last one. Hence, they can be omitted. See appendix A of [Laf03]. So we have $1 \Leftrightarrow 2$.

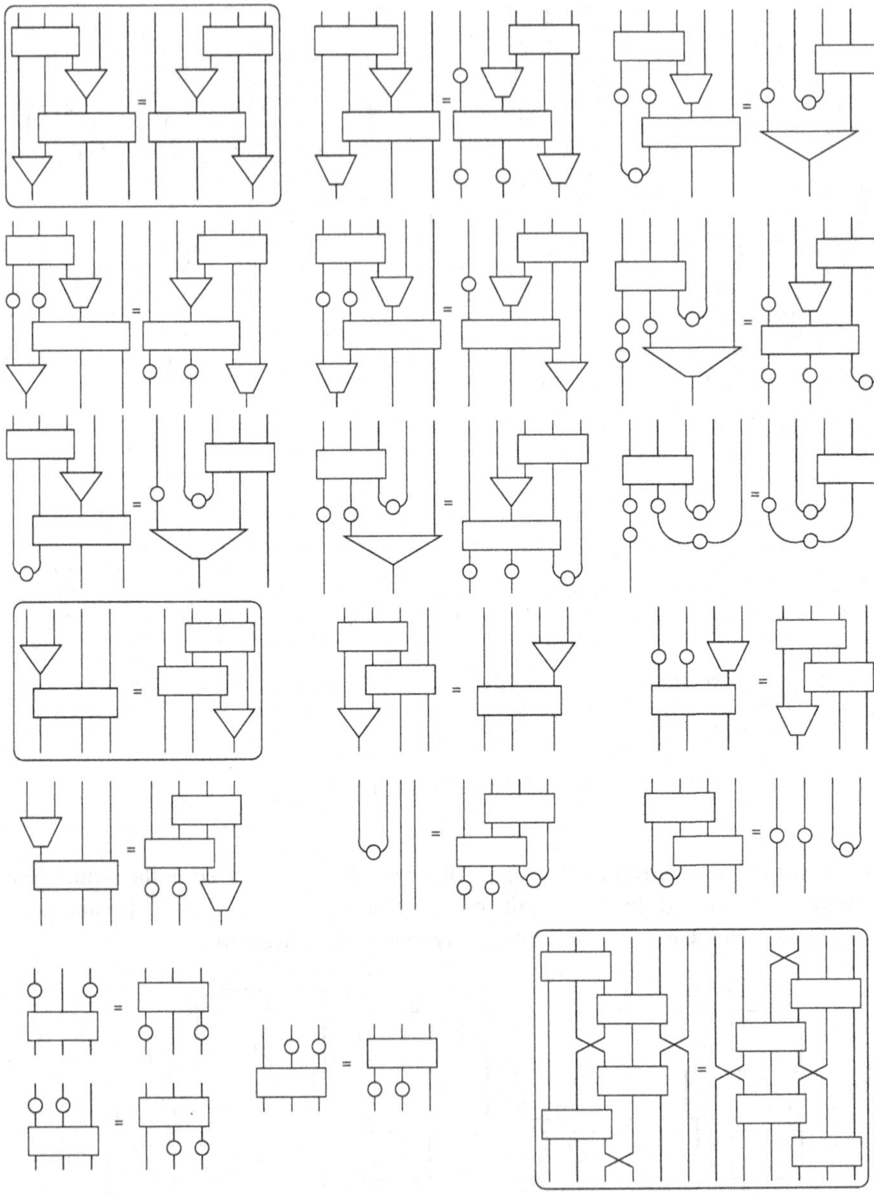

**Fig. 8.** Parametric identities 2 (the three examples of figure 6 are circled)

Now, assume that all peaks of Figure 5 are confluent. In each case, we get two reductions leading to the same diagram. Each reduction yields a parametric diagram representing calculation on parameters, and both must give the same result. Hence, we get an identity between the two parametric diagrams.

Three examples are given in Figure 6: In each case, calculations have been drawn over reductions to show how the corresponding parametric diagrams are built. There is one critical peak in dimension 1 which yields a trivial identity between empty diagrams. Moreover, some critical peaks in dimension 2 yield several identities because there are several cases to consider according to the conditions satisfied by the parameters.

By a careful analysis, we get the list of identities of Figures 7 and 8. Conversely, if those identities are satisfied, then all critical peaks are confluent. So we have $2 \Leftrightarrow 3$. Q.E.D.

## 5 Conclusion

The identities of Figure 7 are those which do not involve $h$. In the case of orthogonal diagrams, they are trivially satisfied. For instance, the first one says that $+$ is associative. The second one is pointless, because if we assume that $\alpha \smile \beta \frown \gamma$, then both sides of the identity are undefined, since for instance, we cannot have $\alpha + \beta + \gamma < \pi$ and $\beta + \gamma > \pi$.

Hence, we are mainly interested in the identities of Figure 8, which express the algebraic properties of $h$. For instance, the last one comes from the critical peak in dimension 4. The confluence of this critical peak, which is also the last example of Figure 6, is strongly related to *Zamolodchikov equation*. See [Cra04] or [Str95].

There are many identities of Figure 8. In the future, we plan to reduce this list by using *undirected parametric diagrams*, which represent predicates rather than partial maps.

## References

[Bur93] Burroni, A.: Higher-dimensional word problems with applications to equational logic. Theoretical Computer Science 115, 43–62 (1993)

[Cra04] Crans, A.S.: Higher linear algebra. In: n-Categories: Foundations and Applications (2004)

[Gui06a] Guiraud, Y.: Termination orders for 3-dimensional rewriting. Journal of Pure and Applied Algebra 207(2), 341–371 (2006)

[Gui06b] Guiraud, Y.: The three dimensions of proofs. Annals of Pure and Applied Logic 141(1–2), 266–295 (2006)

[Laf03] Lafont, Y.: Towards an algebraic theory of boolean circuits. Journal of Pure and Applied Algebra 184, 257–310 (2003)

[Laf07] Lafont, Y.: Algebra and geometry of rewriting. Applied Categorical Structures 15, 415–437 (2007)

[Ran07] Rannou, P.: Théorie algébrique des circuits quantiques, circuits orthogonaux, et circuits paramétriques. Master thesis (2007)

[Str76] Street, R.: Limits indexed by category-valued 2-functors. Journal of Pure and Applied Algebra 8, 149–181 (1976)

[Str95] Street, R.: Higher categories, strings, cubes and simplex equations. Applied Categorical Structures 3, 29–77 & 303 (1995)

# Nominal Unification from a Higher-Order Perspective*

Jordi Levy[1] and Mateu Villaret[2]

[1] Artificial Intelligence Research Institute (IIIA), Spanish Council for Scientific Research (CSIC), Barcelona, Spain
http://www.iiia.csic.es/~levy
[2] Departament d'Informàtica i Matemàtica Aplicada (IMA), Universitat de Girona (UdG), Girona, Spain
http://ima.udg.edu/~villaret

**Abstract.** Nominal Logic is an extension of first-order logic with equality, name-binding, name-swapping, and freshness of names. Contrarily to higher-order logic, bound variables are treated as atoms, and only free variables are proper unknowns in nominal unification. This allows "variable capture", breaking a fundamental principle of lambda-calculus. Despite this difference, nominal unification can be seen from a higher-order perspective. From this view, we show that nominal unification can be reduced to a particular fragment of higher-order unification problems: higher-order patterns unification. This reduction proves that nominal unification can be decided in quadratic deterministic time.

## 1 Introduction

*Nominal Logic* is a version of first-order many-sorted logic with equality and mechanisms for renaming via name-swapping, for name-binding, and for freshness of names. It also provides a *new-quantifier* [GP99], to modeling name generation and locality. It was introduced at the beginning of this decade by Pitts [Pit01,Pit03]. These first works have inspired a sequel of papers where bindings and freshness are introduced in other topics, like equational logic [CP07], rewriting [FG05,FG07], unification [UPG03,UPG04], Prolog [CU04,UC05].

This paper is concerned with *Nominal Unification* [UPG03,UPG04], an extension of first-order unification where terms can contain binders and unification is performed modulo $\alpha$-equivalence. Moreover, (first-order) variables (*unknowns*) are allowed to "capture" bound variables (*atoms*). [UPG03,UPG04] describe a sound and complete, but inefficient (exponential), algorithm for nominal unification. Later this algorithm was extended to deal with the new-quantifier and locality in [FG05]. In [CF07] there is a description of a direct but exponential implementation in Maude, and a polynomial implementation in OCAML based on termgraphs.

* This research has been partially founded by the CICYT research project TIN2007-68005-C04-01/02/03.

A. Voronkov (Ed.): RTA 2008, LNCS 5117, pp. 246–260, 2008.

The use of $\alpha$-equivalence and binders in nominal logic immediately suggests to view nominal unification from a higher-order perspective, the one that we adopt in this paper. Some intuitions about this relation are already roughly described in [UPG04]; and in [Che05] there is a reduction from higher-order pattern unification to nominal unification (here we prove the opposite reduction).

The main benefit of nominal logic, compared to a higher-order logic, is that it allows the use of binding and $\alpha$-equivalence without the other difficulties associated with the $\lambda$-calculus. In particular, with respect to unification, we have that nominal unification is unitary (most general unifiers are unique) and decidable [UPG03,UPG04], whereas higher-order unification is undecidable and infinitary [Luc72,Gol81,Lev98,LV00].

In this paper we fully develop the study of nominal unification from the higher-order view. We show that full higher-order unification is not needed but only a fragment: *Higher-order Pattern Unification* [Mil91,Nip93,Qia96]. This subclass of problems were proposed by Miller [Mil91]. Contrarily to general higher-order unification, higher-order pattern unification is decidable and unitary [Mil91,Nip93]. Moreover, the problem can be solved in linear time [Qia96]. All this will lead us to show how to reduce nominal unification to higher-order pattern unification, and to conclude its decidability in quadratic time as well as the uniqueness of most general unifiers.

From a higher-order perspective, nominal unification can be seen as a variant of higher-order unification where:

1. variables are all first-order typed, and constants are of order at most three (therefore, nominal unification is a fragment of second-order unification),
2. unification is performed modulo $\alpha$-equivalence, instead of the usual $\alpha$ and $\beta$-equivalence,
3. instances of variables (unknowns) are allowed to capture bound variables (atoms), contrarily to the standard higher-order definition, and
4. apart from the usual term-equality predicate, we use a "freshness" predicate $a \# t$ with the intended meaning: bound variable $a$ does not occur free in the instance of term $t$.

The first requirement does not suppose a difficulty. On the contrary, in the reduction to higher-order unification we will add capturable variables as arguments of free variables. The fact that original variables do not have arguments will allow us to reduce nominal unification to higher-order pattern unification.

The second requirement is not a difficulty, either. As all variables are first-order typed, their instantiation can not introduce $\beta$-redexes, and $\beta$-reduction is not really necessary.

The third requirement is the key point that makes nominal unification an interesting subject of research. Variable capture is always a trouble spot. Roughly speaking, the main idea of this paper is to translate (first-order) nominal variables to higher-order variables with the list of bound variables that it can "capture" as arguments. This implies that the arguments of free variables will be lists of pairwise distinct bound variables, hence higher-order patterns.

The fourth requirement can also be overcome by translating freshness predicates into equality predicates.

We structure the paper as follows: after some preliminaries in Section 2, in Section 3, we illustrate by examples the main ideas of the reduction at the same time that we show the main features of nominal unification. In Section 4 we show how to translate a nominal unification problem into a higher-order patterns unification problem, and we prove that this translation is effectively a quadratic time reduction in Section 5. In Section 6 we conclude.

## 2 Preliminaries

### 2.1 Nominal Unification

In nominal logic we talk about *variables* and *atoms.* Only variables may be instantiated, and only atoms may be bounded. They roughly correspond to the higher-order notions of free and bound variables, respectively, but are considered as completely different entities. Therefore, contrarily to the higher-order perspective, the distinction between free and bound variables does not only depend on the occurrences, i.e. in the existence of a binder above them.

In *nominal signatures* we have *sorts of atoms* (typically $\nu$) and *sorts of data* (typically $\delta$) as disjoint sets. *Atoms* (typically $a, b, \ldots$) have one of the sorts of atoms. *Variables*, also called *unknowns*, (typically $X, Y, \ldots$) have a sort of atom or sort of data, i.e. of the form $\nu \,|\, \delta$. *Nominal function symbols* (typically $f, g, \ldots$) have an *arity* of the form $\tau \to \delta$, where $\delta$ is a sort of data and $\tau$ is a sort given by the grammar $\tau ::= \nu \,|\, \delta \,|\, \tau \times \tau \,|\, \langle \nu \rangle \tau$. Abstractions have sorts of the form $\langle \nu \rangle \tau$.

*Nominal terms* (typically $t, u, \ldots$) are given by the grammar:

$$t ::= \langle t_1, t_2 \rangle \,|\, f t \,|\, a \,|\, a.t \,|\, \pi \cdot X$$

where $f$ is a function symbol, $a$ is an atom, $\pi$ is a permutation (finite list of swappings), and $X$ is a variable. They are called respectively *pairs, application, atom, abstraction* and *suspension.* For simplicity, we do not consider the *unit value.*

A *swapping* $(a\, b)$ is a pair of atoms of the same sort. The effect of a swapping over an atom is defined by $(a\, b) \cdot a = b$ and $(a\, b) \cdot b = a$ and $(a\, b) \cdot c = c$, when $c \neq a, b$. For the rest of terms the extension is straightforward, in particular, $(a\, b) \cdot (c.t) = \big((a\, b) \cdot c\big).\big((a\, b) \cdot t\big)$. A *permutation* is a (possibly empty) sequence of swappings. Its effect is defined by $(a_1\, b_1) \ldots (a_n\, b_n) \cdot t = (a_1\, b_1) \cdot \big((a_2\, b_2) \ldots (a_n\, b_n) \cdot t\big)$. Notice that every permutation $\pi$ naturally defines a bijective function from the set of atoms to the sets of atoms, that we will also represent as $\pi$. *Suspensions* are uses of variables with a permutation of atoms waiting to be applied once the variable is instantiated.

A *nominal unification problem* (typically $P$) is a set of equations of the form $t \stackrel{?}{\approx} u$ or $a \,\#^? t$. A *freshness environment* (typically $\nabla$) is a list of pairs $a \,\#\, X$ stating that the instantiation of $X$ cannot capture $a$.

A *solution* of a nominal problem is given by a substitution $\sigma$ and a freshness environment $\nabla$. Substitutions are like in first-order logic, and allow atom capture, for instance $[X \mapsto a]a.X = a.a$. Formally, the pair $\langle \nabla, \sigma \rangle$ solves $P$ if, $\nabla \vdash a \mathrel{\#} \sigma(t)$, for equations $a \mathrel{\#}^? t \in P$, and $\nabla \vdash \sigma(t) \approx \sigma(u)$, for equations $t \stackrel{?}{\approx} u \in P$. The predicates $\approx$ and $\#$ are defined in [UPG03,UPG04] by means of a theory. Their intended meanings are: $\nabla \vdash a \mathrel{\#} t$ holds if, for every substitution $\sigma$ avoiding the atom captures forbidden by $\nabla$, $a$ is not free in $\sigma(t)$; $\nabla \vdash t \approx u$ holds if, for every substitution $\sigma$ avoiding the atom captures forbidden by $\nabla$, $t$ and $u$ are $\alpha$-convertible.

### 2.2 Higher-Order Pattern Unification

In higher-order signatures we have types constructed from a set of basic types (typically $\alpha, \dots$) using the grammar $\tau ::= \alpha \mid \tau \to \tau$, where $\to$ is associative to the right).

$\lambda$-terms are built using the grammar $t ::= x \mid c \mid \lambda x.t \mid t_1\, t_2$, where $x$ is a variable and $c$ is a constant. Other standard concepts of $\lambda$-calculus, like free variables (noted $FV$), bound and free occurrences of variables, $\alpha$-conversion, $\beta$-reduction, $\eta$-long $\beta$-normal form, substitutions, most general unifiers, etc. are defined as usual (see [Dow01]). The domain of a substitution $\sigma$ is denoted by $Dom(\sigma)$, and we say that $X$ occurs in $\sigma$ if $X$ occurs free in $\sigma(Y)$ for some $Y \in Dom(\sigma)$.

A *higher-order pattern* is a simply typed $\lambda$-term where, when written in normal form, all free variable occurrences are applied to lists of pairwise distinct bound variables. *Higher-order pattern unification* is the problem of deciding if there exists a unifier for a set of equations $t \stackrel{?}{=} u$ between higher-order patterns. The most general unifiers of a pattern unification problem is unique (up to free variable renaming). Moreover, it instantiates variables by higher-order patterns. There is an algorithm that finds these unifiers, if exist, in linear time [Qia96].

## 3 Four Examples

In order to describe the reduction of nominal unification to higher-order pattern unification, we will use the unification problems proposed in [UPG03,UPG04] as a quiz.

*Example 1.* The nominal equation

$$a.b.\langle X_1, b\rangle \stackrel{?}{\approx} b.a.\langle a, X_1\rangle$$

has no nominal unifiers. Notice that, although unification is performed modulo $\alpha$-equivalence, as far as we allow atom capture, we can not $\alpha$-convert terms before instantiating them. Therefore, this problem is *not* equivalent to

$$a.b.\langle X_1, b\rangle \stackrel{?}{\approx} a.b.\langle b, X_1\rangle$$

which is solvable, and must be $\alpha$-converted as

$$a.b.\langle X_1, b\rangle \stackrel{?}{\approx} a.b.\langle b, (a\, b)\cdot X_1\rangle$$

Recall that $(a\, b) \cdot X_1$ means that, after instantiating $X_1$ with a term that possibly contain $a$ or $b$, we have to exchange these variables.

According to the ideas exposed in the introduction, we have to replace every occurrence of $X_1$ by $X_1'\,a'\,b'$, since $a, b$ is the list of atoms (bound variables) that can be captured. We get[1]:

$$\lambda a'.\lambda b'.\langle X_1'\,a'\,b'\,,\,b'\rangle \stackrel{?}{=} \lambda b'.\lambda a'.\langle a'\,,\,X_1'\,a'\,b'\rangle$$

Since this is a higher-order unification problem, we can $\alpha$-convert one of the sides of the equation and get:

$$\lambda a'.\lambda b'.\langle X_1'\,a'\,b'\,,b'\,\rangle \stackrel{?}{=} \lambda a'.\lambda b'.\langle b'\,,\,X_1'\,b'\,a'\rangle$$

which is unsolvable, like the original nominal equation.

*Example 2.* The nominal equation

$$a.b.\langle X_2, b\rangle \stackrel{?}{\approx} b.a.\langle a, X_3\rangle$$

is solvable. Its translation is

$$\lambda a'.\lambda b'.\langle X_2'\,a'\,b'\,,\,b'\rangle \stackrel{?}{=} \lambda b'.\lambda a'.\langle a'\,,\,X_3'\,a'\,b'\rangle$$

The most general unifier of this higher-order pattern unification problem is

$$X_2' \mapsto \lambda x.\lambda y.y$$
$$X_3' \mapsto \lambda x.\lambda y.x$$

Now, taking into account that the first argument corresponds to the bound variable $a'$, and the second one to $b'$, we can reconstruct the most general nominal unifier as:

$$X_2 \mapsto b$$
$$X_3 \mapsto a$$

*Example 3.* In some cases, there are interrelationships between the instances of variables that make reconstruction of unifiers more difficult. This is shown with the following nominal equation:

$$a.b.\langle b, X_4\rangle \stackrel{?}{\approx} b.a.\langle a, X_5\rangle$$

that is solvable. Its translation results on:

$$\lambda a'.\lambda b'.\langle b'\,,\,X_4'\,a'\,b'\rangle \stackrel{?}{=} \lambda b'.\lambda a'.\langle a'\,,\,X_5'\,a'\,b'\rangle$$

and its most general unifier is:[2]

$$X_4' \mapsto \lambda x.\lambda y.X_5'\,y\,x$$

This higher-order unifier can be used to reconstruct the nominal unifier

$$X_4 \mapsto (a\ b)\cdot X_5$$

The swapping $(a\,b)$ comes from the fact that the arguments of $X_5'$ and the lambda abstractions in front have a different order.

1 In this example we allow the use of the binary constant $\langle\ ,\ \rangle$ in $\lambda$-calculus for pairs. Later on we will describe formally the translation algorithm and how pairs are really translated.

2 The unifier $X_5' \mapsto \lambda x.\lambda y.X_4'\,y\,x$ is equivalent modulo variable renaming. In this case we obtain the also equivalent nominal unifier $X_5 \mapsto (a\ b)\cdot X_4$.

*Example 4.* The solution of a nominal unification problem is not just a substitution, but a pair $(\nabla, \sigma)$ where $\sigma$ is a substitution and $\nabla$ is a freshness environment imposing some restrictions on the atoms that can occur free in the fresh variables introduced by $\sigma$. The nominal equation

$$a.b.\langle b, X_6\rangle \overset{?}{\approx} a.a.\langle a, X_7\rangle$$

has as solution

$$\begin{array}{l}\sigma = [X_6 \mapsto (b\ a)\cdot X_7]\\ \nabla = \{b \# X_7\}\end{array}$$

where the freshness environment is not empty and requires instances of $X_7$ to not contain (free) occurrences of $b$. Let us see how this is reflected when we translate the problem into a higher unification problem. The translation of the equation using the translation algorithm results on:

$$\lambda a'.\lambda b'.\langle b'\,,\, X'_6\, a'\, b'\rangle \overset{?}{=} \lambda a'.\lambda a'.\langle a'\,,\, X'_7\, a'\, b'\rangle \tag{1}$$

After a convenient $\alpha$-conversion we get

$$\lambda a'.\lambda c'.\langle c'\,,\, X'_6\, a'\, c'\rangle \overset{?}{=} \lambda a'.\lambda c'.\langle c'\,,\, X'_7\, c'\, b'\rangle$$

The most general unifier is again unique:

$$\begin{array}{l}X'_6 \mapsto \lambda x.\lambda y.X_8\, y\, b'\\ X'_7 \mapsto \lambda x.\lambda y.X_8\, x\, y\end{array}$$

Nevertheless, in this case we cannot reconstruct the nominal unifier. Moreover, by instantiating the free variable $b'$, we get other (non-most general) higher-order unifier without nominal counterpart. The translation does not work in this case because $b'$ occurs free in the right hand side of (1). We translate both atoms and unknowns as variables. Occurrences of unknowns become free occurrences of variables, and occurrences of atoms, if are bounded, become bound occurrences of variables. Therefore, in most cases, after the translation the distinction atom/unknown become a distinction free/bound variable. However, if atoms are not bounded, as in this case, they are translated as free variables, hence are instantiable, whereas atoms are not instantiable.

To avoid this problem, we have to ensure that any occurrence of an atom is translated as a bound variable occurrence. This is easily achievable if we add binders in front of both sides of the equation. Therefore, the correct translation of this problem is:

$$\lambda a'.\lambda b'.\lambda a'.\lambda b'.\langle b'\,,\, X'_6\, a'\, b'\rangle \overset{?}{=} \lambda a'.\lambda b'.\lambda a'.\lambda a'.\langle a'\,,\, X'_7\, a'\, b'\rangle$$

where two new binder $\lambda a'.\lambda b'$ have been introduced in front of both sides of the equation. The most general unifier is now:

$$\begin{array}{l}X'_6 \mapsto \lambda x.\lambda y.X'_8\, y\\ X'_7 \mapsto \lambda x.\lambda y.X'_8\, x\end{array}$$

This can be used to reconstruct the nominal substitution:

$$\begin{array}{l}X_6 \mapsto (a\ b)\cdot X_8\\ X_7 \mapsto X_8\end{array}$$

As far as $X'_8\,x$ is translated back as $X_8$, and $X'_8\,x$ does not uses the second argument (the one corresponding to $b$), we have to add a supplementary condition ensuring that $X_8$ does not contain free occurrences of $b$. This results on the freshness environment $\{b \# X_8\}$. Then, $X'_8\,y$ is translated back as $(a\,b) \cdot X_8$.

## 4 The Translation Algorithm

In this Section we formalize the translation algorithm. We transform nominal unification problems into higher-order unification problems. Both kinds of problems are expressed using distinct kinds of signatures. In nominal unification we have sorts of atoms and sorts of data. In higher-order this distinction is no longer necessary, and we will have a *base type* (typically $\nu'$ and $\delta'$) for every sort of atoms $\nu$ or sort of data $\delta$. We give a *sort to types translation function* that allows us to translate any sort into a type.

**Definition 1.** *The translation function is defined on sorts inductively as follows.*

$$\begin{aligned}
&[\![\delta]\!] = \delta' \\
&[\![\nu]\!] = \nu' \\
&[\![\tau_1 \times \tau_2]\!] = ([\![\tau_1]\!] \to [\![\tau_2]\!] \to \top) \to \top \\
&[\![\tau_1 \to \tau_2]\!] = [\![\tau_1]\!] \to [\![\tau_2]\!] \\
&[\![\langle\nu\rangle\tau]\!] = \nu' \to [\![\tau]\!]
\end{aligned}$$

*where $\top$, $\delta'$ and $\nu'$ are base types.*

*Remark 1.* The translation function for terms depends on *all* the atoms occurring in the nominal unification problem. We assume that there exists a *fixed, finite* and *ordered* list of atoms $\langle a_1, \ldots, a_n\rangle$ used in the problem. This seems to contradict the assumption of a countably *infinite* set of atoms for every sort. However, this does not imply a loss of generality as far as every nominal unification problem only contains a finite set of atoms, and its solutions can be expressed without adding new atoms (see Lemma 5). From now on, we will consider this list given and fixed.

For every function symbol $f, \ldots$, we will use a constant with name $f', \ldots$. Nominal atoms $a, b \ldots$ are translated as (bound) variables, with the names $a', b', \ldots$. The lack of distinction between sorts of atoms and data, after the translation, forces us to ensure that the translation of every atom occurrence will correspond to a *bounded* occurrence of variable. For every variable (unknown) $X, Y, \ldots$, we will use a (free) variable with name $X', Y', \ldots$. Trivially, atom abstractions $a.t$ are translated as lambda abstractions, and data $f\,t$ as applications. The translation of suspensions $\pi \cdot X$ is more complicated, as far as it gets rid of atom capture.

**Definition 2.** *Let $\langle a_1, \ldots, a_n\rangle$ be an ordered list of atoms occurring in the nominal unification problem. The translation function from nominal terms with a freshness environments $\nabla$ into $\lambda$-terms is defined inductively as follows.*

$$
\begin{aligned}
&[\![\langle t_1, t_2\rangle]\!]_\nabla = \lambda p.p\, [\![t_1]\!]_\nabla\, [\![t_2]\!]_\nabla \\
&[\![a]\!]_\nabla = a' \\
&[\![f\, t]\!]_\nabla = f'\, [\![t]\!]_\nabla \\
&[\![a.t]\!]_\nabla = \lambda a'.[\![t]\!]_\nabla \\
&[\![\pi \cdot X]\!]_\nabla = X'\, (\pi \cdot b_1)' \dots (\pi \cdot b_m)' \quad \textit{where } b_j \,\#\, X \notin \nabla, \textit{ for } j = 1, \dots, m
\end{aligned}
$$

*where, if* $a : \nu$ *is an atom, then* $a' : [\![\nu]\!]$ *is a bound variable, if* $f : \tau$ *is a function symbol, then* $f' : [\![\tau]\!]$ *is a constant, and if* $X : \tau$ *is a variable, then* $\langle b_1, \dots, b_m\rangle \subseteq \langle a_1, \dots, a_n\rangle$ *is the sublist of atoms such that* $b_j \,\#\, X \notin \nabla$ *and* $X' : [\![\nu_1]\!] \to \dots \to [\![\nu_m]\!] \to [\![\tau]\!]$ *is a free variable and* $b_j : \nu_j$.[3]

**Lemma 1.** *Let* $\nabla$ *be a freshness environment.*
*For every nominal term* $t$ *of sort* $\tau$*, the* $\lambda$*-term* $[\![t]\!]_\nabla$ *has type* $[\![\tau]\!]$*.*
*Therefore,* $[\![t]\!]_\nabla$ *is a well-typed* $\lambda$*-term, for every nominal term* $t$*.*

*Proof.* The proof is simple by structural induction on $t$. The only point that merits a more detailed explanation is the case of suspensions. Since $a_i : \nu_i$, $X : \tau$, and $X' : [\![\nu_{i_1}]\!] \to \dots \to [\![\nu_{i_m}]\!] \to [\![\tau]\!]$, we have $[\![X]\!]_\nabla = X'[\![a_{i_1}]\!]_\nabla \dots [\![a_{i_m}]\!]_\nabla : [\![\tau]\!]$. When $X$ is affected by a swapping $(a_{i_j}\, a_{i_k})$ we also have $[\![(a_{i_j}\, a_{i_k}) \cdot X]\!]_\nabla = X'[\![a_{i_1}]\!]_\nabla \dots [\![a_{i_{j-1}}]\!]_\nabla [\![a_{i_k}]\!]_\nabla [\![a_{i_{j+1}}]\!]_\nabla \dots [\![a_{i_{k-1}}]\!]_\nabla [\![a_{i_j}]\!]_\nabla [\![a_{i_{k+1}}]\!]_\nabla \dots [\![a_{i_m}]\!]_\nabla : [\![\tau]\!]$ because the suspension is not a valid nominal term unless $a_{i_j}$ and $a_{i_k}$ belong to the same sort. The same applies to arbitrary permutations. □

*Example 5.* Given the nominal term $a.b.c.(c\, a)(a\, b) \cdot X$, after applying the substitution $[X \mapsto \langle\langle a, b\rangle, c\rangle]$ we get the instantiation $a.b.c.\langle\langle b, c\rangle, a\rangle$. Let $\langle a, b, c\rangle$ be the (ordered) list of atoms of our problem. The translation of the term and its instantiation results into $\lambda a'.\lambda b'.\lambda c'.X'b'c'a'$ and $\lambda a'.\lambda b'.\lambda c'.\lambda p.p(\lambda p.p\, b'c')a'$, respectively. There is a $\lambda$-substitution $[X' \mapsto \lambda a'.\lambda b'.\lambda c'.\lambda p.p(\lambda p.p\, a'b')c']$ that when applied to the translation of the original term results into the translation of its instantiation. Graphically this can be represented as the commutation of the following diagram, and is proved in general in Lemma 4.

$$
\begin{array}{ccc}
a.b.c.(c\, a)(a\, b) \cdot X & \xrightarrow{[X \mapsto \langle\langle a, b\rangle, c\rangle]} & a.b.c.\langle\langle b, c\rangle, a\rangle \\
\downarrow [\![\;]\!] & & \downarrow [\![\;]\!] \\
\lambda a'.\lambda b'.\lambda c'.X'b'c'a' & \xrightarrow{[X' \mapsto \lambda a'.\lambda b'.\lambda c'.\lambda p.p(\lambda p.p\, a'b')c']} & \lambda a'.\lambda b'.\lambda c'.\lambda p.p(\lambda p.p\, b'c')a'
\end{array}
$$

As we have said nominal unification problems contains two kinds of judgments: *freshness equations* like $a \,\#^?\, t$, and *equality equations* like $t \stackrel{?}{\approx} u$. Equality equations are trivially translated as higher-order unification problems, adding some $\lambda$-bindings in front of both terms to ensure that all occurrences of atoms are translated as bounded occurrences of variables. Freshness equations $a \,\#^?\, t$ are translated as equations of the form $Y \stackrel{?}{\approx} t$ where $Y$ is a fresh variable that will not be able to capture free occurrences of $a$.

[3] Notice that $b_j$ and $\pi \cdot b_j$ are of the same sort.

**Definition 3.** *Let $\langle a_1, \ldots, a_n \rangle$ be an ordered list of atoms occurring in the nominal unification problem. The translation function is defined on nominal problems inductively as follows*

$$[\![\{a_i \,\#^? t\} \cup P]\!] = \{\lambda a'_1. \ldots .\lambda a'_n . Y' a'_1 \ldots a'_{i-1}\, a'_{i+1} \ldots a'_n \stackrel{?}{=} \lambda a'_1. \ldots .\lambda a'_n . [\![t]\!]_\emptyset\} \cup [\![P]\!]$$
$$[\![\{t \stackrel{?}{\approx} u\} \cup P]\!] = \{\lambda a'_1. \ldots .\lambda a'_n . [\![t]\!]_\emptyset \stackrel{?}{=} \lambda a'_1. \ldots .\lambda a'_n . [\![u]\!]_\emptyset\} \cup [\![P]\!]$$

*where $Y'$ is a fresh variable with the appropriate type.*

*Remark 2.* Alternatively to Definition 3, we could decompose freshness equations into simple pieces until we get a freshness environment, and use it in the translation of the equality equations. This would result into the following inductive definition.

$$\begin{array}{ll}
[\![\{a \,\#^? \langle t_1, t_2 \rangle\} \cup P]\!] = [\![\{a \,\#^? t_1, a \,\#^? t_2\} \cup P]\!] & \\
[\![\{a \,\#^? b\} \cup P]\!] = [\![P]\!] & \text{if } a \neq b \\
[\![\{a \,\#^? f\, t\} \cup P]\!] = [\![\{a \,\#^? t\} \cup P]\!] & \\
[\![\{a \,\#^? b.t\} \cup P]\!] = [\![\{a \,\#^? t\} \cup P]\!] & \text{if } a \neq b \\
[\![\{a \,\#^? a.t\} \cup P]\!] = [\![P]\!] & \\
[\![\{a \,\#^? \pi \cdot X\} \cup P]\!] = [\![\{\pi^{-1} \cdot a \,\#\, X\} \cup P]\!] & \\
[\![\nabla \cup P]\!] = [\![P]\!]_\nabla & \text{if for all } \{a \,\#^? t\} \in \nabla, t \text{ is a variable} \\
[\![\{t \stackrel{?}{\approx} u\} \cup P]\!]_\nabla = \{\lambda a'_1. \ldots .\lambda a'_n . [\![t]\!]_\nabla \stackrel{?}{=} \lambda a'_1. \ldots .\lambda a'_n . [\![u]\!]_\nabla\} \cup [\![P]\!]_\nabla &
\end{array}$$

However, in this case, the type of the free variables $X'$ would depend on the freshness equations, and we would have problems to define the translation of a substitution that would have to instantiate such variables. Therefore, for simplicity we opt for Definition 3.

**Lemma 2.** *Given a nominal unification problem $P$, its translation $[\![P]\!]$ is a higher-order pattern unification problem.*
*Moreover, the size of $[\![P]\!]$ is bounded by the square of the size of $P$.*

Finally, we have to translate solutions of nominal unification problems.

**Definition 4.** *Let $\langle a_1, \ldots, a_n \rangle$ be an ordered list of atoms occurring in the nominal unification problem. The translation function is defined on solutions of nominal unification problems inductively as follows.*

$$[\![\langle \nabla, \sigma \rangle]\!] = \bigcup_{X \in Dom(\sigma)} \Big[ X' \mapsto \lambda a'_1 . \cdots \lambda a'_n . [\![\sigma(X)]\!]_\nabla \Big]$$

*Remark 3.* Notice that, if $P$ contains freshness equations, then the set of free variables of $[\![P]\!]$ is bigger than the set of unknowns in $P$ (we have a new free variable, called $Y'$, for every freshness equation). However, $\sigma$ and $[\![\langle \nabla, \sigma \rangle]\!]$ have equivalent domains. This would imply that, if $\langle \nabla, \sigma \rangle$ is a solution of $P$, we will have to extend $[\![\langle \nabla, \sigma \rangle]\!]$, by instantiating also the variables $Y'$, to get a unifier of $[\![P]\!]$. If we had translated nominal unification problems as described in Remark 2, we would obtain a pattern unification problem with equivalent sets of variables. But, we would lose simplicity in other proofs.

We start by proving the following two technical lemmas.

**Lemma 3.** *For any freshness environment* $\nabla$*, nominal terms* $t$*,* $u$*, and atom* $a$*, we have*

1. $\nabla \vdash a \# t$ *if, and only if,* $a' \notin FV([\![t]\!]_\nabla)$*, and*
2. $\nabla \vdash t \approx u$ *if, and only if,* $[\![t]\!]_\nabla =_\alpha [\![u]\!]_\nabla$*.*

*Proof.* The first statement can be proved by routine induction on $t$ and its translation. Notice that atoms are translated "nominally" into variables and that the binding structure is also identically translated, hence, the freshness of an atom $a$ corresponds to the free occurrence of its variable counterpart $a'$. We here only comment the case $t = \pi \cdot X$, in this case, $[\![\pi \cdot X]\!]_\nabla = X'(\pi \cdot a_1)' \ldots (\pi \cdot a_m)'$, where $a_i \# X \notin \nabla$, for any $i \in \{1..m\}$. Therefore, we can establish the following sequence of equivalences $\nabla \vdash a \# \pi \cdot X$ iff $\pi^{-1} \cdot a \# X \in \nabla$ iff $\pi^{-1} \cdot a \notin \{a_1, \ldots, a_m\}$ iff $a \notin \{\pi \cdot a_1, \ldots, \pi \cdot a_m\}$ iff $a' \notin FV(X'(\pi \cdot a_1)' \ldots (\pi \cdot a_m)')$ iff $a' \notin FV([\![\pi \cdot X]\!])$.

The proof of the second statement can be done by induction on the equivalence $t \approx u$. We only comment the equivalence between suspensions: $\pi \cdot X \approx \pi' \cdot X$. Notice that, $\pi \cdot X \approx \pi' \cdot X$ if, and only if, for all atoms $a$ such that $\pi \cdot a \neq \pi' \cdot a$, we have $a \# X \in \nabla$. This condition is equivalent to: the bound variables $(\pi \cdot a)'$ and $(\pi' \cdot a)'$ are passed as a parameter to $X'$ in $[\![\pi \cdot X]\!]_\nabla$ and $[\![\pi' \cdot X]\!]_\nabla$ only when $\pi \cdot a = \pi' \cdot a$. Finally, this condition is equivalent to $[\![\pi \cdot X]\!]_\nabla = [\![\pi' \cdot X]\!]_\nabla$. □

**Lemma 4.** *For any freshness environment* $\nabla$*, nominal terms* $t$*, and nominal substitution* $\sigma$*, we have* $[\![\langle \nabla, \sigma \rangle]\!]([\![t]\!]_\emptyset) = [\![\sigma(t)]\!]_\nabla$*.*[4]

*Proof.* Again this lemma can be proved by structural induction on $t$. We only sketch the suspension case. Let $t = \pi \cdot X$. We have the equalities:

$$\begin{aligned} [\![\langle \nabla, \sigma \rangle]\!]([\![\pi \cdot X]\!]_\emptyset) &= [\ldots, X' \mapsto \lambda a'_1 \ldots \lambda a'_n . [\![\sigma(X)]\!]_\nabla, \ldots](X'(\pi \cdot a_1)' \ldots (\pi \cdot a_n)') \\ &= (\lambda a'_1 \ldots \lambda a'_n . [\![\sigma(X)]\!]_\nabla)(\pi \cdot a_1)' \ldots (\pi \cdot a_n)' \\ &= [\![[a_1 \mapsto \pi \cdot a_1, \ldots, a_n \mapsto \pi \cdot a_n]\sigma(X)]\!]_\nabla \\ &= [\![\pi \cdot \sigma(X)]\!]_\nabla \\ &= [\![\sigma(\pi \cdot X)]\!]_\nabla \end{aligned}$$

□

From these two lemmas we can prove the following result and corollary.

**Theorem 1.** *For any freshness environment* $\nabla$*, nominal unification problem* $P$*, and nominal substitution* $\sigma$*, we have that:*

$\langle \nabla, \sigma \rangle$ *solves the nominal unification problem* $P$*, if, and only if, there exists an extension of* $[\![\langle \nabla, \sigma \rangle]\!]$*, for the variables of* $[\![P]\!]$ *not occurring in* $P$*, that solves the pattern unification problem* $[\![P]\!]$*.*

*Proof.* The pair $\langle \nabla, \sigma \rangle$ solves $P$ iff

$$\begin{array}{ll} \nabla \vdash a_i \# \sigma(t) & \text{for all } a_i \#^? t \in P \\ \nabla \vdash \sigma(t) \approx \sigma(u) & \text{for all } t \stackrel{?}{\approx} u \in P \end{array}$$

By Lemma 3 this is equivalent to:

$$\begin{array}{ll} a' \notin FV([\![\sigma(t)]\!]_\nabla) & \text{for all } a \#^? t \in P \\ [\![\sigma(t)]\!]_\nabla =_\alpha [\![\sigma(u)]\!]_\nabla & \text{for all } t \stackrel{?}{\approx} u \in P \end{array}$$

[4] When we write $=$ between $\lambda$-terms, we mean that they are $\alpha\beta\eta$-equivalent, i.e. that have the same $\eta$-long $\beta$-normal form.

By Lemma 4 this is equivalent to:

$$\begin{array}{ll} a'_i \notin FV(\llbracket\langle\nabla,\sigma\rangle\rrbracket\llbracket t\rrbracket_\emptyset) & \text{for all } a_i \#^? t \in P \\ \llbracket\langle\nabla,\sigma\rangle\rrbracket\llbracket t\rrbracket_\emptyset = \llbracket\langle\nabla,\sigma\rangle\rrbracket\llbracket u\rrbracket_\emptyset & \text{for all } t \stackrel{?}{\approx} u \in P \end{array}$$

Since we avoid variable capture, this is equivalent to:

$$\Big[Y' \mapsto \lambda a'_1 \ldots a'_{i-1} a'_{i+1} \ldots a'_n.\llbracket\langle\nabla,\sigma\rangle\rrbracket\llbracket t\rrbracket_\emptyset\Big]\Big(\lambda a'_1 \ldots a'_n. Y' a'_1 \ldots a'_{i-1}\, a'_{i+1} \ldots a'_n\Big) \stackrel{?}{=}$$
$$\stackrel{?}{=} \llbracket\langle\nabla,\sigma\rangle\rrbracket\Big(\lambda a'_1 \ldots a'_n.\llbracket t\rrbracket_\emptyset\Big) \qquad \text{for all } a_i \#^? t \in P$$
$$\llbracket\langle\nabla,\sigma\rangle\rrbracket\Big(\lambda a'_1.\ldots.\lambda a'_n.\llbracket t\rrbracket_\emptyset\Big) \stackrel{?}{=} \llbracket\langle\nabla,\sigma\rangle\rrbracket\Big(\lambda a'_1.\ldots.\lambda a'_n.\llbracket u\rrbracket_\emptyset\Big) \quad \text{for all } t \stackrel{?}{\approx} u \in P$$

Finally, this means that the following extension solves $\llbracket P\rrbracket$.

$$\sigma' = \llbracket\langle\nabla,\sigma\rangle\rrbracket \cup \bigcup_{Y'} [Y' \mapsto \lambda a'_1.\ldots.\lambda a'_{i-1}.\lambda a'_{i+1}.\ldots.\lambda a'_n.\llbracket\langle\nabla,\sigma\rangle\rrbracket\llbracket t\rrbracket_\emptyset]$$

□

**Corollary 1.** *If the nominal unification problem $P$ is solvable, then the higher-order pattern unification problem $\llbracket P\rrbracket$ is solvable.*

## 5 The Reverse Translation

Notice that Theorem 1 is not enough to prove that, if $\llbracket P\rrbracket$ is solvable, then $P$ is solvable. We still have to prove that if $\llbracket P\rrbracket$ is solvable, then for some solution $\sigma'$ of $\llbracket P\rrbracket$ we can build a nominal solution $\langle\nabla,\sigma\rangle$ of $P$. This is the main objective of this section. Taking into account that $\llbracket P\rrbracket$ is a higher-order pattern unification problem, and that these problems are unitary, we will prove something stronger: if $\llbracket P\rrbracket$ is solvable, then $\llbracket\sigma'\rrbracket^{-1}$ is defined for some most general unifier $\sigma'$ of $\llbracket P\rrbracket$.

In the following example we note that in the solution of pattern unification problems it is important to save *names* of bound variables.

*Example 6.* Consider the nominal problem $a.X \stackrel{?}{\approx} a.f\,(b.Y)$. Its translation is $\lambda a'.\lambda b'.\lambda a'.X'\,a'\,b' \stackrel{?}{=} \lambda a'.\lambda b'.\lambda a'.f'\,(\lambda b'.Y'\,a'\,b')$. An $\alpha$-conversion results in $\lambda a'.\lambda b'.\lambda c'.X'\,c'\,b' \stackrel{?}{=} \lambda a'.\lambda b'.\lambda c'.f'\,(\lambda d'.Y'\,c'\,d')$ and it shows that the parameters of $X'$ and $Y'$ are in fact different. A most general solution is $[X' \mapsto \lambda c'.\lambda b'.f'(\lambda d'.Y'\,c'\,d')]$. Since $Y$ is translated as $Y'\,a'\,b'$, we have to translate back $Y'\,c'\,d'$ as $(a\,c)(d\,b)\cdot Y$. And, since $[X \mapsto t]$ is translated as $[X' \mapsto \lambda a'.\lambda b'.\llbracket t\rrbracket_\nabla]$, we have to translate back $[X' \mapsto \lambda c'.\lambda b'.\llbracket t\rrbracket_\nabla$ as $[X \mapsto (a\,c)\cdot t]$. Therefore, our pattern unifier can be translated back as $[X \mapsto (a\,c)\cdot f(d.(a\,c)(d\,b)\cdot Y)]$. However, the list of atoms is fixed as the list of atoms occurring in the problem, therefore, we know how to translate $a$ and $b$ as $a'$ and $b'$ and vice versa, but we do not know how to translate back $c'$ and $d'$ (here it is done introducing new atoms).

To avoid the problem discussed in the previous example, we will be cautious with the $\alpha$-conversions. Lemma 5 justifies why we can avoid the use of other atoms but the ones occurring in the original nominal problem. Lemma 6 will be also necessary, and also describes some properties of pattern unifiers.

**Lemma 5.** *For any solvable pattern unification problem, there exists a most general unifier that does not use other names and types for bound variables than the ones already used in the problem.*

*Proof.* It can be proved by inspection of the transformations rules in [Mil91,Nip93], that describe a sound and complete algorithm for pattern unification. These transformations introduce fresh variables and new lambda binders. However, the names of the new bound variables can always be chosen to coincide with names of already existing bound variables with the same type. □

**Lemma 6.** *For every pattern unification problem $P$ and most general unifier $\sigma$, if $X$ occurs free in $\sigma$, then for every type of an argument of $X$, there exists a variable $Y$ in $P$ with an argument of the same type, and there exists a variable $Z$ in $P$ with the same return type as $X$.*

*Proof.* Like for the previous lemma, we can analyze the transformations rules in [Mil91,Nip93]. It can be seen that when we introduce a fresh variable with type $\tau_1 \to \ldots \to \tau_n \to \tau_0$, there exist already another variable with type $\tau'_1 \to \ldots \to \tau'_m \to \tau'_0$, such that $\{\tau_1, \ldots, \tau_n\} \subseteq \{\tau'_1, \ldots, \tau'_m\}$ and $\tau_0 = \tau'_0$. The only exception is in rigid-flexible pairs (imitation rule), where fresh variables not satisfying these properties are introduced. But it is easy to see that they always disappear from the solution. □

Even using a solution of the higher-order pattern unification problem with a restricted use of names of bound variables, we still have some freedom to select the unifier $[\![\sigma']\!]^{-1}$. This is reflected in the way we translate back applications of free variables, i.e. in the definition of the list of variable indices $L_{X'}$ for every free variable $X'$.

**Definition 5.** *Let $\langle a_1, \ldots, a_n\rangle$ be the fixed list of atoms. For every free variable $X' : \tau_1 \to \ldots \tau_m \to \tau_0$ we define the list of indexes of atoms $L_{X'} = \langle i_1, \ldots i_m\rangle$ such that $a_{i_j}$ has sort $[\![\tau_j]\!]^{-1}$,* [5] *for $j = 1, \ldots, m$, and we also define the corresponding freshness environment $\nabla_{X'} = \{a_i \# X \mid i \notin L_{X'}\}$.*

**Definition 6.** *Let $\langle a_1, \ldots, a_n\rangle$ be the fixed list of atoms. The back-translation function is defined on $\lambda$-terms in $\eta$-long $\beta$-normal form as follows:*

$$\begin{array}{ll}
[\![\lambda p.\, p\, t_1\, t_2]\!]^{-1} = \langle [\![t_1]\!]^{-1}, [\![t_2]\!]^{-1}\rangle & \textit{if } p : \tau_1 \to \tau_2 \to \top \\
[\![a']\!]^{-1} = a & \\
[\![f'\, t]\!]^{-1} = f\, [\![t]\!]^{-1} & \\
[\![\lambda a'.t]\!]^{-1} = a\, .\, [\![t]\!]^{-1} & \textit{if } a' \textit{ is base typed} \\
[\![X'\, a'_{j_1} \ldots a'_{j_m}]\!]^{-1} = \begin{pmatrix} a_{i_1} & \cdots & a_{i_m} \\ a_{j_1} & \cdots & a_{j_m} \end{pmatrix} \cdot X & \begin{array}{l}\textit{where } L_{X'} = \langle i_1, \ldots i_m\rangle \\ \textit{and } \{a_{j_1}, \cdots, a_{j_m}\} \subseteq \{a_1, \ldots, a_n\}\end{array}
\end{array}$$

*where $a'$ is a bound variable with name $a$, $f'$ is the constant associated to the function symbol $f$, either $X'$ is the free variable associated to $X$, or if $X'$ is a fresh variable then $X$ is a fresh unknown, and the permutation is supposed to be decomposed in terms of transpositions (swappings).*

[5] Notice that Lemma 6 ensures that, even for the introduced fresh variables, for every type $\tau_j$ there exists at least one atom $a_{i_j}$ satisfying $a_{i_j} : [\![\tau_j]\!]^{-1}$. Notice also that for every $X'$, we freely choose one among many possible lists $L_{X'}$.

Notice that the back-translation function is not defined for all λ-terms, even for all higher-order patterns. In particular, $[\![\lambda x.t]\!]^{-1}$ is not defined when $x$ is not base typed nor returns something of type $\top$, or $[\![x\,t]\!]^{-1}$ is not defined when $x$ is a bound variable.

**Definition 7.** *The back-translation function is defined on substitutions inductively as follows.*

$$[\![\sigma']\!]^{-1} = \left\langle \bigcup_{\substack{X' \in Dom(\sigma') \\ X' : \nu'_1 \to \ldots \to \nu'_n \to \delta'}} \left[X \mapsto [\![\sigma'(X')\,a'_1 \ldots a'_n]\!]^{-1}\right], \bigcup_{Z' \text{ occurs in } \sigma'} \nabla_{Z'} \right\rangle$$

Notice that the back-translation only translates those instantiations affecting variables with type $\nu'_1 \to \ldots \to \nu'_n \to \delta'$. When $\sigma'$ is a unifier of a problem $[\![P]\!]$, then it contains in $Dom(\sigma')$ variables $X'$ associated to a nominal variable $X$, that satisfies this condition, and variables $Y'$ resulting from the translation of freshness equations, that do not satisfy the condition because they have type $\nu'_1 \to \ldots \to \nu'_{i-1} \to \nu'_{i+1} \to \ldots \to \nu'_n \to \delta'$. These second type of variables have not back-translation and do not occur in the domain of $[\![\sigma']\!]^{-1}$.

*Example 7.* The nominal unification problem

$$P = \{a.a.X \stackrel{?}{\approx} c.a.X \;,\; a.b.X \stackrel{?}{\approx} b.a.(a\,b) \cdot X \;,\; a \#^? X\}$$

is translated as

$$\begin{aligned}[\![P]\!] = \{\ & \lambda a'.\lambda b'.\lambda c'.\lambda a'.\lambda a'.\ X'\,a'\,b'\,c' \stackrel{?}{=} \lambda a'.\lambda b'.\lambda c'.\lambda c'.\lambda a'.\ X'\,a'\,b'\,c' \;, \\ & \lambda a'.\lambda b'.\lambda c'.\lambda a'.\lambda b'.\ X'\,a'\,b'\,c' \stackrel{?}{=} \lambda a'.\lambda b'.\lambda c'.\lambda b'.\lambda a'.\ X'\,b'\,a'\,c' \;, \\ & \lambda a'.\lambda b'.\lambda c'.\ Y'\,b'\,c' \stackrel{?}{=} \lambda a'.\lambda b'.\lambda c'.\ X'\,a'\,b'\,c'\}\end{aligned}$$

The pattern unifier is

$$\sigma' = [X' \mapsto \lambda a'.\lambda b'.\lambda c'.\ Z'\,b' \;,\; Y' \mapsto \lambda b'.\lambda c'.\ Z'\,b']$$

Fixed $\langle a, b, c\rangle$ as the fixed list of atoms, and taking $L_{Z'} = \langle 2 \rangle$, the nominal solution is

$$\langle \nabla, \sigma \rangle = [\![\sigma']\!]^{-1} = \langle \{a \# Z, c \# Z\} \;,\; [X \mapsto Z] \rangle$$

They satisfy the relation

$$\sigma' = [\![\langle \nabla, \sigma \rangle]\!] \cup [Y' \mapsto \lambda b'.\lambda c'.\ Z'\,b']$$

Notice that $a$ and $b$ are of the same sort. Therefore, we could take $L_{Z'} = \langle 1 \rangle$. Then, we would obtain another "equivalent" nominal unifier:

$$\langle \nabla, \sigma \rangle = [\![\sigma']\!]^{-1} = \langle \{b \# Z, c \# Z\} \;,\; [X \mapsto (a\,b) \cdot Z] \rangle$$

**Lemma 7.** *For every λ-substitution $\sigma'$, if $[\![\sigma']\!]^{-1}$ exists, then $[\![[\![\sigma']\!]^{-1}]\!]$ is extensible to $\sigma'$.*

*Proof.* Straightforward from the definition of $[\![\ ]\!]$ and $[\![\ ]\!]^{-1}$. □

**Lemma 8.** *For every nominal unification problem $P$, if the pattern unification problem $[\![P]\!]$ is solvable, then it has a most general unifier $\sigma'$ such that $[\![\sigma']\!]^{-1}$ is defined.*

*Proof.* By Lemma 5, there exists a most general unifier that does not use bound variables with other names and types than the ones already used in the original problem. This ensures that we can always translate back bound variables $a'$ as the atom with the same name $a$. For the same reason, in all $\lambda$-expressions $\lambda x.t$ the bound variable $x$ will have type $\tau_1 \to \tau_2 \to \top$ (for pairs) or base type $\nu_i'$, which will ensure that its translation back is possible.

By Lemma 6, since all free variables in the original problem $[\![P]\!]$ have type of the form $\nu_1 \to \ldots \to \nu_n \to \delta$, all free variables in the unifier will have type of the form $X' : \nu_{i_i} \to \ldots \to \nu_{i_m} \to \delta$. This ensures the existence of $L_{X'} = \{i_1, \ldots, i_m\}$, as well as the translation back of suspensions.

Finally, we will never be forced to translate back terms of the form $at_1 \ldots t_m$ where $a$ is a bound variable, because in $[\![P]\!]$, hence in $\sigma'$, all bound variables are base typed. □

**Theorem 2.** *For every nominal unification problem $P$, if the pattern unification problem $[\![P]\!]$ is solvable, then $P$ is solvable.*

*Proof.* By Lemma 8, if $[\![P]\!]$ is solvable then there exist a most general unifier $\sigma'$ of such that $\langle\nabla, \sigma\rangle = [\![\sigma']\!]^{-1}$ is defined. By Lemma 7, we have $[\![\langle\nabla, \sigma\rangle]\!]$ is extensible (by instantiating all the variables of $[\![P]\!]$ not corresponding to any variable in $P$) to $\sigma'$, which solves $[\![P]\!]$. Hence, by Theorem 1, $\langle\nabla, \sigma\rangle$ solves $P$. □

**Corollary 2.** *Nominal Unification is quadratic reducible to Higher-Order Pattern Unification.*

*Nominal Unification can be decided in quadratic deterministic time.*

## 6 Conclusion

The paper describes a precise quadratic reduction from Nominal Unification to Higher-order Pattern Unification. This helps to better understand the semantics of the nominal binding and permutations in comparison with $\lambda$-binding and $\alpha$-conversion. Moreover, using the result of linear time decidability for Higher-Order Patterns Unification [Qia96], we prove that Nominal Unification can be decided in quadratic time. It seems not difficult to prove that the translation and the back-translation function that we present transform most general nominal unifiers into most general higher-order patter unifiers and vice versa.

## References

[CF07] Calvès, C., Fernández, M.: Implementing nominal unification. ENTCS 176(1), 25–37 (2007)

[Che05] Cheney, J.: Relating higher-order pattern unification and nominal unification. In: Proc. of the 19th Int. Work on Unification, UNIF 2005, pp. 104–119 (2005)

[CP07] Clouston, R.A., Pitts, A.M.: Nominal equational logic. ENTCS 1496, 223–257 (2007)

[CU04] Cheney, J., Urban, C.: $\alpha$-prolog: A logic programming language with names, binding and $\alpha$-equivalence. In: Demoen, B., Lifschitz, V. (eds.) ICLP 2004. LNCS, vol. 3132, pp. 269–283. Springer, Heidelberg (2004)

[Dow01] Dowek, G.: Higher-order unification and matching. Handbook of automated reasoning, pp. 1009–1062 (2001)

[FG05] Fernández, M., Gabbay, M.: Nominal rewriting with name generation: abstraction vs. locality. In: Proc. of the 7th Int. Conf. on Principles and Practice of Declarative Programming, PPDP 2005, pp. 47–58 (2005)

[FG07] Fernández, M., Gabbay, M.: Nominal rewriting. Information and Computation 205(6), 917–965 (2007)

[GC04] Gabbay, M., Cheney, J.: A sequent calculus for nominal logic. In: Proc. of the 19th Symp. on Logic in Computer Science, LICS 2004, pp. 139–148 (2004)

[Gol81] Goldfarb, W.D.: The undecidability of the second-order unification problem. Theoretical Computer Science 13, 225–230 (1981)

[GP99] Gabbay, M., Pitts, A.M.: A new approach to abstract syntax involving binders. In: Proc. of the 14th Symp. on Logic in Computer Science, LICS 1999, pp. 214–224 (1999)

[Lev98] Levy, J.: Decidable and undecidable second-order unification problems. In: Nipkow, T. (ed.) RTA 1998. LNCS, vol. 1379, pp. 47–60. Springer, Heidelberg (1998)

[Luc72] Lucchesi, C.L.: The undecidability of the unification problem for third-order languages. Technical Report CSRR 2059, Dept. of Applied Analysis and Computer Science, Univ. of Waterloo (1972)

[LV00] Levy, J., Veanes, M.: On the undecidability of second-order unification. Information and Computation 159, 125–150 (2000)

[Mil91] Miller, D.: A logic programming language with lambda-abstraction, function variables, and simple unification. J. of Logic and Computation 1(4), 497–536 (1991)

[Nip93] Nipkow, T.: Functional unification of higher-order patterns. In: Proc. of the 8th Symp. on Logic in Computer Science, LICS 1993, pp. 64–74 (1993)

[Pit01] Pitts, A.M.: Nominal logic: A first order theory of names and binding. In: Kobayashi, N., Pierce, B.C. (eds.) TACS 2001. LNCS, vol. 2215, pp. 219–242. Springer, Heidelberg (2001)

[Pit03] Pitts, A.M.: Nominal logic, a first order theory of names and binding. Information and Computation 186, 165–193 (2003)

[Qia96] Qian, Z.: Unification of higher-order patterns in linear time and space. J. of Logic and Computation 6(3), 315–341 (1996)

[UC05] Urban, C., Cheney, J.: Avoiding equivariance in alpha-prolog. In: Urzyczyn, P. (ed.) TLCA 2005. LNCS, vol. 3461, pp. 401–416. Springer, Heidelberg (2005)

[UPG03] Urban, C., Pitts, A.M., Gabbay, M.J.: Nominal unification. In: Baaz, M., Makowsky, J.A. (eds.) CSL 2003. LNCS, vol. 2803, pp. 513–527. Springer, Heidelberg (2003)

[UPG04] Urban, C., Pitts, A.M., Gabbay, M.J.: Nominal unification. Theoretical Computer Science 323, 473–497 (2004)

# Functional-Logic Graph Parser Combinators

Steffen Mazanek and Mark Minas

Universität der Bundeswehr, München, Germany
{steffen.mazanek,mark.minas}@unibw.de

**Abstract.** Parser combinators are a popular technique among functional programmers for writing parsers. They allow the definition of parsers for string languages in a manner quite similar to BNF rules. In recent papers we have shown that the combinator approach is also beneficial for graph parsing. However, we have noted as well that certain graph languages are difficult to describe in a purely functional way.

In this paper we demonstrate that functional-logic languages can be used to conveniently implement graph parsers. Therefore, we provide a direct mapping from hyperedge replacement grammars to graph parsers. As in the string setting, our combinators closely reflect the building blocks of this grammar formalism. Finally, we show by example that our framework is strictly more powerful than hyperedge replacement grammars.

We make heavy use of key features of both the functional and the logic programming approach: Higher-order functions allow the treatment of parsers as first class citizens. Non-determinism and logical variables are beneficial for dealing with errors and incomplete information. Parsers can even be applied backwards and thus be used as generators or for graph completion.

## 1 Introduction

Declarative languages are known to be exceptionally well-suited for building string parsers. Among functional programmers, the probably most popular approach in this domain are parser combinators. Thereby, some primitive parsers are defined that can be combined into more advanced parsers using a set of powerful combinators. These combinators are higher-order functions that can be used to make parsers resemble a grammar very closely [1,2].

Parser combinators integrate seamlessly into the rest of the program, hence the full power of the host language can be used. Unlike parser generators as Yacc, no extra formalism is needed to specify a grammar. Another benefit is that parsers are first-class values within the language. For example, we can construct lists of parsers or pass them as function parameters. The possibilities are only restricted by the potential of the host language.

Due to these benefits we have started to carry over this approach to the domain of graph parsing recently [3,4]. Graph languages are widely-used nowadays, e.g., for modeling and specification. For instance, we have specified visual languages [5] using so-called hyperedge replacement grammars [6]. There, graphs

A. Voronkov (Ed.): RTA 2008, LNCS 5117, pp. 261–275, 2008.

are used as a model for diagrams and a graph parser can be used to check whether a given diagram is syntactically correct.

Hyperedge replacement grammars, HRG for short, are a well-known way of describing languages of hypergraphs, i.e., graphs where edges are allowed to visit an arbitrary number of nodes. Although restricted in power, this formalism comprises several beneficial properties: It is context-free and still quite powerful. Grammars are comprehensible, and reasonably efficient parsers can be defined for practical languages (in general parsing is NP-complete, though). In this context, rewriting means the replacement of a non-terminal hyperedge of a given hypergraph with a new hypergraph that is glued to the remaining graph by fusing particular nodes (cf. [6]).

In [4] we have discussed how HRGs can be translated to parsers using purely functional combinators. The resulting parsers indeed closely resemble the grammar. They are similar to top-down recursive descent parsers known from string parsing where non-terminal symbols are mapped to functions. In addition, the nodes actually visited by a particular non-terminal edge have to be given as function parameters to ensure the proper embedding of the graph the non-terminal is replaced by. However, these parsers suffer from an inherent problem: Inner nodes occurring in the right-hand side of a production are not known in advance, but have to be guessed in order to establish a match. It is a nontrivial task to realize this guessing efficiently in a purely functional language.

In contrast, logic languages excel at dealing with incomplete information. Free variables can be introduced that are instantiated automatically in order to find solutions. Backtracking is the default behavior and does not need to be implemented by hand. Unfortunately, purely logic languages like Prolog do not support the straightforward definition of higher-order functions like our combinators. Thus, the "remaining input" would have to be passed more explicitly resulting in a lot of boilerplate code.[1]

Having this in mind, graph parsing appears to be a domain asking for multi-paradigm declarative programming languages [8]. Those are already known to be well-suited for string parsing [9]. In the domain of graph parsing their benefits stand out even more. The functional-logic framework of graph parser combinators presented in this paper offers the following striking features:

- Straightforward translation of HRGs to reasonably efficient parsers.
- Application-specific results due to a powerful attribution concept.
- Usable context information. This allows the convenient description of several languages that cannot be defined with a HRG.
- Robust against errors. Valid subgraphs can be extracted.
- Bidirectionality. Besides syntax analysis parsers can be used to construct or complete graphs with respect to the language they describe.

[1] Prolog provides Definite Clause Grammars, syntactic sugar to hide the *difference list* mechanism needed to build efficient string parsers in logic languages. However, a graph is not linearly structured, so this notation cannot be used here. Tanaka's Definite Clause Set Grammars [7] are not supported by common Prolog systems.

This paper is organized as follows: In Sect. 2 we introduce HRGs. We continue with the presentation of an excerpt from the actual framework implemented in the functional-logic programming language Curry (Sect. 3). Thereafter, we discuss the parsing of HRGs and provide some examples (Sect. 4). Finally, we sketch the related work (Sect. 5) and conclude (Sect. 6).

## 2 Hypergraphs and HRGs

In this section we introduce hypergraphs, the notion of graphs our framework is based on, and the HRG formalism [6].

Let $C$ be a set of labels and $type : C \to \mathbb{N}$ a typing function for $C$. In the following, a hypergraph $H$ over $C$ is a finite multiset of (hyper-)edges[2] $e = (lab, ns)$, where $lab \in C$ is an edge label and $ns$ is a sequence of attachment nodes such that $type(lab) = |ns|$, the length of the sequence. The nodes in $ns$ are called *incident* to (or *visited* by) the edge $e$. The position of a particular node $n$ in $ns$ represents the so-called tentacle of $e$ that $n$ is attached to. Hence the order of nodes in $ns$ matters.

Note that our notion of hypergraphs is slightly more restrictive than the more common definition given by [6], because we cannot directly represent isolated nodes. Rather the nodes of $H$ are implicitly given as the union of all nodes incident to its edges. In fact, in many hypergraph application areas isolated nodes simply do not occur. For example, in the context of visual languages, diagram components can be represented by hyperedges, and nodes just represent their connection points, i.e., each node is visited by at least one edge [5].

Throughout this paper we use structured flowcharts as a running example, i.e., flowcharts that have a unique entry and a unique exit point. In Fig. 1a a structured flowchart is given. Here, syntax analysis means to identify the represented structured program (if any).

Flowcharts can be represented by hypergraphs that we call flowgraphs in the following. In Fig. 1b the hypergraph model of the exemplary flowchart is given. Edges are represented by a rectangular box marked with a particular label. For instance, the statement `n:=0` is mapped to an edge labeled "text". The filled black circles represent nodes that we have additionally marked with numbers. A line between an edge and a node indicates that the node is visited by that edge.

The small numbers close to the edges are the tentacle numbers representing the index of a particular node in $ns$. Without these numbers the image may be ambiguous. For instance, the tentacle with number 0 of "text" edges always has to be attached to the node the previous statement ends at whereas the tentacle 1 links the statement to its successor.

The language of flowgraphs can be described using a hyperedge replacement grammar in a straightforward way. Formally, such a HRG $G$ is a quadruple $G = (N, T, P, S)$ that consists of a set of non-terminals $N \subset C$, a set of terminals

[2] We call hyperedges just edges and hypergraphs just graphs if it is clear from the context that we are talking about hypergraphs.

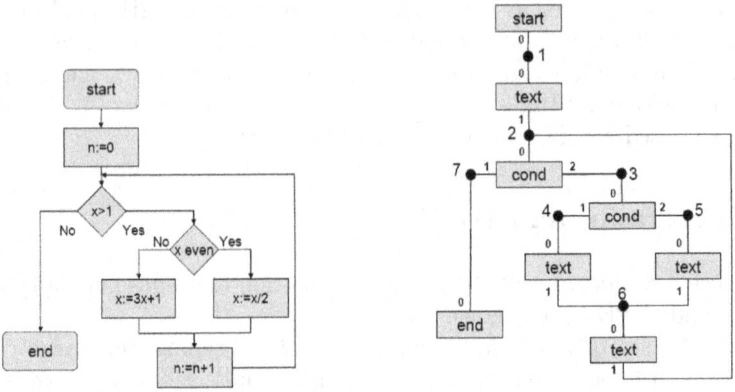


**Fig. 1.** An exemplary flowchart a) and its hypergraph representation b)

$T \subset C$ with $T \cap N = \emptyset$, a finite set of context-free productions $P$ over $N$ and a start symbol $S \in N$.

The HRG for flowgraphs then can be defined as $G_{FC} = (N_{FC}, T_{FC}, P_{FC}, FC)$ where $N_{FC} = \{FC, Stmts, Stmt\}$, $T_{FC} = \{start, end, text, cond\}$ and $P_{FC}$ contains the productions given in Fig. 2a. The notation is similar to BNF rules as known from string grammars. Nodes in a production act as variables. In order to apply a production they have to be instantiated with nodes actually occurring in the graph. We use labels to identify corresponding nodes.

As usual, a language defined by a HRG consists of all graphs whose edges are labeled only with terminal labels and that can be derived in an arbitrary number of steps from the start symbol. Given a HRG and a graph, a graph parser constructs a derivation tree of this graph with respect to the grammar. This can be done, for instance, in a way similar to the algorithm of Cocke, Younger and Kasami well-known from string parsing. How this algorithm actually can be adapted to HRGs is discussed in [10,5]. The (unique) derivation tree of the exemplary flowgraph introduced in Fig. 1b is given in Fig. 2b. Its leaves represent the terminal edges occurring in the graph whereas its inner nodes are marked with non-terminal edge labels indicating the application of a production. The direct descendants of an inner node represent the edges the non-terminal is replaced by. Thereby, the numbers in parentheses identify the nodes actually visited by the particular edge.

## 3 A Basic Combinator-Framework for Graph Parsing

We now introduce the framework as realized in the functional-logic programming language Curry[3]. As we progress, we briefly review some important aspects of Curry to make this paper self-contained.

[3] `http://www.informatik.uni-kiel.de/~curry/report.html`

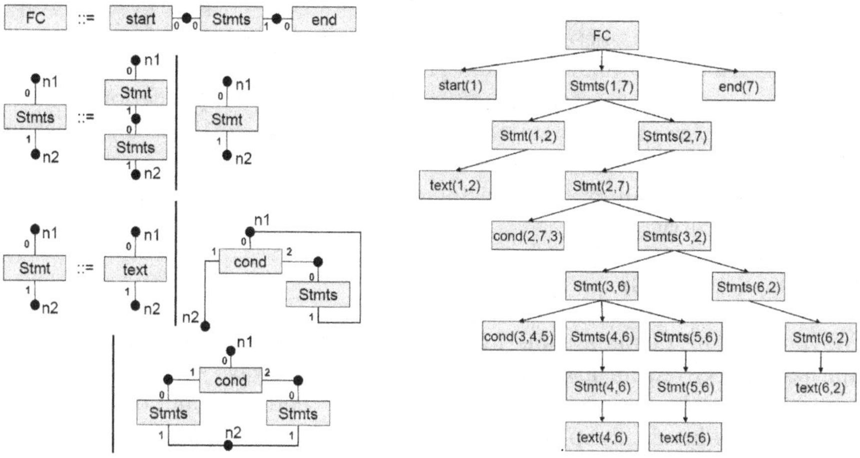


**Fig. 2.** Flowgraphs, a) grammar and b) sample derivation tree

Curry is a declarative multi-paradigm language combining interesting features from both functional and logic programming [8]. The Curry syntax is very close to Haskell[4]. The main addition are free (logic) variables in conditions and right-hand sides of defining rules. A Curry program consists of definitions of functions and data types on which these functions operate. Functions are defined by conditional equations with constraints in the conditions. They are evaluated lazily and can be called with partially instantiated arguments, a feature we make use of heavily. Function calls with free variables are evaluated by a possibly non-deterministic instantiation of the required arguments, i.e. arguments whose values are necessary to decide the applicability of a rule. This mechanism is called *narrowing* [11].

The following Curry code introduces the basic data structures for representing graphs. For the sake of simplicity, we represent nodes by integer numbers and edge labels by strings (although we do not rely on any particular type at all). Corresponding to the definition in Sect. 2 we declare a graph as a list of labeled edges each with its incident nodes. The actual order of edges does not matter.

```
type Node = Int
type Edge = (String, [Node])
type Graph = [Edge]
```

The flowgraph given in Fig. 1b can be represented as follows using the previous declarations:

```
ex = [("start",[1]),("text",[1,2]),("cond",[2,7,3]),("cond",[3,4,5]),
      ("text",[4,6]),("text",[5,6]),("text",[6,2]),("end",[7])]
```

[4] http://www.haskell.org/onlinereport/

Next, we provide the declaration of the type Grappa representing a graph parser. This type is parameterized over the type res of the result. Graph parsers are (non-deterministic) functions from graphs to pairs consisting of the parsing result and the graph that remains after successful parser application. In contrast to Haskell, we do not have to deal with parsing errors and backtracking explicitly (no need for "lists of successes"). Instead, similar to [9], we rely on the non-deterministic notion of functions inherent to functional-logic programming languages like Curry.

```
type Grappa res = Graph -> (res, Graph)
```

We proceed by defining some important primitives for the construction of graph parsers. Given an arbitrary value, pSucceed always succeeds returning this particular value as a result. In contrast, eoi (end of input) only succeeds if the graph is already completely consumed. In this case, as a result we simply return (), the only value of the so-called unit type. Note that in Curry it does not need to be stated explicitly that eoi fails on non-empty input – the absence of a rule is enough.

```
pSucceed::res -> Grappa res
pSucceed v g = (v, g)
```

```
eoi::Grappa ()
eoi [] = ((), [])
```

An especially important primitive parser is edge. It only succeeds if the given edge e is part of the particular graph g. It is implemented in a logic programming style making use of an equational constraint indicated by =:=.

```
edge::Edge -> Grappa ()
edge e g | g=:=(g1++e:g2) = ((), g1++g2)
           where g1, g2 free
```

A constraint $e_1$ =:= $e_2$ is satisfiable if both sides $e_1$ and $e_2$ are reducible to unifiable terms. Here, this means that the edge e indeed is contained in the graph g. In this case, the edge has to be consumed. This is realized by returning just g1++g2 as the remaining graph.[5] Note that, in contrast to Prolog, free variables like g1 and g2 need to be declared explicitly (to make their scopes clear).

In Fig. 3 we provide some important parser combinators. They are defined in a fairly standard way (cf., e.g., [2,9,12]). The choice operator <|> takes two parsers and succeeds if either the first or the second one succeeds. In fact, it is a special case of the standard Curry operator (?)::a->a->a. Two parsers can also be combined via <*>, the successive application where the result is constructed by function application (as in [12]).[6] The second parser thereby starts with the input the first parser has left. For convenience we also define *> and <* that

[5] (++) is the standard operator for list concatenation. In contrast, (:) is the list constructor that can be used to add a single element to the front of a list.

[6] In previous versions of the framework [3,4] we have composed parsers using monads to make use of context. This does not seem to be necessary with the functional-logic approach as we see later. In fact, type classes are not supported in Curry yet.

```
(<|>)::Grappa res -> Grappa res -> Grappa res
p1 <|> _  = p1
_  <|> p2 = p2

(<*>)::Grappa (res1->res2) -> Grappa res1 -> Grappa res2
(p1 <*> p2) g = case p1 g of
                (pv, g') -> case p2 g' of
                            (qv, g'') -> (pv qv, g'')

(<*)::Grappa res1 -> Grappa res2 -> Grappa res1
p <* q = (\x _ -> x) <$> p <*> q
(*>)::Grappa res1 -> Grappa res2 -> Grappa res2
p *> q = (\_ x -> x) <$> p <*> q

(<$>)::(res1->res2) -> Grappa res1 -> Grappa res2
f <$> p = pSucceed f <*> p
(<$)::res1 -> Grappa res2 -> Grappa res1
f <$ p = const f <$> p
```

**Fig. 3.** Standard parser combinators

throw away one of the results. Finally, the parser transformers `<$>` and `<$` can be used to either apply a function to the result of a parser or to just replace it by another value.

On top of these basic combinators we can define various other useful combinators. For instance, we provide the combinator `many` to deal with simple repetition (the graph equivalent to the Kleene star) as:

```
many::Grappa a -> Grappa [a]
many p = pSucceed []
many p = (:) <$> p <*> many p
```

This definition can be read as: "`many p` always succeeds returning nothing (`[]`). It may also succeed by applying `p`, and thereafter `many p` again. In this case their results are combined using the list constructor `(:)`." Note, however, that this definition causes a lot of backtracking. In the string setting a combinator for simple repetition normally returns $n+1$ different results where $n$ is the number of successive occurrences of $p$ **at the beginning of the string**. If a graph contains $n$ occurrences of $p$, altogether $\sum_{i=0}^{n}\binom{n}{i}i!$ results are possible, since any number of occurrences can be chosen in any order. It is possible to disregard "redundant" results by using encapsulated search, but this way we lose some nice properties of our parsers. The problem with `many` is not inherent to our graph parsing approach, though. In fact, `many` is only needed to parse HR languages, which contain either highly disconnected graphs, or graphs which have vertices with high degree – both properties are known to be indicators for high parsing complexity [10].

We provide another typical combinator that we need later. The combinator `chain1Betw p (n1,n2)` can be used to identify a non-empty chain of graphs that can be parsed with `p`. This chain has to be anchored between the nodes `n1` and `n2`. Later we also need a parser `exactChain1Betw` that forces this chain to be of a particular length. We omit its declaration, since it can be defined very similar to `chain1Betw`:[7]

```
chain1Betw::((Node,Node)->Grappa a) -> (Node,Node) -> Grappa [a]
chain1Betw p (n1,n2) = (:[]) <$> p (n1,n2)
chain1Betw p (n1,n2) = (:)   <$> p (n1,n) <*> chain1Betw p (n,n2)
                       where n free
```

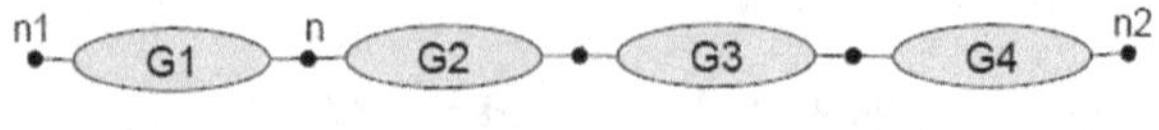


`chain1Betw` can be conveniently defined, because we do not need to know the inner node `n` in advance. We simply define it as a free variable, which can be instantiated according to the Curry narrowing semantics. Representing graph nodes as free variables actually is a functional-logic design pattern [13] that we here exploit in a novel way.

## 4 Parsing of HRGs

In this section we provide a direct mapping from HRGs to parsers based on the previously introduced framework. We exemplify the translation by means of the grammar given in Fig. 2a. We further provide some additional examples to demonstrate interesting properties of our parsers.

In Fig. 4 the parser for flowgraphs is presented. The type annotations are just for convenience and can also be omitted. For each non-terminal edge label $l$ we have defined a parser function that takes a tuple of nodes $(n_1, ..., n_k)$ as a parameter such that $k = \mathit{type}(l)$. For each production over $l$ we insert a new function body. Each terminal edge in the right-hand side of the production is matched and consumed using the primitive parser `edge`, each non-terminal one is translated to a call of the function representing this non-terminal. A free variable is introduced for each inner node of a production.

In contrast to string parsing the order of parsers in a successive composition via `*>` is not that important as long as left recursion is avoided. Nevertheless, the chosen arrangement might have an impact on the performance. Usually, it is advisable to deal with the terminal edges first.

Parsers defined in such a way are quite robust. For instance, they ignore redundant components, i.e., those just remain at the end. However, complete input consumption can be enforced easily by a subsequent application of `eoi`. Thus, instead of `fc` we can use the extended parser `fc <* eoi`.

[7] Actually, `chain1Betw = exactChain1Betw k where k free`, i.e., we can also define `chain1Betw` in terms of `exactChain1Betw`.

```
fc::Grappa ()
fc = edge ("start", [sn]) *> stmts (sn,en) *> edge ("end", [en])
     where sn, en free

stmts::(Node,Node) -> Grappa ()
stmts (n1,n2) = stmt (n1,n2)
stmts (n1,n2) = stmt (n1,n) *>
                stmts (n,n2)
                where n free

stmt::(Node,Node) -> Grappa ()
stmt (n1,n2) = edge ("text", [n1,n2])
stmt (n1,n2) = edge ("cond", [n1,nno,nyes]) *>
               stmts (nno,n2) *> stmts (nyes,n2)
               where nno, nyes free
stmt (n1,n2) = edge ("cond", [n1,n2,nbody]) *>
               stmts (nbody,n1)
               where nbody free
```

**Fig. 4.** A parser for flowgraphs

One problem is still left with our flowgraph parser: As it is, it accepts too many graphs. Further conditions have to be enforced to ensure correctness [6]:

- identification condition: matches have to be injective, i.e., involved nodes have to be pairwise distinct.
- dangling edge condition: there must not be other edges in the remaining graph visiting inner nodes of a match.

For instance, the flowgraphs shown in Fig. 5 can also be parsed successfully with the parser given in Fig. 4 although they are no members of the language defined by $G_{FC}$. In the context of visual languages it often is convenient to relax the dangling edge condition (cf. [5]). This allows for easier specifications. However, from a theoretical point of view, this is not satisfactory. In fact, both conditions can be ensured by additional checks. For instance, we can use inequality constraints on node variables to ensure that they are pairwise distinct, i.e., that a particular match is injective. However, these constraints cannot be globally set, but rather have to be added to the parsers for every single production making them less readable.

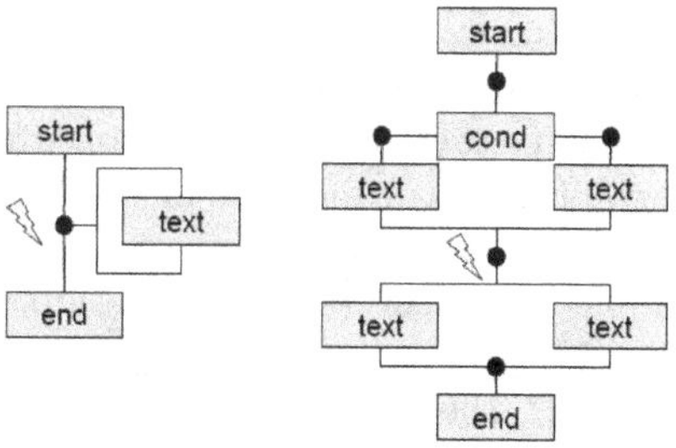


**Fig. 5.** Violation of identification condition and dangling edge condition

**Semantics**

So far we only have checked if the given graph is, or at least contains, a valid flowgraph. However, a major benefit of the combinator approach is that language-specific results can be computed in a flexible way [2,14]. Say, we want to map a flowgraph to its underlying program represented by the recursively defined type Program:

```
type Program = [Stmt]
data Stmt = Text | IfElse Program Program | While Program
```

We do not provide the complete mapping here. Rather we use the translation of the branching production as an example to show how easily parsers can be enriched with attribution.

```
stmt::(Node,Node) -> Grappa Stmt
stmt (n1,n2) = edge ("cond", [n1,nno,nyes]) *>
               IfElse <$> stmts (nno,n2) <*> stmts (nyes,n2)
               where nno, nyes free
```

The result type has to be changed to `Stmt`. Further, we can use the combinator `<$>` to directly construct a statement from the two subprograms. Note that in Curry (as in Haskell) the data constructor `IfElse` is implicitly typed `Program->Program->Stmt`. In this situation we can make the parser definition even more concise, because `stmts` now really is just `chain1Betw stmt`.

### Not just a Parser

Parsing is not the only thing we can do with these functions. We can also apply them backwards to construct graphs of the language. For instance, we can enumerate all graphs in the language up to a particular size. As a result we know that there are only 2 flowgraphs (up to isomorphism) of size 4, 6 of size 5 and 21 of size 6.

We can further use the parser to perform a kind of auto-completion. Say, the edge `text(1,2)` in the graph given in Fig. 1b is missing, such that the flowgraph is not a member of the language anymore. We can try inserting an edge `e` as a free variable and see how `e` is instantiated by the parser. For our example we get several possible completions. For instance, we could add an edge `start(2)`. However, there is only one completion that consumes the whole input: `text(1,2)`, the one we deleted.

This approach could be the starting point for the realization of advanced error correction for graphs. For the error correction of strings a sophisticated and powerful Haskell parser combinator framework has already been proposed [12]. However, a functional-logic approach may be more understandable and easier to adapt to graphs. Such graph completion could be very useful in order to realize powerful content assist for graph grammar based diagram editors like the ones generated with DiaGen [5].

In certain cases we can also perform the mapping of semantics back to a graph, e.g., given a particular program we can construct a corresponding flowgraph. Thereby, nodes are not instantiated, but left as free variables. Indeed the particular node numbers do not matter as long as equal nodes can be identified.

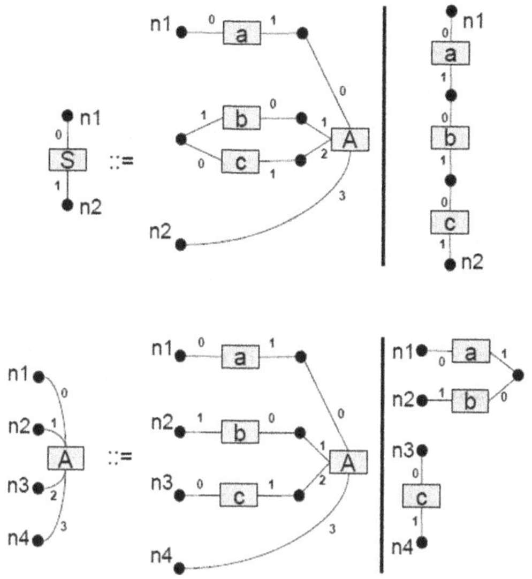

```
s (n1,n2) =
  edge ("a", [n1,n3]) *>
  edge ("b", [n4,n5]) *>
  edge ("c", [n5,n6]) *>
  (+1) <$> a (n3,n4,n6,n2)
  where n3, n4, n5, n6 free
s (n1,n2) =
  edge ("a", [n1,n3]) *>
  edge ("b", [n3,n4]) *>
  edge ("c", [n4,n2]) *>
  pSucceed 1
  where n3, n4 free
a (n1,n2,n3,n4) =
  edge ("a", [n1,n5]) *>
  edge ("b", [n6,n2]) *>
  edge ("c", [n3,n7]) *>
  (+1) <$> a (n5,n6,n7,n4)
  where n5, n6, n7 free
a (n1,n2,n3,n4) =
  edge ("a", [n1,n5]) *>
  edge ("b", [n5,n2]) *>
  edge ("c", [n3,n4]) *>
  pSucceed 1
  where n5 free
```

**Fig. 6.** Graph grammar a) and corresponding parser b) for the graph language $a^k b^k c^k$

## Another Example: $a^k b^k c^k$

We give another example to demonstrate that the readability of a language description can also be improved by using graph parser combinators.

In a string setting the language $\{a^k b^k c^k | k > 0\}$ is not context-free. In contrast, there is a context-free string generating hypergraph grammar that defines a corresponding graph language [6,15]. A hypergraph grammar is string generating, if all graphs in its language have a linear structure, i.e., are a chain of directed edges; below we provide the graph representation of the string "aabbcc" as an example.

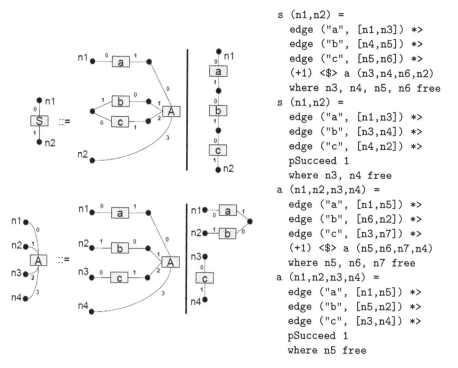

The grammar for the graph language $a^k b^k c^k$ as introduced in [6] is given in Fig. 6a. It is quite complex and hard to grasp despite the structural simplicity of the language. In Fig. 6b a nearly straightforward translation of this grammar to a parser is given. We only have added some attribution to compute the particular value of $k$.

With our framework much more readable descriptions are possible. One of them is given below. Basically it states that there have to be two nodes `n3` and `n4` and a number $k > 0$ such that there is a chain of $k$ "a"-edges between `n1` and n3, a chain of $k$ "b"-edges between n3 and n4 and finally a chain of $k$ "c"-edges between n4 and n2.[8]

```
abc (n1,n2) = exactChain1Betw k (dirEdge "a") (n1,n3) *>
              exactChain1Betw k (dirEdge "b") (n3,n4) *>
              exactChain1Betw k (dirEdge "c") (n4,n2) *>
              pSucceed k
              where k, n3, n4 free
```

The crucial point of this solution is that logic variables like $k$ can be used to share results across parsers. This way context information can be exploited. This approach enables us to not only describe languages more conveniently, but also to describe languages that cannot be defined with a HRG at all. For instance, there is no HRG for Sierpinski triangles that are regular, i.e. equally deep unfolded.[9] In contrast, this can easily be done in our system just by introducing an additional parameter `depth`.

**Performance**

The operational semantics of Curry is based on an optimal evaluation strategy. As an extension to lazy functional programming its behavior is demand-driven. Thus, it ensures optimal evaluation on well-defined classes of programs [11]. However, in [6] it is proved that there are (even context-free) graph languages where parsing is NP-complete. These languages, of course, cannot be parsed with our combinators efficiently. Most practically relevant languages, however, are quite efficient to parse.

To give an impression we provide some performance data for the language $a^k b^k c^k$ in Fig. 7. The measurement has been executed on standard hardware using the Münster Curry Compiler[10]. We see that the more readable parser `abc` is even more efficient. Both parsers have a polynomial runtime behavior (mainly because the first node of the string graph is not known in advance). Since our parsers follow a top-down approach with backtracking we cannot completely avoid that partial results are computed more than once. Bottom-up parsers, which exploit dynamic programming techniques are usually more efficient.

Note, however, that the presented framework has not been optimized with respect to performance. For instance, a more efficient graph representations could be used, e.g., as a mapping `String->[[Node]]` so that all edges with a particular label can be queried much faster.

A good thing with parser combinators is that performance optimizations specific to the particular graph language can be incorporated easily. For instance, a basic improvement for flowgraphs would be to first decompose the given graph

[8] We make use of the primitive `dirEdge lab (n1,n2) = edge (lab, [n1,n2])` to make the combinator `exactChain1Betw` directly applicable.

[9] However, this can be realized with a special kind of parallel replacement mechanism as described in [16].

[10] `http://danae.uni-muenster.de/~lux/curry/`

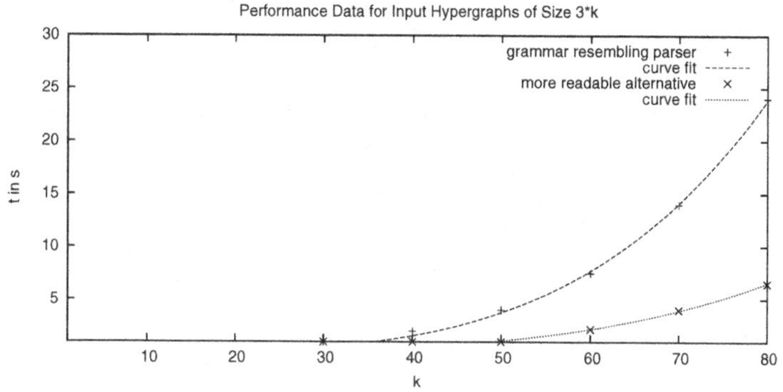


**Fig. 7.** Performance comparison of both parsers for the language $a^k b^k c^k$

into connected components and apply the parser to each of them successively. We provide the combinator `connComp::Grappa a->Grappa [a]` for this task. So, for a broad range of languages, we can start with a parser, which is easy to build and read, and which can be further improved if necessary. Providing additional information also can boost performance, e.g., if we know the first node in our language of string graphs, both parsers need less than a second for $k = 80$.

And even backwardly applied as generators our parsers are reasonably efficient. Compared to other graph transformation tools [17] they seem to be in the center-field. For instance, we have generated a Sierpinski triangle of generation 11 with nearly 200.000 edges in about a minute.

## 5 Related Work

In principle our graph parser combinators are quite similar to conventional string parser combinators like [2,14,9] to name just a few. Many ideas can be carried over straightforwardly. The main differences emerge from the non-linear structure of graphs and the appearance of nodes as connection points between tokens.

From all parser combinator approaches the UU library [12] is special in the sense that it probably provides the most powerful mechanisms to correct all kinds of errors in strings. There, a parser does never fail, but rather constructs a minimal sequence of correction steps. We have shown how our library can be used for restricted kinds of error handling. Redundant edges, for instance, may just remain at the end. This is already quite powerful, since in contrast to strings graphs are sets of components, i.e., there is no particular order imposed. Thus, it does not matter where the redundant components are placed. Furthermore, due to its logic nature, we can conveniently deal with errors that are fixable by embedding additional edges. However, other correction actions like edge relabeling or the gluing of distinct nodes cannot be computed in such a convenient manner.

Another interesting related observation is that parsing of visual languages can be modeled (and even executed) in linear logic [18], a resource-oriented refinement of classical logic. For instance, in [19] the embedding of constraint multiset grammars into linear logic is discussed. However, it seems that hypergraph parsing can be modeled even more straightforwardly. Here, the edges of a hypergraph can be mapped to facts that can be fed into a parser via so-called linear implication ($\multimap$). During the proof the parser consumes these facts and at the end none of them must be left. For instance, to parse a simple flowgraph using Lolli [20] we can write `(start 1, text 1 2, end 2) -o fc`. The combinators presented in this paper also hide the remaining resources from the user. But their major benefit is the flexibility they can be applied with.

Also related are approaches to parsing of particular, restricted kinds of graph grammar formalisms. For instance, in [15] an Earley parser for string generating graph languages has been proposed. The diagram editor generator DiaGen [5] incorporates an HRG parser that is an adaptation of the algorithm of Cocke, Younger and Kasami. And the Visual Language Compiler-Compiler VLCC [21] is based on the methodology of positional grammars that allows to parse restricted kinds of flex grammars (which are essentially HRGs) even in linear time. These approaches have in common that a restricted graph grammar formalism can be parsed efficiently. However, they cannot be generalized straightforwardly to a broader range of languages like our combinators.

## 6 Conclusion

In this paper we have discussed functional-logic graph parser combinators, an extensible framework supporting the flexible construction of special-purpose graph parsers even for (some) context-sensitive graph languages. It has turned out that functional-logic languages are exceptionally well-suited for graph parsing.

In particular we have demonstrated that hyperedge replacement grammars can be mapped to parsers straightforwardly. We have also noted that these grammars are sometimes not the most readable means to describe graph languages; several graph languages cannot even be defined with a HRG at all. In contrast, using our framework we can easily define readable parsers for languages like the string graphs $a^k b^k c^k$ or regular Sierpinski triangles. The resulting parsers are sufficiently efficient for many practical graph languages.

Functional-logic parsers can also be applied backwards. This way they can be used to enumerate a graph language or for graph-completion. Since graphs are well-suited as a model for visual languages, such graph-completion can be very beneficial in the context of diagram editors. We plan to connect our framework with the diagram editor generator DiaGen [5] to provide powerful content assist to the user.

## References

1. Hutton, G.: Higher-order functions for parsing. Journal of Functional Programming 2(3), 323–343 (1992)
2. Fokker, J.: Functional parsers. In: Advanced Functional Programming, First Intl. Spring School on Advanced Functional Programming Techniques-Tutorial Text, London, UK, pp. 1–23. Springer, Heidelberg (1995)

3. Mazanek, S., Minas, M.: Graph parser combinators. In: Proc. of 19th Intl. Symp. on the Impl. and Appl. of Functional Languages. Springer, Heidelberg (2008)
4. Mazanek, S., Minas, M.: Parsing of hyperedge replacement grammars with graph parser combinators. In: Proc. of 7th Intl. Workshop on Graph Transf. and Visual Modeling Techniques. Electronic Communications of the EASST (to appear, 2008)
5. Minas, M.: Concepts and realization of a diagram editor generator based on hypergraph transformation. Science of Computer Programming 44(2), 157–180 (2002)
6. Drewes, F., Habel, A., Kreowski, H.J.: Hyperedge replacement graph grammars. In: Rozenberg, G. (ed.) Handbook of Graph Grammars and Computing by Graph Transformation. Foundations, vol. I, pp. 95–162. World Scientific, Singapore (1997)
7. Tanaka, T.: Definite-clause set grammars: a formalism for problem solving. J. Log. Program. 10(1), 1–17 (1991)
8. Hanus, M.: Multi-paradigm declarative languages. In: Dahl, V., Niemelä, I. (eds.) ICLP 2007. LNCS, vol. 4670, pp. 45–75. Springer, Heidelberg (2007)
9. Caballero, R., López-Fraguas, F.J.: A functional-logic perspective of parsing. In: Middeldorp, A. (ed.) FLOPS 1999. LNCS, vol. 1722, pp. 85–99. Springer, Heidelberg (1999)
10. Lautemann, C.: The complexity of graph languages generated by hyperedge replacement. Acta Inf. 27(5), 399–421 (1989)
11. Antoy, S., Echahed, R., Hanus, M.: A needed narrowing strategy. J. ACM 47(4), 776–822 (2000)
12. Swierstra, S.D., Azero Alcocer, P.R.: Fast, error correcting parser combinators: a short tutorial. In: Pavelka, J., Tel, G., Bartosek, M. (eds.) SOFSEM 1999. LNCS, vol. 1725, pp. 111–129. Springer, Heidelberg (1999)
13. Antoy, S., Hanus, M.: Functional logic design patterns. In: Proc. of the 6th Intl. Symposium on Functional and Logic Programming. LNCS, vol. 2441, pp. 67–87. Springer, Heidelberg (2002)
14. Hutton, G., Meijer, E.: Monadic parser combinators. Technical Report NOTTCS-TR-96-4, Department of Computer Science, University of Nottingham (1996)
15. Seifert, S., Fischer, I.: Parsing string generating hypergraph grammars. In: Ehrig, H., Engels, G., Parisi-Presicce, F., Rozenberg, G. (eds.) ICGT 2004. LNCS, vol. 3256, pp. 352–367. Springer, Heidelberg (2004)
16. Habel, A., Kreowski, H.J.: Pretty patterns produced by hyperedge replacement. In: Göttler, H., Schneider, H.-J. (eds.) WG 1987. LNCS, vol. 314, pp. 32–45. Springer, Heidelberg (1988)
17. Taentzer, G., et al.: Generation of sierpinski triangles: A case study for graph transformation tools. In: Proc. of AGTIVE 2007. LNCS. Springer, Heidelberg (2008)
18. Girard, J.Y.: Linear logic. Theoretical Computer Science 50, 1–102 (1987)
19. Bottoni, P., Meyer, B., Marriott, K., Parisi-Presicce, F.: Deductive parsing of visual languages. In: Proc. of the 4th Intl. Conference on Logical Aspects of Computational Linguistics, London, UK, pp. 79–94. Springer, Heidelberg (2001)
20. Hodas, J.S., Miller, D.: Logic programming in a fragment of intuitionistic linear logic. Inf. Comput. 110(2), 327–365 (1994)
21. Costagliola, G., Lucia, A.D., Orefice, S., Tortora, G.: A parsing methodology for the implementation of visual systems. IEEE Trans. Softw. Eng. 23(12), 777–799 (1997)

# Proving Quadratic Derivational Complexities Using Context Dependent Interpretations*

Georg Moser and Andreas Schnabl

Institute of Computer Science
University of Innsbruck, Austria
{georg.moser,andreas.schnabl}@uibk.ac.at

**Abstract.** In this paper we study *context dependent interpretations*, a semantic termination method extending interpretations over the natural numbers, introduced by Hofbauer. We present two subclasses of context dependent interpretations and establish tight upper bounds on the induced derivational complexities. In particular we delineate a class of interpretations that induces *quadratic derivational complexity*. Furthermore, we present an algorithm for mechanically proving termination of rewrite systems with context dependent interpretations. This algorithm has been implemented and we present ample numerical data for the assessment of the viability of the method.

## 1 Introduction

In order to assess the complexity of a (terminating) term rewrite system (TRS for short) it is natural to look at the maximal length of derivation sequences, as suggested by Hofbauer and Lautemann in [1]. To be precise, let $\mathcal{R}$ denote a finitely branching and terminating TRS over a finite signature. The *derivational complexity function* with respect to $\mathcal{R}$ (denoted as $\mathsf{dc}_{\mathcal{R}}$) relates the length of the longest derivation sequence to the size of the initial term. For *direct* termination techniques it is often possible to infer an upper bound on $\mathsf{dc}_{\mathcal{R}}(n)$ from the termination proof of $\mathcal{R}$, cf. [1,2,3,4,5]. (Currently it is unknown how to estimate the *derivational complexity* of a TRS $\mathcal{R}$, if termination of $\mathcal{R}$ has been shown via transformation methods like the dependency pair method or semantic labelling, but see [4,6] for partial results in this direction.) For example *linear* derivational complexity can be verified by the use of automata techniques: linear match-bounded TRSs induce linear derivational complexity, see [5]. Unfortunately such a feasible growth rate is not typical. Already termination proofs by polynomial interpretations imply a double-exponential upper bound on the derivational complexity, cf. [1]. In both cases the upper bounds are tight.

However, the tightness of the mentioned bounds does not imply that the upper bounds are *always* optimal. In particular polynomial interpretations typically overestimate the derivational complexity. In [7] Hofbauer introduced so-called *context dependent interpretations* as a remedy. These interpretations extend traditional interpretations by introducing an additional parameter. The parameter

* This research is supported by FWF (Austrian Science Fund) project P20133.

A. Voronkov (Ed.): RTA 2008, LNCS 5117, pp. 276–290, 2008.

changes in the course of evaluating a term, which makes the interpretation dependent on the context. The crucial advantage is that context dependent interpretations typically improve the induced bounds on the derivational complexity of TRSs. Furthermore this technique allows the handling of non-simple terminating systems. (See [7] and Section 2 for further details.)

In this paper, we establish theoretical and practical extensions of Hofbauer's approach. As theoretic contributions, we present two subclasses of context dependent interpretations, i.e., we introduce *$\Delta$-linear* and *$\Delta$-restricted interpretations.* We show that $\Delta$-linear interpretations induce *exponential* derivational complexity, while $\Delta$-restricted interpretations induce *quadratic* derivational complexity. Furthermore, we provide examples showing that these bounds are tight. In [7] it is shown that context dependent interpretations are expressive enough to show termination of TRSs that are not simply terminating. We improve upon this and show that $\Delta$-restricted interpretations suffice here. On the practical side, we design an algorithm that automatically searches for $\Delta$-linear interpretations and $\Delta$-restricted interpretations, which shows that the technique can be mechanised. This answers a question posed by Hofbauer in [7]. The procedure has been implemented and we provide ample numerical data to assess its viability. TRSs with *polynomial* derivational complexity appear to be of special interest. Thus, we finally compare the applicability of our method to other termination techniques that also induce polynomial derivational complexity.

The remainder of this paper is organised as follows. In the next section we recall basic notions and starting points of this paper. In Section 3 we introduce the class of $\Delta$-linear interpretations and describe the algorithm that mechanises the search for $\Delta$-linear and $\Delta$-restricted interpretations. In Section 4, we obtain the mentioned results on the derivational complexities induced by either of these interpretations. Furthermore, we show in this section that already $\Delta$-restricted interpretations allow the treatment of non-simple terminating TRSs. Section 5 provides experimental data and finally in Section 6 we conclude and mention future work.

## 2 Context Dependent Interpretations

We assume familiarity with the basics of term rewriting, see [8,9]. Knowledge of context dependent interpretations [7] will be helpful. Below we recall the basic results from the latter paper in a slightly different, but equivalent way, compare [7,10]. See [7] for the motivation and intuition underlying the introduced concepts.

Let $\mathcal{F}$ be a finite signature, let $\mathcal{V}$ be a set of variables and let $\mathcal{R}$ denote a terminating TRS over $\mathcal{F}$. The induced relation $\rightarrow_{\mathcal{R}}$ is assumed to be finitely branching. We simply write $\rightarrow$ for $\rightarrow_{\mathcal{R}}$ if $\mathcal{R}$ is clear from context. The *derivation length* of a term $t$ with respect to $\mathcal{R}$ is defined as follows: $\mathsf{dl}_{\mathcal{R}}(t) = \max\{n \mid \exists u\ t \rightarrow^n u\}$. The *derivational complexity* (with respect to $\mathcal{R}$) is defined as: $\mathsf{dc}_{\mathcal{R}}(n) = \max\{\mathsf{dl}_{\mathcal{R}}(t) \mid |t| \leqslant n\}$, where $|t|$ denotes the size of $t$, i.e., the number of symbols of $t$ as usual. (For example the size of the term $\mathsf{f}(\mathsf{a}, x)$ is 3.) We say the derivational complexity of $\mathcal{R}$ is linear, quadratic, double-exponential, if $\mathsf{dc}_{\mathcal{R}}(n)$ is bounded by a linear, quadratic, double-exponential function in $n$, respectively. A *context dependent $\mathcal{F}$-algebra* (*CDA* for short) $\mathcal{C}$ is a family of $\mathcal{F}$-algebras over the reals parametrised by a set $D \subseteq \mathbb{R}^+$ of positive reals. A CDA $\mathcal{C}$ associates to each function symbol $f \in \mathcal{F}$ of arity $n$, a collection of $n+1$ mappings: $f_{\mathcal{C}}\colon D \times (\mathbb{R}_0^+)^n \to \mathbb{R}_0^+$ and $f_{\mathcal{C}}^i\colon D \to D$ for all $1 \leqslant i \leqslant n$. As usual $f_{\mathcal{C}}$ is called *interpretation function*, while the mappings $f_{\mathcal{C}}^i$ are called *parameter functions*. In addition $\mathcal{C}$ is equipped with a set $\{>_{\Delta} \mid \Delta \in D\}$ of proper orders, where we define: $z >_{\Delta} z'$ if and only if $z - z' \geqslant \Delta$.

Let $\mathcal{C}$ be a CDA and let a *$\Delta$-assignment* denote a mapping: $\alpha : D \times \mathcal{V} \to \mathbb{R}_0^+$. We inductively define a mapping $[\alpha, \Delta]_{\mathcal{C}}$ from the set of terms into the set $\mathbb{R}_0^+$ of non-negative reals:

$$[\alpha, \Delta]_{\mathcal{C}}(t) := \begin{cases} \alpha(\Delta, t) & \text{if } t \in \mathcal{V} \\ f_{\mathcal{C}}(\Delta, [\alpha, f_{\mathcal{C}}^1(\Delta)]_{\mathcal{C}}(t_1), \dots, [\alpha, f_{\mathcal{C}}^n(\Delta)]_{\mathcal{C}}(t_n)) & \text{if } t = f(t_1, \dots, t_n) . \end{cases}$$

We fix some notational conventions: Due to the special role of the additional variable $\Delta$, we often write $f_{\mathcal{C}}[\Delta](z_1, \dots, z_n)$ instead of $f_{\mathcal{C}}(\Delta, z_1, \dots, z_n)$. Furthermore, we usually denote the evaluation of $t$ as $[\alpha, \Delta](t)$, if the respective algebra is clear from context.

We say that a CDA $\mathcal{C}$ is *$\Delta$-monotone* if for all $\Delta \in D$ and for all $a_1, \dots, a_n, b \in \mathbb{R}_0^+$ with $a_i >_{f_{\mathcal{C}}^i(\Delta)} b$ for some $i \in \{1, \dots, n\}$, we have

$$f_{\mathcal{C}}[\Delta](a_1, \dots, a_i, \dots, a_n) >_{\Delta} f_{\mathcal{C}}[\Delta](a_1, \dots, b, \dots, a_n) .$$

Note that if all interpretation functions $f_{\mathcal{C}}[\Delta]$ are weakly monotone with respect to the standard ordering on $\mathbb{R}_0^+$, then validity of the inequalities

$$f_{\mathcal{C}}[\Delta](z_1, \dots, z_i + f_{\mathcal{C}}^i(\Delta), \dots, z_n) - f_{\mathcal{C}}[\Delta](z_1, \dots, z_i, \dots, z_n) \geqslant \Delta ,$$

suffices in order to conclude $\Delta$-monotonicity of $\mathcal{C}$, cf. [7].

A CDA $\mathcal{C}$ is *compatible* with a TRS $\mathcal{R}$ (or $\mathcal{R}$ is compatible with $\mathcal{C}$) if for every rewrite rule $l \to r \in \mathcal{R}$, every $\Delta \in D$, and any assignment $\alpha$: $[\alpha, \Delta](l) >_{\Delta} [\alpha, \Delta](r)$ holds.

*Example 1 ([7]).* As running example, we consider the TRS $\mathcal{R}_1$ with the single rewrite rule $\mathsf{a}(\mathsf{b}(x)) \to \mathsf{b}(\mathsf{a}(x))$. We assume $D = \mathbb{R}^+$. The following interpretation and parameter functions

$$\mathsf{a}_{\mathcal{C}}[\Delta](z) = (1 + \Delta)z \qquad \mathsf{a}_{\mathcal{C}}^1(\Delta) = \frac{\Delta}{1 + \Delta}$$

$$\mathsf{b}_{\mathcal{C}}[\Delta](z) = z + 1 \qquad \mathsf{b}_{\mathcal{C}}^1(\Delta) = \Delta ,$$

define a CDA $\mathcal{C}$ that is $\Delta$-monotone and compatible with $\mathcal{R}_1$, compare [7].

**Theorem 2 ([7]).** *Let $\mathcal{R}$ be a TRS and suppose that there exists a $\Delta$-monotone and compatible CDA $\mathcal{C}$. Then $\mathcal{R}$ is terminating and*

$$\mathsf{dl}_{\mathcal{R}}(t) \leqslant \inf_{\Delta \in D} \frac{[\alpha, \Delta](t)}{\Delta} \tag{1}$$

*holds for all terms $t \in \mathcal{T}(\mathcal{F}, \mathcal{V})$.*

The next example clarifies the impact of Theorem 2, compare [7].

*Example 3.* Consider the TRS $\mathcal{R}_1$ together with the CDA $\mathcal{C}$ in Example 1. Suppose $\mathsf{c} \in \mathcal{F}$ is a constant and $\mathsf{c}_{\mathcal{C}}[\Delta] = 0$. We assert $D = \mathbb{R}^+$. Then we obtain $[\alpha, \Delta](\mathsf{a}^n(\mathsf{b}^m(\mathsf{c}))) = (1 + \Delta n)m$ and hence:

$$\inf_{\Delta>0} \frac{[\alpha, \Delta](\mathsf{a}^n(\mathsf{b}^m(\mathsf{c})))}{\Delta} = \inf_{\Delta>0} (\frac{1}{\Delta} + n)m = nm \geqslant \mathsf{dl}_{\mathcal{R}_1}(\mathsf{a}^n(\mathsf{b}^m(\mathsf{c}))) .$$

Furthermore, an easy inductive argument reveals: $\mathsf{dl}_{\mathcal{R}_1}(\mathsf{a}^n(\mathsf{b}^m(\mathsf{c}))) = nm$. Hence with respect to the term $\mathsf{a}^n(\mathsf{b}^m(\mathsf{c}))$, compatibility with $\mathcal{C}$ entails an optimal upper bound on the derivation length of $\mathcal{R}_1$. This is also true for all ground terms. A proof of $\inf_{\Delta>0} \frac{[\alpha,\Delta](t)}{\Delta} = \mathsf{dl}_{\mathcal{R}_1}(t)$ for all $t \in \mathcal{T}(\mathcal{F})$ can be found in [7].

**Definition 4.** *A $\Delta$-*quotient *is an expression of the form*

$$\frac{\Delta}{a + b\Delta} ,$$

*where $a, b \in \mathbb{N}$ and either $a > 0$ or $b > 0$. A $\Delta$-quotient $d$ is* nontrivial, *if $d \neq \Delta$.*

**Lemma 5.** *Let $d_1$, $d_2$ be $\Delta$-quotients and let $d = d_1[\Delta := d_2]$ denote the result of substituting $d_2$ for $\Delta$ in $d_1$. Then $d$ is a $\Delta$-quotient.*

As usual a *polynomial* $P$ in the variables $z_1, \ldots, z_n$ (over the reals) is a finite sum $\sum_{i=1}^{m} c_i z_1^{i_1} \ldots z_n^{i_n}$. To accommodate $\Delta$-quotients we slightly generalise polynomials.

**Definition 6.** *An* extended monomial *$M$ in the variables $\Delta$ and $z_1, \ldots, z_n$ is a finite product $c \cdot \prod_i v_i$ such that $c$ is an integer and $v_i$ is $x^n$, $x \in \{\Delta, z_1, \ldots, z_n\}$ or $v_i$ is a $\Delta$-quotient. The integer $c$ is called the* coefficient *and the expression $v_i$ a* literal. *Finally, an* extended polynomial *$P$ over $\Delta \in D$ and $z_1, \ldots, z_n \in \mathbb{R}_0^+$ is a finite sum $\sum_i M_i$ of extended monomials $M_i$ (in $\Delta$ and $z_1, \ldots, z_n$).*

Note that the coefficients of an extended polynomial are integers. If the context clarifies what is meant, we will drop the qualifier "extended". Examples 1 and 3 as well as the examples studied in [7] suggest a restricted notion of context dependent algebras. This is the subject of the next definition.

**Definition 7.** *A* polynomial context dependent interpretation *of $\mathcal{F}$ is a CDA $(\mathcal{C}, \{>_\Delta \mid \Delta \in D\})$ satisfying the following properties:*

- *the interpretation function $f_{\mathcal{C}}$ is an extended polynomial,*
- *the parameter set $D$ equals $\mathbb{R}^+$, and*
- *for each $f \in \mathcal{F}$ the parameter functions $f_{\mathcal{C}}^i$ are $\Delta$-quotients.*

**Lemma 8.** *Let $\mathcal{C}$ denote a polynomial context dependent interpretation, let $\alpha$ be a $\Delta$-assignment, and let $t$ be a term. Then $[\alpha, \Delta](t)$ is an extended polynomial.*

*Proof.* The lemma is a direct consequence of the definitions and Lemma 5. □

*Remark 9.* Hofbauer showed in [7] that for any monotone polynomial interpretation compatible with a TRS $\mathcal{R}$, there exists a polynomial context dependent interpretation which is $\Delta$-monotone and compatible with $\mathcal{R}$ and induces at least the same upper bound on the derivational complexity as the polynomial interpretation.

## 3 Automated Search for Context Dependent Interpretations

One approach to find context dependent interpretations (semi-)automatically was already mentioned in Hofbauer's paper [7]. A given polynomial interpretation is suitably lifted to a context dependent interpretation such that monotonicity and compatibility are preserved, but the upper bound on the derivational complexity is often improved. Unfortunately, experimental evidence suggests that the applicability of this heuristics is limited, if one is interested in automatically finding complexity bounds, see Section 5 for further details. However, the standard approach for automatically proving termination via polynomial interpretations as stipulated by Contejean et al. [11] can be adapted. The description of this adaption is the topic of this section. We restrict the form of parametric interpretations that we consider.

**Definition 10.** *A* (parametric) $\Delta$-linear interpretation *is a polynomial context dependent interpretation $C$ whose interpretation functions and parameter functions have the following form:*

$$f_C(\Delta, z_1, \ldots, z_n) = \sum_{i=1}^{n} a_{(f,i)} z_i + \sum_{i=1}^{n} b_{(f,i)} z_i \Delta + c_f \Delta + d_f$$

$$f_C^i(\Delta) = \frac{\Delta}{a_{(f,i)} + b_{(f,i)} \Delta}$$

*where the occurring coefficients are supposed to be natural numbers. For a* parametric *$\Delta$-linear interpretation, $a_{(f,i)}$, $b_{(f,i)}$, $c_f$, and $d_f$ ($f \in \mathcal{F}$, $1 \leqslant i \leqslant n$) are called* coefficient variables.

Note that for any $\Delta$-linear interpretation, we have $a_{(f,i)} > 0$ or $b_{(f,i)} > 0$ ($f \in \mathcal{F}$, $1 \leqslant i \leqslant n$): Any $\Delta$-linear interpretation is a polynomial context dependent interpretation by definition. And hence the parameter functions have to be $\Delta$-quotients, cf. Definition 7. Moreover the coefficients $a_{(f,i)}$, $b_{(f,i)}$ are used in the interpretation function *and* the parameter functions. This is necessary for the correctness of Lemma 12 below.

*Example 11.* Consider the TRS $\mathcal{R}_1$ from Example 1. The parametric interpretation and parameter functions have the form:

$$\mathsf{a}_C[\Delta](z) = az + bz\Delta + c\Delta + d \qquad \mathsf{a}_C^1(\Delta) = \frac{\Delta}{a + b\Delta}$$

$$\mathsf{b}_C[\Delta](z) = ez + fz\Delta + g\Delta + h \qquad \mathsf{b}_C^1(\Delta) = \frac{\Delta}{e + f\Delta} .$$

The following lemma is a direct consequence of the definitions.

**Lemma 12.** *Let $C$ be an $\Delta$-linear interpretation. Then $C$ is $\Delta$-monotone.*

Due to Lemma 12, in order to prove termination of a given TRS $\mathcal{R}$, it suffices to find a $\Delta$-linear interpretation compatible with $\mathcal{R}$. This observation is reflected in the following definition.

**Definition 13.** *Let $\mathcal{R}$ be a TRS and let $\mathcal{C}$ be a parametric $\Delta$-linear interpretation. The* compatibility constraints *of $\mathcal{R}$ with respect to $\mathcal{C}$ are defined as*

$$\begin{aligned}\mathsf{CC}(\mathcal{R},\mathcal{C}) =& \{[\alpha,\Delta](l) - [\alpha,\Delta](r) - \Delta \geqslant 0 \mid l \to r \in \mathcal{R}\} \cup \\ & \cup \{a_{(f,i)} + b_{(f,i)} - 1 \geqslant 0 \mid f \in \mathcal{F}, 1 \leqslant i \leqslant \mathrm{ar}(f)\} .\end{aligned}$$

*Here* $\mathrm{ar}(f)$ *denotes the arity of $f$ and $\alpha$ refers to a* symbolic $\Delta$-assignment*: Expressions of the form $[\alpha,\Delta](x)$ for $x \in \mathcal{V}$ remain unevaluated.*

While the first half of $\mathsf{CC}(\mathcal{R},\mathcal{C})$ represents compatibility with $\mathcal{R}$, the second set of constraints guarantees that the denominators of the occurring $\Delta$-quotients are different from 0. Thus any solution to $\mathsf{CC}(\mathcal{R},\mathcal{C})$, instantiating coefficients with natural numbers, represents a polynomial context dependent interpretation compatible with $\mathcal{R}$.

*Example 14.* Consider the (parametric) CDA $\mathcal{C}$ from Example 11 and set $\Delta_1 = \mathsf{a}^1_{\mathcal{C}}(\Delta)$ and $\Delta_2 = \mathsf{b}^1_{\mathcal{C}}(\Delta)$. Let $\alpha_1 = [\alpha, \Delta_2[\Delta := \Delta_1]](x)$ and let $\alpha_2 = [\alpha, \Delta_1[\Delta := \Delta_2]](x)$. Then the constraint $[\alpha,\Delta](\mathsf{a}(\mathsf{b}(x))) - [\alpha,\Delta](\mathsf{b}(\mathsf{a}(x))) - \Delta \geqslant 0$ becomes:

$$\begin{aligned}&\big(ae\alpha_1 + af\alpha_1\Delta_1 + ag\Delta_1 + be\alpha_1\Delta + bf\alpha_1\Delta_1\Delta + bg\Delta_1\Delta + (bh+c)\Delta + \\ &+ah+d\big) - \big(ae\alpha_2 + be\alpha_2\Delta_2 + ce\Delta_2 + af\alpha_2\Delta + bf\alpha_2\Delta_2\Delta + cf\Delta_2\Delta + \\ &+(df+g)\Delta + de + h\big) - \Delta \geqslant 0 .\end{aligned}$$

For all constraints $(P \geqslant 0) \in \mathsf{CC}(\mathcal{R},\mathcal{C})$, $P$ is an extended polynomial, cf. Lemma 8. It is easy to see how an extended polynomial (over $\Delta, z_1, \dots, z_n$) is transferable into a (standard) polynomial (over $\Delta, z_1, \dots, z_n$): Multiply (symbolically) with denominators of (nontrivial) $\Delta$-quotients till all (nontrivial) $\Delta$-quotients are eliminated. This simple procedure is denoted as $\mathsf{A}$. Correctness and termination of the procedure follow trivially.

**Definition 15.** *Let $\mathcal{R}$ be a TRS and let $\mathcal{C}$ be a parametric $\Delta$-linear interpretation. The* polynomial compatibility constraints *of $\mathcal{R}$ with respect to $\mathcal{C}$ are defined as follows:* $\mathsf{PCC}(\mathcal{R},\mathcal{C}) := \{P' \geqslant 0 \mid P \geqslant 0 \in \mathsf{CC}(\mathcal{R},\mathcal{C})$ *and* $P' := \mathsf{A}(P)\}$.

*Example 16.* Consider the constraint $P \geqslant 0$ depicted in Example 14. To apply the algorithm $\mathsf{A}$ we first have to symbolically multiply with the expression $a+b\Delta$ and later with $e+f\Delta$. The resulting constraint $P' \geqslant 0$ (with the polynomial $P'$ in the "variables" $\Delta$, $\alpha_1$, and $\alpha_2$) has the form:

$$\begin{aligned}&\big((b^2ef + bf^2)\alpha_1\Delta^3 + (2abef + af^2 + b^2e^2 + bef)\alpha_1\Delta^2 \\ &+(2abe^2 + a^2ef + aef)\alpha_1\Delta + (a^2e^2)\alpha_1\big) \\ &-\big((abf^2 + b^2f)\alpha_2\Delta^3 + (a^2f^2 + 2abef + abf + b^2e)\alpha_2\Delta^2 \\ &+(2a^2ef + abe^2 + aeb)\alpha_2\Delta + (a^2e^2)\alpha_2\big) + \big((b^2fh - bdf^2 - bf)\Delta^3 \\ &+(2abfh + b^2eh + bdf - adf^2 - 2bdef - bfh - be - af)\Delta^2 \\ &+(a^2fh + 2abeh + adf + bde - 2adef - afh - bde^2 - beh - ae)\big)\Delta \\ &+(a^2eh + ade - ade^2 - aeh)\big) \geqslant 0 .\end{aligned}$$

We obtain $\mathsf{PCC}(\mathcal{R}_1, \mathcal{C}) = \{P' \geqslant 0, a + b - 1 \geqslant 0, e + f - 1 \geqslant 0\}$, where the last two constraints reflect that all denominators of $\Delta$-quotients are non-zero.

Let $P \geqslant 0$ be a constraint in $\mathsf{PCC}(\mathcal{R}, \mathcal{C})$ such that $n$ distinct symbolic assignments $[\alpha, d](x)$ occur in $P$ ($x \in \mathcal{V}$, $d$ a $\Delta$-quotient). (In Example 16 two symbolic assignments occur: $\alpha_1$ and $\alpha_2$.) Then $P$ is conceivable as a polynomial in $\mathbb{Z}[\Delta, z_1, \ldots, z_n]$. It remains to verify that (a suitable instance of) $P$ is *positive*, i.e., we have to prove that $P(\Delta, z_1, \ldots, z_n) \geqslant 0$ for any values $\Delta > 0$, $z_i \geqslant 0$. This is achieved by testing for absolute positivity instead of positivity, compare [11].

A polynomial $P$ is *absolutely positive* if $P$ has non-negative coefficients only. A parametric polynomial $P$ is called *absolutely positive* if there exists an instance $P'$ of $P$ such that $P'$ is absolutely positive. Clearly any absolutely positive polynomial is positive. Thus for a given constraint $P \geqslant 0 \in \mathsf{PCC}(\mathcal{R}, \mathcal{C})$ it suffices to find instantiations of the coefficient variables such that all coefficients are natural numbers. This is achieved through the construction of suitable Diophantine inequalities over the coefficients.

**Lemma 17.** *Let $\mathcal{R}$ be a TRS and let $\mathcal{C}$ denote a parametric $\Delta$-linear interpretation. If for all $P \geqslant 0 \in \mathsf{PCC}(\mathcal{R}, \mathcal{C})$, $P$ is absolutely positive then there exists an instantiation of $\mathcal{C}$ compatible with $\mathcal{R}$.*

*Proof.* If $P$ is absolutely positive, there exist natural numbers that can be substituted to the coefficient variables in $P$ such that the resulting polynomial $P'$ is absolutely positive and thus positive. By definition this implies that the constraints in $\mathsf{CC}(\mathcal{R}, \mathcal{V})$ are fulfilled. We define an instantiation $\mathcal{C}'$ of $\mathcal{C}$ by applying the same substitution to the coefficient variables in $\mathcal{C}$. Then $\mathcal{C}'$ is compatible with $\mathcal{R}$. □

As an immediate consequence of Lemmata 12, 17, and Theorem 2 we obtain the following theorem.

**Theorem 18.** *Let $\mathcal{R}$ be a TRS and let $\mathcal{C}$ denote a parametric $\Delta$-linear interpretation. Suppose for all $P \geqslant 0 \in \mathsf{PCC}(\mathcal{R}, \mathcal{C})$, $P$ is absolutely positive. Then $\mathcal{R}$ is terminating and property* (1) *holds for $D = \mathbb{R}^+$.*

It is easy to see that the Diophantine inequalities induced by Example 16 cannot be solved, if the symbolic assignments $\alpha_1$ and $\alpha_2$ are treated as different variables. This motivates the next definition.

**Definition 19.** *Given a TRS $\mathcal{R}$ and a $\Delta$-linear interpretation $\mathcal{C}$, the* equality constraints *of $\mathcal{R}$ with respect to $\mathcal{C}$ are defined as follows:*

$$\mathsf{EC}(\mathcal{R}, \mathcal{C}) = \{(a + b\Delta) - (c + d\Delta) = 0 \mid \textit{Property } (*) \textit{ is fulfilled}\}$$

(∗) *There exists $P \geqslant 0 \in \mathsf{PCC}(\mathcal{R}, \mathcal{C})$, $x \in \mathcal{V}$ such that $[\alpha, d_1](x)$ and $[\alpha, d_2](x)$ occur in $P$ and $d_1 = \frac{\Delta}{a+b\Delta} \neq \frac{\Delta}{c+d\Delta} = d_2$.*

*Example 20.* Consider Example 16. Property (∗) is applicable to the $\Delta$-quotients $d_1$, $d_2$ in the $\Delta$-assignments $\alpha_1 = [\alpha, d_1]$ and $\alpha_2 = [\alpha, d_2]$ as

$$d_1 = \frac{\Delta}{ae + (be + f)\Delta} \neq \frac{\Delta}{ae + (af + b)\Delta} = d_2 .$$

Thus the constraint $(ae + (be + f)\Delta) - (ae + (af + b)\Delta) = 0$ occurs in $\mathsf{EC}(\mathcal{R}_1, \mathcal{C})$. This is the only constraint in $\mathsf{EC}(\mathcal{R}_1, \mathcal{C})$.

Let $P \geqslant 0 \in \mathsf{PCC}(\mathcal{R}, \mathcal{C})$, assume the equality constraints in $\mathsf{EC}(\mathcal{R}, \mathcal{C})$ are fulfilled and assume we want to test for absolute positivity of $P$. By assumption distinct symbolic assignments can be treated as equal, which may change the coefficients we need to consider in $P$. This is expressed by writing $P \geqslant 0 \in \mathsf{PCC}(\mathcal{R}, \mathcal{C}) \cup \mathsf{EC}(\mathcal{R}, \mathcal{C})$. Furthermore, we call a parametric polynomial a *zero polynomial* if there exists an instance $P'$ of $P$ such that $P' = 0$.

**Corollary 21.** *Let $\mathcal{R}$ be a TRS and let $\mathcal{C}$ denote a parametric $\Delta$-linear interpretation. Suppose for all $P \geqslant 0$ ($P = 0$) $\in \mathsf{PCC}(\mathcal{R}, \mathcal{C}) \cup \mathsf{EC}(\mathcal{R}, \mathcal{C})$, $P$ is absolutely positive ($P$ is a zero polynomial). Then $\mathcal{R}$ is terminating and property* (1) *holds for $D = \mathbb{R}^+$.*

Corollary 21 opens the way to efficiently search for CDAs: Finding a $\Delta$-monotone and compatible CDA $\mathcal{C}$ amounts to solving the Diophantine constraints in $\mathsf{PCC}(\mathcal{R}, \mathcal{C}) \cup \mathsf{EC}(\mathcal{R}, \mathcal{C})$. It is well-known that solvability of Diophantine constraints is undecidable [12]. However, there is an easy remedy for this: we restrict the domain of the coefficient variables to a finite one.

*Example 22.* Consider the TRS $\mathcal{R}_1$ from Example 1 and the $\Delta$-linear interpretation $\mathcal{C}$ from Example 11. Applying the above described algorithm, the following Diophantine (in)equalities need to be solved.

$$
\begin{array}{ll}
b^2ef + bf^2 - abf^2 - b^2f \geqslant 0 & abe^2 + a^2ef + aef - 2a^2ef - aeb \geqslant 0 \\
b^2fh - bdf^2 - bf \geqslant 0 & a^2eh + ade - ade^2 - aeh \geqslant 0 \\
a + b - 1 \geqslant 0 & e + f - 1 \geqslant 0 \\
be + f - af - b = 0 & af^2 + b^2e^2 + bef - a^2f^2 - abf - b^2e \geqslant 0 \\
\multicolumn{2}{l}{2abfh + b^2eh + bdf - adf^2 - 2bdef - bfh - be - af \geqslant 0} \\
\multicolumn{2}{l}{a^2fh + 2abeh + adf + bde - 2adef - afh - bde^2 - beh - ae \geqslant 0 \,.}
\end{array}
$$

Here the constraints $a + b - 1 \geqslant 0$, $e + f - 1 \geqslant 0$ guarantee that the denominators of occurring $\Delta$-quotients are positive, and the equality $be + f - af - b = 0$ expresses the equality constraint in $\mathsf{EC}(\mathcal{R}_1, \mathcal{C})$. Our below discussed implementations of the algorithm presented in this section find the following satisfying assignments for the coefficient variables fully automatically:

$$a = b = e = h = 1 \qquad c = d = f = g = 0 \,.$$

## 4 Derivational Complexities Induced by Polynomial Context Dependent Interpretations

In this section we show that the derivational complexity induced by $\Delta$-linear interpretations is exponential and that this bound is tight. Furthermore, we introduce a restricted subclass of $\Delta$-linear interpretations that induces (tight) quadratic derivational complexity.

Recall the TRS $\mathcal{R}_1$ considered in Example 1. This TRS belongs to a family of TRSs $\mathcal{R}_k$ for $k > 0$: $\mathsf{a}(\mathsf{b}(x)) \to \mathsf{b}^k(\mathsf{a}(x))$ and it is not difficult to see that for $k \geqslant 2$ the derivational complexity of $\mathcal{R}_k$ is exponential. In [7] $\Delta$-linear interpretations $\mathcal{C}_k$ were introduced such that

$$\inf_{\Delta > 0} \frac{[\alpha, \Delta]_{\mathcal{C}_k}(t)}{\Delta} = \mathsf{dl}_{\mathcal{R}_k}(t) ,$$

holds for any ground term. I.e., for all $k > 0$ there exist $\Delta$-linear interpretations that optimally bound the derivational complexities of $\mathcal{R}_k$. This triggers the question whether we can find such context dependent interpretations automatically. The next example answers this question affirmatively, for $k = 2$.[1]

*Example 23.* Consider the TRSs $\mathcal{R}_2$: $\mathsf{a}(\mathsf{b}(x)) \to \mathsf{b}(\mathsf{b}(\mathsf{a}(x)))$.[2] To find a $\Delta$-linear interpretation, we employ the same parametric interpretation $\mathcal{C}$, as in Example 11 and build the set of constraints $\mathsf{CC}(\mathcal{R}_2, \mathcal{C})$ and consecutively the polynomial compatibility constraints $\mathsf{PCC}(\mathcal{R}_2, \mathcal{C})$ together with the equality constraints $\mathsf{EC}(\mathcal{R}_2, \mathcal{C})$. We only state the (automatically) obtained interpretation and parameter functions:

$$\mathsf{a}_{\mathcal{C}}[\Delta](z) = (2 + 2\Delta)z \qquad \mathsf{a}^1_{\mathcal{C}}(\Delta) = \frac{1}{2 + 2\Delta}$$

$$\mathsf{b}_{\mathcal{C}}[\Delta](z) = z + 1 \qquad \mathsf{b}^1_{\mathcal{C}}(\Delta) = \Delta .$$

As a consequence of Example 23 we see the existence of TRSs, compatible with $\Delta$-linear interpretations, whose derivational complexity function is exponential. Moreover, we have the following lemma.

**Lemma 24.** *Let $\mathcal{C}$ denote a $\Delta$-linear interpretation and let $K$ denote the maximal coefficient occurring in $\mathcal{C}$. Further let $t$ be a ground term, $\alpha$ a $\Delta$-assignment and $\Delta > 0$. Then $[\alpha, \Delta](t) \leqslant (K + 2)^{|t|}(\Delta + 1)$.*

*Proof.* Straightforward induction on $t$. □

**Theorem 25.** *Let $\mathcal{R}$ be a TRS and let $\mathcal{C}$ denote a $\Delta$-linear interpretation compatible with $\mathcal{R}$. Then $\mathcal{R}$ is terminating and $\mathsf{dc}_{\mathcal{R}}(n) = 2^{\mathcal{O}(n)}$. Moreover there exists a TRS $\mathcal{R}$ such that $\mathsf{dc}_{\mathcal{R}}(n) = 2^{\Omega(n)}$.*

*Proof.* The proof of the upper bound follows the pattern of the proof of Theorem 29 below. To show that this upper bound is tight, we consider the TRS $\mathcal{R}_2$ from Example 23. It is easy to see that $\mathsf{dc}_{\mathcal{R}_2}(n) = 2^{\Omega(n)}$ holds. □

In order to establish a termination method that induces *polynomial* derivational complexity, we restrict the class of $\Delta$-linear interpretations.

[1] The answer remains positive for $k = 3$. Detailed experimental evidence and additional information on the considered constraints are available at `http://cl-informatik.uibk.ac.at/~aschnabl/experiments/cdi/`

[2] This is Example 2.50 in Steinbach and Kühler's collection [13].

**Definition 26.** *A* $\Delta$-restricted interpretation *is a* $\Delta$*-linear interpretation. In addition we require that for the interpretation functions and parameter functions*

$$f_{\mathcal{C}}(\Delta, z_1, \dots, z_n) = \sum_{i=1}^{n} a_{(f,i)} z_i + \sum_{i=1}^{n} b_{(f,i)} z_i \Delta + c_f \Delta + d_f$$
$$f_{\mathcal{C}}^i(\Delta) = \frac{\Delta}{a_{(f,i)} + b_{(f,i)} \Delta} ,$$

*we have* $a_{(f,i)} \in \{0,1\}$ *for all* $1 \leqslant i \leqslant n$.

*Example 27.* Consider the TRS $\mathcal{R}_1$ from Example 1. The assignment of coefficient variables as defined in Example 22 induces a $\Delta$-restricted interpretation.

**Lemma 28.** *Let* $\mathcal{C}$ *denote a* $\Delta$*-restricted interpretation with coefficients* $a_{(f,i)}$, $b_{(f,i)}$, $c_f$, $d_f$ $(f \in \mathcal{F},\ 1 \leqslant i \leqslant \operatorname{ar}(f))$ *and we set*

$$M := \max(\{c_f, d_f \mid f \in \mathcal{F}\} \cup \{1\})$$
$$N := \max(\{b_{(f,i)} \mid f \in \mathcal{F}, 1 \leqslant i \leqslant \operatorname{ar}(f)\} \cup \{1\}) .$$

*Further let* $t$ *be a ground term,* $\alpha$ *a* $\Delta$*-assignment and let* $\Delta > 0$*. Then* $[\alpha, \Delta](t) \leqslant M(|t| + N|t|^2 \Delta)$.

*Proof.* We proceed by induction on $t$. As $t \in \mathcal{T}(\mathcal{F})$, the evaluation is independent of the assignment. Hence we write $[\Delta](t)$ instead of $[\alpha, \Delta](t)$. If $t = f \in \mathcal{F}$, then

$$[\Delta](t) = c_f \Delta + d_f \leqslant M(\Delta + 1) \leqslant M(|t| + N|t|^2 \Delta) .$$

If on the other hand $t = f(t_1, \dots, t_n)$, then

$$\begin{aligned}
[\Delta](t) &= \sum_i (a_{f_i} + b_{f_i}\Delta)[f_{\mathcal{C}}^i(\Delta)](t_i) + c_f\Delta + d_f && (2)\\
&\leqslant \sum_i (a_{f_i} + b_{f_i}\Delta)\big(M(|t_i| + N|t_i|^2 \frac{\Delta}{a_{f_i} + b_{f_i}\Delta}\big) + c_f\Delta + d_f && (3)\\
&= \sum_i \big((a_{f_i} + b_{f_i}\Delta)M|t_i| + MN|t_i|^2\Delta\big) + c_f\Delta + d_f && (4)\\
&\leqslant \sum_i \big((1 + N\Delta)M|t_i| + MN|t_i|^2\Delta\big) + M(\Delta + 1) && (5)\\
&\leqslant \sum_i |t_i|\big((1 + N\Delta)M + MN(|t| - 1)\Delta\big) + M(\Delta + 1) && (6)\\
&= (|t| - 1)\big((1 + N\Delta)M + MN(|t| - 1)\Delta\big) + M(\Delta + 1) && (7)\\
&= M\big((|t| - 1)(1 + N\Delta) + N(|t| - 1)^2\Delta + (\Delta + 1)\big) && (8)\\
&\leqslant M(|t| + N|t|^2\Delta) . && (9)
\end{aligned}$$

In line (3) we employ the induction hypothesis, in (6) we use $|t_i| \leqslant |t| - 1$ and for (9) a simple calculation reveals: $(|t| - 1)(1 + N\Delta) + N\Delta(|t| - 1)^2 + (\Delta + 1) = |t| + N|t|^2\Delta + \Delta - N|t|\Delta \leqslant |t| + N|t|^2\Delta$. □

**Theorem 29.** *Let $\mathcal{R}$ be a TRS and let $\mathcal{C}$ denote a $\Delta$-restricted interpretation compatible with $\mathcal{R}$. Then $\mathcal{R}$ is terminating and* $\mathsf{dc}_{\mathcal{R}}(n) = \mathcal{O}(n^2)$. *Moreover there exists a TRS $\mathcal{R}$ such that* $\mathsf{dc}_{\mathcal{R}}(n) = \Omega(n^2)$.

*Proof.* By Theorem 2 $\mathcal{R}$ is terminating and by Lemma 28, there exists $K \in \mathbb{N}$, such that for any ground term $t$: $[\Delta](t) \leqslant K(|t| + K|t|^2\Delta) \leqslant K^2|t|^2(\Delta + 1)$ and hence

$$\mathsf{dl}_{\mathcal{R}}(t) \leqslant \inf_{\Delta>0} \frac{[\Delta](t)}{\Delta} \leqslant \inf_{\Delta>0} \frac{K^2|t|^2(\Delta+1)}{\Delta} = K^2|t|^2 \ .$$

We obtain $\mathsf{dl}_{\mathcal{R}}(t) = \mathcal{O}(|t|^2)$ for any $t \in \mathcal{T}(\mathcal{F}, \mathcal{V})$ and thus $\mathsf{dc}_{\mathcal{R}}(n) = \mathcal{O}(n^2)$. The tightness of the bound follows by Example 1. □

By definition the constant employed in Theorem 29 depends only on the employed interpretation functions. Moreover this dependence is linear. In concluding this section, we want to stress that $\Delta$-restricted interpretation are even strong enough to handle non-simple terminating TRSs.

*Example 30 ([7]).* Consider the TRS $\mathcal{R}$ with the one rule $\mathsf{a}(\mathsf{a}(x)) \to \mathsf{a}(\mathsf{b}(\mathsf{a}(x)))$. By applying the algorithm described in Section 3, we find the below given $\Delta$-restricted interpretation $\mathcal{C}$ automatically:

$$\mathsf{a}_{\mathcal{C}}[\Delta](z) = 2z\Delta + 2 \qquad \mathsf{b}_{\mathcal{C}}[\Delta](z) = z\Delta \qquad \mathsf{a}^1_{\mathcal{C}}(\Delta) = \frac{1}{2} \qquad \mathsf{b}^1_{\mathcal{C}}(\Delta) = 1 \ .$$

By Theorem 18, $\mathcal{C}$ is compatible with $\mathcal{R}$. Hence Theorem 29 implies that the derivational complexity of $\mathcal{R}$ is (at most) quadratic.

## 5 Experimental Results

In this section we describe the programs $\mathsf{cdi}_1$, $\mathsf{cdi}_2$, and $\mathsf{cdi}_3$. These programs provide search procedures for context dependent interpretations. The program $\mathsf{cdi}_1$ implements the heuristics of Hofbauer in [7], mentioned in Section 3 above. On the other hand, programs $\mathsf{cdi}_2$ and $\mathsf{cdi}_3$ implement the algorithm presented in Section 3 and incorporate constraint solvers for Diophantine (in)equalities. The program $\mathsf{cdi}_1$ searches for $\Delta$-linear interpretations, while $\mathsf{cdi}_2$ and $\mathsf{cdi}_3$ can search for $\Delta$-linear and $\Delta$-restricted interpretations. We summarise further differences below:

$\mathsf{cdi}_1$ Firstly, the program searches for a polynomial interpretation compatible with a TRS $\mathcal{R}$. This interpretation is then lifted to a polynomial context dependent interpretation $\mathcal{C}$ as follows: Coefficients of the form $k + 1$ are replaced by $k + \Delta$. Finally MATHEMATICA[3] is invoked to verify that the resulting CDA $\mathcal{C}$ is $\Delta$-monotone and compatible with $\mathcal{R}$.

$\mathsf{cdi}_2$ This programs employs a *constraint propagation procedure* to solve the Diophantine constraints in $\mathsf{PCC}(\mathcal{R}, \mathcal{C}) \cup \mathsf{EC}(\mathcal{R}, \mathcal{C})$. Essentially the implementation follows the technique suggested in [11].

[3] `http://www.wolfram.com/products/mathematica/`

$\mathsf{cdi}_3$ The Diophantine (in)equalities in $\mathsf{PCC}(\mathcal{R},\mathcal{C}) \cup \mathsf{EC}(\mathcal{R},\mathcal{C})$ are translated into *propositional logic* and suitable assignments are found by employing a SAT solver, in our case MiniSat[4]. The implementation follows ideas presented in [14] and employs the `plogic` library of $\mathsf{T_TT_2}$.[5]

The implementation of the transformation steps as described in Section 3, is the same for $\mathsf{cdi}_2$ and $\mathsf{cdi}_3$. The programs $\mathsf{cdi}_1$, $\mathsf{cdi}_2$, and $\mathsf{cdi}_3$ are written in OCaml[6] (and parts of $\mathsf{cdi}_1$ in C). All three programs are fairly small: $\mathsf{cdi}_1$ consists of about 2000 lines of code, while $\mathsf{cdi}_2$ and $\mathsf{cdi}_3$ use roughly 3000 lines of code each.In Table 1 we summarise the comparison between the different programs $\mathsf{cdi}_1$, $\mathsf{cdi}_2$, and $\mathsf{cdi}_3$. The numbers in the third line of the table refer to the number of bits maximally used in $\mathsf{cdi}_3$ to encode coefficients. Correspondingly for $\mathsf{cdi}_2$ we used 32 as strict bound on the coefficients. We are interested in automatically verifying the complexity of terminating TRSs. Consequentially, as testbed we employ those 957 TRSs from the version 4.0 of the Termination Problem Data Base (TPDB for short) that can be shown terminating with at least one of the tools that participated in the termination competition 2007.[7] The presented tests were performed single-threaded on a 2.40 GHz Intel® Core™ 2 Duo with 2 GB of memory. For each system we used a timeout of 60 seconds, the times in the tables are given in milliseconds.

**Table 1.** 957 terminating TRSs

| | $\mathsf{cdi}_1$ | $\mathsf{cdi}_2$ | | $\mathsf{cdi}_3$ | | | | | | | |
|---|---|---|---|---|---|---|---|---|---|---|---|
| | $\Delta$-linear | $\Delta$-restr. | $\Delta$-linear | $\Delta$-restricted | | | | $\Delta$-linear | | | |
| | | | | 2 | 3 | 4 | 5 | 2 | 3 | 4 | 5 |
| # success | 19 | 61 | 62 | 83 | 86 | 86 | 86 | 82 | 82 | 82 | 83 |
| average time | - | 3132 | 3595 | 3652 | 4041 | 4008 | 3986 | 5496 | 4981 | 5010 | 5527 |
| # timeout | - | 276 | 782 | 144 | 189 | 222 | 238 | 525 | 687 | 751 | 797 |

Observe that the heuristic proposed in [7] is not suitable as an automatic procedure. (We have not indicated the time spent by $\mathsf{cdi}_1$ as the timing is incomparable to the stand-alone approach of $\mathsf{cdi}_2$ or $\mathsf{cdi}_3$.) With respect to the comparison between $\mathsf{cdi}_2$ and $\mathsf{cdi}_3$, the latter outperforms the former, if at least 2 bits are used. Perhaps surprisingly the performance of $\mathsf{cdi}_2$ and $\mathsf{cdi}_3$ on $\Delta$-restricted and $\Delta$-linear is almost identical. This can be explained by the strong impact of larger bounds for the coefficients $a_{(f,i)}$ ($f \in \mathcal{F}, 1 \leqslant i \leqslant \mathsf{ar}(f)$) in the complexity of the issuing Diophantine (in)equalities. However, for both programs $\mathsf{cdi}_2$ and $\mathsf{cdi}_3$, the stronger technique gains one crucial system: Example 23.

---

[4] `http://minisat.se/`

[5] `http://colo6-c703.uibk.ac.at/ttt2/`

[6] `http://www.caml.inria.fr/`

[7] These 957 systems and full experimental evidence can be found at `http://cl-informatik.uibk.ac.at/~aschnabl/experiments/cdi/`

**Table 2.** Termination Methods as Complexity Analysers

| | SL | T$_\mathsf{T}$Tbox | cdi$_3$—$\Delta$-restricted | cdi$^+$—$\Delta$-restricted |
|---|---|---|---|---|
| # success | 41 | 125 | 86 | 87 |
| average time | 20 | 577 | 3986 | 3010 |
| # timeout | 0 | 225 | 238 | 237 |

Table 2 relates existing methods that induce polynomial derivational complexities of TRSs to cdi$_3$. SL refers to *strongly linear* interpretations, i.e., only interpretation functions of the form $f_\mathcal{A}(x_1, \ldots, x_n) = \sum_i x_i + c$, $c \in \mathbb{N}$ are allowed. Clearly compatibility with strongly linear interpretations induces *linear* derivational complexity. Secondly, T$_\mathsf{T}$Tbox refers to the implementation of the match-bound technique as in [15]: Linear TRSs are tested for match-boundedness, non-linear, but non-duplicating TRSs are tested for match-raise-boundedness. This technique again implies *linear* derivational complexity. (Employing [16] (as in [5]) one sees that any match-raise bounded TRS has linear derivational complexity. Then the claim follows from Lemma 8 in [15].) Note that the restriction to non-duplicating TRS is harmless, as any duplicating TRS induces at least exponential derivational complexity. No further termination methods that induce at most polynomial derivational complexities for TRSs have previously been known. In particular related work on implicit complexity (for example [17,18,19,20,21]) does not provide methods that induce polynomial *derivational complexities*, even if sometimes the derivation length can be bounded polynomially, if the set of start terms is suitably restricted. Finally cdi$^+$ denotes our standard strategy: First, we search for a strongly linear interpretation. If such an interpretation cannot be found, then a $\Delta$-restricted interpretation is sought (with 5 bits as bound).

Some comments on the results reported in Table 2: By definition the set of TRSs compatible with a strongly linear interpretation is a (strict) subset of those treatable with cdi$^+$. On the other hand the comparison between T$_\mathsf{T}$Tbox and cdi$^+$ (or cdi$_3$) may appear not very favourable for our approach. However, cdi$^+$ (and cdi$_3$) can handle TRSs that cannot be handled by T$_\mathsf{T}$Tbox. More precisely with respect to $\Delta$-restricted interpretations cdi$^+$ (and cdi$_3$) can handle 38 (37) TRSs that cannot be handled with T$_\mathsf{T}$Tbox. For instance the following example can only be handled with cdi$^+$ (and cdi$_3$).

*Example 31.* Consider the following rewrite system $\mathcal{R}_{+,-}$. (This is Example 2.11 in Steinbach and Kühler's collection [13].)

$$
\begin{array}{lll}
\mathsf{0}+y \to y & \mathsf{0}-y \to x & \mathsf{s}(x)-\mathsf{s}(y) \to x-y \\
\mathsf{s}(x)+y \to \mathsf{s}(x+y) & x-\mathsf{0} \to x &
\end{array}
$$

It is easy to see that $\mathcal{R}_{+,-}$ is compatible with the following (automatically generated) $\Delta$-restricted interpretation $\mathcal{C}$.

$$
\begin{array}{ll}
-_\mathcal{C}[\Delta](x,y) = x + y + 3y\Delta + 2\Delta & \mathsf{0}_\mathcal{C}[\Delta] = 0 \\
+_\mathcal{C}[\Delta](x,y) = x + y + x\Delta + \Delta & \mathsf{s}_\mathcal{C}[\Delta](x) = x + 2 \,,
\end{array}
$$

with parameter functions: $-^1_{\mathcal{C}}(\Delta) = +^2_{\mathcal{C}}(\Delta) = \mathsf{s}^1_{\mathcal{C}}(\Delta =)\Delta$, $-^2_{\mathcal{C}}(\Delta) = \frac{\Delta}{1+3\Delta}$, and $+^1_{\mathcal{C}}(\Delta) = \frac{\Delta}{1+\Delta}$. Due to Theorem 29 we conclude *quadratic* derivational complexity, while the standard polynomial interpretation would only allow to conclude an *exponential* upper bound. Note that the deduced quadratic derivational complexity provides an optimal upper bound.

Another issue is the high average yes time (and the higher number of timeouts) of $\mathsf{cdi}_3$ and $\mathsf{cdi}^+$ in relation to existing techniques. Although a closer look reveals that the total times spent by $\mathsf{T_TTbox}$ and $\mathsf{cdi}^+$ (or $\mathsf{cdi}_3$) is relatively equal, an improvement of the efficiency of the introduced tools seems worthwhile.

*Remark 32.* Note that $\mathsf{cdi}^+$ in conjunction with $\mathsf{T_TTbox}$ can automatically verify that 163 TRSs in the testbed are of at most *quadratic* derivational complexity. Put differently more than 10% of all 1381 TRSs (and more than a third of the 445 non-duplicating TRSs) in version 4.0 of the TPDB are of quadratic derivational complexity.

## 6 Conclusion

In this paper we have presented two subclasses of context dependent interpretations, and established tight upper bounds on the induced derivational complexities. More precisely, we have delineated two subclasses: $\Delta$-linear and $\Delta$-restricted context dependent interpretations that induce exponential and quadratic derivational complexity, respectively. Further, we introduced an algorithm for mechanically proving termination of rewrite systems with context dependent interpretations. As a consequence we established a technique to automatically verify quadratic derivational complexity of TRSs. Finally, we reported on different implementations of this algorithm and presented numerical data to compare these implementations with existing methods that allow to automatically verify polynomial derivational complexity of TRSs.

We believe the here presented approach can be extended further. A starting point for future work would be to decide whether it is possible to define additional subclasses of context dependent interpretations inducing polynomial derivational complexities that grow faster than quadratic. One possible approach is to drop the restriction to *integer* coefficients and thus generalise the notion of polynomial context dependent interpretations. By Tarski's quantifier elimination method, such an extension turns the undecidable *positivity problem* for Diophantine (in)equalities into a decidable problem. Further research will clarify the impact of this extension. A crucial problem in practical considerations is the known ineffectivity of quantfier elimination, see for example [22].

## References


1. Hofbauer, D., Lautemann, C.: Termination proofs and the length of derivations. In: Dershowitz, N. (ed.) RTA 1989. LNCS, vol. 355, pp. 167–177. Springer, Heidelberg (1989)

2. Hofbauer, D.: Termination proofs by multiset path orderings imply primitive recursive derivation lengths. TCS 105, 129–140 (1992)
3. Weiermann, A.: Termination proofs for term rewriting systems with lexicographic path orderings imply multiply recursive derivation lengths. TCS 139, 355–362 (1995)
4. Moser, G.: Derivational complexity of Knuth Bendix orders revisited. In: Hermann, M., Voronkov, A. (eds.) LPAR 2006. LNCS (LNAI), vol. 4246, pp. 75–89. Springer, Heidelberg (2006)
5. Geser, A., Hofbauer, D., Waldmann, J., Zantema, H.: On tree automata that certify termination of left-linear term rewriting systems. IC 205, 512–534 (2007)
6. Hirokawa, N., Moser, G.: Automated complexity analysis based on the dependency pair method. In: Proc. 4th IJCAR. Springer, Heidelberg (accepted for publication, 2008)
7. Hofbauer, D.: Termination proofs by context-dependent interpretations. In: Middeldorp, A. (ed.) RTA 2001. LNCS, vol. 2051, pp. 108–121. Springer, Heidelberg (2001)
8. Baader, F., Nipkow, T.: Term Rewriting and All That. Cambridge University Press, Cambridge (1998)
9. Terese: Term Rewriting Systems. Cambridge Tracts in Theoretical Computer Science, vol. 55. Cambridge University Press, Cambridge (2003)
10. Schnabl, A.: Context Dependent Interpretations. Master's thesis, Universität Innsbruck (2007), `http://cl-informatik.uibk.ac.at/~aschnabl/`
11. Contejean, E., Marché, C., Tomás, A.P., Urbain, X.: Mechanically proving termination using polynomial interpretations. JAR 34, 325–363 (2005)
12. Matiyasevich, Y.: Enumerable sets are diophantine. Soviet Mathematics (Dokladi) 11, 354–357 (1970)
13. Steinbach, J., Kühler, U.: Check your ordering - termination proofs and open problems. Technical Report SEKI-Report SR-90-25, University of Kaiserslautern (1990)
14. Fuhs, C., Giesl, J., Middeldorp, A., Schneider-Kamp, P., Thiemann, R., Zankl, H.: SAT solving for termination analysis with polynomial interpretations. In: Marques-Silva, J., Sakallah, K.A. (eds.) SAT 2007. LNCS, vol. 4501, pp. 340–354. Springer, Heidelberg (2007)
15. Korp, M., Middeldorp, A.: Proving termination of rewrite systems using bounds. In: Baader, F. (ed.) RTA 2007. LNCS, vol. 4533, pp. 273–287. Springer, Heidelberg (2007)
16. Hofbauer, D., Waldmann, J.: Deleting string rewriting systems preserve regularity. TCS 327, 301–317 (2004)
17. Bonfante, G., Cichon, A., Marion, J.Y., Touzet, H.: Algorithms with polynomial interpretation termination proof. JFP 11, 33–53 (2001)
18. Marion, J.Y.: Analysing the implicit complexity of programs. IC 183, 2–18 (2003)
19. Bonfante, G., Marion, J.Y., Moyen, J.Y.: Quasi-intepretations and small space bounds. In: Giesl, J. (ed.) RTA 2005. LNCS, vol. 3467, pp. 150–164. Springer, Heidelberg (2005)
20. Marion, J.Y., Péchoux, R.: Resource analysis by sup-interpretation. In: Hagiya, M., Wadler, P. (eds.) FLOPS 2006. LNCS, vol. 3945, pp. 163–176. Springer, Heidelberg (2006)
21. Avanzini, M., Moser, G.: Complexity analysis by rewriting. In: Proc.9th FLOPS. LNCS, vol. 4989, pp. 130–146 (2008)
22. Caviness, B., Johnson, J. (eds.): Quantifier Elimination and Cylindrical Algebraic Decomposition. Springer, Heidelberg (2004)

# Tree Automata for Non-linear Arithmetic

Naoki Kobayashi[1] and Hitoshi Ohsaki[2]

[1] Tohoku University, Japan
koba@ecei.tohoku.ac.jp
[2] National Institute of Advanced Industrial Science and Technology, Japan
ohsaki@ni.aist.go.jp

**Abstract.** Tree automata modulo associativity and commutativity axioms, called *AC tree automata*, accept trees by iterating the transition modulo equational reasoning. The class of languages accepted by *monotone* AC tree automata is known to include the solution set of the inequality $x \times y \geqslant z$, which implies that the class properly includes the AC closure of regular tree languages. In the paper, we characterize more precisely the expressiveness of monotone AC tree automata, based on the observation that, in addition to polynomials, a class of exponential constraints (called *monotone exponential Diophantine inequalities*) can be expressed by monotone AC tree automata with a minimal signature. Moreover, we show that a class of arithmetic logic consisting of monotone exponential Diophantine inequalities is definable by monotone AC tree automata. The results presented in the paper are obtained by applying our novel tree automata technique, called *linearly bounded projection*.

## 1 Introduction

When reasoning about system properties of complex software automatically, analysis tools are often required to deal with arithmetic constraints together with system transitions. Safety-critical software used in automobiles, airplanes, and spacecrafts are examples whose internal transitions are triggered according to whether a certain arithmetic condition is satisfied. Hybrid automata ([8]) is a typical framework that provides a formalism for modeling such hybrid systems. However, interesting decision problems about hybrid automata are undecidable in most cases.

Equational tree automata were proposed in [17] as an extension of tree automata, in which equational reasoning is allowed at each transition step. The idea of the transition *modulo* equivalence looks simple, but the flexibility introduced in the framework turns out to define an enriched classification of formal tree languages. It is known that tree automata are closely related to context-free grammars: the leaf language of trees accepted by a regular tree automaton is a context-free language, and conversely, the set of derivation trees generated by a context-free grammar is a tree language accepted by a regular tree automaton [20]. But therefore, tree encoding by regular tree automata is no longer powerful enough to express linear arithmetic. On the other hand, tree automata

A. Voronkov (Ed.): RTA 2008, LNCS 5117, pp. 291–305, 2008.

with associativity and commutativity (AC) axioms are expressive enough to encode Presburger formulas. In fact, there are several papers discussing, based on equational tree automata or a related framework, how to embed linear arithmetic feature in XML type checking [4] and query processing [2].

Tree automata with AC axioms are called AC tree automata (AC-TA for short). Depending on which types of transition rules are equipped, we distinguish two classes of AC-TA ([17]): *regular* AC tree automata and *monotone* AC tree automata. The former is allowed to have transition rules of the form $f(\alpha_1, \ldots, \alpha_n) \to \beta$ or $\alpha \to \beta$. The latter class may additionally have rules of the form $f(\alpha_1, \ldots, \alpha_n) \to f(\beta_1, \ldots, \beta_n)$. One can easily observe that monotone AC tree automata are a super-class of regular AC tree automata, though it is not obvious that the language hierarchy of the two classes is strict. Indeed, in the absence of AC axioms, the additional type of transition rules does not increase the expressive power of tree automata.

In [15], however, Ohsaki *et al.* showed that tree languages accepted by monotone AC tree automata are not closed under complement. This implies that the class of monotone AC tree automata is a proper super-class of regular AC tree automata, as tree languages accepted by regular AC tree automata are closed under all Boolean operations. They also showed that there exists a monotone AC tree automaton whose accepted language $L$ satisfies $t \in L$ iff $|t|_{\mathsf{a}} \times |t|_{\mathsf{b}} \geqslant |t|_{\mathsf{c}}$ $\wedge |t|_{\mathsf{d}} = 1 \wedge |t|_{\mathsf{e}} = 2$, where $|t|_{\mathsf{a}}$ denotes the number of occurrences of a constant $\mathsf{a}$ in a tree $t$. This example reveals that monotone AC tree automata are a candidate framework for representing non-linear arithmetic constraints.

The goal of this paper is to investigate the expressive power of monotone AC tree automata. Although the closure properties of this class have been studied extensively [15,17], little is known about the expressiveness. So we devote our attentions to the problem of defining the class of arithmetic reducible to monotone AC tree automata. Arithmetic constraints of particular interest in the paper are of the form of $E(\boldsymbol{x}_n) \geqslant L(\boldsymbol{x}_n)$, such that $E(\boldsymbol{x}_n)$ is an arithmetic expression formed of non-negative integers, first-order variables, addition, multiplication, and exponentiation. The right-hand side $L(\boldsymbol{x}_n)$ is a linear polynomial with integer coefficients. We call the above constraint a *monotone exponential Diophantine inequality*. And a formula consisting of monotone exponential Diophantine inequalities is called a *monotone exponential Diophantine formula*. The syntax of the formulas is given later.

The paper is organized as follows. The remainder of the section reviews some background and related work. The next section introduces equational tree automata. The previous results and related properties are also presented. In Section 3, we demonstrate two monotone AC tree automata for multiplication and exponentiation. That is, we define an AC-TA $\mathcal{A}_{\mathcal{E}}^{\mathsf{mult}}$ such that $\mathcal{A}_{\mathcal{E}}^{\mathsf{mult}}$ accepts a tree $t$ if and only if $|t|_{\mathsf{a}} \times |t|_{\mathsf{b}} \geqslant |t|_{\mathsf{c}}$. Similarly, we define $\mathcal{A}_{\mathcal{E}}^{\mathsf{exp}}$ such that $\mathcal{A}_{\mathcal{E}}^{\mathsf{exp}}$ accepts a tree $t$ if and only if $|t|_{\mathsf{a}}^{|t|_{\mathsf{b}}} \geqslant |t|_{\mathsf{c}}$. Unlike the monotone AC tree automaton shown in the previous paper [15], $\mathcal{A}_{\mathcal{E}}^{\mathsf{mult}}$ does not require extra symbols to interpret $|t|_{\mathsf{a}} \times |t|_{\mathsf{b}} \geqslant |t|_{\mathsf{c}}$. In Section 4, we discuss a decidable sub-class of exponential Diophantine formulas. In particular, we show through tree automata techniques that the satisfiability of monotone Diophantine formulas is decidable, in contrast to the undecidability of Hilbert's 10th problem for non-negative solutions. In Section 5, we show that every monotone exponential Diophantine inequality is *monotone AC tree automata definable*. This means that one can construct a monotone AC tree automaton over a minimal signature whose accepted language $M$ satisfies $t \in M$ iff $E(|t|_{\mathsf{a}_1}, \ldots, |t|_{\mathsf{a}_n}) \geqslant L(|t|_{\mathsf{a}_1}, \ldots, |t|_{\mathsf{a}_n})$. Our proof is based on a special projection, called a *linearly bounded projection*. This result is *not* an immediate consequence of the previous observation that $x \times y \geqslant z$ and $x^y \geqslant z$ are monotone AC-TA definable. For instance, $x^3 \geqslant z$ is equivalent to $\exists y( x^2 \geqslant y \wedge x \times y \geqslant z)$; however, we do not know whether the class of monotone AC tree automata is closed under projection in general. Finally, in Section 6, we conclude the paper by summarizing the results and remaining questions.

### 1.1 Related Work

Given a signature $F$ that contains only an AC symbol $\otimes$ and constants, one can regard Petri nets as ground AC rewriting systems over the signature $F$. A configuration of a net $\mathcal{N}$ is a tree $t$ over $F$, such that $|t|_{\mathsf{a}}$ is the number of *tokens* on a *place* $\mathsf{a}$ of $\mathcal{N}$. In the setting, transition of $\mathcal{N}$ with the configuration $t$ is performed by ground AC rewrite rules. For instance, the transition consuming "a token on the place $\mathsf{a}$ and another token on $\mathsf{b}$" and producing "two tokens on $\mathsf{a}$ and a token on $\mathsf{c}$" is represented by the rewrite rule $\mathsf{a} \otimes \mathsf{b} \to \mathsf{a} \otimes \mathsf{a} \otimes \mathsf{c}$. For an initial configuration $t_0$ of the net $\mathcal{N}$, the set of all reachable trees from $t_0$ is called a *reachability set*.

Monotone AC tree automata over the above signature $F$ are a sub-class of Petri nets. So far we do not know whether Petri nets are properly more expressive than monotone AC tree automata, but it is worth investigating a reasonable subclass of Petri nets for automated reasoning purposes. In fact, the membership problem for monotone AC tree automata is PSPACE-complete, while the upper bound complexity of its Petri net counterpart (the reachability problem) is not known. In [7] Hack showed that for a given polynomial $P(\boldsymbol{x}_n)$ with non-negative integer coefficients, there effectively exists a Petri net such that the projection of its reachability set onto the first $n+1$ places is $\{(x_1, \ldots, x_n, y) \mid P(\boldsymbol{x}_n) \geqslant y\}$. This observation is, however, not directly applied to the case of monotone AC tree automata, because the transition rules of monotone AC tree automata are more restricted than those of Petri nets.

## 2 Preliminaries

We assume the reader is familiar with term rewriting [1] and tree automata [3]. An *equational theory* is a pair $\mathcal{E} = (F, E)$ of a signature $F$ (a finite set of function symbols, each with an associated unique *arity*) and a finite set $E$ of orientation-sensitive axioms over function symbols in $F$ possibly with some variables. The binary relation $\to_{\mathcal{E}}$ induced by $\mathcal{E}$ is the rewrite relation, i.e. $s \to_{\mathcal{E}} t$ if there exist an axiom $l \approx r$ in $E$, a context $C[\,]$ and a substitution $\sigma$ such that $s = C[l\sigma]$ and

$t = C[r\sigma]$. The equivalence closure and the reflexive-transitive of $\rightarrow_\mathcal{E}$ are denoted $=_\mathcal{E}$ and $\rightarrow^*_\mathcal{E}$, respectively. For a binary function symbol $f \in F$, the associativity axiom is written as $f(f(x,y),z) \approx f(x,f(y,z))$, and the commutativity axiom is $f(x,y) \approx f(y,x)$. The associative and commutative theory (AC-theory) is an equational theory whose axioms are the associativity and commutativity for some of the binary function symbols.

A *(monotone) equational tree automaton* (ETA) is a 4-tuple $(\mathcal{E}, Q, Q_{\mathsf{fin}}, \Delta)$, that consists of an equational theory $\mathcal{E}$, a finite set $Q$ of states disjoint from symbols in $F$, a subset $Q_{\mathsf{fin}}$ of $Q$, and a finite set $\Delta$ of transition rules whose shapes are in the following forms:

| (Regular) | (Epsilon) | (Monotone) |
|---|---|---|
| $f(\alpha_1,\dots,\alpha_n) \rightarrow \beta_1$ | $\alpha_1 \rightarrow \beta_1$ | $f(\alpha_1,\dots,\alpha_n) \rightarrow f(\beta_1,\dots,\beta_n)$ |

such that $f \in F$, $\mathsf{arity}(f) = n$ and $\alpha_1,\dots,\alpha_n,\beta_1,\dots,\beta_n \in Q$. State symbols occurring at different positions, $\alpha_i, \alpha_j$ or $\beta_i, \beta_j$ $(i \neq j)$, in transition rules can be the same. In the paper we write $\mathcal{A}_\mathcal{E}$ to denote an ETA $\mathcal{A}$ equipped with an equational theory $\mathcal{E}$.

The move relation $\rightarrow_{\mathcal{A}_\mathcal{E}}$ is the equational rewrite relation of $\mathcal{A}_\mathcal{E}$, that is, $s \rightarrow_{\mathcal{A}_\mathcal{E}} t$ iff $s =_\mathcal{E} C[l]$ and $t =_\mathcal{E} C[r]$ for a transition rule $l \rightarrow r$ in $\Delta$ and a context $C[\,]$. A tree $t$ is *accepted* by $\mathcal{A}_\mathcal{E}$ if $t \in \mathcal{T}_F$ and $t \rightarrow^*_{\mathcal{A}_\mathcal{E}} \alpha$ for some $\alpha \in Q_{\mathsf{fin}}$, and the set of trees accepted by $\mathcal{A}_\mathcal{E}$ is denoted $\mathcal{L}(\mathcal{A}_\mathcal{E})$ in the paper. A tree language accepted by $\mathcal{A}_\mathcal{E}$ is called $\mathcal{E}$*-monotone.* In particular, if $\mathcal{E}$ is the AC-theory, the accepted tree language is called AC-monotone.

**Proposition 1 ([15,17]).** *The class of monotone AC-TA is effectively closed under union and intersection, but is not closed under complement. Moreover, the membership and emptiness problems are decidable, but the inclusion problem is undecidable.* □

If transition rules in $\Delta$ of $\mathcal{A}_\mathcal{E}$ are all in the REGULAR form, $\mathcal{A}_\mathcal{E}$ is called a *regular* ETA. A regular ETA with the AC-theory $\mathcal{E}$ is a regular AC-TA. A tree language accepted by the regular ETA (resp. the regular AC-TA) is called $\mathcal{E}$-regular (AC-regular). If $\mathcal{E}$ is the free theory ($E = \varnothing$), a regular ETA is called, as is followed by customary, a *regular tree automaton*, and a tree language accepted by the regular tree automaton is called *regular.* The above definition of "regularity" (regular tree automata and regular tree languages) is identical to the standard notion, e.g. found in [3].

*Leaves* of a tree $t$, denoted $\mathsf{leaf}(t)$, are the sequence in left-to-right order of constants occurring in the tree: $\mathsf{leaf}(f(t_1,\dots,t_n)) = \mathsf{leaf}(t_1)\cdots\mathsf{leaf}(t_n)$ if $n \geqslant 1$; $\mathsf{leaf}(a) = a$, otherwise. The *leaf language* associated to a tree language $L$ is the set of leaves obtained from $L$. The *commutative image* of a tree language $L$ is the commutative closure of a leaf language of $L$. Parikh mapping $\Psi_F$ ([18]) is the mapping from tree languages to the commutative image of the languages: Given a tree language $L$ whose signature $F$ contains the set of constants $F_0 = \{\mathsf{a}_1,\dots,\mathsf{a}_k\}$, then $\Psi_F(L) = \{(|t|_{\mathsf{a}_1},\dots,|t|_{\mathsf{a}_k}) \mid \exists t \in L\}$. A subset of vectors in $\mathbb{N}^k$ is *linear* if the set is $\{v \mid \exists x_1,\dots,x_n \in \mathbb{N}\colon v = c + x_1 v_1 + \cdots + x_n v_n\}$ for some vectors $c, v_1, \ldots, v_n$ in $\mathbb{N}^k$. We call the vector expression $c + x_1v_1 + \cdots + x_nv_n$ a *non-negative vector addition system*[1] (NNVAS). The finite union of linear sets is a *semi-linear* set. By definition, every linear set is non-empty. However, the empty set is semi-linear, being the union of zero linear sets. A finite subset of vectors is also semi-linear, while the finite subset is not linear if it contains more than one element.

The Parikh image $\Psi_F(L)$ of a regular tree language $L$ is effectively semi-linear, meaning that: Given a regular tree automaton $\mathcal{A}$, one can construct a finite sequence of NNVAS's $V_1, \ldots, V_n$ ($n \geqslant 0$) such that the union of linear sets generated by $V_1, \ldots, V_n$ coincides with $\Psi_F(\mathcal{L}(\mathcal{A}))$.

We introduce the two special operations for subsets of vectors. The $i$-th *projection* $\mathsf{pr}_i$ of a subset $W$ of $\mathbb{N}^k$ ($1 \leqslant i \leqslant k$) is the mapping from $\mathbb{N}^k$ to $\mathbb{N}^{k-1}$ such that $\mathsf{pr}_i(W) = \{ (v(1), \ldots, v(i-1), v(i+1), \ldots, v(k)) \mid \exists v \in W \}$. Similarly, the $i$-th *cylindrification* $\mathsf{cy}_i$ of $W$ is the mapping from $\mathbb{N}^k$ to $\mathbb{N}^{k+1}$ such that $\mathsf{cy}_i(W) = \{ (v(1), \ldots, v(i-1), x, v(i), \ldots, v(k)) \mid \exists v \in W, \ x \in \mathbb{N} \}$. Here we denote $v(i)$ for the $i$-th element of a vector $v$.

**Proposition 2 ([6]).** *The class of semi-linear sets is effectively closed under Boolean operations, projection, and cylindrification. Moreover, the emptiness (and thus, the membership and inclusion problems also) are decidable.* □

Using the property, one can show that the class of AC-regular tree languages is closed under Boolean operations. Similarly, Dal Zilio and Lugies for multitree automata [4,13] and Verma and Goubault-Larrecq for two-way AC-tree automata [21] showed that their equationally extended tree automata enjoy the closure properties of Boolean operations.

As the benefit from the closure properties of Boolean operations and the positive decidability results, several fundamental decision problems are translated to language problems in the class of regular AC-TA. Satisfiability of monadic second-order logic with Presburger arithmetic (Presburger MSO) [19] is one of the examples.

What about then the class of monotone AC-TA?

## 3 Monotone AC Tree Automata for Multiplication and Exponentiation

Hereafter in the following sections, we consider a special class of monotone AC-TA over a *flat signature.* A signature $F$ is flat if $F$ consists of one AC symbol $\mathsf{f}$ and constants only. We write $F_0$ for the set of all constants in $F$. So $F = \{\mathsf{f}\} \cup F_0$.

Ohsaki *et al.* showed in [15] that there exists a monotone AC-TA whose language $L$ over $F$ satisfies $\Psi_F(L) = \{ (x, y, z, 1, 2) \in \mathbb{N}^5 \mid x \times y \geqslant z \wedge x + y + z > 0 \}$. The class of regular AC-TA or the related automata does not satisfy this property due to the expressiveness limited by linear arithmetic. This example thus reveals that there exists the strict hierarchy between monotone AC-TA and regular AC-TA.

[1] Vector addition systems (e.g. [12]) are equipped with vectors $v_1, \ldots, v_n$ from $\mathbb{Z}^k$.

$F$ : f a b c

$E$ : $\mathsf{f}(\mathsf{f}(x,y),z) \approx \mathsf{f}(x,\mathsf{f}(y,z)) \quad \mathsf{f}(x,y) \approx \mathsf{f}(y,x)$

$\Delta_1$ : $\mathsf{a} \to \lambda \quad \mathsf{b} \to \lambda \quad \mathsf{f}(\lambda,\lambda) \to \lambda$

$\Delta_2$ : $\mathsf{a} \to \alpha \quad \mathsf{b} \to \beta \quad \mathsf{c} \to \gamma \quad \beta \to \beta_1 \quad \beta_1 \to \beta_2 \quad \beta_2 \to \delta \quad \mathsf{f}(\alpha,\delta) \to \delta$

$\mathsf{f}(\alpha,\beta_1) \to \mathsf{f}(\alpha_1,\beta_1) \quad \mathsf{f}(\alpha_1,\gamma) \to \alpha_2 \quad \mathsf{f}(\alpha_2,\beta_2) \to \mathsf{f}(\alpha,\beta_2) \quad \mathsf{f}(\beta_2,\beta) \to \beta_1$

$Q$ : $\alpha \ \alpha_1 \ \alpha_2 \ \beta \ \beta_1 \ \beta_2 \ \gamma \ \delta \ \lambda$

$Q_{\mathsf{fin}}$ : $\lambda, \ \delta$

**Fig. 1.** $\mathcal{A}^{\mathsf{mult}} = (\mathcal{E}, Q, Q_{\mathsf{fin}}, \Delta_1 \cup \Delta_2)$ and $\mathcal{E} = (F, E)$

In this section, we show that there exist monotone AC tree automata $\mathcal{A}^{\mathsf{mult}}_{\mathcal{E}}$ and $\mathcal{A}^{\mathsf{exp}}_{\mathcal{E}}$ such that

$$\Psi_F(\mathcal{L}(\mathcal{A}^{\mathsf{mult}}_{\mathcal{E}})) = \{\,(x,y,z) \in \mathbb{N}^3 \mid x \times y \geqslant z \wedge x+y+z > 0\,\}$$
$$\Psi_F(\mathcal{L}(\mathcal{A}^{\mathsf{exp}}_{\mathcal{E}})) = \{\,(x,y,z) \in \mathbb{N}^3 \mid x^y \geqslant z \wedge x+y > 0\,\}$$

Note that the condition $x+y+z>0$ is necessary since our tree encoding does not allow to have a tree whose Parikh image is $(0,0,0)$.

In the previous paper ([15]), it was unknown whether there is an automaton with a minimal signature (i.e., an automaton without any extra constants). As far as we know, it was also unknown whether there exists a monotone AC tree automaton for exponentiation even when extra symbols are allowed.

### 3.1 AC Tree Automaton for "$x \times y \geqslant z$"

**Lemma 1.** *Let $F = \{\mathsf{f},\mathsf{a},\mathsf{b},\mathsf{c}\}$ with one AC-symbol $\mathsf{f}$ and three constants $\mathsf{a},\mathsf{b},\mathsf{c}$. There exists a monotone AC-TA $\mathcal{A}^{\mathsf{mult}}_{\mathcal{E}}$ over $F$ such that*

$$\Psi_F(\mathcal{L}(\mathcal{A}^{\mathsf{mult}}_{\mathcal{E}})) = \{\,(x,y,z) \in \mathbb{N}^3 \mid x \times y \geqslant z \ \wedge \ x+y+z > 0\,\}.$$

□

Our example is exhibited in Fig. 1. Based on case analysis of the number of occurrences of c, the AC-TA $\mathcal{A}^{\mathsf{mult}} = (\mathcal{E}, Q, Q_{\mathsf{fin}}, \Delta_1 \cup \Delta_2)$ is defined as the union of the two automata $\mathcal{A}^{\mathsf{mult}}_1 = (\mathcal{E}, \{\lambda\}, \{\lambda\}, \Delta_1)$ and $\mathcal{A}^{\mathsf{mult}}_2 = (\mathcal{E}, Q - \{\lambda\}, Q_{\mathsf{fin}} - \{\lambda\}, \Delta_2)$.

If $|t|_{\mathsf{c}} = 0$, obviously $t \to^*_{\Delta_1/\mathcal{E}} \lambda$, and thus $\mathcal{A}^{\mathsf{mult}}_{1\mathcal{E}}$ accepts $t$. Otherwise $|t|_{\mathsf{c}} > 0$, we observe for the recursive computation that $|t|_{\mathsf{a}} \times |t|_{\mathsf{b}} \geqslant |t|_{\mathsf{c}}$ iff $|t|_{\mathsf{c}} = m \times |t|_{\mathsf{a}} + n$ for some $m, n$ such that $0 \leqslant m < |t|_{\mathsf{b}}$ and $0 < n \leqslant |t|_{\mathsf{a}}$. That means, a tree $t = \mathsf{f}(\mathsf{a}^x, \mathsf{b}^y, \mathsf{c}^z)$ with $x \times y \geqslant z$ is considered to be the tree with the following leaves:

$$\mathsf{f}(\,\mathsf{a}^x, \mathsf{b}, \underbrace{\mathsf{b}, \mathsf{c}^x, \ \ldots\ , \mathsf{b}, \mathsf{c}^x}_{m \,\times\, (\mathsf{b},\mathsf{c}^x)}, \mathsf{b}^{y-m-1}, \mathsf{c}^n\,).$$

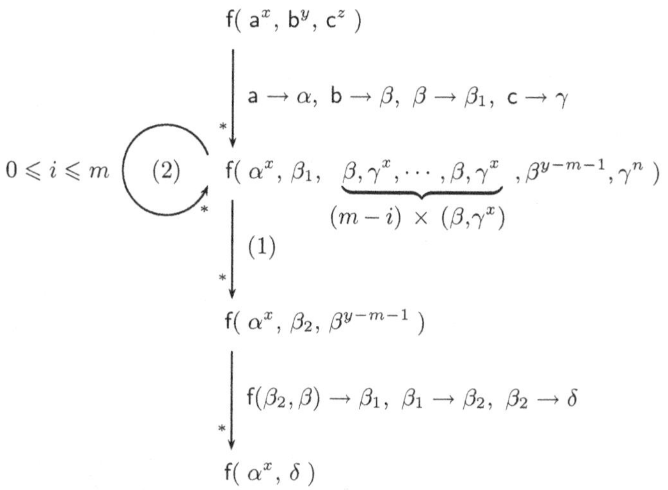


**Fig. 2.** A derivation of $\mathsf{f}(\,\mathsf{a}^x,\ \mathsf{b}^y,\ \mathsf{c}^z)$

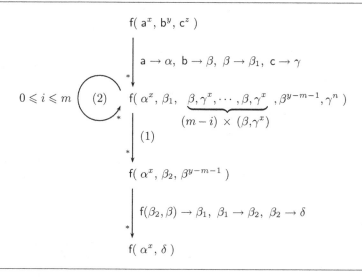

Here $\mathsf{f}(\mathsf{a}^x, \mathsf{b}^y, \mathsf{c}^z)$ represents that it has $x$ occurrences of the constant $\mathsf{a}$, $y$ occurrences of $\mathsf{b}$, and $z$ occurrences of $\mathsf{c}$, respectively.

Regarding the derivation from $\mathsf{f}(\mathsf{a}^x, \mathsf{b}^y, \mathsf{c}^z)$, we have the two properties

(1) $\mathsf{f}(\,\alpha^x, \beta_1, \gamma^n\,) \to^*_{\Delta_2/\mathcal{E}} \mathsf{f}(\,\alpha^x, \beta_2\,)$ if $n \leqslant x$,

(2) $\mathsf{f}(\,\alpha^x, \beta_1, \gamma^n, \beta\,) \to^*_{\Delta_2/\mathcal{E}} \mathsf{f}(\,\alpha^x, \beta_1\,)$ if $n \leqslant x$.

The derivation in the property (1) is obtained from $\mathsf{f}(\alpha^x, \beta_1, \gamma^n)$ by applying $\mathsf{f}(\alpha, \beta_1) \to \mathsf{f}(\alpha_1, \beta_1)$ and $\mathsf{f}(\alpha_1, \gamma) \to \alpha_2$ repeatedly $n$ times, and then by applying $\beta_1 \to \beta_2$ and $\mathsf{f}(\alpha_2, \beta_2) \to \mathsf{f}(\alpha, \beta_2)$. Furthermore, when $n \leqslant x$, we have

$$\mathsf{f}(\,\alpha^x, \beta_1, \gamma^n, \beta\,) \to^*_{\Delta_2/\mathcal{E}} \mathsf{f}(\,\alpha^x, \beta_2, \beta\,) \to_{\Delta_2/\mathcal{E}} \mathsf{f}(\,\alpha^x, \beta_1\,)$$

that is the derivation in the property (2). According to this second property, $t = f(\mathsf{a}^x, \mathsf{b}^y, \mathsf{c}^z)$ reaches $\mathsf{f}(\alpha^x, \beta_1, \beta^{y-m-1}, \gamma^n)$ by $\Delta_2/\mathcal{E}$. Thus, by the first property, $t$ reaches $\mathsf{f}(\alpha^x, \beta_2, \beta^{y-m-1})$. That means, we have the derivation

$$t \to^*_{\mathcal{A}^{\mathsf{mult}}_{2\,\mathcal{E}}} \mathsf{f}(\alpha^x, \beta^y, \gamma^z) \to^*_{\mathcal{A}^{\mathsf{mult}}_{2\,\mathcal{E}}} \mathsf{f}(\alpha^x, \beta_1, \beta^{y-m-1}, \gamma^n) \to^*_{\mathcal{A}^{\mathsf{mult}}_{2\,\mathcal{E}}} \mathsf{f}(\alpha^x, \beta_2, \beta^{y-m-1}).$$

Since $\mathsf{f}(\beta_2, \beta) \to^*_{\mathcal{A}^{\mathsf{mult}}_{\mathcal{E}}} \beta_2$ and $\beta_2 \to_{\mathcal{A}^{\mathsf{mult}}_{\mathcal{E}}} \delta$, the derivation from $t$ reaches the final state $\delta$:

$$\mathsf{f}(\alpha^x, \beta_2, \beta^{y-m-1}) \to^*_{\mathcal{A}^{\mathsf{mult}}_{2\,\mathcal{E}}} \mathsf{f}(\alpha^x, \beta_2) \to^*_{\mathcal{A}^{\mathsf{mult}}_{2\,\mathcal{E}}} \mathsf{f}(\alpha^x, \delta) \to^*_{\mathcal{A}^{\mathsf{mult}}_{2\,\mathcal{E}}} \delta.$$

Therefore, every tree $t$ over $F = \{\,\mathsf{f}, \mathsf{a}, \mathsf{b}, \mathsf{c}\,\}$ is accepted if $|t|_{\mathsf{a}} \times |t|_{\mathsf{b}} \geqslant |t|_{\mathsf{c}}$. The derivation from $t$ to $\mathsf{f}(\alpha^x, \delta)$ by $\mathcal{A}^{\mathsf{mult}}_{2\,\mathcal{E}}$ is illustrated in Fig. 2.

See Appendix A.1 in [10] for the completeness proof of $\mathcal{A}^{\mathsf{mult}}_{\mathcal{E}}$.

### 3.2 AC Tree Automaton for Exponentiation

Let us first illustrate the AC tree automaton $\mathcal{A}^{\mathsf{sqr}} = (\mathcal{E}, Q, \{\lambda, \delta\}, \Delta_1 \cup \Delta_2)$ that accepts $\{t \mid 2^{|t|_{\mathsf{a}}} \geqslant |t|_{\mathsf{b}}\}$.

$$\Delta_1 : \mathsf{a} \to \lambda \qquad \mathsf{f}(\lambda, \lambda) \to \lambda$$

$$\Delta_2 : \mathsf{a} \to \alpha \qquad \mathsf{b} \to \beta \qquad \mathsf{b} \to \delta \qquad \alpha \to \gamma_1 \qquad \gamma_1 \to \gamma_3$$

$$\left.\begin{array}{l} \mathsf{f}(\beta,\gamma_1) \to \mathsf{f}(\beta_1,\gamma_2) \\ \mathsf{f}(\beta,\gamma_2) \to \gamma_1 \end{array}\right\} (1) \quad \left.\begin{array}{l} \mathsf{f}(\beta_1,\gamma_3) \to \mathsf{f}(\beta,\gamma_3) \\ \mathsf{f}(\alpha,\gamma_3) \to \gamma_1 \end{array}\right\} (2) \quad \left.\begin{array}{l} \mathsf{f}(\beta,\gamma_3) \to \delta \\ \mathsf{f}(\alpha,\delta) \to \delta \end{array}\right\} (3)$$

Regarding the above transition rules, we note that for every $t \in \mathcal{T}_F$, if $|t|_{\mathsf{b}} = 0$, $t$ is accepted by $\mathcal{A}^{\mathsf{sqr}}_{\mathcal{E}}$ using the transition rules in $\Delta_1$; otherwise, $t$ is accepted by the rules in $\Delta_2$. If $|t|_{\mathsf{a}} = 0$ and $|t|_{\mathsf{b}} = 1$, it is obvious. If $|t|_{\mathsf{a}} > 0$, the derivation from $t$ to $\delta$ by $\Delta_2$ is illustrated in Fig. 3.

The derivation in Fig. 3. is separated to the three phases: First, $\mathsf{a}, \mathsf{b}$ in $t$ are replaced by $\alpha, \beta$, and one of $\alpha$'s is replaced by $\gamma_1$. This is for initializing the computation. Next, we apply the computation for "reducing by half" the number of occurrences of $\beta$. At most the half of $\beta$'s are replaced by $\beta_1$, and the same number of $\beta$'s are removed. This transition is performed by applying the rules in (1). In the third phase, $\beta_1$'s are replaced back to $\beta$'s by the rule $\mathsf{f}(\beta_1, \gamma_3) \to \mathsf{f}(\beta, \gamma_3)$ in (2). Therefore, we have the derivation

$$\mathsf{f}(\,\alpha^m, \gamma_1, \beta^n\,) \to^*_{\mathcal{A}^{\mathsf{sqr}}_{\mathcal{E}}} \mathsf{f}(\alpha^m, \gamma_3, \beta^{n-k}) \text{ such that } 0 \leqslant k \leqslant \frac{1}{2}n.$$

If $n - k = 1$, the transition rule $\mathsf{f}(\beta, \gamma_3) \to \delta$ in (3) is applied to $\mathsf{f}(\alpha^m, \gamma_3, \beta^{n-k})$, and the remaining $\alpha$'s are eliminated by using the rule $\mathsf{f}(\alpha, \delta) \to \delta$. Otherwise, $\mathsf{f}(\alpha, \gamma_3) \to \gamma_1$ is applied for iterative computation, and it restarts the computation from the reduction-by-half phase.

Generalizing the above observation, we obtain the next lemma.

**Lemma 2.** *Let $F = \{\mathsf{f}, \mathsf{a}, \mathsf{b}, \mathsf{c}\}$ with one AC-symbol $\mathsf{f}$ and three constants $\mathsf{a}, \mathsf{b}, \mathsf{c}$. There exists a monotone AC-TA $\mathcal{A}^{\mathsf{exp}}_{\mathcal{E}}$ over $F$ such that*

$$\Psi_F(\mathcal{L}(\mathcal{A}^{\mathsf{exp}}_{\mathcal{E}})) = \{\,(x, y, z) \in \mathbb{N}^3 \mid x^y \geqslant z \,\wedge\, x + y > 0\,\}$$

*Proof.* Appendix A.2 in [10]. □

## 4 Monotone Exponential Diophantine Formulas

An arithmetic constraint over a finite set of variables $x_1, \ldots, x_n$ $(n \geqslant 0)$ is an *exponential Diophantine formula* if the formula is in $D$ of the following syntax:

$$D ::= A \mid \neg(\,D\,) \mid D \vee D \mid D \wedge D$$

$$A ::= \exists x_i\,(\,D\,) \mid \sum_{i \in I} a_i x_i \geqslant b \mid x_i \times x_j \geqslant x_k \mid x_i^{x_j} \geqslant x_k$$

such that $a_i, b \in \mathbb{Z}$ for all $i \in I$. If a formula $\psi$ does not contain an atomic formula $x_i^{x_j} \geqslant x_k$, $\psi$ is simply called a *Diophantine formula*.

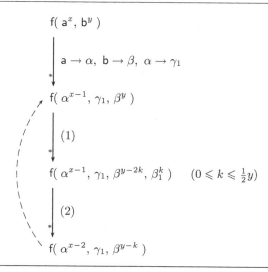

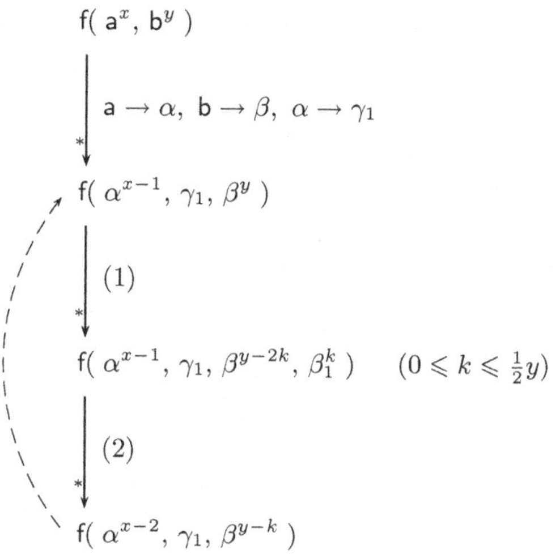


**Fig. 3.** Three phases in the derivation from $\mathsf{f}(\,\mathsf{a}^x,\ \mathsf{b}^y\,)$

A formula $\psi$ is *satisfiable* if there exists an assignment $\theta$ to free-variables in $\psi$ such that $\psi\theta$ is true. Similarly, $\psi$ is *valid* if $\psi\theta$ is true for any assignment $\theta$. For instance, the formula $x \geqslant 1$ is satisfiable but not valid, because $(x \geqslant 1)\{x \mapsto 0\}$ is not true. In the paper we write $[\![\, \psi \,]\!]_{\mathbb{N}}$ for the set of all solution vectors of $\psi$, that is, $v \in [\![\, \psi \,]\!]_{\mathbb{N}}$ iff $v \in \mathbb{N}^n$ and $\psi\{x_1 \mapsto v(1), \ldots, x_n \mapsto v(n)\}$ is true. Note that $\psi$ is satisfiable iff $[\![\, \psi \,]\!]_{\mathbb{N}} \neq \varnothing$; $\psi$ is valid iff $[\![\, \psi \,]\!]_{\mathbb{N}} = \mathbb{N}^n$.

A formula $\psi$ is *logically equivalent* to $\phi$, denoted $\psi \Leftrightarrow \phi$, if for every assignment $\theta$ to free-variables in formulas, $\psi\theta$ is true iff $\phi\theta$ is true. Obviously, $\Leftrightarrow$ is an equivalence relation. See the reference, e.g. [9], for more logical background.

We define a new class of arithmetic constraints.

> A formula $\psi$ is *monotone* if none of the negative sub-formulas $\neg(\phi)$ of $\psi$ contains an atomic formula of the form $x_i \times x_j \geqslant x_k$ or $x_i^{x_j} \geqslant x_k$.

For instance, $x^2 \geqslant y$ is logically equivalent to the formula $\exists z(\,x \times z \geqslant y \wedge x \geqslant z\,)$. However, there is no monotone (exponential) Diophantine formula equivalent to $x^2 = y$, because every formula equivalent to $x^2 = y$ has a negative sub-formula containing a non-linear constraint, e.g. $(x^2 \geqslant y) \wedge \exists z(\neg(x^2 \geqslant z) \wedge (z = y + 1))$.

Next we define arithmetic expressions over variables $x_1, \ldots, x_n$ $(n \geqslant 0)$.

$$E(\boldsymbol{x}_n) ::= a \mid x_i \mid E(\boldsymbol{x}_n) + E(\boldsymbol{x}_n) \mid E(\boldsymbol{x}_n) \times E(\boldsymbol{x}_n) \mid E(\boldsymbol{x}_n)^{E(\boldsymbol{x}_n)}$$

where $a \in \mathbb{N}$. Elements in $E(\boldsymbol{x}_n)$ are called *exponentials*. We write $P(\boldsymbol{x}_n)$ instead of $E(\boldsymbol{x}_n)$ for a *polynomial* with non-negative integer coefficients. A *linear polynomial* (with arbitrary integer coefficients) is denoted $L(\boldsymbol{x}_n)$.

Note that for every $E(\boldsymbol{x}_n)$ and $L(\boldsymbol{x}_n)$, one can find a monotone exponential Diophantine formula equivalent to $E(\boldsymbol{x}_n) \geqslant L(\boldsymbol{x}_n)$, because:

$$E_1(\boldsymbol{x}_n) + E_2(\boldsymbol{x}_n) \geqslant y \Leftrightarrow \exists z_1, \exists z_2 (E_1(\boldsymbol{x}_n) \geqslant z_1 \wedge E_2(\boldsymbol{x}_n) \geqslant z_2 \wedge z_1 + z_2 \geqslant y)$$
$$E_1(\boldsymbol{x}_n) \times E_2(\boldsymbol{x}_n) \geqslant y \Leftrightarrow \exists z_1, \exists z_2 (E_1(\boldsymbol{x}_n) \geqslant z_1 \wedge E_2(\boldsymbol{x}_n) \geqslant z_2 \wedge z_1 \times z_2 \geqslant y)$$
$$E_1(\boldsymbol{x}_n)^{E_2(\boldsymbol{x}_n)} \geqslant y \Leftrightarrow \exists z_1, \exists z_2 (E_1(\boldsymbol{x}_n) \geqslant z_1 \wedge E_2(\boldsymbol{x}_n) \geqslant z_2 \wedge z_1^{z_2} \geqslant y)$$

Given a polynomial $P(\boldsymbol{x}_n)$ with *arbitrary* integer coefficients, the question $[\![ P(\boldsymbol{x}_n) = 0 ]\!]_{\mathbb{N}} \stackrel{?}{=} \varnothing$ (HILBERT'S 10TH PROBLEM for non-negative solutions) is undecidable [14]. Therefore, the satisfiability question for (exponential) Diophantine formulas is undecidable.

A formula $\psi$ is *linear Diophantine*, on the other hand, if every atomic formula of $\psi$ is a linear constraint $\sum_{i \in I} a_i x_i \geqslant b$ such that $a_i, b \in \mathbb{Z}$ for all $i \in I$. In the literature a linear Diophantine formula is often called a *Presburger formula.* The solution set of a given Presburger formula is effectively semi-linear. And thus, the satisfiability of Presburger formulas is decidable [5].

Every linear Diophantine formula is monotone, but a monotone Diophantine formula may not be linear. Moreover, $x^2 = y$ is not a monotone formula, while the formula is Diophantine. So we have the following relation among the classes of arithmetic constraints:

$$\text{Presburger} \subsetneq \text{monotone Diophantine} \subsetneq \text{Diophantine}$$

The following decidability result follows immediately from the results of the previous section and the closure properties of monotone AC tree automata.

**Theorem 1 (Satisfiability).** *Satisfiability of monotone exponential Diophantine formulas is decidable.*

*Proof.* We observe that $\exists x\,(\psi) \vee \phi \Leftrightarrow \exists x\,(\psi \vee \phi)$ and $\exists x\,(\psi) \wedge \phi \Leftrightarrow \exists x\,(\psi \wedge \phi)$ if $x$ does not freely occur in $\phi$. This implies that given a monotone exponential Diophantine formula $\delta$, one can move existential quantifiers outside of $\delta$, so that $\delta$ is logically equivalent to a formula $\exists \boldsymbol{x}\,(\psi)$ for some $\psi$ in $S$.

$$S ::= x \times y \geqslant z \mid x^y \geqslant z \mid C \mid S \wedge S \mid S \vee S$$

Here $C$ ranges over the set of Presburger formulas. By Lemmas 1 and 2 and Proposition 1, one can construct $\mathcal{A}_\mathcal{E}$ such that $\Psi_F(\mathcal{L}(\mathcal{A}_\mathcal{E})) = [\![ \psi ]\!]_{\mathbb{N}} - \mathbf{0}$. Obviously, $\exists \boldsymbol{x}\,(\psi)$ is satisfiable if and only if $\mathcal{L}(\mathcal{A}_\mathcal{E}) \neq \varnothing$ or $\mathbf{0}$ is a solution of $\psi$. By Proposition 2, $\mathcal{L}(\mathcal{A}_\mathcal{E}) \neq \varnothing$ is decidable. Moreover, the question if $\mathbf{0}$ is a solution of $\psi$ is decidable. Therefore, the satisfiability of $\exists \boldsymbol{x}\,(\psi)$ is so. □

## 5 Linearly Bounded Projection and Definable Formulas

In this section, we focus on the main question, that is, the expressiveness of monotone AC tree automata relative to the first-order theory of arithmetic.

According to the examples in Section 3 (Lemmas 1 and 2), the class of tree languages accepted by monotone AC tree automata includes the solution sets of $x \times y \geqslant z$ and $x^y \geqslant z$. We generalize this result, so that the class covers more complex arithmetic constraints, such as $(x^2 + y)^x \geqslant z$.

A formula $\psi$ is *monotone AC-TA definable* if there effectively exists a monotone AC-TA $\mathcal{A}_\mathcal{E}$ over a flat signature $F$ such that the Parikh image of the accepted language is the set of all solution vectors except zero, i.e. $\Psi_F(\mathcal{L}(\mathcal{A}_\mathcal{E})) = [\![\,\phi\,]\!]_\mathbb{N} - \mathbf{0}$. In the following of the section, we attempt to characterize the expressive power of monotone AC tree automata in comparison with the class of arithmetic formulas. In particular, we show that every formula in the following sub-class of monotone exponential Diophantine formulas is monotone AC-TA definable.

$$M \;::=\; E(\boldsymbol{x}_n) \geqslant L(\boldsymbol{x}_n) \;\mid\; C \;\mid\; M \vee M \;\mid\; M \wedge M$$

As appeared previously, $C$ ranges over the set of Presburger formulas.

*Remark 1.* Our discussion in the sequel does not lose generality, though $\mathbf{0}$ is excluded in the above definition. Indeed, there are several ways to deal with the zero vector in tree automata. For instance, one can introduce the special constant $\dot{0}$ such that the solution set $[\![\,\phi\,]\!]_\mathbb{N}$ contains $\mathbf{0}$ iff $\mathcal{A}_\mathcal{E}$ accepts $\dot{0}$. Note that the notion of definability is invariant in the treatment of $\mathbf{0}$.

We note that the class of languages of *regular* AC-TA over a flat signature (which is a *commutative context-free grammar* in formal languages) coincides with the class of solution sets of Presburger formulas, called Presburger sets, [5]. For Petri nets, the set $\{(x_1,\ldots,x_n,y) \mid P(\boldsymbol{x}_n) \geqslant y\}$ is a Petri net language if $P(\boldsymbol{x}_n)$ is a polynomial with non-negative integer coefficients [7], while the transition rules of monotone AC tree automata are more restricted.

According to the above observation about regular AC-TA together with the fact that the class of monotone AC-TA subsumes regular AC-TA and is closed under union and intersection (Proposition 1), we have the following lemma.

**Lemma 3.** *Every Presburger formula is monotone AC-TA definable. Moreover, if formulas $\psi_1$ and $\psi_2$ are monotone AC-TA definable, so are $\psi_1 \wedge \psi_2$ and $\psi_1 \vee \psi_2$.* □

As noted in the proof of Theorem 1, every monotone exponential Diophantine formula is logically equivalent to some formula $\exists \boldsymbol{x}(\,\psi\,)$ such that $\psi$ is a monotone AC-TA definable formula. Therefore, the definability of monotone exponential Diophantine formulas could follow immediately if the class of monotone AC-TA *were* closed under projection. However, it is unknown so far whether the class of monotone AC-TA is closed under *arbitrary* projection.

Our key observation is that the language class of monotone AC-TA is closed under a special projection, called *linearly bounded projection*, discussed below. It turns out that the linearly bounded projection suffices to show that the above sub-class $M$ of monotone exponential Diophantine formulas is monotone AC-TA definable.

**Lemma 4** (Linearly Bounded Projection: special case). *Given a monotone AC-TA $\mathcal{A}_\mathcal{E}$ over a flat signature $F$ containing constants* $\mathsf{a}$ *and* $\mathsf{b}$, *one can construct a monotone AC-TA $\mathcal{B}_\mathcal{E}$ over $F - \{\mathsf{a}\}$ such that*

$$\mathcal{L}(\mathcal{B}_\mathcal{E}) = \{\, u \mid \exists t \in \mathcal{L}(\mathcal{A}_\mathcal{E})\colon |t|_\mathsf{a} \leqslant |t|_\mathsf{b} \text{ and } |t|_\mathsf{c} = |u|_\mathsf{c} \text{ for all } \mathsf{c} \in F_0 - \{\mathsf{a}\}\,\}.$$

*Namely, if a formula $\psi$ is monotone AC-TA definable, so is $\exists x\,(\,x \leqslant y \wedge \psi\,)$.*

*Proof (Outline).* We suppose, without loss of generality, that $\mathcal{A}{=}(\mathcal{E}, Q_1, \{\gamma\}, \Delta_1)$ such that $\Delta_1$ contains no transition rule for $\mathsf{a}, \mathsf{b}$ except $\mathsf{a} \to \alpha$ and $\mathsf{b} \to \beta$. The set

$Q_2$ of state symbols of $\mathcal{B}_\mathcal{E}$ are the union of $Q_1$ and the set $\{ M \in \mathsf{mul}(Q_1) \mid |M| \leqslant 2 \}$ of multisets of $Q_1$ whose size is at most 2. (Here, $\mathsf{mul}(S)$ stands for the set of multisets of a set $S$.) Moreover, to distinguish multisets from sets, we write $\{\!\!\{\cdot\}\!\!\}$ to denote a multiset, and $\uplus$ for the multiset union. The set $\Delta_2$ of transition rules of $\mathcal{B}_\mathcal{E}$ consists of rules from $(\Delta_1 - \{ \mathsf{a} \to \alpha,\ \mathsf{b} \to \beta \}) \cup \{ \mathsf{b} \to \{\!\!\{\alpha, \beta\}\!\!\},\ \mathsf{b} \to \{\!\!\{\beta\}\!\!\} \}$ and the rules defined below:

$$\begin{array}{rcll} M \uplus \{\!\!\{p\}\!\!\} & \to & M \uplus \{\!\!\{q\}\!\!\} & \text{if } p \to^*_{\mathcal{A}_\mathcal{E}} q \\ \{\!\!\{p, q\}\!\!\} & \to & \{\!\!\{r\}\!\!\} & \text{if } \mathsf{f}(p, q) \to^*_{\mathcal{A}_\mathcal{E}} r \\ \mathsf{f}(M \uplus \{\!\!\{p\}\!\!\}, q) & \to & M \uplus \{\!\!\{r\}\!\!\} & \\ \mathsf{f}(M \uplus \{\!\!\{p\}\!\!\}, \{\!\!\{q\}\!\!\}) & \to & M \uplus \{\!\!\{r\}\!\!\} & \\ \mathsf{f}(M \uplus \{\!\!\{p\}\!\!\}, N \uplus \{\!\!\{q\}\!\!\}) & \to & \mathsf{f}(M \uplus \{\!\!\{r\}\!\!\}, N) & (N \neq \varnothing) \\ \{\!\!\{p, q\}\!\!\} & \to & \{\!\!\{r, s\}\!\!\} & \text{if } \mathsf{f}(p, q) \to^*_{\mathcal{A}_\mathcal{E}} \mathsf{f}(r, s) \\ \mathsf{f}(M \uplus \{\!\!\{p\}\!\!\}, q) & \to & \mathsf{f}(M \uplus \{\!\!\{r\}\!\!\}, s) & \\ \mathsf{f}(M \uplus \{\!\!\{p\}\!\!\}, N \uplus \{\!\!\{q\}\!\!\}) & \to & \mathsf{f}(M \uplus \{\!\!\{r\}\!\!\}, N \uplus \{\!\!\{s\}\!\!\}) & \end{array}$$

such that $p, q, r, s \in Q_1$.

Let $\mathcal{B}_\mathcal{E}$ be the automaton equipped with the above $Q_2$ and $\Delta_2$ such that the final states are $\gamma$ and $\{\!\!\{\gamma\}\!\!\}$. For instance, the transition move $C[\mathsf{a}, \mathsf{b}] \to_{\mathcal{A}_\mathcal{E}} C[\alpha, \mathsf{b}] \to_{\mathcal{A}_\mathcal{E}} C[\alpha, \beta]$ is simulated in $\mathcal{B}_\mathcal{E}$ by the single transition $D[\mathsf{b}] \to_{\mathcal{B}_\mathcal{E}} D[\{\!\!\{\alpha, \beta\}\!\!\}]$. It is not difficult to show that for all $t \in \mathcal{T}_F$ and $u \in \mathcal{T}_{F - \{\mathsf{a}\}}$, if

- $|t|_\mathsf{a} \leqslant |t|_\mathsf{b}$
- $|u|_\mathsf{c} = |t|_\mathsf{c}$ for all $c \in F_0 - \{\mathsf{a}\}$,

then $\mathcal{A}_\mathcal{E}$ accepts $t$ iff there exists a derivation $u \to^*_{\mathcal{B}_\mathcal{E}} \gamma$ or $u \to^*_{\mathcal{B}_\mathcal{E}} \{\!\!\{\gamma\}\!\!\}$. □

This lemma can be generalized as follows.

**Lemma 5** (LINEARLY BOUNDED PROJECTION). *If $L(\boldsymbol{x}_n)$ is a linear polynomial with non-negative integer coefficients, every monotone AC-TA $\mathcal{A}_\mathcal{E}$ over the flat signature $F = \{\mathsf{f}\} \cup \{\mathsf{a}\} \cup \{\mathsf{b}_1, \dots, \mathsf{b}_n\}$ can be transformed to $\mathcal{B}_\mathcal{E}$ over $F - \{\mathsf{a}\}$ such that*

$$\mathcal{L}(\mathcal{B}_\mathcal{E}) = \{ u \mid \exists t \in \mathcal{L}(\mathcal{A}_\mathcal{E}) : |t|_\mathsf{a} \leqslant L(|t|_{\mathsf{b}_1}, \dots, |t|_{\mathsf{b}_n}) \text{ and } |t|_{\mathsf{b}_i} = |u|_{\mathsf{b}_i} \text{ for all } i \}.$$

*Proof.* Appendix A.3 in [10]. □

The following theorem is an immediate consequence.

**Theorem 2.** *Let $L(\boldsymbol{y}_n)$ be a linear polynomial (with integer coefficients) such that free-variables $\boldsymbol{y}_n$ in $L$ do not contain $x$. Then, if $\psi$ is monotone AC-TA definable, so is $\exists x\, ( x \leqslant L(\boldsymbol{y}_n) \wedge \psi )$.* □

Let us demonstrate the linearly bounded projection as a tool for translating exponential constraints by monotone AC tree automata.

*Example 1.* Consider the super-exponential inequality $x^{x^x} \geqslant y$. This constraint is equivalent to $( x = 1 \wedge 1 \geqslant y ) \vee \exists z\, ( x \geqslant 2 \wedge x^z \geqslant y \wedge x^x \geqslant z)$. Let $\psi$ be the formula $x \geqslant 2 \wedge x^z \geqslant y \wedge x^x \geqslant z$. Then $\psi$ is monotone AC-TA definable

(Lemma 2). Note that the definability of $x^x \geqslant y$ follows from Theorem 2, meaning that, $x^x \geqslant y$ is logically equivalent to $\exists w\,(\,w \leqslant x \,\wedge\, x^w \geqslant y \wedge w \geqslant x\,)$ and the sub-formula $x^w \geqslant y \wedge w \geqslant x$ is monotone AC-TA definable.

Observe that $x \geqslant 2$ implies $x^z \geqslant z$. Thus we have

$$\begin{aligned}\exists z\,(\psi) &\Leftrightarrow \exists z\,(\,z \leqslant y \wedge \psi\,) \vee \exists z\,(\,z \geqslant y \wedge \psi\,)\\ &\Leftrightarrow \exists z\,(\,z \leqslant y \wedge \psi\,) \vee \exists z\,(\,z \geqslant y \,\wedge\, x \geqslant 2 \,\wedge\, x^x \geqslant z)\\ &\Leftrightarrow \exists z\,(\,z \leqslant y \wedge \psi\,) \vee (\,x \geqslant 2 \,\wedge\, x^x \geqslant y\,)\end{aligned}$$

The left sub-formula $\exists z\,(\,z \leqslant y \wedge \psi\,)$ is monotone AC-TA definable (Theorem 2), and the right sub-formula is so according to the above observation. Therefore, $x^{x^x} \geqslant y$ is monotone AC-TA definable. □

It is not difficult to generalize the technique we have used in the example, so that we state the next lemma. See Appendix A.4 in [10] for a proof.

**Lemma 6.** *Every monotone exponential Diophantine inequality* $E(\boldsymbol{x}_n) \geqslant L(\boldsymbol{x}_n)$ *is monotone AC-TA definable.* □

Accordingly, we have the main result of the paper.

**Theorem 3 (Definability of exponential Diophantine formulas).** *Every arithmetic constraint in* $M$ *of the following syntax is monotone AC-TA definable:*

$$M \;::=\; E(\boldsymbol{x}_n) \geqslant L(\boldsymbol{x}_n) \;\mid\; C \;\mid\; M \vee M \;\mid\; M \wedge M$$

*where* $C$ *denotes a Presburger formula.* □

*Remark 2.* We have the following inclusion of formulas:

$$\begin{array}{ccccc}\textit{quantifier-free}\text{ monotone} & & & & \text{monotone}\\ \text{exponential Diophantine} & \subsetneq & M & \subsetneq & \text{exponential Diophantine}\end{array}$$

For the left (strict) inclusion, we consider $x^2 + y^x \geqslant z$ in $M$. This constraint requires $\exists$-quantifiers when it is expressed as an exponential Diophantine formula. For the right strict inclusion, consider $\exists z\,(z^3 \geqslant y \wedge x \geqslant z+1)$, which is equivalent to $(x-1)^3 \geqslant y \wedge x \geqslant 1$. It does not belong to $M$, since the co-efficient of $x^2$ is negative.

## 6 Concluding Remarks

In the paper we have discussed the expressiveness of monotone AC-TA. First we demonstrated in Lemmas 1 and 2 that the non-linear inequalities $x \times y \geqslant z$ and $x^y \geqslant z$ over natural numbers are interpretive by monotone AC-TA. These examples also refine the previous results in [15]. Next we proposed the class of monotone exponential Diophantine formulas, which is a decidable fragment of exponential Diophantine formulas (Theorem 1). Using the transformation by linearly bounded projection, we have shown that every monotone exponential Diophantine inequality $E(\boldsymbol{x}_n) \geqslant L(\boldsymbol{x}_n)$ is monotone AC-TA definable (Lemma 6), and therefore, a sub-class of monotone exponential Diophantine formulas is monotone AC-TA definable (Theorem 3). As an example of the main result, we have presented that the super-exponential $x^{x^{x}} \geqslant y$ is monotone AC-TA definable.

The definability results in terms of monotone AC-TA obtained in the paper are based on our original technique, LINEARLY BOUNDED PROJECTION. Due to the unsolved question about the closedness under projection of AC-monotone tree languages, this special projection plays an essential role overall in the paper.

As a remark of this projection, one can apply it for showing that $x \times y \geqslant z$ is monotone AC-TA definable, using the previous result ([15]) that $x \times y \geqslant z \wedge u = 1 \wedge v = 2$ is monotone AC-TA definable, because $x \times y \geqslant z$ is logically equivalent to $\exists u, v\,(\,u \leqslant 1 \,\wedge\, v \leqslant 2 \,\wedge\, x \times y \geqslant z \,\wedge\, u = 1 \,\wedge\, v = 2\,)$. This is an alternative proof for Lemma 1.

There are two interesting questions remaining open:

1. *Closedness under projection*: Exponential monotone Diophantine formulas are monotone AC-TA definable if the class of monotone AC-TA is closed under projection. Moreover, under the same condition, one can show that the language classes of Petri nets and monotone AC-TA are identical, which seems to be negatively observed, without a clear proof, in [11].

As we have seen in the paper, using linearly bounded projection, one can eliminate a certain type of ∃-quantifiers in the formulas. Moreover, it may be possible to transform every monotone exponential Diophantine formula into an equivalent formula formed of linearly bounded monotone AC-TA definable sub-formulas. If so, regardless of the answer to the above question 1, one can have the positive answer to the definability question about monotone exponential Diophantine formulas—whether every monotone exponential Diophantine formula is monotone AC-TA definable. Furthermore, the validity of monotone exponential Diophantine formulas is an interesting question, though the decidability of the universality problem of monotone AC-TA is not known.

2. *Complete characterization of monotone AC-TA*: The definability theorem for exponential Diophantine formulas (Theorem 3) appeals that this class of formulas is related to monotone AC-TA. However, we do not know exactly which class of the arithmetic can be the counterpart of monotone AC-TA.

One can observe that the class of monotone AC-TA is a proper sub-class of monotone A-TA. *Monotone A-TA* are the class of ETA whose axioms are associativity only [16]. This hierarchy stems from the observation that every monotone AC-TA can be simulated by a monotone A-TA which additionally has transition rules $f(\alpha,\beta) \to f(\beta,\alpha)$ for all states $\alpha,\beta$ instead of the commutativity axioms of $f$. Tree languages accepted by monotone A-TA are closely related to context-sensitive languages, which is a sub-class of primitive recursive sets. Thus, monotone AC-TA may relate to the first-order theory of arithmetic with a certain type of, e.g. monotonic, primitive recursions.

**Acknowledgments.** The authors would like to thank Jun-ichi Abo for participating in our discussions with enthusiasm. The completeness proof of Lemma 1 is partly owing to him. We also thank for many fruitful comments from four anonymous reviewers.

## References


1. Baader, F., Nipkow, T.: Term Rewriting and All That. Cambridge University Press, Cambridge (1998)
2. Boneva, I., Talbot, J.-M., Tison, S.: Expressiveness of a Spatial Logic for Trees. In: Proc. of 20th LICS, Chicago (USA), pp. 280–289. IEEE Computer Society, Los Alamitos (2005)
3. Comon-Lundh, H., Dauchet, M., Gilleron, R., Jacquemard, F., Lugiez, D., Tison, S., Tommasi, M.: Tree Automata Techniques and Applications, draft (2005), `http://www.grappa.univ-lille3.fr/tata`
4. Dal Zilio, S., Lugiez, D.: XML Schema, Tree Logic and Sheaves Automata. Applicable Algebra in Engineering, Communication and Computing 17, 337–377 (2006)
5. Ginsburg, S., Spanier, E.H.: Semigroups, Presburger Formulas, and Languages. Pacific Journal of Mathematics 16, 285–296 (1966)
6. Ginsburg, S.: The Mathematical Theory of Context-Free Languages. McGraw-Hill, New York (1966)
7. Hack, M.H.T.: Decidability Questions for Petri Nets, Ph.D. thesis, Massachusetts Institute of Technology, USA (1976)
8. Henzinger, T.A.: The Theory of Hybrid Automata. In: Proc. of 11th LICS, New Brunswick (USA). IEEE Computer Society, Los Alamitos (1996) (Extended version), `http://mtc.epfl.ch/~tah/Publications`
9. Hinman, P.G.: Fundamentals of Mathematical Logic. A K Peters (2005)
10. Kobayashi, N., Ohsaki, H.: Tree Automata for Non-Linear Arithmetic, draft (February 2008), `http://staff.aist.go.jp/hitoshi.ohsaki/`
11. Kudlek, M., Mitrana, V.: Normal Forms of Grammars, Finite Automata, Abstract Families, and Closure Properties of Multiset Languages. In: Calude, C.S., Pun, G., Rozenberg, G., Salomaa, A. (eds.) Multiset Processing. LNCS, vol. 2235, pp. 135–146. Springer, Heidelberg (2001)
12. Landweber, L.H.: Properties of Vector Addition Systems, Technical Report 258, University of Wisconsin-Madison, USA (1975)
13. Lugiez, D.: Multitree Automata That Count. TCS 333, 225–263 (2005)
14. Matiyasevich, Y.: Hilbert's Tenth Problem. MIT Press, Cambridge (1993)
15. Ohsaki, H., Talbot, J.-M., Tison, S., Roos, Y.: Monotone AC-Tree Automata. In: Sutcliffe, G., Voronkov, A. (eds.) LPAR 2005. LNCS (LNAI), vol. 3835, pp. 337–351. Springer, Heidelberg (2005)
16. Ohsaki, H., Seki, H., Takai, T.: Recognizing Boolean Closed A-Tree Languages with Membership Conditional Rewriting Mechanism. In: Nieuwenhuis, R. (ed.) RTA 2003. LNCS, vol. 2706, pp. 483–498. Springer, Heidelberg (2003)
17. Ohsaki, H.: Beyond Regularity: Equational Tree Automata for Associative and Commutative Theories. In: Fribourg, L. (ed.) CSL 2001 and EACSL 2001. LNCS, vol. 2142, pp. 539–553. Springer, Heidelberg (2001)
18. Parikh, R.: On Context-Free Languages. JACM 13, 570–581 (1966)
19. Seidl, H., Schwentick, T., Muscholl, A.: Numerical Document Queries. In: Proc. of 22nd PODS, SanDiego (USA), pp. 155–166. ACM, New York (2003)
20. Thatcher, J.W.: Characterizing Derivation Trees of Context-Free Grammars Through a Generalization of Automata Theory. Journal of Computer and System Sciences 1, 317–322 (1967)
21. Verma, K.N., Goubault-Larrecq, J.: Alternating Two-Way AC-Tree Automata. Information and Computation 205, 817–869 (2007)

# Confluence by Decreasing Diagrams

## Converted

Vincent van Oostrom*

Utrecht University, Dept. of Philosophy
Heidelberglaan 6, 2584 CS, Utrecht, The Netherlands
Vincent.vanOostrom@phil.uu.nl

**Abstract.** The decreasing diagrams technique is a complete method to reduce confluence of a rewrite relation to local confluence. Whereas previous presentations have focussed on the *proof* the technique is correct, here we focus on *applicability*. We present a simple but powerful generalisation of the technique, requiring peaks to be closed only by conversions instead of valleys, which is demonstrated to further ease applicability.

## 1 Introduction

The decreasing diagrams technique [1,2] is a method to reduce the problem of showing confluence of a rewrite relation to showing its local confluence. In exchange for localisation, the confluence diagrams need to be *decreasing* with respect to some labelling. The method is complete in the sense that any (countable) confluent rewrite relation *can be* equipped with such a labelling. But by undecidability of confluence completeness also entails that *finding* such a labelling is hard. The goal of this paper is to ease the latter, thus enhancing applicability of the technique. We try to achieve this in two ways.

First, in Sect. 3, we relax the local confluence constraint. Instead of requiring that for every pair of diverging steps a *pair of reductions* exists such that the resulting diagram is decreasing, we show it suffices that a *conversion* exists such that the resulting diagram is decreasing, by analogy with the way in which Winkler & Buchberger's confluence criterion [3, Lemma 3.1] relaxes Newman's Lemma [4, Theorem 3].

Next, in Sect. 4, we provide heuristics for finding appropriate labellings, illustrated by many examples from the literature, ranging from abstract rewriting via term rewriting and $\lambda$-calculi to process algebra.

In the examples we use results from the literature. Other than that, we assume only basic rewriting knowledge, which is recapitulated in Sect. 2. That section serves also to recapitulate from [1,2] the core of the decreasing diagrams technique. Those not yet familiar with that technique are advised to consult one of its textbook accounts [5,6] first.

* This research was supported by a two-month grant from LIX, Paris, France.

A. Voronkov (Ed.): RTA 2008, LNCS 5117, pp. 306–320, 2008.

## 2 Preliminaries

A *rewrite* relation is a binary relation on a set of objects. To stress we are interested in the direction of rewrite relations we use arrow-like notations like $\rightarrow$, $\succ$, $\triangleright$, and $\blacktriangleright$ to denote them. For a rewrite relation $\rightarrow$, we inductively define an object $a$ to be *terminating*, if for all objects $b$ such that $a \rightarrow b$, $b$ is terminating. The rewrite relation $\rightarrow$ is *terminating* if all its objects are. For a rewrite relation denoted by an arrow-like notation $\rightarrow$, its converse is denoted by the converse $\leftarrow$ of the notation. We denote the union of two rewrite relations by the union of their notations, e.g. $\triangleleft\!\blacktriangleright$ denotes $\triangleleft \cup \blacktriangleright$, and $\leftrightarrow$ denotes $\leftarrow \cup \rightarrow$, the symmetric closure of $\rightarrow$. We use $\rightarrow \cdot \triangleright$ to denote the composition of $\rightarrow$ and $\triangleright$, and $\rightarrow^=$ and $\rightarrow^+$ to denote respectively the reflexive and transitive closure of $\rightarrow$. To denote the reflexive–transitive closure of $\rightarrow$, i.e. its 'repetition', we employ the 'repetition' $\twoheadrightarrow$ of its notation. When both $\triangleright$,$\blacktriangleright$ are defined, we abbreviate $\triangleright \cup \blacktriangleright$ to $\rightarrow$. Further notions and notations will be introduced on a by-need basis. We now state the decreasing diagrams theorem, illustrating it by means of a running example.

**Definition 1.** *A pair* $(\triangleright, \blacktriangleright)$ *of rewrite relations* commutes *if* $\lll \cdot \blacktriangleright\!\!\blacktriangleright \subseteq \blacktriangleright\!\!\blacktriangleright \cdot \lll$, *and commutes* locally *if* $\triangleleft \cdot \blacktriangleright \subseteq \blacktriangleright\!\!\blacktriangleright \cdot \lll$. *A rewrite relation* $\rightarrow$ *is* confluent *if* $(\rightarrow, \rightarrow)$ *commutes, and* locally *confluent if* $(\rightarrow, \rightarrow)$ *commutes locally.*

*Example 1.* A rewrite relation is confluent if locally confluent and terminating.

The decreasing diagrams technique generalises this example, that is, Newman's Lemma, by weakening the termination assumption to *decreasingness.*

**Definition 2.** *A pair* $((\triangleright_\ell)_{\ell \in L}, (\blacktriangleright_m)_{m \in M})$ *of families of rewrite relations is* decreasing *if the union* $L \cup M$ *of their sets* $L$,$M$ *of labels comes equipped with a terminating and transitive rewrite relation* $\succ$. *A pair of rewrite relations* $(\triangleright, \blacktriangleright)$ *is* decreasing, *if* $\triangleright = \bigcup_{\ell \in L} \triangleright_\ell$, $\blacktriangleright = \bigcup_{m \in M} \blacktriangleright_m$ *for such a decreasing pair of families, such that for all* $\ell \in L, m \in M$, $\triangleleft_\ell \cdot \blacktriangleright_m \subseteq \blacktriangleright\!\!\blacktriangleright_{\curlyvee\ell} \cdot \blacktriangleright^=_m \cdot \blacktriangleright\!\!\blacktriangleright_{\curlyvee\{\ell,m\}} \cdot \lll_{\curlyvee\{\ell,m\}} \cdot \triangleleft^=_\ell \cdot \lll_{\curlyvee m}$ *(see Fig. 1), where* $\curlyvee N = \{n \in L \cup M \mid \exists k \in N\ k \succ n\}$, *and* $\curlyvee n$ *abbreviates* $\curlyvee\{n\}$.

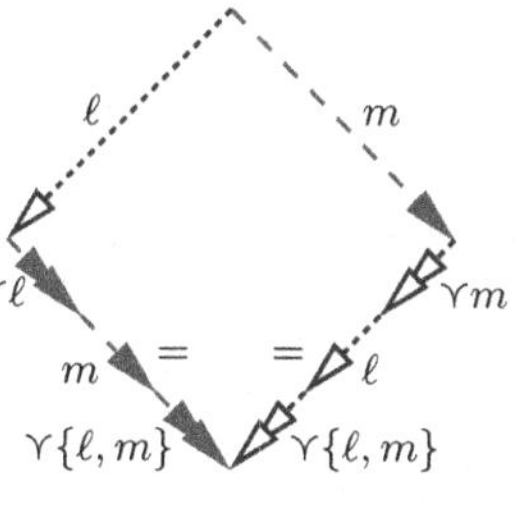


**Fig. 1.** Decreasingness

*A family* $(\rightarrow_\ell)_{\ell \in L}$ *of rewrite relations is* decreasing *if* $((\rightarrow_\ell)_{\ell \in L}, (\rightarrow_\ell)_{\ell \in L})$ *is. A rewrite relation* $\rightarrow$ *is* decreasing, *if* $(\rightarrow, \rightarrow)$ *is.*

*Example 2.* A rewrite relation $\to$ on $A$ as in Example 1 is decreasing: The family $(\to_a)_{a \in A}$, with $\to_a$ defined as $\to$ with domain restricted to $\{a\}$, and with the set of labels $A$ ordered by $\to^+$, is decreasing by the assumption that $\to$ is terminating. Clearly $\to = \bigcup_{a \in A} \to_a$, and as for any peak $b \leftarrow_a a \to_a c$, there is a valley between $b$ and $c$ by local confluence and since for any object, i.e. label, $d$ in this valley it trivially holds $a \to^+ d$, we conclude to decreasingness of $\to$.

**Theorem 1 ([1]).** *A pair of rewrite relations commutes if it is decreasing. A rewrite relation is confluent if it is decreasing.*

*Proof.* We recapitulate the core of the proof in [1] for easy adaptation later on. Instead of proving commutation one proves the stronger property:

(∗) Every peak $b \twoheadleftarrow_\sigma \cdot \twoheadrightarrow_\tau c$ can be completed by a valley $b \twoheadrightarrow_{\tau'} \cdot \twoheadleftarrow_{\sigma'} c$, into a so-called *decreasing* diagram, i.e. such that $|\sigma\tau'| \preceq_{mul} |\sigma| \uplus |\tau| \succeq_{mul} |\tau\sigma'|$.

where $|\sigma|$ is the *lexicographic maximum measure* of the string of labels $\sigma$, i.e. the multiset inductively defined by: $|\varepsilon| = \emptyset$ and $|\ell\sigma| = [\ell] \uplus |\sigma| - \curlyvee\ell$, and $\succ_{mul}$ is the (terminating) multiset extension of the (terminating) relation $\succ$ on the labels.

For such a peak, completability into a decreasing diagram is proved by $\succ_{mul}$-induction on its *measure* $|\sigma| \uplus |\tau|$. The proof being trivial in case either of the reductions in the assumption is empty, the interesting cases are seen to be of shape $\twoheadleftarrow_\sigma \cdot \lhd_\ell \cdot \blacktriangleright_m \cdot \twoheadrightarrow_\tau$, for which one concludes by the following three steps corresponding to the three components of Fig. 2:

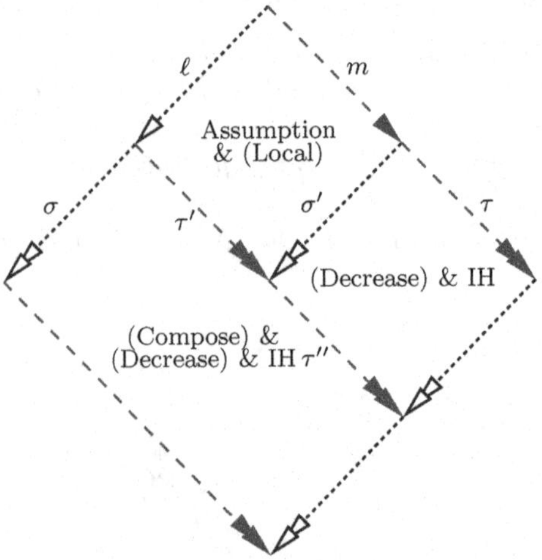


**Fig. 2.** Decreasing ⇒ Commutes

1. by the decreasingness assumption on $(\rhd, \blacktriangleright)$ the local peak $\lhd_\ell \cdot \blacktriangleright_m$ can be completed by a valley $\twoheadrightarrow_{\tau'} \cdot \twoheadleftarrow_{\sigma'}$ yielding a decreasing diagram by (Local);
2. by the induction hypothesis, which applies by (Decrease) and (1), the peak $\twoheadleftarrow_{\sigma'} \cdot \twoheadrightarrow_\tau$ can be completed by a valley $\twoheadrightarrow_{\tau''} \cdot \twoheadleftarrow_{\sigma''}$ into a decreasing diagram;

3. by the induction hypothesis, which applies by (Decrease) and (Compose) applied to (1),(2), the peak $\lhd\!\!\lhd_{\sigma} \cdot \blacktriangleright\!\!\blacktriangleright_{\tau'} \cdot \blacktriangleright\!\!\blacktriangleright_{\tau''}$, can be completed into a decreasing diagram, which by another application of (Compose) proves the result;

where the following facts haven been used for diagrams $D_i$ with $i \in \{1,2\}$, where a *diagram* $D_i$ consists of a peak $\lhd\!\!\lhd_{\sigma_i} \cdot \blacktriangleright\!\!\blacktriangleright_{\tau_i}$ completed by a valley $\blacktriangleright\!\!\blacktriangleright_{\tau'_i} \cdot \lhd\!\!\lhd_{\sigma'_i}$:

(Local) If $D_1$ is a local diagram, i.e. if its peak is of shape $\lhd_{\ell} \cdot \blacktriangleright_m$ for some labels $\ell$,$m$, then $D_1$ is decreasing if and only if its valley is of shape $\blacktriangleright\!\!\blacktriangleright_{\Upsilon\ell} \cdot \blacktriangleright^{=}_{m} \cdot \blacktriangleright\!\!\blacktriangleright_{\Upsilon\{\ell,m\}} \cdot \lhd\!\!\lhd_{\Upsilon\{\ell,m\}} \cdot \lhd^{=}_{\ell} \cdot \lhd\!\!\lhd_{\Upsilon m}$ [1, Proposition 2.3.16].

(Decrease) If $D_1$ is a non-empty decreasing diagram, i.e. if its reduction $\blacktriangleright\!\!\blacktriangleright_{\tau_1}$ is not empty, then filling the peak $\lhd\!\!\lhd_{\sigma_1} \cdot \blacktriangleright\!\!\blacktriangleright_{\tau_1} \cdot \blacktriangleright\!\!\blacktriangleright_{\tau_2}$ with $D_1$ decreases the measure, i.e. $|\sigma_1| \uplus |\tau_1\tau_2| \succ_{mul} |\sigma'_1| \uplus |\tau_2|$ [1, Lemma 2.3.19].

(Compose) If diagrams $D_1$,$D_2$ are decreasing and can be composed, i.e. if their respective reductions $\lhd\!\!\lhd_{\sigma'_1}$ and $\lhd\!\!\lhd_{\sigma_2}$ coincide, then this composition, consisting of the peak $\lhd\!\!\lhd_{\sigma_1} \cdot \blacktriangleright\!\!\blacktriangleright_{\tau_1} \cdot \blacktriangleright\!\!\blacktriangleright_{\tau_2}$ completed by the valley $\blacktriangleright\!\!\blacktriangleright_{\tau'_1} \cdot \blacktriangleright\!\!\blacktriangleright_{\tau'_2} \cdot \lhd\!\!\lhd_{\sigma'_2}$, is decreasing again [1, Lemma 2.3.17]. □

*Example 3.* Theorem 1 applied to Example 2 yields a proof of Example 1.

*Remark 1.* Conversely, any *countable* confluent relation is decreasing [1, Corollary 2.3.30]. It is an open problem whether countability can be dropped, and also whether any pair of commuting relations, countable or not, is decreasing.

Theorem 1 provides a method to prove properties stronger than commutation.

**Theorem 2.** *Let $P$ be a property of diagrams which is closed under composition (defined as in the proof of Theorem 1). If $(\rhd, \blacktriangleright)$ is a decreasing pair of rewrite relations such that every local peak can be completed into a decreasing diagram having property $P$, then every peak can be so completed.*

*Proof.* Require the diagram in $(*)$ in the proof of Theorem 1 to satisfy $P$. □

*Example 4.* Consider the property $P$ expressing that in a diagram $D$ with peak $b \twoheadleftarrow a \twoheadrightarrow c$ and valley $b \twoheadrightarrow d \twoheadleftarrow c$, its 'left-reduction' $a \twoheadrightarrow b \twoheadrightarrow d$ is not longer than its 'right-reduction' $a \twoheadrightarrow b \twoheadrightarrow d$. As $P$ is preserved under composition of diagrams, it suffices under the assumptions of Newman's Lemma to check that $P$ holds for all local diagrams. If it does, then all maximal reductions from a given object end in the same normal form, *reached in the same number of steps* [7].

*Example 5.* The *strict* commutation property expressing that in a diagram with peak $b \lhd\!\!\lhd a \blacktriangleright\!\!\blacktriangleright c$ and valley $b \blacktriangleright\!\!\blacktriangleright d \lhd\!\!\lhd c$, if $a \blacktriangleright\!\!\blacktriangleright c$ is non-empty then so is $b \blacktriangleright\!\!\blacktriangleright d$, is easily seen to be preserved under composition. Hence, it suffices to verify that local decreasing diagrams are strict.

## 3 Conversion

We generalise the decreasing diagrams technique as presented in the previous section, by allowing local peaks to be completed by conversions instead of valleys.

**Definition 3.** *A pair of rewrite relations* $(\triangleright, \blacktriangleright)$ *is* decreasing with respect to conversions, *if* $\triangleright = \bigcup_{\ell \in L} \triangleright_\ell$, $\blacktriangleright = \bigcup_{m \in M} \blacktriangleright_m$ *for a decreasing pair of families* $((\triangleright_\ell)_{\ell \in L}, (\triangleright_m)_{m \in M})$ *such that for all* $\ell \in L, m \in M$, $\triangleleft_\ell \cdot \blacktriangleright_m \subseteq \triangleleft\!\blacktriangleright^*_{\curlyvee \ell} \cdot \blacktriangleright^=_m \cdot \triangleleft\!\blacktriangleright^*_{\curlyvee \{\ell,m\}} \cdot \triangleleft^=_\ell \cdot \triangleleft\!\blacktriangleright^*_{\curlyvee m}$. *A rewrite relation* $\to$ *is* decreasing with respect to conversions, *if* $(\to, \to)$ *is.*

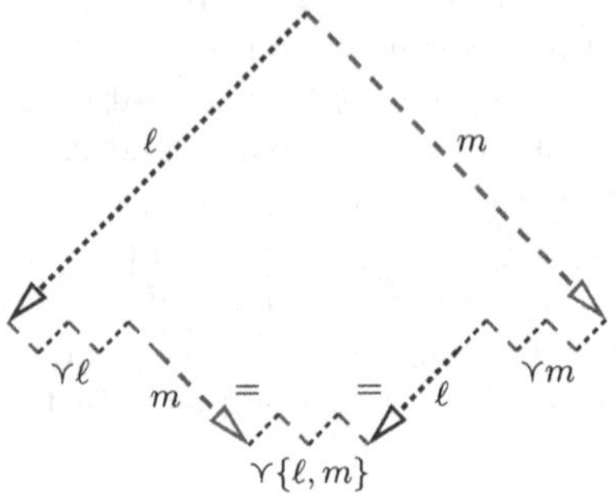


**Fig. 3.** Decreasingness with respect to conversions

Decreasingness is illustrated in Fig. 3, where, to avoid clutter, arrowheads in the conversions have been elided. We will henceforth refer to decreasingness in the sense of Definition 2 as *decreasing with respect to valleys*, abbreviated to $\diamondsuit$, and we abbreviate decreasingness in the present sense of Definition 3 to $\triangle$.

*Example 6.* Let $\to$ be a terminating rewrite relation such that for every local peak $b \leftarrow a \to c$ the objects $b$ and $c$ are *convertible below* $a$, i.e. $b = a_1 \leftrightarrow \ldots \leftrightarrow a_n = c$ with $a \to^+ a_i$ for all $1 \le i \le n$. From Example 2 we already know that labelling steps by their source and ordering the labels by $\to^+$ yields a decreasing labelling, and it is easy to see that the requirement that every local peak $b \leftarrow a \to c$ be convertible below $a$, entails that the rewrite relation $\to$ is decreasing with respect to conversions for this labelling.

**Theorem 3.** *A pair of rewrite relations commutes if it is decreasing with respect to conversions, and idem for confluence of a single rewrite relation.*

*Proof.* We adapt the proof of Theorem 1, keeping the same invariant and induction. Observe that the only difference arises in case the peak is of shape $b \blacktriangleleft\!\!\blacktriangleleft_\sigma b' \blacktriangleleft_\ell \cdot \triangleright_m c' \triangleright\!\!\triangleright_\tau c$ for some $\ell \in L, m \in M$. By the assumption that $(\triangleright, \blacktriangleright)$ is decreasing with respect to conversions, the local peak $b' \blacktriangleleft_\ell \cdot \triangleright_m c'$ can be transformed into $b' \triangleleft\!\blacktriangleright^*_{\curlyvee \ell} \cdot \blacktriangleright^=_m \cdot \triangleleft\!\blacktriangleright^*_{\curlyvee \{\ell,m\}} \cdot \triangleleft^=_\ell \cdot \triangleleft\!\blacktriangleright^*_{\curlyvee m} c'$, see (1) in Fig. 4. We show decreasingness with respect to valleys, by transforming the conversion into a valley of shape $b' \blacktriangleright\!\!\blacktriangleright_{\curlyvee \ell} \cdot \blacktriangleright^=_m \cdot \blacktriangleright\!\!\blacktriangleright_{\curlyvee \{\ell,m\}} \cdot \triangleleft\!\!\triangleleft_{\curlyvee \{\ell,m\}} \cdot \triangleleft^=_\ell \cdot \triangleleft\!\!\triangleleft_{\curlyvee m} c'$, and conclude.

First observe that if the peak of a decreasing diagram consists only of labels in $\curlyvee M$ then so does its valley. Thus by repeatedly applying the induction hypothesis, the peaks in the conversions $\triangleleft\!\blacktriangleright^*_{\curlyvee \ell}, \triangleleft\!\blacktriangleright^*_{\curlyvee m}$ can be transformed into valleys smaller than $\ell, m$, yielding $b' \blacktriangleright\!\!\blacktriangleright_{\curlyvee \ell} \cdot \triangleleft\!\!\triangleleft_{\curlyvee \ell} \cdot \blacktriangleright^=_m \cdot \triangleleft\!\blacktriangleright^*_{\curlyvee \{\ell,m\}} \cdot \triangleleft^=_\ell \cdot \blacktriangleright\!\!\blacktriangleright_{\curlyvee m} \cdot \triangleleft\!\!\triangleleft_{\curlyvee m} c'$,

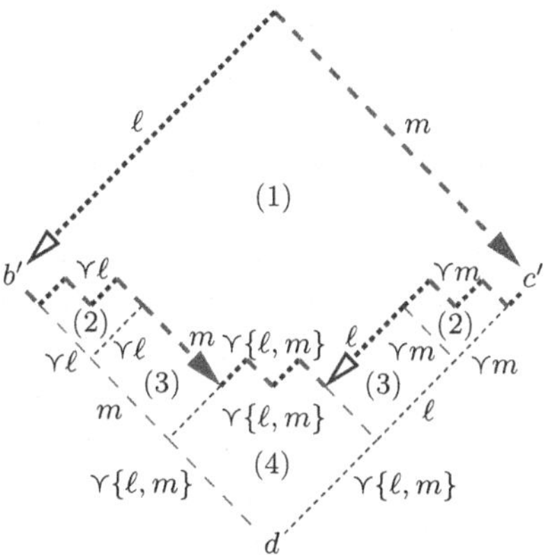


**Fig. 4.** Decreasing with respect to conversions ⇒ Commutes

see (2) in Fig. 4. Applying the induction hypothesis to the peaks $\twoheadleftarrow_{\curlyvee\ell} \cdot \blacktriangleright^{=}_{m}$ and $\triangleleft^{=}_{\ell} \cdot \twoheadrightarrow_{\curlyvee m}$ gives by analogous reasoning, valleys of shapes $\twoheadrightarrow_{\curlyvee\ell} \cdot \blacktriangleright^{=}_{m} \cdot \twoheadrightarrow_{\curlyvee\{\ell,m\}} \cdot \twoheadleftarrow_{\curlyvee\{\ell,m\}}$ and $\twoheadrightarrow_{\curlyvee\{\ell,m\}} \cdot \twoheadleftarrow_{\curlyvee\{\ell,m\}} \cdot \triangleleft^{=}_{\ell} \cdot \twoheadleftarrow_{\curlyvee m}$, see (3) in Fig. 4, giving $b' \twoheadrightarrow_{\curlyvee\ell} \cdot \blacktriangleright^{=}_{m} \cdot \triangleleft\!\blacktriangleright^{*}_{\curlyvee\{\ell,m\}} \cdot \triangleleft^{=}_{\ell} \cdot \twoheadleftarrow_{\curlyvee m} c'$. Finally, repeatedly applying the induction hypothesis to the peaks in the conversion $\triangleleft\!\blacktriangleright^{*}_{\curlyvee\{\ell,m\}}$ transforms it into a valley $\twoheadrightarrow_{\curlyvee\{\ell,m\}} d \twoheadleftarrow_{\curlyvee\{\ell,m\}}$, see (4) in Fig. 4, resulting in a decreasing diagram with respect to valleys. □

*Example 7.* Theorem 3 applied to Example 6 yields Winkler & Buchberger's result [3, Lemma 3.1] stating that any terminating rewrite relation such that the targets of any local peak are convertible below its source, is confluent.

*Remark 2.* It would be interesting to see whether the proofs of confluence by decreasing diagrams as presented in [8,9] can be adapted in a similar way.

Observe that from the proof of Theorem 3 it follows that, for any *given* labelling, decreasingness with respect to valleys is equivalent to decreasingness with respect to conversions, Although equivalent, the latter is in principle easier to check: one 'just' has to find an appropriate conversion instead of an appropriate valley for each local peak $\leftarrow_{\ell} \cdot \rightarrow_{m}$. Of course, the 'search space' for conversions is in general much larger than for valleys. To keep searching feasible nonetheless, observe first that one only needs to search (forward or backward) rewrite steps having labels smaller than or equal to $\ell$ or $m$, and second that one may opt to linearly bound the amount of time spent on searching decreasing conversions by that spent on valleys, where the latter restriction is complete by the observation above. That searching for conversions instead of valleys can be advantageous is witnessed by the following example, where searching conversions for all local peaks takes time linear in $n$ whereas searching for valleys takes quadratic time.

*Example 8.* Consider for every natural number $n$, the confluent rewrite relation given by $b_i \leftarrow a_i \rightarrow c_i$, $b_i \rightarrow b_{i+1}$, and $c_i \rightarrow c_{i+1}$, for $1 \leq i \leq n$, with $b_{n+1} = c_{n+1}$. Completing a local peak $b_i \leftarrow a_i \rightarrow c_i$ by a valley takes time $n - i$, whereas completing it by a conversion takes constant time, say 4. This proves our claim since $\sum_{i=1}^{n} n-i$ is quadratic in $n$ and $\sum_{i=1}^{n} 4$ is linear.

*Remark 3.* In his dissertation Geser argues [10, p. 38] that checking for valleys is less complex than checking for conversions. The above shows on the contrary that both can be combined fruitfully without changing the worst case $O$-behaviour. Whether such combinations are useful, i.e. less complex on average, in practice or in theory, for all or some labellings, remains to be investigated.

Like Theorem 1, also Theorem 3 provides a method to prove properties stronger than commutation. However, compared to Theorem 2, the property now has to apply to all *conversion* diagrams, i.e. diagrams consisting of a peak completed by a conversion instead of a valley, and be *closed* not only under composition, but also under filling a peak of the conversion by another conversion diagram.

**Theorem 4.** *Let $P$ be a property of conversion diagrams which is closed. If $(\triangleright, \blacktriangleright)$ is a decreasing pair of rewrite relations such that every local peak can be completed into a decreasing diagram with respect to conversions, having property $P$, then every peak can be completed into a decreasing diagram with respect to valleys, having property $P$.*

*Proof.* Load the induction hypothesis in the proof of Theorem 3 with $P$. □

We generalise Examples 4 and 5 to conversion diagrams.

*Example 9.* Consider the property $P$ expressing that the *distance* $d(D)$ of the diagram $D$ with peak $b \twoheadleftarrow a \twoheadrightarrow c$ and conversion $b \leftrightarrow^* c$, is not positive, where $d(D)$ is the integer defined as the number of forward steps $(\rightarrow)$ minus the number of backward steps $(\leftarrow)$ on the cycle $a \twoheadrightarrow b \leftrightarrow^* c \twoheadleftarrow a$. The property $P$ is closed, since the distance of a conversion diagram obtained by 'glueing' two such diagrams together is the sum of their distances (note that shared steps contribute oppositely). Hence under the assumptions of Winkler & Buchberger's Lemma it suffices to verify that $P$ holds for local conversion diagrams. If it does, then all maximal reductions from a given object end in the same normal form, reached in the same number of steps, generalizing Example 4.

*Example 10.* Strictness as in Example 5 can easily be extended to a closed property of conversion diagrams, by requiring in a diagram with peak $b \lhd\!\!\lhd a \blacktriangleright\!\!\blacktriangleright c$ and conversion $b \lhd\!\!\blacktriangleright^* c$, if $a \blacktriangleright\!\!\blacktriangleright c$ is non-empty then $b \lhd\!\!\blacktriangleright^* c$ contains some $\blacktriangleright$-step. Again, it suffices to verify that local decreasing conversion diagrams are strict.

## 4 Application

We apply the results of the previous section, providing heuristics for finding decreasing labellings along the way. First, we present the 'self-labelling' heuristic and show that it can be used to deal with several known commutation and confluence results for abstract rewrite relations. More generally, we cover and systematise *all* such results in [6, Chapter 1] and [11]. Finally, we present the 'rule-labelling' and 'self-duplication' heuristics and show they can be used to deal with commutation and confluence problems in term rewriting.

## 4.1 Abstract Rewriting

The labelling employed in case of Newman's Lemma and Winkler & Buchberger's Lemma (Examples 2 and 6) may seem like sorcery, but is in fact an instance of a general idea: self-labelling. Let us try to explain this by showing what *fails* if one would try to devise a decreasing labelling in case of Kleene's standard counterexample to the implication 'local-confluence $\Rightarrow$ confluence'.

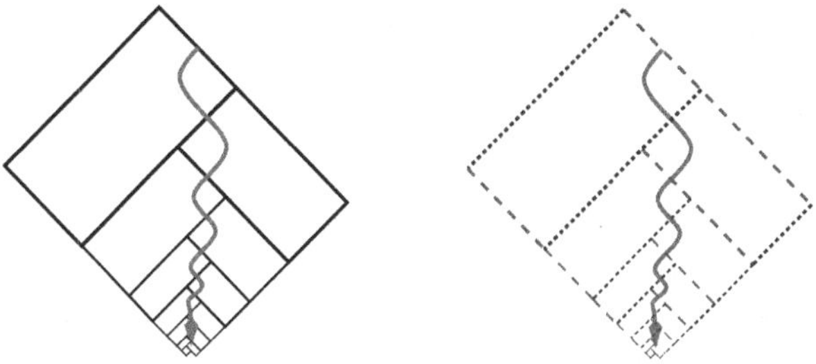

**Fig. 5.** Failure of local confuence (commutation) $\Rightarrow$ confluence (commutation)

*Example 11.* Consider the rewrite relation $\rightarrow$ given by $b \leftarrow a \leftrightarrows a' \rightarrow c$. Since every local peak can be completed by means of a valley, e.g. $b \leftarrow a \rightarrow a'$ can be completed by means of $b \leftarrow a \leftarrow a'$, the rewrite relation is locally confluent. However it is not confluent, as e.g. the peak $b \leftarrow a \rightarrow a' \rightarrow c$ cannot be completed. Fig. 5 (left) illustrates what goes wrong when trying a proof of confluence for this peak by means of tiling: the tiling process never terminates so does not lead to a completed confluence diagram. The (green) curved downward arrow, intersecting steps all of shape $a \leftrightarrow a'$, shows how such an infinite regress must *transfer* to an infinitely decreasing sequence of labels, preventing the construction of a decreasing labelling. Vice versa, in case of Newman's Lemma such an infinite decreasing sequence of labels cannot occur since it would, by concatenating the $a \leftrightarrow a'$-steps intersected by the arrow, immediately *transfer* into an infinite $\rightarrow$-reduction, contradicting the assumed termination of $\rightarrow$.

*Remark 4.* In [12] it is noted that the commuting version of Kleene's counterexample, Fig. 5 (right), plays a similar obstructive rôle in process algebra in proving similarity: Taking $\triangleright$ as reduction, $\mathbf{0} \triangleleft a \overset{\triangleleft}{\blacktriangleright} \tau.a \blacktriangleright \mathbf{0}$ witnesses that despite $\blacktriangleright$ being a weak simulation modulo transitivity, it is not contained in weak similarity.

The example suggests one may *transfer* termination of a (rewrite) relation on the objects, to a decreasing labelling by means of the following heuristic:

($H_1$: **Self-labelling**) Given a terminating relation on the objects, try using steps (or objects) *themselves* as labels, ordered by the transitive closure of the relation, and label a step $a \rightarrow b$ by itself (or its source $a$ or target $b$).

One may think that this heuristic is so much geared towards Newman's Lemma that it doesn't apply to any other interesting cases. But in fact it was inspired by self-labelling as used in proving termination by means of monotone algebras, and below we will see several other important instances of this heuristic.

We proceed by systematically treating *all* the abstract confluence and commutation results by analysis of local peaks as found in [6, Chapter 1] and [11] and some more. On the one hand, the systematisation is based on relating results based on decreasing *valleys* and *conversions* (the rows in Fig. 6) for a *given* labelling. On the other hand, it is based on the different *trace patterns* which arise when labelling the diagrams (the columns in Fig. 6). On the gripping hand, we relate *commutation* to *confluence* results (the tables below).

**Unqualified references in tables are to [6, Chapter 1].**

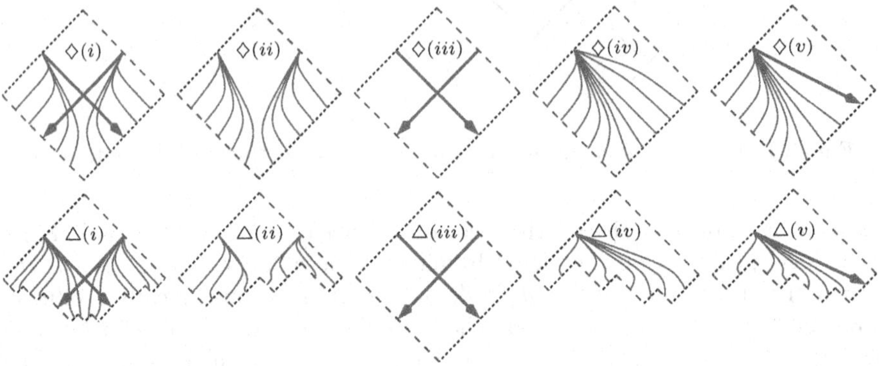


**Fig. 6.** Valleys (top) vs. conversions (bottom) for various trace patterns

Trace patterns [8,9,6] as displayed in Fig. 6 are the *patterns* obtained by *tracing* each label in the conclusion of a decreasing (valley or conversion) diagram back to a label in its hypothesis (the peak), which is either the same (the thick arrows in the figure) or greater (the thin lines) in concordance with the requirements imposed by decreasingness. E.g. Fig. 6 $\triangle(i)$ displays the general trace pattern for decreasing conversions of Theorem 3, and the patterns to its right are special cases of interest to us.

*Example 12.* Consider the labelling src which labels each step $a \to b$ by its *source* as $a \to_a b$. Then, if $\to$ is terminating, we may order the labels via the transitive closure $\to^+$, giving rise to a decreasing labelling. The trace pattern corresponding to the commutation version of Newman's Lemma (Example 2) is Fig. 6 $\Diamond(ii)$, and the one corresponding to the commutation version of Winkler & Buchberger's Lemma (Examples 6) is Fig. 6 $\triangle(ii)$. This, together with corresponding references to [6], is summarised in Table 1. Note that $\to^+$ could be replaced by any transitive and terminating relation $\succ$ such that $\to \subseteq \succ$, as in the usual presentation of Winkler & Buchberger's result.

**Table 1.** Source self-labelling with termination of $\rightarrow$

| src | confluence | commutation |
|---|---|---|
| $\Diamond$ | (Theorem 1.2.1) Newman | (Exercise 1.3.2) folklore |
| $\triangle$ | (Exercise 1.3.12) Winkler & Buchberger | new? |

**Table 2.** The empty order $\emptyset$

| $\emptyset$ | confluence | commutation |
|---|---|---|
| $\Diamond = \triangle$ | (Theorem 1.2.2(iii)) Hindley & Rosen | folklore |

*Example 13.* Any family can be made decreasing simply by equipping it with the empty order. However, a local peak $\triangleleft_\ell \cdot \blacktriangleright_m$ can then only be turned into a decreasing diagrams by means of $\blacktriangleright_m^= \cdot \triangleleft_\ell^=$, so decreasing valleys (Fig. 6 $\Diamond(iii)$) and conversions (Fig. 6 $\triangle(iii)$) coincide, yielding Table 2. For a singleton set of labels the result (subcommutativity implies confluence) goes back to [4].

Based on the previous two examples one could be led to believe the decreasing diagrams technique is just the 'sum of Newman's Lemma and the Lemma of Hindley–Rosen'. The following examples show it is much more powerful than a simple 'sum'; it is also our second instance of the self-labelling heuristic.

*Example 14.* Consider the labelling tgt which labels each step $a \rightarrow b$ by its *target* as $a \rightarrow_b b$. If the relation $\blacktriangleright$ relative *to* $\triangleleft$, defined by $\blacktriangleright/\triangleleft = \triangleleft\!\!\triangleleft \cdot \blacktriangleright \cdot \triangleleft\!\!\triangleleft$, is terminating, then ordering the labels via the transitive closure $(\blacktriangleright/\triangleleft)^+$ gives a decreasing labelling. If we also require that only *smaller* labels be used, then to have decreasing valleys coincides with strict commutation (see Example 5), to have decreasing conversions with *quasi*-commutation, i.e. $\triangleleft \cdot \blacktriangleright \subseteq \blacktriangleright \cdot \triangleleft\!\blacktriangleright^*$, and trace patterns are as in Fig. 6(iv), since labelling $b \triangleleft a \blacktriangleright c$ yields $b \triangleleft_b a \blacktriangleright_c c$. Since $b \blacktriangleright/\triangleleft\ c$, if $c$ is greater than some label, $b$ is so as well. Although the requirement that labels be *smaller* allows to reduce ([6, Exercise 1.3.19] Bachmair & Dershowitz) termination of $\blacktriangleright/\triangleleft$ to that of $\blacktriangleright$ as in the usual presentation of these results, cf. Table 3, it is not a necessary requirement. To wit, $b \triangleleft a \blacktriangleright c \triangleright b$ *is* decreasing although $c \triangleright_b b$ is equal to not smaller than $a \triangleright_b b$, with trace pattern as in Fig. 6 $\Diamond(v)$. For confluence this result is not interesting as $\rightarrow/\leftarrow$ is never terminating, for non-empty $\rightarrow$.

*Example 15.* Extending Example 13, a pair $(\triangleright, \blacktriangleright)$ can be made decreasing by letting $\triangleright$ be *stronger* than $\blacktriangleright$, i.e. by ordering $\triangleright$ above $\blacktriangleright$. A local peak $\triangleleft \cdot \blacktriangleright$ can

**Table 3.** Target self-labelling with termination of $\blacktriangleright/\triangleleft$

| tgt | commutation |
|---|---|
| $\Diamond$ | (Exercise 1.3.15) Geser |
| $\triangle$ | [10, Sect. 3.3] Geser |

**Table 4.** The stronger-than order $\rhd \succ \blacktriangleright$

| $\rhd \succ \blacktriangleright$ | confluence | commutation |
|---|---|---|
| $\Diamond = \triangle$ | (Exercise 1.3.11) Huet | (Exercise 1.3.6) Hindley |

be completed into a decreasing valley or conversion (only) by means of $\blacktriangleright\!\!\blacktriangleright \cdot \lhd^{=}$, its trace pattern being a special case of Fig. 6 $\Diamond(v)$. Interestingly, of the ensuing results in Table 4, Huet's result that *strong* confluence, i.e. $\leftarrow \cdot \rightarrow \subseteq \twoheadrightarrow \cdot \leftarrow^{=}$, implies confluence, is more recent than Hindley's that *strong* commutation, i.e. $\lhd \cdot \blacktriangleright \subseteq \blacktriangleright\!\!\blacktriangleright \cdot \lhd^{=}$, implies commutation, despite being an instance. In fact, also Staples' later result [6, Exercise 1.3.7] that $\lhd\!\!\lhd \cdot \blacktriangleright \subseteq \blacktriangleright\!\!\blacktriangleright \cdot \lhd\!\!\lhd$ implies commutation is seen to be an instance of Hindley's result, noting that $\rhd\!\!\rhd^{*} = \rhd\!\!\rhd^{=} = \rhd\!\!\rhd$.

Whereas *stronger-than* as in the previous example orders one family above another, *requests* as in the next example orders within families.

*Example 16.* If both $\rhd, \blacktriangleright$ are $\{1,2\}$-labelled families and all diagrams are decreasing with respect to $1 \succ 2$, then 1 *requests* 2. The reason for this terminology becomes clear when considering the most general shapes concrete decreasing diagrams, say for decreasing valleys, may have: $\lhd_1 \cdot \blacktriangleright_1 \subseteq \blacktriangleright\!\!\blacktriangleright_2 \cdot \blacktriangleright_1^{=} \cdot \blacktriangleright\!\!\blacktriangleright_2 \cdot \lhd\!\!\lhd_2 \cdot \lhd_1^{=} \cdot \lhd\!\!\lhd_2$ with trace pattern $\Diamond(i)$, or $\lhd_1 \cdot \blacktriangleright_2 \subseteq \blacktriangleright\!\!\blacktriangleright_2 \cdot \lhd\!\!\lhd_2 \cdot \lhd_1^{=}$ or its symmetric version $\lhd_2 \cdot \blacktriangleright_1 \subseteq \blacktriangleright_1^{=} \cdot \blacktriangleright\!\!\blacktriangleright_2 \cdot \lhd\!\!\lhd_2$ both with trace pattern $\Diamond(v)$, or $\lhd_2 \cdot \blacktriangleright_2 \subseteq \blacktriangleright_2 \cdot \lhd_2$ with trace pattern $\Diamond(iii)$; for commutation a 1-step may *request* 2-steps to find the common reduct, but not the other way around. This generalises the classical notion of request as employed in the results of Table 5, strengthening these.

**Table 5.** Requests for ordering $1 \succ 2$ within a family $\{1,2\}$

| $1 \succ 2$ | confluence | commutation |
|---|---|---|
| $\Diamond$ | (Exercise 1.3.8) Rosen & Staples | new? |
| $\triangle$ | (Exercise 1.3.10) van Oostrom | new? |

A third instance of the self-labelling heuristic is obtained by noting that in the commuting version of Example 12, it is in fact not necessary to have termination of $\rhd \cup \blacktriangleright$, but termination of $\rhd^{+} \cdot \blacktriangleright^{+}$ suffices; translated to Fig. 5 (right) this reads: It suffices that there's no infinite *zigzag*, alternating $\rhd$ and $\blacktriangleright$ reductions.

*Example 17.* If $(\rhd, \blacktriangleright)$ commutes locally and $\rhd^{+} \cdot \blacktriangleright^{+}$ is terminating, then $(\rhd, \blacktriangleright)$ commutes [12, Corollary 4.6]. To see this, consider the **stp**-labelling which labels a step of either type, say, $a \rhd b$, by *itself* as $a \rhd_{a \rhd b} b$, which is ordered above a step of the *other* type $c \blacktriangleright_{c \blacktriangleright d} d$ if $b \twoheadrightarrow c$. Note that the **stp**-labelling is decreasing since an infinite decreasing sequence of labels would entail an infinitely zigzagging reduction, contradicting termination of $\rhd^{+} \cdot \blacktriangleright^{+}$. As a local peak $b \lhd_{a \rhd b} a \blacktriangleright_{a \blacktriangleright c} c$ can by assumption be completed by a commuting valley the steps of which are either reachable by a zig from $b$ or a zag from $c$, we conclude.

Variations on this example like e.g. [12, Lemma 4.5] are easily accomodated by our techniques as well, and even in a modular way by choosing an appropriate property to load the induction in Theorem 2.

The above examples cover the results in [6, Chapter 1], which in turn subsume those in [11], except for [11, Exercise 2.0.8(11)] which we deal with now.

*Example 18.* Let $\rightarrow\,=\,\rhd \cup \blacktriangleright$ and consider the labelling $\mathsf{pred}$ which may label any step $a \rightarrow b$ as $a \rightarrow_{c,i} b$, where $c$ is any *predecessor* of $a$, i.e. $c \twoheadrightarrow a$, and $i$ is set to 1 for a backward step and to 2 for a forward step. If $\rhd/\blacktriangleright$ is terminating, then the lexicographic product of $(\rhd/\blacktriangleright)^+$ and $1 \succ 2$ gives a decreasing labelling. We verify by case analysis that then the combination of $\rhd$ being locally confluent and $\blacktriangleright$ being *non-splitting*, i.e. $\leftarrow \cdot \blacktriangleright \subseteq \twoheadrightarrow \cdot \leftarrow^=$, results in decreasing diagrams:

for a peak $\lhd_{a,1} \cdot \rhd_{b,2}$, local confluence of $\rhd$ yields a valley $\rhd\!\!\rhd \cdot \lhd\!\!\lhd$ the steps of which can be labelled simply by their sources to result in a decreasing diagram;

for a peak $\lhd_{a,1} \cdot \blacktriangleright_{b,2}$, non-splittingness of $\blacktriangleright$ yields a valley $\twoheadrightarrow \cdot \leftarrow^=$, which after labelling the steps in $\twoheadrightarrow$ by their source and the $\leftarrow$-step as $\leftarrow_{a,1}$, results in a decreasing diagram;

a peak $\blacktriangleleft_{a,1} \cdot \rhd_{b,2}$ is dealt with symmetric to the previous case;

for a peak $\blacktriangleleft_{a,1} \cdot \blacktriangleright_{b,2}$ non-splittingness of $\blacktriangleright$ yields a valley $\twoheadrightarrow \cdot \leftarrow^=$, which after labelling as in the second item except for labelling the steps of $\twoheadrightarrow$ up to and including the first $\rhd$-step (if any) as $\rightarrow_{a,2}$, results in a decreasing diagram.

Although a bit more involved this labelling and case analysis directly cover the Full Localisation Lemma, the most complex (p.71–73) confluence result in [10], stating $\rightarrow\,=\,\rhd \cup \blacktriangleright$ is confluent, if $\rhd/\blacktriangleright$ is terminating and $\rightarrow$ is locally confluent, with the condition that in case a local peak $b \leftarrow a \blacktriangleright c$ needs to be completed by a valley of shape $b \twoheadrightarrow d \leftarrow^+ a' \blacktriangleleft c$ then $a\ (\rhd/\blacktriangleright)^+\ a'$.

*Remark 5.* The final two examples both are covered by the original decreasing diagrams technique. It would be interesting to consider their conversion versions.

## 4.2 Term Rewriting

We show the usefulness of the rule-labelling heuristic in first- or higher-order term rewriting systems.

($H_2$: **Rule-labelling**) Try labelling steps by the rule applied.

*Example 19.* Consider the TRS with rules [13, Example 2]: (1) $\mathsf{nats} \rightarrow 0 : \mathsf{inc}(\mathsf{nats})$, (2) $\mathsf{inc}(x : y) \rightarrow \mathsf{s}(x) : \mathsf{inc}(y)$, (3) $\mathsf{hd}(x : y) \rightarrow x$, (4) $\mathsf{tl}(x : y) \rightarrow y$, (5) $\mathsf{inc}(\mathsf{tl}(\mathsf{nats})) \rightarrow \mathsf{tl}(\mathsf{inc}(\mathsf{nats}))$. The rule-labelling heuristic which labels every step by the rule applied, yields, by left- and right-linearity of the rules, that $\leftarrow_i \cdot \rightarrow_j \subseteq \rightarrow_j^= \cdot \leftarrow_i^=$ for all $i, j \in \{1, \ldots, 5\}$ and non-overlapping steps. The only critical pair arises from the local peak $\mathsf{tl}(\mathsf{inc}(\mathsf{nats})) \leftarrow_5 \mathsf{inc}(\mathsf{tl}(\mathsf{nats})) \rightarrow_1 \mathsf{inc}(\mathsf{tl}(0 : \mathsf{inc}(\mathsf{nats})))$ which can be completed by $\mathsf{tl}(\mathsf{inc}(\mathsf{nats})) \rightarrow_1 \mathsf{tl}(\mathsf{inc}(0 : \mathsf{inc}(\mathsf{nats}))) \rightarrow_2 \mathsf{tl}(\mathsf{s}(0) : \mathsf{inc}(\mathsf{inc}(\mathsf{nats}))) \rightarrow_4 \mathsf{inc}(\mathsf{inc}(\mathsf{nats})) \leftarrow_4 \mathsf{inc}(\mathsf{tl}(0 : \mathsf{inc}(\mathsf{nats})))$. As the latter diagram can be easily made decreasing, e.g. by ordering $5 \succ 1, 2, 4$, we conclude confluence.

More generally, for any finite left- and right-linear term rewriting system, it is decidable whether the rule-labelling entails decreasingness, simply by trying all possible orderings of the rules,[1] refuting the claim of [13, Footnote 1] that this requires 'careful and smart design choices'. Of course, this does not allow to deal with non-right-linear rules:

*Example 20.* Consider the TRS with rules [13, Example 1]: (1) $g(a) \to f(g(a))$, (2) $g(b) \to c$, (3) $a \to b$, (4) $f(x) \to h(x,x)$, (5) $h(x,y) \to c$. Since the TRS is not right-linear, the above observation does not apply. In particular, the rule-labelling cannot work as rule 4 can *self-duplicate*, consider e.g. the term $f(f(c))$! Still, it is easy to find a decreasing labelling noting the duplicated variable has on the right-hand side less $f$-symbols above it than on its left-hand side: labelling steps by first the number of $f$-symbols above it and then the rule, and ordering these lexicographically by first $>$ and then $\succ$ given by $3 \succ 2,4,5$ does the job.

The trick in the example fails if variables occur 'deeper' in the right-hand side than in the left-hand side of a rule. Even in such cases the heuristic might be applicable, by solving the problem of self-duplication by brute force:

($H_3$: **Self-duplication**) First try to separate out self-duplicating rules, and then switch for these to 'multi' steps in which an arbitrary number of redexes for that rule may be contracted.

This technique may be applied to prove confluence of orthogonal term rewriting systems, but also of some term rewriting systems *with* critical pairs, as is nicely illustrated in the following example.

Extending Bloo and Rose's $\lambda$x-calculus with explicit substitutions, with a rule encoding the substitution lemma of the $\lambda$-calculus, yields the $\lambda$xc-calculus.

**Theorem 5.** *The following CRS [6, Chapter 11]) for* $\lambda$xc*-calculus is confluent.*

$$\begin{aligned}
(\lambda y.X(y))Y &\to X(y)[y{:=}Y] \\
X[y{:=}Y] &\to X \\
y[y{:=}Y] &\to Y \\
(X_1(y)X_2(y))[y{:=}Y] &\to (X_1(y)[y{:=}Y])(X_2(y)[y{:=}Y]) \\
(\lambda x.X(y))[y{:=}Y] &\to \lambda x.X(y)[y{:=}Y] \\
X(y,z)[y{:=}Y(z)][z{:=}Z] &\to X(y,z)[z{:=}Z][y{:=}Y(z)[z{:=}Z]]
\end{aligned}$$

*Proof.* It will turn out handy to split the set of rules as follows. The first rule is the `Beta`-rule, the next four are the `x`-rules, and the final rule is the `c`-rule.

Let $\twoheadrightarrow\!\!\!\!\!\circ_{\texttt{Beta}}$ denote the contraction of any number of `Beta`-redexes, let $\to_{\texttt{x}}$ denote contracting an arbitrary (possibly garbage collecting) `x`-redex, and let $\twoheadrightarrow\!\!\!\!\!\circ_{\texttt{c}}$ denote a `c`-reduction which is a prefix of contracting all $c$-redexes in the term (see [14]). To show confluence of $\lambda$xc, it then suffices to prove confluence of these relations, since $\to_{\lambda\texttt{xc}} \subseteq \twoheadrightarrow\!\!\!\!\!\circ_{\texttt{Beta}} \cup \to_{\texttt{x}} \cup \twoheadrightarrow\!\!\!\!\!\circ_{\texttt{c}} \subseteq \twoheadrightarrow_{\lambda\texttt{xc}}$. In the rest of the proof, the three types of rules are referred to simply as `Beta`, `x`, and `c`. It suffices to show that they are decreasing with respect to the order $\succ$ given by `Beta` $\succ$ `c` $\succ$ `x`,

[1] This is analogous to the way one may proceed when checking whether a TRS is terminating via recursive path orders.

using the src-labelling with respect to (terminating) x-reduction to order x-steps among each other. Distinguish cases on the types of the rules in a local peak.

- If both are Beta, then the result follows from Beta being *linear orthogonal*, in the sense that a common reduct is found in either zero or one Beta-steps on both sides.
- If $s \leftarrow_{\mathtt{x}} t \mathrel{-\!\!\circ\!\!\rightarrow}_{\mathtt{Beta}} r$, then $s \rightarrow_{\mathtt{x}} \cdot \mathrel{-\!\!\circ\!\!\rightarrow}_{\mathtt{Beta}} q$ and $r \mathrel{-\!\!\circ\!\!\rightarrow}_{\mathtt{c}} q$ in case the x-step overlaps one of the Beta-redexes, and $s \mathrel{-\!\!\circ\!\!\rightarrow}_{\mathtt{Beta}} q \twoheadleftarrow_{\mathtt{x}} r$ otherwise.
- Beta is orthogonal to c, and they commute in a single step on either side.
- If both are x, a common reduct is found in at most two further x-steps having smaller sources.
- If $s \leftarrow_{\mathtt{x}} t \mathrel{-\!\!\circ\!\!\rightarrow}_{\mathtt{c}} r$, then $s \twoheadrightarrow_{\mathtt{x}} \cdot \mathrel{-\!\!\circ\!\!\rightarrow}_{\mathtt{c}} q \twoheadleftarrow_{\mathtt{x}} r$ in case the x-step overlaps one of the c-redexes (needing a number of garbage collection steps). Otherwise the steps simply commute (c may duplicate x, but not *vice versa*).
- If both rules are c, a common reduct is found in at most one further c-step, which holds since the c-rule is an instance of self-distributivity [14]. □

Note that a common reduct is found in an amount of work which is linear in the diverging steps, measuring each step by the number of steps performed by it. This is not that good, but still better than always reducing to x-normal form as is done in proofs relying on the so-called interpretation method.

## 5 Conclusion

We have improved upon our earlier decreasing diagrams technique. It was shown that in many cases it is not difficult to find a labelling showing decreasingness. The heuristics presented could be a stepping stone for constructing an automatic confluence prover.

We conclude by noting that the generalization does straightforwardly extend to Ohlebusch's confluence by decreasing diagrams *modulo an equivalence relation* results [5, Chapter 2].

*Acknowledgments.* We thank the anonymous referees, Alfons Geser, and Jean-Pierre Jouannaud for useful discussions, and the latter and Delia Kesner for their hospitality during my research stay in Paris, where this paper was conceived.

## References

1. van Oostrom, V.: Confluence for Abstract and Higher-Order Rewriting. PhD thesis. Vrije Universiteit, Amsterdam (March 1994)
2. van Oostrom, V.: Confluence by decreasing diagrams. Theoretical Computer Science 126(2), 259–280 (1994)
3. Winkler, F., Buchberger, B.: A criterion for eliminating unnecessary reductions in the Knuth–Bendix algorithm. In: Algebra, Combinatorics and Logic in Computer Science, Györ (Hungary) (1983) Colloquia Mathematica Societatis János Bolyai, 42, vol. II, North-Holland, Amsterdam (1986)

4. Newman, M.: On theories with a combinatorial definition of "equivalence". Annals of Mathematics 43(2), 223–243 (1942)
5. Ohlebusch, E.: Advanced Topics in Term Rewriting. Springer, Heidelberg (2002)
6. Terese: Term Rewriting Systems. Cambridge Tracts in Theoretical Computer Science, vol. 55. Cambridge University Press, Cambridge (2003)
7. van Oostrom, V.: Random descent. In: Baader, F. (ed.) RTA 2007. LNCS, vol. 4533, pp. 314–328. Springer, Heidelberg (2007)
8. Bezem, M., Klop, J., van Oostrom, V.: Diagram techniques for confluence. Information and Computation 141(2), 172–204 (1998)
9. Klop, J., van Oostrom, V., de Vrijer, R.: A geometric proof of confluence by decreasing diagrams. Journal of Logic and Computation 10(3), 437–460 (2000)
10. Geser, A.: Relative Termination. PhD thesis, Universität Passau (1990)
11. Klop, J.: Term rewriting systems. In: Handbook of Logic in Computer Science. Background: Computational Structures, vol. 2, pp. 1–116. Oxford Science Publications (1992)
12. Pous, D.: New up-to techniques for weak bisimulation. Theoretical Computer Science 380(1-2), 164–180 (2007)
13. Gramlich, B., Lucas, S.: Generalizing Newman's lemma for left-linear rewrite systems. In: Pfenning, F. (ed.) RTA 2006. LNCS, vol. 4098, pp. 66–80. Springer, Heidelberg (2006)
14. Dehornoy, P.: Braids and Self-Distributivity. Progress in Mathematics, vol. 192. Birkhäuser, Basel (2000)

# A Finite Simulation Method in a Non-deterministic Call-by-Need Lambda-Calculus with Letrec, Constructors, and Case

Manfred Schmidt-Schauss[1] and Elena Machkasova[2]

[1] Dept. Informatik und Mathematik, Inst. Informatik, J.W. Goethe-University, PoBox 11 19 32, D-60054 Frankfurt, Germany
schauss@ki.informatik.uni-frankfurt.de
[2] Division of Science and Mathematics, University of Minnesota, Morris, MN 56267-2134, U.S.A.
elenam@morris.umn.edu

**Abstract.** The paper proposes a variation of simulation for checking and proving contextual equivalence in a non-deterministic call-by-need lambda-calculus with constructors, case, seq, and a letrec with cyclic dependencies. It also proposes a novel method to prove its correctness. The calculus' semantics is based on a small-step rewrite semantics and on may-convergence. The cyclic nature of letrec bindings, as well as non-determinism, makes known approaches to prove that simulation implies contextual preorder, such as Howe's proof technique, inapplicable in this setting. The basic technique for the simulation as well as the correctness proof is called pre-evaluation, which computes a set of answers for every closed expression. If simulation succeeds in finite computation depth, then it is guaranteed to show contextual preorder of expressions.

## 1 Introduction and Related Work

The construction of compilers and the compilation of programs in higher level, expressive programming languages is an important process in computer science that is a highly sophisticated engineering task. Unfortunately there remains a gap between theory and practice. Usually compilers incorporating lots of complicated transformations and optimizations are built with only a partial knowledge about correctness issues. This gap increases with the number of features, such as higher-order functions, concurrency, store, and system- or user-interaction. The ability to reason about program equivalence in the presence of non-determinism opens a door to a rigorous handling of these features.

We study these issues using a call-by-need lambda-calculus $L$ with data structures and non-determinism that in addition has letrec allowing cyclic binding dependencies. This may have applications for concurrent Haskell [Pey03, PGF96], and also for other functional programming languages [Han96].

Our language $L$ comes with a rewrite semantics (a small-step semantics) that is more appropriate to investigate non-determinism than a big-step semantics,

A. Voronkov (Ed.): RTA 2008, LNCS 5117, pp. 321–335, 2008.

since it explicitly models interleaving and atomicity. On top of the operational semantics we define a contextual semantics with may-convergence, which is maximal, since all expressions that cannot be distinguished by observations are identified.

This follows an approach pioneered in [Plo75] of considering two small-step rewrite relations in a calculus: a normal order reduction which represents evaluation of a term by some evaluation engine, such as an interpreter, and transformation steps performed by a compiler to optimize a program. The latter steps may include reductions from the calculus as well as other transformations. The goal is to prove contextual equivalence of the original and the transformed expressions, i.e. that any transformation step performed anywhere in a term does not change the term's convergence behavior.

Unfortunately the approach in [Plo75] cannot be applied to systems with cyclic dependencies (such as `letrec`) or with non-determinism, since it requires confluence of transformations which fails in such systems (see [AW96]). Some alternative approaches include restrictions on cyclic substitution [AK97] or considering terms up to infinite unwindings of cycles [AB02].

Investigations of correctness (also called meaning-preservation) for a call-by-value system of mutually recursive components with applications to modules and linking were undertaken in [MT00, Mac02] where a proof method based on diagrams called *lift* and *project* was introduced. The diagram approach was later extended and generalized in [WDK03] in an abstract setting. Another approach based on multihole contexts was used for a call-by-name system of mutually recursive components in [Mac07]. However, the diagram-based and context-based approaches above require that all normal-order reductions preserve behavior of a term, which is not the case for (choice)-reductions. Contextual equivalence for non-deterministic call-by-need calculi was investigated in [KSS98, SSS07a, SSSS08] using the method of forking and commuting diagrams, and in [MSC99] using abstract machine-reductions.

An important tool to prove contextual equivalence of concrete expressions is simulation-based (it is also called applicative simulation) since it allows to show contextual equivalence of expressions $s, t$ based only on the analysis of the reductions of $s, t$, in contrast to the definition, which requires checking reduction in infinitely many contexts. This method was used for variants of lambda calculi, see e.g [Abr90, How89, Gor99]. An extension of simulation to a non-deterministic call-by-need calculus was investigated in [Man04] (and generalized in [SSM07]), where Howe's [How89, How96] proof technique is extended to call-by-need non-determinism by using an intermediate approximation calculus. Unfortunately, these proof methods based on the approach of Howe appear not to be adaptable to call-by-need non-deterministic calculi with letrec, since cyclic dependencies cannot be treated: the proof technique of Howe fails in a subtle way.

In this paper we propose a method of finite simulation, i.e., a simulation with a finite depth, and prove its correctness by a new proof technique that uses approximations. The simulation method constructs a set of answer-terms for a given expression. These sets are then compared for various closed expressions in order to show contextual equivalence. An answer is either an abstraction or an constructor-expression, built from constructors, $\Omega$, and abstractions, such as partial lists. Consider the two (non-convertible) expressions $s, t$ where $s = \texttt{repeat True}$ is a nonending list and $t$ is recursively defined by $t = \texttt{choice} \perp (\texttt{Cons True}\ t)$. The latter will evaluate, depending on the choices, to $\perp$, $(\texttt{Cons True} \perp)$, $(\texttt{Cons True} (\texttt{Cons True} \perp))$, .... The expressions $s, t$ are equivalent w.r.t. observational equivalence based on may-convergence. Our simulation method permits to show their equivalence solely on the basis of the (approximative) answers that can be derived from each expression.

The proof of validity of the proposed method for our calculus $L$ (see section 2) requires several steps. The first step is to investigate the correctness of several reductions and transformations in $L$. Note that the normal-order reduction in the language $L$ treats chains of variable-variable bindings as transparent for several reductions. This is crucial for constructing correctness proofs which otherwise may not even be possible, since the measures for inductions are insufficient. A context lemma and standardization of reductions are proved. The second step is a transfer from $L$ to the calculus $L_S$ (see section 4), which has the same contextual equivalence as $L$, but simpler reduction rules, e.g. variable-variable bindings are now opaque. The third step (see Sections 5 – 7) is to define the computation of answers from a closed expression, and to prove criteria for contextual equivalence on the basis of the answer sets. We also provide a method to analyze contextual equivalence and preorder of answers. In particular, we show that abstractions can be compared based on applying them to all closed answers or to $\Omega$. As an application of this technique, we show that `choice` (see Section 8) has useful algebraic properties, such as idempotency, commutativity and associativity, for all expressions, including open ones.

Missing proofs can be found in [SSM08].

## 2 The Calculus L

### 2.1 Syntax and Reductions of the Functional Core Language L

We define the calculus L consisting of a language $\mathcal{L}(\mathrm{L})$, its reduction rules, the normal order reduction strategy, and contextual equivalence. $L$ is the calculus considered in [SSSS04] and an extension by `choice` of the one in [SSSS08]. The rules of the calculus limit copying of abstractions and prohibit copying of constructor expressions, thus limiting the level of complexity of proofs. There are finitely many types, and for every type $T$ there are finitely many, say $\#(T)$, constants called constructors $c_{T,i}, i = 1, \dots, \#(T)$, each with an arity $\mathrm{ar}(c_{T,i}) \geq 0$. The syntax for expressions $E$ is as follows:

$$\begin{aligned} E ::= {} & V \mid (c\ E_1 \dots E_{\mathrm{ar}(c)}) \mid (\texttt{seq}\ E_1\ E_2) \mid (\texttt{case}_T\ E\ \mathit{Alt}_1 \dots \mathit{Alt}_{\#(T)}) \mid (E_1\ E_2) \\ & (\texttt{choice}\ E_1\ E_2) \mid (\lambda\ V.E) \mid (\texttt{letrec}\ V_1 = E_1, \dots, V_n = E_n\ \texttt{in}\ E) \\ \mathit{Alt} ::= {} & (\mathit{Pat}\ \rightarrow\ E) \qquad \mathit{Pat} ::= (c\ V_1 \dots V_{\mathrm{ar}(c)}) \end{aligned}$$

where $E, E_i$ are expressions, $V, V_i$ are variables, and $c$ denotes a constructor. Expressions $(\texttt{case}_T \dots)$ have exactly one alternative for every constructor of type

$T$. We assign the names *application, abstraction, constructor application*, `seq`-*expression*, `case`-*expression*, or `letrec`-*expression* to the expressions $(E_1\ E_2)$, $(\lambda V.E)$, $(c\ E_1 \ldots E_{\mathrm{ar}(c)})$, $(\mathtt{seq}\ E_1\ E_2)$, $(\mathtt{case}_T\ E\ Alt_1 \ldots Alt_{\#(T)})$, $(\mathtt{letrec}\ V_1 = E_1, \ldots, V_n = E_n\ \mathtt{in}\ E)$, respectively. A *value* $v$ is defined as either an abstraction or a constructor application (with any subexpressions).

We assume that variables $V_i$ in `letrec`-bindings are all distinct, that the bindings can be interchanged, and that there is at least one binding. `letrec` is recursive, i.e., the scope of $x_j$ in $(\mathtt{letrec}\ x_1 = E_1, \ldots, x_j = E_j, \ldots\ \mathtt{in}\ E)$ is $E$ and all expressions $E_i$. Free and bound variables in expressions and $\alpha$-renaming are defined using the usual conventions. The set of free variables in $t$ is denoted as $FV(t)$. For simplicity we use the distinct variable convention, i.e., all bound variables in expressions are assumed to be distinct, and free variables are distinct from bound variables. The reduction rules are assumed to implicitly rename bound variables in the result by $\alpha$-renaming if necessary. We will use some obvious abbreviations of the syntax. E.g. $\{x_i = x_{i+1}\}_{i=m}^{n}$ abbreviates $x_m = x_{m+1}, \ldots, x_n = x_{n+1}$.

**Definition 2.1.** *The class $\mathcal{C}$ of all* contexts *is defined as the set of expressions $C$ from* L, *where the symbol $[\cdot]$, the* hole, *is a predefined context, treated as an atomic expression, such that $[\cdot]$ occurs exactly once in $C$.*

*Given a term $t$ and a context $C$, we will write $C[t]$ for the expression constructed from $C$ by plugging $t$ into the hole, i.e, by replacing $[\cdot]$ in $C$ by $t$, where this replacement is meant syntactically, i.e., a variable capture is permitted.*

**Definition 2.2 (Reduction Rules of the Calculus L).** *The (base) reduction rules for the calculus and language* L *are defined in figures 1 and 2, where the labels $S, V$ are to be ignored in this subsection, but will be used in subsection 2.2. The abbreviation Env means a set of bindings. The reduction rules can be applied in any context. Several reduction rules are denoted by their name prefix, e.g. the union of (llet-in) and (llet-e) is called (llet). The union of (llet), (lcase), (lapp), (lseq) is called (lll).*

*Reductions (and transformations) are denoted using an arrow with super and/or subscripts: e.g.* $\xrightarrow{llet}$. *Transitive closure of reductions is denoted by a $+$, reflexive transitive closure by a $*$. E.g.* $\xrightarrow{*}$ *is the reflexive, transitive closure of* $\rightarrow$.

## 2.2 Normal Order Reduction and Contextual Equivalence

The normal order reduction strategy of the calculus L is a call-by-need strategy, which is a call-by-name strategy adapted to sharing. The labeling algorithm in figure 3 will detect the position to which a reduction rule is applied according to the normal order. It uses the labels: $S$ (subterm), $T$ (top term), $V$ (visited), $W$ (visited, no copy-target). For a term $s$ the labeling algorithm starts with $s^T$, where no other subexpression in $s$ is labeled, and exhaustively applies the rules in figure 3. The algorithm may terminate with a failure if a relabeling occurs, and otherwise with success, which indicates a potential normal-order redex, usually as the direct superterm of the $S$-marked subexpression.

| | |
|---|---|
| (lbeta) | $((\lambda x.s)^S\ r) \to (\texttt{letrec}\ x = r\ \texttt{in}\ s)$ |
| (cp-in) | $(\texttt{letrec}\ x_1 = v^S, \{x_i = x_{i-1}\}_{i=2}^m, Env\ \ \texttt{in}\ C[x_m^V])$<br>$\to (\texttt{letrec}\ x_1 = v, \{x_i = x_{i-1}\}_{i=2}^m, Env\ \ \texttt{in}\ C[v])$<br>where $v$ is an abstraction |
| (cp-e) | $(\texttt{letrec}\ x_1 = v^S, \{x_i = x_{i-1}\}_{i=2}^m, Env, y = C[x_m^V]\ \texttt{in}\ r)$<br>$\to (\texttt{letrec}\ x_1 = v, \{x_i = x_{i-1}\}_{i=2}^m, Env, y = C[v]\ \texttt{in}\ r)$<br>where $v$ is an abstraction |
| (llet-in) | $(\texttt{letrec}\ Env_1\ \texttt{in}\ (\texttt{letrec}\ Env_2\ \texttt{in}\ r)^S)$<br>$\to (\texttt{letrec}\ Env_1, Env_2\ \texttt{in}\ r)$ |
| (llet-e) | $(\texttt{letrec}\ Env_1, x = (\texttt{letrec}\ Env_2\ \texttt{in}\ s_x)^S\ \texttt{in}\ r)$<br>$\to (\texttt{letrec}\ Env_1, Env_2, x = s_x\ \texttt{in}\ r)$ |
| (lapp) | $((\texttt{letrec}\ Env\ \texttt{in}\ t)^S\ s) \to (\texttt{letrec}\ Env\ \texttt{in}\ (t\ s))$ |
| (lcase) | $(\texttt{case}_T\ (\texttt{letrec}\ Env\ \texttt{in}\ t)^S\ alts) \to (\texttt{letrec}\ Env\ \texttt{in}\ (\texttt{case}_T\ t\ alts))$ |
| (seq-c) | $(\texttt{seq}\ v^S\ t) \to t$ if $v$ is a value |
| (seq-in) | $(\texttt{letrec}\ x_1 = v^S, \{x_i = x_{i-1}\}_{i=2}^m, Env\ \texttt{in}\ C[(\texttt{seq}\ x_m^V\ t)])$<br>$\to (\texttt{letrec}\ x_1 = v, \{x_i = x_{i-1}\}_{i=2}^m, Env\ \texttt{in}\ C[t])$ if $v$ is a value |
| (seq-e) | $(\texttt{letrec}\ x_1 = v^S, \{x_i = x_{i-1}\}_{i=2}^m, Env, y = C[(\texttt{seq}\ x_m^V\ t)]\ \texttt{in}\ r)$<br>$\to (\texttt{letrec}\ x_1 = v, \{x_i = x_{i-1}\}_{i=2}^m, Env, y = C[t]\ \texttt{in}\ r)$ if $v$ is a value |
| (lseq) | $(\texttt{seq}\ (\texttt{letrec}\ Env\ \texttt{in}\ s)^S\ t) \to (\texttt{letrec}\ Env\ \texttt{in}\ (\texttt{seq}\ s\ t))$ |
| (choice-l) | $(\texttt{choice}\ s\ t)^{S \vee T} \to s$ |
| (choice-r) | $(\texttt{choice}\ s\ t)^{S \vee T} \to t$ |

**Fig. 1.** Reduction rules, part a

**Definition 2.3 (Normal Order Reduction of L).** *Let $t$ be an expression. Then a single normal order reduction step $\xrightarrow{no}$ is defined by first applying the labeling algorithm to $t$. If the labeling algorithm terminates successfully, then one of the rules in figures 1 and 2 has to be applied, if possible, where the labels $S, V$ must match the labels in the expression $t$. The* normal order redex *is defined as the subexpression to which the reduction rule is applied.*

**Definition 2.4.** *A reduction context $R$ is any context, such that its hole will be labeled with $S$ or $T$ by the labeling algorithm. A* surface context, *denoted as $\mathcal{S}$, is a context where the hole is not contained in an abstraction. An* application surface context, *denoted as $\mathcal{AS}$, is a surface context where the hole is neither contained in an abstraction nor in an alternative of a case-expression.*

Note that the normal order redex is unique, and that a normal-order reduction is unique with the only exception of (choice).

A *weak head normal form (WHNF)* is either a value $v$ or an expression $(\texttt{letrec}\ Env\ \texttt{in}\ v)$, or $(\texttt{letrec}\ x_1 = (c\ \overrightarrow{t}), \{x_i = x_{i-1}\}_{i=2}^m, Env\ \texttt{in}\ x_m)$.

**Definition 2.5.** *A normal order reduction sequence is called an* evaluation *if the last term is a WHNF. For a term $t$, we write $t\downarrow$ iff there is an evaluation starting from $t$. We also say that $t$ is* converging *(or* terminating*). Otherwise, if there is no evaluation of $t$, we write $t\Uparrow$. A specific representative of non-converging*

| | |
|---|---|
| (case-c) | $(\texttt{case}_T\ (c_i\ \overrightarrow{t})^S\ \ldots((c_i\ \overrightarrow{y}) \to t)\ldots) \to (\texttt{letrec}\ \{y_i = t_i\}_{i=1}^n\ \texttt{in}\ t)$ where $n = \text{ar}(c_i) \geq 1$ |
| (case-c) | $(\texttt{case}_T\ c_i^S\ \ldots\ (c_i \to t)\ldots) \to t$ if $\text{ar}(c_i) = 0$ |
| (case-in) | $\texttt{letrec}\ x_1 = (c_i\ \overrightarrow{t})^S, \{x_i = x_{i-1}\}_{i=2}^m, Env$ $\texttt{in}\ C[\texttt{case}_T\ x_m^V\ \ldots((c_i\ \overrightarrow{z})\ldots \to t)\ldots]$ $\to \texttt{letrec}\ x_1 = (c_i\ \overrightarrow{y}), \{y_i = t_i\}_{i=1}^n, \{x_i = x_{i-1}\}_{i=2}^m, Env$ $\texttt{in}\ C[(\texttt{letrec}\ \{z_i = y_i\}_{i=1}^n\ \texttt{in}\ t)]$ where $n = \text{ar}(c_i) \geq 1$ and $y_i$ are fresh variables |
| (case-in) | $\texttt{letrec}\ x_1 = c_i^S, \{x_i = x_{i-1}\}_{i=2}^m, Env\ \texttt{in}\ C[\texttt{case}_T\ x_m^V\ \ldots\ (c_i \to t)\ldots]$ $\to \texttt{letrec}\ x_1 = c_i, \{x_i = x_{i-1}\}_{i=2}^m, Env\ \texttt{in}\ C[t]$ if $\text{ar}(c_i) = 0$ |
| (case-e) | $\texttt{letrec}\ x_1 = (c_i\ \overrightarrow{t})^S, \{x_i = x_{i-1}\}_{i=2}^m,$ $u = C[\texttt{case}_T\ x_m^V\ \ldots((c_i\ \overrightarrow{z}) \to r_1)\ldots], Env\ \ \texttt{in}\ r_2$ $\to \texttt{letrec}\ x_1 = (c_i\ \overrightarrow{y}), \{y_i = t_i\}_{i=1}^n, \{x_i = x_{i-1}\}_{i=2}^m,$ $u = C[(\texttt{letrec}\ z_1 = y_1, \ldots, z_n = y_n\ \texttt{in}\ r_1)], Env\ \ \texttt{in}\ r_2$ where $n = \text{ar}(c_i) \geq 1$ and $y_i$ are fresh variables |
| (case-e) | $\texttt{letrec}\ x_1 = c_i^S, \{x_i = x_{i-1}\}_{i=2}^m, u = C[\texttt{case}_T\ x_m^V\ \ldots\ (c_i \to r_1)\ldots], Env\ \ \texttt{in}\ r_2$ $\to \texttt{letrec}\ x_1 = c_i, \{x_i = x_{i-1}\}_{i=2}^m \ldots, u = C[r_1], Env\ \texttt{in}\ r_2$ if $\text{ar}(c_i) = 0$ |

**Fig. 2.** Reduction rules, part b

*expressions is* $\Omega := (\lambda z.(z\ z))\ (\lambda x.(x\ x))$, *i.e.* $\Omega\Uparrow$. *For consistency with our earlier work (e.g. [SSS07b]) the must-divergence notation* $\Uparrow$ *is used.*

As an example for normal-order reduction, some reductions of $\Omega$: $(\lambda z.(z\ z))\ (\lambda x.(x\ x)) \xrightarrow{no,lbeta} (\texttt{letrec}\ z = \lambda x.(x\ x)\ \texttt{in}\ (z\ z)) \xrightarrow{no,cp} (\texttt{letrec}\ z = \lambda x.(x\ x)\ \texttt{in}\ ((\lambda x'.(x'\ x'))\ z)) \xrightarrow{no,lbeta} (\texttt{letrec}\ z = \lambda x.(x\ x)\ \texttt{in}\ (\texttt{letrec}\ x_1 = z\ \texttt{in}\ (x_1\ x_1))) \xrightarrow{no,llet} (\texttt{letrec}\ z = \lambda x.(x\ x), x_1 = z\ \texttt{in}\ (x_1\ x_1)) \longrightarrow \ldots$.

**Definition 2.6 (contextual preorder and equivalence).** *Let* $s, t$ *be terms. Then:*

$$s \leq_c t \text{ iff } \forall C[\cdot]:\quad C[s]{\downarrow} \Rightarrow C[t]{\downarrow}$$
$$s \sim_c t \text{ iff } s \leq_c t \wedge t \leq_c s$$

By standard arguments, we see that $\leq_c$ is a precongruence and that $\sim_c$ is a congruence, where a *precongruence* $\leq$ is a preorder on expressions, such that $s \leq t \Rightarrow C[s] \leq C[t]$ for all contexts $C$, and a *congruence* is a precongruence that is also an equivalence relation.

## 3 Correctness of Reductions and Transformations

**Theorem 3.1.** *All the reductions (viewed as transformations) in the base calculus* L *with the exception of (choice) maintain contextual equivalence, i.e., whenever* $t \xrightarrow{a} t'$, *with* $a \in \{cp, lll, case, seq, lbeta\}$, *then* $t \sim_c t'$. *The same holds for all transformations in figure 4. Moreover,* $s \xrightarrow{choice} t$ *implies* $t \leq_c s$.

$$
\begin{array}{ll}
(\texttt{letrec}\ Env\ \texttt{in}\ t)^T & \to (\texttt{letrec}\ Env\ \texttt{in}\ t^S)^V \\
(s\ t)^{S\vee T} & \to (s^S\ t)^V \\
(\texttt{seq}\ s\ t)^{S\vee T} & \to (\texttt{seq}\ s^S\ t)^V \\
(\texttt{case}_T\ s\ alts)^{S\vee T} & \to (\texttt{case}_T\ s^S\ alts)^V \\
(\texttt{letrec}\ x = s, Env\ \texttt{in}\ C[x^S]) & \to (\texttt{letrec}\ x = s^S, Env\ \texttt{in}\ C[x^V]) \\
(\texttt{letrec}\ x = s, y = C[x^S], Env\ \texttt{in}\ t) & \to (\texttt{letrec}\ x = s^S, y = C[x^V], Env\ \texttt{in}\ t) \\
 & \quad \text{if } C[x] \neq x \\
(\texttt{letrec}\ x = s, y = x^S, Env\ \texttt{in}\ t) & \to (\texttt{letrec}\ x = s^S, y = x^W, Env\ \texttt{in}\ t)
\end{array}
$$

The labeling rules can be applied in any context

**Fig. 3.** Labeling algorithm for $L$

We define non-reduction transformations in Figure 4. Some transformations have two or more forms, e.g. (ve1) and (ve2). The side condition for (abs2) guarantees finiteness of (abs2) sequences. The transformations are used later either for the pre-evaluation or to aid correctness proofs.

**Proposition 3.2.** *The expression $\Omega$ is the least element w.r.t. $\leq_c$, and for every closed expression $s$ with $s\Uparrow$, the equation $s \sim_c \Omega$ holds.*

We summarize correctness of transformations and decreasing property of (choice) in the *standardization* result, which shows that reduction sequences can be standardized using normal-order reduction.

**Theorem 3.3 (Standardization).** *If $t \xrightarrow{*} t'$ where $t'$ is a WHNF and the sequence $\xrightarrow{*}$ consists of any reduction from L in figures 1 and 2 and of transformation steps from figure 4, then $t\Downarrow$.*

## 4 A Simpler Calculus

We define a simpler calculus $L_S$ that is used to produce a set of values of any closed expression. It is formulated such that a so-called *pre-evaluation* can be defined and shown to be a correct tool to prove contextual preorder and contextual equivalence of expressions in almost the same way as the simulation method would do it. The calculus $L_S$ does not use variable-binding chains for reduction steps, and permits also copying expressions of the form $(c\ x_1 \ldots x_n)$, where $x_i$ are variables. Such expressions are called *cv-expression.*

The rules of the calculus $L_S$ are defined in figure 6. We use labels $S, T, V$ indicating the normal order redex The labeling algorithm in 5 starts with $t^T$, where no subexpression of $t$ is labeled, and uses the rules exhaustively, which can be applied in any context.

An $L_S$-*WHNF* is defined as $v$ or $(\texttt{letrec}\ Env\ \texttt{in}\ v)$, where $v$ is an abstraction or a cv-expression. It is easy to see that every $L_S$-WHNF is also an $L$-WHNF, and that for every $L$-WHNF $t$, there is an $L_S$-WHNF $t'$ with: $t \xrightarrow{L_S,no,*} t'$ using only (*abs*), (*lll*), and (*cp*).

| | |
|---|---|
| (ve1) | $(\texttt{letrec}\ x = y, x_1 = t_1, \ldots, x_n = t_n\ \texttt{in}\ r) \to (\texttt{letrec}\ x_1 = t'_1, \ldots, x_n = t'_n\ \texttt{in}\ r')$ where $t'_i = t_i[y/x], r' = r[y/x], n \geq 0$ and if $x \neq y$ |
| (ve2) | $(\texttt{letrec}\ x = y\ \texttt{in}\ s) \to s[y/x]$ if $x \neq y$ |
| (abs1) | $(\texttt{letrec}\ x = c\ \overrightarrow{t}, Env\ \ \texttt{in}\ s) \to (\texttt{letrec}\ x = c\ \overrightarrow{x}, \{x_i = t_i\}_{i=1}^{\mathrm{ar}(c)}, Env\ \ \texttt{in}\ s)$ where $\mathrm{ar}(c) \geq 1$ and for $1 \leq i \leq \mathrm{ar}(c)$ $x_i \in FV(Env)$ |
| (abs2) | $(c\, t_1 \ldots t_n) \to (\texttt{letrec}\ x_1 = t_1, \ldots, x_n = t_n\ \texttt{in}\ (c\, x_1 \ldots x_n))$ where at least one of $t_i$ is not a variable |
| (cpcx-in) | $(\texttt{letrec}\ x = c\ \overrightarrow{t}, Env\ \texttt{in}\ C[x])$ $\to (\texttt{letrec}\ x = c\ \overrightarrow{y}, \{y_i = t_i\}_{i=1}^{\mathrm{ar}(c)}, Env\ \texttt{in}\ C[c\ \overrightarrow{y}])$ |
| (cpcx-e) | $(\texttt{letrec}\ x = c\ \overrightarrow{t}, z = C[x], Env\ \texttt{in}\ t)$ $\to (\texttt{letrec}\ x = c\ \overrightarrow{y}, \{y_i = t_i\}_{i=1}^{\mathrm{ar}(c)}, z = C[c\ \overrightarrow{y}], Env\ \texttt{in}\ t)$ |
| (gc1) | $(\texttt{letrec}\ \{x_i = s_i\}_{i=1}^{n}, Env\ \texttt{in}\ t) \to (\texttt{letrec}\ Env\ \texttt{in}\ t)$ if for all $i : x_i$ does not occur in $Env$ nor in $t$ |
| (gc2) | $(\texttt{letrec}\ \{x_i = s_i\}_{i=1}^{n}\ \texttt{in}\ t) \to t$ if for all $i : x_i$ does not occur in $t$ |
| (ucp1) | $(\texttt{letrec}\ Env, x = t\ \texttt{in}\ S[x]) \to (\texttt{letrec}\ Env\ \texttt{in}\ S[t])$ |
| (ucp2) | $(\texttt{letrec}\ Env, x = t, y = S[x]\ \texttt{in}\ r) \to (\texttt{letrec}\ Env, y = S[t]\ \texttt{in}\ r)$ |
| (ucp3) | $(\texttt{letrec}\ x = t\ \texttt{in}\ S[x]) \to S[t]$ where in the (ucp)-rules, $x$ has at most one occurrence in $S[x]$ and no occurrence in $Env, t, r$; and $S$ is a surface context |
| (cpbot1) | $(\texttt{letrec}\ x = \Omega, Env\ \texttt{in}\ C[x]) \to (\texttt{letrec}\ x = \Omega, Env\ \texttt{in}\ C[\Omega])$ |
| (cpbot2) | $(\texttt{letrec}\ x = \Omega, y = C[x], Env\ \texttt{in}\ r) \to (\texttt{letrec}\ x = \Omega, y = C[\Omega], Env\ \texttt{in}\ r)$ |

**Fig. 4.** Transformations in $L$ calculus

$$(\texttt{letrec}\ x = s, Env\ \texttt{in}\ C[x^S]) \to (\texttt{letrec}\ x = s^S, Env\ \texttt{in}\ C[x^V])$$
$$(\texttt{letrec}\ x = s, y = C[x^S], Env\ \texttt{in}\ r) \to (\texttt{letrec}\ x = s^S, y = C[x^V], Env\ \texttt{in}\ r)$$
$$\text{if } C \neq [.]$$

The rules for $(\texttt{letrec}\ Env\ \texttt{in}\ t)^T$, $(s\ t)$, $(\texttt{seq}\ s\ t)$ and $(\texttt{case}\ s\ alts)$ are as for $L$

**Fig. 5.** Labeling rules of $L_S$

Using diagrams and an induction on the length of a reduction sequence, the equivalence is shown in [SSM08]:

**Theorem 4.1.** *Let $s$ be an expression. Then* $s{\downarrow}_{L_S} \Leftrightarrow s{\downarrow}_L$.

## 5 Pre-evaluation of Expressions

In the following we will use the technical observation that during a normal-order reduction of $t$ we can trace the bindings $x_i = r_i$ of a closed subexpression $r = (\texttt{letrec}\ x_1 = r_1, \ldots, x_n = r_n\ \texttt{in}\ s')$ of $t$, if $r$ occurs on the surface of $t$. The application of this observation in proofs allows us to draw several nice and important conclusions.

We will use evaluation in $L_S$ to reduce closed expressions in all possible ways, where reduction takes place in surface contexts. The intention is to have a means to compare closed expressions by their sets of results, even perhaps infinite sets.

| | |
|---|---|
| (cp-in) | $(\texttt{letrec}\ x = v^S, Env\ \ \texttt{in}\ C[x^V]) \rightarrow (\texttt{letrec}\ x = v, Env\ \ \texttt{in}\ C[v])$ <br> where $v$ is an abstraction or a cv-expression |
| (cp-e) | $(\texttt{letrec}\ x = v^S, Env, y = C[x^V]\ \texttt{in}\ r) \rightarrow (\texttt{letrec}\ x = v, Env, y = C[v]\ \texttt{in}\ r)$ <br> where $v$ is an abstraction or a cv-expression |
| (abs) | $(c\ t_1 \dots t_n)^{S \vee T} \rightarrow (\texttt{letrec}\ x_1 = t_1, \dots, x_n = t_n\ \texttt{in}\ (c\ x_1 \dots x_n))$ <br> if $(c\ t_1 \dots t_n)$ is not a cv-expression |
| (lbeta), (seq-c), (case), (choice), (lll) are as in $L$ | |

**Fig. 6.** Reduction rules of $L_S$

We use the additional constant ⊚ (called stop) in order to indicate stopped reductions. Its semantical value is $\bot$, but it is clearer if there is a notational distinction between them.

**Definition 5.1.** *A* pseudo-value *is an expression built from ⊚, constructors, and abstractions, and an* answer *is a pseudo-value not equal to ⊚.*

We show the intention of the pre-evaluation by an example. The idea is to first obtain by reduction all possible WHNFs, and then to apply normal-order reductions locally to the bindings. Since this in general does not terminate, we stop the reduction at any point and then fill the results into the in-expression: the bindings that are cv-expressions or abstractions are copied sufficiently often into the in-expression. Due to recursive bindings, this may also be a non-terminating process that has to be stopped. We strip away the top letrec-environment and replace the occurrences of the previously let-bound variables by ⊚.

*Example 5.2.* The expression ($\texttt{letrec}\ x =$ ($\texttt{Cons True}\ x$) $\texttt{in}\ x$) has the following resulting answers: (`Cons` ⊚ ⊚), (`Cons True` ⊚), (`Cons` ⊚ (`Cons True` ⊚)), (`Cons True` (`Cons True` ⊚)), ....

The approximation reduction $\xrightarrow{A}$ is based upon $L_S$-reduction:

**Definition 5.3.** *Let $s$ be a closed expression. We define the approximation reduction $\xrightarrow{A}$ as follows:*
*Then $s \xrightarrow{A} v$ holds for some closed answer $v$ iff there is a reduction starting from* ($\texttt{letrec}\ x = s\ \texttt{in}\ x$) *to $v$ using the following intermediate steps.*

1. ($\texttt{letrec}\ x = s\ \texttt{in}\ x$) $\xrightarrow{*} s'$ *using an $L_S$-evaluation to a WHNF $s'$. Continuing from $s'$, we perform any number of $L_S$-reductions in application surface contexts (non-deterministically), where the target variables of (cp) are also in application surface contexts.*
2. *Perform any number of copy-reductions into the "in"-expression. Here the target variable of (cp) may be in any context $C$.*
3. *The last step is to remove the top-letrec-environment, and to replace all remaining let-bound variables in the "in"-expression by ⊚. The resulting expression is now either ⊚ or one of the desired answers $v$.*

*The set of answers reachable from $s$ by this procedure is defined as $ans(s)$.*

**Lemma 5.4.** *Let $s$ be a closed expression and $v \in ans(s)$. Then $v \leq_c s$.*

*Proof.* This follows from the correctness of the transformations proved in the previous sections, from decreasingness of (choice) (see Theorem 3.1) and from the fact that $\circledcirc \sim_c \Omega$ is the least element w.r.t. $\leq_c$ (see Proposition 3.2). □

Now we prove that sufficiently many answers are reached by these reductions.

**Theorem 5.5.** *Let $R$ be a reduction context, $s$ be a closed expression such that $R[s]\downarrow$. Then there is an answer $v$ with $(\texttt{letrec}\ x = s\ \texttt{in}\ x) \xrightarrow{A} v$, such that $R[v]\downarrow$. Note that $s \sim_c (\texttt{letrec}\ x = s\ \texttt{in}\ x)$ (see Theorem 3.1).*

*Proof.* In the proof we always refer to the calculus $L_S$.
Let $R$ be a reduction context and $s$ be a closed expression. Let *Red* be a normal-order reduction of $R[(\texttt{letrec}\ x = s\ \texttt{in}\ x)] \xrightarrow{no} r_1 \ldots \xrightarrow{no} r_n$, where $r_n$ is a WHNF, and $n$ is the number of normal-order reductions. In every expression of *Red*, the bindings inherited from $x = s$ can be identified in every $r_i$ by labeling them with †. Thus we label letrec-bound variables and the bound expression in surface positions that are derived from $s$. An important invariant is that for all †-labeled bindings $y_i = a_i$, and all free variables $y$ in $a_i$, $y$ is also a †-labeled variable, which follows by induction on the length of the reduction from the fact that $s$ is closed. If a WHNF $w$ of $R[(\texttt{letrec}\ x = s\ \texttt{in}\ x)]$ is reached, then from the WHNF we can gather all the †-labeled bindings in the top level letrec environment of $w$, and construct the expression $s' := (\texttt{letrec}\ \mathit{Env}\ \texttt{in}\ x)$, where we denote $x_1 = s_1, \ldots, x_m = s_m$ by *Env* and where $x_1 = x$ for convenience. Now we compute one possible answer $v$ from $s'$ as required by our claim as follows. We perform $n+1$ of the following macro-copy-steps within the environment *Env* into the "in"-expression:

One step consists of replacing all occurrences of $x_i$ by $s_i$ in the "in"-expression (initially $x$) for all $x_i = s_i$ in *Env* s.t. $s_i$ is an abstraction or a cv-expression. We do this in parallel for every letrec-bound variable, which is the same as applying the substitution $\sigma$ that is formed from *Env*. This is repeated $n+1$ times. The last step is to remove the top-environment, and to replace all letrec-bound variables in the in-expression by $\circledcirc$. This may produce either $\circledcirc$, or the desired answer $v$, and we have $s' \xrightarrow{A} v$ according to Definition 5.3. Since we assumed that a WHNF is reached, and $s$ was in a reduction context before, it is not possible that only $\circledcirc$ is reached, since the initial variable $x$ was in a reduction context and there must be at least one normal-order copy into $x$. Thus at least one of the macro-copy steps will replace $x$ by a constructor-expression or an abstraction.

Now we have to show that $R[v]\downarrow$. We start by rearranging the normal-order reduction *Red* of $R[(\texttt{letrec}\ x = s\ \texttt{in}\ x)]$, such that all the reductions that are within the †-labeled *Env* are performed first, i.e., $R[(\texttt{letrec}\ x = s\ \texttt{in}\ x)] \xrightarrow{*} R[s'] \xrightarrow{no,*} r_n$. It is easy to see that the $R[(\texttt{letrec}\ x = s\ \texttt{in}\ x)] \xrightarrow{*} R[s']$ starts with an $L_S$-normal-order reduction to a WHNF, since $x$ is "demanded" first. The subsequent reductions remain in application surface positions. The reduction $R[s'] \xrightarrow{no,*} r_n$ is normal-order, and has length at most $n$. The reduction sequence

*Red* is a mixture of reduction steps within †-labeled components, or reduction steps that modify the non-†-labeled components. All reductions are in surface contexts. Hence the †-reductions can be shifted to the left over non-†-reductions, since they are independent.

Now we focus on $R[s'] \xrightarrow{no,*} r_n$ of length at most $n$. We have to show that for $s' \xrightarrow{*} v$, we also have $R[v] \xrightarrow{no,*} u$, where $u$ is a WHNF. The term $v$ and its descendents can be represented using $\phi_{\geq k}$, which is defined as follows: $\phi_{\geq k}(r)$ denotes $r$ modified by the following operations: first $k$ applications of $\sigma$ are performed (the substitution corresponding to the $s$-environment $Env$), then any number of (cp)-steps using $Env$ and variables in $r$ as target variables, and as a final step $[\circledcirc/x_i]$-replacements in $r$ for all let-bound variables in $Env$. Now we have to show that $(\phi_{\geq n+1} R[s'])\downarrow$. The steps $\xleftarrow{no,a} \cdot \xrightarrow{\phi_{\geq k}}$ can be switched, i.e. replaced by $\xrightarrow{\phi_{\geq k-1}} \cdot \xleftarrow{\rho,*}$, where $\rho = \{(no,a), (abs), (lll), (cp), (cpbot), (ve)\}$. Using this commutation, it is easily shown by induction that, finally, we obtain a WHNF that is the result of a macro-copy reduction using $\phi_{\geq i}$, where $i \geq 1$. The argument now is that the replaced positions do not contribute to the WHNF $w$, hence it remains a WHNF after applying $\phi_{\geq 1}$. This means there is a reduction sequence $R[v] \xrightarrow{*} w'$, where $w'$ is a WHNF. Finally, the standardization theorem 3.3 shows that $R[v]\downarrow$. □

## 6 Least Upper Bounds and Sets of Answers

**Definition 6.1.** *Let $W$ be a set of expressions, and let $t$ be an expression. Then $t$ is a* lub *of $W$ iff $\forall u \in W : u \leq_c t$, and for every $s$ with $\forall u \in W : u \leq_c s$, it is $t \leq_c s$.*

*The expression $t$ is called a* contextual lub (club) *of $W$, iff for all contexts $C$: $C[t]$ is a lub of $\{C[r] \mid r \in W\}$. The notation is $t \in club(W)$. An expression $t$ is called a* linear club (lclub) *of $W$ if the set $W$ is a $\leq_c$-ascending chain of expressions. The notation is $t \in lclub(W)$. The set of all $t$ such that $t \in lclub(A)$ for some $A \subseteq W$ is denoted as $sublclub(W)$.*

*Example 6.2.* The following $\leq_c$-ascending chain $\lambda x_1.\Omega$, $\lambda x_1, x_2.\Omega$, $\ldots \lambda x_1, \ldots, x_n.\Omega$ has $YK$ as lclub, which is equivalent to the value $\lambda x.(Y\ K)$. The combinators are defined as $Y = \lambda f.(\lambda x.f(x\ x))\ (\lambda x.f(x\ x))$, and $K = \lambda x, y.x$.

Easy arguments show that the following holds:

**Lemma 6.3.** *For any closed expression $s$: $s \sim_c \Omega$ iff $ans(s) = \emptyset$. Otherwise, if $s \not\sim_c \Omega$, then $s \in club(ans(s))$.*

This yields an immediate criterion for contextual preorder:

**Corollary 6.4.** *Let $s, t$ be closed expressions. If for all $w \in ans(s)$ we also have $w \leq_c t$, then $s \leq_c t$.*

The following useful sufficient condition immediately follows from the corollary:

**Theorem 6.5.** *Let $s, t$ be closed expressions. If for all $v \in ans(s)$ there is some $w \in sublclub(ans(t))$ with $v \leq_c w$, then $s \leq_c t$.*

Note that a simplistic subset-condition for answer-sets is insufficient for $s \leq_c t$:

*Example 6.6.* Let $s := \lambda x.Y\ K$, $f\ z :=$ `choice` $z$ (`letrec` $u = f\ z$ `in` $\lambda x.u$) and $t := f\ \Omega$, where an explicit definition of $f$ is $f = Y\ (\lambda g.\lambda z.$`choice` $\Omega$ (`letrec` $u = g\ z$ `in` $\lambda x.u$)). Then for every $v \in ans(t)$, we have $v <_c s$. However, it is easy to see that $s \sim_c t$, since $\lambda x.Y\ K$ is the club of the ascending chain of values in $ans(t)$.

The following is obvious using contexts:

**Proposition 6.7.** $(c\ s_1 \dots s_n) \leq_c (c\ t_1 \dots t_n) \Leftrightarrow s_i \leq_c t_i$ *for all* $i$.

## 7 Criteria for Abstractions

Besides the trivial method to compare two abstractions $\lambda x.s$ and $\lambda x.t$ by $\alpha$-equivalence, perhaps combined with other correct transformations, we give a stronger condition for $\lambda x.s \leq_c \lambda x.t$ that is based on applying the abstractions to all possible pseudo-value arguments not using the criteria for all contexts. The following is proved in [SSM08].

**Lemma 7.1.** *[Context Lemma for Closing Reduction Contexts] Let $s, t$ be expressions. Then $s \leq_c t$ iff for all reduction contexts $R$: if $R[s], R[t]$ are closed and $R[s]\downarrow$, then also $R[t]\downarrow$,*

In a *pseudo-value environment Env* every bound term is a (closed) pseudo-value.

**Proposition 7.2.** *Let $s, t$ be two expressions. Then $s \leq_c t$ iff for all pseudo-value environments Env: if* (`letrec` *Env* `in` $s$), (`letrec` *Env* `in` $t$) *are closed then* (`letrec` *Env* `in` $s$) $\leq_c$ (`letrec` *Env* `in` $t$).

*Proof.* In order to show the non-trivial direction, we will use Lemma 7.1. Let $R$ be a reduction context such that $R[s], R[t]$ are closed and such that $R[s]\downarrow$. It is no restriction to assume that $R[\cdot]$ is of the form (`letrec` $Env_1, Env_2$ `in` $R'[\cdot]$), where $Env_1$ binds all the variables in $FV(s, t)$, and (`letrec` $Env_1$ `in` $[\cdot]$) is closed. Since $s' :=$ (`letrec` $Env_1, Env_2$ `in` $R'[s]$)$\downarrow$, there is a normal-order reduction *Red* of $s'$. In the same way as in the proof of Theorem 5.5, we can evaluate all bindings in $Env_1$ first, obtaining $Env_1'$ such that $s'' :=$ (`letrec` $Env_1', Env_2$ `in` $R'[s]$)$\downarrow$. The environment $Env_1'$ will be further modified into $Env_1''$ as follows: every binding $x = r$, where $r$ is not an abstraction and not a cv-expression is changed into $x = \circledcirc$. Again we have $s^{(3)} :=$ (`letrec` $Env_1'', Env_2$ `in` $R'[s]$)$\downarrow$, since the $\circledcirc$-bindings do not influence the normal-order reduction. Let $n$ be the length of a normal-order reduction of $s^{(3)}$. We further modify $Env_1''$ into $Env_1^{(3)}$ by applying the substitution $\sigma$ corresponding to $Env_1''$ at least

$n$ times to the environment, and then replacing all remaining occurrences of variables by $\circledcirc$. Similar as in the proof of Theorem 5.5, we argue that $(\texttt{letrec}\ \mathit{Env}_1^{(3)}, \mathit{Env}_2\ \texttt{in}\ R'[s])\downarrow$. Using the knowledge about correct transformations, it can be proved using induction that $(\texttt{letrec}\ \mathit{Env}_1^{(3)}, \mathit{Env}_2\ \texttt{in}\ R'[s]) \sim_c (\texttt{letrec}\ \mathit{Env}_1^{(3)}, \mathit{Env}_2\ \texttt{in}\ R'[(\texttt{letrec}\ \mathit{Env}_1^{(3)}\ \texttt{in}\ s)])$, hence $(\texttt{letrec}\ \mathit{Env}_1^{(3)}, \mathit{Env}_2\ \texttt{in}\ R'[(\texttt{letrec}\ \mathit{Env}_1^{(3)}\ \texttt{in}\ s)])\downarrow$.

Now we argue the reverse way for $t$: By the assumption, we have $(\texttt{letrec}\ \mathit{Env}_1^{(3)}, \mathit{Env}_2\ \texttt{in}\ R'[(\texttt{letrec}\ \mathit{Env}_1^{(3)}\ \texttt{in}\ s)]) \leq_c (\texttt{letrec}\ \mathit{Env}_1^{(3)}, \mathit{Env}_2\ \texttt{in}\ R'[(\texttt{letrec}\ \mathit{Env}_1^{(3)}\ \texttt{in}\ t)])$, hence $(\texttt{letrec}\ \mathit{Env}_1^{(3)}, \mathit{Env}_2\ \texttt{in}\ R'[(\texttt{letrec}\ \mathit{Env}_1^{(3)}\ \texttt{in}\ t)])\downarrow$. The same argument as above shows that also $(\texttt{letrec}\ \mathit{Env}_1^{(3)}, \mathit{Env}_2\ \texttt{in}\ R'[t])\downarrow$. Since $\circledcirc \sim_c \bot$ is the $\leq_c$-least element, and (cp) does not change the $\sim_c$ equivalence class, we also have $(\texttt{letrec}\ \mathit{Env}_1'', \mathit{Env}_2\ \texttt{in}\ R'[t])\downarrow$. Since $(\texttt{letrec}\ \mathit{Env}_1'', \mathit{Env}_2\ \texttt{in}\ R'[t])$ can be reached from $(\texttt{letrec}\ \mathit{Env}_1, \mathit{Env}_2\ \texttt{in}\ R'[t])$ by reductions that only decrease by $\leq_c$ due to Theorem 3.1, we finally have $(\texttt{letrec}\ \mathit{Env}_1, \mathit{Env}_2\ \texttt{in}\ R'[t])\downarrow$. Since $R$ was arbitrary, we can apply Lemma 7.1 and obtain that $s \leq_c t$ □

**Theorem 7.3.** $\lambda x.s \leq_c \lambda x.t$ *iff for all pseudo-values* $v$: $(\lambda x.s)\ v \leq_c (\lambda x.t)\ v$.

*Proof.* Follows from Proposition 7.2, since $(\lambda x.s)\ v \sim_c (\texttt{letrec}\ x = v\ \texttt{in}\ s)$ □

## 8 Finite Simulation Method and Examples

Now we have several criteria to prove $s \leq_c t$ for closed expressions $s, t$.

1. If $\mathit{ans}(s) \subseteq \mathit{ans}(t)$, then $s \leq_c t$.
2. If for every $v \in \mathit{ans}(s)$, there is some $w \in \mathit{ans}(t)$ with $v \leq_c w$, then $s \leq_c t$.
3. If for every $v \in \mathit{ans}(s)$, there is some $w \in \mathit{ans}(t)$ with $v \leq_c w$ or some $w \in \mathit{sublclub}(\mathit{ans}(t))$ with $v \leq_c w$, then $s \leq_c t$.
4. if $s = c\ s_1 \dots s_n$, $t = c\ t_1 \dots t_n$, and $s_i \leq_c t_i$ for all $i$, then $s \leq_c t$.
5. if $s = \lambda x.s'$, $t = \lambda x.t'$, and for all pseudo-values $v$: $s\ v \leq_c t\ v$, then $s \leq_c t$.

The following (non-effective) procedure is a prototype of "finite simulation" for testing two closed expressions $s, t$ whether they are in a relation $s \leq_c t$:

1. Compute the answer-sets $\mathit{ans}(s)$ and $\mathit{ans}(t)$.
2. For every value $v \in \mathit{ans}(s)$, find a value $w \in \mathit{ans}(t)$ such that $v \leq_c w$.
3. For the $v \leq_c w$-test, use the following tests, recursively:
   (a) If $v = (c\ s_1 \dots s_m)$, $w = (c\ t_1 \dots t_m)$, make sure that $v_i \leq_c w_i$ for all $i$.
   (b) If $v = \lambda x.s'$, $w = \lambda x.t'$, make sure that for all pseudo-values $a$: $(v\ a) \leq_c (w\ a)$, again using this procedure.

If answers-sets are finite, the recursion depth is bounded, and all the involved tests are decidable, the procedure becomes effective. Proving $s \sim t$ for expressions $s, t$ can be done by checking $s \leq_c t$ and $t \leq_c s$.

*Example 8.1.* Let $s := \mathit{repeat}\ \texttt{True}$, $t := Y\ (\lambda a.\texttt{choice}\ \Omega\ (\texttt{Cons True}\ a)$ where $\mathit{repeat} := Y\ (\lambda r.\lambda x.\texttt{Cons}\ x\ (r\ x))$. Then $s$ can be reduced to the answers $(\texttt{Cons True}\ (\texttt{Cons True}\ (\ldots(\texttt{Cons True}\ \Omega))))$ and $t$ can be reduced to the same answers, where we use $\circledcirc \sim_c \Omega$. This implies that $s \sim_c t$.

*Example 8.2.* Finite simulation can distinguish expressions that differ only by sharing: Let $s := (\texttt{letrec}\ x = \texttt{choice True False in}\ \lambda y.x)$ and let $t = \lambda y.(\texttt{letrec}\ x = \texttt{choice True False in}\ x)$. These expressions are contextually different, using the context $C[\cdot] := (\texttt{letrec}\ z = [\cdot]\ \texttt{in if}\ (z\ \bot)\ \texttt{then}\ (\texttt{if}\ (z\ \bot)\ \texttt{then True else}\ \bot)\ \texttt{else True})$. The answer-sets are: $\mathit{ans}(t) = \{t\}$, $\mathit{ans}(s) = \{\lambda y.\texttt{True}, \lambda y.\texttt{False}\}$.

*Example 8.3.* We are able to prove idempotency, commutativity and associativity for `choice` as a binary operator or all expressions, and hence these identities can be used as program transformation using Proposition 7.2: For instance, for commutativity: every pseudo-value environment *Env* that closes $s$ and $t$, we consider $(\texttt{letrec}\ \mathit{Env}\ \texttt{in choice}\ s\ t)$ and $(\texttt{letrec}\ \mathit{Env}\ \texttt{in choice}\ t\ s)$. The first step of the approximation is the choice-reduction. Then the right and left hand side have the same set of answers, hence they are equivalent, which follows from Corollary 6.4. The same can be done for the other identities.

## 9 Conclusion and Further Research

We have shown that in a call-by-need non-deterministic lambda calculus with letrec, where the proof method of Howe fails to prove correctness of co-inductive simulation, the correctness of finite simulation can be established as a tool with almost the same practical power. Further research is to adapt and extend the methods to an appropriately defined simulation, and to investigate an extension of the tools and methods to a combination of may- and must-convergence.

## Acknowledgements

We thank David Sabel for reading several versions of the paper.

## References

[AB02] Ariola, Z.M., Blom, S.: Skew confluence and the lambda calculus with letrec. Annals of Pure and Applied Logic 117, 95–168 (2002)

[Abr90] Abramsky, S.: The lazy lambda calculus. In: Turner, D.A. (ed.) Research Topics in Functional Programming, pp. 65–116. Addison-Wesley, Reading (1990)

[AK97] Ariola, Z.M., Klop, J.W.: Lambda calculus with explicit recursion. Inform. and Comput. 139(2), 154–233 (1997)

[AW96] Ariola, Z.M., Klop, J.W.: Equational term graph rewriting. Fundamentae Informaticae 26(3,4), 207–240 (1996)

[Gor99] Gordon, A.D.: Bisimilarity as a theory of functional programming. Theoret. Comput. Sci. 228(1-2), 5–47 (1999)

[Han96] Hanus, M.: A unified computation model for functional and logic programming. In: POPL 1997, pp. 80–93. ACM, New York (1996)

[How89] Howe, D.: Equality in lazy computation systems. In: 4th IEEE Symp. on Logic in Computer Science, pp. 198–203 (1989)

[How96] Howe, D.: Proving congruence of bisimulation in functional programming languages. Inform. and Comput. 124(2), 103–112 (1996)

[KSS98] Kutzner, A., Schmidt-Schauß, M.: A nondeterministic call-by-need lambda calculus. In: ICFP 1998, pp. 324–335. ACM Press, New York (1998)

[Mac02] Machkasova, E.: Computational Soundness of Non-Confluent Calculi with Applications to Modules and Linking. PhD thesis, Boston University (2002)

[Mac07] Machkasova, E.: Computational soundness of a call by name calculus of recursively-scoped records. In: 7th WRS. ENTCS (2007)

[Man04] Mann, M.: Congruence of bisimulation in a non-deterministic call-by-need lambda calculus. In: SOS 2004, BRICS NS-04-1, pp. 20–38 (2004)

[MSC99] Moran, A.K.D., Sands, D., Carlsson, M.: Erratic fudgets: A semantic theory for an embedded coordination language. In: Ciancarini, P., Wolf, A.L. (eds.) COORDINATION 1999. LNCS, vol. 1594, pp. 85–102. Springer, Heidelberg (1999)

[MT00] Machkasova, E., Turbak, F.A.: A calculus for link-time compilation. In: Smolka, G. (ed.) ESOP 2000 and ETAPS 2000. LNCS, vol. 1782, pp. 260–274. Springer, Heidelberg (2000)

[Pey03] Jones, S.P.: Haskell 98 language and libraries: the Revised Report. Cambridge University Press, Cambridge (2003), `http://www.haskell.org`

[PGF96] Peyton Jones, S., Gordon, A., Finne, S.: Concurrent Haskell. In: Proc. 23th Principles of Programming Languages (1996)

[Plo75] Plotkin, G.D.: Call-by-name, call-by-value, and the lambda-calculus. Theoret. Comput. Sci. 1, 125–159 (1975)

[SSM07] Schmidt-Schauß, M., Mann, M.: On equivalences and standardization in a non-deterministic call-by-need lambda calculus. Frank report 31, Inst. f. Informatik, J.W.Goethe-University, Frankfurt (August 2007)

[SSM08] Schmidt-Schauß, M., Machkasova, E.: A finite simulation method in a non-deterministic call-by-need calculus with letrec, constructors and case. Frank 32, Inst. f. Informatik, J.W.Goethe-University, Frankfurt (2008)

[SSS07a] Sabel, D., Schmidt-Schauß, M.: A call-by-need lambda-calculus with locally bottom-avoiding choice: Context lemma and correctness of transformations. Math. Structures Comput. Sci (accepted for publication, 2007)

[SSS07b] Schmidt-Schauß, M., Sabel, D.: On generic context lemmas for lambda calculi with sharing. Frank 27, Inst. Informatik, J.W.G-Univ., Frankfurt (2007)

[SSSS04] Schmidt-Schauß, M., Schütz, M., Sabel, D.: On the safety of Nöcker's strictness analysis. Frank 19, Inst. Informatik, J.W.G-Univ., Frankfurt (2004)

[SSSS08] Schmidt-Schauß, M., Schütz, M., Sabel, D.: Safety of Nöcker's strictness analysis. J. Funct. Programming (accepted for publication, 2008)

[WDK03] Wells, J.B., Plump, D., Kamareddine, F.: Diagrams for meaning preservation. In: Nieuwenhuis, R. (ed.) RTA 2003. LNCS, vol. 2706, pp. 88–106. Springer, Heidelberg (2003)

# Root-Labeling*

Christian Sternagel and Aart Middeldorp

Institute of Computer Science
University of Innsbruck
Austria

**Abstract.** In 2006 Jambox, a termination prover developed by Endrullis, surprised the termination community by winning the string rewriting division and almost beating AProVE in the term rewriting division of the international termination competition. The success of Jambox for strings is partly due to a very special case of semantic labeling. In this paper we integrate this technique, which we call root-labeling, into the dependency pair framework. The result is a simple processor with help of which $\mathsf{T_TT_2}$ surprised the termination community in 2007 by producing the first automatically generated termination proof of a string rewrite system with non-primitive recursive complexity (Touzet, 1998). Unlike many other recent termination methods, the root-labeling processor is trivial to automate and completely unsuitable for producing human readable proofs.

## 1 Introduction

Semantic labeling is a complete method for proving the termination of term rewrite systems (TRSs), introduced by Zantema [24]. It transforms a given TRS into a termination equivalent TRS by labeling function symbols based on their semantics. The challenge when applying and automating semantic labeling is to choose the labeling functions in such a way that the resulting TRS is easier to prove terminating. Koprowski and Zantema [13,15] showed how this can be done when algebras over the natural numbers are used together with the lexicographic path order to deal with the resulting infinite TRSs over infinite signatures. In [14] Koprowski and Middeldorp combined predictive labeling—a version of semantic labeling with less constraints [11]—with dependency pairs and modeled the search space as a SAT problem.

A very special version of semantic labeling for string rewrite systems (SRSs), due to Johannes Waldmann (Jörg Endrullis, personal communication), in which the semantic and labeling components are completely determined by the SRS at hand was used by Matchbox/SatELite [23] and Jambox [4] in the string rewriting division of the 2006 international termination competition[1] with remarkable success. This special version, which we call root-labeling, is extended to TRSs in this paper. More importantly but equally straightforward, we present root-labeling

* This research is supported by FWF (Austrian Science Fund) project P18763.

[1] `www.lri.fr/~marche/termination-competition/2006/`

A. Voronkov (Ed.): RTA 2008, LNCS 5117, pp. 336–350, 2008.

as a processor in the dependency pair framework [8,19]. Due to this new root-labeling processor, in 2007 $\mathsf{T_TT_2}$ [16] could prove the termination of exactly one SRS that had eluded all termination tools (many of which use highly specialized techniques for SRSs) before, resulting in the first automatic termination proof of an SRS whose derivational complexity is not primitive recursive (Touzet [21]).

The remainder of this paper is organized as follows. In the next section we recall basic definitions and results concerning semantic labeling and dependency pairs. In Section 3, semantic labeling is specialized to root-labeling. Incorporating dependency pairs is the topic of Section 4 and in Section 5 we present our main example in some detail. Experimental results are presented in Section 6 and we conclude in Section 7 with suggestions for future research.

## 2 Preliminaries

We assume basic knowledge of term rewriting [2,18]. Let $\mathcal{R}$ be a TRS over a signature $\mathcal{F}$ and let $\mathcal{A} = (A, \{f_\mathcal{A}\}_{f\in\mathcal{F}})$ be an $\mathcal{F}$-algebra. Let $\mathcal{V}$ be the set of variables. We say that $\mathcal{A}$ is a model of $\mathcal{R}$ if $[\alpha]_\mathcal{A}(l) = [\alpha]_\mathcal{A}(r)$ for every rule $l \to r \in \mathcal{R}$ and every assignment $\alpha\colon \mathcal{V} \to A$. A labeling $\ell$ for $\mathcal{A}$ consists of sets of labels $L_f$ for every $f \in \mathcal{F}$ together with mappings $\ell_f\colon A^n \to L_f$ for every $n$-ary function symbol $f \in \mathcal{F}$ with $L_f \neq \varnothing$. The labeled signature $\mathcal{F}_\mathsf{lab}$ consists of $n$-ary function symbols $f_a$ for every $n$-ary function symbol $f \in \mathcal{F}$ and label $a \in L_f$ together with all function symbols $f \in \mathcal{F}$ such that $L_f = \varnothing$. The mapping $\ell_f$ determines the label of the root symbol $f$ of a term $f(t_1, \ldots, t_n)$ based on the values of the arguments $t_1, \ldots, t_n$. For every assignment $\alpha\colon \mathcal{V} \to A$ the mapping $\mathsf{lab}_\alpha\colon \mathcal{T}(\mathcal{F}, \mathcal{V}) \to \mathcal{T}(\mathcal{F}_\mathsf{lab}, \mathcal{V})$ is inductively defined as follows: $\mathsf{lab}_\alpha(t) = t$ if $t$ is a variable, $\mathsf{lab}_\alpha(f(t_1, \ldots, t_n)) = f(\mathsf{lab}_\alpha(t_1), \ldots, \mathsf{lab}_\alpha(t_n))$ if $L_f = \varnothing$, and $\mathsf{lab}_\alpha(f(t_1, \ldots, t_n)) = f_a(\mathsf{lab}_\alpha(t_1), \ldots, \mathsf{lab}_\alpha(t_n))$ if $L_f \neq \varnothing$ where $a$ denotes the label $\ell_f([\alpha]_\mathcal{A}(t_1), \ldots, [\alpha]_\mathcal{A}(t_n))$. The *labeled* TRS $\mathcal{R}_\mathsf{lab}$ over the signature $\mathcal{F}_\mathsf{lab}$ consists of the rewrite rules $\mathsf{lab}_\alpha(l) \to \mathsf{lab}_\alpha(r)$ for all rules $l \to r \in \mathcal{R}$ and assignments $\alpha\colon \mathcal{V} \to A$.

**Theorem 1** (Zantema [24]). *Let $\mathcal{R}$ be a TRS. Let the algebra $\mathcal{A}$ be a non-empty model of $\mathcal{R}$ and let $\ell$ be a labeling for $\mathcal{A}$. The TRS $\mathcal{R}$ is terminating if and only if the TRS $\mathcal{R}_\mathsf{lab}$ is terminating.* □

*Example 2 ([24]).* Consider the TRS $\mathcal{R}$ (Toyama [22]) consisting of the single rule $\mathsf{f}(\mathsf{a}, \mathsf{b}, x) \to \mathsf{f}(x, x, x)$. To ease the termination proof, we label function symbol $\mathsf{f}$ such that its occurrence on the left gets a different label from the one on the right. This is achieved by taking the algebra $\mathcal{A}$ with carrier $\{0, 1\}$ and interpretations $\mathsf{a}_\mathcal{A} = 0$, $\mathsf{b}_\mathcal{A} = 1$, $\mathsf{f}_\mathcal{A}(x, y, z) = 0$ for all $x, y, z \in \{0, 1\}$, together with $L_\mathsf{a} = L_\mathsf{b} = \varnothing$, $L_\mathsf{f} = \{0, 1\}$ and $\ell_\mathsf{f}(x, y, z) = 0$ if $x = y$ and $\ell_\mathsf{f}(x, y, z) = 1$ if $x \neq y$. The algebra $\mathcal{A}$ is a model of $\mathcal{R}$ and $\mathcal{R}_\mathsf{lab}$ consists of the rule $\mathsf{f}_1(\mathsf{a}, \mathsf{b}, x) \to \mathsf{f}_0(x, x, x)$. Termination of $\mathcal{R}_\mathsf{lab}$ is obvious as there are no dependency pairs.

A stronger version of semantic labeling is obtained by equipping the carrier of the algebra and the label sets with a well-founded order such that all algebra functions and all labeling functions are weakly monotone in all coordinates. The model condition is then weakened to $[\alpha]_{\mathcal{A}}(l) \geqslant [\alpha]_{\mathcal{A}}(r)$ for every rule $l \to r \in \mathcal{R}$ and every assignment $\alpha\colon \mathcal{V} \to A$. Further, all rules of the form $f_a(x_1,\ldots,x_n) \to f_b(x_1,\ldots,x_n)$ with $a, b \in L_f$ such that $a > b$ have to be added to $\mathcal{R}_{\mathsf{lab}}$ in order to obtain a sound transformation (Zantema [24]). This version of semantic labeling is capable of transforming any terminating TRS into a TRS whose termination proof is particularly simple, see [17]. This result is however only of theoretical interest. Recent variants inspired by the need for automation are presented in [11,20].

The dependency pair method [1] is a powerful approach for proving termination of TRSs. It is used in most termination tools for term rewriting. The dependency pair framework [8,19] is a modular reformulation and improvement of this approach. We present a simplified version which is sufficient for our purposes. Let $\mathcal{R}$ be a TRS over a signature $\mathcal{F}_{\mathcal{R}}$. The signature $\mathcal{F}_{\mathcal{R}}$ is extended with symbols $f^\sharp$ for every symbol $f \in \{\mathsf{root}(l) \mid l \to r \in \mathcal{R}\}$, where $f^\sharp$ has the same arity as $f$. In examples we write $F$ for $f^\sharp$. If $t \in \mathcal{T}(\mathcal{F},\mathcal{V})$ with $\mathsf{root}(t)$ defined then $t^\sharp$ denotes the term that is obtained from $t$ by replacing its root symbol with $\mathsf{root}(t)^\sharp$. If $l \to r \in \mathcal{R}$ and $t$ is a subterm of $r$ with a defined root symbol that is not a proper subterm of $l$ then the rule $l^\sharp \to t^\sharp$ is a *dependency pair* of $\mathcal{R}$. The set of dependency pairs of $\mathcal{R}$ is denoted by $\mathsf{DP}(\mathcal{R})$. A *DP problem* is a pair of TRSs $(\mathcal{P},\mathcal{R})$ such that symbols in $\mathcal{F}^\sharp = \{\mathsf{root}(l), \mathsf{root}(r) \mid l \to r \in \mathcal{P}\}$ do neither occur in $\mathcal{R}$ nor in proper subterms of the left and right-hand sides of rules in $\mathcal{P}$. Writing $\mathcal{F}_{\mathcal{P}}$ for the signature of $\mathcal{P}$, the signature $\mathcal{F}_{\mathcal{R}} \cup (\mathcal{F}_{\mathcal{P}} \setminus \mathcal{F}^\sharp)$ is denoted by $\mathcal{F}$. The problem $(\mathcal{P},\mathcal{R})$ is said to be *finite* if there is no infinite sequence $s_1 \xrightarrow{\epsilon}_{\mathcal{P}} t_1 \to_{\mathcal{R}}^* s_2 \xrightarrow{\epsilon}_{\mathcal{P}} t_2 \to_{\mathcal{R}}^* \cdots$ such that all terms $t_1$, $t_2$, ... are terminating with respect to $\mathcal{R}$. Such an infinite sequence is said to be *minimal*. Here the $\epsilon$ in $\xrightarrow{\epsilon}_{\mathcal{P}}$ denotes that the application of the rule in $\mathcal{P}$ takes place at the root position. The main result underlying the dependency pair approach states that a TRS $\mathcal{R}$ is terminating if and only if the DP problem $(\mathsf{DP}(\mathcal{R}),\mathcal{R})$ is finite.

In order to prove finiteness of a DP problem a number of so-called *DP processors* have been developed. DP processors are functions that take a DP problem as input and return a set of DP problems as output. In order to be employed to prove termination they need to be *sound*, that is, if all DP problems in a set returned by a DP processor are finite then the initial DP problem is finite. *Complete* DP processors, which are those processors with the property that if one of the returned DP problems is not finite then the original DP problem is not finite, can be used to prove non-termination.

## 3 Plain Root-Labeling[2]

The challenge when implementing semantic labeling is to find an appropriate carrier and suitable interpretation and labeling functions such that the labeled

[2] As stated in the introduction, the results in this section for SRSs are due to Johannes Waldmann; they are mentioned in [26].

system is easier to prove terminating. This issue has been addressed in several recent papers ([13,15,14]).

In the special version of semantic labeling defined in this section, everything is fixed. This has the disadvantage of reducing the power of semantic labeling significantly but the advantage of making automation a trivial issue.

**Definition 3.** *Let $\mathcal{R}$ be a TRS over a signature $\mathcal{F}$. The algebra $\mathcal{A}_\mathcal{F}$ has carrier $\mathcal{F}$ and interpretation functions $f_{\mathcal{A}_\mathcal{F}}(x_1, \ldots, x_n) = f$ for every $n$-ary $f \in \mathcal{F}$ and all $x_1, \ldots, x_n \in \mathcal{F}$. The labeling $\ell$ is defined as follows: $L_f = \mathcal{F}^n$ if the arity $n$ of $f$ is at least 1 and $L_f = \varnothing$ otherwise, and $\ell_f(x_1, \ldots, x_n) = (x_1, \ldots, x_n)$ for all $f$ with $L_f \neq \varnothing$ and all $x_1, \ldots, x_n \in \mathcal{F}$. In examples we write $x_1$ for $(x_1)$. The resulting labeled TRS $\mathcal{R}_\mathsf{lab}$ is denoted by $\mathcal{R}_\mathsf{rl}$.*

*Example 4.* Consider the TRS $\mathcal{R}$ from Example 2, extended with the two rules $\mathsf{c} \to \mathsf{a}$ and $\mathsf{c} \to \mathsf{b}$. The TRS $\mathcal{R}_\mathsf{rl}$ consists of the six rules

$$\begin{array}{lll} \mathsf{f}_{(\mathsf{a},\mathsf{b},\mathsf{a})}(\mathsf{a},\mathsf{b},x) \to \mathsf{f}_{(\mathsf{a},\mathsf{a},\mathsf{a})}(x,x,x) & \mathsf{f}_{(\mathsf{a},\mathsf{b},\mathsf{c})}(\mathsf{a},\mathsf{b},x) \to \mathsf{f}_{(\mathsf{c},\mathsf{c},\mathsf{c})}(x,x,x) & \mathsf{c} \to \mathsf{a} \\ \mathsf{f}_{(\mathsf{a},\mathsf{b},\mathsf{b})}(\mathsf{a},\mathsf{b},x) \to \mathsf{f}_{(\mathsf{b},\mathsf{b},\mathsf{b})}(x,x,x) & \mathsf{f}_{(\mathsf{a},\mathsf{b},\mathsf{f})}(\mathsf{a},\mathsf{b},x) \to \mathsf{f}_{(\mathsf{f},\mathsf{f},\mathsf{f})}(x,x,x) & \mathsf{c} \to \mathsf{b} \end{array}$$

and is terminating because there are no dependency pairs. The TRS $\mathcal{R}$, however, admits an infinite rewrite sequence starting from the term $\mathsf{f}(\mathsf{a},\mathsf{b},\mathsf{c})$.

The problem in the previous example is that $\mathcal{A}_\mathcal{F}$ is not a model of $\mathcal{R}$. In order to solve this problem we close every rule $l \to r$ where $l$ and $r$ have different root symbols under flat contexts, before performing the root-labeling operation.

**Definition 5.** *Let $\mathcal{R}$ be a TRS over a signature $\mathcal{F}$. The rules in $\mathcal{R}_\mathsf{p} = \{l \to r \in \mathcal{R} \mid \mathsf{root}(l) = \mathsf{root}(r)\}$ are* root-preserving. *The rules in $\mathcal{R}_\mathsf{a} = \mathcal{R} \setminus \mathcal{R}_\mathsf{p}$ are* root-altering. *The set $\{f(x_1, \ldots, x_{i-1}, \Box, x_{i+1}, \ldots, x_n) \mid f \in \mathcal{F}$ has arity $n \geqslant 1$ and $1 \leqslant i \leqslant n\}$ of* flat *contexts is denoted by $\mathcal{FC}$. The* flat context closure *of $\mathcal{R}$ is defined as $\mathcal{FC}(\mathcal{R}) = \mathcal{R}_\mathsf{p} \cup \{C[l] \to C[r] \mid l \to r \in \mathcal{R}_\mathsf{a}$ and $C \in \mathcal{FC}\}$.*

*Example 6.* For the TRS $\mathcal{R}$ from Example 4, the TRS $\mathcal{FC}(\mathcal{R})$ consisting of the rules

$$\begin{array}{lll} \mathsf{f}(\mathsf{a},\mathsf{b},x) \to \mathsf{f}(x,x,x) & \mathsf{f}(\mathsf{c},x,y) \to \mathsf{f}(\mathsf{a},x,y) & \mathsf{f}(\mathsf{c},x,y) \to \mathsf{f}(\mathsf{b},x,y) \\ & \mathsf{f}(x,\mathsf{c},y) \to \mathsf{f}(x,\mathsf{a},y) & \mathsf{f}(x,\mathsf{c},y) \to \mathsf{f}(x,\mathsf{b},y) \\ & \mathsf{f}(x,y,\mathsf{c}) \to \mathsf{f}(x,y,\mathsf{a}) & \mathsf{f}(x,y,\mathsf{c}) \to \mathsf{f}(x,y,\mathsf{b}) \end{array}$$

is obtained. Note that like $\mathcal{R}$, $\mathcal{FC}(\mathcal{R})$ is non-terminating for $\mathsf{f}(\mathsf{a},\mathsf{b},\mathsf{c})$.

**Lemma 7.** *The transformation $\mathcal{FC}(\cdot)$ on TRSs is termination preserving and reflecting, i.e., $\mathcal{R}$ is terminating if and only if $\mathcal{FC}(\mathcal{R})$ is terminating.*

*Proof.* By construction, every rewrite step in $\mathcal{FC}(\mathcal{R})$ can be simulated by a rewrite step in $\mathcal{R}$. This proves the "only if" direction. For the "if" direction we reason as follows. Suppose $\mathcal{R}$ is not terminating. Then there exists an infinite sequence $t_1 \to_\mathcal{R} t_2 \to_\mathcal{R} \cdots$. Let $C \in \mathcal{FC}$ be an arbitrary flat context. Since rewriting is closed under contexts, we obtain $C[t_1] \to_{\mathcal{R}} C[t_2] \to_{\mathcal{R}} \cdots$. We claim that this sequence is a rewrite sequence in $\mathcal{FC}(\mathcal{R})$. Fix $i \geqslant 1$ and consider the step $C[t_i] \to_{\mathcal{R}} C[t_{i+1}]$. Let $l \to r \in \mathcal{R}$ be the employed rewrite rule. We distinguish two cases.

1. If $l \to r$ is root-preserving then $l \to r$ belongs to $\mathcal{FC}(\mathcal{R})$ and the result is clear.
2. Suppose $l \to r$ is root-altering. If the rule was applied below the root position in $t_i \to_{\mathcal{R}} t_{i+1}$, then $t_i \to_{\mathcal{FC}(\mathcal{R})} t_{i+1}$ by applying the rule $C'[l] \to C'[r]$ for the flat context $C' \in \mathcal{FC}$ uniquely determined by the function symbol in $t_i$ directly above the redex and the position of the redex. Formally, if $\pi \cdot j$ is the redex position in $t_i$ then $C' = f(x_1, \ldots, x_{j-1}, \Box, x_{j+1}, \ldots, x_n)$ with $f = \mathsf{root}(t_i|_\pi)$. Hence also $C[t_i] \to_{\mathcal{FC}(\mathcal{R})} C[t_{i+1}]$. If the rule $l \to r$ was applied at the root position in $t_i \to_{\mathcal{R}} t_{i+1}$ then $C[t_i] \to_{\mathcal{FC}(\mathcal{R})} C[t_{i+1}]$ by applying the rule $C[l] \to C[r] \in \mathcal{FC}(\mathcal{R})$.

So $C[t_1] \to_{\mathcal{FC}(\mathcal{R})} C[t_2] \to_{\mathcal{FC}(\mathcal{R})} \cdots$ and thus $\mathcal{FC}(\mathcal{R})$ is non-terminating. □

To pave the way for the developments in the next section, we need the observation that Lemma 7 remains true if we allow an arbitrary extension $\mathcal{G}$ of the signature $\mathcal{F}$ of $\mathcal{R}$ when building flat contexts. We write $\mathcal{FC}_\mathcal{G}$ when we want to make the signature of the flat contexts clear.

**Theorem 8.** *The transformation $\mathcal{FC}(\cdot)_{\mathsf{rl}}$ on TRSs is termination preserving and reflecting, i.e., $\mathcal{R}$ is terminating if and only if $\mathcal{FC}(\mathcal{R})_{\mathsf{rl}}$ is terminating.*

*Proof.* According to Lemma 7, termination of $\mathcal{R}$ is equivalent to termination of $\mathcal{FC}(\mathcal{R})$. By construction, all rules in $\mathcal{FC}(\mathcal{R})$ are root-preserving. Hence $\mathcal{A}_\mathcal{F}$ is a model for $\mathcal{FC}(\mathcal{R})$ and Theorem 1 yields the termination equivalence of $\mathcal{FC}(\mathcal{R})$ and $\mathcal{FC}(\mathcal{R})_{\mathsf{rl}}$. Combining the two equivalences yields the desired result. □

We conclude this section with two more examples. A string rewrite system (SRS) is a TRS over a signature consisting of unary function symbols. We write strings $\mathsf{a}(\mathsf{b}(\mathsf{c}(x)))$ as $\mathsf{abc}$ (the variable is implicit).

*Example 9.* Consider the SRS $\mathcal{R} = \{\mathsf{aa} \to \mathsf{aba}\}$. Since the rule is root-preserving, $\mathcal{FC}(\mathcal{R}) = \mathcal{R}$. The SRS $\mathcal{FC}(\mathcal{R})_{\mathsf{rl}}$ consists of the two rules $\mathsf{a_a a_a} \to \mathsf{a_b b_a a_a}$ and $\mathsf{a_a a_b} \to \mathsf{a_b b_a a_b}$, and is terminating because its rules are oriented from left to right by the polynomial interpretation $[\mathsf{a_a}](x) = x + 1$ and $[\mathsf{a_b}](x) = [\mathsf{b_a}](x) = x$.

*Example 10.* Consider the TRS $\mathcal{R}$ (`teparla3.trs`) consisting of the two rules $\mathsf{f}(y, \mathsf{f}(x, \mathsf{f}(\mathsf{a}, x))) \to \mathsf{f}(\mathsf{f}(\mathsf{a}, \mathsf{f}(x, \mathsf{a})), \mathsf{f}(\mathsf{a}, y))$ and $\mathsf{f}(x, \mathsf{f}(x, y)) \to \mathsf{f}(\mathsf{f}(\mathsf{f}(x, \mathsf{a}), \mathsf{a}), \mathsf{a})$. None of the tools participating in the 2007 international competition could prove its termination. Like in the preceding example we have $\mathcal{R} = \mathcal{R}_\mathsf{p}$ and hence $\mathcal{FC}(\mathcal{R}) = \mathcal{R}$. The TRS $\mathcal{FC}(\mathcal{R})_{\mathsf{rl}}$ consists of the following eight rules

$$\mathsf{f}_{(\mathsf{a},\mathsf{f})}(y, \mathsf{f}_{(\mathsf{a},\mathsf{f})}(x, \mathsf{f}_{(\mathsf{a},\mathsf{a})}(\mathsf{a}, x))) \to \mathsf{f}_{(\mathsf{f},\mathsf{f})}(\mathsf{f}_{(\mathsf{a},\mathsf{f})}(\mathsf{a}, \mathsf{f}_{(\mathsf{a},\mathsf{a})}(x, \mathsf{a})), \mathsf{f}_{(\mathsf{a},\mathsf{a})}(\mathsf{a}, y))$$
$$\mathsf{f}_{(\mathsf{f},\mathsf{f})}(y, \mathsf{f}_{(\mathsf{a},\mathsf{f})}(x, \mathsf{f}_{(\mathsf{a},\mathsf{a})}(\mathsf{a}, x))) \to \mathsf{f}_{(\mathsf{f},\mathsf{f})}(\mathsf{f}_{(\mathsf{a},\mathsf{f})}(\mathsf{a}, \mathsf{f}_{(\mathsf{a},\mathsf{a})}(x, \mathsf{a})), \mathsf{f}_{(\mathsf{a},\mathsf{f})}(\mathsf{a}, y))$$

$$
\begin{aligned}
\mathsf{f}_{(\mathsf{a},\mathsf{f})}(y, \mathsf{f}_{(\mathsf{f},\mathsf{f})}(x, \mathsf{f}_{(\mathsf{a},\mathsf{f})}(\mathsf{a}, x))) &\to \mathsf{f}_{(\mathsf{f},\mathsf{f})}(\mathsf{f}_{(\mathsf{a},\mathsf{f})}(\mathsf{a}, \mathsf{f}_{(\mathsf{f},\mathsf{a})}(x, \mathsf{a})), \mathsf{f}_{(\mathsf{a},\mathsf{a})}(\mathsf{a}, y)) \\
\mathsf{f}_{(\mathsf{f},\mathsf{f})}(y, \mathsf{f}_{(\mathsf{f},\mathsf{f})}(x, \mathsf{f}_{(\mathsf{a},\mathsf{f})}(\mathsf{a}, x))) &\to \mathsf{f}_{(\mathsf{f},\mathsf{f})}(\mathsf{f}_{(\mathsf{a},\mathsf{f})}(\mathsf{a}, \mathsf{f}_{(\mathsf{f},\mathsf{a})}(x, \mathsf{a})), \mathsf{f}_{(\mathsf{a},\mathsf{f})}(\mathsf{a}, y)) \\
\mathsf{f}_{(\mathsf{a},\mathsf{f})}(x, \mathsf{f}_{(\mathsf{a},\mathsf{a})}(x, y)) &\to \mathsf{f}_{(\mathsf{f},\mathsf{a})}(\mathsf{f}_{(\mathsf{f},\mathsf{a})}(\mathsf{f}_{(\mathsf{a},\mathsf{a})}(x, \mathsf{a}), \mathsf{a}), \mathsf{a}) \\
\mathsf{f}_{(\mathsf{a},\mathsf{f})}(x, \mathsf{f}_{(\mathsf{a},\mathsf{f})}(x, y)) &\to \mathsf{f}_{(\mathsf{f},\mathsf{a})}(\mathsf{f}_{(\mathsf{f},\mathsf{a})}(\mathsf{f}_{(\mathsf{a},\mathsf{a})}(x, \mathsf{a}), \mathsf{a}), \mathsf{a}) \\
\mathsf{f}_{(\mathsf{f},\mathsf{f})}(x, \mathsf{f}_{(\mathsf{f},\mathsf{a})}(x, y)) &\to \mathsf{f}_{(\mathsf{f},\mathsf{a})}(\mathsf{f}_{(\mathsf{f},\mathsf{a})}(\mathsf{f}_{(\mathsf{f},\mathsf{a})}(x, \mathsf{a}), \mathsf{a}), \mathsf{a}) \\
\mathsf{f}_{(\mathsf{f},\mathsf{f})}(x, \mathsf{f}_{(\mathsf{f},\mathsf{f})}(x, y)) &\to \mathsf{f}_{(\mathsf{f},\mathsf{a})}(\mathsf{f}_{(\mathsf{f},\mathsf{a})}(\mathsf{f}_{(\mathsf{f},\mathsf{a})}(x, \mathsf{a}), \mathsf{a}), \mathsf{a})
\end{aligned}
$$

and can be proved terminating by the 2007 competition versions of AProVE [7], Jambox, and $\mathsf{T_TT_2}$.

## 4 Root-Labeling with Dependency Pairs

The performance of Jambox and Matchbox/SatELite [23] in the 2006 international termination competition revealed that root-labeling is a powerful transformation on SRSs. So it is not a surprise that other tools adopted this technique, too. In 2007, MultumNonMulta [12] and Torpa [25] followed suit. So did $\mathsf{T_TT_2}$, with one important difference: the combination with dependency pairs.

In the previous section we defined root-labeling as a transformation on TRSs. In order to benefit from the numerous termination techniques that are available in connection with the dependency pair framework, it is worthwhile to extend root-labeling to dependency pair problems. (For the same reason, in [14] semantic labeling with respect to quasi-models is extended to dependency pair problems.) In this section we present two different approaches that achieve this.

In the first approach, which is the one implemented in the 2007 competition version of $\mathsf{T_TT_2}$, the insertion of a fresh unary function symbol below dependency pair symbols ensures that the strict separation of rules in $\mathcal{P}$ and in $\mathcal{R}$ is maintained.

**Definition 11.** *Let $(\mathcal{P}, \mathcal{R})$ be a DP problem. Let $\mathcal{F}_\mathcal{R}$ be the signature of $\mathcal{R}$ and let $\mathcal{F}_\mathcal{P}$ be the signature of $\mathcal{P}$. We denote $\{\mathsf{root}(l), \mathsf{root}(r) \mid l \to r \in \mathcal{P}\}$ by $\mathcal{F}^\sharp$ and $\mathcal{F}_\mathcal{R} \cup (\mathcal{F}_\mathcal{P} \setminus \mathcal{F}^\sharp)$ by $\mathcal{F}$. Let $\triangle$ be a fresh unary function symbol. The function* block *inserts $\triangle$ between the root symbol $f$ and the arguments $t_1, \ldots, t_n$ of a term $f(t_1, \ldots, t_n)$ with $n \geqslant 1$: $\mathsf{block}(t) = t$ if $t$ is a variable or a constant and $\mathsf{block}(t) = f(\triangle(t_1), \ldots, \triangle(t_n))$ if $t = f(t_1, \ldots, t_n)$ with $n \geqslant 1$. We define $\mathcal{FC}_1(\mathcal{P}, \mathcal{R})$ as the pair $(\mathsf{block}(\mathcal{P}), \mathcal{FC}_1(\mathcal{R}))$ where $\mathsf{block}(\mathcal{P}) = \{\mathsf{block}(l) \to \mathsf{block}(r) \mid l \to r \in \mathcal{P}\}$ and $\mathcal{FC}_1(\mathcal{R}) = \mathcal{FC}_{\mathcal{F} \cup \{\triangle\}}(\mathcal{R})$.*

**Lemma 12.** *The pair $\mathcal{FC}_1(\mathcal{P}, \mathcal{R})$ is a DP problem.*

*Proof.* The set $\{\mathsf{root}(l), \mathsf{root}(r) \mid l \to r \in \mathsf{block}(\mathcal{P})\}$ coincides with $\mathcal{F}^\sharp$ and function symbols in $\mathcal{F}^\sharp$ do occur neither in $\mathcal{FC}_1(\mathcal{R})$ (as $\mathcal{F}^\sharp \cap (\mathcal{F} \cup \{\triangle\}) = \varnothing$) nor in proper subterms of terms in $\mathsf{block}(\mathcal{P})$. □

**Lemma 13.** *The DP problem $(\mathcal{P}, \mathcal{R})$ is finite if and only if the DP problem $\mathcal{FC}_1(\mathcal{P}, \mathcal{R})$ is finite.*

*Proof.* First assume that $(\mathcal{P},\mathcal{R})$ is not finite. Then there exists a minimal sequence $s_1 \xrightarrow{\epsilon}_{\mathcal{P}} t_1 \rightarrow^*_{\mathcal{R}} s_2 \xrightarrow{\epsilon}_{\mathcal{P}} t_2 \rightarrow^*_{\mathcal{R}} \cdots$. We may assume without loss of generality that all function symbols occurring at non-root positions belong to $\mathcal{F}$. (By replacing all maximal proper subterms with a root symbol that belongs to $\mathcal{F}^\sharp$ by the same variable, we obtain a minimal sequence that satisfies this property.) We claim that

$$\mathsf{block}(s_1) \xrightarrow{\epsilon}_{\mathsf{block}(\mathcal{P})} \mathsf{block}(t_1) \rightarrow^*_{\mathcal{FC}_1(\mathcal{R})} \mathsf{block}(s_2) \xrightarrow{\epsilon}_{\mathsf{block}(\mathcal{P})} \mathsf{block}(t_2) \rightarrow^*_{\mathcal{FC}_1(\mathcal{R})} \cdots$$

is a minimal sequence with respect to $\mathcal{FC}_1(\mathcal{P},\mathcal{R})$. Fix $i \geqslant 1$. By construction of $\mathsf{block}(\mathcal{P})$, the step $s_i \xrightarrow{\epsilon}_{\mathcal{P}} t_i$ gives rise to the step $\mathsf{block}(s_i) \xrightarrow{\epsilon}_{\mathsf{block}(\mathcal{P})} \mathsf{block}(t_i)$. Next we consider the sequence $t_i \rightarrow^*_{\mathcal{R}} s_{i+1}$. Let $s \rightarrow_{\mathcal{R}} t$ be an arbitrary step in this sequence, using the rewrite rule $l \rightarrow r \in \mathcal{R}$ at position $\pi$. Since $\pi > \epsilon$ we may write $s = F(u_1,\dots,u_n)$ and $t = F(u_1,\dots,u_{j-1},v_j,u_{j+1},\dots,u_n)$ with $u_j \rightarrow_{\mathcal{R}} v_j$ and $\pi \geqslant j$. If $l \rightarrow r$ is root-preserving then $l \rightarrow r \in \mathcal{FC}_1(\mathcal{R})$ and thus $u_j \rightarrow_{\mathcal{FC}_1(\mathcal{R})} v_j$, which implies $\triangle(u_j) \rightarrow_{\mathcal{FC}_1(\mathcal{R})} \triangle(v_j)$. Similar as in the proof of Lemma 7, if $l \rightarrow r$ is root-altering then we obtain $\triangle(u_j) \rightarrow_{\mathcal{FC}_1(\mathcal{R})} \triangle(v_j)$ by using an appropriate flat context from $\mathcal{FC}_{\mathcal{F}}$ when $\pi > j$ and the flat context $\triangle(\Box)$ when $\pi = j$. So in all cases we have $\triangle(u_j) \rightarrow_{\mathcal{FC}_1(\mathcal{R})} \triangle(v_j)$ and hence also $\mathsf{block}(s) \rightarrow_{\mathcal{FC}_1(\mathcal{R})} \mathsf{block}(t)$. Since the step $s \rightarrow_{\mathcal{R}} t$ was an arbitrary step in the sequence $t_i \rightarrow^*_{\mathcal{R}} s_{i+1}$, we obtain $\mathsf{block}(t_i) \rightarrow^*_{\mathcal{FC}_1(\mathcal{R})} \mathsf{block}(s_{i+1})$. Next we show that $\mathsf{block}(t_i)$ is terminating with respect to $\mathcal{FC}_1(\mathcal{R})$. By construction of $\mathcal{FC}_1(\mathcal{R})$, every application of a rule in $\mathcal{FC}_1(\mathcal{R})$ can be performed by a rule in $\mathcal{R}$. Hence if $\mathsf{block}(t_i)$ is not terminating with respect to $\mathcal{FC}_1(\mathcal{R})$ then $\mathsf{block}(t_i)$ is not terminating with respect to $\mathcal{R}$ and this implies in turn that $t_i$ is not terminating with respect to $\mathcal{R}$ as $\triangle$ does not appear in the rules of $\mathcal{R}$. This, however, contradicts the minimality of the initial sequence $s_1 \xrightarrow{\epsilon}_{\mathcal{P}} t_1 \rightarrow^*_{\mathcal{R}} s_2 \xrightarrow{\epsilon}_{\mathcal{P}} t_2 \rightarrow^*_{\mathcal{R}} \cdots$. It follows that the sequence displayed above is a minimal sequence with respect to $\mathcal{FC}_1(\mathcal{P},\mathcal{R})$. Therefore, $\mathcal{FC}_1(\mathcal{P},\mathcal{R})$ is not finite, which concludes the proof of the "if" direction.

For the "only if" direction, suppose that

$$s_1 \xrightarrow{\epsilon}_{\mathsf{block}(\mathcal{P})} t_1 \rightarrow^*_{\mathcal{FC}_1(\mathcal{R})} s_2 \xrightarrow{\epsilon}_{\mathsf{block}(\mathcal{P})} t_2 \rightarrow^*_{\mathcal{FC}_1(\mathcal{R})} \cdots$$

is a minimal sequence with respect to $\mathcal{FC}_1(\mathcal{P},\mathcal{R})$. Let the mapping $\mathsf{unblock}$ erase all occurrences of $\triangle$ from terms:

$$\mathsf{unblock}(t) = \begin{cases} t & \text{if } t \text{ is a variable,} \\ \mathsf{unblock}(t_1) & \text{if } t = \triangle(t_1), \\ f(\mathsf{unblock}(t_1),\dots,\mathsf{unblock}(t_n)) & \text{if } t = f(t_1,\dots,t_n) \text{ with } f \neq \triangle. \end{cases}$$

Using similar reasoning as in the "if" direction, we easily obtain

$$\mathsf{unblock}(s_1) \xrightarrow{\epsilon}_{\mathcal{P}} \mathsf{unblock}(t_1) \rightarrow^*_{\mathcal{R}} \mathsf{unblock}(s_2) \xrightarrow{\epsilon}_{\mathcal{P}} \mathsf{unblock}(t_2) \rightarrow^*_{\mathcal{R}} \cdots$$

To show minimality, suppose that $\mathsf{unblock}(t_i)$ is non-terminating with respect to $\mathcal{R}$. Using the special structure of $t_i$, it readily follows that $t_i$ is non-terminating with respect to $\mathcal{R}$ and with respect to $\mathcal{FC}_1(\mathcal{R})$. The latter provides the desired contradiction. □

When applying root-labeling to $\mathcal{FC}_1(\mathcal{P},\mathcal{R})$, it is not useful to label the root symbols of $\mathsf{block}(\mathcal{P})$, since identical symbols will always have identical labels consisting solely of $\triangle$'s. This is reflected in the following definition.

**Definition 14.** *Let $(\mathcal{P},\mathcal{R})$, $\mathcal{F}^\sharp$, $\mathcal{F}$, and $\triangle$ be as in Definition 11. The first root-labeling transformation $\mathcal{FC}_1(\mathcal{P},\mathcal{R})_{\mathsf{rl}}$ is defined as the pair $(\mathsf{block}(\mathcal{P})_{\mathsf{rl}}, \mathcal{FC}_1(\mathcal{R})_{\mathsf{rl}})$ with $L_f = \varnothing$ if $f \in \mathcal{F}^\sharp$ or if $f$ is a constant in $\mathcal{F}$ and $L_f = \mathcal{F}^n$ if $f \in \mathcal{F} \cup \{\triangle\}$ has arity $n \geqslant 1$, and $f_{\mathcal{A}_{\mathcal{F}\cup\mathcal{F}^\sharp\cup\{\triangle\}}}(x_1,\ldots,x_n) = g$ for every $f \in \mathcal{F}^\sharp$ and arbitrary but fixed $g \in \mathcal{F}^\sharp$.*

The modification of the algebra $\mathcal{A}_{\mathcal{F}\cup\mathcal{F}^\sharp\cup\{\triangle\}}$ in the last line of the above definition ensures that the model condition is trivially satisfied for the rules in $\mathcal{P}$. Hence these rules need not be closed under flat contexts, even if they are root-altering.

**Theorem 15.** *The DP problem $(\mathcal{P},\mathcal{R})$ is finite if and only if the DP problem $\mathcal{FC}_1(\mathcal{P},\mathcal{R})_{\mathsf{rl}}$ is finite.*

In other words, the mapping $(\mathcal{P},\mathcal{R}) \mapsto \{\mathcal{FC}_1(\mathcal{P},\mathcal{R})_{\mathsf{rl}}\}$ is a sound and complete DP processor.

*Proof.* According to Lemma 13 finiteness of $(\mathcal{P},\mathcal{R})$ is equivalent to finiteness of $\mathcal{FC}_1(\mathcal{P},\mathcal{R})$. The latter is equivalent to finiteness of $\mathcal{FC}_1(\mathcal{P},\mathcal{R})_{\mathsf{rl}}$. The proof is standard. Starting from a minimal sequence in $\mathcal{FC}_1(\mathcal{P},\mathcal{R})$, a minimal sequence in $\mathcal{FC}_1(\mathcal{P},\mathcal{R})_{\mathsf{rl}}$ is obtained by applying $\mathsf{lab}_\alpha$ to every term in the sequence in $\mathcal{FC}_1(\mathcal{P},\mathcal{R})$, where $\alpha$ assigns an arbitrary element of $\mathcal{F}$ to every variable. Conversely, a minimal sequence in $\mathcal{FC}_1(\mathcal{P},\mathcal{R})_{\mathsf{rl}}$ is transformed into a minimal sequence in $\mathcal{FC}_1(\mathcal{P},\mathcal{R})$ by simply erasing all labels. □

In the second approach for incorporating root-labeling into the dependency pair framework, we preserve the model condition by closing the rules of $\mathcal{R}$ also under flat contexts with a root symbol from $\mathcal{F}^\sharp$. To avoid problems by mixing up dependency pair symbols with symbols of $\mathcal{R}$, those additional closure rules are moved to the first component of dependency pair problems.

**Definition 16.** *Let $(\mathcal{P},\mathcal{R})$, $\mathcal{F}^\sharp$, and $\mathcal{F}$ be as in Definition 11. Let $\mathcal{FC}_2(\mathcal{P},\mathcal{R})$ be the pair $(\mathcal{P} \cup \mathcal{FC}_{\mathcal{F}^\sharp}(\mathcal{R}_{\mathsf{a}}), \mathcal{FC}_{\mathcal{F}}(\mathcal{R}))$. The second root-labeling transformation $\mathcal{FC}_2(\mathcal{P},\mathcal{R})_{\mathsf{rl}}$ is defined as the pair $(\mathcal{P}_{\mathsf{rl}} \cup \mathcal{FC}_{\mathcal{F}^\sharp}(\mathcal{R}_{\mathsf{a}})_{\mathsf{rl}}, \mathcal{FC}_{\mathcal{F}}(\mathcal{R})_{\mathsf{rl}})$ with $L_f = \varnothing$ if $f$ is a constant in $\mathcal{F} \cup \mathcal{F}^\sharp$ and $L_f = \mathcal{F}^n$ if $f \in \mathcal{F} \cup \mathcal{F}^\sharp$ has arity $n \geqslant 1$, and $f_{\mathcal{A}_{\mathcal{F}\cup\mathcal{F}^\sharp}}(x_1,\ldots,x_n) = g$ for every $f \in \mathcal{F}^\sharp$ and arbitrary but fixed $g \in \mathcal{F}^\sharp$.*

It is obvious that the pair $\mathcal{FC}_2(\mathcal{P},\mathcal{R})$ is a DP problem.

**Lemma 17.** *The DP problem $(\mathcal{P},\mathcal{R})$ is finite if and only if the DP problem $\mathcal{FC}_2(\mathcal{P},\mathcal{R})$ is finite.*

*Proof.* We abbreviate $\mathcal{P} \cup \mathcal{FC}_{\mathcal{F}^\sharp}(\mathcal{R}_{\mathsf{a}})$ to $\mathcal{FC}_2(\mathcal{P})$. Assume that $(\mathcal{P},\mathcal{R})$ is not finite. Hence there exists a minimal sequence $s_1 \xrightarrow{\epsilon}_{\mathcal{P}} t_1 \to_{\mathcal{R}}^* s_2 \xrightarrow{\epsilon}_{\mathcal{P}} t_2 \to_{\mathcal{R}}^* \cdots$.

Without loss of generality we assume that all function symbols occurring at non-root positions belong to $\mathcal{F}$. We claim that

$$s_1 \xrightarrow{\epsilon}_{\mathcal{P}} t_1 \rightarrow^*_{\mathcal{FC}_{\mathcal{F}}(\mathcal{R}) \cup \mathcal{FC}_{\mathcal{F}^\sharp}(\mathcal{R}_\mathsf{a})} s_2 \xrightarrow{\epsilon}_{\mathcal{P}} t_2 \rightarrow^*_{\mathcal{FC}_{\mathcal{F}}(\mathcal{R}) \cup \mathcal{FC}_{\mathcal{F}^\sharp}(\mathcal{R}_\mathsf{a})} \cdots$$

is a minimal sequence with respect to $\mathcal{FC}_2(\mathcal{P}, \mathcal{R})$. Fix $i \geqslant 1$. We obviously have $s_i \xrightarrow{\epsilon}_{\mathcal{P}} t_i$. Let $s \rightarrow_{\mathcal{R}} t$ be an arbitrary step in the sequence $t_i \rightarrow^*_{\mathcal{R}} s_{i+1}$, using the rewrite rule $l \rightarrow r \in \mathcal{R}$ at position $\pi$. Since $\pi > \epsilon$ we may write $s = F(u_1, \ldots, u_n)$ and $t = F(u_1, \ldots, u_{j-1}, v_j, u_{j+1}, \ldots, u_n)$ with $u_j \rightarrow_{\mathcal{R}} v_j$ and $\pi \geqslant j$. If $l \rightarrow r$ is root-preserving then $l \rightarrow r \in \mathcal{FC}_{\mathcal{F}}(\mathcal{R})$ and thus $s \rightarrow_{\mathcal{FC}_{\mathcal{F}}(\mathcal{R})} t$. Similar as in the proof of Lemma 7, if $l \rightarrow r$ is root-altering and $\pi > j$ then we obtain $u_j \rightarrow_{\mathcal{FC}_{\mathcal{F}}(\mathcal{R})} v_j$ and thus $s \rightarrow_{\mathcal{FC}_{\mathcal{F}}(\mathcal{R})} t$ by using an appropriate flat context from $\mathcal{FC}_{\mathcal{F}}$. If $\pi = j$ then $s \rightarrow_{\mathcal{FC}_{\mathcal{F}^\sharp}(\mathcal{R}_\mathsf{a})} t$ by using the flat context $F(x_1, \ldots, x_{j-1}, \Box, x_{j+1}, \ldots, x_n) \in \mathcal{FC}_{\mathcal{F}^\sharp}$. So in all cases we have $s \rightarrow_{\mathcal{FC}_{\mathcal{F}}(\mathcal{R}) \cup \mathcal{FC}_{\mathcal{F}^\sharp}(\mathcal{R}_\mathsf{a})} t$. Hence the sequence displayed above exists. By pinpointing the steps from $\mathcal{P} \cup \mathcal{FC}_{\mathcal{F}^\sharp}(\mathcal{R}_\mathsf{a})$, this sequence can be written as

$$s_1 \xrightarrow{\epsilon}_{\mathcal{FC}_2(\mathcal{P})} t'_1 \rightarrow^*_{\mathcal{FC}_{\mathcal{F}}(\mathcal{R})} s'_2 \xrightarrow{\epsilon}_{\mathcal{FC}_2(\mathcal{P})} t'_2 \rightarrow^*_{\mathcal{FC}_{\mathcal{F}}(\mathcal{R})} \cdots$$

where for every $i \geqslant 1$ there exists a $j \geqslant i$ such that $t_i = t'_j$. We need to show that every $t'_j$ is terminating with respect to $\mathcal{FC}_{\mathcal{F}}(\mathcal{R})$. Let $j \geqslant 1$. We distinguish two cases. If $t'_j = t_i$ for some $i$ then $t'_j$ is terminating with respect to $\mathcal{R}$, due to the minimality of the initial sequence in $(\mathcal{P}, \mathcal{R})$. According to Lemma 7, more precisely the extension of Lemma 7 mentioned in the paragraph after the proof, $t'_j$ is terminating with respect to $\mathcal{FC}_{\mathcal{F}}(\mathcal{R})$. In the other case we have $t_i \rightarrow^*_{\mathcal{FC}_{\mathcal{F}}(\mathcal{R}) \cup \mathcal{FC}_{\mathcal{F}^\sharp}(\mathcal{R}_\mathsf{a})} t'_j$ for some $i$. If $t'_j$ is not terminating with respect to $\mathcal{FC}_{\mathcal{F}}(\mathcal{R})$ then it is also not terminating with respect to $\mathcal{FC}_{\mathcal{F} \cup \mathcal{F}^\sharp}(\mathcal{R})$ and hence $t_i$ is not terminating with respect $\mathcal{FC}_{\mathcal{F} \cup \mathcal{F}^\sharp}(\mathcal{R})$. This contradicts Lemma 7 because $t_i$ is terminating with respect to $\mathcal{R}$, due to minimality. This concludes the proof of minimality. We conclude that $\mathcal{FC}_2(\mathcal{P}, \mathcal{R})$ is not finite, which settles the "if" direction.

For the "only if" direction, suppose that

$$s_1 \xrightarrow{\epsilon}_{\mathcal{FC}_2(\mathcal{P})} t_1 \rightarrow^*_{\mathcal{FC}_{\mathcal{F}}(\mathcal{R})} s_2 \xrightarrow{\epsilon}_{\mathcal{FC}_2(\mathcal{P})} t_2 \rightarrow^*_{\mathcal{FC}_{\mathcal{F}}(\mathcal{R})} \cdots$$

is a minimal sequence with respect to $\mathcal{FC}_2(\mathcal{P}, \mathcal{R})$. Without loss of generality we assume that symbols from $\mathcal{F}^\sharp$ occur exclusively at root positions. Using the termination equivalence of $\mathcal{FC}_{\mathcal{F}}(\mathcal{R})$ and $\mathcal{FC}_{\mathcal{F} \cup \mathcal{F}^\sharp}(\mathcal{R})$, which is a consequence of Lemma 7, and the fact that every $t_i$ is terminating with respect to $\mathcal{FC}_{\mathcal{F}}(\mathcal{R})$, it follows that this sequence contains infinitely many steps in $\mathcal{P}$. Since every rule in $\mathcal{FC}_{\mathcal{F} \cup \mathcal{F}^\sharp}(\mathcal{R})$ is simulated by a rule in $\mathcal{R}$, the sequence is a sequence in $(\mathcal{P}, \mathcal{R})$. After dropping the (possibly empty) initial steps using rules from $\mathcal{R}_\mathsf{a}$, we obtain a sequence in $(\mathcal{P}, \mathcal{R})$ which is easily shown to be minimal. □

**Theorem 18.** *The DP problem* $(\mathcal{P}, \mathcal{R})$ *is finite if and only if the DP problem* $\mathcal{FC}_2(\mathcal{P}, \mathcal{R})_{\mathsf{rl}}$ *is finite.*

*Proof.* According to Lemma 17 finiteness of $(\mathcal{P}, \mathcal{R})$ is equivalent to finiteness of $\mathcal{FC}_2(\mathcal{P}, \mathcal{R})$. The equivalence of the latter with finiteness of $\mathcal{FC}_2(\mathcal{P}, \mathcal{R})_{\mathsf{rl}}$ follows as in the proof of Theorem 15. □

## 5 Touzet's SRS

The first approach detailed in the preceding section, $\mathcal{FC}_1(\cdot)_{\mathsf{rl}}$, was implemented and incorporated into the 2007 competition version of $\mathsf{T_TT_2}$. Together with linear polynomial interpretations with coefficients from $\{0,1\}$ and standard dependency pair refinements (usable rules with argument filtering [9], recursive SCC [10]), $\mathsf{T_TT_2}$ could automatically prove termination of `z090.srs`, which is an example from Touzet [21] of a simply terminating SRS whose derivational complexity is not primitive recursive. This prompted Johannes Waldmann to write

> "I find this astonishing: (link to url omitted)
> To my knowledge, this would be the first automatic proof
> for an SRS with non-primitive-recursive complexity"

on the termtools[3] mailing list (7 June 2007).

The SRS $\mathcal{T}$ consists of the rules

$$\begin{array}{llll} \mathsf{bu} \to \mathsf{bs} & \mathsf{sbs} \to \mathsf{bt} & \mathsf{tb} \to \mathsf{bs} & \mathsf{ts} \to \mathsf{tt} \\ \mathsf{sb} \to \mathsf{bsss} & \mathsf{su} \to \mathsf{ss} & \mathsf{tbs} \to \mathsf{utb} & \mathsf{tu} \to \mathsf{ut} \end{array}$$

and simulates the following process on fixed-length lists of natural numbers (s denotes successor and b separates numbers):

$$\begin{aligned} (\cdots, n+1, m, \cdots) &\to (\cdots, n, m+3, \cdots) \\ (\cdots, n+1, m+1, k, \cdots) &\to (\cdots, n, k, m+1, \cdots) \end{aligned}$$

Moreover, the function

$$\phi\colon (x,y) \mapsto \max\{\, z \mid (\, y+1, \overbrace{0, \cdots, 0}^{2x+1}\,) \to^* (\,\overbrace{0, \cdots, 0}^{2x+1}, z+1\,)\,\}$$

dominates the Ackermann function (Touzet [21]), which proves that the derivational complexity of $\mathcal{T}$ is not primitive recursive. The (simple) termination of $\mathcal{T}$ is shown in [21] by a complicated ad-hoc argument.

Below we present some details of the termination proof generated by $\mathsf{T_TT_2}$. The SRS $\mathcal{T}$ has the following 17 dependency pairs:

| | | | | | | | |
|---|---|---|---|---|---|---|---|
| Bu → Bs | (1) | Sb → S | (6) | Su → S | (10) | Tbs → B | (14) |
| Bu → S | (2) | Sbs → Bt | (7) | Tb → Bs | (11) | Ts → Tt | (15) |
| Sb → Bsss | (3) | Sbs → T | (8) | Tb → S | (12) | Ts → T | (16) |
| Sb → Sss | (4) | Su → Ss | (9) | Tbs → Tb | (13) | Tu → T | (17) |
| Sb → Ss | (5) | | | | | | |

In the first step of the proof, the ensuing DP problem $(\mathsf{DP}(\mathcal{T}), \mathcal{T})$ is subjected to the interpretations $[\mathsf{B}](x) = [\mathsf{S}](x) = [\mathsf{T}](x) = [\mathsf{s}](x) = [\mathsf{t}](x) = [\mathsf{u}](x) =$

[3] `http://lists.lri.fr/pipermail/termtools/`

$x$ and $[\mathsf{b}](x) = x + 1$. which causes the pairs (3)–(8), (11), (12), (14) to be eliminated. The eight remaining dependency pairs give rise to three SCCs: $\{(1)\}$, $\{(9), (10)\}$, and $\{(13), (15)\text{–}(17)\}$. The first two are easily handled. The last one is the problematic one. Dependency pairs (15) and (16), subsequently followed by pair (17), are removed by using the following interpretations (and considering the induced usable rules):

- $[\mathsf{T}](x) = [\mathsf{u}](x) = x$, $[\mathsf{b}](x) = [\mathsf{t}](x) = 0$, and $[\mathsf{s}](x) = x + 1$,
- $[\mathsf{T}](x) = x$, $[\mathsf{u}](x) = x + 1$, and $[\mathsf{b}](x) = [\mathsf{s}](x) = [\mathsf{t}](x) = 0$.

The remaining DP problem $(\{(13)\}, \mathcal{T})$ is very resistant against automatic termination proof methods, even though it has just one dependency pair. This is the point where root-labeling comes into play. Since $\mathcal{T}_\mathsf{a}$ contains five of the eight rules of $\mathcal{T}$, the second component of the flat context closure

$$\mathcal{FC}_1(\{(13)\}, \mathcal{T}) = (\{\mathsf{T}\triangle\mathsf{bs} \to \mathsf{T}\triangle\mathsf{b}\}, \mathcal{FC}_{\{\mathsf{b},\mathsf{s},\mathsf{t},\mathsf{u},\triangle\}}(\mathcal{T}))$$

consists of 28 rewrite rules. As there are four symbols in the carrier for the labeling step, $\mathcal{FC}_1(\{(13)\}, \mathcal{T})_{\mathsf{rl}} = (\mathcal{P}, \mathcal{R})$ with 112 rules in $\mathcal{R}$, which the reader will be spared, and $\mathcal{P}$ consisting of the rules

$$\mathsf{T}\triangle_\mathsf{b}\mathsf{b}_\mathsf{s}\mathsf{s}_\mathsf{b} \to \mathsf{T}\triangle_\mathsf{b}\mathsf{b}_\mathsf{b} \quad (a) \qquad \mathsf{T}\triangle_\mathsf{b}\mathsf{b}_\mathsf{s}\mathsf{s}_\mathsf{t} \to \mathsf{T}\triangle_\mathsf{b}\mathsf{b}_\mathsf{t} \quad (c)$$
$$\mathsf{T}\triangle_\mathsf{b}\mathsf{b}_\mathsf{s}\mathsf{s}_\mathsf{s} \to \mathsf{T}\triangle_\mathsf{b}\mathsf{b}_\mathsf{s} \quad (b) \qquad \mathsf{T}\triangle_\mathsf{b}\mathsf{b}_\mathsf{s}\mathsf{s}_\mathsf{u} \to \mathsf{T}\triangle_\mathsf{b}\mathsf{b}_\mathsf{u} \quad (d)$$

Rule $(a)$ is eliminated as it is not part of any SCC. By counting function symbols, the 112 rules in $\mathcal{R}$ are successively reduced to 104 ($\triangle_\mathsf{s}$ is counted), 92 ($\triangle_\mathsf{t}$ is counted), and 78 ($\mathsf{u}_\mathsf{s}$ is counted) rules. Now all rules of $\mathcal{R}$ that hindered the automatic orientation of the rules $(b)$–$(d)$ were removed and the termination proof of $\mathcal{T}$ is concluded by using the following interpretations:

$$\begin{aligned}
&[\mathsf{s}_\mathsf{s}](x) = [\mathsf{s}_\mathsf{t}](x) = [\mathsf{s}_\mathsf{u}](x) = x + 1 \\
&[\mathsf{T}](x) = [\triangle_\mathsf{b}](x) = [\mathsf{b}_\mathsf{s}](x) = [\mathsf{b}_\mathsf{t}](x) = [\mathsf{b}_\mathsf{u}](x) = x \\
&[\mathsf{b}_\mathsf{b}](x) = [\mathsf{s}_\mathsf{b}](x) = [\mathsf{s}_\mathsf{s}](x) = [\mathsf{s}_\mathsf{t}](x) = [\mathsf{s}_\mathsf{u}](x) = [\mathsf{t}_\mathsf{b}](x) = [\mathsf{t}_\mathsf{s}](x) \\
&\qquad = [\mathsf{t}_\mathsf{t}](x) = [\mathsf{t}_\mathsf{u}](x) = [\mathsf{u}_\mathsf{b}](x) = [\mathsf{u}_\mathsf{t}](x) = [\mathsf{u}_\mathsf{u}](x) = 0
\end{aligned}$$

Inspired by the success of $\mathsf{T_TT_2}$ on Touzet's SRS, Hans Zantema announced on the termtools mailing list (16 August 2007) a much simpler example of a simply terminating SRS whose derivational complexity is not primitive recursive:

$$\mathsf{ab} \to \mathsf{baa} \qquad \mathsf{abb} \to \mathsf{bc} \qquad \mathsf{ca} \to \mathsf{ac} \qquad \mathsf{cb} \to \mathsf{bb}$$

This SRS can be automatically proved terminating by Torpa [25] (and several other tools as well), without using root-labeling.

## 6 Experiments

Extensive tests were conducted to evaluate the usefulness of the root-labeling processors. We used the rewrite systems in version 4.0 of the Termination

**Table 1.** Experimental results for SRSs

| | | | $\mathcal{FC}_1(\cdot)_{\mathsf{rl}}$ | $\mathcal{FC}_2(\cdot)_{\mathsf{rl}}$ | $\mathcal{FC}_2(\cdot)^*_{\mathsf{rl}}$ |
|---|---|---|---|---|---|
| P(1) | | 25 (1.12) | 94 ( 2.30) | 98 (2.31) | 119 (3.03) |
| P(1) | with usable rules | 37 (0.06) | 112 ( 1.84) | 118 (2.32) | 138 (3.00) |
| P(2) | | 49 (0.63) | 144 ( 4.14) | 148 (4.05) | 172 (6.84) |
| P(2) | with usable rules | 57 (0.53) | 155 ( 4.01) | 169 (4.40) | 188 (6.33) |
| P(1;2) | | 50 (0.80) | 160 ( 3.40) | 165 (3.36) | 198 (5.28) |
| P(1;2) | with usable rules | 57 (3.35) | 171 ( 3.10) | 183 (3.54) | 209 (4.92) |
| M(2,1) | with usable rules | 87 (0.73) | 181 ( 3.61) | 181 (4.12) | 193 (4.36) |
| M(3,1) | with usable rules | 109 (2.99) | 151 (10.67) | 158 (7.98) | 163 (7.50) |
| total number of proofs | | 118 | 214 | 220 | 241 |

Problems Data Base,[4] extended with the secret systems of the 2007 termination competition, which amount to 1381 TRSs and 724 SRSs. All tests were performed on a workstation equipped with an Intel® Pentium™ M processor running at a CPU rate of 2 GHz on 1 GB of system memory and with a time limit of 60 seconds.

Our results for SRSs are summarized in Table 1. Numbers in parentheses indicate the average time (in seconds) to prove termination. For every entry in the table the dependency pair framework with common processors based on an estimation of the dependency graph, the recursive SCC algorithm [10], and the rule removal processor [8] (which in the case of P(1) amounts to counting certain function symbols) together with some reduction pair processor are used. Different rows in the table correspond to different reduction pair processors based on polynomial (P) and matrix (M) interpretations. For polynomial interpretations the numbers in parentheses indicate how many bits are used for coefficients in the SAT-encoding described in [6]. Rows with P(1;2) indicate that 2 bits are used only when no progress can be made with 1 bit. This is faster (and thus more powerful) than P(2). For matrix interpretations [5] the first number in parentheses provides the dimension of the matrices and the second number denotes the number of bits used for the matrix elements in the SAT-encoding. In rows containing 'with usable rules' the respective processor is used to orient the usable rules of $\mathcal{R}$ with respect to the implicit argument filtering obtained from the 0 coefficients in the interpretations [9], as opposed to all rules of the $\mathcal{R}$ component of a DP problem $(\mathcal{P}, \mathcal{R})$.

In columns $\mathcal{FC}_1(\cdot)_{\mathsf{rl}}$ and $\mathcal{FC}_2(\cdot)_{\mathsf{rl}}$ the two root-labeling processors are executed as soon as the processors described in the preceding paragraph no longer make progress. In the final column an optimization of the $\mathcal{FC}_2(\cdot)_{\mathsf{rl}}$ processor is used which takes effect when root-labeling an already labeled SRS is attempted. This is described at the end of this section and goes back to the implementation of root-labeling in Jambox. We now comment on the obtained data.

- First of all, when increasing the number of bits for polynomial or matrix interpretations, it can happen that systems which can be proved using smaller

[4] `www.lri.fr/~marche/tpdb`

**Table 2.** Experimental results for TRSs

| | | | $\mathcal{FC}_1(\cdot)_{\mathsf{rl}}$ | $\mathcal{FC}_2(\cdot)_{\mathsf{rl}}$ |
|---|---|---|---|---|
| P(1) | with usable rules | 549 (0.16) | 636 (0.61) | 631 (0.39) |
| P(2) | with usable rules | 584 (0.52) | 663 (0.92) | 665 (0.93) |
| M(2,1) | with usable rules | 666 (0.57) | 697 (0.91) | 700 (0.88) |
| M(3,1) | with usable rules | 662 (1.58) | 680 (1.94) | 684 (2.05) |
| M(2,2) | with usable rules | 648 (2.17) | 664 (2.62) | 668 (2.74) |
| M(3,2) | with usable rules | 608 (4.29) | 613 (4.49) | 614 (4.51) |
| total number of proofs | | 686 | 716 | 719 |

values, are no longer handled due to a timeout in the SAT solver (in our case Minisat [3]).

- Due to the nature of root-labeling, every run that does not succeed to prove termination results in a timeout or a memory overflow.
- On first sight it seems that $\mathcal{FC}_2(\cdot)_{\mathsf{rl}}$ is strictly stronger than $\mathcal{FC}_1(\cdot)_{\mathsf{rl}}$. However, what is not apparent from the table is that for every row there are a number of systems which can be proved terminating using $\mathcal{FC}_1(\cdot)_{\mathsf{rl}}$ but not using $\mathcal{FC}_2(\cdot)_{\mathsf{rl}}$. These numbers are 1, 1, 5, 2, 4, 2, 9, and 6 (top to bottom).

Our results for TRSs are summarized in Table 2. Most of the remarks for SRSs also hold for TRSs. The optimization for SRSs detailed below is however not implemented for TRSs (which is apparent from the missing fourth column).

We conclude this experimental section with a brief description of the optimized root-labeling processor $\mathcal{FC}_2(\cdot)^*_{\mathsf{rl}}$. During the execution of $\mathcal{FC}_1(\cdot)_{\mathsf{rl}}$ and $\mathcal{FC}_2(\cdot)_{\mathsf{rl}}$, function symbols of an already labeled DP problem do not have structure, which entails that both closure under flat contexts as well as the actual root-labeling steps create many more new rules due the increased size of the labeled signature. The implementation of plain root-labeling for SRSs in Jambox reduces the number of new rules by using the original signature in subsequent root-labeling steps. This is best explained on a concrete example. When labeling the rule $\mathsf{a_b b_a} \to \mathsf{a_c c_a}$, the unlabeled rule $\mathsf{ab} \to \mathsf{ac}$ is closed under flat contexts

$$\mathsf{aab} \to \mathsf{aac} \qquad \mathsf{bab} \to \mathsf{bac} \qquad \mathsf{cab} \to \mathsf{cac}$$

and subsequently labeled by propagating all one symbol extensions (aa, ab, ac) of the original starting assignment a, resulting in the nine rules:[5]

$$\begin{array}{lll}
\mathsf{a_{ab}a_{ba}b_{aa}} \to \mathsf{a_{ac}a_{ca}c_{aa}} & \mathsf{a_{ab}a_{ba}b_{ab}} \to \mathsf{a_{ac}a_{ca}c_{ab}} & \mathsf{a_{ab}a_{ba}b_{ac}} \to \mathsf{a_{ac}a_{ca}c_{ac}} \\
\mathsf{b_{ab}a_{ba}b_{aa}} \to \mathsf{b_{ac}a_{ca}c_{aa}} & \mathsf{b_{ab}a_{ba}b_{ab}} \to \mathsf{b_{ac}a_{ca}c_{ab}} & \mathsf{b_{ab}a_{ba}b_{ac}} \to \mathsf{b_{ac}a_{ca}c_{ac}} \\
\mathsf{c_{ab}a_{ba}b_{aa}} \to \mathsf{c_{ac}a_{ca}c_{aa}} & \mathsf{c_{ab}a_{ba}b_{ab}} \to \mathsf{c_{ac}a_{ca}c_{ab}} & \mathsf{c_{ab}a_{ba}b_{ac}} \to \mathsf{c_{ac}a_{ca}c_{ac}}
\end{array}$$

With $\mathcal{FC}_2(\cdot)_{\mathsf{rl}}$ we would obtain at least sixteen rules. Although the numbers in Table 1 may suggest otherwise, $\mathcal{FC}_2(\cdot)^*_{\mathsf{rl}}$ does not subsume $\mathcal{FC}_2(\cdot)_{\mathsf{rl}}$. For instance,

[5] The transformation $\mathcal{FC}_2(\cdot)^*_{\mathsf{rl}}$ is easily formalized, starting from the $\mathcal{F}$-algebra $\mathcal{A}_{\mathcal{F}\times\mathcal{F}}$ in Definition 3 with $f_{\mathcal{A}_{\mathcal{F}\times\mathcal{F}}}(a, b) = (f, a)$.

three (`x02.srs`, `z108.srs`, and `z114.srs`) of the 181 systems in the $\mathcal{FC}_2(\cdot)_{\mathsf{rl}}$ entry of the row 'M(2,1) with usable rules' are not included in the 193 systems in the $\mathcal{FC}_2(\cdot)^*_{\mathsf{rl}}$ entry.

## 7 Summary and Future Work

In this paper we introduced the technique of root-labeling for TRSs into the dependency pair framework, resulting in two different root-labeling processors. Touzet's example showed the usefulness of these processors for SRSs. Although root-labeling is trivial to automate, further research is needed to determine when to apply it. Besides, it is unclear whether one of the two introduced processors is (at least in theory) strictly stronger than the other. As explained in the experimental section, the current implementation in $\mathsf{T_TT_2}$ applies root-labeling as soon as the other processors do not make progress. In particular, we never undo the effect of root-labeling. Since the root-labeling processors are always applicable, the DP problems quickly grow too large. This is especially a problem for TRSs, where a single application of root-labeling typically blows up the system. A related problem is the possibility to verify the correctness of the produced termination proofs.

**Acknowledgments.** We are grateful to Johannes Waldmann for inventing root-labeling (for string rewrite systems) and for drawing our attention to the termination proof of `z090.srs` that $\mathsf{T_TT_2}$ produced during the 2007 international termination competition. Without these contributions, this paper would not have been written. Discussions with Jörg Endrullis and Johannes Waldmann improved our results.

## References

1. Arts, T., Giesl, J.: Termination of term rewriting using dependency pairs. Theoretical Computer Science 236, 133–178 (2000)
2. Baader, F., Nipkow, T.: Term Rewriting and All That. Cambridge University Press, Cambridge (1998)
3. Eén, N., Sörensson, N.: An extensible SAT-solver. In: Giunchiglia, E., Tacchella, A. (eds.) SAT 2003. LNCS, vol. 2919, pp. 502–518. Springer, Heidelberg (2004)
4. Endrullis, J.: Jambox (2005), `http://joerg.endrullis.de`
5. Endrullis, J., Waldmann, J., Zantema, H.: Matrix interpretations for proving termination of term rewriting. In: Furbach, U., Shankar, N. (eds.) IJCAR 2006. LNCS (LNAI), vol. 4130, pp. 574–588. Springer, Heidelberg (2006)
6. Fuhs, C., Giesl, J., Middeldorp, A., Schneider-Kamp, P., Thiemann, R., Zankl, H.: SAT solving for termination analysis with polynomial interpretations. In: Marques-Silva, J., Sakallah, K.A. (eds.) SAT 2007. LNCS, vol. 4501, pp. 340–354. Springer, Heidelberg (2007)
7. Giesl, J., Schneider-Kamp, P., Thiemann, R.: AProVE 1.2: Automatic termination proofs in the dependency pair framework. In: Furbach, U., Shankar, N. (eds.) IJCAR 2006. LNCS (LNAI), vol. 4130, pp. 281–286. Springer, Heidelberg (2006)

8. Giesl, J., Thiemann, R., Schneider-Kamp, P.: The dependency pair framework: Combining techniques for automated termination proofs. In: Baader, F., Voronkov, A. (eds.) LPAR 2004. LNCS (LNAI), vol. 3452, pp. 301–331. Springer, Heidelberg (2005)
9. Giesl, J., Thiemann, R., Schneider-Kamp, P., Falke, S.: Mechanizing and improving dependency pairs. Journal of Automated Reasoning 37(3), 155–203 (2006)
10. Hirokawa, N., Middeldorp, A.: Automating the dependency pair method. Information and Computation 199(1,2), 172–199 (2005)
11. Hirokawa, N., Middeldorp, A.: Predictive labeling. In: Pfenning, F. (ed.) RTA 2006. LNCS, vol. 4098, pp. 313–327. Springer, Heidelberg (2006)
12. Hofbauer, D.: MultumNonMulta (2006), `www.theory.informatik.uni-kassel.de/~dieter/multum/`
13. Koprowski, A.: TPA: Termination proved automatically. In: Pfenning, F. (ed.) RTA 2006. LNCS, vol. 4098, pp. 257–266. Springer, Heidelberg (2006)
14. Koprowski, A., Middeldorp, A.: Predictive labeling with dependency pairs using SAT. In: Pfenning, F. (ed.) CADE 2007. LNCS (LNAI), vol. 4603, pp. 410–425. Springer, Heidelberg (2007)
15. Koprowski, A., Zantema, H.: Recursive path ordering for infinite labelled rewrite systems. In: Furbach, U., Shankar, N. (eds.) IJCAR 2006. LNCS (LNAI), vol. 4130, pp. 332–346. Springer, Heidelberg (2006)
16. Korp, M., Sternagel, C., Zankl, H., Middeldorp, A.: Tyrolean termination tool 2 (2007), `http://colo6-c703.uibk.ac.at/ttt2`
17. Middeldorp, A., Ohsaki, H., Zantema, H.: Transforming termination by self-labelling. In: McRobbie, M.A., Slaney, J.K. (eds.) CADE 1996. LNCS, vol. 1104, pp. 373–386. Springer, Heidelberg (1996)
18. Terese: Term Rewriting Systems. Cambridge Tracts in Theoretical Computer Science, vol. 55. Cambridge University Press, Cambridge (2003)
19. Thiemann, R.: The DP Framework for Proving Termination of Term Rewriting. PhD thesis, RWTH Aachen, Technical report AIB-2007-17 (2007)
20. Thiemann, R., Middeldorp, A.: Innermost termination of rewrite systems by labeling. In: Giesl, J. (ed.) WRS 2007. ENTCS, vol. 204, pp. 3–19 (2008)
21. Touzet, H.: A complex example of a simplifying rewrite system. In: Larsen, K.G., Skyum, S., Winskel, G. (eds.) ICALP 1998. LNCS, vol. 1443, pp. 507–517. Springer, Heidelberg (1998)
22. Toyama, Y.: Counterexamples to the termination for the direct sum of term rewriting systems. Information Processing Letters 25, 141–143 (1987)
23. Waldmann, J.: Matchbox: A tool for match-bounded string rewriting. In: van Oostrom, V. (ed.) RTA 2004. LNCS, vol. 3091, pp. 85–94. Springer, Heidelberg (2004)
24. Zantema, H.: Termination of term rewriting by semantic labelling. Fundamenta Informaticae 24, 89–105 (1995)
25. Zantema, H.: TORPA: Termination of rewriting proved automatically. In: van Oostrom, V. (ed.) RTA 2004. LNCS, vol. 3091, pp. 95–104. Springer, Heidelberg (2004)
26. Zantema, H., Waldmann, J.: Termination by quasi-periodic interpretations. In: Baader, F. (ed.) RTA 2007. LNCS, vol. 4533, pp. 404–418. Springer, Heidelberg (2007)

# Combining Rewriting with Noetherian Induction to Reason on Non-orientable Equalities

Sorin Stratulat

LITA, Université Paul Verlaine-Metz, 57000, France
stratulat@univ-metz.fr

**Abstract.** We propose a new (Noetherian) induction schema to reason on equalities and show how to integrate it into implicit induction-based inference systems. Non-orientable conjectures of the form $lhs = rhs$ and their instances can be soundly used as induction hypotheses in rewrite operations. It covers the most important rewriting-based induction proof techniques: i) *term rewriting induction* if $lhs = rhs$ is orientable, ii) *enhanced rewriting induction* if $lhs$ and $rhs$ are comparable, iii) *ordered rewriting induction* if the instances of $lhs = rhs$ are orientable, and iv) *relaxed rewriting induction* if the instances of $lhs = rhs$ are not comparable. In practice, it helps to automatize the (rewrite-based) reasoning on a larger class of non-orientable equalities, like the permutative and associativity equalities.

## 1 Introduction

Since the publication of the seminal paper of Knuth and Bendix on completion [12], rewriting has become a difficult to circumvent tool in the automation of reasoning on equalities. Today, modern theorem provers incorporate sophisticated rewriting techniques for replacing equals by equals.

The most common rewrite operations rest on term rewrite systems (TRS), representing sets of orientable equalities, also known as rewrite rules. The equalities are oriented according to well-founded orderings that, in addition, ensure the termination of the rewriting process. On the other hand, well-founded orderings are paramount to establish the soundness of induction principles. Lankford [13] was among the first to remark that the (ordered) saturation process during completion-like rewriting procedures contains induction reasoning. He called this kind of induction 'inductionless'; compared to conventional (explicit) induction, the tests for identifying the equalities that may serve as induction hypotheses are embodied in the proof method.

Other rewrite-based, this time goal-oriented, proof techniques fall in the category of 'implicit induction'. Equalities from TRSs are distinguished from the equalities to be proved, called *conjectures*. In a proof step, a conjecture from the current proof state is replaced by a (potentially empty) set of new conjectures. Starting with the *term rewriting induction* proof technique proposed by Reddy [14], the states from any derivation (or proof) contain not only conjectures, but also previously processed equalities, called *premises* in [16]. Some new conjectures are resulted from the rewriting with rules from i) TRS and/or ii) smaller instances of orientable premises and conjectures from the

A. Voronkov (Ed.): RTA 2008, LNCS 5117, pp. 351–365, 2008.

current state, considered as induction hypotheses. *Ordered rewriting induction* [4] improves the previous technique since it allows the use of non-orientable induction hypotheses as long as they simplify conjectures. Recent improvements are the *enhanced* and *incremental rewriting induction* [1] as a solution to prove equivalent-sides equalities, like many associativity/commutativity equalities.

In this paper, we propose a new Noetherian induction schema that allows for *relaxed rewriting* [7] to deal with non-orientable equalities: compared to incremental rewriting induction techniques, some instances of a processing conjecture can be soundly used as (explicit) induction hypotheses during the current step, even if they are not smaller or have incomparable sides. The Noetherian induction schema is based on a well-founded ordering defined over terms and can be seamlessly integrated into implicit induction inference systems.

The paper is organised in 6 sections, as follows: after the introduction of the basic notions and notations in Section 2, we reproduce and analyse in Section 3 the iRI system, an incremental rewriting induction inference system from [1]. The relaxed rewriting and the new induction schema are introduced in Section 4 and integrated into iRI. The resulted system is then shown sound using a methodology based on the instantiation of the inference system A, that allowed in the past to build and analyse several ordered saturation-based and implicit induction inference systems [16,17,18]. A abstracts the computation and points out the conjectures and premises to be considered as induction hypotheses in any proof step. The *soundness* property states that A preserves particular minimal false conjectures (*counterexamples*) in any A-derivation. To achieve this, it implements the Fermat's 'Descente Infinie' (DI) induction principle [19] using formula-based well-founded orderings. Compared to other similar systems, A provides maximal sets of induction hypotheses at every derivation step. In Section 5, we describe the implementation of the induction schema into a recent version of the SPIKE prover [5]. Finally, we prove its soundness and compare its performances w.r.t. iRI when proving, for example, some permutative equalities. The last section gives the conclusions and outlines the directions for future work.

## 2 Preliminaries

Most of the notions and notations from this section are standard; some related to iRI and A are imported from [1] and [18], respectively. For details on term rewriting, the reader may consult [3].

**Terms and substitutions.** We denote by $\mathcal{F}$ and $\mathcal{V}$ the set of (arity-fixed) function symbols and (universally quantified) variables, respectively, and by $\mathcal{T}(\mathcal{F},\mathcal{V})$ the set of terms over them. $V(t)$ represents the set of variables of the term $t$. A term is *ground* if it has no variables. New terms can be obtained by replacing variables from existing terms with terms, by means of mappings from variables to terms called *substitutions*. Given a substitution $\sigma$, the term $t\sigma$ is an *instance* of the term $t$. If $s$ and $t$ are two terms, we denote by $mgu(s,t)$ their most general unifier. The subterm $s$ of a (non-variable) term $t$ is identified by its position $p$, denoted by $t[s]_p$. $\epsilon$ is the head position of a term. If $p \neq \epsilon$ in $t[s]_p$, then $s$ is a *strict* subterm of $t$. $t\,|_p$ is the subterm of $t$ at the position $p$ and $t[s]$ indicates that $s$ is a subterm of $t$.

**Orderings over terms.** The reflexive transitive closure, transitive closure, equivalence closure and the inverse of a binary relation $\rightarrow$ are denoted by $\stackrel{*}{\rightarrow}$, $\stackrel{+}{\rightarrow}$, $\stackrel{*}{\leftrightarrow}$ and $\rightarrow^{-1}$, respectively. The composition of two binary relations $A$ and $B$ is represented by $A \circ B$. Given two binary relations $\rightarrow_i$ and $\rightarrow_j$, $\rightarrow_{i/j}$ abbreviates $\stackrel{*}{\leftrightarrow}_j \circ \rightarrow_i \circ \stackrel{*}{\leftrightarrow}_j$.

A *quasi-ordering* $\geq$ is a reflexive and transitive binary relation. The *equivalent* part $\sim$ of a quasi-ordering $\geq$ is defined by $x \sim y$ iff $x \geq y$ and $y \geq x$. Its *strict* part $>$, called *ordering*, is denoted by $x > y$ iff $x \geq y$ and $y \not\geq x$. A binary relation $R$ is *stable by substitution* if whenever $s\,R\,t$ then $(s\sigma)\,R\,(t\sigma)$ for any substitution $\sigma$. A quasi-ordering $\geq$ defined over the elements of a nonempty set $A$ is *well-founded* if any nonempty subset of $A$ contains a minimal element. Moreover, it is stable if its strict and equivalent parts are stable. A *reduction* ordering is a transitive and irreflexive relation that is well-founded, stable by substitution and stable by context (i.e. $s\,R\,t$ implies $u[s]\,R\,u[t]$). The reduction quasi-orderings $\succeq$, have their strict ($\succ$) and equivalent ($\approx$) parts stable by context and substitution; moreover, $\succ$ has to be a reduction ordering. We write $s \lhd t$ if $s$ is a strict subterm of $t$ and $t \succ s$. We consider *simplification* quasi-orderings $\succeq$ as reduction quasi-orderings satisfying the subterm property, i.e. $t \succ s$ whenever $s$ is a strict subterm of $t$. Two terms $s$ and $t$ are *incomparable*, denoted by $s \not\asymp t$, if neither $s \succ t$, nor $t \succ s$, nor $s \approx t$, otherwise they are *comparable*. $\equiv$ is the syntactic identity of terms.

**Rewriting.** The formulas of interest in the paper are *equalities* of the form $s = t$, where $s$ and $t$ are two terms. The equality $s = t$ is *equivalent-sides* if $s \approx t$, *incomparable* if $s \not\asymp t$, and *permutative* if $s$ can be obtained from $t$ by permuting variables. It can be transformed into the *rewrite rule* $s \rightarrow t$ if $s \succ t$. A *term rewriting system* (TRS) is a set of rewrite rules. The rewrite relation of a TRS $\mathcal{R}$ is denoted by $s \rightarrow_{\mathcal{R}} t$. We say that a term $s$ is in *$\mathcal{R}$-normal form* if there is no $t$ such that $s \rightarrow_{\mathcal{R}} t$, otherwise $s$ is *$\mathcal{R}$-reducible*. Any equality that can be transformed into a rewrite rule is *orientable*, otherwise it is *non-orientable*. Given a substitution $\sigma$, an equality $s = t$ and a term $l$ such that $l[s\sigma]_u$, a *rewrite operation* replaces $l[s\sigma]_u$ by $l[t\sigma]_u$. Examples of rewrite operations are: the *ordered rewriting* [4] if $s\sigma \succ t\sigma$, and the *relaxed rewriting* [7] if $s\sigma \not\asymp t\sigma$.

Substitutions can also be applied to equalities: $(s = t)\sigma$ is defined as $s\sigma = t\sigma$.

**Orderings over formulas.** Generally, if $\leq$ is a quasi-ordering over formulas, we denote by $\Phi_{\leq\phi}$ (resp. $\Phi_{<\phi}$ and $\Phi_{\sim\phi}$) the set $\{\psi\gamma \mid \psi \in \Phi, \gamma$ a substitution and $\psi\gamma \leq$ (resp. $<$ and $\sim$)$\phi\}$ of instances of formulas from a set $\Phi$ that are 'smaller or equal' to (resp. 'smaller than' and 'equivalent' to) a formula $\phi$. A reduction quasi-ordering over terms can be extended to a quasi-ordering over equalities, denoted by $\leq_e$, by comparing the multisets of their sides using the multiset extension of $\prec$ [11]. We write $s = t <_e l = r$ if $\{s, t\} \prec\!\!\prec \{l, r\}$, where $\prec\!\!\prec$ is the multiset extension of $\prec$ as follows. Given two multisets $A_1$ and $A_2$, we write $A_1 \prec\!\!\prec A_2$ if, after eliminating pairwisely the equivalent terms from $A_1$ and $A_2$, for each term from $A_1$ there exists a greater term in $A_2$. In addition, $s = t \sim_e l = r$ if the sets $\{s, t\}$ and $\{l, r\}$ are empty after eliminating pairwisely from $A_1$ and $A_2$ the equivalent terms. It can be shown that $\leq_e$ is a well-founded and stable by substitution quasi-ordering if $\preceq$ is, too.

**Inductive theorems.** $s = t$ is an *inductive theorem* of a TRS $\mathcal{R}$, denoted by $\mathcal{R} \models_{ind} s = t$, if for any of its ground (i.e. variable-free) instances $(s = t)\sigma$, we have $s\sigma \overset{*}{\leftrightarrow}_{\mathcal{R}} t\sigma$. A ground equality is a *counterexample* if it is not an inductive theorem. $s = t$ is *false* and denoted by $\mathcal{R} \not\models_{ind} s = t$ if it contains (i.e. one of its instances is) a counterexample.

In this paper, we will consider inductive theorems of *constructor-based* TRS; in this case, $\mathcal{F}$ is the disjoint union between the set of defined functions symbols $\mathcal{D}$ and the set of constructor symbols $\mathcal{C}$. $\mathcal{D}$ represents the set of root symbols of the lhs of rewrite rules from the TRS and $\mathcal{T}(\mathcal{C}, \mathcal{V})$ denotes the set of *constructor terms*. A *constructor substitution* replaces variables with constructor terms. A set of instances $\{s\sigma_1 = t\sigma_1, \ldots, s\sigma_n = t\sigma_n\}$, denoted by $\Psi(s = t)$, is a *cover set* of $s = t$ if for any ground constructor substitution $\tau$, there is a ground constructor substitution $\tau'$ and $i \in [1..n]$ for which $(s = t)\tau \equiv (s = t)\sigma_i\tau'$. Such instances are also called *cover-instances* and the corresponding substitutions *cover-substitutions*. A term of the form $f(c_1, \ldots, c_n)$, where $f \in \mathcal{D}$ and $c_1, \ldots, c_n$ are constructor terms, is *basic*. The set of basic subterms of a term $s$ is denoted by $\mathcal{B}(s)$. The lhs of the rewrite rules from a TRS are expected to be basic. A TRS $\mathcal{R}$ is *quasi-reducible* if no basic ground term is in $\mathcal{R}$-normal form.

**Noetherian induction for equalities.** Inductive theorems can be proved by induction-based proof techniques, as those based on *Noetherian induction* [9], according to *induction schemas*. Given an equality $s = t$, $\Psi(s = t)$ one of its cover sets and $\prec$ a well-founded ordering over terms, an induction schema can be defined as follows: $s = t$ is an inductive theorem of a TRS $\mathcal{R}$ if, for any cover-instance $s\sigma = t\sigma$ from $\Psi(s = t)$, we have $\mathcal{R} \cup \Theta \models_{ind} (s = t)\sigma$, where $\Theta$ is the set of instances $(s = t)\theta$ such that $s\theta \prec s\sigma$. The elements of $\Theta$ are *Noetherian induction hypotheses.*

**Reasoning systems.** An *inference system* consists of a set of *inference rules* that verify whether a set of (quantifier-free first order) formulas, called *conjectures*, are consequences of another set of (quantifier-free first order) formulas, called *axioms*. The inference rules define transitions between two *states* containing conjectures in which the application of an inference rule replaces one conjecture with a set of new conjectures. A *derivation* is built by successive applications of inference rules.

## 3 Rewriting Induction Inference Systems

Recently, Aoto [1] proposed several extensions of the *rewriting induction* (RI, for short), firstly introduced by Reddy in [14] under the name of *term rewriting induction*, as a solution to rewrite non-orientable equalities with previously processed orientable equalities. In this section, we firstly present the most powerful of his inference systems, iRI, reproduced in Fig. 1 and based on *incremental rewriting induction*, then discuss some of its drawbacks.

W.r.t. [14], equivalent-sides equalities are also stored in iRI-states and rewriting operations with them are allowed. More exactly, $(E, H, G)$ is an iRI-state, where $E$ is the set of current conjectures, while $H$ and $G$ are two sets of previously processed conjectures

$$\textsc{Expand}: (E \cup \{s = t\},\ H,\ G) \vdash^{iRI} (E \cup Expd_u(s,t),\ H \cup \{s \to t\},\ G)$$
if $u \in \mathcal{B}(s)$, $s \succ t$

$$\textsc{Expand2}: (E \cup \{s = t\},\ H,\ G) \vdash^{iRI} (E \cup Expd2_{u,v}(s,t),\ H,\ G \cup \{s = t\})$$
if $u \in \mathcal{B}(s)$, $v \in \mathcal{B}(t)$, $s \approx t$

$$\textsc{Simplify}: (E \cup \{s = t\},\ H,\ G) \vdash^{iRI} (E \cup \{s' = t\},\ H,\ G)$$
if $s \to_{(\mathcal{R} \cup H)/G} s'$

$$\textsc{Simplify2}: (E \cup \{s = t\},\ H,\ G) \vdash^{iRI} (E \cup \{s' = t\},\ H,\ G)$$
if $s \overset{*}{\leftrightarrow}_{\mathcal{R} \cup \mathcal{L}} s'$, $s \succeq s'$

$$\textsc{Delete}: (E \cup \{s = t\},\ H,\ G) \vdash^{iRI} (E,\ H,\ G)$$
if $s \overset{*}{\leftrightarrow}_G t$

$$\textsc{Delete2}: (E \cup \{s = t\},\ H,\ G) \vdash^{iRI} (E,\ H,\ G)$$
if $s \overset{*}{\leftrightarrow}_{\mathcal{R} \cup \mathcal{L}} t$

**Fig. 1.** iRI - an incremental RI system

containing orientable and equivalent-sides equalities, respectively. The iRI system allows for rewriting with lemmas, denoted by $\mathcal{L}$, which are previously proved conjectures. An i*RI-proof* of a set of conjectures $E$ is an iRI-derivation that starts with the state $(E, \emptyset, \emptyset)$ and ends in a 'current conjecture'-free state. iRI has been shown sound (see Theorem 2 in [1]): if $\mathcal{R}$ is a quasi-reducible TRS, $E$ a set of conjectures for which there exists an iRI-proof, then all conjectures in $E$ are inductive theorems of $\mathcal{R}$. The reduction quasi-ordering $\succeq$ must be compatible with $\mathcal{R}$, i.e. $\mathcal{R} \subseteq \succ$. The preservation property of minimal counterexamples is an alternative solution to show soundness when the ordering over conjectures is well-founded: by contradiction, let's assume that there is a proof of a false conjecture using an inference system satisfying this property. So, there is a counterexample and therefore a minimal counterexample in the derivation. On the other hand, by the preservation property, such a minimal counterexample is expected to be found in the set of conjectures from the last proof state. Even if we will not show here that iRI itself satisfies the 'preservation' property, we will do it for a variant of iRI in the next section.

The last four iRI-rules in Fig. 1 apply as long as rewriting operations or equal-by-equal replacements with equivalent-sides equalities can be performed. When none of them apply, the iRI-derivations may still continue if some variables of the conjectures are instantiated by applying one of the first two rules. The instantiation process is based on unification, as described by the functions $Expd$ and $Expd2$:

$$Expd_u(s,t) \quad := \{s[r]\sigma = t\sigma \mid s \equiv s[u],\ \sigma \text{ is } mgu(u,l),\ l \to r \in \mathcal{R},\ l: \text{basic}\}$$
$$Expd2_{u,v}(s,t) := \bigcup \{Expd_{v\sigma}(t\sigma, s') \mid s' = t\sigma \in Expd_u(s,t)\}$$

The function $Expd$ returns the equalities resulted from the narrowing operations with $\mathcal{R}$-rules on its first argument. $Expd2$ performs a second narrowing operation, this time on the instances of the second argument computed by $Expd$. Thanks to the quasi-reducibility property of $\mathcal{R}$, it can be shown that the set of mgu-resulted instances of $s = t$ from the definition of $Expd_u(s,t)$ (and implicitly those in $Expd2_{u,v}(s,t)$, too) are cover sets of $s = t$ (see Lemma 1 and Lemma 2 from [1]).

A major drawback of iRI is its restriction to treat only the equivalent-sides equalities from the class of non-orientable $\mathcal{R}$-irreducible equalities. It is therefore important to find reducible quasi-orderings that transform non-orientable conjectures into equivalent-sides equalities. In [1], the recursive path ordering (rpo) [10] is used to handle the commutativity equalities but not the associativity equalities, while the lexicographic path ordering (lpo) is used to handle the associativity equalities but not the commutativity equalities. Another consequence of this restriction is the need for changing the kind of such orderings during successive proofs. For example, the iRI-proof of the commutativity of the multiplication over the naturals requires as lemmas the (previously proved) commutativity and associativity properties of the addition over the naturals. Therefore, during the proof of the lemmas, the ordering should change from lpo to rpo (or vice-versa, depending on which lemma is firstly proved). Notice that it is generally difficult to find a reduction quasi-ordering that simultaneously orients the axioms and makes comparable the side terms of an arbitrary equality conjecture. Sometimes it is even impossible, as shown by the following example.

*Example 1.* Let's consider the reverse $rev2$ and length $len$ functions over lists:

$$len(Nil) = 0 \tag{1}$$
$$len(Cons(x,l)) = S(len(l)) \tag{2}$$
$$rev2(Nil,l) = l \tag{3}$$
$$rev2(Cons(a,l),l') = rev2(l,Cons(a,l')) \tag{4}$$

and the conjecture $len(rev2(x,y)) = len(rev2(y,x))$. We will show that there is no reduction quasi-ordering that simultaneously orients the axiom (4) from left to right and makes the conjecture an equivalent-sides equality.[1] By contradiction, assume that there is such a reduction quasi-ordering $\succeq$. Since $\approx$ is stable by substitution, we have $len(rev2(Cons(a,Nil),Nil)) \approx len(rev2(Nil,Cons(a,Nil)))$ by instantiating the conjecture with the substitution $\{x \mapsto Cons(a,Nil), y \mapsto Nil\}$. For recursive and lexicographic path orderings, this is possible only if $rev2(Cons(a,Nil),Nil) \approx rev2(Nil,Cons(a,Nil))$. On the other hand, $\succ$ is also stable by substitution, so $rev2(Cons(a,Nil),Nil) \succ rev2(Nil,Cons(a,Nil))$, by instantiating (4) with the substitution $\{l \mapsto Nil, l' \mapsto Nil\}$. ◇

## 4 Proving Incomparable Equalities

In the following, we introduce a Noetherian induction-based technique that instantiates variables from a given conjecture $e$ and allows to (ordered or relaxed) rewrite some of its $\mathcal{R}$-reduced instances with (instances of) $e$.

[1] A similar example has been given by Frank Pfenning in a private communication.

### 4.1 The Noetherian Induction-Based Technique

The idea behind the proposed technique is to choose one side of an equality conjecture $s = t$, w.l.o.g. be the lhs, and rewrite $\mathcal{R}$-reduced cover-instances of $s = t$ with instances of $s = t$ representing Noetherian induction hypotheses. More exactly, if $(s = t)\theta$ is the Noetherian induction hypothesis for the cover-instance $(s = t)\sigma$, we allow to replace $s\theta$ with $t\theta$ into any $\mathcal{R}$-reduced equality derived from $(s = t)\sigma$. Both the $\mathcal{R}$-reduction and replacement operations should happen on the lhs of the equalities in order to guarantee that $s\theta \prec s\sigma$. Formally, if $(s = t)\sigma$ is a cover-instance of $s = t$, the function $\rho(s\sigma = t\sigma)$ assigns one equality from the set

$$\{a' = b' \mid t\sigma \xrightarrow{*}_{\mathcal{R}} b' \text{ and either } s\sigma \xrightarrow{+}_{\mathcal{R}} a_1[s\theta]_u \text{ and } a_1[t\theta]_u \xrightarrow{*}_{\mathcal{R}} a', \text{ or } s\sigma \xrightarrow{*}_{\mathcal{R}} a'\}.$$

The technique is captured by the function $\Phi$, which applies the function $\rho$ on any element from the cover set $\Psi(s = t)$ of $s = t$:

$$\Phi(s = t) := \{\rho(a = b) \mid a = b \in \Psi(s = t)\}$$

We can show that $\Phi(s = t)$ and $s = t$ are inductively $\mathcal{R}$-equivalent:

**Theorem 1.** *Let $\mathcal{R}$ be a quasi-reducible TRS and $\succeq$ a simplification quasi-ordering such that $\mathcal{R} \subseteq \succ$. Then $\mathcal{R} \models_{ind} s = t$ iff for any equality $e \in \Phi(s = t)$, $\mathcal{R} \models_{ind} e$.*

The technique allows for i) term rewriting induction if $s \succ t$, ii) enhanced and incremental rewriting induction when, in addition, the case $s \approx t$ is accepted, iii) ordered rewriting induction if $s\theta \succ t\theta$, and iv) *relaxed rewriting induction* when $s\theta$ and $t\theta$ are not comparable. Unusually, it allows to replace $s\theta$ by $t\theta$ even if $s\theta \prec t\theta$.

*Example 2.* Let us assume a simplification quasi-ordering $\preceq$ that orients from left to right the axioms $\mathcal{R}$ of $plus$, defining the addition over the naturals:

$$plus(0, x) = x \tag{5}$$

$$plus(S(x), y) = S(plus(x, y)) \tag{6}$$

For example, let $\prec$ be the rpo with status, based on the precedence $>_F$ over the function symbols $plus >_F S >_F 0$, and assume that $plus$ has a multiset status. Then, the conjecture $(s = t \equiv) plus(x, plus(y, z)) = plus(y, plus(x, z))$ is incomparable but the function $\Phi$ can be applied using the cover set $\{plus(0, plus(y, z)) = plus(y, plus(0, z)),\ plus(S(xs), plus(y, z)) = plus(y, plus(S(xs), z))\}$ to yield two equalities:

1. the cover-instance $plus(0, plus(y, z)) = plus(y, plus(0, z))$ is $\mathcal{R}$-reduced by (5), firstly on the lhs, then on the rhs, to yield $plus(y, z) = plus(y, z)$;
2. when $\sigma$ is $\{x \mapsto S(xs)\}$, $(s\sigma = t\sigma \equiv) plus(S(xs), plus(y, z)) = plus(y, plus(S(xs), z))$ is $\mathcal{R}$-reduced on the both sides by (6) to $S(\underline{plus(xs, plus(y, z))}) = plus(y, S(plus(xs, z)))$. The underlined term is a $\theta$-instance of the lhs of the conjecture when $\theta$ is $\{x \mapsto xs\}$. The relaxed rewriting operation will produce $S(plus(y, plus(xs, z))) = plus(y, S(plus(xs, z)))$. Notice that $s\theta \prec s\sigma$. ◊

In the next section, we will detail its integration into the iRI system.

### 4.2 Integration into the iRI Inference System

The first step towards the integration of any induction reasoning over incomparable equalities in iRI is to store incomparable equalities inside the iRI-states. Instead of adding a new component to the iRI-states, we will mix them with the equalities from the $H$ and $G$ components into only one $\mathcal{H}$ component. Its elements are called *premises*.[2] The set of rewrite rules (resp. the equivalent-sides equalities) from $\mathcal{H}$ is denoted by $\mathcal{H}^{\rightarrow}$ (resp. $\mathcal{H}^{\approx}$).

We define two derived operations, $\Phi^{Expd}$ and $\Phi^{Expd2}$, based on $Expd$ and $Expd2$, respectively:

$$\Phi^{Expd}(s = t) := \{\rho(a = b) \mid a = b \in Expd_u(s, t), u \in \mathcal{B}(s)\} \text{ and}$$
$$\Phi^{Expd2}(s = t) := \{\rho(a = b) \mid a = b \in Expd2_{u,v}(s, t), u \in \mathcal{B}(s), v \in \mathcal{B}(t)\}.$$

The new system, integrating them and denoted by iRI$'$, is depicted in Fig. 2. The substitution $\theta$ from EXPAND$'$ and EXPAND2$'$ may be used optionally by $\Phi^{Expd}$ and $\Phi^{Expd2}$, respectively. When $a \equiv a'$ and $b \equiv b'$, these two iRI-rules represent EXPAND$'$ and EXPAND2$'$ in the $(E, \mathcal{H})$ state form; in this case, the cover-substitution $\sigma$ from the definition of EXPAND$'$ is the mgu substitution used in the definition of $Expd$ to compute $a = b$. The applicability condition $(a' \equiv)a = b(\equiv b') <_e (s = t)\sigma$ holds because $b \equiv t\sigma$ and $a \prec s\sigma$ (since $s\sigma \rightarrow_{\mathcal{R}} a$). On the other hand, the substitution $\sigma$ from the definition of EXPAND2$'$ is the composition of the mgu substitutions used to instantiate successively the lhs and rhs of $s = t$ when computing $a = b$ by $Expd2$. Again, a similar reasoning can prove that $(a' \equiv)a = b(\equiv b') <_e (s = t)\sigma'$. The next rules are the equivalent representation of the corresponding iRI-rules in the $(E, \mathcal{H})$ state form, excepting the rules SIMPLIFY$'$ and DELETE$'$. They allow to rewrite with equivalent-sides premises only if their instances used during the rewriting process are smaller or equal than the processed conjecture. Otherwise, one can derive unsound inference systems, as presented in Fig. 2 from [1].

An iRI$'$-proof of a set $E$ of conjectures starts with $(E, \emptyset)$ and ends in a conjecture-free state.

*Example 3 (iRI$'$-proof based on Example 2).* In [1], the iRI-proof of the conjecture $plus(x, plus(y, z)) = plus(y, plus(x, z))$ requires two additional lemmas. However, a completely automatic proof can be done with an $Expd$-based variable instantiation schema, using the iRI$'$ system instead.

As in Example 2, EXPAND$'$ instantiates the variable $x$ and yields two conjectures: i) $plus(y, z) = plus(y, z)$, which is smaller than the cover-instance $plus(0, plus(y, z)) = plus(y, plus(0, z))$, and ii) $S(plus(y, plus(xs, z))) = plus(y, S(plus(xs, z)))$. Notice that both the last conjecture and the corresponding $\theta$-instance (recall that $\theta$ is $\{x \mapsto xs\}$ in Example 2) are smaller than the cover-instance $plus(S(xs), plus(y, z)) = plus(y, plus(S(xs), z))$, too.

The identity can be eliminated either by DELETE$'$ or DELETE2$'$. The rule EXPAND$'$ is again applied on the last conjecture, which is orientable: $plus(y, S(plus(xs, z))) \rightarrow S(plus(y, plus(xs, z)))$. The variable $y$ is instantiated this time:

[2] We will show later that they have the same properties as the A-premises (see Section 4.3).

1. $\sigma$ is $\{y \mapsto 0\}$: $plus(0, S(plus(xs, z))) = S(plus(0, plus(xs, z)))$ is $\mathcal{R}$-reduced by (5) on the both sides to the identity $S(plus(xs, z)) = S(plus(xs, z))$;
2. $\sigma$ is $\{y \mapsto S(ys)\}$: $plus(S(ys), S(plus(xs, z))) = S(plus(S(ys), plus(xs, z)))$ is $\mathcal{R}$-reduced by (6) on its both sides to $S(\underline{plus(ys, S(plus(xs, z)))}) = S(S(plus(ys, plus(xs, z))))$. The underlined term is an instance of the lhs of the remaining conjecture, so it will be replaced by its corresponding rhs to yield another identity. In both cases, it can be easily noticed that the applicability conditions of EXPAND$'$ are satisfied.

Finally, the two identities are deleted to obtain an empty set of conjectures. $\diamond$

In the following, we will prove the soundness of the iRI$'$ inference system, i.e. that the initial set of conjectures of any iRI$'$-proof are inductive consequences of the axioms, using a 'Descente Infinie' (DI, for short) induction-based approach.

EXPAND$'$: $(E \cup \{s = t\}, \mathcal{H}) \vdash^{\text{iRI}'} (E \cup \Phi^{Expd}(s = t), \mathcal{H} \cup \{s = t\})$,
if for any new conjecture $a' = b'$ computed with $\theta$ (optional) and $\sigma$:
$(s = t)\theta <_e (s = t)\sigma$ (optional) and $(a' = b') <_e (s = t)\sigma$

EXPAND2$'$: $(E \cup \{s = t\}, \mathcal{H}) \vdash^{\text{iRI}'} (E \cup \Phi^{Expd2}(s = t), \mathcal{H} \cup \{s = t\})$,
if for any new conjecture $a' = b'$ computed with $\theta$ (optional) and $\sigma$:
$(s = t)\theta <_e (s = t)\sigma$ (optional) and $(a' = b') <_e (s = t)\sigma$

SIMPLIFY$'$: $(E \cup \{s = t\}, \mathcal{H}) \vdash^{\text{iRI}'} (E \cup \{s' = t\}, \mathcal{H})$
if $s \rightarrow_{(\mathcal{R} \cup \mathcal{H}^{\rightarrow})/\mathcal{H}^{\approx}_{\leq_e s=t}} s'$

SIMPLIFY2$'$: $(E \cup \{s = t\}, \mathcal{H}) \vdash^{\text{iRI}'} (E \cup \{s' = t\}, \mathcal{H})$
if $s \overset{*}{\leftrightarrow}_{\mathcal{R} \cup \mathcal{L}} s'$, $s \succeq s'$

DELETE$'$: $(E \cup \{s = t\}, \mathcal{H}) \vdash^{\text{iRI}'} (E, \mathcal{H})$
if $s \overset{*}{\leftrightarrow}_{\mathcal{H}^{\approx}_{\leq_e s=t}} t$

DELETE2$'$: $(E \cup \{s = t\}, \mathcal{H}) \vdash^{\text{iRI}'} (E, \mathcal{H})$
if $s \overset{*}{\leftrightarrow}_{\mathcal{R} \cup \mathcal{L}} t$

**Fig. 2.** iRI$'$ - the iRI-system extended to reason on incomparable equalities

### 4.3 The DI Induction-Based Soundness Proving Approach

The DI inference systems implement the Fermat's DI induction principle in order to prove inductive theorems [17]. Their soundness is ensured by the existence of a well-founded and stable by substitution quasi-ordering over the formulas. It guarantees that whenever there is a false formula in the states of a derivation, there is also a minimal (w.r.t. the quasi-ordering) counterexample. A DI inference system is *sound* if the minimal counterexamples are preserved in the derivations.

In [16], we proposed an abstract DI inference system, denoted by A, that allows for automatic reasoning. It defines explicitly the set of induction hypotheses at any derivation step, only with information gathered from the current state. In order to do this, the states of a derivation are allowed to contain not only conjectures but also previously processed conjectures not containing minimal counterexamples, called *premises*. Therefore, a DI inference rule of an inference system $P$ has the form $(E \cup \{\phi\}, H) \vdash^{P} (E \cup \Phi, H')$, where $E, \{\phi\}, \Phi$ and $H, H'$ are sets of conjectures and premises, respectively; $\phi$ is the *processed* conjecture, $\Phi$ the set of new conjectures, and $H'$ is $H \cup \{\phi\}$ if $\phi$ does not contain minimal counterexamples, otherwise is $H$.

ADDPREMISE
$(E \cup \{\phi\},\ H) \ \vdash^{\mathsf{A}} \ (E \cup \Phi,\ H \cup \{\phi\})$
  if $\{\phi\} \ \ \sqsubset_{(H,\,E)} \ \Phi$
SIMPLIFY
$(E \cup \{\phi\},\ H) \ \vdash^{\mathsf{A}} \ (E \cup \Phi,\ H)$
  if $\{\phi\} \ \ \sqsubseteq_{(H \cup E,\,\emptyset)} \ \Phi$

**Fig. 3.** The A-inference system

The soundness is ensured if, whenever a processed conjecture contains a minimal counterexample, there is a conjecture in the next state that contains an equivalent (w.r.t. the quasi-ordering) minimal counterexample. Formally, we denote by $Ax$ the set of axioms. Then, $\phi$ is allowed to be replaced by $\Phi$ in the rule $(E \cup \{\phi\}, H) \vdash^{P} (E \cup \Phi, H')$ if $Ax \cup \mathcal{C}^1_{\leq \phi\gamma} \cup \mathcal{C}^2_{<\phi\gamma} \cup \Phi_{\leq \phi\gamma} \models_{ind} \phi\gamma$, for any ground instance $\phi\gamma$, i.e. when $\Phi$ is a (*general* labelled) *contextual cover set* (CCS for short) of $\phi$ in the context $(\mathcal{C}^1, \mathcal{C}^2)$ [16]. The formulas from the context are instances of formulas from the current state, in our case $E$ and $H$. Particular CCSs can be labelled differently and non-exclusively: i) *cover set*[3] if $\mathcal{C}^1 = \mathcal{C}^2 = \emptyset$, ii) *strict* if $\Phi_{<\phi\gamma}$ instead of $\Phi_{\leq \phi\gamma}$, and iii) *empty* if $\Phi = \emptyset$. As can be noticed from the CCS's definition, the formulas from $\mathcal{C}^1_{\leq \phi\gamma}$ and $\mathcal{C}^2_{<\phi\gamma}$ are allowed to deduce $\phi\gamma$, even if they are false or not yet proved to be true. They play the role of *induction hypotheses*. Contexts can be compared: a context $\mathcal{C}^1 = (\mathcal{C}^{11}, \mathcal{C}^{12})$ is smaller or equal to (or included into) another context $\mathcal{C}^2 = (\mathcal{C}^{21}, \mathcal{C}^{22})$ if $(\mathcal{C}^{11}_{\leq \phi} \cup \mathcal{C}^{12}_{<\phi}) \subseteq (\mathcal{C}^{21}_{\leq \phi} \cup \mathcal{C}^{22}_{<\phi})$, for any ground formula $\phi$. The 'contextually cover' relation can be extended to sets of conjectures: $\Psi \sqsubseteq_{\mathcal{C}}$ (resp. $\sqsubset_{\mathcal{C}}$) $\Phi$ iff $\Phi$ is a general (resp. strict) CCS of any $\phi \in \Psi$ in the context $\mathcal{C}$. The contexts are defined by the inference rules of A.

A simplified and sound version of A [17] is given in Fig. 3. It consists of two inference rules. ADDPREMISE adds the processed conjecture to the set of premises if the set of new conjectures $\Phi$ is a strict CCS in the context $(H, E)$. SIMPLIFY does not allow such addition but, in exchange, the set of new conjectures is a general CCS and the context is bigger.

A is at the heart of a methodology to analyse the soundness of concrete DI inference systems. Mainly, an inference system $P$ is sound if it is an instance of A, i.e. any $P$-rule $p$ is an instance of an A-rule $r$. To show this, we have to i) represent the $P$-derivation states under the form of $(E, H)$, ii) identify the reasoning techniques that may build $P$-CCSs, iii) identify the label and context of each $P$-CCS, and iv) verify the label matching for each $P$-CCS of $p$ and that their context is included into that of the

[3] Notice that this notion of cover set generalises that introduced in Section 3. From now on, we will use this notion, unless otherwise stated.

corresponding CCSs from $r$. Only the strict $P$-CCS and the empty $P$-CCSs can match strict A-CCSs, and any $P$-CCS can match general A-CCSs.

### 4.4 Soundness Proof and Sound Extensions of iRI′

The soundness of iRI′ is the consequence of the theorem that shows iRI′ as an A-instance.

**Theorem 2** **(iRI′ as A-instance).** *Any iRI′-rule is an instance of an A-rule.*

**Corollary 1.** *Let $\mathcal{R}$ be a quasi-reducible TRS and $\succeq$ a simplification quasi-ordering such that $\mathcal{R} \subseteq \succ$. Then, the minimal counterexamples are persistent in any iRI′-derivation starting with an empty set of premises.*

**Proof:** According to Theorem 1 from [17], the minimal counterexamples are persistent in any A-derivation starting with an empty set of premises. This is also true for iRI′, since iRI′ is an instance of A, according to Theorem 2. ■

**Theorem 3** **(soundness of iRI′).** *Let $\mathcal{R}$ be a quasi-reducible TRS and $\succeq$ a simplification quasi-ordering such that $\mathcal{R} \subseteq \succ$. For any iRI′-proof $(E, \emptyset) \vdash^{iRI'} \ldots \vdash^{iRI'} (\emptyset, \mathcal{H})$, we have $\mathcal{R} \models_{ind} e$ for any $e \in E$.*

**Proof:** By contradiction, assume an iRI′-proof $(E, \emptyset) \vdash^{iRI'} \ldots \vdash^{iRI'} (\emptyset, \mathcal{H})$ such that there exists $e \in E$ with $\mathcal{R} \not\models_{ind} e$. The conjectures encountered along the proof contain a minimal counterexample which, according to Corollary 1, is persistent. On the other hand, the proof finishes with an empty set of conjectures. ■

The instantiation result allows for sound extensions of iRI′: the strict cover sets built by the expand rules may be generalized to any strict CCS having the context $(\mathcal{H}, E)$ allowed by AddPremise. For example, the $\mathcal{R}$-rewrite operations $\rightarrow_{\mathcal{R}}$ can be replaced by $\rightarrow_{\mathcal{R} \cup (\mathcal{H} \cup E)^{\rightarrow}}$ in the definitions of $\Phi^{Expd}$ and $\Phi^{Expd2}$. Similarly, the other rules can use induction hypotheses from the context $(\mathcal{H} \cup E, \emptyset)$ as Simplify instances: the applicability conditions for Simplify′, Simplify2′, Delete′ and Delete2′ can be relaxed to $s \rightarrow_{(\mathcal{R} \cup (E \cup \mathcal{H})^{\rightarrow})/(E \cup \mathcal{H})^{\approx}_{\leq_{e\, s=t}}} s'$, $s \overset{*}{\leftrightarrow}_{\mathcal{R} \cup \mathcal{L} \cup (E \cup \mathcal{H})^{\approx}_{\leq_{e\, s=t}}} s'$ with $s \succeq s'$, $s \overset{*}{\leftrightarrow}_{(E \cup \mathcal{H})^{\approx}_{\leq_{e\, s=t}}} t$ and $s \overset{*}{\leftrightarrow}_{\mathcal{R} \cup \mathcal{L} \cup (E \cup \mathcal{H})^{\approx}_{\leq_{e\, s=t}}} t$, respectively.

The applicability condition for the Expand2 and Expand2′ rules, $(a' = b') <_e (s = t)\sigma$, can also be relaxed to $(a' = b') \leq_e (s = t)\sigma$ if the current conjecture is no longer added to the set of premises. The resulted (compactly written) two rules:

SimpExpd(2): $(E \cup \{s = t\}, \mathcal{H}) \vdash^{iRI'} (E \cup \Phi^{Expd(2)}(s = t), \mathcal{H})$,
if for any new conjecture $a' = b'$ computed with $\theta$ (optional) and $\sigma$:
$(s = t)\theta <_e (s = t)\sigma$ (optional) and $(a' = b') \leq_e (s = t)\sigma$

can be easily proved as instances of A-Simplify. Therefore, the extended iRI′ integrating them is sound.

In practice, it may happen that the applicability conditions related to Expand2 and Expand2′ be too restrictive. In the extreme case when they are not satisfied, our induction-based technique can still be used if the set of premises becomes empty, as follows:

RESETPREMISES(2): $(E \cup \{s = t\},\ \mathcal{H}) \vdash^{\mathrm{iRI}'} (E \cup \Phi^{Expd(2)}(s = t),\ \emptyset)$

**Theorem 4 (soundness of the extended iRI′).** *The iRI′ system extended with the* RESETPREMISES(2) *rules is sound.*

The systems iRI and (the extensions of) iRI′ are not refutationally complete. For example, the derivation of $app(xs, ys) = app(ys, xs)$ (see [1] for the definition of $app$) is blocked when dealing with a derived constructor conjecture of the form $cons(x, \ldots) = cons(y, \ldots)$. They lack mechanisms for identifying and refuting false constructor conjectures.

## 5 Implementation into the SPIKE Inference System

SPIKE [7,2,5] is a DI induction-based theorem prover adapted for automated reasoning on conditional theories. It has been successfully used to verify real-size applications like the JavaCard Platform [5] and a non-trivial telecommunications algorithm [15]. It integrates powerful reasoning techniques, based on (conditional) rewriting, subsumption and decision procedures.

Previous versions of SPIKE have already proved non-orientable equalities, like the commutativity of the addition over naturals [7]. However, as [1] points out, SPIKE [7,6] failed to prove other interesting non-orientable equalities, for example the commutativity of the multiplication over naturals. To augment even more its proving power and degree of automation, we have integrated the proposed induction-based technique into the most recent version of SPIKE [5]. GENERATE, the only inference rule that deals with variable instantiations, computes by unification (from the mgus) and simultaneously the cover-substitutions and the corresponding rewrite rules that simplify the cover-instances of the processed conjecture. When dealing with unconditional equational conjectures, the heuristics for instantiating variables are a mixture between the instantiation schemes of EXPAND′ and EXPAND2′.

In SPIKE, we have added a new rule that integrates our induction-based technique. It employs the variable instantiation schema of GENERATE and, therefore, implements both EXPAND′ and EXPAND2′ rules; more exactly, the heuristics decide which of the two rules to be used for a given conjecture. Their applicability conditions have been implemented such that the induction technique becomes more 'inductionless' for particular cases. Mainly, we took into account the definition of $<_e$, the stability by substitution and subterm properties of $\preceq$. Firstly, we conclude that whenever Noetherian induction hypotheses are applied, i) $s\sigma \succ s\theta$ since $s\sigma \xrightarrow{+}_{\mathcal{R}} a[s\theta]$, and ii) $t\sigma \succeq b'$ since $t\sigma \xrightarrow{*}_{\mathcal{R}} b'$. As shown in Table 1, for the particular case when $s \succ t$, the applicability conditions $(s = t)\theta <_e (s = t)\sigma$ and $(a' = b') <_e (s = t)\sigma$ are implicitly verified since $\{s\theta, t\theta, a', b'\} \lll \{s\sigma\}$. Similarly, when $s\theta \succ t\theta$ or $s\theta \approx t\theta$, we have $\{s\theta, t\theta, a'\} \lll \{s\sigma\}$. Finally, when $s\theta \not\asymp t\theta$, the condition $(s = t)\theta <_e (s = t)\sigma$ is satisfied whenever either $t\theta \prec s\sigma$ or $t\theta \prec t\sigma$, while the condition $(a' = b') <_e (s = t)\sigma$ holds if either $a' \prec s\sigma$ or $a' \prec t\sigma$. We are interested to keep $a'$ as small as possible

**Table 1.** The EXPAND$(2)'$ applicability conditions for particular cases

| Case \ Condition | $(s = t)\theta <_e (s = t)\sigma$ | $(a' = b') <_e (s = t)\sigma$ |
|---|---|---|
| $s \succ t$ | Verified | Verified |
| $s\theta \succ t\theta$ or $s\theta \approx t\theta$ | Verified | Verified |
| $s\theta \not\succ t\theta$ | $t\theta \prec s\sigma$ or $t\theta \prec t\sigma$ | $a' \prec s\sigma$ or $a' \prec t\sigma$ |

in order to satisfy the condition $(a' = b') <_e (s = t)\sigma$; in the implementation, the rewriting operations $\xrightarrow{+}_{\mathcal{R}}$ and $\xrightarrow{*}_{\mathcal{R}}$ are performed up to normalization.

The new inference system has been tested on several conjectures, most of them unsuccessfully attempted by previous versions of SPIKE. For lack of space, in Table 2, we have compared the SPIKE and iRI proofs only on some non-trivial permutative equalities from [1] (the boxed equalities in the table). The first conjecture has been proved completely automatically while the second requires only one lemma (which can be proved completely automatically, too). The last one failed to be proved by the old version of SPIKE, mainly because its inference system is not able to rewrite with unorientable premises in simplification steps, thus producing proof divergence. In the new version, it needs two lemmas: i) the first conjecture and ii) a new equality that needs the first conjecture as lemma. Compared to iRI, SPIKE used only one (rpo) ordering for all proofs. The prover and the full SPIKE specification of these examples and other permutative and associativity equalities can be found at the SPIKE section of the web page `http://lita.sciences.univ-metz.fr/~stratula`.

**Table 2.** Comparison between the lemmas used in some iRI and SPIKE proofs

| iRI lemmas | SPIKE lemmas |
|---|---|
| $plus(x, plus(y, z)) = plus(y, plus(x, z))$ | |
| $plus(x, y) = plus(y, x)$<br>$plus(x, plus(y, z)) = plus(plus(x, y), z)$ | no lemmas |
| $sum(app(xs, ys)) = sum(app(ys, xs))$ | |
| $plus(x, y) = plus(y, x)$<br>$plus(x, plus(y, z)) = plus(plus(x, y), z)$ | $plus(x, S(y)) = S(plus(x, y))$ |
| $times(x, y) = times(y, x)$ | |
| $plus(x, y) = plus(y, x)$<br>$plus(x, plus(y, z)) = plus(plus(x, y), z)$ | $plus(x, plus(y, z)) = plus(y, plus(x, z))$<br>$times(x, S(y)) = plus(x, times(x, y))$ |

The SPIKE inference system has already been represented as an instance of A in [16]. The augmented inference system is therefore sound because the new rule behaves as either EXPAND$'$ or EXPAND2$'$ (recall that both of them have been proved instances of ADDPREMISE in the proof of Theorem 2). As shown for previous versions, SPIKE can also refute conjectures under the conditions explained in [16]. The new rule can safely replace GENERATE without affecting the refutational completeness property because it always succeeds, as GENERATE does, on conjectures containing basic non-ground terms when dealing with quasi-reducible TRSs.

## 6 Conclusions and Future Work

We have proposed a new Noetherian induction technique to reason on (pure) equational logic, adapted for DI inference systems. It instantiates variables of equalities and allows to rewrite their instances using any existing rewriting technique. We have advantageously integrated it into the iRI incremental rewriting induction inference system. The resulted system, iRI′, was able to produce more automated proofs on some concrete examples in a setting that no longer requires that i) the non-orientable conjectures be considered as equivalent-sides equalities, and ii) the proof orderings change in successive proofs. To prove its soundness, the iRI′ system has been represented as an instance of the abstract system A. This witnesses that the rewriting induction systems are members of the DI family of inference systems, like those based on implicit induction and saturation. Therefore, they share the same underlying logical principles. The instantiation result also allowed for some easy and sound iRI′ extensions.

We have also integrated the proposed induction technique into SPIKE and implemented it. The resulted system was tested on the conjectures presented in the conclusion part of [1]. The proofs of some of them successfully performed completely automatically or with a higher degree of automation than the iRI-proofs. Thanks to the generality of the DI induction-based approach, we expect that these results be applicable to other implicit induction and saturation-based systems [18].

Rewriting with non-orientable equalities is a common and successful technique specific to Noetherian, explicit induction-based provers like ACL2, RRL and CLAM. Its integration into explicit induction schemes is made easier because there is *a priori* no ordering restriction over the resulted conjectures. On the other hand, DI-based techniques are sometimes more effective (see [8] for a comparison between different proofs of the Gilbreath card trick problem done with SPIKE and some explicit induction provers), mainly because they do not require any hierarchy between the intermediate lemmas to be proved and, therefore, manage better the inductive hypotheses. From this point of view, the presented work is a step forward to combining the advantages of the two approaches.

An interesting direction for future work is to study the conditions for the sound replacement of the induction schema used by our technique with different Noetherian induction schemas such that the applicability conditions of rules like EXPAND(2)' and SIMPEXPD(2) be satisfied and, in another direction, to study its applicability to arbitrary (quantifier-free first order) formulas. We also intend to study the unsound integrations of Noetherian induction schemas into RESETPREMISES(2)-like rules. However, such rules have to be employed parsimoniously because all the current premises, representing potential induction hypotheses in the implicit induction setting, are lost. Therefore, we intend to identify the cases when current premises can be soundly saved.

## References

1. Aoto, T.: Dealing with non-orientable equations in rewriting induction. In: Pfenning, F. (ed.) RTA 2006. LNCS, vol. 4098, pp. 242–256. Springer, Heidelberg (2006)
2. Armando, A., Rusinowitch, M., Stratulat, S.: Incorporating decision procedures in implicit induction. J. Symb. Comput. 34(4), 241–258 (2002)

3. Baader, F., Nipkow, T.: Term Rewriting and All That. Cambridge University Press, Cambridge (1998)
4. Bachmair, L., Dershowitz, N., Plaisted, D.A.: Completion without failure. In: Resolution of Equations in Algebraic Structure, vol. 2, pp. 1–30. Academic Press, London (1989)
5. Barthe, G., Stratulat, S.: Validation of the JavaCard platform with implicit induction techniques. In: Nieuwenhuis, R. (ed.) RTA 2003. LNCS, vol. 2706, pp. 337–351. Springer, Heidelberg (2003)
6. Bouhoula, A.: Automated theorem proving by test set induction. Journal of Symbolic Computation 23, 47–77 (1997)
7. Bouhoula, A., Kounalis, E., Rusinowitch, M.: Automated mathematical induction. Journal of Logic and Computation 5(5), 631–668 (1995)
8. Bouhoula, A., Rusinowitch, M.: Implicit induction in conditional theories. Journal of Automated Reasoning 14(2), 189–235 (1995)
9. Bundy, A.: The automation of proof by mathematical induction. In: Robinson, J.A., Voronkov, A. (eds.) Handbook of Automated Reasoning, pp. 845–911. Elsevier and MIT Press (2001)
10. Dershowitz, N.: Orderings for term-rewriting systems. Theoretical Computer Science 17(3), 279–301 (1982)
11. Dershowitz, N., Manna, Z.: Proving termination with multiset orderings. Communications of the ACM 22(8), 465–476 (1979)
12. Knuth, D., Bendix, P.: Simple word problems in universal algebras. In: Leech (ed.) Computational problems in abstract algebra, pp. 263–297. Pergamon Press, Oxford (1970)
13. Lankford, D.S.: Some remarks on inductionless induction. Technical Report MTP-11, Louisiana Tech University, Ruston, LA (1980)
14. Reddy, U.: Term rewriting induction. In: 10th International Conference on Automated Deduction. LNCS, vol. 814, pp. 162–177 (1990)
15. Rusinowitch, M., Stratulat, S., Klay, F.: Mechanical verification of an ideal incremental ABR conformance algorithm. J. Autom. Reasoning 30(2), 53–177 (2003)
16. Stratulat, S.: A general framework to build contextual cover set induction provers. Journal of Symbolic Computation 32(4), 403–445 (2001)
17. Stratulat, S.: Automatic 'Descente Infinie' induction reasoning. In: Beckert, B. (ed.) TABLEAUX 2005. LNCS (LNAI), vol. 3702, pp. 262–276. Springer, Heidelberg (2005)
18. Stratulat, S.: 'Descente Infinie' induction-based saturation procedures. In: SYNASC 2007: Proceedings of the Ninth International Symposium on Symbolic and Numeric Algorithms for Scientific Computing (SYNASC 2007), Washington, DC, USA, pp. 17–24. IEEE Computer Society, Los Alamitos (2007)
19. Wirth, C.-P.: Descente infinie + deduction. Logic Journal of the IGPL 12(1), 1–96 (2004)

# Deciding Innermost Loops*

René Thiemann[1], Jürgen Giesl[2], and Peter Schneider-Kamp[2]

[1] Institute of Computer Science, University of Innsbruck, Austria
rene.thiemann@uibk.ac.at

[2] LuFG Informatik 2, RWTH Aachen University, Germany
{giesl,psk}@informatik.rwth-aachen.de

**Abstract.** We present the first method to disprove *innermost* termination of term rewrite systems automatically. To this end, we first develop a suitable notion of an innermost loop. Second, we show how to detect innermost loops: One can start with any technique amenable to find loops. Then our novel procedure can be applied to decide whether a given loop is an innermost loop. We implemented and successfully evaluated our method in the termination prover AProVE.

## 1 Introduction

Termination is an important property of term rewrite systems (TRSs). Therefore, much effort has been spent on developing and automating powerful techniques for showing (innermost) termination of TRSs. An important application area for these techniques is termination analysis of functional programs. Since the evaluation mechanism of functional languages is mainly term rewriting, one can transform functional programs into TRSs and prove termination of the resulting TRSs to conclude termination of the functional programs [9]. Although "full" rewriting does not impose any evaluation strategy, this approach is sound even if the underlying programming language has an innermost evaluation strategy.

But in order to detect bugs in programs, it is at least as important to prove *non-termination* of programs or of the corresponding TRSs. Here, the evaluation strategy cannot be ignored, because a non-terminating TRS may still be innermost terminating. Thus, in order to disprove termination of programming languages with an innermost strategy, it is important to develop techniques to disprove *innermost* termination of TRSs automatically.

Only a few techniques for showing non-termination of TRSs have been introduced so far [7,12,17,18,20]. Nevertheless, there already exist several tools that are able to prove non-termination of TRSs automatically by finding loops (e.g., AProVE [8], Jambox [5], Matchbox [23], NTI [20], TORPA [24], TTT [14]). But up to now, all of these techniques and tools only disprove full and not innermost termination. So they can only be applied to disprove innermost termination if the TRS belongs to a known class where termination and innermost termination coincide [11]. In this paper, we demonstrate how to extend all of these techniques such that they can be directly used for disproving innermost termination for any

* Supported by the Deutsche Forschungsgemeinschaft (DFG) under grant GI 274/5-2.

A. Voronkov (Ed.): RTA 2008, LNCS 5117, pp. 366–380, 2008.

kind of TRS. For instance, this is needed for the following program where the resulting TRS is not confluent and hence, does not fall into a known class where innermost and full termination are the same.

*Example 1 (Factorial function).* The following ACL2 program [15] computes the factorial function where $x$ is increased from 0 to $y - 1$ and in every iteration the result is multiplied by $1 + x$.

```
(defun factorial (y) (fact 0 y))
(defun fact (x y)
  (if (== x y)
    1
    (× (+ 1 x) (fact (+ 1 x) y))))
```

Using a translation to TRSs suggested by [22], we obtain the following TRS $\mathcal{R}$ where the rules (5) – (12) are needed to handle the built-in functions of ACL2.

| | | | |
|---|---|---|---|
| $\mathsf{factorial}(y) \to \mathsf{fact}(\mathsf{0}, y)$ | (1) | $\mathsf{0} \times y \to \mathsf{0}$ | (7) |
| $\mathsf{fact}(x, y) \to \mathsf{if}(x == y, x, y)$ | (2) | $\mathsf{suc}(x) \times y \to y + (x \times y)$ | (8) |
| $\mathsf{if}(\mathsf{true}, x, y) \to \mathsf{suc}(\mathsf{0})$ | (3) | $x == y \to \mathsf{chk}(\mathsf{eq}(x, y))$ | (9) |
| $\mathsf{if}(\mathsf{false}, x, y) \to \mathsf{suc}(x) \times \mathsf{fact}(\mathsf{suc}(x), y)$ | (4) | $\mathsf{eq}(x, x) \to \mathsf{true}$ | (10) |
| $\mathsf{0} + y \to y$ | (5) | $\mathsf{chk}(\mathsf{true}) \to \mathsf{true}$ | (11) |
| $\mathsf{suc}(x) + y \to \mathsf{suc}(x + y)$ | (6) | $\mathsf{chk}(\mathsf{eq}(x, y)) \to \mathsf{false}$ | (12) |

Here, it is crucial to use *innermost* instead of full rewriting. Otherwise, it would always be possible to rewrite $s == t \to_{\mathcal{R}} \mathsf{chk}(\mathsf{eq}(s, t)) \to_{\mathcal{R}} \mathsf{false}$, i.e., terms like $\mathsf{0} == \mathsf{0}$ could then be evaluated to both true and false. In contrast, for innermost rewriting one has to apply rule (10) first if $s$ and $t$ are equal.

Note that in this TRS, $s == t$ is indeed evaluated to false whenever $s$ and $t$ are *any* terms that are syntactically different. This is essential to model the semantics of ACL2 correctly, since here there are – like in term rewriting – no types. At the same time, all functions in ACL2 must be "completely defined".

So to perform non-termination proofs for languages like ACL2, we need a way to disprove innermost termination. This problem is harder than disproving termination since one has to take care of the evaluation strategy.

In this paper we investigate *looping* reductions. These are specific kinds of infinite reductions which can be represented in a finite way. To disprove innermost termination of TRSs, we develop an automatic method which in case of success, presents the innermost loop to the user as a counterexample.

For the TRS of Ex. 1, there is indeed an innermost loop. It corresponds to the non-terminating reduction of the ACL2 program when calling $\mathsf{fact}(n, m)$ for natural numbers $n > m$. The reason is that the first argument is increased over and over again, and it will never become equal to $m$.

The paper is organized as follows. In Sect. 2, we extend the notion of a loop to innermost rewriting. Then as the main contribution of the paper, we describe a novel decision procedure in Sect. 3 which detects whether a loop for full rewriting is still a loop in the innermost case. How to combine our work with dependency pairs is discussed in Sect. 4. Finally, Sect. 5 summarizes our results and describes their empirical evaluation with the termination prover AProVE.

## 2 Loops

We only regard finite signatures and TRSs and refer to [2] for the basics of rewriting. An obvious approach to find infinite reductions is to search for a term $s$ which rewrites to a term $t$ containing an instance of $s$, i.e., $s \to_{\mathcal{R}}^{+} t = C[s\mu]$ for some context $C$ and substitution $\mu$. The corresponding infinite reduction is

$$s \to_{\mathcal{R}}^{+} C[s\mu] \to_{\mathcal{R}}^{+} C[C\mu[s\mu^2]] \to_{\mathcal{R}}^{+} C[C\mu[C\mu^2[s\mu^3]]] \to_{\mathcal{R}}^{+} \cdots$$

Equivalently, one can also represent it as an infinite reduction w.r.t. $\to_{\mathcal{R}}^{+} \circ \trianglerighteq$, where $\trianglerighteq$ is the weak subterm relation:

$$s \to_{\mathcal{R}}^{+} \circ \trianglerighteq s\mu \to_{\mathcal{R}}^{+} \circ \trianglerighteq s\mu^2 \to_{\mathcal{R}}^{+} \circ \trianglerighteq s\mu^3 \to_{\mathcal{R}}^{+} \circ \trianglerighteq \ldots \qquad (\star)$$

Here, for every $s\mu^n$ the same rules are applied at the same positions to obtain $s\mu^{n+1}$. A reduction of the form $s \to_{\mathcal{R}}^{+} t \trianglerighteq s\mu$ is called a *loop* and a TRS which admits a loop is called *looping*.

*Example 2.* The TRS of Ex. 1 admits the following loop where $s = \mathsf{fact}(x, y)$ and $\mu = \{x/\mathsf{suc}(x)\}$.

$$\begin{aligned} s &\to_{\mathcal{R}} \mathsf{if}(x == y, x, y) \\ &\to_{\mathcal{R}} \mathsf{if}(\mathsf{chk}(\mathsf{eq}(x, y)), x, y) \\ &\to_{\mathcal{R}} \mathsf{if}(\mathsf{false}, x, y) \\ &\to_{\mathcal{R}} \mathsf{suc}(x) \times \mathsf{fact}(\mathsf{suc}(x), y) \\ &\trianglerighteq \mathsf{fact}(\mathsf{suc}(x), y) \\ &= s\mu \end{aligned}$$

Clearly, a naive search for looping terms is very costly. Therefore, in current non-termination provers the techniques of forward closures [3,18], unfoldings [20], ancestor graphs [17], forward- or backward-narrowing [7], and overlap closures [12] are used, where all these techniques are special forms of overlap closures. As all mentioned techniques essentially perform narrowing steps, one can modify them by also allowing narrowings into variables. This is proposed in [7] and [20]. For example, the loop of the TRS $\{\mathsf{f}(x, y, x, y, z) \to \mathsf{f}(0, 1, z, z, z), \mathsf{a} \to 0, \mathsf{a} \to 1\}$ of [25] cannot be detected by overlap closures if one does not permit narrowings into variables. Nevertheless, most of these techniques are able to detect the loop of Ex. 2. Another alternative to detect loops (at least for string rewriting) could be based on specialized unification procedures, cf. [4].

However, if one does not consider full rewriting but innermost rewriting, then loopingness does not imply non-termination,[1] since the innermost rewrite relation $\xrightarrow{\mathsf{i}}_{\mathcal{R}}$ is not stable under substitutions. More precisely, one should not define

[1] As usual, a TRS is *innermost non-terminating* iff there is a (possibly non-ground) term starting an infinite innermost reduction.

any TRS with a reduction $s \xrightarrow{\mathsf{i}}^+_{\mathcal{R}} \circ \trianglerighteq s\mu$ to be "innermost looping", because then an "innermost looping" TRS could still be innermost terminating as shown by Ex. 3. The reason is that $s \xrightarrow{\mathsf{i}}^+_{\mathcal{R}} \circ \trianglerighteq s\mu$ does not imply $s\mu \xrightarrow{\mathsf{i}}^+_{\mathcal{R}} \circ \trianglerighteq s\mu^2$. And even if $s\mu \xrightarrow{\mathsf{i}}^+_{\mathcal{R}} \circ \trianglerighteq s\mu^2$ is true, then it could be that later on for some larger $n$ there is no reduction $s\mu^n \xrightarrow{\mathsf{i}}^+_{\mathcal{R}} \circ \trianglerighteq s\mu^{n+1}$.

*Example 3.* Consider the TRS $\mathcal{R}$ consisting of the following rules.

$$\begin{aligned} \mathsf{f}(\mathsf{g}(x)) &\to \mathsf{f}(\mathsf{g}(\mathsf{g}(x))) \\ \mathsf{g}(\mathsf{g}(\mathsf{g}(x))) &\to \mathsf{a} \end{aligned}$$

This TRS would be "innermost looping" according to the definition discussed above, e.g., $\mathsf{f}(\mathsf{g}(x)) \to_{\mathcal{R}} \mathsf{f}(\mathsf{g}(\mathsf{g}(x))) = \mathsf{f}(\mathsf{g}(x))\{x/\mathsf{g}(x)\}$, but it is innermost terminating. The reason is that the first rule is applicable at most twice. Afterwards, one has to use the second rule and no reduction is possible afterwards.

To solve this problem, one might define that a TRS $\mathcal{R}$ is "innermost looping" iff there are a term $s$ and a substitution $\mu$ such that $s\mu^n \xrightarrow{\mathsf{i}}^+_{\mathcal{R}} \circ \trianglerighteq s\mu^{n+1}$ *for every natural number $n$*. A similar definition was already used in [7, Footnote 6]. Then indeed, innermost loopingness implies innermost non-termination. However, the following example shows that this definition does not correspond to a loop in the intuitive way where the reduction $s\mu^n \xrightarrow{\mathsf{i}}^+_{\mathcal{R}} \circ \trianglerighteq s\mu^{n+1}$ always has the same form and length. Consequently, it would be undecidable whether a known loop is also an innermost loop.

*Example 4.* Consider the TRS $\mathcal{R}$ with the following rules.

$$\begin{aligned} \mathsf{f}(x, y) &\to \mathsf{f}(\mathsf{suc}(x), \mathsf{g}(\mathsf{h}(x, \mathsf{0}))) \\ \mathsf{h}(\mathsf{suc}(x), y) &\to \mathsf{h}(x, \mathsf{suc}(y)) \\ \mathsf{g}(\mathsf{h}(x, y)) &\to \mathsf{j}(y) \end{aligned}$$

$\mathcal{R}$ is "innermost looping", as for $s = \mathsf{f}(\mathsf{suc}(x), \mathsf{g}(\mathsf{h}(x, \mathsf{0})))$ and $\mu = \{x/\mathsf{suc}(x)\}$ there is the following reduction for every $n \in \mathbb{N}$.

$$\begin{aligned} s\mu^n &= \mathsf{f}(\mathsf{suc}^{n+1}(x), \mathsf{g}(\mathsf{h}(\mathsf{suc}^n(x), \mathsf{0}))) \\ &\xrightarrow{\mathsf{i}}^n_{\mathcal{R}} \mathsf{f}(\mathsf{suc}^{n+1}(x), \mathsf{g}(\mathsf{h}(x, \mathsf{suc}^n(\mathsf{0})))) \\ &\xrightarrow{\mathsf{i}}_{\mathcal{R}} \mathsf{f}(\mathsf{suc}^{n+1}(x), \mathsf{j}(\mathsf{suc}^n(\mathsf{0}))) \\ &\xrightarrow{\mathsf{i}}_{\mathcal{R}} \mathsf{f}(\mathsf{suc}^{n+2}(x), \mathsf{g}(\mathsf{h}(\mathsf{suc}^{n+1}(x), \mathsf{0}))) \\ &= s\mu^{n+1} \end{aligned}$$

The problem is that the form and the *length* of the reduction from $s\mu^n$ to $s\mu^{n+1}$ depend on $n$. Therefore, with this definition of "innermost looping", it is not even semi-decidable whether a known loop is an innermost loop.

To see this, recall that it is not semi-decidable whether a (computable) function $\mathsf{j}$ over the naturals is total. Since term rewriting is Turing-complete, we can assume that there are confluent rules which compute $\mathsf{j}$ by innermost rewriting. But then we can add the three rules of $\mathcal{R}$ and totality of $\mathsf{j}$ is equivalent to the question whether the reduction above is an innermost loop, since we obtain an innermost loop iff all terms $\mathsf{j}(\mathsf{suc}^n(\mathsf{0}))$ are innermost terminating.

So the problem with the requirement $s\mu^n \xrightarrow{\mathrm{i}}{}^+_{\mathcal{R}} \circ \trianglerighteq s\mu^{n+1}$ is that for every $n$, the reduction from $s\mu^n$ to $s\mu^{n+1}$ may be completely different. In contrast, in the infinite reduction ($\star$) that corresponds to a loop for full rewriting, the reductions from $s\mu^n$ to $s\mu^{n+1}$ always have the same form. For every $n$, one can apply exactly the same rules in exactly the same order at exactly the same positions. Hence, one only has to give the reduction $s \to^+_{\mathcal{R}} t \trianglerighteq s\mu$. Then one immediately knows how to continue for $s\mu, s\mu^2, \ldots$. This gives rise to our final definition of "innermost looping".

**Definition 5 (Innermost Looping TRS).** *A TRS $\mathcal{R}$ is* innermost looping *iff there are a substitution $\mu$, a number $m \geq 1$, terms $s_1, \ldots, s_m, t$, rules $\ell_1 \to r_1, \ldots, \ell_m \to r_m \in \mathcal{R}$, and positions $p_1, \ldots, p_m$ such that for all $n \in \mathbb{N}$ all steps in the following looping reduction*[2] *are innermost steps.*

$$s_1\mu^n \to_{\ell_1 \to r_1, p_1} s_2\mu^n \to_{\ell_2 \to r_2, p_2} \ldots s_m\mu^n \to_{\ell_m \to r_m, p_m} t\mu^n \trianglerighteq s_1\mu^{n+1} \qquad (\star\star)$$

Note that ($\star\star$) is the same as the looping reduction in ($\star$), which is just written down in a more detailed way. Hence, one can represent an innermost loop in the same way as a loop for termination: by just giving the reduction $s_1 \to_{\mathcal{R}} s_2 \to_{\mathcal{R}} \ldots s_m \to_{\mathcal{R}} t \trianglerighteq s_1\mu$, i.e., $s_1 \to^+_{\mathcal{R}} t \trianglerighteq s_1\mu$.

## 3 Detecting Innermost Loops

It is clear that with Def. 5, every innermost looping TRS is innermost non-terminating. Moreover, there exist several techniques and tools to find ordinary loops (for full rewriting). Such loops are good starting points when searching for innermost loops because an innermost loop is a loop which satisfies the additional requirements of Def. 5. The only remaining problem is to check whether such an ordinary loop is also an innermost loop.

*Example 6.* Consider the looping reduction of Ex. 2. To check whether this is an innermost loop we have to check for $\mu = \{x/\mathsf{suc}(x)\}$ and for all $n \in \mathbb{N}$ whether the corresponding steps are innermost steps when instantiating the terms with $\mu^n$. The problem in this example is the reduction $\mathsf{if}(\mathsf{chk}(\mathsf{eq}(x,y)), x, y)\mu^n \to_{\mathcal{R}} \mathsf{if}(\mathsf{false}, x, y)\mu^n$ at position 1 since the redex contains the subterm $\mathsf{eq}(x,y)\mu^n$ which might not be a normal form for some $n$ due to rule $\mathsf{eq}(x,x) \to \mathsf{true}$.

In the remainder of this section we will show the main result that it is decidable whether a given loop is an innermost loop. For example, it will turn out that the loop in Ex. 2 is an innermost loop whereas the one of Ex. 3 is not. We show this result in 4 steps, corresponding to the sections 3.1 – 3.4.

### 3.1 From Innermost Loops to Redex Problems

Note that ($\star\star$) is an innermost loop iff every direct subterm of every redex $s_i\mu^n|_{p_i}$ is in normal form. Since a term $t$ is in normal form iff $t$ does not contain a redex w.r.t. $\mathcal{R}$, we can reformulate the question about innermost loopingness in terms of so-called *redex problems*.

[2] Here, $\to_{\ell \to r, p}$ denotes a rewrite step with the rule $\ell \to r$ at position $p$.

**Definition 7 (Redex, Matching, and Identity Problems).** *Let $s$ and $\ell$ be terms, let $\mu$ be a substitution (with finite domain). Then a* redex problem *is a triple $(s \mathrel{|\!\!>} \ell, \mu)$, a* matching problem *is a triple $(s > \ell, \mu)$, and an* identity problem *is a triple $(s \approxeq \ell, \mu)$.*

*A redex problem $(s \mathrel{|\!\!>} \ell, \mu)$ is* solvable *iff there are a position $p$, a substitution $\sigma$, and an $n \in \mathbb{N}$ such that $s\mu^n|_p = \ell\sigma$. A matching problem is* solvable *iff there are a substitution $\sigma$ and an $n \in \mathbb{N}$ such that $s\mu^n = \ell\sigma$. An identity problem is* solvable *iff there is an $n \in \mathbb{N}$ such that $s\mu^n = \ell\mu^n$.*

**Theorem 8 (Setting up Redex Problems).** *In the reduction $(\star\star)$ all steps are innermost steps iff for all direct subterms $s$ of the $s_i|_{p_i}$ and all left-hand sides $\ell$ of rules from $\mathcal{R}$, the redex problem $(s \mathrel{|\!\!>} \ell, \mu)$ is not solvable.*

*Proof.* Some reduction $s_i\mu^n \to_{\ell_i \to r_i, p_i} u$ is not an innermost step iff for some direct subterm $s$ of $s_i|_{p_i}$, the term $s\mu^n$ is not in normal form, since $s_i\mu^n|_{p_i} = s_i|_{p_i}\mu^n$. (Note that even for $n = 0$ we have a reduction at position $p_i$ in $(\star\star)$. Hence, $p_i$ is a position of $s_i$ and moreover, $s_i|_{p_i}$ cannot be a variable. Thus, the "direct subterms of $s_i|_{p_i}$" are indeed properly defined.) Equivalently, there are some rule $\ell \to r$ and position $p$ such that $s\mu^n|_p = \ell\sigma$. But then the redex problem $(s \mathrel{|\!\!>} \ell, \mu)$ is solvable. □

*Example 9.* The loop of Ex. 2 is an innermost loop iff for $\mu = \{x/\mathsf{suc}(x)\}$ all redex problems $(s \mathrel{|\!\!>} \ell, \mu)$ are not solvable where $s$ is from the set $\{x, y, \mathsf{eq}(x, y), \mathsf{false}\}$ of direct subterms of redexes in the loop and $\ell$ is a left-hand side of $\mathcal{R}$.

The loop of Ex. 3 is an innermost loop iff both $(\mathsf{g}(x) \mathrel{|\!\!>} \mathsf{f}(\mathsf{g}(x)), \mu')$ and $(\mathsf{g}(x) \mathrel{|\!\!>} \mathsf{g}(\mathsf{g}(\mathsf{g}(x))), \mu')$ are not solvable where $\mu' = \{x/\mathsf{g}(x)\}$.

To find out whether a redex problem $(s \mathrel{|\!\!>} \ell, \mu)$ is solvable, we search for three unknowns: the position $p$, the substitution $\sigma$, and the number $n$. We will now eliminate these unknowns one by one and start with the position $p$. This will result in matching problems. Then in a second step we will further transform matching problems into identity problems where only the number $n$ is unknown. Finally, we will present an algorithm to decide identity problems. Therefore, at the end of this section we will have a decision procedure for redex problems, and thus also for the question whether a given loop is an innermost loop.

## 3.2 From Redex Problems to Matching Problems

To start with simplifying a redex problem $(s \mathrel{|\!\!>} \ell, \mu)$ into a finite disjunction of matching problems, note that since the position $p$ can be chosen freely within any of the terms $s, s\mu, s\mu^2, \ldots$, it is not feasible to just try out all possibilities. But the following theorem shows that it is indeed possible to reduce redex problems to finitely many matching problems. Essentially, it states that it suffices to consider all subterms of $s$ and all subterms of terms that are introduced by $\mu$. Here, $\mathcal{V}$ is the set of all variables and for any term $t$, $\mathcal{V}(t)$ is the set of its variables and $Pos(t)$ is the set of its positions.

**Theorem 10 (Solving Redex Problems).** *Let $(s \mathrel{|\!\!>} \ell, \mu)$ be a redex problem. Let $\mathcal{W} = \bigcup_{i \in \mathbb{N}} \mathcal{V}(s\mu^i)$. Then $(s \mathrel{|\!\!>} \ell, \mu)$ is solvable iff $\ell$ is a variable or if one of the matching problems $(u \gtrdot \ell, \mu)$ is solvable for some non-variable subterm $u$ of a term in $\{s\} \cup \{x\mu \mid x \in \mathcal{W}\}$.*

*Proof.* If $\ell$ is a variable then the redex problem is obviously solvable, so let $\ell \notin \mathcal{V}$. We consider both directions separately.

First, let $(u \gtrdot \ell, \mu)$ be solvable, i.e., there are $\sigma$ and $n$ such that $u\mu^n = \ell\sigma$. If $u$ is a subterm of $s$, i.e., $u = s|_p$ for some $p$, then $s\mu^n|_p = s|_p\mu^n = u\mu^n = \ell\sigma$ proves that $(s \mathrel{|\!\!>} \ell, \mu)$ is solvable. Otherwise, if $u$ is a subterm of some $x\mu$ with $x \in \mathcal{W}$ then there is some $i$ such that $x \in \mathcal{V}(s\mu^i)$. Hence, there is a position $p$ such that $s\mu^{i+1}|_p = u$. Again, $s\mu^{i+1+n}|_p = u\mu^n = \ell\sigma$ proves that $(s \mathrel{|\!\!>} \ell, \mu)$ is solvable.

For the other direction of the equivalence we assume that $(s \mathrel{|\!\!>} \ell, \mu)$ is solvable, so let $s\mu^n|_p = \ell\sigma$ for some $p$, $\sigma$, and $n$. If $p \in Pos(s)$ and $s|_p \notin \mathcal{V}$, then we are done as the matching problem $(u \gtrdot \ell, \mu)$ for the corresponding subterm $u = s|_p$ is obviously solvable.

Otherwise, there must be an $0 \leq i < n$ such that $p \in Pos(s\mu^{i+1})$ with $s\mu^{i+1}|_p \notin \mathcal{V}$ (as $\ell \notin \mathcal{V}$) and either $s\mu^i|_p \in \mathcal{V}$ or $p \notin Pos(s\mu^i)$. In both cases there must be a variable $x$ and a position $p'$ such that $x \in \mathcal{V}(s\mu^i) \subseteq \mathcal{W}$ and $x\mu|_{p'} = s\mu^{i+1}|_p$. We choose the non-variable subterm $u = x\mu|_{p'}$ of $x\mu$. Then indeed the matching problem $(u \gtrdot \ell, \mu)$ is solvable since

$$u\mu^{n-(i+1)} = x\mu|_{p'}\mu^{n-(i+1)} = s\mu^{i+1}|_p\mu^{n-(i+1)} = s\mu^n|_p = \ell\sigma. \qquad \square$$

Note that the set $\mathcal{W}$ is a subset of the finite set $\mathcal{V}(s) \cup \bigcup_{x \in Dom(\mu)} \mathcal{V}(x\mu)$. Thus, one can compute $\mathcal{W}$ by adding $\mathcal{V}(s\mu^i)$ for larger and larger $i$ until one reaches an $i$ where the set does not increase anymore. Hence, Thm. 10 can easily be automated.

*Example 11.* We use Thm. 10 for the redex problems of Ex. 9. We first consider the redex problems resulting from the loop of Ex. 2. Since there are no new variables occurring when applying $\mu$ we obtain $\mathcal{W} = \{x, y\}$. Thus, the loop is an innermost loop iff none of the matching problems $(s \gtrdot \ell, \mu)$ is solvable where $s$ is now chosen from $\{\mathsf{suc}(x), \mathsf{eq}(x, y), \mathsf{false}\}$. (So the variables $x$, $y$ do not have to be regarded anymore, but now one has to consider the new term $\mathsf{suc}(x)$ from the substitution.)

In the same way, the loop of Ex. 3 is an innermost loop iff none of the matching problems $(\mathsf{g}(x) \gtrdot \mathsf{f}(\mathsf{g}(x)), \mu')$ and $(\mathsf{g}(x) \gtrdot \mathsf{g}(\mathsf{g}(\mathsf{g}(x))), \mu')$ is solvable.

### 3.3 From Matching Problems to Identity Problems

Now the question remains whether a given matching problem is solvable. This amounts to detecting the matcher $\sigma$ and the unknown number $n$. Our next aim is to reduce this problem to a conjunction of identity problems, i.e., to eliminate the need to search for matchers $\sigma$. However, we first have to generalize the notion of matching problems $(s \gtrdot \ell, \mu)$ which contain one pair of terms $s \gtrdot \ell$ to matching problems which allow a set of pairs of terms.

**Definition 12 (General Matching Problem).** *A* general matching problem *$(\mathcal{M}, \mu)$ consists of a set $\mathcal{M}$ of pairs $\{s_1 \gg \ell_1, \dots, s_k \gg \ell_k\}$ together with a substitution $\mu$. A general matching problem $(\mathcal{M}, \mu)$ is* solvable *iff there are a substitution $\sigma$ and an $n \in \mathbb{N}$ such that for all $1 \le j \le k$ the equality $s_j \mu^n = \ell_j \sigma$ is valid.*

*If $\mathcal{M}$ only contains one pair $s \gg \ell$ then we identify $(\mathcal{M}, \mu)$ with $(s \gg \ell, \mu)$, and if $\mu$ is clear from the context we write $\mathcal{M}$ as an abbreviation for $(\mathcal{M}, \mu)$.*

We now give a set of four transformation rules which either detect that a matching problem is not solvable (indicated by $\bot$), or which transform a matching problem into solved form. Here, a (general) matching problem $(\{s_1 \gg \ell_1, \dots, s_k \gg \ell_k\}, \mu)$ is in *solved form* iff all $\ell_1, \dots, \ell_k$ are variables. Once we have reached a matching problem in solved form, it is easily possible to translate it into identity problems.

**Definition 13 (Transformation of Matching Problems).** *We define the following transformation $\Rightarrow$ on general matching problems. If $(\mathcal{M}, \mu)$ is a general matching problem with $\mathcal{M} = \mathcal{M}' \uplus \{s \gg \ell\}$ where $\ell \notin \mathcal{V}$, and if $\mathcal{V}_{incr} = \{x \in \mathcal{V} \mid$ there is some $n \in \mathbb{N}$ with $x\mu^n \notin \mathcal{V}\}$ is the set of* increasing variables, *then*

*(i)* $\mathcal{M} \Rightarrow \{s'\mu \gg \ell' \mid s' \gg \ell' \in \mathcal{M}\}$, *if* $s \in \mathcal{V}_{incr}$
*(ii)* $\mathcal{M} \Rightarrow \bot$, *if* $s \in \mathcal{V} \setminus \mathcal{V}_{incr}$
*(iii)* $\mathcal{M} \Rightarrow \bot$, *if* $s = f(\dots)$, $\ell = g(\dots)$, *and* $f \neq g$
*(iv)* $\mathcal{M} \Rightarrow \mathcal{M}' \cup \{s_1 \gg \ell_1, \dots, s_k \gg \ell_k\}$, *if* $s = f(s_1, \dots, s_k)$, $\ell = f(\ell_1, \dots, \ell_k)$

Rule (iv) just decomposes terms and Rule (iii) handles a symbol-clash. These rules are standard for classical matching algorithms. However, if the left-hand side is a variable $x$ and the right-hand side is not, then a matching problem may still be solvable. If $x$ is increasing then we just have to apply $\mu$ until a new symbol is produced on the left-hand side. This is done by Rule (i) and will be illustrated in more detail when solving the matching problems of the loop in Ex. 3. However, if $x$ is not increasing then the matching problem is not solvable since $x\mu^n$ will always remain a variable. Hence, $\bot$ is obtained by Rule (ii). The following theorem shows that every matching problem $(s \gg \ell, \mu)$ can be automatically reduced to a finite conjunction of identity problems.

**Theorem 14 (Solving Matching Problems).** *Let $(\mathcal{M}, \mu)$ be a general matching problem.*

*(i) The transformation rules of Def. 13 are confluent and terminating.*
*(ii) If $\mathcal{M} \Rightarrow \bot$ then $\mathcal{M}$ is not solvable.*
*(iii) If $\mathcal{M} \Rightarrow \mathcal{M}'$ with $\mathcal{M}' \neq \bot$, then $\mathcal{M}$ is solvable iff $\mathcal{M}'$ is solvable.*
*(iv) $\mathcal{M}$ is solvable iff $\mathcal{M} \Rightarrow^* \mathcal{M}'$ for some matching problem $\mathcal{M}' = \{s_1 \gg x_1, \dots, s_k \gg x_k\}$ in solved form, such that for all $i \neq j$ with $x_i = x_j$ the identity problem $(s_i \approxeq s_j, \mu)$ is solvable.*

*Proof.* (i) To prove confluence one can show that $\Rightarrow$ is strongly confluent by a simple case analysis.
To show termination of $\Rightarrow$ first note that no transformation rule increases the size of the terms in the right-hand sides of a matching problem. Thus, Rule (iv) can only be applied finitely often. But since every sequence of transformations with Rule (i) eventually triggers an application of Rule (iii) or (iv), Rule (i) cannot be used infinitely often either.

(ii) If $\mathcal{M} \Rightarrow \bot$ due to Rule (iii) then $s \gg \ell \in \mathcal{M}$ with $s = f(\ldots)$ and $\ell = g(\ldots)$ where $f \neq g$. But then for every $n \in \mathbb{N}$ the terms $s\mu^n = f(\ldots)$ and $\ell\sigma = g(\ldots)$ are different. Hence, $\mathcal{M}$ is not solvable.
If $\mathcal{M} \Rightarrow \bot$ due to Rule (ii) then $x \gg \ell \in \mathcal{M}$ with $x \in \mathcal{V} \setminus \mathcal{V}_{incr}$ and $\ell = f(\ldots)$. But since $x$ is not an increasing variable we know that $x\mu^n \in \mathcal{V}$ for all $n \in \mathbb{N}$. Thus, the terms $x\mu^n$ and $\ell\sigma = f(\ldots)$ are different for all $n$. Hence, $\mathcal{M}$ is not solvable.

(iii) We first consider Rule (i) for $\mathcal{M} = \{s_1 \gg \ell_1, \ldots, s_k \gg \ell_k\}$.

$$\begin{array}{ll} & \mathcal{M} \text{ is solvable} \\ \text{iff} & \exists \sigma, n : s_1\mu^n = \ell_1\sigma \wedge \cdots \wedge s_k\mu^n = \ell_k\sigma \\ \text{iff} & \exists \sigma', n : s_1\mu^{n+1} = \ell_1\sigma' \wedge \cdots \wedge s_k\mu^{n+1} = \ell_k\sigma' \quad \text{(as } s_i \in \mathcal{V} \text{ for some } i\text{)} \\ \text{iff} & \mathcal{M}' = \{s_1\mu \gg \ell_1, \ldots, s_k\mu \gg \ell_k\} \text{ is solvable} \end{array}$$

For Rule (iv) the result follows from the fact that $f(s_1, \ldots, s_k)\mu^n = f(\ell_1, \ldots, \ell_k)\sigma$ iff $s_i\mu^n = \ell_i\sigma$ for all $1 \leq i \leq k$.

(iv) If $\mathcal{M}$ is solvable then due to (ii) and (iii), $\mathcal{M}$ cannot be transformed to $\bot$. So let $\mathcal{M}'$ be a normal form of $\mathcal{M}$ w.r.t. $\Rightarrow$. Then, obviously $\mathcal{M}'$ has the form $\{s_1 \gg x_1, \ldots, s_k \gg x_k\}$ and $\mathcal{M}'$ is solvable due to (iii). Thus, there are a substitution $\sigma$ and a number $n$ such that for all $1 \leq i \leq k$ the equality $s_i\mu^n = x_i\sigma$ is valid. Hence, for all $i \neq j$ with $x_i = x_j$ the identity problem $(s_i \approxeq s_j, \mu)$ is solvable.
For the other direction let $\mathcal{M} \Rightarrow^* \mathcal{M}' = \{s_1 \gg x_1, \ldots, s_k \gg x_k\}$ where for every $i \neq j$ with $x_i = x_j$ there is some $n_{ij}$ with $s_i\mu^{n_{ij}} = s_j\mu^{n_{ij}}$. Let $n$ be the maximum of all $n_{ij}$. Then, obviously $s_i\mu^n = s_j\mu^n$ for all these $i$ and $j$. We define $\sigma = \{x_1/s_1\mu^n, \ldots, x_k/s_k\mu^n\}$. First note that $\sigma$ is well defined by construction. But as then $s_i\mu^n = x_i\sigma$ is valid for all $1 \leq i \leq k$ we know that $\mathcal{M}'$ is solvable. Using (iii) we finally conclude that $\mathcal{M}$ is solvable. □

*Example 15.* We illustrate the transformation rules by continuing Ex. 11.

For the loop of Ex. 2 we can reduce all but one matching problem to $\bot$ by Rule (iii). Only the matching problem $(\mathsf{eq}(x,y) \gg \mathsf{eq}(x,x), \mu)$ is transformed by Rule (iv) into its solved form $\{x \gg x, y \gg x\}$. Hence, by Thm. 14 the loop is an innermost loop iff the identity problem $(x \approxeq y, \mu)$ is not solvable.

For the loop of Ex. 3, we had to find out whether $(\mathsf{g}(x) \gg \mathsf{g}(\mathsf{g}(\mathsf{g}(x))), \mu')$ is solvable. Applying Rule (iv) yields $(x \gg \mathsf{g}(\mathsf{g}(x)), \mu')$. Since $x$ is an increasing variable for $\mu'$, we now have to apply Rule (i) and obtain $(\mathsf{g}(x) \gg \mathsf{g}(\mathsf{g}(x)), \mu')$ as $x\mu' = \mathsf{g}(x)$. Repeated application of Rules (iv) and (i) results in the solved form $(x \gg x, \mu')$. Hence, by Thm. 14 the matching problem $(\mathsf{g}(x) \gg \mathsf{g}(\mathsf{g}(\mathsf{g}(x))), \mu')$ is solvable as no identity problems are created. Thus, we have detected that the loop of Ex. 3 is not an innermost loop.

### 3.4 Deciding Identity Problems

Note that for left-linear TRSs, identity problems are never created, since there the right-hand sides of a general matching problem are always variable disjoint.

Input: An identity problem $(s \approxeq t, \mu)$.
Output: "Yes", if the identity problem is solvable, and "No", if it is not.

(i) While $\mu$ contains a cycle of length $n > 1$ do $\mu := \mu^n$.
(ii) $S := \emptyset$
(iii) If $s = t$ then stop with result "Yes".
(iv) If there is a shared position $p$ of $s$ and $t$ such that $s|_p = f(\dots)$ and $t|_p = g(\dots)$ and $f \neq g$ then stop with result "No".
(v) If there is a shared position $p$ of $s$ and $t$ such that $s|_p = x$, $t|_p = g(\dots)$, and $x$ is not an increasing variable then stop with result "No".
Repeat this step with $s$ and $t$ exchanged.
(vi) If there is a shared position $p$ of $s$ and $t$ such that $s|_p = x$, $t|_p = y$, $x \neq y$, and $x, y \notin \mathit{Dom}(\mu)$ then stop with result "No".
(vii) Add the triple $(x, p, t|_p)$ to $S$ for all shared positions $p$ of $s$ and $t$ such that $x = s|_p \neq t|_p$ where $x$ is an increasing variable.
Repeat this step with $s$ and $t$ exchanged.
(viii) If $(x, p_1, u_1) \in S$ and $(x, p_2, u_2) \in S$ where
(a) $u_1$ and $u_2$ are not unifiable or where
(b) $u_1 = u_2$ and $p_1 < p_2$,
then stop with result "No".
(ix) $s := s\mu, t := t\mu$
(x) Continue with Step (iii).

**Fig. 1.** An algorithm to decide solvability of identity problems

However, in order to handle also non-left-linear TRSs, it remains to give an algorithm which decides solvability of an identity problem.[3] This algorithm is presented in Fig. 1, and we now explain its steps one by one.

First we replace the substitution $\mu$ by $\mu^n$ such that $\mu^n$ does not contain cycles. Here, a substitution $\delta$ contains a *cycle* of length $n$ iff $\delta = \{x_1/x_2,\ x_2/x_3,\ \dots,\ x_n/x_1, \dots\}$ where the $x_i$ are pairwise different variables. Obviously, if $\delta$ contains a cycle of length $n$ then in $\delta^n$ all variables $x_1, \dots, x_n$ do not belong to the domain any more. Thus, Step (i) terminates and afterwards, $\mu$ does not contain cycles of length 2 or more.

Note that the identity problem $(s \approxeq t, \mu)$ is solvable iff $(s \approxeq t, \mu^n)$ is solvable. Hence, after Step (i) we still have to decide solvability of $(s \approxeq t, \mu)$ for the modified $\mu$. The advantage is that now $\mu$ has a special structure. For all $x \in \mathit{Dom}(\mu)$, either $x$ is an increasing variable or for some $n$ the term $x\mu^n$ is a variable which is not in $\mathit{Dom}(\mu)$. For such substitutions $\mu$, the terms $s, s\mu, s\mu^2, \dots$ finally become *stationary* at each position, i.e., for every position $p$ there is some $n$ such that either all terms $s\mu^n|_p, s\mu^{n+1}|_p, s\mu^{n+2}|_p, \dots$ are of the form $f(\dots)$, or all these terms are the same variable $x \notin \mathit{Dom}(\mu)$. Therefore, it is possible to define

[3] It could also be possible to express identity problems as primal unification problems and to use an algorithm for primal unification [13] instead. But then one would have to extend the results of [13] to allow arbitrary dependencies of function symbols. Moreover, our algorithm has the advantage of being very easy to implement.

$s\mu^\infty$ as the (possibly infinite) term where $\mathrm{root}(s\mu^\infty|_p) = f$ iff $\mathrm{root}(s\mu^n|_p) = f$ for some $n$, and $s\mu^\infty|_p = x$ iff there is some $n$ such that $s\mu^m|_p = x$ for all $m \geq n$.

If the identity problem is solvable then there is some $n$ such that $s\mu^n = t\mu^n$ which will be detected in Step (iii). The reason is that with Steps (ix) and (x) one iterates over all pairs $(s,t)$, $(s\mu, t\mu)$, $(s\mu^2, t\mu^2)$, ....

If the identity problem is not solvable, then this could be due to a *stationary conflict*, i.e., $s\mu^\infty \neq t\mu^\infty$. Then the identity problem is unsolvable since $s\mu^n = t\mu^n$ would imply $s\mu^\infty = t\mu^\infty$. If the terms $s\mu^\infty$ and $t\mu^\infty$ differ, then there is some position $p$ such that the symbols at position $p$ in $s\mu^\infty$ and $t\mu^\infty$ differ, or $s\mu^\infty|_p$ is a variable and $t\mu^\infty|_p$ is not a variable (or vice versa), or both $s\mu^\infty|_p$ and $t\mu^\infty|_p$ are different variables. Recall that the terms $s, s\mu, s\mu^2, \ldots$ and the terms $t, t\mu, t\mu^2, \ldots$ finally become stationary. Hence, if we choose $n$ high enough, then the conflict at position $p$ can already be detected by inspecting $s\mu^n|_p$ and $t\mu^n|_p$. Thus, then one of three cases in Steps (iv)–(vi) will hold.

With the steps described up to now, we can detect all solvable identity problems and all identity problems which are not solvable due to a stationary conflict. However, there remain other identity problems which are not solvable, but which do not have a stationary conflict. As an example consider $(x \approxeq y, \{x/\mathsf{f}(x), y/\mathsf{f}(y)\})$. Then $s\mu^\infty = \mathsf{f}(\mathsf{f}(\mathsf{f}(\ldots))) = t\mu^\infty$ but this identity problem is not solvable since $x\mu^n = \mathsf{f}^n(x) \neq \mathsf{f}^n(y) = y\mu^n$ for all $n \in \mathbb{N}$. We call such identity problems *infinite*.

The remaining steps (ii), (vii), and (viii) are used to detect infinite identity problems. In the set $S$ we store sub-problems $(x, p, u)$ such that whenever the identity problem is solvable, then $x\mu^m = u\mu^m$ must hold for some $m$ to make the terms $s\mu^n$ and $t\mu^n$ equal at position $p$.

We give some intuition why the two abortion criteria in Step (viii) are correct. For (viii–a), note that if $u_1$ and $u_2$ are not unifiable then $x\mu^m$ cannot be both $u_1\mu^m$ and $u_2\mu^m$, which means that the sub-problems $(x, p_1, u_1)$ and $(x, p_2, u_2)$ (resp. $(x \approxeq u_1, \mu)$ and $(x \approxeq u_2, \mu)$) are not solvable. For (viii–b), in order to make $x\mu^m$ equal to $u_1\mu^m$, we again produced the same problem at a lower position. Then the original identity problem is again not solvable, since this repeated generation of the same sub-problem would continue forever. As usual, $p_1 < p_2$ denotes that position $p_1$ is strictly above $p_2$.

The following theorem shows that all answers of the algorithm are indeed correct and it also shows that it always returns an answer. The termination proof is quite involved since we have to show that the criteria in Step (viii) suffice to detect all infinite identity problems.

**Theorem 16 (Solving Identity Problems).** *The algorithm in Fig. 1 to decide solvability of identity problems is correct and it terminates.*

*Proof.* One can easily show that in the $k$-th iteration, $S$ is the following set $S_k$.

$$S_k = \{(x, p, u) \mid x \in \mathcal{V}_{incr} \wedge x \neq u \wedge \exists m \leq k : \\ (s\mu^m|_p = x \wedge u = t\mu^m|_p) \vee (t\mu^m|_p = x \wedge u = s\mu^m|_p)\}$$

Since the correctness of Steps (i)–(vi) was already illustrated in the explanation of the algorithm, we only prove the correctness of Step (viii) formally. So let $(x, p_1, u_1)$ and $(x, p_2, u_2)$ be elements of some $S_k$. Hence, there exist $m_1 \leq k$ and $m_2 \leq k$ such that w.l.o.g. for both $i = 1$ and $i = 2$, we have $s\mu^{m_i}|_{p_i} = x$ and $t\mu^{m_i}|_{p_i} = u_i$ where $x$ is a variable with $x \neq u_i$. If the identity problem $(s \approxeq t, \mu)$ is not solvable then there is nothing to show. Otherwise, there is some $n$ with $s\mu^n = t\mu^n$. Since $s\mu^{m_i}|_{p_i} = x \neq u_i = t\mu^{m_i}|_{p_i}$, we know that $n > m_i$ for both $i$. Hence, we can conclude the following equalities for both $i \in \{1, 2\}$:

$$x\mu^{n-m_i} = s\mu^{m_i}|_{p_i}\mu^{n-m_i} = s\mu^n|_{p_i} = t\mu^n|_{p_i} = t\mu^{m_i}|_{p_i}\mu^{n-m_i} = u_i\mu^{n-m_i} \quad (13)$$

Assume that we have applied (viii–a) and the algorithm wrongly returned "No". This directly leads to a contradiction since by (13), $u_1\mu^n = x\mu^n = u_2\mu^n$ proves that $u_1$ and $u_2$ are unifiable.

Now assume that we applied (viii–b) and wrongly obtained "No". W.l.o.g. let $p_1 < p_2$. Since $s\mu^{m_1}|_{p_1}$ is the variable $x$, we must apply $\mu$ at least one more time to obtain a term with the position $p_2$ and thus, $m_1 < m_2$. As $x\mu^{n-m_1} = u_1\mu^{n-m_1}$ by (13), there must be some smallest number $n' \leq n$ such that $x\mu^{n'-m_1} = u_1\mu^{n'-m_1}$ is valid. From $x \neq u_1$ we conclude $n' > m_1$ and from $s\mu^{n'}|_{p_1} = x\mu^{n'-m_1} = u_1\mu^{n'-m_1} = t\mu^{n'}|_{p_1}$ we derive that also the subterms $s\mu^{n'}|_{p_2}$ of $s\mu^{n'}|_{p_1}$ and $t\mu^{n'}|_{p_2}$ of $t\mu^{n'}|_{p_1}$ are identical. Again, $n' > m_2$ must hold and we obtain $x\mu^{n'-m_2} = u_2\mu^{n'-m_2}$. But this is a contradiction to the minimality of $n'$ since $u_1 = u_2$ and $n' - m_2 < n' - m_1$.

To prove termination, we have already argued in the explanation of the algorithm why we can detect all solvable identity problems and all those problems which have a stationary conflict. So it remains to prove that all infinite problems can be detected. To this end, we start with three observations on infinite identity problems, i.e., unsolvable identity problems $(s \approxeq t, \mu)$ where $s\mu^\infty = t\mu^\infty$.

First, if $(s \approxeq t, \mu)$ is infinite then $(s\mu \approxeq t\mu, \mu)$ is infinite as well.

Second, if $(s \approxeq t, \mu)$ is infinite then there is no position $p$ where $(s|_p \approxeq t|_p, \mu)$ has a stationary conflict (i.e., $s|_p\mu^\infty \neq t|_p\mu^\infty$). Otherwise there would also be a stationary conflict for $(s \approxeq t, \mu)$ which contradicts the infinity of $(s \approxeq t, \mu)$.

And third, whenever $(s \approxeq t, \mu)$ is infinite then there is some position $p$ such that $s|_p \neq t|_p$, at least one of the terms $s|_p$ or $t|_p$ is an increasing variable, and $(s|_p\mu \approxeq t|_p\mu, \mu)$ is infinite, too. This can be proved as follows. Let $p$ be one of the longest (i.e., lowest) shared positions of $s$ and $t$ such that $(s|_p \approxeq t|_p, \mu)$ is not solvable. (Such positions must exist since $(s \approxeq t, \mu)$ is not solvable.) Due to the second observation we know that $(s|_p \approxeq t|_p, \mu)$ again is infinite. Moreover, using the maximality of $p$ we conclude that at least one of the terms $s|_p$ or $t|_p$ is a variable. Since $(s|_p \approxeq t|_p, \mu)$ is infinite, this variable must be increasing. Finally, by the first observation, $(s\mu|_p \approxeq t\mu|_p, \mu)$ is infinite as well.

Now we show that if there were an infinite run of the algorithm, we would insert an infinite number of triples into $S$ where the corresponding positions $p_0$, $p_0\,p_1$, $p_0\,p_1\,p_2$, ... are getting longer and longer: Since $(s \approxeq t, \mu)$ is infinite, due to the third observation there is a position $p_0$ such that a triple $(x_0, p_0, s|_{p_0})$ or $(x_0, p_0, t|_{p_0})$ is added to $S$. Moreover, $(s\mu|_{p_0} \approxeq t\mu|_{p_0}, \mu)$ is infinite. Hence, again using the third observation we obtain a position $p_1$ such that $(s\mu|_{p_0}\mu|_{p_1} \approxeq t\mu|_{p_0}\mu|_{p_1}, \mu) = (s\mu^2|_{p_0p_1} \approxeq t\mu^2|_{p_0p_1}, \mu)$ is infinite where $s\mu^2|_{p_0p_1}$ and $t\mu^2|_{p_0p_1}$ are different terms, one of them being an increasing variable $x_1$. Thus, again the corresponding triple $(x_1, p_0\, p_1, s\mu|_{p_0\, p_1})$ or $(x_1, p_0\, p_1, t\mu|_{p_0\, p_1})$ is added to $S$. By iterating this reasoning, we obtain the desired infinite sequence of triples in $S$.

As there exist only finitely many increasing variables, there must be some $x$ which occurs infinitely often in this sequence. Thus, we obtain an infinite subsequence $(x, p_0 \ldots p_{i_1}, u_{i_1})$, $(x, p_0 \ldots p_{i_2}, u_{i_2})$, ... where $i_1 < i_2 < \ldots$ and $p_0 \ldots p_{i_1} < p_0 \ldots p_{i_2} < \ldots$. Due to Kruskal's tree theorem [16], there must be some $i_j$ and $i_k$ such that $i_j < i_k$ and $u_{i_j}$ is embedded in $u_{i_k}$. If $u_{i_j} = u_{i_k}$ then this is a contradiction to an infinite run of the algorithm since then the criterion in Step (viii–b) would hold and the algorithm would be stopped. Otherwise, $u_{i_j}$ is strictly embedded in $u_{i_k}$. But then $u_{i_j}$ cannot be unified with $u_{i_k}$ since the embedding relation is stable under substitutions. Hence in that case, the criterion in Step (viii–a) will stop the algorithm. □

*Example 17.* We illustrate the algorithm with the identity problem $(x \approxeq y, \mu)$ where $\mu = \{x/\mathsf{f}(y, u_0), y/\mathsf{f}(z, u_0), z/\mathsf{f}(x, u_0), u_0/u_1, u_1/u_0\}$.

As $\mu$ contains a cycle of length 2 we replace $\mu$ by $\mu^2 = \{x/\mathsf{f}(\mathsf{f}(z, u_0), u_1)$, $y/\mathsf{f}(\mathsf{f}(x, u_0), u_1)$, $z/\mathsf{f}(\mathsf{f}(y, u_0), u_1)\}$. Since $x\mu^\infty = \mathsf{f}(\mathsf{f}(\mathsf{f}(\ldots, u_1), u_0), u_1) = y\mu^\infty$, we know that the problem is either solvable or infinite. Hence, the criteria in Steps (iv)–(vi) will never apply. We start with $s = x$ and $t = y$. Since the terms are different we add $(x, \varepsilon, y)$ and $(y, \varepsilon, x)$ to $S$. In the next iteration we have $s = \mathsf{f}(\mathsf{f}(z, u_0), u_1)$ and $t = \mathsf{f}(\mathsf{f}(x, u_0), u_1)$. Again, the terms are different and we add $(x, 11, z)$ and $(z, 11, x)$ to $S$. The next iteration yields the new triples $(y, 1111, z)$ and $(z, 1111, y)$, and after having applied $\mu$ three times, we obtain the two last triples $(x, 111111, y)$ and $(y, 111111, x)$. Then due to the criterion (viii–b), the algorithm terminates with "No".

By simply combining all theorems of Sect. 3, we finally obtain a decision procedure which solves the question whether a loop is also an innermost loop.

**Corollary 18 (Deciding Innermost Loops).** *For every loop*

$$s_1 \to_{\mathcal{R}} s_2 \to_{\mathcal{R}} \ldots \to_{\mathcal{R}} s_m \to_{\mathcal{R}} t \trianglerighteq s_1\mu$$

*of a TRS $\mathcal{R}$, it is decidable whether that loop is also an innermost loop.*

*Example 19.* In Ex. 15 we observed that the loop of Ex. 2 is an innermost loop iff $(x \approxeq y, \mu)$ is not solvable where $\mu = \{x/\mathsf{suc}(x)\}$. We apply the algorithm of Fig. 1 to show that this identity problem is not solvable. Hence, we show that the loop is an innermost loop and thus, the TRS of Ex. 1 is not innermost terminating.

Since $\mu$ only contains cycles of length 1, we skip Step (i). So, let $s = x$ and $t = y$. Then none of the steps (iii)–(vi) is applicable. Hence, we add $(x, \varepsilon, y)$ to $S$ and continue with $s = \mathsf{suc}(x)$ and $t = y$. Then, in Step (v) the algorithm is stopped with the answer "No" due to a stationary conflict.

## 4 Integration into the Dependency Pair Framework

In [7], we showed that in order to find loops automatically, it is advantageous to use the *dependency pair framework* [1,6,10] because of a reduced search space. There are two main reasons for this: First, one can drop the contexts when looking for loops, i.e., one can drop the $\trianglerighteq$ in "$\to_{\mathcal{R}}^{+} \circ \trianglerighteq$" and will still be able to detect every looping TRS [7, Thm. 23]. Second and more important, by using dependency pairs one can often prove termination of large parts of the TRS, and hence only has to search for loops for a small subsystem of the original TRS.

While the results of this paper have only been presented for TRSs, it is easy to extend our notion of "innermost looping" (Def. 5) to DP problems – the basic data structure within the dependency pair framework. Then the methods of Sect. 3 can again be used to decide whether a looping DP problem is innermost looping. Moreover, one can extend [7, Thm. 23] to the innermost case, i.e., a TRS is innermost looping iff the corresponding DP problem is innermost looping. The details of these extensions can be found in [21, Chapter 8].

## 5 Conclusion

To prove non-termination of innermost rewriting, we first extended the notion of a loop to the innermost case. An innermost loop is an innermost reduction with a strong regularity which admits the same infinite reduction as an ordinary loop does for full rewriting. Afterwards, we developed a novel procedure to decide whether a given loop is also an innermost loop. Our procedure can be combined with any method to detect loops for full rewriting, regardless whether it directly searches for loops of the TRS or whether it performs this search within the dependency pair framework.

We have implemented our procedure in combination with dependency pairs in our termination prover AProVE [8] which already featured a method to detect loops, cf. [7]. Note that while proving the soundness and the termination of our novel decision procedure is non-trivial, the procedure itself is very easy to implement. To evaluate its usefulness empirically, we tested it on the *termination problem data base* (TPDB). This is the collection of examples used in the annual *International Competition of Termination Tools* [19]. Currently, the TPDB contains 129 TRSs where at least one tool has been able to disprove termination in the competition in 2007. With the results of this paper, AProVE now also disproves *innermost* termination for 93 of these TRSs (where we use a time limit of 1 minute per example). In contrast, we are not aware of any other existing tool for disproving innermost termination. The fact that from the remaining 36 TRSs at least 30 are innermost terminating demonstrates the power of our approach. Moreover, of course AProVE can also disprove innermost termination of Ex. 1. Concerning efficiency, the check whether a loop that was found is also an innermost loop needs less than 8 seconds in total for all TRSs of the TPDB. For further details on our experiments and to run this new version of AProVE via a web interface, we refer to `http://aprove.informatik.rwth-aachen.de/eval/decidingLoops`.

## References


1. Arts, T., Giesl, J.: Termination of term rewriting using dependency pairs. Theoretical Computer Science 236, 133–178 (2000)
2. Baader, F., Nipkow, T.: Term Rewriting and All That. Cambridge (1998)

3. Dershowitz, N.: Termination of rewriting. J. Symb. Comp. 3, 69–116 (1987)
4. Diekert, V.: Makanin's algorithm. In: Lothaire, M. (ed.) Combinatorics on Words, pp. 387–442. Cambridge University Press, Cambridge (2002)
5. Endrullis, J.: Jambox, `http://joerg.endrullis.de`
6. Giesl, J., Thiemann, R., Schneider-Kamp, P.: The dependency pair framework: Combining techniques for automated termination proofs. In: Baader, F., Voronkov, A. (eds.) LPAR 2004. LNCS (LNAI), vol. 3452, pp. 301–331. Springer, Heidelberg (2005)
7. Giesl, J., Thiemann, R., Schneider-Kamp, P.: Proving and disproving termination of higher-order functions. In: Gramlich, B. (ed.) FroCos 2005. LNCS (LNAI), vol. 3717, pp. 216–231. Springer, Heidelberg (2005)
8. Giesl, J., Schneider-Kamp, P., Thiemann, R.: AProVE 1.2: Automatic termination proofs in the DP framework. In: Furbach, U., Shankar, N. (eds.) IJCAR 2006. LNCS (LNAI), vol. 4130, pp. 281–286. Springer, Heidelberg (2006)
9. Giesl, J., Swiderski, S., Schneider-Kamp, P., Thiemann, R.: Automated termination analysis for Haskell: From term rewriting to programming languages. In: Pfenning, F. (ed.) RTA 2006. LNCS, vol. 4098, pp. 297–312. Springer, Heidelberg (2006)
10. Giesl, J., Thiemann, R., Schneider-Kamp, P., Falke, S.: Mechanizing and improving dependency pairs. Journal of Automated Reasoning 37(3), 155–203 (2006)
11. Gramlich, B.: Abstract relations between restricted termination and confluence properties of rewrite systems. Fundamenta Informaticae 24, 3–23 (1995)
12. Guttag, J., Kapur, D., Musser, D.: On proving uniform termination and restricted termination of rewriting systems. SIAM J. Computation 12, 189–214 (1983)
13. Hermann, M., Galbavý, R.: Unification of infinite sets of terms schematized by primal grammars. Theoretical Computer Science 176(1-2), 111–158 (1997)
14. Hirokawa, N., Middeldorp, A.: Tyrolean Termination Tool: Techniques and features. Information and Computation 205(4), 474–511 (2007)
15. Kaufmann, M., Manolios, P., Moore, J.S.: Computer-Aided Reasoning: An Approach. Kluwer, Dordrecht (2000)
16. Kruskal, J.B.: Well-quasi-orderings, the Tree Theorem, and Vazsonyi's conjecture. Transactions of the American Mathematical Society 95, 210–223 (1960)
17. Kurth, W.: Termination und Konfluenz von Semi-Thue-Systemen mit nur einer Regel. PhD thesis, Technische Universität Clausthal, Germany (1990)
18. Lankford, D., Musser, D.: A finite termination criterion. Unpublished Draft. USC Information Sciences Institute (1978)
19. Marché, C., Zantema, H.: The termination competition. In: Baader, F. (ed.) RTA 2007. LNCS, vol. 4533, pp. 303–313. Springer, Heidelberg (2007)
20. Payet, É.: Detecting non-termination of term rewriting systems using an unfolding operator. In: Puebla, G. (ed.) LOPSTR 2006. LNCS, vol. 4407, pp. 194–209. Springer, Heidelberg (2007)
21. Thiemann, R.: The DP Framework for Proving Termination of Term Rewriting. PhD thesis, RWTH Aachen University. Technical Report AIB-2007-17 (2007), `http://aib.informatik.rwth-aachen.de/2007/2007-17.pdf`
22. Vroon, D.: Personal communication (2007)
23. Waldmann, J.: Matchbox: A tool for match-bounded string rewriting. In: van Oostrom, V. (ed.) RTA 2004. LNCS, vol. 3091, pp. 85–94. Springer, Heidelberg (2004)
24. Zantema, H.: Termination of string rewriting proved automatically. Journal of Automated Reasoning 34, 105–139 (2005)
25. Zhang, X.: Overlap closures do not suffice for termination of general term rewriting systems. Information Processing Letters 37(1), 9–11 (1991)

# Termination Proof of S-Expression Rewriting Systems with Recursive Path Relations

Yoshihito Toyama

RIEC, Tohoku University
Katahira 2-1-1, Aoba-ku, Sendai 980-8577, Japan
toyama@nue.riec.tohoku.ac.jp

**Abstract.** S-expression rewriting systems were proposed by the author (RTA 2004) for termination analysis of Lisp-like untyped higher-order functional programs. This paper presents a short and direct proof for the fact that every finite S-expression rewriting system is terminating if it is compatible with a recursive path relation with status. By considering well-founded binary relations instead of well-founded orders, we give a much simpler proof than the one depending on Kruskal's tree theorem.

## 1 Introduction

An important syntactical method to prove termination of a first-order term rewriting system is the one using recursive path orders [1,4,5,8,16] relying on Kruskal's tree theorem [9]. Higher-order rewriting systems are rewriting systems to accommodate higher-order functions and several syntactical methods for proving termination of them are presented [7,10,11,15,17].

In the previous paper [17] the author proposed Lisp-like untyped higher-order rewriting systems without $\lambda$-abstraction, called S-expression rewriting systems, and proved that every finite S-expression rewriting system is terminating if it is compatible with a lexicographic path order, by using the notion of the simplification order based on Kruskal's tree theorem. However, our proof in [17] is neither direct nor simple because of *unbounded-variadic* and *higher-order* feature of S-expressions, and it is not easy to extend it to more expressive orders like recursive path orders with lexicographic and multiset status.

The purpose of this paper is to present a short and direct proof for the main result in [17] and to extend it to recursive path relations with lexicographic and multiset status on S-expressions. Like modern proofs in Buchholz [2], Dawson and Goré [3], Goubault-Larrecq [6], Jouannaud and Rubio [7], our termination proof does not use Kruskal's tree theorem, and it uses well-founded binary relations instead of well-founded orders as in Goubault-Larrecq [6], Mellies [12]. Thus the recursive path relation is not required to be well-ordered nor transitive for proving well-foundedness. Our proof technique is based on Goubault-Larrecq's work [6] in which he presents a general termination proof of abstract recursive path relations, independent on term-structure. However, since Goubault-Larrecq

A. Voronkov (Ed.): RTA 2008, LNCS 5117, pp. 381–391, 2008.

[6] defines the recursive path relation as the greatest fixed point of the monotonic operator by the co-induction principle, his proof of the stability does not work for the least fixed point. We modify his proof for treating the recursive path relation defined by the induction principle in common use and for adapting it to *unbounded-variadic* and *higher-order* feature of S-expressions.

The remainder of this paper is organized as follows. After a preliminary section, in Section 3 we give a general termination proof of abstract recursive path relations. Section 4 studies lexicographic path relation on S-expressions and discusses termination of S-expression rewriting systems compatible with lexicographic path relations. Section 5 extends our result to recursive path relations with lexicographic and multiset status on S-expressions.

## 2 Preliminaries

We mainly follow the notation of [1,16,17]. Let $\mathcal{C}$ be a (finite or infinite) set of *constants*, denoted by $f, g, h, \cdots$, and $\mathcal{V}$ a countably infinite set of *variables*, denoted by $x, y, z, \cdots, \alpha, \beta, \gamma, \cdots$, where $\mathcal{C} \cap \mathcal{V} = \phi$.

The set $\mathcal{S}(\mathcal{C}, \mathcal{V})$ of terms, called *symbolic expressions* (*S-expressions* for short) [17], is recursively defined by: (i) $\mathcal{C} \cup \mathcal{V} \subseteq \mathcal{S}(\mathcal{C}, \mathcal{V})$, and (ii) $(s_1\ s_2\ \cdots\ s_n) \in \mathcal{S}(\mathcal{C}, \mathcal{V})$ for $s_1, \cdots, s_n \in \mathcal{S}(\mathcal{C}, \mathcal{V})$ $(n \geq 0)$. Terms are denoted by $s, t, r, \cdots$. The set of variables in a term $t$ is denoted by $\mathcal{V}(t)$.

A *substitution* $\theta$ is a mapping from $\mathcal{V}$ into $\mathcal{S}(\mathcal{C}, \mathcal{V})$. Substitutions are extended into homomorphisms from $\mathcal{S}(\mathcal{C}, \mathcal{V})$ into $\mathcal{S}(\mathcal{C}, \mathcal{V})$. Following common usage, we write $t\theta$ instead of $\theta(t)$.

Consider an extra constant $\Box$ called a hole. Then $C \in \mathcal{S}(\mathcal{C} \cup \{\Box\}, \mathcal{V})$ is called a context. We use the notation $C[\ ]$ for the context containing precisely one hole, and $C[t]$ denotes the result of placing a term $t$ in the hole of $C[\ ]$. We use the notation $(\cdots t \cdots)$ for $C[t]$ if $C[\ ] = (s_1 \cdots s_{k-1}\ \Box\ s_{k+1} \cdots s_n)$. A term $s$ is called a subterm of $t$ if $t = C[s]$ and an immediate subterm if $t = (\cdots s \cdots)$.

A rewrite rule is a pair $\langle l, r \rangle$ of terms such that $l \notin \mathcal{V}$ and $\mathcal{V}(r) \subseteq \mathcal{V}(l)$. We write $l \to r$ for $\langle l, r \rangle$. A *S-expression rewriting system* (SRS for short) $\mathcal{R}$ is a set of rewrite rules [17]. The rewrite rules of a SRS $\mathcal{R}$ define a reduction relation $\to_{\mathcal{R}}$ on $\mathcal{S}(\mathcal{C}, \mathcal{V})$ as follows: $t \to_{\mathcal{R}} s$ if and only if there exist a rewrite rule $l \to r \in \mathcal{R}$, a context $C[\ ]$ and a substitution $\theta$ such that $t = C[l\theta]$ and $s = C[r\theta]$. A SRS $\mathcal{R}$ is *terminating* if there exists no infinite reduction sequence $t_0 \to_{\mathcal{R}} t_1 \to_{\mathcal{R}} t_2 \to_{\mathcal{R}} \cdots$.

*Example 1 (SRS).* The higher-order function *map* is presented by the following SRS $\mathcal{R}$, where *map*, *cons*, *nil*, *plus*, *s*, 0 are constants and $\alpha$, $x$, $y$ variables.

$$\mathcal{R} \begin{cases} (map\ \alpha\ nil) \to nil \\ (map\ \alpha\ (cons\ x\ y)) \to (cons\ (\alpha\ x)\ (map\ \alpha\ y)) \\ ((plus\ (s\ x))\ y) \to (s\ ((plus\ x)\ y)) \\ ((plus\ 0)\ y) \to y \end{cases}$$

Then we have the reduction sequence:
$(map\ (plus\ (s\ 0))\ (cons\ (s\ 0)\ (cons\ (s\ (s\ 0))\ nil))) \xrightarrow{+}_{\mathcal{R}}$

$(cons\ (s\ (s\ 0))\ (map\ (plus\ (s\ 0))\ (cons\ (s\ (s\ 0))\ nil))) \xrightarrow{+}_{\mathcal{R}}$
$(cons\ (s\ (s\ 0))\ (cons\ (s\ (s\ (s\ 0)))\ nil))$. □

Let $\succ$ be a binary relation on $\mathcal{S}(\mathcal{C}, \mathcal{V})$. For short, we write $\prec$ for its inverse, $\succeq$ for its reflexive closure, $\succ^+$ for its transitive closure, $\succ^*$ for its transitive-reflexive closure, $\succ^n$ for its $n$-fold composition.

A term $t \in \mathcal{S}(\mathcal{C}, \mathcal{V})$ is strongly normalizing with respect to $\succ$ if there is no infinite decreasing sequence $t = t_0 \succ t_1 \succ t_2 \succ \cdots$. $SN(\succ)$ is the set of all terms that are strongly normalizing with respect to $\succ$. $\succ$ is well-founded if $SN(\succ) = \mathcal{S}(\mathcal{C}, \mathcal{V})$. A binary relation $\succ$ is well-founded on a term set $T \subseteq \mathcal{S}(\mathcal{C}, \mathcal{V})$ if there is no infinite decreasing sequence $t_0 \succ t_1 \succ t_2 \succ \cdots$ on $T$.

A binary relation $\succ$ is monotonic if $s \succ t$ implies $(\cdots s \cdots) \succ (\cdots t \cdots)$ (i.e., closed under context), and it is stable if $s \succ t$ implies $s\theta \succ t\theta$ (i.e., closed under substitution).

A binary relation $\succ$ is called a rewrite relation if it is monotonic and stable. A SRS $\mathcal{R}$ is compatible with a binary relation $\succ$ if $l \succ r$ for every rule $l \to r \in \mathcal{R}$. It is easy to show that a SRS $\mathcal{R}$ is terminating if and only if it is compatible with a well-founded rewrite relation.

## 3 Termination Theorem

Goubault-Larrecq [6] gives a general constructive termination proof of abstract recursive path relations. We adapt his general proof to recursive path relations on S-expressions. Our termination proof is non-constructive, different from [6], but simpler.

Let $\succ, \succ^{\Lambda}, \rhd$ be three binary relations on $\mathcal{S}(\mathcal{C}, \mathcal{V})$ and $\rhd$ well-founded. Assume that $\succ$ has the following property [6].

*Property 1.* $s \succ t$ if and only if

(P1) $\exists u.[s \rhd u \wedge u \succeq t]$, or
(P2) $\forall u.[t \rhd u \Rightarrow s \succ u] \ \wedge \ s \succ^{\Lambda} t$.

Let $\overline{SN(\succ)} = \{ s \mid \forall u \lhd s.\ u \in SN(\succ)\}$ and a term set $S \subseteq \mathcal{S}(\mathcal{C}, \mathcal{V})$ be closed with respect to $\rhd$ (i.e., if $s \in S$ and $s \rhd t$ then $t \in S$). Inspired by Ferreira [5] we say that $\succ^{\Lambda}$ is a *term lifting* on $S$ (with respect to $\succ$ and $\rhd$) if $\succ^{\Lambda}$ is well-founded on $\overline{SN(\succ)} \cap S$. (Note that in [5] a term lifting is defined as a specific lifting, i.e., $\succ^{\Lambda}$ is well-founded on first order terms if $\succ$ is well-founded on their arguments. Our definition is essentially equivalent to that in [5] if $\rhd$ means the immediate subterm relation.) Then we have the following termination theorem.

**Theorem 1 (Well-Foundedness).** Let $\succ^{\Lambda}$ be a term lifting on $S$. Then $\succ$ is well-founded on $S$.

*Proof.* We prove the claim by contradiction. The proof is similar to a *minimal bad sequence argument* [14]. Suppose $\succ$ is not well-founded on $S$ and consider a minimal infinite sequence $s_0 \succ s_1 \succ s_2 \succ \cdots$ on $S$ with respect to $\rhd$ in the following sense:

(i) if $s_0 \rhd t$ then $t$ is strongly normalizing with respect to $\succ$;
(ii) if $s_{i+1} \rhd t$ and $s_i \succ t$ then $t$ is strongly normalizing with respect to $\succ$.

We show that every $s_k \succ s_{k+1}$ satisfies (P2). Suppose $s_0 \succ s_1$ satisfies (P1). Then $u \succeq s_1$ for some $u \lhd s_0$; contradicting minimality of the infinite sequence on $S$. Thus $s_0 \succ s_1$ must satisfy (P2). Suppose $s_0 \succ s_1 \succ s_2 \succ \cdots \succ s_k$ $(k > 0)$ satisfy (P2) and $s_k \succ s_{k+1}$ satisfies (P1). Then $u \succeq s_{k+1}$ for some $u \lhd s_k$. Since $s_{k-1} \succ s_k$ satisfies (P2), we have $s_{k-1} \succ u \succeq s_{k+1}$; again contradicting minimality of $s_k$.

We next show every $s_k$ in $\overline{SN(\succ)} \cap S$. Suppose $s_m \notin \overline{SN(\succ)} \cap S$. Then we have some $u \lhd s_m$ such that $u$ is not in $SN(\succ)$. Since $s_{m-1} \succ s_m$ satisfies (P2), $s_{m-1} \succ u$. Thus we have an infinite sequence $s_0 \succ s_1 \succ s_2 \succ \cdots \succ s_{m-1} \succ u \cdots$; contradicting minimality of $s_m$.

Since every $s_k \succ s_{k+1}$ satisfies (P2), we have an infinite sequence $s_0 \succ^{\Lambda} s_1 \succ^{\Lambda} s_2 \succ^{\Lambda} \cdots$. It contradicts the assumption that $\succ^{\Lambda}$ is well-founded on $\overline{SN(\succ)} \cap S$. □

**Lemma 1 (Monotonicity).** Let $\rhd$ be the immediate subterm relation defined as $(\cdots u \cdots) \rhd u$. Assume that if $s \succ t$ then $(\cdots s \cdots) \succ^{\Lambda} (\cdots t \cdots)$. Then the relation $\succ$ is monotonic, i.e., $s \succ t$ implies $(\cdots s \cdots) \succ (\cdots t \cdots)$.

*Proof.* Let $s \succ t$. From the assumption we have $(\cdots s \cdots) \succ^{\Lambda} (\cdots t \cdots)$. Thus, in order to obtain $(\cdots s \cdots) \succ (\cdots t \cdots)$ by (P2), we show that for every $u$, $(\cdots t \cdots) \rhd u$ implies $(\cdots s \cdots) \succ u$. Let $(\cdots t \cdots) \rhd u$. If $u \neq t$ then $(\cdots s \cdots) \rhd u$. Since $u \succeq u$, we have $(\cdots s \cdots) \succ u$ by (P1). If $u = t$ then $s \succ u$. Thus we have $(\cdots s \cdots) \rhd s \succ u$. By (P1), it holds that $(\cdots s \cdots) \succ u$. □

**Lemma 2 (Subterm).** $s \rhd^+ t$ implies $s \succ t$.

*Proof.* By induction on $n \geq 1$ we prove the claim that $s \rhd^n t$ implies $s \succ t$. Base step: Let $s \rhd t$. Then, by (P1) and $t \succeq t$, we have $s \succ t$. Induction step: Let $s \rhd u \rhd^n t$. From induction hypothesis it follows that $u \succ t$. By (P1), we have $s \succ t$. (Note that the claim holds without the assumption that $\rhd$ is the immediate subterm relation.) □

We say that a term pair $\langle s, t \rangle \in \mathcal{S}(\mathcal{C}, \mathcal{V})$ is stable with respect to $\succ$ if $s \succ t$ implies $s\theta \succ t\theta$ for every substitution $\theta$. A term pair set $P \subseteq \mathcal{S}(\mathcal{C}, \mathcal{V})^2$ is stable with respect to $\succ$ if every $\langle s, t \rangle \in P$ is stable with respect to $\succ$. $ST(\succ)$ denotes the set of all term pairs that are stable with respect to $\succ$. Thus, $\succ$ is stable (i.e., closed under substitution) if and only if $ST(\succ) = \mathcal{S}(\mathcal{C}, \mathcal{V})^2$. We define $\langle s, t \rangle \rhd_2 \langle s', t' \rangle$ as $s \trianglerighteq s'$, $t \trianglerighteq t'$ and $\langle s, t \rangle \neq \langle s', t' \rangle$; $\lhd_2$ as its inverse. It is clear that $\rhd_2$ is well-founded. Let $\overline{ST(\succ)} = \{ \langle s, t \rangle \mid \forall \langle u, v \rangle \lhd_2 \langle s, t \rangle .\ \langle u, v \rangle \in ST(\succ) \}$.

**Lemma 3 (Stability).** Let $\rhd$ be the immediate subterm relation defined as $(\cdots u \cdots) \rhd u$. Let $P \subseteq \mathcal{S}(\mathcal{C}, \mathcal{V})^2$ be closed with respect to $\rhd_2$. Assume that $\overline{ST(\succ)} \cap P$ is stable with respect to $\succ^{\Lambda}$ (i.e., every $\langle s, t \rangle \in \overline{ST(\succ)} \cap P$ is stable with respect to $\succ^{\Lambda}$) and that $s \succ x$ implies $s \rhd^+ x$ for every variable $x$. Then $P$ is stable with respect to $\succ$.

*Proof.* We prove the claim by contradiction. Suppose $P$ is not stable with respect to $\succ$ and consider a minimal $\langle s,t\rangle \in P$ with respect to $\rhd_2$ such that $\langle s,t\rangle \notin ST(\succ)$. Then there exist some $\theta$ such that $s \succ t$ and $s\theta \not\succ t\theta$. If $t = x \in \mathcal{V}$ then by the assumption we have $s \rhd^+ x$. Since $\rhd$ is the immediate subterm relation, $s\theta \rhd^+ x\theta$. From Lemma 2 it follows that $s\theta \succ x\theta$; contradiction. Thus we suppose $t \notin \mathcal{V}$. If $s \succ t$ satisfies (P1) then $s \rhd u$ and $u \succeq t$ for some $u$. Since $\rhd$ is the immediate subterm relation, we have $s\theta \rhd u\theta$. From minimality of $\langle s,t\rangle$ and $\langle s,t\rangle \rhd_2 \langle u,t\rangle$, $\langle u,t\rangle$ is stable with respect to $\succ$. Thus $u\theta \succeq t\theta$. By (P1) we have $s\theta \succ t\theta$; contradiction. If $s \succ t$ satisfies (P2) then $t \rhd u$ implies $s \succ u$ for every $u$ and $s \succ^{\Lambda} t$. Let $t\theta \rhd v$. Since $t$ is not a variable, there exists some $t'$ such that $t \rhd t'$ and $v = t'\theta$. From (P2), $t \rhd t'$ implies $s \succ t'$. From minimality of $\langle s,t\rangle$ and $\langle s,t\rangle \rhd_2 \langle s,t'\rangle$, $\langle s,t'\rangle$ is stable with respect to $\succ$; we have $s\theta \succ t'\theta = v$. Thus it follows that $t\theta \rhd v$ implies $s\theta \succ v$ for every $v$. From minimality of $\langle s,t\rangle$, $\langle s,t\rangle$ is in $\overline{ST(\succ)} \cap P$. Thus, from the assumption, $s\theta \succ^{\Lambda} t\theta$. By (P2), it follows that $s\theta \succ t\theta$; contradiction. □

**Remark.** Goubault-Larrecq [6] proves the stability of $\succ$ under the weaker condition that $\succ^{\Lambda}$ is stable whenever $\succ$ is stable, but $\succ$ is defined as the greatest fixed point of the monotonic operator induced from (P1) and (P2) (See Remark 4 in [6]). Thus his proof does not work for $\succ$ recursively defined as the least fixed point in common use, differently from ours.

## 4 Lexicographic Path Relation

In this section we discuss lexicographic path relation on S-expressions and prove termination of finite S-expression rewriting systems compatible with lexicographic path relations. Thanks to the abstract results presented in the previous section, our proof is much simpler than that in [17]. For a binary relation $\succ$ on $\mathcal{S}(\mathcal{C},\mathcal{V})$, $\succ^{lex}$ denotes the lexicographic extension of $\succ$. The lexicographic path relation on $\mathcal{S}(\mathcal{C},\mathcal{V})$ is defined as follows [17].

**Definition 1 (Lexicographic Path Relation).** Let $>$ be a well-founded relation on $\mathcal{C}$. The lexicographic path relation $\succ_{lp}$ on $\mathcal{S}(\mathcal{C},\mathcal{V})$ is recursively defined as follows: $s \succ_{lp} t$ if and only if

(L0) $s, t \in \mathcal{C}$ and $s > t$ , or
(L1) $s = (s_1 \cdots s_m)$ $(m \geq 1)$ and $\exists s_i.\ s_i \succeq_{lp} t$, or
(L2) $t = (t_1 \cdots t_n)$ $(n \geq 0)$ and $\forall t_i.\ s \succ_{lp} t_i$, and either
(L2-1) $s \in \mathcal{C}$, or
(L2-2) $s = (s_1 \cdots s_m)$ $(m \geq 1)$ and $\langle s_1, \cdots, s_m\rangle \succ_{lp}^{lex} \langle t_1, \cdots, t_n\rangle$.

Note that the lexicographic path relation $\succ_{lp}$ is not well-founded due to *unbounded-variadic* feature of S-expressions; for example, we can have the infinite descending sequence $(1) \succ_{lp} (0\ 1) \succ_{lp} (0\ 0\ 1) \succ_{lp} (0\ 0\ 0\ 1) \succ_{lp} \cdots$ if $0 < 1$. The problem arises from the fact that lexicographic sequences of unbounded size are not well-founded. To overcome this difficulty we introduce the notion of the maximum degree of terms.

The maximum degree $d(t)$ of a term $t \in \mathcal{S}(\mathcal{C},\mathcal{V})$ is defined by: (i) $d(t)=0$ for $t \in \mathcal{C} \cup \mathcal{V}$, and (ii) $d(t) = max\{n, d(t_1), \cdots, d(t_n)\}$ for $t = (t_1 \cdots t_n)$ $(n \geq 0)$. The set of terms having the maximum degree not more than $p$ is defined by $\mathcal{S}_p(\mathcal{C},\mathcal{V}) = \{t \mid d(t) \leq p\}$.

**Theorem 2.** The lexicographic path relation $\succ_{lp}$ is well-founded on $\mathcal{S}_p(\mathcal{C},\mathcal{V})$ for every $p \geq 0$, and it is monotonic and stable.

*Proof.* Let $\rhd$ be the immediate subterm relation. From the definition of $\succ_{lp}$ it is obvious that $s \succ_{lp} x$ implies $s \rhd^+ x$ for every variable $x$. We define $s \succ_{lp}^{\Lambda} t$ as (L0) $s, t, \in \mathcal{C}$ and $s > t$, or (L2-1) $s \in \mathcal{C}$ and $t = (t_1 \cdots t_n)$ $(n \geq 0)$, or (L2-2) $s = (s_1 \cdots s_m)$ $(m \geq 1)$, $t = (t_1 \cdots t_n)$ $(n \geq 0)$ and $\langle s_1, \cdots, s_m\rangle \succ_{lp}^{lex} \langle t_1, \cdots, t_n\rangle$. Then $\succ_{lp}$ obviously satisfies Property 1. Monotonicity and stability of $\succ_{lp}$ directly follow from Lemma 1 and Lemma 3 by taking $P = \mathcal{S}(\mathcal{C},\mathcal{V})^2$ respectively. Thus, it remains to prove well-foundedness of $\succ_{lp}$ on $\mathcal{S}_p(\mathcal{C},\mathcal{V})$. From Theorem 1, it is enough to show that $\succ_{lp}^{\Lambda}$ is actually a term lifting on $\mathcal{S}_p(\mathcal{C},\mathcal{V})$ with respect to $\succ_{lp}$ and $\rhd$, i.e., it is well-founded on $\overline{SN(\succ_{lp})} \cap \mathcal{S}_p(\mathcal{C},\mathcal{V})$. Suppose $\succ_{lp}^{\Lambda}$ is not well-founded on $\overline{SN(\succ_{lp})} \cap \mathcal{S}_p(\mathcal{C},\mathcal{V})$ and consider an infinite sequence $s_0 \succ_{lp}^{\Lambda} s_1 \succ_{lp}^{\Lambda} s_2 \succ_{lp}^{\Lambda} \cdots$ on $\overline{SN(\succ_{lp})} \cap \mathcal{S}_p(\mathcal{C},\mathcal{V})$. Since $>$ is well-founded on $\mathcal{C}$, there exists some $k$ such that for every $i \geq k$, $s_i \succ_{lp}^{\Lambda} s_{i+1}$ satisfies (L2-2). Thus, we have an infinite sequence $\langle s_{k,1}, \cdots, s_{k,n}\rangle \succ_{lp}^{lex} \langle s_{k+1,1}, \cdots, s_{k+1,n'}\rangle \succ_{lp}^{lex} \langle s_{k+2,1}, \cdots, s_{k+2,n''}\rangle \succ_{lp}^{lex} \cdots$ where all $s_{i,j}$ are in $SN(\succ_{lp})$ and $n, n', n'', \cdots \leq p$. However, since $\succ_{lp}$ on the set of all $s_{i,j}$ is well-founded, its lexicographic extension over the sequences of length bounded by $p$ is well-founded. Thus, we get a contradiction. □

A SRS $\mathcal{R}$ is *bounded* [17] if for any infinite reduction sequence $t_0 \rightarrow_{\mathcal{R}} t_1 \rightarrow_{\mathcal{R}} t_2 \rightarrow_{\mathcal{R}} t_3 \rightarrow_{\mathcal{R}} \cdots$ there exists some natural number $p$ such that $d(t_i) \leq p$.

**Lemma 4.** Every finite SRS $\mathcal{R}$ is bounded.

*Proof.* Let $t_0 \rightarrow_{\mathcal{R}} t_1 \rightarrow_{\mathcal{R}} t_2 \rightarrow_{\mathcal{R}} t_3 \rightarrow_{\mathcal{R}} \cdots$ be an infinite reduction sequence and let $c = max\{d(r) \mid l \rightarrow r \in \mathcal{R}\}$. Consider a reduction $s \rightarrow_{\mathcal{R}} t$ where $s = C[l\theta]$, $t = C[r\theta]$ and $l \rightarrow r \in \mathcal{R}$. Then we have $d(t) = d(C[r\theta]) =$
$max\{d(C[\,]), d(r), d(x\theta) \mid x \in \mathcal{V}(r)\} \leq$
$max\{d(C[\,]), d(l), d(r), d(x\theta) \mid x \in \mathcal{V}(r)\} \leq$
$max\{d(C[l\theta]), d(r)\} = max\{d(s), d(r)\} \leq$
$max\{d(s), c\}$. Thus it holds that $d(t_{i+1}) \leq max\{d(t_i), c\}$ for every $i \geq 0$. By taking $p = max\{d(t_0), c\}$ we conclude the claim. □

**Theorem 3.** Let $\mathcal{R}$ be a finite SRS compatible with a lexicographic path relation $\succ_{lp}$. Then $\mathcal{R}$ is terminating.

*Proof.* The claim follows from Theorem 2 and Lemma 4. □

By replacing Conditions (L2), (L2-1) and (L2-2) in Definition 1 of $\succ_{lp}$ with Conditions (L2-a), (L2-b) and (L2-c), we obtain the following practical definition in [17], which is a bit easier to apply. It is easy to show that the two definitions of $\succ_{lp}$ are equivalent.

**Definition 2 (Lexicographic Path Relation).** Let $>$ be a well-founded relation on $\mathcal{C}$. The lexicographic path relation $\succ_{lp}$ on $\mathcal{S}(\mathcal{C}, \mathcal{V})$ is recursively defined as follows: $s \succ_{lp} t$ if and only if

(L0) $s, t, \in \mathcal{C}$ and $s > t$ , or
(L1) $s = (s_1 \cdots s_m)$ $(m \geq 1)$ and $\exists s_i.\ s_i \succeq_{lp} t$, or
(L2-a) $s \in \mathcal{C}$, $t = (t_1 \cdots t_n)$ $(n \geq 0)$ and $\forall t_i.\ s \succ_{lp} t_i$, or
(L2-b) $s = (s_1 \cdots s_m)$, $t = (t_1 \cdots t_n)$ $(m > n \geq 0)$, and $s_1 = t_1, \cdots, s_n = t_n$, or
(L2-c) $s = (s_1 \cdots s_m)$, $t = (t_1 \cdots t_n)$ $(m, n \geq 1)$ and $\exists i.\ s_1 = t_1, \cdots, s_{i-1} = t_{i-1}, s_i \succ_{lp} t_i, s \succ_{lp} t_{i+1}, \cdots, s \succ_{lp} t_n$.

Now we verify termination of S-expression rewriting systems by Theorem 3. Various higher-order rewriting systems in the literature [7,10,11,15] are naturally expressed in SRSs and termination of them is easily proven as that of SRSs without using the notion of *type*. In the following examples, we denote variables acting as higher-order variables by $\alpha, \beta, \gamma, \cdots$ in distinction from $x, y, z, \cdots$ for readability.

*Example 2 (map).* Let $\mathcal{C} = \{map, cons, nil\}$ with $map > cons > nil$ and consider the following SRS $\mathcal{R}$.

$$\mathcal{R}\begin{cases} ((map\ \alpha)\ nil) \to nil \\ ((map\ \alpha)\ (cons\ x\ y)) \to (cons\ (\alpha\ x)\ ((map\ \alpha)\ y)) \end{cases}$$

For the first rule $((map\ \alpha)\ nil) \succ_{lp} nil$ is trivial by the subterm property. For the second rule we have (i) $(map\ \alpha) \succ_{lp} cons$ by (L1) because $map \succ_{lp} cons$ by (L0), (ii) $((map\ \alpha)\ \ (cons\ x\ y)) \succ_{lp} (\alpha\ x)$ by (L2-c) because $(map\ \alpha) \succ_{lp} \alpha$ and $((map\ \alpha)\ (cons\ x\ y)) \succ_{lp} x$ by the subterm property, and (iii) $((map\ \alpha)\ (cons\ x\ y)) \succ_{lp} ((map\ \alpha)\ y)$ by (L2-c) because $(cons\ x\ y) \succ_{lp} y$ by the subterm property. From (i), (ii) and (iii) we have $((map\ \alpha)\ (cons\ x\ y)) \succ_{lp} (cons\ (\alpha\ x)\ ((map\ \alpha)\ y))$ by (L2-c). Thus $\mathcal{R}$ is terminating since it is compatible with $\succ_{lp}$. □

In the above example, *map* is represented as a currying notation $((map\ s)\ t)$ by the extra parentheses as in [10,17], instead of the usual flat notation $(map\ s\ t)$. This currying is necessary for guaranteeing the compatibility with $\succ_{lp}$ because of $(map\ \alpha\ (cons\ x\ y)) \not\succ_{lp} (cons\ (\alpha\ x)\ (map\ \alpha\ y))$ for the flat notation of the second rewrite rule. For a currying-transformation technique to strengthen the power of the lexicographic path relation, see [17].

*Example 3 (maplist).* Let $\mathcal{C} = \{fmap, cons, nil\}$ with $fmap > cons > nil$ and consider the following SRS $\mathcal{R}$.

$$\mathcal{R}\begin{cases} ((fmap\ nil)\ x) \to nil \\ ((fmap\ (cons\ \alpha\ \beta))\ \ x) \to (cons\ (\alpha\ x)\ ((fmap\ \beta)\ x)) \end{cases}$$

Then $\mathcal{R}$ is terminating since it is compatible with $\succ_{lp}$. □

*Example 4 (twice).* Let $\mathcal{C} = \{twice, dapply\}$ with $twice > dapply$ and consider the following SRS $\mathcal{R}$.

$$\mathcal{R}\begin{cases}((dapply\ \alpha\ \beta)\ x) \to (\alpha\ (\beta\ x))\\ (twice\ \alpha) \to (dapply\ \alpha\ \alpha)\end{cases}$$

Then $\mathcal{R}$ is terminating since it is compatible with $\succ_{lp}$. □

*Example 5 (applylist).* Let $\mathcal{C} = \{lapply, cons, nil\}$ with $lapply > cons > nil$ and consider the following SRS $\mathcal{R}$.

$$\mathcal{R}\begin{cases}((lapply\ nil)\ x) \to x\\ ((lapply\ (cons\ \alpha\ \beta))\ x) \to (\alpha\ ((lapply\ \beta)\ x))\end{cases}$$

Then $\mathcal{R}$ is terminating since it is compatible with $\succ_{lp}$. □

*Example 6 (recursor).* Let $\mathcal{C} = \{rec, s, 0\}$ with $rec > s > 0$ and consider the following SRS $\mathcal{R}$.

$$\mathcal{R}\begin{cases}((rec\ \alpha\ x)\ 0) \to x\\ ((rec\ \alpha\ x)\ (s\ y)) \to (\alpha\ (s\ y)\ ((rec\ \alpha\ x)\ y))\end{cases}$$

Then $\mathcal{R}$ is terminating since it is compatible with $\succ_{lp}$. □

## 5 Recursive Path Relation with Status

In this section we discuss the recursive path relation with status on S-expressions, which is a higher-order extension of the well-known one on the first-order terms [1,5,8,13,16]. For a binary relation $\succ$ on $\mathcal{S}(\mathcal{C},\mathcal{V})$, $\succ^{mul}$ denotes the multiset extension of $\succ$. We assume a status (function) $\tau$ as a mapping from $\mathcal{S}(\mathcal{C},\mathcal{V})$ to $\{lex, mul\}$. Then, the recursive path relation with status on $\mathcal{S}(\mathcal{C},\mathcal{V})$ is defined as follows.

**Definition 3 (Recursive Path Relation with Status).** Let $>$ be a well-founded relation on $\mathcal{C}$. The recursive path relation $\succ_{rp}$ with status on $\mathcal{S}(\mathcal{C},\mathcal{V})$ is recursively defined as follows: $s \succ_{rp} t$ if and only if

(R0) $s, t, \in \mathcal{C}$ and $s > t$ , or
(R1) $s = (s_1 \cdots s_m)$ $(m \geq 1)$ and $\exists s_i.\ s_i \succeq_{rp} t$, or
(R2) $t = (t_1 \cdots t_n)$ $(n \geq 0)$ and $\forall t_i.\ s \succ_{rp} t_i$, and either
(R2-1) $s \in \mathcal{C}$, or
(R2-2) $s = (s_1 \cdots s_m)$ $(m \geq 1)$ and $t = ()$, or
(R2-3) $s = (s_1 \cdots s_m)$ and $t = (t_1 \cdots t_n)$ $(m, n \geq 1)$, and
(a) $s_1 \succ_{rp} t_1$, or
(b) $s_1 = t_1$ and $\langle s_2, \cdots, s_m\rangle \succ_{rp}^{\tau(s_1)} \langle t_2, \cdots, t_n\rangle$.

**Theorem 4.** The recursive path relation $\succ_{rp}$ with status is well-founded on $\mathcal{S}_p(\mathcal{C},\mathcal{V})$ for every $p \geq 0$, and it is monotonic.

*Proof.* Let $\rhd$ be the immediate subterm relation and $s \succ_{rp}^{\Lambda} t$ defined as (R0) $s, t, \in \mathcal{C}$ and $s > t$, or (R2-1) $s \in \mathcal{C}$ and $t = (t_1 \cdots t_n)$ $(n \geq 0)$, or (R2-2) $s = (s_1 \cdots s_m)$ $(m \geq 1)$ and $t = ()$, or (R2-3a) $s = (s_1 \cdots s_m)$, $t = (t_1 \cdots t_n)$ $(m, n \geq 1)$ and $s_1 \succ_{rp} t_1$, or (R2-3b) $s = (s_1 \cdots s_m)$, $t = (t_1 \cdots t_n)$ $(m, n \geq 1)$, $s_1 =$

$t_1$ and $\langle s_2, \cdots, s_m\rangle \succ_{rp}^{\tau(s_1)} \langle t_2, \cdots, t_n\rangle$. Then $\succ_{rp}$ obviously satisfies Property 1. Monotonicity of $\succ_{rp}$ directly follows from Lemma 1. Thus, it remains to prove well-foundedness of $\succ_{rp}$ on $\mathcal{S}_p(\mathcal{C},\mathcal{V})$. From Theorem 1, it is enough to show that $\succ_{rp}^{\Lambda}$ is well-founded on $\overline{SN(\succ_{rp})} \cap \mathcal{S}_p(\mathcal{C},\mathcal{V})$. Suppose $\succ_{rp}^{\Lambda}$ is not well-founded on $\overline{SN(\succ_{rp})} \cap \mathcal{S}_p(\mathcal{C},\mathcal{V})$ and consider an infinite sequence $s_0 \succ_{rp}^{\Lambda} s_1 \succ_{rp}^{\Lambda} s_2 \succ_{rp}^{\Lambda} \cdots$ on $\overline{SN(\succ_{rp})} \cap \mathcal{S}_p(\mathcal{C},\mathcal{V})$. Since $>$ is well-founded on $\mathcal{C}$, there exists some $k$ such that for every $i \geq k$, $s_i \succ_{rp}^{\Lambda} s_{i+1}$ satisfies (R2-3). Thus, we have an infinite sequence $(s_{k,1} \cdots s_{k,n}) \succ_{rp} (s_{k+1,1} \cdots s_{k+1,n'}) \succ_{rp} (s_{k+2,1} \cdots s_{k+2,n''}) \succ_{rp} \cdots$, where all $s_{i,j}$ are in $SN(\succ_{rp})$, $n, n', n'', \cdots \leq p$ and $s_{k,1} \succeq_{rp} s_{k+1,1} \succeq_{rp} s_{k+2,1} \succeq_{rp} \cdots$. Thus there exists some $q \geq k$ such that $s_{q,1} = s_{q+1,1} = s_{q+2,1} = \cdots$. Then, we have an infinite sequence $\langle s_{q,1}, \cdots, s_{q,m}\rangle \succ_{rp}^{\tau(s_{q,1})} \langle s_{q+1,1}, \cdots, s_{q+1,m'}\rangle \succ_{rp}^{\tau(s_{q,1})} \langle s_{q+2,1}, \cdots, s_{q+2,m''}\rangle \succ_{rp}^{\tau(s_{q,1})} \cdots$. However, since $\succ_{rp}$ on the set of all $s_{i,j}$ is well-founded, its lexicographic (multiset) extension over the sequences of length bounded by $p$ is well-founded. Thus, we get a contradiction. □

We note that the recursive path relation $\succ_{rp}$ with status on $\mathcal{S}(\mathcal{C},\mathcal{V})$ is not stable due to *higher-order* feature of S-expressions, different from the one on the first-order terms. For example, assume that $\tau(x) = lex$, $\tau(0) = mul$, and let $s = (x\ 1\ 0\ 0)$, $t = (x\ 0\ 1\ 1)$. Then $s \succ_{rp} t$ but $t\theta \succ_{rp} s\theta$ when $\theta = [x := 0]$. This problem arises whenever status $\tau$ is not trivial, i.e., $\tau(x) \neq \tau(u)$ for some term $u$. However, stability of all term pairs is not necessary for proving termination of a SRS $\mathcal{R}$; indeed, it is enough to guarantee that $\mathcal{R}$ is stable with respect to $\succ_{rp}$. Hence, we present criteria for stability of $\mathcal{R}$.

For status $\tau$, we define $(s_1 \cdots s_m) \succ_{rp}^{\tau} (t_1 \cdots t_n)$ $(m, n \geq 1)$ by (R2-3b) $s_1 = t_1$ and $\langle s_2, \cdots, s_m\rangle \succ_{rp}^{\tau(s_1)} \langle t_2, \cdots, t_n\rangle$.

**Lemma 5 (Stability of $\succ_{rp}$).** Let $\rhd$ be the immediate subterm relation and $P \subseteq \mathcal{S}(\mathcal{C},\mathcal{V})^2$ closed with respect to $\rhd_2$. Let $\overline{ST(\succ_{rp})} \cap P$ be stable with respect to $\succ_{rp}^{\tau}$. Then $P$ is stable with respect to $\succ_{rp}$.

*Proof.* From the definition of $\succ_{lp}$ it is obvious that $s \succ_{rp} x$ implies $s \rhd^+ x$ for every variable $x$. Similarly to the proof of Theorem 4, we define $\succ_{rp}^{\Lambda}$ as (R0), (R2-1), (R2-2), (R2-3a), or (R2-3b). Then the claim directly follows from Lemma 3. □

For a term pair set $Q \subseteq \mathcal{S}(\mathcal{C},\mathcal{V})^2$, we define the closure of $Q$ with respect to $\rhd_2$ as $Q^* = \{ \langle s,t\rangle \mid \langle u,v\rangle \rhd_2^* \langle s,t\rangle \text{ for some } \langle u,v\rangle \in Q \}$.

**Theorem 5.** Let $\mathcal{R}$ be a finite SRS compatible with a recursive path relation with status $\tau$. Let $\overline{ST(\succ_{rp})} \cap \mathcal{R}^*$ be stable with respect to $\succ_{rp}^{\tau}$. Then $\mathcal{R}$ is terminating.

*Proof.* The claim follows from Theorem 4 and Lemma 5. □

We say status $\tau$ is stable on a term set $S \subseteq \mathcal{S}(\mathcal{C},\mathcal{V})$ if for every $s \in S$, $\tau(s) = \tau(s\theta)$ for any $\theta$. The head term of $s = (s_1 \cdots s_m)$ $(m \geq 1)$ is denoted by $hd(s) = s_1$. For $Q \subseteq \mathcal{S}(\mathcal{C},\mathcal{V})^2$ we define $H_{eq}(Q) = \{ hd(s) \mid \langle s,t\rangle \in Q \text{ and } hd(s) = hd(t) \}$.

**Corollary 1.** Let $\mathcal{R}$ be a finite SRS compatible with a recursive path relation with status $\tau$. Let status $\tau$ be stable on $H_{eq}(\mathcal{R}^*)$. Then $\mathcal{R}$ is terminating.

*Proof.* The claim follows from Theorem 5. □

*Example 7 (folding: adapted from [11]).* Let $\mathcal{C} = \{sum, prod, fold, *, +, s, 0, cons, nil\}$ with $sum, prod > fold > * > + > s > 0 > cons > nil$ and consider the following SRS $\mathcal{R}$.

$$\mathcal{R} \begin{cases} ((fold\ \alpha\ x)\ nil) \to x \\ ((fold\ \alpha\ x)\ (cons\ y\ z)) \to (\alpha\ y\ ((fold\ \alpha\ x)\ z)) \\ sum \to (fold\ +\ 0) \\ prod \to (fold\ *\ (s\ 0)) \\ (+\ 0\ y) \to y \\ (+\ (s\ x)\ y) \to (s\ (+\ y\ x)) \\ (*\ 0\ y) \to 0 \\ (*\ (s\ x)\ y) \to (+\ y\ (*\ y\ x)) \end{cases}$$

Note that $\mathcal{R}$ is not compatible with $\succ_{lp}$ because $(+\ (s\ x)\ y) \not\succ_{lp} (s\ (+\ y\ x))$ and $(*\ (s\ x)\ y) \not\succ_{lp} (+\ y\ (*\ y\ x))$. We take status $\tau$ as $\tau(t) = mul$ if $t = +, *$ and $\tau(t) = lex$ otherwise. Then it is easily shown that $\mathcal{R}$ is compatible with $\succ_{rp}$. As $H_{eq}(R^*) = \{(fold\ \alpha\ x), fold, *, +, s\}$, $\tau$ is stable on $H_{eq}(R^*)$. From Corollary 1, it follows that $\mathcal{R}$ is terminating. □

*Example 8 (sorting: adapted from [7]).* Let $\mathcal{C} = \{asort, dsort, sort, ins, max, min, s, 0, cons, nil\}$ with $asort, dsort > sort > ins > max, min > s > 0 > cons > nil$ and consider the following SRS $\mathcal{R}$.

$$\mathcal{R} \begin{cases} ((sort\ \alpha\ \beta)\ nil) \to nil \\ ((sort\ \alpha\ \beta)\ (cons\ x\ y)) \to ((ins\ \alpha\ \beta)\ ((sort\ \alpha\ \beta)\ y)\ x) \\ ((ins\ \alpha\ \beta)\ nil\ y) \to (cons\ y\ nil) \\ ((ins\ \alpha\ \beta)\ (cons\ x\ z)\ y) \to (cons\ (\alpha\ x\ y)\ ((ins\ \alpha\ \beta)\ z\ (\beta\ x\ y))) \\ (max\ x\ x) \to x \\ (max\ 0\ y) \to y \\ (max\ x\ 0) \to x \\ (max\ (s\ x)\ (s\ y)) \to (s\ (max\ y\ x)) \\ (min\ x\ x) \to x \\ (min\ 0\ y) \to 0 \\ (min\ x\ 0) \to 0 \\ (min\ (s\ x)\ (s\ y)) \to (s\ (min\ y\ x)) \\ (asort\ z) \to ((sort\ min\ max)\ z) \\ (dsort\ z) \to ((sort\ max\ min)\ z) \end{cases}$$

Since $(max\ (s\ x)\ (s\ y)) \not\succ_{lp} (s\ (max\ y\ x))$ and $(min\ (s\ x)\ (s\ y)) \not\succ_{lp} (s\ (min\ y\ x))$, $\mathcal{R}$ is not compatible with $\succ_{lp}$. We take status $\tau$ as $\tau(t) = mul$ if $t = max, min$ and $\tau(t) = lex$ otherwise. Then, $\mathcal{R}$ is compatible with $\succ_{rp}$. As $H_{eq}(R^*) = \{(sort\ \alpha\ \beta), sort, (ins\ \alpha\ \beta), ins, cons, max, min, s\}$, $\tau$ is stable on $H_{eq}(R^*)$. Thus, from Corollary 1, it follows that $\mathcal{R}$ is terminating. □

**Acknowledgment.** I would like to thank Jeroen Ketema for his suggestions on improving an early version of this article and the anonymous referees for their helpful comments. This work was partially supported by grants from Japan Society for the Promotion of Science, No. 19500003.

## References

1. Baader, F., Nipkow, T.: Term Rewriting and All That. Cambridge University Press, Cambridge (1998)
2. Buchholz, W.: Proof-theoretic analysis of termination proofs. Ann. Pure Appl. Logic 75, 57–65 (1995)
3. Dawson, J.E., Goré, R.: A general theorem on termination of rewriting. In: Marcinkowski, J., Tarlecki, A. (eds.) CSL 2004. LNCS, vol. 3210, pp. 100–114. Springer, Heidelberg (2004)
4. Dershowitz, N.: Ordering for term-rewriting systems. Theoretical Comput. Sci. 17, 279–301 (1982)
5. Ferreira, M.C.F.: Termination of term rewriting: Well-foundedness, totality and transformation, PhD thesis, Dep. of Comput. Sci., Utrecht University (1995)
6. Goubault-Larrecq, J.: Well-founded recursive relations. In: Fribourg, L. (ed.) CSL 2001 and EACSL 2001. LNCS, vol. 2142, pp. 484–497. Springer, Heidelberg (2001)
7. Jouannaud, J.-P., Rubio, A.: Polymorphic higher-order recursive path orderings. J. ACM 54, Article 2, 1–48 (2007)
8. Kamin, S., Levy, J.-J.: Two generalizations of the recursive path ordering, University of Illinois (unpublished manuscript) (1980)
9. Kruskal, B.: Well quasi ordering, the tree theorem and Vazsonyi's conjecture. Trans. Am. Math. Soc. 95, 210–225 (1960)
10. Linfantsev, M., Bachmair, L.: An LPO-based termination ordering for higher-order terms without $\lambda$-abstraction. In: Grundy, J., Newey, M. (eds.) TPHOLs 1998. LNCS, vol. 1479, pp. 277–293. Springer, Heidelberg (1998)
11. Lysne, O., Piris, J.: A termination ordering for higher order rewrite systems. In: Hsiang, J. (ed.) RTA 1995. LNCS, vol. 914, pp. 26–40. Springer, Heidelberg (1995)
12. Mellies, P.-A.: On a duality between Kruskal and Dershowitz theorems. In: Larsen, K.G., Skyum, S., Winskel, G. (eds.) ICALP 1998. LNCS, vol. 1443, pp. 518–529. Springer, Heidelberg (1998)
13. Middeldorp, A., Zantema, H.: Simple termination of rewrite systems. Theoretical Comput. Sci. 175, 127–158 (1997)
14. Nash-Williams, C.St.J.A.: On well-quasi-ordering finite trees. Proc. of the Cambridge Phil. Soc. 59, 833–835 (1963)
15. van Raamsdonk, F.: On termination of higher-order rewriting. In: Middeldorp, A. (ed.) RTA 2001. LNCS, vol. 2051, pp. 261–275. Springer, Heidelberg (2001)
16. Terese: Term Rewriting Systems. Cambridge University Press, Cambridge (2003)
17. Toyama, Y.: Termination of S-expression rewriting systems: Lexicographic path ordering for higher-order terms. In: van Oostrom, V. (ed.) RTA 2004. LNCS, vol. 3091, pp. 40–54. Springer, Heidelberg (2004)

# Encoding the Pure Lambda Calculus into Hierarchical Graph Rewriting

Kazunori Ueda

Dept. of Computer Science and Engineering, Waseda University
3-4-1, Okubo, Shinjuku-ku, Tokyo 169-8555, Japan
ueda@ueda.info.waseda.ac.jp

**Abstract.** Fine-grained reformulation of the lambda calculus is expected to solve several difficulties with the notion of substitutions—definition, implementation and cost properties. However, previous attempts including those using explicit substitutions and those using Interaction Nets were not ideally simple when it came to the encoding of the pure (as opposed to weak) lambda calculus. This paper presents a novel, fine-grained, and highly asynchronous encoding of the pure lambda calculus using LMNtal, a hierarchical graph rewriting language, and discusses its properties. The major strength of the encoding is that it is significantly simpler than previous encodings, making it promising as an alternative formulation, rather than just the encoding, of the pure lambda calculus. The membrane construct of LMNtal plays an essential role in encoding colored tokens and operations on them. The encoding has been tested using the publicly available LMNtal implementation.

## 1 Introduction

The $\lambda$-calculus and $\lambda$-terms play fundamental roles not only in functional languages but in the treatment of variable binding and scoping that appear in various formalisms including programming languages, mathematics and logic.

The core of the $\lambda$-calculus is $\beta$-reduction, $(\lambda x.M)N \to M[x \mapsto N]$, but the definition of substitutions used here is far from simple and provoked various alternative formulations. In particular, "to replace all the free occurrences of $x$ by copies of $N$" does not necessarily reflect actual implementation, which may share the representation of $N$ whenever possible but must sometimes make copies of $N$ (e.g., when applying another $\lambda$-term to $N$).

One of the formalisms aiming at the precise representation of the $\lambda$-calculus is the $\lambda\sigma$-calculus [1], which provides two syntactic categories, $\lambda$-terms and explicit substitutions, and gives rewrite rules to both.

Another approach to formalizing the $\lambda$-calculus is to adopt graph representation of $\lambda$-terms; a bound variable can most naturally be represented as an edge (or a hyperedge) that connects the defining and applied occurrences of the same variable. Most previous work in this approach adopted Interaction Nets [6] to represent and manipulate graphs ([7][9][10][12], to name a few). Many of the encodings of the $\lambda$-calculus into Interaction Nets pursued optimal sharing

A. Voronkov (Ed.): RTA 2008, LNCS 5117, pp. 392–408, 2008.

or efficiency, and resulted in more or less involved representation of $\lambda$-terms to achieve the objective. One notable exception is the encoding by Sinot [12], which addressed the simplicity of the encoding, but it focused on the weak $\lambda$-calculus that did not evaluate the body of $\lambda$-abstractions. Indeed, as KCLE [10] suggests, encoding of the pure calculus can be much less concise (in terms of the number of rules involved) than the encoding of the weak calculus. Weak $\lambda$-calculi may be appropriate for the foundations of functional languages, but the applications of the $\lambda$-calculus as a whole call for *strong* (or *pure*) $\lambda$-calculi as well.

This raises one question: Is there any *concise* graph-based encoding of the *pure* $\lambda$-calculus? There can be several criteria for conciseness, but the number of rewrite rules is clearly important and sufficiently objective. With Interaction Nets, YALE [9] proposes a relatively simple solution but still needed to simulate "boxes" for scope management. So the next question is: To obtain a more concise encoding (appropriate, say, for an undergraduate text), what additional constructs should be included to the graph rewriting framework?

The purpose and the contribution of this paper are to give a concrete answer to these questions by proposing a fine-grained and highly nondeterministic encoding of the pure $\lambda$-calculus (with open terms) and discussing its properties. Specifically, the paper employs LMNtal [14], a hierarchical graph rewriting language that uses logical variables to represent connectivity and membranes to represent hierarchy. LMNtal shares some of the ideas (such as membranes) with Chemical Abstract Machine (CHAM) [3], but it is better considered as extending Interaction Nets with membranes that can enclose graph nodes, rewrite rules and other membranes. To the best of the author's knowledge, work on the encoding of the $\lambda$-calculus into CHAM (such as [3]) dealt with lazy or weak calculus, and hence has little technical relation to the present work unlike encodings into Interaction Nets.

LMNtal aims at a substrate language of various computational models, especially those for concurrency, and the $\pi$-calculus and the ambient calculus have been encoded before [15]. An implementation is publicly available on the web[1]. The membrane construct of LMNtal provides powerful functionalities such as the copying of the graph structure enclosed by a membrane, but our encoding uses membranes only to represent and manipulate fresh local names, called *colors*, so that each rewrite step can be executed in (almost) constant amortized time. Thus our encoding is not too specific to a particular graph rewriting framework; rather, it is to give insights on what constructs are necessary or useful for concise encoding.

A graph representation was employed also in the graph reduction system [13] for combinators. This approach avoids difficulties arising from scoping by focusing on a form of closed reduction and compiling $\lambda$-terms into combinator expressions, but it cannot be regarded as a direct formulation of the operational semantics of the $\lambda$-calculus.

Our method decomposes graph copying into microsteps that may proceed asynchronously with $\beta$-reductions. Our method shares its granularity and asyn-

[1] `http://www.ueda.info.waseda.ac.jp/lmntal/`

$$
\begin{array}{rl}
\text{(process)} & P ::= \mathbf{0} \mid p(X_1,\ldots,X_m) \mid P,P \mid \{P\} \mid T\texttt{:-}T \\
\text{(process template)} & T ::= \mathbf{0} \mid p(X_1,\ldots,X_m) \mid T,T \mid \{T\} \mid T\texttt{:-}T \\
& \quad \mid \texttt{@}p \mid \$p[X_1,\ldots,X_m \,|\, A] \mid p(*X_1,\ldots,*X_n) \\
\text{(residual)} & A ::= \texttt{[]} \mid *X
\end{array}
$$

**Fig. 1.** Syntax of LMNtal

$$(\text{E1})\quad \mathbf{0},P \equiv P \qquad (\text{E2})\quad P,Q \equiv Q,P \qquad (\text{E3})\quad P,(Q,R) \equiv (P,Q),R$$

$$(\text{E4})\quad P \equiv P[Y/X] \qquad \text{if } X \text{ is a local link of } P$$

$$(\text{E5})\quad P \equiv P' \Rightarrow P,Q \equiv P',Q \qquad (\text{E6})\quad P \equiv P' \Rightarrow \{P\} \equiv \{P'\}$$

$$(\text{E7})\quad X\texttt{=}X \equiv \mathbf{0} \qquad (\text{E8})\quad X\texttt{=}Y \equiv Y\texttt{=}X$$

$$(\text{E9})\quad X\texttt{=}Y,\ P \equiv P[Y/X] \qquad \text{if } P \text{ is an atom and } X \text{ occurs free in } P$$

$$(\text{E10})\quad \{X\texttt{=}Y,\ P\} \equiv X\texttt{=}Y,\ \{P\} \qquad \text{if exactly one of } X \text{ and } Y \text{ occurs free in } P$$

$$(\text{R1})\ \frac{P \longrightarrow P'}{P,Q \longrightarrow P',Q} \qquad (\text{R2})\ \frac{P \longrightarrow P'}{\{P\} \longrightarrow \{P'\}} \qquad (\text{R3})\ \frac{Q \equiv P \quad P \longrightarrow P' \quad P' \equiv Q'}{Q \longrightarrow Q'}$$

$$(\text{R4})\quad \{X\texttt{=}Y,P\} \longrightarrow X\texttt{=}Y,\{P\} \qquad \text{if } X \text{ and } Y \text{ occur free in } \{X\texttt{=}Y,P\}$$

$$(\text{R5})\quad X\texttt{=}Y,\{P\} \longrightarrow \{X\texttt{=}Y,P\} \qquad \text{if } X \text{ and } Y \text{ occur free in } P$$

$$(\text{R6})\quad T\theta,(T\texttt{:-}U) \longrightarrow U\theta,(T\texttt{:-}U)$$

**Fig. 2.** Structural congruence and reduction relation of LMNtal

chrony with the $\lambda\sigma$-calculus, though they employ very different representations of $\lambda$-terms. Since each of the proposed rewrite rules is simple and well-motivated, the proposed method is expected to serve not only as an encoding but as a fine-grained reformulation of the pure $\lambda$-calculus.

## 2 LMNtal: A Hierarchical Graph Rewriting Language

This section briefly introduces the hierarchical graph rewriting model and language LMNtal. For further details and a diverse range of related work of LMNtal, the readers are referred to [14].

### 2.1 Syntax

The syntax of LMNtal is given in Fig. 1, where two syntactic categories, *link names* (denoted by $X$) and *atom names* (denoted by $p$), are presupposed.

Being a model of concurrency, LMNtal uses the terms *(hierarchical) graphs* and *processes* interchangeably. **0** is an inert process, $p(X_1,\ldots,X_m)$ $(m \geq 0)$ is an atom with the arity (valency) $m$, $P,P$ is parallel composition, $\{P\}$ is a *cell* formed by enclosing $P$ by a *membrane* { }, and $T$ :- $T$ is a rewrite rule. The built-in, binary atom name = represents a *connector* for interconnecting its argument links.

Occurrences of a link name represent endpoints of a one-to-one link between atoms. For this purpose, each link name in a process $P$ is allowed to occur *at most twice* (*Link Condition*). A link whose name occurs only once in $P$ is called a *free link* of $P$. Links may cross membranes and connect atoms located at different "places" of the membrane hierarchy.

Process templates on the both sides of rewrite rules allow *process contexts*, *rule contexts*, and *aggregates* [14]. Here we only describe process contexts used in our encoding. A process context, denoted $\$p\texttt{[}X_1\texttt{,}\ldots\texttt{,}X_m\,\texttt{|}\,A\texttt{]}$ $(m \geq 0)$, is to match "the rest of the processes" (except rewrite rules) within the membrane in which it appears. The arguments specify what free link names may or must occur. $X_1,\ldots,X_m$ are the link names that must occur free in $\$p$. When the *residual* $A$ is of the form $*X$, links other than $X_1,\ldots,X_m$ may occur free, and $*X$ stands for the sequence of those optional free links. When $A$ is of the form `[]`, no other free links may occur.

Rewrite rules must observe several syntactic conditions so that the Link Condition is preserved in the course of program execution [14]. Most importantly, link names in a rewrite rule must occur exactly twice, and each process context must occur exactly once at the toplevel of distinct cells in the LHS of a rule. There are no constraints on the number of occurrences of a process context on the RHS of a rule, but our encoding belongs to the most standard case where each process context occurs exactly once in the RHS.

### 2.2 Operational Semantics

The operational semantics of LMNtal (Fig. 2) consists of structural congruence defined by (E1)–(E10) and the reduction relation defined by (R1)–(R6). (E4) stands for $\alpha$-conversion. (E9)–(E10) are the interaction rules between atoms/cells and connectors.

Computation proceeds by rewriting processes using rules collocated in the same place of the nested membrane structure. (R1)–(R3) are standard structural rules, while (R4)–(R5) are the mobility rules of `=`. The central rule of LMNtal is (R6), in which $\theta$ is to map process contexts into actual processes.

### 2.3 Syntactic Conventions

LMNtal provides several syntactic conventions to allow concise description of processes, of which our encoding shown in the next section uses the following:

1. `+`$(X)$ can be written as `+`$X$, where `+` is a unary atom.
2. $p(s_1,\ldots,s_{m-1},X), q(t_1,\ldots,t_{k-1},X,t_{k+1},\ldots,t_n)$ (atoms connected by the link $X$ occurring as the final argument of $p$) can be written as $q(t_1,\ldots,t_{k-1}, p(s_1,\ldots,s_{m-1}),t_{k+1},\ldots,t_n)$.
3. When a process context of the form $\$p\texttt{[|}*X\texttt{]}$ occurs exactly twice in a rule (i.e., once in the RHS), they both can be abbreviated to $\$p$.
4. Each rule may be prefixed by a rule name and two `@`'s (Fig. 4).

A notable consequence of the above convention is that a process `(f(A,B,L), g(L))` can be written as `f(A,B,g)` and as `g(f(A,B))`. Furthermore, `(f(A,B,L), g(L))` can be expanded to `(f(A,B,L),g(M),L=M)` or `(f(A,B,L),g(M),M=L)` by (E9), from which we obtain `f(A,B)=g` and `g=f(A,B)`, respectively. They all stand for the same graph consisting of a ternary atom and a unary atom connected by a single link. The connector `=` is heavily used as the symbol representing interconnection (rather than equality) between atoms.

## 3 Encoding the λ-Calculus into LMNtal

Now we describe our encoding of the λ-calculus into hierarchical graph rewriting. Our starting point was the encoding into Interaction Nets. Interaction Nets is a non-hierarchical graph rewriting formalism with strong syntactic conditions, and LMNtal can be considered as a model and a language that extends Interaction Nets by alleviating their syntactic conditions and introducing the notions of membranes and contexts. Of various encodings into Interaction Nets, Sinot's encoding [12] is one of the simplest in the sense that it dispenses with the explicit management of free variables in each λ-abstraction. However, the method is to compute weak head normal forms (terms of the form $xM_0 \ldots M_n$ $(n \geq 0)$ or $\lambda x.M$, where $M$ and $M_i$ are not necessarily in normal form) and the computation is serialized using a control token navigating over the λ-graph. Our goal, in contrast, is to encode the basic reduction semantics of the pure λ-calculus, preserving and manifesting nondeterminism inherent in the formalism.

First of all, we define the encoding from a λ-term $L$ into an LMNtal process. The result must have exactly one free link (say $R$), which is connected to the atom referring to $L$. So the translation function $\mathcal{T}$ receives as arguments the λ-term $L$ and the free link name $R$.

- When $L$ is a variable $x$, it is represented as a unary atom with the name $x$ which is connected to $R$ via a binary atom `fv` indicating a free variable:

$$\mathcal{T}(x,R) \stackrel{\text{def}}{=} \mathtt{fv}(x,R) \quad (= R\,\mathtt{=}\,\mathtt{fv}(x)).$$

- When $L$ is a λ-abstraction $\lambda x.M$, let $k$ $(\geq 0)$ be the number of free occurrences of $x$ in $M$, and $\mathcal{T}'(M,R,R_1,\ldots,R_k)$ be a process obtained from $\mathcal{T}(M,R)$ by removing all unary atoms $x$ and their tags `fv` and changing them into free links $R_1,\ldots,R_k$. Then

$$\begin{aligned}\mathcal{T}(\lambda x.M,R) \stackrel{\text{def}}{=} {} & \mathtt{lambda}(R_0,R',R),\ \mathcal{T}'(M,R',R_1,\ldots,R_k),\\ & \mathit{connect}[R_0,R_1,\ldots,R_k]\,,\end{aligned}$$

where $\mathit{connect}[R_0,R_1,\ldots,R_k]$ is a process with free links $R_0$, $R_1$, ..., $R_k$ defined as follows:

$$\begin{aligned}\mathit{connect}[R_0] &\stackrel{\text{def}}{=} \mathtt{rm}(R_0)\\ \mathit{connect}[R_0,R_1] &\stackrel{\text{def}}{=} R_0\,\mathtt{=}\,R_1\\ \mathit{connect}[R_0,R_1,\ldots,R_n] &\stackrel{\text{def}}{=} \mathtt{cp}(R_1,R_0',R_0),\ \mathit{connect}[R_0',R_2,\ldots,R_n] \quad (n \geq 2).\end{aligned}$$

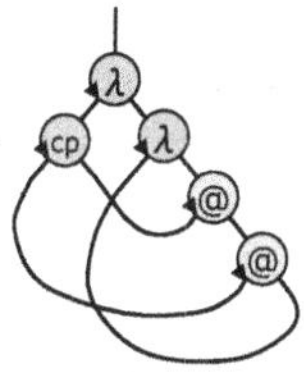


**Fig. 3.** Graph representation of the Church numeral 2

– When $L$ is an application $MN$:

$$\mathcal{T}(MN, R) \stackrel{\text{def}}{=} \texttt{apply}(R_1, R_2, R),\ \mathcal{T}(M, R_1),\ \mathcal{T}(N, R_2)\,.$$

Bound variables are encoded into LMNtal links, but because of the Link Condition of LMNtal, bound variables not occurring exactly twice requires the branching or termination of links. We employ a unary atom `rm` (remove) to terminate unused bound variables and a ternary atom `cp` (copy) to bifurcate links. The encoding of a bound variable with more than two occurrences forms a tree of `cp`'s, but the form of the tree does not count for our encoding and its properties. For example, a combinator $\mathbf{I} = \lambda x.x$ is represented as

`lambda(X,X,Result)` (= `Result = lambda(X,X)`)

where `Result` is the free link name representing the result. The Church encodings of natural numbers, $\lambda fx.f^n x$ ($n \geq 0$), can be represented as

```
0: lambda(rm,lambda(X,X),Result)
1: lambda(F,lambda(X,apply(F,X)),Result)
2: lambda(cp(F0,F1),lambda(X,apply(F0,apply(F1,X))),Result)
```

and so on, where '`@`' in Fig. 3 stands for `apply` and the arrowheads indicate the atoms' first arguments and the ordering of arguments.

Figure 4 shows a complete set of rules that encodes the pure λ-calculus using our λ-term representation. As will be discussed later in this section, only eight of those rules are essential.

The first rule, `beta`, performs "bare" β-reduction, that is, performs parameter passing without copying the argument even when it is referenced more than once.

Rule `beta` alone is sufficient if all formal parameters are used exactly once; otherwise we need reaction rules for the atoms `cp` and `rm` (13 rules following `beta`) to destroy or copy graph structures incrementally. The final three rules are for the color management of `cp`'s described next.

The ternary `cp`'s in λ-terms are first converted to quaternary `cp`'s by Rule `c2c`. The additional third argument is to distinguish between `cp`'s with different origins when copying nested λ-abstractions. The additional information is called a *color* after the Petri Net terminology. Each color is represented using an LMNtal cell, where links entering the same cell are interpreted as referring to the same color.

```
beta@@ H=apply(lambda(A, B), C) :- H=B, A=C.

l_c@@  lambda(A,B)=cp(C,D,L), {+L,$q} :-
         C=lambda(E,F), D=lambda(G,H), A=cp(E,G,L1), B=cp(F,H,L2),
         {{+L1},+L2,sub(S)}, {super(S),$q}.
a_c@@  apply(A,B)=cp(C,D,L), {+L,$q} :-
         C= apply(E,F), D= apply(G,H), A=cp(E,G,L1), B=cp(F,H,L2),
         {+L1,+L2,$q}.

c_c1@@ cp(A,B,L1)=cp(C,D,L2), {{+L1,$p},+L2,$q} :- A=C, B=D, {{$p},$q}.
c_c2@@ cp(A,B,L1)=cp(C,D,L2), {{+L1,$p},$q}, {+L2,top,$r}
          :- C=cp(E,F,L3), D=cp(G,H,L4), {{+L3,+L4,$p},$q},
             A=cp(E,G,L5), B=cp(F,H,L6), {+L5,+L6,top,$r}.
f_c@@  fv($u)=cp(A,B,L), {+L,$q} :- unary($u) |
          A=fv($u), B=fv($u), {$q}.
l_r@@  lambda(A,B)=rm :- A=rm, B=rm.
a_r@@  apply(A,B)=rm :- A=rm, B=rm.
c_r1@@ cp(A,B,L)=rm, {+L,$q} :- A=rm, B=rm, {$q}.
c_r2@@ cp(A,B,L)=rm, {{+L,$p},$q} :- A=rm, B=rm, {{$p},$q}.
c_r3@@ A=cp(B,rm,L), {+L,$p} :- A=B, {$p}.
c_r4@@ A=cp(rm,B,L), {+L,$p} :- A=B, {$p}.
r_r@@  rm=rm :- .
f_r@@  fv($u)=rm :- unary($u) | .

promote@@ {{},$p,sub(S)}, {$q,super(S)} :- {$p,$q}.
c2c@@  A=cp(B,C) :- A=cp(B,C,L), {+L,top}.
gc@@   {top} :- .
```

**Fig. 4.** The LMNtal encoding of the pure λ-calculus

Colors form tree-shaped partial order. Color cells in the parent-child relationship are interconnected by a link terminated by a unary `super` atom on one end and a `sub` on the other. Colors without parents hold a nullary `top` instead of a `sub`. Figure 5 represents a tree of colors with one top color and two subordinate sibling colors. Each quaternary `cp` is given a top color initially. Rule `c2c` creates an independent top color cell for each `cp`, but whether to merge top color cells or not does not affect the correctness of our encoding.

Graph copying starts when links representing formal and actual parameters are interconnected and a `cp` on the formal side meets `lambda`, `apply`, or `fv`. When `cp` meets `apply`, it copies the partner, splits itself, and proceeds to the copying of the `apply`'s two arguments, but the color of the split `cp`'s remains unchanged (Rule `a_c`). When `cp` meets `lambda`, it copies the partner and splits itself in the same manner, but the split `cp`'s are made to have a fresh, subordinate color (Rule `l_c`). The cell `{+L,$q}` on the LHS of `l_c` stands for the current color and is inherited as `{super(S),$q}` on the RHS (`$q` stands for incident links from other `cp`'s). The RHS creates another color cell `{{+L1},+L2,sub(S)}` referred to by the

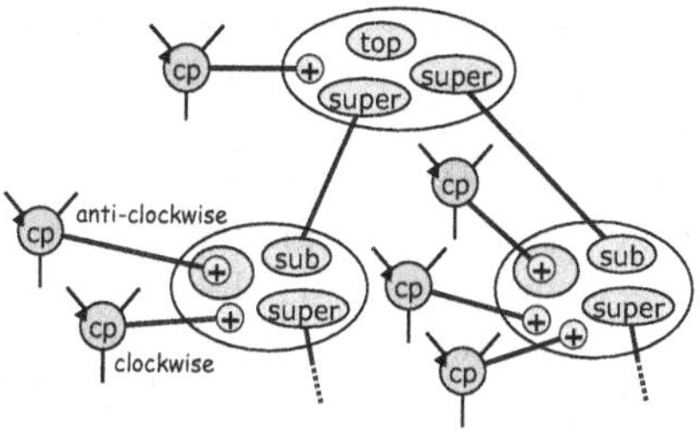


**Fig. 5.** Coloring cp atoms

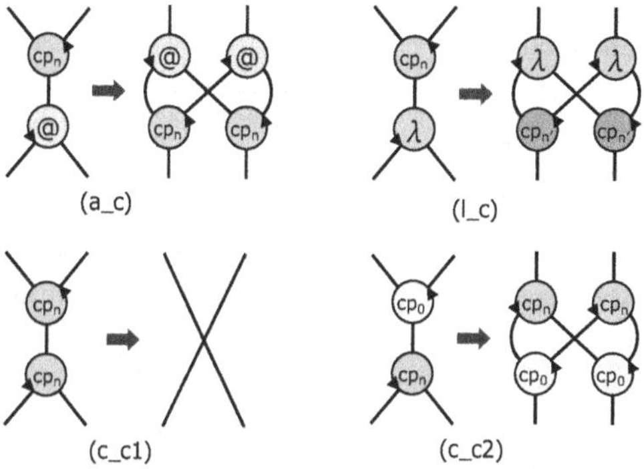


**Fig. 6.** Reaction rules for cp atoms

color links of the new cp's. In a color cell, a cp atom moving anticlockwise (in the representation of Fig. 3) from the $x$ side of a $\lambda$-term $\lambda x.M$ is distinguished from a cp moving clockwise from the $M$ side by letting the former's color link enter a nested membrane inside the color cell. A clockwise cp and an anticlockwise cp (hereafter denoted as $\mathtt{cp}^+$ and $\mathtt{cp}^-$ for exposition) are considered to have *positive* and *negative polarities*, respectively. Figure 5 shows a graph structure consisting of a $\mathtt{cp}^+$ with a top color, a $\mathtt{cp}^+$ and a $\mathtt{cp}^-$ with a subordinate color, and two $\mathtt{cp}^+$'s and a $\mathtt{cp}^-$ with another subordinate color.

As can be seen from the Church numerals example, the link representing the bound variable of a $\lambda$-abstraction is either terminated by rm or is split using zero or more $\mathtt{cp}^+$'s and connected to some places in the body. Accordingly, each $\mathtt{cp}^-$ will eventually meet, and is annihilated by, either a rm or a $\mathtt{cp}^+$ with the same color (possibly after crossing and copying $\mathtt{cp}^+$'s with the top color). When a $\mathtt{cp}^-$ meets a $\mathtt{cp}^+$ with the top color, it copies the partner using c_c2, splits itself, and proceeds.

In contrast, a $\mathtt{cp}^+$ may not meet a $\mathtt{cp}^-$ with the same color, because it may escape the scope of the $\lambda$-abstraction through a link representing nonlocal variables. The color of a $\mathtt{cp}^+$ that has escaped must be changed back to the original parent color. This is done using promote, which merges some color into its parent color when all the $\mathtt{cp}^-$'s of that color disappear. The cell {{},$p,sub(S)} on the LHS represents a color that has no incident links from cp$^-$'s, while the cell {$p,$q} on the RHS represents a single, fused color.

Rule promote is applied asynchronously; it is not necessarily applied immediately after all cp$^-$'s of some color disappear. However, the delay of promote simply delays the reaction between cp$^+$'s and cp$^-$'s (using c_c1 and c_c2), and does not cause wrong reactions by affecting the applicability of other rules.

Figure 6 depicts rewrite rules related to cp's, where the difference of colors is shown using suffixes for simplicity and the suffix 0 indicates the topmost color.

Of the remaining rules, f_c copies global free variables. This rule contains an side condition, unary($u), that specifies that the first argument of fv is connected to some unary atom, which will be copied in the RHS because $u occurs twice there. Rules l_r, a_r, c_r1, c_r2, r_r, and f_r are to delete any partner that an rm may encounter. c_r3 and c_r4 are to remove a cp with one branch terminated by an rm. Rule gc is to delete a topmost color not referenced any more.

Most previous encodings into Interaction Nets used two kinds (i.e., two colors) of copying tokens. Two colors sufficed in [12] because it did not evaluate bodies of λ-abstractions. YALE and KCLE computed normal forms, but did so by explicitly managing nonlocal variables, which added certain complexity. Although not for computing normal forms, Lang's encoding [8] employed many colors, where colors were represented as sequences of fresh names. Color comparison was based on whether one color was a prefix of the other, whose efficient implementation was yet to be studied. Lamping's optimal sharing [7] also employed many colors (called *levels*), and further employed tokens called croissants and brackets (both coming with many colors as well) to achieve sharing and complicated level management. Our encoding pursues the simplicity of the rewrite system: Color hierarchy implemented using membranes lead to a rewrite system that added only a few rules to the rules for handling all possible pairs of atoms that may encounter.

Our rewrite system could be slimmed down further. Rule c2c can be dispensed with by starting with colored cp atoms. Furthermore, Rules l_r, a_r, c_r1, c_r3, c_r4, r_r, f_r (i.e., all rules involving rm except c_r2), plus gc are just for garbage collection and tidying up the tree of cp's, and could be dispensed with. (Rule c_r2 cannot be removed because it kills cp$^-$'s whose numbers are counted.) This leaves us only eight essential rules, beta, l_c, a_c, c_c1, c_c2, f_c, c_r2 and promote, which suffice for the full evaluation (without the need of any further 'read-back') of colored λ-term representation.

### 3.1 An Example

Of numerous examples we have run using our LMNtal system, the exponentiation of Church numerals seems to be an important test of λ-calculus encodings because the encoding of $m^n$, $\lambda mn.nm$, is extremely simple and yet involves exponential amount of graph copying. It is important also because it requires the evaluation of the bodies of λ-abstractions.

```
N=n(2) :- N=lambda(cp(F0,F1),lambda(X,apply(F0,apply(F1,X)))).
N=n(3) :- N=lambda(cp(F0,cp(F1,F2)),
                   lambda(X,apply(F0,apply(F1,apply(F2,X))))).
result=apply(apply(apply(apply(n(2),n(2)),n(3)),fv(s)),fv(0)).
H=apply(fv(s), fv(I)) :- int(I) | H=fv(I+1).
```

**Fig. 7.** Church numerals and their exponentiation

The program in Fig. 7 reduces to `result=fv(81)` ($81 = 3^{2^2}$). As illustrated in the example, the encoding into LMNtal allows $\delta$-reduction rules, namely rules for rewriting free names such as `s` and `0`, to be added to the pure $\lambda$-calculus.

# 4 Properties of the Encoding

The encoding described above decomposes $\beta$-reduction into many small microsteps that allow asynchronous, out-of-order execution. The adequacy of the encoding is therefore not obvious; recall that the confluence and termination of the $\lambda\sigma$-calculus was not obvious, either [4][11]. Furthermore, because of the asynchrony, the "meaning" of an intermediate state of graph reduction broken into microsteps is far from obvious. To address the above problems, we interpret hierarchical graphs using $\lambda$-terms with extended binder syntax and establish the properties of the encoding through well-known properties of the $\lambda$-calculus.

One might wonder why we translate graphs back to (extended) $\lambda$-terms rather than leveraging the properties of LMNtal and of the rules in Fig. 4. However, LMNtal itself is not equipped with strong properties such as confluence (unlike Interaction Nets) because they may limit the expressive power of LMNtal as a unifying computational model. Indeed, using the experimental model-checking mode of our LMNtal implementation, we have found that the evaluation of $2^2$ contains 940 reachable states and two final states, while $2^3$ contains 63118 reachable states and three final states. These numbers demonstrate the highly nondeterministic and asynchronous nature of our encoding. Of course, the above final states are identical up to the associativity and commutativity of `cp` (Section 3).

Henceforth we refer to extended $\lambda$-terms as *$\lambda\gamma\delta$-terms.*

## 4.1 $\lambda\gamma\delta$-Terms

The purpose of extending the syntax of $\lambda$-terms here is to be able to distinguish between copied and shared subterms. Rather than taking the $\lambda\sigma$-calculus approach, we take an approach (similar in spirit to the one found in [5]) of enriching binders of the $\lambda$-calculus and restricting the use of each binder.

Specifically, we employ several binder symbols that are semantically equivalent to $\lambda$ but have different roles and restrictions. First, we continue to use $\lambda$ to construct $\lambda$-abstractions, but allows the bound variable $x$ of $\lambda x.M$ to occur free in $M$ *exactly once*. Instead, we introduce a new binder $\gamma$ to represent sharing, namely intervening cp's. In the following, we interpret each $\gamma$ to represent a cp with some color. When colors matter, we distinguish between $\gamma$'s using '$'$'s and write $\gamma^0$ for a $\gamma$ with the top color. By $(\gamma x.M)N$ we mean a term in which the two places indicated by the free $x$'s *share* the term $N$, while as usual, we mean by $M[x \mapsto N]$ a term obtained by replacing the two free $x$'s by *copies* of $N$. To be an interpretation of cp, $\gamma x.M$ must have exactly two free occurrences of $x$ in $M$, and conversely, when a bound variable occurs twice or more in a body, they must be bound by one or more $\gamma$'s.

For instance, $\lambda f.\lambda x.f(fx)$ is represented either as $\lambda f.(\gamma f'.\lambda x.f'(f'x))f$ or as $\lambda f.\lambda x.(\gamma f'.f'(f'x))f$. This example indicates that there may be more than one right place of $\gamma$, but we regard two $\lambda\gamma\delta$-terms representing the same graph structure as structurally equivalent (written $\equiv_{\lambda\gamma\delta}$), that is, convertible in zero steps in either direction. Formally, $\equiv_{\lambda\gamma\delta}$ enjoys (as in the $\lambda$-calculus with `let`):

$$L((\gamma x.M)N) \equiv_{\lambda\gamma\delta} (\gamma x.LM)N \quad (x \notin fv(L)) \tag{1}$$

$$((\gamma x.L)N)M \equiv_{\lambda\gamma\delta} (\gamma x.LM)N \quad (x \notin fv(M)) \tag{2}$$

$$\star y.(\gamma x.M)N \equiv_{\lambda\gamma\delta} (\gamma x.\star y.M)N \quad (x \not\equiv y \text{ and } \star \text{ is any binder}) \tag{3}$$

Another binder we need is $\delta$ to discard a bound variable, which corresponds to rm. For instance, $\lambda f.\lambda x.x$ is represented as $\lambda f.(\delta f'.\lambda x.x)f$ or $\lambda f.\lambda x.(\delta f'.x)f$. Due to space limitations, this paper focuses on $\gamma$ and cp; the treatment of $\delta$ and rm is analogous and much simpler. Likewise, we omit reactions involving global free variables, which is even more straightforward.

We define another equivalence relation, $\equiv_\lambda$. $M_1 \equiv_\lambda M_2$ means $\ell(M_1) \equiv \ell(M_2)$, where $\ell(M)$ stands for a $\lambda$-term obtained by a sequence of $\beta$-reductions that eliminate any binder (including $\gamma$ and $\delta$) *except* $\lambda$. For instance, $\lambda f.(\delta f'.\lambda x.x)f$ $\equiv_\lambda \lambda f.\lambda x.(\delta f'.x))f$ because both are reduced to the same $\lambda$-term, $\lambda f.\lambda x.x$.

## 4.2 Relating Reduction Relations and $\lambda\gamma\delta$-Terms

This section illustrates how our rewrite rules in Fig. 4 can be interpreted as the rewriting of $\lambda\gamma\delta$-terms.

First, the $\beta$-reduction of the $\lambda$-calculus directly gives a $\lambda\gamma\delta$ representation of Rule beta of Fig. 4:

$$(\lambda x.M)N \rightsquigarrow M[x \mapsto N]\,. \tag{4}$$

We use $\rightsquigarrow$ to mean the reduction relation between encoded graphs (rather than the reduction relation over $\lambda$-terms they represent). Because $x$ is supposed to occur exactly once in $M$, the occurrence conditions of the bound variables of any $\lambda$-term containing $(\lambda x.M)N$ as a subterm is preserved.

Rule a_c expresses interaction between cp and application, and its $\lambda\gamma\delta$ representation varies depending on how many upper-level binders the copied subterm refers to. When an application $MN$ is a closed term, Rule a_c corresponds to the rewriting

$$(\gamma y.L)(MN) \rightsquigarrow (\gamma m.\gamma n.L[y \mapsto mn])MN\,. \tag{5}$$

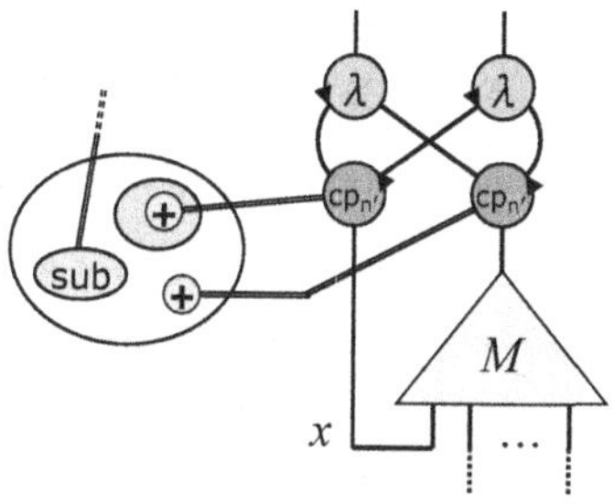


**Fig. 8.** Copying $\lambda x.M$

Note that the order of the two $\gamma$'s in the RHS is inessential and that the RHS is equivalent (modulo $\equiv_{\lambda\gamma\delta}$) to $(\gamma n.\gamma m.L[y \mapsto mn])NM$. The RHS of (5) is regarded as the result of copying the toplevel of application only. Since both $m$ and $n$ occur twice in $L$ after rewriting, the occurrence condition of $\gamma$ is preserved.

Rule `l_c` copies a term containing $\lambda$-abstractions, so its $\lambda\gamma\delta$ representation is far less obvious. Again we start with cases without upper-level binders:

$$(\gamma y.L)(\lambda x.M) \rightsquigarrow (\gamma' g.L[y \mapsto \lambda x.gx])(\overline{\gamma}' x.M)\,. \tag{6}$$

This represents the copying of the binding constructor, $\lambda$, of $\lambda x.M$. The body $M$ is still shared by the two copies, but the two bound variables of the two $\lambda$-abstractions must be connected to the bound variable $x$ in the original $M$ via a `cp` with a fresh color. The resulting configuration is represented using the pair of $\gamma'$ and $\overline{\gamma}'$, where $\gamma'$ stands for a `cp` with a fresh color immediately beneath $\gamma$ in the color ordering, while $\overline{\gamma}'$ (to be used with $\gamma'$) stands for the pair of $x$ (occurring *exactly once* in $M$) and $M$. Thus, when the argument of $\gamma'$ is of the form $\overline{\gamma}' x.M$, it means that $M$ is shared by two $\lambda$-abstractions, and the whole term stands for a graph in which two `cp`'s with the same color and inverse polarities are inserted at the result side and the $x$ side of $M$ as shown in Fig. 8. The free variables of $M$ other than $x$ in Fig. 8 stand for global free variables or variables bound at upper levels ((6) is the case where there are no such variables).

When a reaction using `a_c` or `l_c` occurs within the (still) shared part of a $\lambda$-abstraction being copied, we must handle the bound variable of that $\lambda$-abstraction properly. For instance, the $\lambda\gamma\delta$ representation of an `l_c` reaction that copies the second $\lambda$ of $\lambda x.\lambda z.M$ (whose first $\lambda$ has already been copied) is represented as

$$(\gamma' g.L[y \mapsto \lambda x.gx])(\overline{\gamma}' x.\lambda z.M) \rightsquigarrow (\gamma'' g.L[y \mapsto \lambda x.\lambda z.gxz])(\overline{\gamma}' x.\overline{\gamma}'' z.M). \tag{7}$$

where the $\overline{\gamma}'$ stands for the pair of $x$ and $\lambda z.M$ containing $x$ as the sole free variable. In general, more $\overline{\gamma}$'s will be used to represent the copying of deeply nested $\lambda$-abstractions, but the full inductive definition that generalizes (6) and (7) is omitted here.

When applying a_c inside a $\lambda$-abstraction being copied and $x$ occurs in $M$ or $N$ (but not both) of $\lambda x.MN$, we have either of the following:

$$(\gamma' g.L[y \mapsto \lambda x.gx])(\overline{\gamma}' x.MN) \rightsquigarrow (\gamma' mn.L[y \mapsto \lambda x.(mx)n])(\overline{\gamma}' x.M)N \quad (8)$$
$$(\gamma' g.L[y \mapsto \lambda x.gx])(\overline{\gamma}' x.MN) \rightsquigarrow (\gamma' mn.L[y \mapsto \lambda x.m(nx)])M(\overline{\gamma}' x.N). \quad (9)$$

When $x$ occurs free in both $M$ and $N$, the $\lambda\gamma\delta$ representation of the application takes the form $\overline{\gamma}' x.(\gamma^0 x.MN)x$, where the $\gamma^0$ represents cp to feed $x$ into both $M$ and $N$. This case requires temporary translation to a combinator form, and we allow this $\lambda\gamma\delta$-term to be written also as $\mathbf{AS}(\overline{\gamma}' x.M)(\overline{\gamma}' x.N)$ (i.e., they are considered equivalent modulo $\equiv_{\lambda\gamma\delta}$), where $\mathbf{S}$ is the standard $S$-combinator $\lambda mn.\gamma^0 x.(mx)(nx)$ that corresponds to cp for distributing $x$ to $M$ and $N$, and $\mathbf{A}$ is $\lambda smn.smn$ that corresponds to apply. Then we allow the four "components", $\mathbf{A}$, $\mathbf{S}$, $(\overline{\gamma}' x.M)$, and $(\overline{\gamma}' x.N)$, to be brought independently by repeatedly using the equivalence

$$(\gamma y.L)(MN) \equiv_{\lambda\gamma\delta} (\gamma m.\gamma n.L[y \mapsto mn])MN \quad (10)$$

which we apply only when $M$ and $N$ are the components of a combinator representation (cf. (5) for real apply). As a consequence, we have

$$(\gamma' g.L[y \mapsto \lambda x.gx])(\overline{\gamma}' x.(\gamma^0 x.MN)x) \equiv_{\lambda\gamma\delta}$$
$$(\gamma' asmn.L[y \mapsto \lambda x.asmnx])\mathbf{AS}(\overline{\gamma}' x.M)(\overline{\gamma}' x.N)$$

and finally Rule a_c for this (awkward) case is represented as

$$(\gamma' a.L)\mathbf{A} \rightsquigarrow L[a \mapsto \mathbf{A}]\,. \quad (11)$$

(The copying of $\mathbf{S}$ (for cp) will be discussed later.)

The $\lambda\gamma\delta$ representation of Rule c_c1 to handle the annihilation of cp's is written as

$$(\gamma' g.M)(\overline{\gamma}' x.x) \rightsquigarrow M[g \mapsto \mathbf{I}] \quad (12)$$

where $\mathbf{I}$ stands for the identity combinator. The LHS stands for the pair of a $cp^+$ and a $cp^-$ with the same color, which both disappear on the RHS. The $\mathbf{I}$ combinator corresponds to the LMNtal connector = and annihilates itself as connectors do.

Rule c_c2 to deal with the reaction of a $cp^-$ and a toplevel $cp^+$ is another case where a combinator representation should be employed, and is considered dual to a_c:

$$(\gamma' s.M)\mathbf{S} \rightsquigarrow M[s \mapsto \mathbf{S}]. \quad (13)$$

Finally, the $\lambda\gamma\delta$ representation of promote for updating the colors of binders is written as

$$\gamma' x.M \rightsquigarrow \gamma x.M \quad (14)$$

where the color of $\gamma'$ is immediately beneath the color of $\gamma$. According to the semantics of promote, this rule is applicable only when there is no term of the form $\overline{\gamma}' x.M$ inside the $\lambda\gamma\delta$-term being evaluated.

### 4.3 Properties

Now we are in a position to give several properties of our encoding through the reduction relations given in the previous subsection.

We say that a graph (and a $\lambda\gamma\delta$-term representing the graph) is in a *cp-normal form* (representing a $\lambda$-term $M$) if it is equivalent, modulo the associativity and commutativity of `cp`, to the result of translating $M$ using the translation function $\mathcal{T}$ (Section 3).

Because the reduction of a $\lambda$-term may not terminate, we cannot expect the termination of our graph rewriting in general. However, we can establish the termination of graph rewriting in the absence of `beta`.

**Proposition 1** *Suppose a graph in a cp-normal form is rewritten finitely many times using the rules in Fig. 4 and then using any rules except* `beta` *after that. Then the rewriting terminates with a cp-normal form.*

*Proof.* See Appendix. □

**Proposition 2** *Suppose $M \longrightarrow M'$ (in the $\lambda$-calculus) and let a $\lambda\gamma\delta$-term $G$ be a cp-normal form representing $M$. Then there exists a cp-normal form $G'$ representing $M'$ such that $G \rightsquigarrow^* G'$.*

*Proof.* Use `beta` (i.e., (4)) once to rewrite the $\beta$-redex in $G$ corresponding to the $\beta$-redex of $M$ to obtain $G''$, and then use the rules other than `beta` to obtain a $\lambda\gamma\delta$-term $G'$ representing its cp-normal form. It is easy to check that each rewrite rule in Section 4.2, except (4) for $\beta$-reduction, rewrites a $\lambda\gamma\delta$-term $L$ to $L'$ such that $L \equiv_\lambda L'$. Therefore $G'' \equiv_\lambda G'$. It is also easy to check that $\ell(G'') = M'$. Therefore $G'$ is a cp-normal form representing $M'$. □

**Theorem 3 (completeness)** *Suppose $M_0 \longrightarrow^* M_n$ (in the $\lambda$-calculus), and let $G_0$ be a cp-normal form representing $M_0$. Then there is a $G_n$ in a cp-normal form representing $M_n$ such that $G_0 \rightsquigarrow^* G_n$ .*

*Proof.* By repeated application of Proposition 2. □

**Theorem 4 (confluence modulo)** *Suppose $G_0$ is in a cp-normal form, $G_0 \rightsquigarrow^* G_1$, and $G_0 \rightsquigarrow^* G_2$. Then there exists a $\lambda$-term $M'$ and $\lambda\gamma\delta$-terms $G_3$, $G_4$ such that $G_1 \rightsquigarrow^* G_3$, $G_2 \rightsquigarrow^* G_4$ and both $G_3$ and $G_4$ are cp-normal forms representing $M'$.*

*Proof.* We first reduce $G_1$ and $G_2$ using rules other than (4) to obtain their cp-normal forms $G_1'$ and $G_2'$, which must represent some $\lambda$-terms $N$ and $N'$, respectively. From the confluence property of the $\lambda$-calculus, there exists an $M'$ satisfying $N \longrightarrow^* M'$ and $N' \longrightarrow^* M'$. Along the reduction paths from $N$ to $M'$ and from $N'$ to $M'$, apply Proposition 2 repeatedly to obtain cp-normal forms representing $M'$. □

**Theorem 5 (preservation of strong normalization)** *Suppose a λ-term $M$ is strongly normalizing (i.e., does not have infinite reduction sequences). Then a λγδ-term in a cp-normal form representing $M$ is strongly normalizing in the λγδ-calculus.*

*Proof.* First we discuss the cases not involving `rm`. Suppose the λγδ-term has an infinite reduction sequence. Because reduction without Rule `beta` is terminating (Proposition 1), an infinite sequence must have infinite applications of `beta`. Let $G \leadsto G'$ be the $i$th application of `beta`. This reduction may be done in a shared part (i.e., $N'$ of the form $(\gamma x.N)N'$) or elsewhere, but in any case there are λγδ-terms $L$ and $L'$ such that (i) $L$ is a cp-normal form of $G$, (ii) $L'$ is a cp-normal form of $G'$, and (iii) $\ell(L) \rightarrow^k \ell(L')$ for some $k\,(\geq 1)$ determined by the degree of sharing. This contradicts the assumption of strong normalization.

Therefore an infinite reduction sequence should contain at least one application of `beta` that "activates" `rm`, i.e., derives a λγδ-term $L$ containing $(\delta x.N)N'$. Let $L'$ be $L$ with $(\delta x.N)N'$ replaced by $N$. Either $N'$ or $L'$ should have an infinite reduction sequence. However, both $\ell(N')$ and $\ell(L')$ are strongly normalizing in the λ-calculus with the maximum length of reductions strictly less than that of the original λ-term. Therefore the same argument does not apply to $N'$ or $L'$ indefinitely. □

**Theorem 6 (soundness)** *Suppose $G \leadsto^* G'$. Then $\ell(G) \longrightarrow^* \ell(G')$.*

*Proof Sketch.* By an argument analogous to the first half of the proof of Theorem 5. □

Thus we have made sure that our encoding is quite well-behaved.

Finally, a brief comment on the cost of each reduction. By using a stack to maintain initial atoms and atoms newly created by rewriting, we can achieve O(1) amortized time complexity for all the rewrite rules of Fig. 4 except the three rules that compare or fuse colors: `c_c1`, `c_c2` and `promote`. The three rules are essentially operations on sets for which the well-known UNION-FIND algorithm can achieve almost constant (inverse Ackermann) amortized time complexity.

## 5 Discussions and Conclusion

This paper described an encoding of the pure λ-calculus into hierarchical graph encoding and showed that it had many desirable properties. The proposed rewrite rules allow highly asynchronous and concurrent execution of parameter passing, incremental copying, and scope control using color hierarchy. The set of rules we have proposed consists of just eight essential rules, plus rules for tidying up and initial coloring, and allows graphical understanding. This makes it promising as an alternative formulation, rather than just the encoding, of the pure λ-calculus. The encoding has been implemented and tested using LMNtal. To justify our encoding, we introduced λγδ-terms with additional binders, and illustrated how intermediate states of graph rewriting could be interpreted as λγδ-terms. The

$\lambda\gamma\delta$ interpretation of graphs made it rather straightforward to show the adequacy of our encoding.

The key idea of our encoding was the color management of copy tokens, where the "colors" are essentially local names and local names are essentially equivalence classes of local name occurrences. In other words, we have shown how a very generic notion of local names can be smoothly integrated into graph rewriting and contributes to its expressive power. Although LMNtal doesn't feature an explicit construct for local names, it allows us to *encode* local name and their operations that consist of

1. ordering between names,
2. checking of name equality,
3. detection of garbage names (names with no references), and
4. fusion of two names,

all of which were essential in our encoding but not all of which may be available in other computational models featuring local name creation.
This paper focused on the pure $\lambda$-calculus with no particular reduction strategies. It is a subject of future work to achieve the sharing of computation. Previous work on optimal sharing and optimal reduction is quite sophisticated, and there remains a broad spectrum between optimality and simplicity.

**Acknowledgments.** The author is indebted to anonymous referees for their useful comments and pointers (especially one in French) to the literature. This work was partially supported by Grant-In-Aid for Scientific Research (B)(2)16300009, (C)(2)13324050 (priority area), (B)(2)14085205 (priority area) and 04560009 (priority area), MEXT and JSPS.

## References

1. Abadi, M., Cardelli, L., Curien, P.-L., Lévy, J.-J.: Explicit Substitutions. Journal of Functional Programming 1(4), 375–416 (1991)
2. Bloo, R., Geuvers, H.: Explicit Substitution: on the Edge of Strong Normalization. Theoretical Computer Science 211(1–2), 375–395 (1999)
3. Berry, G., Boudol, G.: The Chemical Abstract Machine. In: Proc. POPL 1990, pp. 81–94. ACM, New York (1990)
4. Curien, P.-L., Hardin, T., Lévy, J.-J.: Confluence Properties of Weak and Strong Calculi of Explicit Substitutions. J. ACM 43(2), 362–397 (1996)
5. Fernández, M., Mackie, I., Sinot, F.-R.: Closed Reduction: Explicit Substitutions without $\alpha$-conversion. Mathematical Structures in Computer Science 15(2), 343–381 (2005)
6. Lafont, Y.: Interaction Nets. In: Proc. POPL 1990, pp. 95–108. ACM, New York (1990)
7. Lamping, J.: An Algorithm for Optimal Lambda-Calculus Reductions. In: Proc. POPL 1990, pp. 16–30. ACM, New York (1990)
8. Lang, F.: Modèles de la $\beta$-réduction pour les implantations. Ph.D. Thesis, Ècole Normale Supérieure de Lyon (1998)
9. Mackie, I.: YALE: Yet Another Lambda Evaluator Based on Interaction Nets. In: Proc. ICFP 1998, pp. 117–128. ACM, New York (1998)

10. Mackie, I.: Efficient λ-Evaluation with Interaction Nets. In: van Oostrom, V. (ed.) RTA 2004. LNCS, vol. 3091, pp. 155–169. Springer, Heidelberg (2004)
11. Melliès, P.-A.: Typed λ -Calculi with Explicit Substitutions May Not Terminate. In: Dezani-Ciancaglini, M., Plotkin, G. (eds.) TLCA 1995. LNCS, vol. 902, pp. 328–334. Springer, Heidelberg (1995)
12. Sinot, F.-R.: Call-by-Name and Call-by-Value as Token-Passing Interaction Nets. In: Urzyczyn, P. (ed.) TLCA 2005. LNCS, vol. 3461, pp. 386–400. Springer, Heidelberg (2005)
13. Peyton Jones, S.L.: The Implementation of Functional Programming Languages. Prentice-Hall, Englewood Cliffs (1987)
14. Ueda, K., Kato, N.: LMNtal: a Language Model with Links and Membranes. In: Mauri, G., Păun, G., Jesús Pérez-Jímenez, M., Rozenberg, G., Salomaa, A. (eds.) WMC 2004. LNCS, vol. 3365, pp. 110–125. Springer, Heidelberg (2005)
15. Ueda, K.: Encoding Distributed Process Calculi into LMNtal. Electronic Notes in Theoretical Computer Science 209, 187–200 (2008)

# A Appendix

**Proposition 1.** *Suppose a graph in a cp-normal form is rewritten finitely many times using the rules in Fig. 4 and then using any rules except* `beta` *after that. Then the rewriting terminates with a cp-normal form.*

*Proof Sketch.* Let $a_k$ denote the $k$th argument of an atom $a$. We first establish invariant properties on how `cp`'s are connected to other atoms:

(i) $\mathtt{cp}^+_3$ is connected *only* to $\mathtt{fv}_3$, $\mathtt{lambda}_3$, $\mathtt{apply}_3$, $\mathtt{lambda}_1$, $\mathtt{cp}^-_3$, $\mathtt{cp}^+_1$ or $\mathtt{cp}^+_2$.
(ii) $\mathtt{cp}^-_3$ is connected *only* to $\mathtt{apply}_1$, $\mathtt{apply}_2$, $\mathtt{lambda}_2$, $\mathtt{cp}^-_1$, $\mathtt{cp}^-_2$ or $\mathtt{cp}^+_3$.

Clearly a graph in a cp-normal form satisfies these two properties, and it is easy to see that each rewrite rule preserves them.

Reactions involving a $\mathtt{cp}^+$ starts when a $\beta$-reduction eliminates an atom `lambda` connected to $\mathtt{cp}^+_3$ somewhere in a graph. Then the $\mathtt{cp}^+$ descends the tree formed by `apply`, `lambda`, and `fv` until it is annihilated by `fv` or meets no more `apply`'s or `lambda`'s. This descent is eventually finished because $\beta$-reduction is not performed except in the first $n$ steps, at that stage the depth of the tree is fixed.

When the descent of a $\mathtt{cp}^+$ terminates, because of the invariant, $\mathtt{cp}^+_3$ is connected either to $\mathtt{lambda}_1$, $\mathtt{cp}^-_3$, $\mathtt{cp}^+_1$ or $\mathtt{cp}^+_2$. Because $\mathtt{cp}^+_3$ and $\mathtt{cp}^-_3$ either cross each other or annihilate themselves, any remaining $\mathtt{cp}^+_3$ ends up with connected to $\mathtt{lambda}_1$, $\mathtt{cp}^+_1$ or $\mathtt{cp}^+_2$.

A $\mathtt{cp}^+$ entering the left branch of `lambda` is changed to $\mathtt{cp}^-$, but because the links from the left branch of `lambda` are connected to its right branch via `cp`, `lambda` or `apply`, the $\mathtt{cp}^-$ will eventually meet a $\mathtt{cp}^+$ with the same color and disappears. This implies also that all the newly created colors will be changed back to their parent colors and eventually become the top color.

Therefore, the reaction involving `cp`'s will terminate, and when it terminates, all the $\mathtt{cp}^+_3$'s will have the top color and will be connected to $\mathtt{lambda}_1$, $\mathtt{cp}^+_1$ or $\mathtt{cp}^+_2$, satisfying the cp-normal form condition. □

# Revisiting Cut-Elimination: One Difficult Proof Is Really a Proof

Christian Urban[1] and Bozhi Zhu[2]

[1] Technical University of Munich
[2] North China Electric Power University

**Abstract.** Powerful proof techniques, such as logical relation arguments, have been developed for establishing the strong normalisation property of term-rewriting systems. The first author used such a logical relation argument to establish strong normalising for a cut-elimination procedure in classical logic. He presented a rather complicated, but informal, proof establishing this property. The difficulties in this proof arise from a quite subtle substitution operation, which implements proof transformation that permute cuts over other inference rules. We have formalised this proof in the theorem prover Isabelle/HOL using the Nominal Datatype Package, closely following the informal proof given by the first author in his PhD-thesis. In the process, we identified and resolved a gap in one central lemma and a number of smaller problems in others. We also needed to make one informal definition rigorous. We thus show that the original proof is indeed a proof and that present automated proving technology is adequate for formalising such difficult proofs.

## 1 Introduction

Proofs about syntax are often not very deep; rather the difficulties arise from the huge amount of details. Human reasoners seem to be ill-equipped to cope with such amounts of details. This observation is based on the experience obtained with a formalisation [18] of a paper on LF by Harper and Pfenning [6]. Their paper contained many informal proofs spread over more than 30 pages. The formalisation revealed a gap in one of the proofs and a small number of minor lacunae in others. Also in the present paper we describe a formalisation of an informal 20-page proof given by the first author [14] (see also [17]). This proof claims to establish a strong normalisation result of cut-elimination in classical logic. However, this formalisation, too, uncovers a number of errors in the informal proof, including one that required to restate two central lemmas.

In the literature there are numerous informal proofs for the termination of various cut-elimination procedures. One of the main applications of these procedures is to ensure consistency of sequent-calculi, that means that there is no proof for the sequent $\vdash \bot$. Gentzen [5] was the first who proved in this way the consistency of a sequent-calculus for intuitionistic and classical logic. Most of such cut-elimination procedures, including Gentzen's original, are weakly normalising, i.e., they employ a particular cut-elimination procedure strategy. While for establishing consistency a weakly normalising procedure is usually sufficient, if one wants to do computations with sequent proofs then strong normalisation is a more useful property. One reason for this is that

A. Voronkov (Ed.): RTA 2008, LNCS 5117, pp. 409–424, 2008.

cut-elimination in classical logic is *not* confluent and therefore one might reach different cut-free proofs by reducing cuts in a different order or applying different reduction rules.

For the purpose of calculating the collection of all cut-free proofs reachable from a classical proof, the first author introduced in [14,17] a strongly normalising procedure for cut-elimination. Note that simply taking an unrestricted version of Gentzen's cut-elimination procedure, that is removing the strategy, leads to infinite reduction sequences. Therefore the strongly-normalising cut-elimination procedure in [14,17] uses Gentzen's original rules for logical cuts, but modifies the rules for commuting cuts. (An instance of the cut-rule is said to be a *logical* cut when both cut-formulae are introduced by axioms or logical inference rules; otherwise the cut is said to be a *commuting* cut.) An interesting feature of this procedure is that it allows commuting cuts to pass over other cuts. It achieves strong normalisation by restricting the rules for commuting cuts so that they must "transport" in *one* step a commuting cut to *all* places where the corresponding cut-formula is introduced (Gentzen defined for this process local reduction rules, which only rewrite neighboring inference rules in a proof). As a result one ends up with a quite general reduction system for cut-elimination: for example it can simulate $\beta$-reduction in the $\lambda$-calculus [14].

Unfortunately, the generality of the reduction system means also that strong normalisation is much more difficult to prove. Our proof establishing this property is based on symmetric reducibility candidates [2], a powerful proof technique from the term-rewriting literature. To present the proof in a convenient form, sequent proofs are annotated with terms and the cut-elimination procedure is defined as a term-rewriting system. In particular, the proof transformation for commuting cuts is expressed as a special sort of proof substitution.

The disadvantage of using terms is that in order to deal with them in a convenient manner they nearly always need to be quotiented modulo $\alpha$-equivalence—for example in order to have capture-avoiding substitution being definable as a total function. However, this quotienting makes (formal) reasoning much harder: inductions and recursions over the structure of $\alpha$-equated terms are not immediately defined concepts; that means one has to spend some effort to derive them (in contrast with unquotient, or raw, terms where these concepts are for "free"). Moreover function definitions need to respect $\alpha$-equivalence. This precludes, for example, the definition of the function that returns the immediate subterms of an $\alpha$-equated term [15]. When working with such terms, one also often employs an informal variable convention [3] without giving a proper justification for its validity. By using this convention, one does not consider truly arbitrary bound variables, as *required* by the induction principles, but rather bound variables about which various freshness assumptions are made. Such reasoning is in general however unsound (see [16] for an example).

In informal "pencil-and-paper" proofs such problems are usually ignored. While this is harmless in easy proofs of simple properties, in difficult ones ignoring such problems carries the danger of overlooking errors (see [8, Page 16] for one overlooked by Kleene). Since the proof given by the first author for the strong-normalisation property is quite difficult and since a number of researchers have built their results directly on the strong-normalisation property (for example the *lemuridæ* system [4] and the typed version of the $\mathcal{X}$-calculus [20]) or adapted the same proof-technique to other rewrite systems [21], it seems prudent to reconsider whether the original informal proof is actually a proof.

The Nominal Datatype Package [19] provides an infrastructure for reasoning conveniently about datatypes with a built-in notion of $\alpha$-equivalence: it allows to specify such datatypes, provides appropriate recursion combinators and derives strong induction principles that have the usual variable convention already built-in. The latter comes with safeguards that make the variable convention a safe reasoning principle.

The main contribution of this paper is a complete formalisation of a difficult strong normalisation proof.[1] This formalisation uncovers a number of errors in the informal proof and makes one informal definition from the infromal proof more rigorous. The techniques used for the latter are applicable also in other calculi where non-trivial operation need to be defined over terms with involving binders. In the formalisation we also encounter some difficulties with a standard formulation for notion of strong normalisation. The rest of the paper is organised as follows: Sec. 2 reviews the informal proof, that is the definitions for terms, typing-rules and cut-elimination reductions given in [14,17]. The details about the formalisation are given in Sec. 3; Sec. 4 concludes and gives suggestions for further work.

## 2 Sequent Proofs and Cut-Elimination

The main idea behind the cut-elimination procedure presented in [14,17] is to transport one subderivation of a commuting cut to the place(s) where the cut-formula is introduced. To specify this operation, we used terms to annotate sequent proofs, whose inference rules are inspired by Kleene's sequent calculus G3a [7] and the sequent calculus G3c of [13]. These terms encode the structure of a proof and are defined as:

$$\begin{array}{rlll}
M, N ::= & \mathsf{Ax}(x,a) & \text{Axiom} & \\
\mid & \mathsf{Cut}(\langle a\rangle M, (x)N) & \text{Cut} & \\
\mid & \mathsf{And}_R(\langle a\rangle M, \langle b\rangle N, c) & \text{And-R} & \\
\mid & \mathsf{And}^i_L((x)M, y) & \text{And-L}_i & (i=1,2) \\
\mid & \mathsf{Or}^i_R(\langle a\rangle M, b) & \text{Or-R}_i & (i=1,2) \\
\mid & \mathsf{Or}_L((x)M, (y)N, z) & \text{Or-L} & \\
\mid & \mathsf{Imp}_R((x)\langle a\rangle M, b) & \text{Imp-R} & \\
\mid & \mathsf{Imp}_L(\langle a\rangle M, (x)N, y) & \text{Imp-L} &
\end{array} \qquad (1)$$

where $x, y, z$ are taken from a set of *names* and $a, b, c$ from a set of *co-names*. We use round brackets to signify that a name becomes bound and angle brackets that a co-name becomes bound.

Our sequents, or *typing judgements*, are of the form $\Gamma \triangleright M \triangleright \Delta$, where $\Gamma$ is a left-context, $M$ a term and $\Delta$ a right-context. The inference rules for those typing judgements are given in Fig. 1. One distinguishing feature of this term-calculus is that the structural rules, weakening and contraction, are completely implicit in the form of the inference rules. Thus we regard contexts as sets of (label,formula)-pairs, as in type theory, and *not* as multisets, as in LK or LJ. A label is either a name (for left-contexts) or a co-name (for right-contexts).

[1] Available at `http://isabelle.in.tum.de/nominal`.

$$\overline{x:B,\Gamma \triangleright \mathsf{Ax}(x,a) \triangleright \Delta, a:B}$$

$$\frac{x:B_i,\Gamma \triangleright M \triangleright \Delta}{y:B_1{\wedge}B_2,\Gamma \triangleright \mathsf{And}^i_L(\langle x\rangle M, y) \triangleright \Delta}\ \wedge_{L_i} \qquad \frac{\Gamma \triangleright M \triangleright \Delta, a:B \quad \Gamma \triangleright N \triangleright \Delta, b:C}{\Gamma \triangleright \mathsf{And}_R(\langle a\rangle M, \langle b\rangle N, c) \triangleright \Delta, c:B{\wedge}C}\ \wedge_R$$

$$\frac{x:B,\Gamma \triangleright M \triangleright \Delta \quad y:C,\Gamma \triangleright N \triangleright \Delta}{z:B{\vee}C,\Gamma \triangleright \mathsf{Or}_L(\langle x\rangle M, \langle y\rangle N, z) \triangleright \Delta}\ \vee_L \qquad \frac{\Gamma \triangleright M \triangleright \Delta, a:B_i}{\Gamma \triangleright \mathsf{Or}^i_R(\langle a\rangle M, b) \triangleright \Delta, b:B_1{\vee}B_2}\ \vee_{R_i}$$

$$\frac{\Gamma \triangleright M \triangleright \Delta, a:B \quad x:C,\Gamma \triangleright N \triangleright \Delta}{y:B{\supset}C,\Gamma \triangleright \mathsf{Imp}_L(\langle a\rangle M, \langle x\rangle N, y) \triangleright \Delta}\ \supset_L \qquad \frac{x:B,\Gamma \triangleright M \triangleright \Delta, a:C}{\Gamma \triangleright \mathsf{Imp}_R(\langle x\rangle\langle a\rangle M, b) \triangleright \Delta, b:B{\supset}C}\ \supset_R$$

$$\frac{\Gamma_1 \triangleright M \triangleright \Delta_1, a:B \quad x:B,\Gamma_2 \triangleright N \triangleright \Delta_2}{\Gamma_1,\Gamma_2 \triangleright \mathsf{Cut}(\langle a\rangle M, \langle x\rangle N) \triangleright \Delta_1,\Delta_2}\ \mathit{Cut}$$

**Fig. 1.** The inference, or typing, rules of our sequent calculus.

To see how our terms encode sequent proofs, suppose a sequent $\ldots A \vdash B \ldots$ can be proved. Then in our judgments, $A$ and $B$ have labels (say $x : A$ and $a : B$), and $M$ would be an encoding of the proof of $\ldots A \vdash B \ldots$, with these labels, so denoted $\ldots x : A \triangleright M \triangleright a : B \ldots$ Then, where the sequent proof is extended further downwards, $x : A$ and $a : B$ might disappear from the contexts. At the point where they disappear, the corresponding proof-term includes the binding $\langle x\rangle\_$ or $\langle a\rangle\_$, reflecting the fact that the choice of label ($x$ or $a$) is not relevant to the proof as a whole.

To form contexts we have the following conventions: a context can only include a single association for each name (similarly for co-names); a comma in a conclusion stands for the set union and a comma in a premise stands for the *disjoint* set union. Consider for example the $\supset_R$-rule. This rule introduces the (co-name,formula)-pair $b : B{\supset}C$ in the conclusion, and consequently, $b$ is a free co-name in the term $\mathsf{Imp}_R(\langle x\rangle\langle a\rangle M, b)$. However, $b$ can already be free in the subterm $M$, in which case $b : B{\supset}C$ belongs to $\Delta$. Thus the conclusion of the $\supset_R$-rule is of the form

$$\Gamma \triangleright \mathsf{Imp}_R(\langle x\rangle\langle a\rangle M, b) \triangleright \Delta \oplus b : B{\supset}C$$

where $\oplus$ denotes set union. Note that $x : B$ and $a : C$ in the premise are *not* part of the conclusion because they are intended to become bound. Hence the premise must be of the form

$$x : B \otimes \Gamma \triangleright M \triangleright \Delta \otimes a : C$$

where $\otimes$ denotes disjoint set union. Our cut-rule requires that two contexts are joined on each side of the conclusion. Thus we take this rule to be of the following form:

$$\frac{\Gamma_1 \vdash \Delta_1 \otimes a : B \quad x : B \otimes \Gamma_2 \vdash \Delta_2}{\Gamma_1 \oplus \Gamma_2 \vdash \Delta_1 \oplus \Delta_2}\ \mathit{Cut}$$

Next we focus on the cut-elimination rules. For this consider the following logical cut:

$$\dfrac{\dfrac{\Gamma_1 \vdash \Delta_1, B \quad \Gamma_1 \vdash \Delta_1, C}{\Gamma_1 \vdash \Delta_1, B \wedge C}\wedge_R \quad \dfrac{B, \Gamma_2 \vdash \Delta_2}{B \wedge C, \Gamma_2 \vdash \Delta_2}\wedge_{L_1}}{\Gamma_1, \Gamma_2 \vdash \Delta_1, \Delta_2}\mathit{Cut}$$

where we omitted for better readability the labels and term-annotations. We expect this cut to reduce to

$$\dfrac{\Gamma_1 \vdash \Delta_1, B \quad B, \Gamma_2 \vdash \Delta_2}{\Gamma_1, \Gamma_2 \vdash \Delta_1, \Delta_2}\mathit{Cut}\ .$$

However because of our implicit treatment of the structural rules, some care is needed: we have to ensure that the cut-formula $B \wedge C$ does not occur in $\Delta_1$ or $\Gamma_2$. If it does, then we have not a logical cut, but a commuting cut. In order to distinguish between both kinds of cuts, we introduce the notion when a term introduces freshly a name or a co-name (this corresponds to the usual definition of a main formula in an inference rule).

**Definition 1.** *A term, $M$, introduces the name $z$ or co-name $c$, if and only if $M$ is of the form*

| | | | |
|---|---|---|---|
| *for z:* | $\mathsf{Ax}(z, c)$ | *for c:* | $\mathsf{Ax}(z, c)$ |
| | $\mathsf{And}^i_L(\langle x\rangle S, z)$ | | $\mathsf{And}_R(\langle a\rangle S, \langle b\rangle T, c)$ |
| | $\mathsf{Or}_L(\langle x\rangle S, \langle y\rangle T, z)$ | | $\mathsf{Or}^i_R(\langle a\rangle S, c)$ |
| | $\mathsf{Imp}_L(\langle a\rangle S, \langle x\rangle T, z)$ | | $\mathsf{Imp}_R(\langle x\rangle\langle a\rangle S, c)$ |

*A term freshly introduces a name, if and only if none of its proper subterms introduces this name. In other words, the name must not be free in a proper subterm. Similarly for co-names.*

Armed with this definition we can state the cut-reduction rules for dealing with logical cuts (they correspond to Gentzen's rules for logical cuts):

**Definition 2 (Reductions for Logical Cuts, $i = 1, 2$)**

$\mathsf{Cut}(\langle b\rangle \mathsf{And}_R(\langle a_1\rangle M_1, \langle a_2\rangle M_2, b), \langle y\rangle \mathsf{And}^i_L(\langle x\rangle N, y)) \longrightarrow \mathsf{Cut}(\langle a_i\rangle M_i, \langle x\rangle N)$
*if* $\mathsf{And}_R(\langle a_1\rangle M_1, \langle a_2\rangle M_2, b)$ *and* $\mathsf{And}^i_L(\langle x\rangle N, y)$ *freshly introduce $b$ and $y$, resp.*

$\mathsf{Cut}(\langle b\rangle \mathsf{Or}^i_R(\langle a\rangle M, b), \langle y\rangle \mathsf{Or}_L(\langle x_1\rangle N_1, \langle x_2\rangle N_2, y)) \longrightarrow \mathsf{Cut}(\langle a\rangle M, \langle x_i\rangle N_i)$
*if* $\mathsf{Or}^i_R(\langle a\rangle M, b)$ *and* $\mathsf{Or}_L(\langle x_1\rangle N_1, \langle x_2\rangle N_2, y)$ *freshly introduce $b$ and $y$, resp.*

$\mathsf{Cut}(\langle b\rangle \mathsf{Imp}_R(\langle x\rangle\langle a\rangle M, b), \langle z\rangle \mathsf{Imp}_L(\langle c\rangle N, \langle y\rangle P, z))$
$\longrightarrow \mathsf{Cut}(\langle a\rangle \mathsf{Cut}(\langle c\rangle N, \langle x\rangle M), \langle y\rangle P)$ *or*
$\longrightarrow \mathsf{Cut}(\langle c\rangle N, \langle x\rangle \mathsf{Cut}(\langle a\rangle M, \langle y\rangle P))$
*if* $\mathsf{Imp}_R(\langle x\rangle\langle a\rangle M, b)$ *and* $\mathsf{Imp}_L(\langle c\rangle N, \langle y\rangle P, z)$ *freshly introduce $b$ and $z$, resp.*

$\mathsf{Cut}(\langle a\rangle M, \langle x\rangle \mathsf{Ax}(x, b)) \longrightarrow M[a \mapsto b]$ *if $M$ freshly introduces $a$*

$\mathsf{Cut}(\langle a\rangle \mathsf{Ax}(y, a), \langle x\rangle M) \longrightarrow M[x \mapsto y]$ *if $M$ freshly introduces $x$*

In this definition we use $M[a \mapsto b]$ to stand for capture-avoiding renaming of $a$ to $b$ in $M$ (similarly $M[x \mapsto y]$ for names).

The definition of the reduction rules for dealing with commuting cuts is more subtle. Consider the following proof (where again we left out the labels and annotations):

$$\pi_1\left\{\dfrac{\dfrac{\dfrac{}{A,B\vdash C,A^\bullet}}{A\vdash B\supset C,A^\bullet}\supset_R \quad \dfrac{}{A\vdash B\supset C,A}}{A\vee A\vdash B\supset C,A}\vee_L\right. \qquad \left.\dfrac{\dfrac{\dfrac{}{A^\star\vdash D,A}\quad\dfrac{}{A^\star\vdash D,A}}{A\vdash D,A\wedge A}\wedge_R \quad \dfrac{\dfrac{}{A^\star,E\vdash A}\quad\dfrac{}{A^\star,E\vdash A}}{A,E\vdash A\wedge A}\wedge_R}{A,D\supset E\vdash A\wedge A}\supset_L\right\}\pi_2$$
$$\overline{\qquad\qquad A\vee A,D\supset E\vdash B\supset C,A\wedge A\qquad\qquad}\;Cut$$

The cut-formula $A$ is neither a main formula in the inference rule $\vee_L$, nor in $\supset_L$ (on the term-level that means that the terms for $\pi_1$ and $\pi_2$ do not freshly introduce the name and co-name corresponding for $A$). Therefore the cut is a commuting cut. In $\pi_1$ the cut-formula is a main formula in the axioms marked with a bullet; in $\pi_2$, respectively, in the axioms marked with a star. Eliminating the cut in the proof above means to either transport the derivation $\pi_2$ to the places marked with a bullet and "cut it against" the corresponding axioms, or to transport $\pi_1$ and "cut it against" the axioms marked with a star. In both cases the derivation being transported is duplicated. We realise these operations with two symmetric forms of substitution, which we shall write as $P\{x:=\langle a\rangle Q\}$ and $S\{b:=(y)T\}$.

Whenever such a substitution is "next" to a term in which the cut-formula is introduced, then the substitution becomes an instance of the Cut-term constructor. In the following two examples we shall write $\{\sigma\}$ and $\{\tau\}$ for the substitutions $\{c:=(x)P\}$ and $\{x:=\langle b\rangle Q\}$, respectively.

$$\begin{array}{rcl}\mathsf{And}_R(\langle a\rangle M,\langle b\rangle N,c)\{\sigma\} &=& \mathsf{Cut}(\langle c\rangle\mathsf{And}_R(\langle a\rangle\, M\{\sigma\},\langle b\rangle\, N\{\sigma\},c),(x)P)\\ \mathsf{Imp}_L(\langle a\rangle M,(y)N,x)\{\tau\} &=& \mathsf{Cut}(\langle b\rangle Q,(x)\mathsf{Imp}_L(\langle a\rangle\, M\{\tau\},(y)\, N\{\tau\},x))\end{array}$$

In the first term the formula labelled with $c$ is the main formula and in the second the one labelled with $x$. So in both cases the substitutions "expand" to cuts, and in addition, the substitutions are pushed inside the subterms. This is because there might be several occurrences of $c$ and $x$: both labels need not have been freshly introduced. We are left with specifying the cases where the name or co-name that is being substituted for is not a label of the main formula. In these cases the substitutions are pushed inside the subterms or vanish in case of the axioms. Fig. 2 gives all clauses for the cases where a cut expands and the clauses for when a substitution is pushed inside the terms. We do not need to worry about inserting contraction rules when a term is duplicated, since our contexts are sets of labelled formulae, and thus contractions are made implicitly.

There is one point worth mentioning about the clauses marked with $\star$ in Fig. 2. Although these clauses are not needed for strong normalisation, they are needed to have the property

$$M\{x:=\langle a\rangle P\}\{b:=(y)Q\} \;=\; M\{b:=(y)Q\}\{x:=\langle a\rangle P\}$$

for $b$ not free in $\langle a\rangle P$ and $x$ not free in $(y)Q$. This property is crucial in our strong normalisation proof. However, this property does *not* hold for a slightly simpler definition of the substitution operation where the lines marked with $\star$ are deleted and the first two clauses are replaced by $\mathsf{Ax}(x,c)\{c:=(y)P\} \overset{\text{def}}{=} P[y\mapsto x]$ and $\mathsf{Ax}(y,a)\{y:=$

$$\begin{array}{rl}
\mathsf{Ax}(x,c)\{c:=(y)P\} & \stackrel{\text{def}}{=} \mathsf{Cut}(\langle c\rangle\mathsf{Ax}(x,c),(y)P)\\
\mathsf{Ax}(y,a)\{y:=\langle c\rangle P\} & \stackrel{\text{def}}{=} \mathsf{Cut}(\langle c\rangle P,(y)\mathsf{Ax}(y,a))\\
\mathsf{And}_R(\langle a\rangle M,\langle b\rangle N,c)\{c:=(y)P\} & \stackrel{\text{def}}{=} \mathsf{Cut}(\langle c\rangle\mathsf{And}_R(\langle a\rangle M\{c:=(y)P\},\langle b\rangle N\{c:=(y)P\},c),(y)P)\\
\mathsf{And}_L^i((x)M,y)\{y:=\langle c\rangle P\} & \stackrel{\text{def}}{=} \mathsf{Cut}(\langle c\rangle P,(y)\mathsf{And}_L^i((x)M\{y:=\langle c\rangle P\},y))\\
\mathsf{Or}_R^i(\langle a\rangle M,c)\{c:=(y)P\} & \stackrel{\text{def}}{=} \mathsf{Cut}(\langle c\rangle\mathsf{Or}_R^i(\langle a\rangle M\{c:=(y)P\},c),(y)P)\\
\mathsf{Or}_L((x)M,(y)N,z)\{z:=\langle c\rangle P\} & \stackrel{\text{def}}{=} \mathsf{Cut}(\langle c\rangle P,(z)\mathsf{Or}_L((x)M\{z:=\langle c\rangle P\},(y)N\{z:=\langle c\rangle P\},z))\\
\mathsf{Imp}_R((x)\langle a\rangle M,b)\{b:=(y)P\} & \stackrel{\text{def}}{=} \mathsf{Cut}(\langle b\rangle\mathsf{Imp}_R((x)\langle a\rangle M\{b:=(y)P\},b),(y)P)\\
\mathsf{Imp}_L(\langle a\rangle M,(x)N,y)\{y:=\langle c\rangle P\} & \stackrel{\text{def}}{=} \mathsf{Cut}(\langle c\rangle P,(y)\mathsf{Imp}_L(\langle a\rangle M\{y:=\langle c\rangle P\},(x)N\{y:=\langle c\rangle P\},y))\\
\mathsf{Cut}(\langle a\rangle\mathsf{Ax}(x,a),(y)M)\{x:=\langle b\rangle P\} & \stackrel{\text{def}}{=} \mathsf{Cut}(\langle b\rangle P,(y)M\{x:=\langle b\rangle P\}) \quad \star\\
\mathsf{Cut}(\langle a\rangle M,(x)\mathsf{Ax}(x,b))\{b:=(y)P\} & \stackrel{\text{def}}{=} \mathsf{Cut}(\langle a\rangle M\{b:=(y)P\},(y)P) \quad \star
\end{array}$$

Otherwise we push the substitution inside the subterms

$$\begin{array}{rl}
\mathsf{Ax}(x,a)\{\sigma\} & \stackrel{\text{def}}{=} \mathsf{Ax}(x,a)\\
\mathsf{Cut}(\langle a\rangle M,(x)N)\{\sigma\} & \stackrel{\text{def}}{=} \mathsf{Cut}(\langle a\rangle\ M\{\sigma\},(x)\ N\{\sigma\})\\
\mathsf{And}_R(\langle a\rangle M,\langle b\rangle N,c)\{\sigma\} & \stackrel{\text{def}}{=} \mathsf{And}_R(\langle a\rangle\ M\{\sigma\},\langle b\rangle\ N\{\sigma\},c)\\
\mathsf{And}_L^i((x)M,y)\{\sigma\} & \stackrel{\text{def}}{=} \mathsf{And}_L^i((x)\ M\{\sigma\},y)\\
\mathsf{Or}_R^i(\langle a\rangle M,b)\{\sigma\} & \stackrel{\text{def}}{=} \mathsf{Or}_R^i(\langle a\rangle\ M\{\sigma\},b)\\
\mathsf{Or}_L((x)M,(y)N,z)\{\sigma\} & \stackrel{\text{def}}{=} \mathsf{Or}_L((x)\ M\{\sigma\},(y)\ N\{\sigma\},z)\\
\mathsf{Imp}_R((x)\langle a\rangle M,b)\{\sigma\} & \stackrel{\text{def}}{=} \mathsf{Imp}_R((x)\langle a\rangle\ M\{\sigma\},b)\\
\mathsf{Imp}_L(\langle a\rangle M,(x)N,y)\{\sigma\} & \stackrel{\text{def}}{=} \mathsf{Imp}_L(\langle a\rangle\ M\{\sigma\},(x)\ N\{\sigma\},y)
\end{array}$$

**Fig. 2.** Definition of the substitution operation. This operation is used in the cut-reduction dealing with commuting cuts. For more details see [14,17].

$\langle c\rangle P\} \stackrel{\text{def}}{=} P[c \mapsto a]$. This simpler definition corresponds to the more familiar method how cuts are eliminated. A consequence of the lines marked with $\star$, as we shall see, is that calculations and properties involving substitution are quite subtle.

However, we are now in a position to complete the definition of our cut-elimination procedure by stating how commuting cuts reduce, namely:

**Definition 3 (Reductions for Commuting Cuts)**

$$\begin{array}{rll}
\mathsf{Cut}(\langle a\rangle M,(x)N) & \longrightarrow M\{a:=(x)N\} & \textit{if } M \textit{ does not freshly introduce } a, \ \textbf{or}\\
& \longrightarrow N\{x:=\langle a\rangle M\} & \textit{if } N \textit{ does not freshly introduce } x
\end{array}$$

and close the reduction relation under term-formation. The important properties of this cut-elimination procedure are subject-reduction and strong normalisation.

**Theorem 1 (Subject Reduction and Strong Normalisation [14,17])**

- *If* $\Gamma \triangleright M \triangleright \Delta$ *and* $M \longrightarrow M'$ *then* $\Gamma \triangleright M' \triangleright \Delta$.
- *If* $\Gamma \triangleright M \triangleright \Delta$ *then* $M$ *is strongly normalising w.r.t.* $\longrightarrow$.

## 3 The Formalisation of the Strong Normalisation Proof

In this section we describe the formalisation of the strong normalisation proof. While the informal description of this proof is already quite detailed—the details are spread over more than 20 pages in [14], we found that several subtle points were overlooked, one central lemma is faulty and has to be restated, and a definition has to be made rigorous.

The definition of the terms given in (1) and formulae (which we omitted in this paper) pose no problem for the Nominal Datatype Package, as it was designed to deal with such definitions. From the definition of terms, the package derives automatically a weak and a strong structural induction principle (the strong one has the variable convention already built in [19]), and provides a recursion combinator for defining functions over the structure of the terms [15]. With this combinator, it is easy to define the capture-avoiding renaming functions $M[a \mapsto b]$ and $M[x \mapsto y]$, although these definitions require that several proof-obligations are discharged by the user (the proof-obligations ensure that the renaming-functions preserve $\alpha$-equivalence).

The typing-system rules given in Fig. 1 can also be formalised with ease, except that we have chosen to represent typing-context as (label,formula)-lists rather than sets. This requires that we add appropriate validity and freshness-constraints to the inference rules. A context is defined to be *valid* provided no name or co-name occurs twice. This can be stated with the rules:

$$\frac{}{valid([])} \qquad \frac{a \# \Delta \quad valid(\Delta)}{valid((a:B)::\Delta)} \qquad \frac{x \# \Gamma \quad valid(\Gamma)}{valid((x:B)::\Gamma)}$$

where $a \# \Delta$ (similarly $x \# \Gamma$) stands for $a$ being fresh for $\Delta$ (i.e. not occurring in $\Delta$). Using this definition and freshness, the axiom and $\wedge_R$-rules, for example, look in the formalisation as follows:

$$\frac{valid(\Gamma) \quad valid(\Delta) \quad (x:B) \in \Gamma \quad (a:B) \in \Delta}{\Gamma \triangleright \mathsf{Ax}(a,b) \triangleright \Delta}$$

$$\frac{\begin{array}{c}\Gamma \triangleright M \triangleright (a:B)::\Delta \quad \Gamma \triangleright N \triangleright (b:C)::\Delta \\ a \# \Delta \quad b \# \Delta \quad \Delta' \approx (c:B \wedge C)::\Delta \quad valid(\Delta')\end{array}}{\Gamma \triangleright \mathsf{And}_R(\langle a\rangle M, \langle b\rangle N, c) \triangleright \Delta'} \wedge_R$$

where $\approx$ stands for two lists being equal if regarded as sets.

Most of the effort during the formalisation we had to invest in defining the substitution operation and proving associated lemmas. One reason for this is that the informal definition given in Fig. 2 makes from a formal point of view only little sense. For example, merging the two clauses given for $\{c := (x)P\}$ and the term-constructor $\mathsf{And}_R(\langle a\rangle M, \langle b\rangle N, d)$ into an *if*-statement (as is required in a formal definition by recursion over the structure of terms) leads to:

$$\begin{array}{l}\mathsf{And}_R(\langle a\rangle M, \langle b\rangle N, d)\{c{:=}(x)P\} \stackrel{\text{def}}{=} \\ \quad \mathit{if}\ c = d \\ \qquad \mathit{then}\ \mathsf{Cut}(\langle d\rangle \mathsf{And}_R(\langle a\rangle(M\{c{:=}(x)P\}), \langle b\rangle(N\{c{:=}(x)P\}), d), (x)P) \\ \qquad \mathit{else}\ \mathsf{And}_R(\langle a\rangle(M\{c{:=}(x)P\}), \langle b\rangle(N\{c{:=}(x)P\}), d)\end{array}$$

where the "true"-branch corresponds to the case where the substitution expands to a cut, and the "false"-branch where the substitution is just pushed inside the subterms $M$ and $N$. The obvious problem is that we attempt to push a substitution under binders (in this example the binders are $\langle a\rangle\_$ and $\langle b\rangle\_$). This is only possible provided $a$ and $b$ do not occur freely in the term $P$. Hence we have to restrict the clause with suitable preconditions, namely:

$$\begin{array}{l}\text{provided } a \# P \text{ and } b \# P \text{ then}\\ \quad \mathsf{And}_R(\langle a\rangle M, \langle b\rangle N, d)\{c{:=}(x)P\} \stackrel{\text{def}}{=}\\ \qquad \textit{if } c = d\\ \qquad\quad \textit{then } \mathsf{Cut}(\langle d\rangle \mathsf{And}_R(\langle a\rangle(M\{c{:=}(x)P\}), \langle b\rangle(N\{c{:=}(x)P\}), d), (x)P)\\ \qquad\quad \textit{else } \mathsf{And}_R(\langle a\rangle(M\{c{:=}(x)P\}), \langle b\rangle(N\{c{:=}(x)P\}), d)\end{array}$$

Since we define substitution over $\alpha$-equivalence classes, we still obtain a total function with this restriction in place. The hope is that we can always rename $\mathsf{And}_R(\langle a\rangle M, \langle b\rangle N, d)$ appropriately so that the preconditions are met. However, this is futile for the proof substitution operation, because in the "true"-branch also the (free) co-name $d$ is bound with the scope of $P$—and we cannot rename (potentially) free co-names in a term without violating $\alpha$-equivalence. The way out is to choose in the "true"-branch explicitly a fresh co-name $d'$ and define the clause formally as

$$\begin{array}{l}\text{provided } a \# P \text{ and } b \# P \text{ then}\\ \quad \mathsf{And}_R(\langle a\rangle M, \langle b\rangle N, d)\{c{:=}(x)P\} \stackrel{\text{def}}{=}\\ \qquad \textit{if } c = d\\ \qquad\quad \textit{then fresh } (\lambda d'.\mathsf{Cut}(\langle d'\rangle \mathsf{And}_R(\langle a\rangle(M\{c{:=}(x)P\}), \langle b\rangle(N\{c{:=}(x)P\}), d'), (x)P))\\ \qquad\quad \textit{else } \mathsf{And}_R(\langle a\rangle(M\{c{:=}(x)P\}), \langle b\rangle(N\{c{:=}(x)P\}), d)\end{array}$$

using the fresh function defined in [10]. Space constraints prevent us to give more details about this function here, except that this function characterises when a construction that picks a fresh (co-)name is independent of which fresh (co-)name is chosen. While this clause (and similar ones for the other term-constructors) give us the properties we expect from the substitution operation (which defines cut-reductions), the corresponding definition leads to quite complicated proofs. One reason is that in the "true"-branches we need to find a fresh (co-)name so that the fresh function provides us with the desired result. At the moment this has to be done by hand, as the Nominal Datatype Package provides only little help for dealing conveniently with the fresh functions.

Having properly defined the substitution operation, we can prove facts about how substitutions interact; for example the following form of the substitution lemma is needed in several places in the proof:

**Lemma 1 (Some Substitution Lemmas)**

- *If* $x \# P$ *then* $M\{x{:=}\langle c\rangle \mathsf{Ax}(y,c)\}\{y{:=}\langle c\rangle P\} = M\{y{:=}\langle c\rangle P\}\{x{:=}\langle c\rangle P\}$
- *If* $x \# (y)Q$ *then* $M\{x{:=}\langle a\rangle P\}\{b{:=}(y)Q\} = M\{b{:=}(y)Q\}\{x{:=}\langle a\rangle(P\{b{:=}(y)Q\})\}$

where the first one is needed in the proof of the second (note that in the second $x \# (y)Q$ stands for $x = y$ or $x \# Q$). We prove such lemmas using the strong structural induction principle for terms, as this minimises the need for renaming bound names and co-names. Still these proofs require some considerable effort due to the sheer number of cases that need to be analysed. In order to appreciate the difficulties involving the substitution operation for terms, note that the symmetric property for the first part of the lemma, namely

$$M\{y:=\langle c\rangle P\}\{x:=\langle c\rangle \mathsf{Ax}(y,c)\} = M\{y:=\langle c\rangle P\}\{x:=\langle c\rangle P\}$$

does not hold. The formalisation of properties about substitution requires approximately 20% of the formalisation code.

In comparison with the definition of the substitution operation, the definition of the cut-reduction relation is relatively simple. It relies on the auxiliary notions for when a term freshly introduces a name or co-name; for example

$$\frac{}{\mathit{fin}(\mathsf{Ax}(z,a),z)} \qquad \frac{z\ \#\ (x)M \quad z\ \#\ (y)N}{\mathit{fin}(\mathsf{Or}_L((x)M,(y)N,z),z)}$$

and so on for the notion of freshly introducing a name (similarly *fic* for freshly introducing a co-name). The formal definition of the cut-reductions look then as follows (the first two are examples for logical cuts; the last for a commuting cut):

$$\frac{\mathit{fic}(M,a)}{\mathsf{Cut}(\langle a\rangle M,(x)\mathsf{Ax}(x,b)) \longrightarrow M[a\mapsto b]}$$

$$\frac{\mathit{fic}(\mathsf{And}_R(\langle a_1\rangle M_1,\langle a_2\rangle M_2,b),b) \qquad \mathit{fin}(\mathsf{And}_L^1((x)N,y),y)}{\mathsf{Cut}(\langle b\rangle\mathsf{And}_R(\langle a_1\rangle M_1,\langle a_2\rangle M_2,b),(y)\mathsf{And}_L^1((x)N,y)) \longrightarrow \mathsf{Cut}(\langle a_1\rangle M_1,(x)N)}$$

$$\frac{\neg\mathit{fic}(M,a)}{\mathsf{Cut}(\langle a\rangle M,(x)N) \longrightarrow M\{a:=(x)N\}}$$

In addition to those rules we specified in the formalisation a slew of congruence rules, such as:

$$\frac{M \longrightarrow M'}{\mathsf{And}_R(\langle a\rangle M,\langle b\rangle N,c) \longrightarrow \mathsf{And}_R(\langle a\rangle M',\langle b\rangle N,c)}$$

$$\frac{N \longrightarrow N'}{\mathsf{And}_R(\langle a\rangle M,\langle b\rangle N,c) \longrightarrow \mathsf{And}_R(\langle a\rangle M,\langle b\rangle N',c)}$$

These rules were not explicitly mentioned in the informal proof. For the cut-reductions and *fin* (similarly *fic*) we need to establish the lemma:

**Lemma 2.** *If $M \longrightarrow M'$ and $\mathit{fin}(M,x)$ then $\mathit{fin}(M',x)$.*

This is relatively easy to prove by a strong induction over $M \longrightarrow M'$. We next establish important properties characterising the interactions between cut-reductions and substitution:

**Lemma 3**

- $M\{x:=\langle c\rangle\mathsf{Ax}(y,c)\} \longrightarrow^* M[x\mapsto y]$
- $M\{c:=(x)\mathsf{Ax}(x,d)\} \longrightarrow^* M[c\mapsto d]$
- *If $M \longrightarrow M'$ then $M\{\sigma\} \longrightarrow^* M'\{\sigma\}$.*

The first two properties are by strong structural inductions over $M$; the third is by strong induction over the reduction $M \longrightarrow M'$. All proofs require many case distinctions and rely on additional proofs relating substitutions and the inductively defined predicates *fin* and *fic*. The third property in this lemma is interesting insofar as it is quite un-intuitive considering a similar property for capture avoiding substitution in the lambda-calculus.[2] In the informal proof [14,17], this lemma was stated and proved as:

**Lemma 4 (Faulty).** *If $M \longrightarrow M'$ then either $M\{\sigma\} = M'\{\sigma\}$ or $M\{\sigma\} \longrightarrow M'\{\sigma\}$.*

The case where $M\{\sigma\} = M'\{\sigma\}$ was correctly analysed. It involves reductions of the form

$$\mathsf{Cut}(\langle a\rangle M, (x)\mathsf{Ax}(x, b)) \longrightarrow M[a \mapsto b]$$

with $M$ being of the form $\mathsf{Ax}(y, a)$ and $\{\sigma\}$ being $\{y := \langle c\rangle P\}$. In this case $M\{y := \langle c\rangle P\}$ is defined as $\mathsf{Cut}(\langle c\rangle P, (x)\mathsf{Ax}(x, b))$, and $M'\{y := \langle c\rangle P\}$ as $\mathsf{Ax}(y, b)\{y := \langle c\rangle P\}$, which in turn is defined as $\mathsf{Cut}(\langle c\rangle P, (y)\mathsf{Ax}(y, b))$. Both terms are equal by $\alpha$-equivalence.

However, the case where $M\{\sigma\}$ needs more than one reduction to reach $M'\{\sigma\}$ was overlooked! Such a case occurs with logical cuts, for example

$$\mathsf{Cut}(\langle b\rangle \mathsf{And}_R(\langle a_1\rangle M_1, \langle a_2\rangle M_2, b), (z)\mathsf{And}^1_L((x)N, z)) \longrightarrow \mathsf{Cut}(\langle a_1\rangle M_1, (x)N)$$

with the proviso that $M_1$ is of the form $\mathsf{Ax}(y, a_1)$. In this case the left-hand side $M\{\sigma\}$ is

$$\mathsf{Cut}(\langle b\rangle \mathsf{And}_R(\langle a_1\rangle \mathsf{Cut}(\langle c\rangle P, (y)\mathsf{Ax}(y, a_1)), \langle a_2\rangle M_2\{\sigma\}, b), (z)\mathsf{And}^1_L((x)N\{\sigma\}, z))$$

which in a single step reduces to

$$\mathsf{Cut}(\langle a_1\rangle \mathsf{Cut}(\langle c\rangle P, (y)\mathsf{Ax}(y, a_1)), (x)N\{\sigma\})$$

and in possibly more than one step reduces to

$$\mathsf{Cut}(\langle a_1\rangle P[c \mapsto a_1], (x)N)$$

which in turn is equal to $M'\{\sigma\}$. Since in the formalisation we have to go through every case one by one, such cases cannot be overlooked there. Fortunately, the proof of the more general lemma goes through. Fortunately, also, the more general property does not destroy the overall proof: the next lemma (Lem. 7 below) that uses this lemma can be modified to deal with the many step-reduction sequence.

Next the informal proof considered the notion for a term being strongly normalising. This notion was stated as all reductions sequences starting from a term must be finite. As the formal definition of a term $M$ being strongly normalising we used the inductive definition:

$$\frac{\forall M'. M \longrightarrow M' \text{ implies } M' \in SN}{M \in SN} \tag{2}$$

This is a standard definition used in many formalisations. Two interesting phenomena arose however with this definition. One was that in the informal proof we stated in a passing (one-sentence) remark that the strong normalisation is preserved under renamings, namely

[2] In the $\lambda$-calculus the property is if $M \longrightarrow_\beta M'$ then $M\{\sigma\} \longrightarrow_\beta M'\{\sigma\}$.

**Lemma 5.** *If $M \in SN$ then $M[a{\mapsto}b] \in SN$ and also $M[x{\mapsto}y] \in SN$.*

This lemma is "obvious" because renaming cannot create any new redexes, or cuts (unlike the proof substitution which might create new cuts). Surprisingly, however, this fact caused us a lot of frustration in the formalisation and resulted in slightly more than 10%(!) of the formalisation code. The problem is that we know by induction hypothesis that $(\forall M'.M \longrightarrow M'$ implies $M' \in SN)$. We can further assume that for an $M'$, $M[a \mapsto b]$ reduces to $M'$, and we have to show that $M' \in SN$. To do so, we have to analyse how $M[a{\mapsto}b]$ reduces w.r.t. $M$. As a result we have to show a fact:

**Lemma 6.** *If $M[a \mapsto b] \longrightarrow M'$, then there exists an $M_0$ such that $M' = M_0[a \mapsto b]$ and $M \longrightarrow M_0$.*

Its proof needs to analyse all the term constructors and all the applicable reductions. This is extremely laborious. The problem is independent of our calculus and would also arise in the $\lambda$-calculus. The fact that the lemma is "obvious", but its proof is hard, seems to indicate that the definition shown in (2) is not the right definition for establishing Lemma 5. We have however not seen any better formal definition for strong normalisation in the term-rewriting literature.

Finally the informal proof establishes that all typable terms are strongly normalising. Surprisingly the symmetric candidates $[\![(B)]\!]$ and $[\![\langle B\rangle]\!]$ defined for this part of the proof do not create any difficulties (the corresponding definitions are therefore omitted here, see [14,17]). We show that the candidates are closed under reductions

**Lemma 7 (Reduction Preserves Candidates)**

- *If $\langle a\rangle M \in [\![\langle B\rangle]\!]$ and $M \longrightarrow^* M'$, then $\langle a\rangle M' \in [\![\langle B\rangle]\!]$.*
- *If $(x)M \in [\![(B)]\!]$ and $M \longrightarrow^* M'$, then $(x)M' \in [\![(B)]\!]$.*

In comparison with the informal proof, however, the assumptions in this lemma had to be strengthened to deal with many-step reductions (i.e. $\longrightarrow^*$) because of the flaw in the third part of Lemma 3. However, this generalisation does not affect the structural induction over $B$ that is employed to establish Lemma 7. Now we can show how the candidates imply the property of strong normalisation, namely

**Lemma 8**

- *If $\langle a\rangle M \in [\![\langle B\rangle]\!]$, then $M \in SN$;*
- *If $(x)M \in [\![(B)]\!]$, then $M \in SN$.*

The last difficult lemma spells out the conditions when a cut is strongly normalising, namely:

**Lemma 9.** *If $M, N \in SN$ and $\langle a\rangle M \in [\![\langle B\rangle]\!]$, $(x)N \in [\![(B)]\!]$ then*

$$\mathsf{Cut}(\langle a\rangle M, (x)N) \in SN\ .$$

The informal proof of this lemma is inspired by a technique of Prawitz [11]. It proceeds by induction over a lexicographically ordered induction value of the form $(\delta, \mu, \nu)$ where $\delta$ is the size of the cut-formula $B$; $\mu$ and $\nu$ are the longest reductions sequences starting from $M$ and $N$. Because of the assumptions that $M \in SN$and $N \in SN$, the informal proof claims without proof that these maximal lengths must be finite and the induction therefore is sensible.

Here arises, however, the second problem with the definition of strong normalisation shown in (2): while this claim is indeed true for the reduction system at hand, it is not true in general. The reason is that a strongly normalising term does not need to have an upper bound for the longest reduction sequence: consider the term $M$ that reduces in one step to the normal form $M_1$, but also in *two* steps to the normal form $M_2$ and so on for any $n$. For this term we have that every reduction sequence starting from $M$ is finite, but there is no upper bound for the length of the longest reduction sequence starting from $M$. This problem does not arise in our reduction system, because $\longrightarrow$ is only finitely branching. Together with the König's lemma one can then infer that a longest reduction sequence indeed exists for every strongly normalising term. The problem with this argument, however, is that establishing that $\longrightarrow$ is only finitely branching is far from trivial and also a formalisation of König's lemma is not readily available in Isabelle.

We were able to completely avoid the work involved with this argument by performing a well-founded induction, not on the triple using the longest reduction sequence, but directly on the predicate $SN$ (which is well-founded). This change in the induction value does not require any changes to the other arguments in the proof. Though we had to supply details for cases which were not present in the informal proof and which were not like the other cases that were shown.

The final proof builds up a closing substitution for a well-typed term $\Gamma \triangleright M \triangleright \Delta$ and shows that $M$ is strongly normalising under this closing substitution. While all these proofs involving candidates are quite laborious, they do not contain any surprises (except the point about the length of the longest reduction sequence of a strongly normalising term). Therefore we omit all details about them. The formalisation is part of the Nominal Datatype Package and can be downloaded from

`http://isabelle.in.tum.de/nominal`

## 4 Conclusion

We have described a formalisation of an informal proof establishing the strong-normalisation property of the cut-elimination procedure for classical logic given in [14,17]. Besides confirming that the informal proof is really a proof (all errors can be fixed), the purpose of this paper is to convey the point that such formalisations are feasible and the formal proving techniques are within reach of being useful in "everyday" reasoning (The distribution of the Nominal Datatype Package contains a number of other formalisations from a wide range of topics). The formalisation of the strong-normalisation proof was still quite demanding. However, given that the time formalising the informal proof (which was however already quite detailed) is roughly equal to the time finding and writing down this informal proof, then this additional effort seems more than acceptable to us. The additional time spent with formalising the proof ensures that no case is overlooked and that the definitions are rigorous. Also, having a formalisation of the proof allows one to "play" with the definitions. This is in contrast with an informal proof where it is rather impractical to change any definition, since checking that the change does not affect the proof is "equal" to re-doing the proof. In contrast, we hope to be able to improve in the future upon the problems we encountered with the definition of strong normalisation. Once we find a more convenient alternative definition for strong normalisation, we can just re-run the formalisation and quickly focus on the places where the proof might break with the new definition.

Our formalisation is at the moment the biggest single-file formalisation in the whole Isabelle distribution. Its size is slightly more than 770 KByte. It took us approximately 5 person-weeks to complete the formalisation (including finding fixes for all the problems). The big size and speed with which the formalisation was completed is due to the fact that in cut-elimination proofs many cases are repetitive and only differ in details. So we were often able to complete one case and then cut-and-paste this case in place of the other cases. The copied code then often only needed tweaking to deal with slightly different assumptions and proof-obligations. The formalisation needs approximately 14 minutes to check on a standard laptop. Our work is now used for benchmarking Isabelle and also has proved to be a very useful testcase for any new features that are implemented in Isabelle.

The Nominal Datatype Package has been invaluable for proving properties about terms involving *simple*, lambda-calculus-like binders (our terms annotated to sequent-proofs are only slightly more complicated than the $\lambda$-terms annotated to natural deduction proofs). We note that de-Bruijn indices can be used in *principle* for such formalisations involving $\alpha$-equated terms; but also note practical difficulties when several kinds of binders need to be treated and some binders even occur iterated in term-constructors (like in our $\mathsf{Imp}_R$). In our opinion, a proof on the scale that we have done here employing de-Bruijn indices is not feasible, because of the complications arising from our substitution operation. Twelf, a system that provides an infrastructure for reasoning about higher-order abstract syntax—another existing technique for dealing with binders, seems not yet streamlined enough to deal conveniently with logical relation arguments on the scale that are used in the informal proofs above (See [12] for an approach about how to perform logical relation arguments in Twelf). The formalisation of a weakly normalising cut-elimination procedure done by Pfenning [9] using higher-order abstract syntax in Twelf does not seem to scale to our strong normalisation proof, as it is impossible to define our notion of symmetric reducibility candidates in Twelf. Also our proof is substantially more complex than the proof underlying the formalisation by Pfenning (he considers only weak normalisation). Aydemir *et al.* have reported recently [1] that a locally nameless representation for terms with binders has been very useful in formalising informal proofs from programming language theory. We have not yet been able to thoroughly compare their results with ours and do not know how their results scale to our quite difficult proof. For example, what helped us to avoid mistakes in our formalisation was that names and co-names have different type. As a result the type-system will immediately complain whenever we mixed up these names. It remains to be seen whether a locally nameless representation of terms can be defined so that formalisations have a similar convenience.

The most annoying aspect in our formalisation is the lack of automated support for dealing with the fresh function. Finding an appropriate fresh name or co-name that meets the conditions can be easily automated. However the verification of the conditions associated with the fresh function seems hard to automate. This and comparing our work with the one by Aydemir *et al.* we leave as future work.

**Acknowledgements.** The first author thanks Jeremy Dawson and Michael Norrish who gave helpful hints to formalise Lemma 9. Markus Wenzel and Stefan Berghofer streamlined Isabelle to cope with the size of the formalisation.

## References

1. Aydemir, B., Charguéraud, A., Pierce, B.C., Pollack, R., Weirich, S.: Engineering Formal Metatheory. In: Proc. of the 35rd Symposium on Principles of Programming Languages (POPL), pp. 3–15. ACM, New York (2008)
2. Barbanera, F., Berardi, S.: A Symmetric Lambda Calculus for "Classical" Program Extraction. In: Hagiya, M., Mitchell, J.C. (eds.) TACS 1994. LNCS, vol. 789, pp. 495–515. Springer, Heidelberg (1994)
3. Barendregt, H.: The Lambda Calculus: Its Syntax and Semantics. Studies in Logic and the Foundations of Mathematics, vol. 103. North-Holland, Amsterdam (1981)
4. Brauner, P., Houtmann, C., Kirchner, C.: Principles of Superdeduction. In: Proc. of the 22nd Annual IEEE Symposium on Logic in Computer Science (LICS), pp. 41–50 (2007)
5. Gentzen, G.: Untersuchungen über das logische Schließen I and II. Mathematische Zeitschrift 39, 176–210, 405–431 (1935)
6. Harper, R., Pfenning, F.: On Equivalence and Canonical Forms in the LF Type Theory. ACM Transactions on Computational Logic 6(1), 61–101 (2005)
7. Kleene, S.C.: Introduction to Metamathematics. North-Holland, Amsterdam (1952)
8. Kleene, S.C.: Disjunction and Existence Under Implication in Elementary Intuitionistic Formalisms. Journal of Symbolic Logic 27(1), 11–18 (1962)
9. Pfenning, F.: Structural Cut Elimination. Information and Computation 157(1–2), 84–141 (2000)
10. Pitts, A.: Alpha-Structural Recursion and Induction. Journal of the ACM 53, 459–506 (2006)
11. Prawitz, D.: Ideas and Results of Proof Theory. In: Proceedings of the 2nd Scandinavian Logic Symposium. Studies in Logic and the Foundations of Mathematics, vol. 63, pp. 235–307. North-Holland, Amsterdam (1971)
12. Schürmann, C., Sarnat, J.: Towards a Judgemental Reconstruction of Logical Relation Proofs. In: Proc. of the 23rd IEEE Symposium on Logic in Computer Science (LICS) (to appear, 2008)
13. Troelstra, A.S., Schwichtenberg, H.: Basic Proof Theory. Cambridge Tracts in Theoretical Computer Science, vol. 43. Cambridge University Press, Cambridge (1996)
14. Urban, C.: Classical Logic and Computation. PhD thesis, Cambridge University (October 2000)
15. Urban, C., Berghofer, S.: A Recursion Combinator for Nominal Datatypes Implemented in Isabelle/HOL. In: Furbach, U., Shankar, N. (eds.) IJCAR 2006. LNCS (LNAI), vol. 4130, pp. 498–512. Springer, Heidelberg (2006)
16. Urban, C., Berghofer, S., Norrish, M.: Barendregt's Variable Convention in Rule Inductions. In: Pfenning, F. (ed.) CADE 2007. LNCS (LNAI), vol. 4603, pp. 35–50. Springer, Heidelberg (2007)

17. Urban, C., Bierman, G.: Strong Normalisation of Cut-Elimination in Classical Logic. Fundamenta Informaticae 45(1–2), 123–155 (2001)
18. Urban, C., Cheney, J., Berghofer, S.: Mechanizing the Metatheory of LF. In: Proc. of the 23rd IEEE Symposium on Logic in Computer Science (LICS). Technical report (to appear, 2008), `http://isabelle.in.tum.de/nominal/LF`
19. Urban, C., Tasson, C.: Nominal Techniques in Isabelle/HOL. In: Nieuwenhuis, R. (ed.) CADE 2005. LNCS (LNAI), vol. 3632, pp. 38–53. Springer, Heidelberg (2005)
20. van Bakel, S., Lengrand, S., Lescanne, P.: The Language X: Circuits, Computations and Classical Logic. In: Coppo, M., Lodi, E., Pinna, G.M. (eds.) ICTCS 2005. LNCS, vol. 3701, pp. 81–96. Springer, Heidelberg (2005)
21. Yoshida, N., Berger, M., Honda, K.: Strong Normalisation in the $\pi$-Calculus. In: Proc. of the 16th IEEE Symposium on Logic in Computer Science (LICS), pp. 311–322 (2001)

# Reduction Under Substitution

Jörg Endrullis and Roel de Vrijer

VU Vrije Universiteit Amsterdam

**Abstract.** The Reduction-under-Substitution Lemma (RuS), due to van Daalen [Daa80], provides an answer to the following question concerning the lambda calculus: given a reduction $M[x := L] \twoheadrightarrow N$, what can we say about the contribution of the substitution to the result $N$. It is related to a not very well-known lemma that was conjectured by Barendregt in the early 70's, addressing the similar question as to the contribution of the argument $M$ in a reduction $FM \twoheadrightarrow N$. The origin of Barendregt's Lemma lies in undefinablity proofs, whereas van Daalen's interest came from its application to the so-called Square Brackets Lemma, which is used in proofs of strong normalization.

In this paper we compare various forms of RuS. We strengthen RuS to multiple substitution and context filling and show how it can be used to give short and perspicuous proofs of undefinability results. Most of these are known as consequences of Berry's Sequentiality Theorem, but some fall outside its scope. We show that RuS can also be used to prove the sequentiality theorem itself. To that purpose we give a further adaptation of RuS, now also involving "bottom" reduction rules, sending unsolvable terms to a bottom element and in the limit producing Böhm trees.

## 1 Introduction

The Reduction-under-Substitution Lemma (RuS) addresses the following question concerning the $\lambda$-calculus: given a reduction $M[x := L] \twoheadrightarrow N$, what is the contribution of the substitution to the result $N$? Or, equivalently: how much of $N$ can be produced already by $M$, independently of the substitution? The answer to the second question will turn out to be: a prefix of $N$. Thus there is a natural inverse correspondence with the so-called prefix property, cf. [BKV00] or [Ter03], Ch. 8.

RuS was formulated by Diederik van Daalen [Daa80] as a slightly strengthened version of an observation of Barendregt [Bar74], addressing the same questions as to the contribution of the argument $M$ in a reduction $FM \twoheadrightarrow N$. We will study Barendregt's Lemma (BL) in Section 2. Because of its more general form, van Daalen's formulation allowed for an easier and more elegant proof than BL. RuS found its way into Barendregt's book on the $\lambda$-calculus [Bar84], where it ended up as Exercise 15.4.8. This literally seemed to be the end of the story, as subsequently little more attention has been paid in the literature to either BL or RuS. Unjustly so, as we hope to make clear in this paper.

The origin of Barendregt's Lemma lies in undefinability. In accordance, Exercise 15.4.8 in [Bar84] is employed there as one of two methods to obtain the

A. Voronkov (Ed.): RTA 2008, LNCS 5117, pp. 425–440, 2008.

undefinability of Church's $\delta$ (using a particular encoding of numerals), the other method using a Böhm-out technique. In [Vri87] Barendregt's Lemma was used for a quick proof of the undefinability of surjective pairing in the $\lambda$-calculus, which was one of the early results of Barendregt in [Bar74], there proved using the technique of underlining.

Van Daalen's interest in reduction under substitution derived from the fact that it implied the so-called Square Brackets Lemma (SqBL), a structural lemma on the contribution of a substitution in a reduction to abstractor form. The SqBL was the key to van Daalen's new and original method for proving strong normalization. Use has been made of this method in [Daa80], [Lév75], [Bar84], Ch. 14, and [Oos97]. It is also discussed in [Vri07], to which we refer for a detailed historical account of Barendregt's Lemma and reduction under substitution.

The aim of this paper is twofold. First, to give a cogent exposition of reduction under substitution. Thereto the first two sections are explanatory in character. The second goal is to explore the potential of RuS for producing new insights, proof methods and results in the $\lambda$-calculus, starting by generalizing RuS to multiple subsitution, and later also extending it to filling holes in contexts. The essential difference is that hole filling may introduce variables that are captured by a binder, whereas substitution may not.

We will present new elementary proofs of undefinability results that are sometimes presented as applications of Berry's Sequentiality Theorem (BST), [Ber78, Ber79], [Bar84]. BST is in terms of Böhm trees, and therefore it is intrinsically infinitary, whereas RuS is just a structural observation on finite reductions. We also use RuS to prove the Perpendicular Lines Theorem for open terms with respect to $\beta$-conversion, thereby confirming a conjecture from [BS99]. Finitary proofs of classic undefinability results have also been obtained in [BKOV99]. In Section 5 we will briefly discuss the relation to our approach.

We will also pay attention to the issue of sequentiality itself. In Section 7 we first prove a new sequentiality result that is purely in terms of $\beta$-reduction. Then we tackle the original BST, adapting RuS to cover also the Böhm-reduction rules, sending unsolvable terms to a bottom element and in the limit producing Böhm trees. We note that some of our results fall outside the scope of BST.

### 1.1 Outline

In Section 2 we start out by a discussion of Barendregt's Lemma. We illustrate the use of BL by giving short proofs for the undefinability of surjective pairing in the $\lambda$-calculus and for the Genericity Lemma. We generalize the Genericity Lemma to a form that is not implied by BST.

Then in Section 3 the Reduction-under-Substitution Lemma (RuS) is introduced and its relation to Barendregt's Lemma indicated. The use of RuS is illustrated by the Square Brackets Lemma.

A proof of RuS will be given in Section 4, at the same time generalizing it to multiple substitutions.

Then in Section 5 a couple of undefinability results are presented, related to the sequential nature of the $\lambda$-calculus.

In Section 6 we indicate how our analysis can be extended from substitutions to subterms within an arbitrary context. As an application we prove a form of the Perpendicular Lines Theorem.

Finally in Section 7 we turn to the theme of sequentiality. First a new sequentiality result is established as a corollary to RuS and then we use RuS in an analysis of Berry's Sequentiality Theorem.

We conclude by assessing our results in Section 8, giving links to relevant related work and pointing out possible lines for further research.

### 1.2 Preliminaries

We are concerned with the pure $\lambda$-calculus, with which we assume familiarity. We adopt the notations and conventions of the standard text [Bar84]. In particular, we use $\rightarrow$ to denote one-step $\beta$-reduction, $\twoheadrightarrow$ for the reflexive, transitive closure of $\rightarrow$, and $=$ for $\beta$-convertibility. Moreover, $\equiv$ stands for syntactive equivalence modulo $\alpha$-conversion.

For terms $t, s$ and a position $p$ in $t$ we use $t|_p$ for the subterm of $t$ at position $p$, and $t[s]_p$ denotes the result of replacing the subterm at position $p$ in $t$ by $s$. The empty context is denoted by $[\,]$. Note that in particular $C[[\,]]_p$ denotes the result of placing a hole at position $p$ in the context $C$.

## 2 Barendregt's Lemma

At the end of [Bar72], a handwritten note of Henk Barendregt on the undefinability of Church's $\delta$ in combinatory logic (CL), one finds a statement that seems to be added just as an afterthought. It is not widely known, probably just by a group of insiders, who refer to it as Barendregt's Lemma (BL). We quote [Bar72] verbatim:

**Theorem 12.** *If $CL \vdash FM \twoheadrightarrow N$, then there are subterm occurrences $A_i$ of $N$ such that $CL \vdash Fx \twoheadrightarrow N'$ where $N'$ is the result of substituting $x\boldsymbol{N_i}$ for the subterm occurrence $A_i$ and such that $CL \vdash [x/M]N' \twoheadrightarrow N$.*

*Proof.* Same method as the proof of 9.

Here "Same method as the proof of 9" refers to the method used earlier in the manuscript, an intricate syntactic analysis using the technique of underlining.

We will now give a rendering of BL for the $\lambda$-calculus that is in several aspects somewhat more explicit.

First, the prefix that remains invariant in passing from $N'$ to $N$ can be specified as a multi-hole context $C$ (with 0 or more holes!), such that we have $N \equiv C[A_1, \ldots, A_n]$ and $N' \equiv C[x\boldsymbol{N_1}, \ldots, x\boldsymbol{N_n}]$, with $n \geq 0$.

Secondly, the notation $x\boldsymbol{N_i}$ should be elucidated. Define an $x$-vector as a term of the form $xP_1 \ldots P_k$ $(k \geq 0)$. Then what is meant is that each $x\boldsymbol{N_i}$ is an $x$-vector, that is, a term $x\boldsymbol{N_i} \equiv xN_{i,1} \ldots N_{i,k_i}$.

Thirdly, we can be more specific about the reduction $N'[x := M] \twoheadrightarrow N$. It takes place below the prefix $C$, so it can be divided into reductions $(x\boldsymbol{N_i})[x := M] \twoheadrightarrow A_i$. Making this explicit rules out syntactic accidents.

**Lemma 1 (Barendregt's Lemma).** *Let $FM \twoheadrightarrow N$ and let $x$ be a variable not occurring in $F$. Then there are a term $N'$, an $n$-hole context $C$ (with $n \geq 0$), $x$-vectors $B_1, \ldots, B_n$ and terms $A_1, \ldots, A_n$, such that $Fx \twoheadrightarrow N' \equiv C[B_1, \ldots, B_n]$, $B_i[x := M] \twoheadrightarrow A_i$ $(1 \leq i \leq n)$ and $N \equiv C[A_1, \ldots, A_n]$. See Fig. 1.*

*Proof.* In the next section we will see that this lemma follows immediately from Lem. 6, the Reduction-under-Substitution Lemma. □

The lemma is depicted in Fig. 1, where we use the notations $B_i^* \equiv B_i[x := M]$ and $B_i \mapsto B_i^*$.

$$\begin{array}{ccccc} Fx & \twoheadrightarrow & N' & \equiv & C[B_1, \ldots, B_n] \\ & & & & \mapsdown \ \cdots \ \mapsdown \\ & & & & C[B_1^*, \ldots, B_n^*] \\ & & & & \twoheaddownarrow \ \cdots \ \twoheaddownarrow \\ & & & & C[A_1, \ldots, A_n] \ \equiv \ N \end{array}$$

**Fig. 1.** Barendregt's Lemma, pictorial

Heuristically, BL describes the contribution of the argument $M$ to the result $N$ in a reduction $FM \twoheadrightarrow N$. Namely, the result $N$ can be decomposed in two parts:

(i) A prefix $C$ of $N$ that is independent of $M$.
(ii) Subterm occurrences $A_i$, immediately below $C$, that depend on $M$ in an essential way, namely as reducts of $x$-vectors in which $M$ has been substituted for $x$.

We now give two typical applications of Barendregt's Lemma.

### 2.1 Undefinability of Surjective Pairing

A *surjective pairing* would consist of a triple of lambda terms $D, D_1, D_2$, such that for arbitrary $M, N$ we have:

$$D_1(DMN) = M \qquad D_2(DMN) = N \qquad D(D_1M)(D_2M) = M$$

The undefinability of surjective pairing in the $\lambda$-calculus is the central result of [Bar74], where it is proved via underlining. Here we present the short proof from [Vri87] using Barendregt's Lemma.

We recall the notion of terms of order 0, see [Bar84], 17.3.2-3.

**Definition 2.** A term $Z$ has order 0 if it does not reduce to a term in abstraction form, that is if $\neg\exists P\colon\ Z \twoheadrightarrow \lambda x.P$

For a term $Z$ of order 0 we have the following implication:

$$ZM_1 \dots M_p \twoheadrightarrow N \;\Rightarrow\; N \equiv Z'M_1' \dots M_p',\;\; Z \twoheadrightarrow Z',\;\; M_i \twoheadrightarrow M_i' \tag{1}$$

The paradigmatic example of a term of order 0 is $\Omega \equiv (\lambda x.xx)\lambda x.xx$, and in this case we even have the stronger implication:

$$\Omega M_1 \dots M_p \twoheadrightarrow N \;\Rightarrow\; N \equiv \Omega M_1' \dots M_p',\;\; M_i \twoheadrightarrow M_i' \tag{2}$$

The same holds for the case that $Z$ is a variable or an $x$-vector.

**Theorem 3.** *In the λ-calculus a surjective pairing does not exist.*

*Proof.* Assume there were $D, D_1, D_2$ satisfying the equations for surjective pairing. Define $F \equiv \lambda x.D(D_1\Omega)(D_2x)$. Then $F\Omega = D(D_1\Omega)(D_2\Omega) = \Omega$ and hence by the Church–Rosser Theorem the terms $F\Omega$ and $\Omega$ have a common reduct, which can only be $\Omega$ itself. So $F\Omega \twoheadrightarrow \Omega$ and BL can be applied to yield an $N'$ with the ascribed properties (taking $M \equiv N \equiv \Omega$). Since for $\Omega$ we have (2), one easily verifies that there are only two possibilities for $N'$, namely either $N' \equiv \Omega$ or $N' \equiv x$. We investigate both cases.

*Case 1* $N' \equiv \Omega$. Then $Fx \twoheadrightarrow \Omega$ and so $Fx = \Omega$ and by substitutivity of conversion $FM = \Omega$ for an arbitrary term $M$. So for any $M$ we have $D_2M = D_2(D(D_1\Omega)(D_2M)) = D_2(FM) = D_2\Omega$ and hence for arbitrary $N$ we have $N = D_2(DNN) = D_2\Omega$. It follows that all terms are equal, contradicting consistency of the λ-calculus.

*Case 2* $N' \equiv x$. Then $Fx \twoheadrightarrow x$ and so $Fx = x$ and we have $FM = M$ for an arbitrary term $M$. Hence $D_1M = D_1(FM) = D_1(D(D_1\Omega)(D_2M)) = D_1\Omega$ for any $M$. From this a contradiction is derived in the same way as in Case 1. □

## 2.2 Genericity

The following theorem is due to Barendregt [Bar84]. As far as we know the observation that it follows from Barendregt's Lemma is new.

**Theorem 4 (Genericity).** *If $F\Omega = I$, then $Fx = I$.*

*Proof.* Apply BL to a reduction $F\Omega \twoheadrightarrow I$, which exists according to the Church–Rosser Theorem. We get the following situation.

$$\begin{array}{rcl} Fx \twoheadrightarrow N' & \equiv & C[\dots,\; xM_1 \dots M_p,\; \dots] \\ & & \quad\quad\downarrow \\ & & C[\dots,\; \Omega M_1^* \dots M_p^*,\; \dots] \\ & & \quad\quad\downarrow \\ & & C[\dots,\; \Omega M_1' \dots M_p',\; \dots] \equiv I \end{array}$$

Since the term $I$ contains no occurrence of $\Omega$, the context $C$ must have zero holes, hence $N' \equiv C \equiv I$. It follows that $Fx = I$. □

By inspecting the proof one sees that the Genericity Lemma can be generalized to arbitrary order-zero terms, if they do not occur in the result of the reduction.

**Theorem 5 (Generalized Genericity).** *If $FZ \twoheadrightarrow N$ for a term $Z$ of order zero and $Z \not\twoheadrightarrow S$ for all subterms $S$ of $N$, then $Fx = N$.*

*Proof.* Applying BL to a reduction $FZ \twoheadrightarrow N$, we get the following situation. Since the term $Z$ is of order zero and does not rewrite to any subterm of $N$, the context $C$ must have zero holes, hence $C \equiv N$. It follows that $Fx = N$. □

It is interesting to note that, in contrast with the original Thm. 4, this generalized Genericity Theorem does not follow from Berry's Sequentiality Theorem. An example of an application of Thm. 5 that is not in the scope of Berry's Sequentiality Theorem can be obtained by taking $Z$ and $N$ to be both unsolvable terms, e.g. $Z \equiv \Omega\Omega$ and $N \equiv \Omega$. If $F(\Omega\Omega) = \Omega$, then $Fx = \Omega$ by Thm. 5, but the Böhm trees of $Z$ as well as $N$ are just $\perp$.

## 3 Reduction Under Substitution

Barendregt's Lemma can be cast in a different way, in terms of substitution instead of function application. This is the form that originates with Diederik van Daalen [Daa80] and that found its way into the book [Bar84], as Exercise 15.4.8. It is slightly stronger than BL and easier to prove.

$$
\begin{array}{ccccc}
M & \twoheadrightarrow & N' & \equiv & C[B_1, \dots, B_n] \\
 & & & & \bar{\downarrow} \quad \cdots \quad \bar{\downarrow} \\
 & & & & C[B_1^*, \dots, B_n^*] \\
 & & & & \downarrow\!\!\downarrow \quad \cdots \quad \downarrow\!\!\downarrow \\
 & & & & C[A_1, \dots, A_n] \;\equiv\; N
\end{array}
$$

**Fig. 2.** Reduction under substitution, pictorial

**Lemma 6 (Reduction under Substitution).** *Let $M[x := L] \twoheadrightarrow N$. Then there are a term $N'$, an $n$-hole context $C$ (with $n \geq 0$), $x$-vectors $B_1, \dots, B_n$ and terms $A_1, \dots, A_n$, such that $M \twoheadrightarrow N' \equiv C[B_1, \dots, B_n]$, $B_i[x := L] \twoheadrightarrow A_i$ for all $1 \leq i \leq n$ and $N \equiv C[A_1, \dots, A_n]$. See Fig. 2.*

*Proof.* In Sec. 4 we will prove the Reduction-under-Substitution Lemma for multiple substitution, Thm. 13, of which the present form is just a special case. □

So the proof will be postponed, but we already point out that Lem. 1 immediately follows from Lem. 6 by taking $Fx$ for $M$ and $M$ for $L$.

It should be remarked that the context $C$ and the $x$-vectors $B_i$ are in general not unique. Consider for example $M \equiv xzx$ with the substitution $[x := \lambda y.y]$ together with the reduction $M[x := \lambda y.y] \twoheadrightarrow z(\lambda y.y)$. Then we have

(i) $M \twoheadrightarrow C_1[B_1]$ with $C_1 \equiv [\,]$, $B_1 \equiv xzx$, $B_1^* \twoheadrightarrow z(\lambda y.y)$
(ii) $M \twoheadrightarrow C_2[B_2, B_3]$ with $C_2 \equiv [\,][\,]$, $B_2 \equiv xz$, $B_3 \equiv x$, $B_2^* \twoheadrightarrow z$, $B_3^* \twoheadrightarrow \lambda y.y$

In the second factorization the context $C_2$ shows more of the stucture of the result $z(\lambda y.y)$ than $C_1$ does, namely that it is an application term. We call $C_2$ *finer* than $C_1$, and $C_1$ *coarser*, $C_1 \lhd C_2$.

Van Daalen's interest in the substitution variant of BL was because of the Square Brackets Lemma, which he used in his proof of strong normalization.[1]

**Lemma 7 (Square Brackets Lemma).** *Let $M[x := L] \twoheadrightarrow \lambda y.P$. Then we have one of the following two cases.*

1. *$M \twoheadrightarrow \lambda y.P'$ for a $P'$ such that $P'[x := L] \twoheadrightarrow P$*
2. *$M \twoheadrightarrow x\boldsymbol{Q}$ and $(x\boldsymbol{Q})[x := L] \twoheadrightarrow \lambda y.P$*

*Proof.* The prefix $C$ found by Lem. 6 can either be of the form $\lambda y.C'$ or it must be the empty context. If $C \equiv \lambda y.C'$ then $N' \equiv \lambda y.P'$ for some $P'$ and we are in Case 1. If $C \equiv [\,]$ then $N'$ is an $x$-vector and we are in Case 2. □

It is noted in [Daa80] that the lemma can be generalized to situations where the outer shape of the reduct is not an abstraction. In [Oos97] a similar lemma is stated for arbitrary patterns, the generalization is called there "Invert".

## 4 Reduction Under Multiple Substitution

We now prove the Reduction-under-Substitution Lemma for multiple substitutions. Throughout this section, and in some of the following ones, we will work with a fixed substitution $[\boldsymbol{x} := \boldsymbol{L}]$ with $\boldsymbol{x} = x_1, \ldots, x_m$ and $\boldsymbol{L} = L_1, \ldots, L_m$ ($m \geq 0$). We tacitly assume that no lambdas binding the variables $x_i$ are used (this can always be achieved by $\alpha$-renaming), so that occurrences of $x_1, \ldots, x_m$ will always be free.

The following definition sums up some technical notions and convenient notations (some of which we already used in the previous sections).

**Definition 8**

1. An $\boldsymbol{x}$*-vector* is a term of the form $x_i P_1 \ldots P_k$ with $1 \leq i \leq m$ and $k \geq 0$.
2. $M^*$ denotes $M[\boldsymbol{x} := \boldsymbol{L}]$.
3. $M \mapsto N$ when $N \equiv M^*$.
4. $C \lhd M$ if context $C$ is a prefix of term (or context) $M$.
5. $C \blacktriangleleft M$ if $C \lhd M$, $x_1, \ldots, x_m \notin FV(C)$ and $M \equiv C[B_1, \ldots, B_n]$ for some $\boldsymbol{x}$-vectors $B_i$.
6. $M \rightsquigarrow_C C[A_1, \ldots, A_n]$ if $C \blacktriangleleft M$ as in 5 and moreover $B_i^* \twoheadrightarrow A_i$ for $i = 1, ..., n$
7. $M \rightsquigarrow N$ if there exists a context $C$ such that $M \rightsquigarrow_C N$

[1] Why "square brackets"? The lemma analyses the contribution of the substitution in a reduction to an abstraction term. In the notation of Automath square brackets were used to denote lambda abstraction.

**Lemma 9.** *We have $M \rightsquigarrow N$ if and only if one of the following four cases applies:*

*(i)* $M$ *is an* $\boldsymbol{x}$*-vector with* $M^* \twoheadrightarrow N$
*(ii)* $M \equiv N \equiv y$ *for some variable* $y$ *with* $y \not\equiv x_1, \ldots, y \not\equiv x_m$
*(iii)* $M \equiv M_1 M_2$ *and* $N = N_1 N_2$ *with* $M_1 \rightsquigarrow N_1$ *and* $M_2 \rightsquigarrow N_2$
*(iv)* $M \equiv \lambda y.M'$ *and* $N \equiv \lambda y.N'$ *with* $M' \rightsquigarrow N'$

*As a consequence, if* $M \rightsquigarrow_C N$ *then* $M|_p \rightsquigarrow_{C|_p} N|_p$ *for every position* $p$ *in* $C$.

*Proof.* Follows directly from the definition. □

**Lemma 10.** *Let* $C \blacktriangleleft M$ *and* $C' \blacktriangleleft M$ *with* $C' \lhd C$, *then* $\rightsquigarrow_C \subseteq \rightsquigarrow_{C'}$.

*Proof.* Let $B_i$, $B'_j$ be $\boldsymbol{x}$-vectors such that $M \equiv C[B_1, \ldots, B_n] \equiv C'[B'_1, \ldots, B'_{n'}]$. Then $C[B_1^*, \ldots, B_n^*] \equiv C'[B_1'^*, \ldots, B_{n'}'^*]$ since $x_1, \ldots, x_m \notin FV(C) \cup FV(C')$ and all occurrences of $x_1, \ldots, x_m$ are free. Now $\rightsquigarrow_C \subseteq \rightsquigarrow_{C'}$ follows since the $B_i^*$ are disjoint, and each of them is a subterm of some $B_j'^*$. □

**Lemma 11.** *If* $y \not\equiv x_1, \ldots, y \not\equiv x_m$, *then*

$$M \rightsquigarrow M',\ N \rightsquigarrow N' \Rightarrow M[y := N] \rightsquigarrow M'[y := N'] .$$

*Proof.* Let $\sigma$ be shorthand for $[y := N]$, $\sigma'$ for $[y := N']$ and $\sigma^*$ for $[y := N^*]$. We use induction over the structure of $M$ according to Lem. 9:

(i) If $M$ is an $\boldsymbol{x}$-vector, then $M\sigma$ is and $(M\sigma)^* \equiv M^*\sigma^*$ since $y \not\equiv x_1, \ldots, x_m$. From $M^* \twoheadrightarrow M'$ and $N^* \twoheadrightarrow N'$ follows $M^*\sigma^* \twoheadrightarrow M'\sigma'$ and $M\sigma \rightsquigarrow M'\sigma'$.
(ii) If $M \equiv M' \equiv z$ for a variable $z$ with $z \not\equiv x_1, \ldots, x_m$, then either $z \equiv y$ and $M\sigma \equiv N \rightsquigarrow N' \equiv M'\sigma'$, or $z \not\equiv y$ and $M\sigma \equiv z \rightsquigarrow z \equiv M'\sigma'$.
(iii) If $M \equiv M_1 M_2$ and $M' \equiv M'_1 M'_2$ with $M_i \rightsquigarrow M'_i$, then $M_i\sigma \rightsquigarrow M'_i\sigma'$ by IH and since $M\sigma \equiv M_1\sigma(M_2\sigma)$ and $M'\sigma' \equiv M'_1\sigma'(M_2\sigma')$ we get $M\sigma \rightsquigarrow M'\sigma'$.
(iv) If $M \equiv \lambda z.M_1$ and $M' \equiv \lambda z.M'_1$ with $M_1 \rightsquigarrow M'_1$, then either $z \equiv y$ and $M\sigma \equiv M \rightsquigarrow M' \equiv M'\sigma'$, or $z \not\equiv y$, $M\sigma \equiv \lambda z.M_1\sigma \rightsquigarrow^{\mathrm{IH}} \lambda z.M'_1\sigma' \equiv M'\sigma'$.

□

**Lemma 12.** $\rightsquigarrow \cdot \twoheadrightarrow \subseteq \twoheadrightarrow \cdot \rightsquigarrow$

*Proof.* By induction it suffices to show $\rightsquigarrow \cdot \rightarrow \subseteq \rightarrow \cdot \rightsquigarrow$. Let $M \rightsquigarrow N \rightarrow O$, then there are a context $C$, $\boldsymbol{x}$-vectors $B_i$ and terms $A_i$ such that: $M \equiv C[B_1, \ldots, B_n]$, $B_i \mapsto B_i^* \twoheadrightarrow A_i$ for all $1 \leq i \leq n$, $N \equiv C[A_1, \ldots, A_n]$, and a step $\rho : N \rightarrow O$ at position $p$. Note that $M^* \twoheadrightarrow_C N$ where $\twoheadrightarrow_C$ means that all steps are below $C$.

Assume $\rho$ is entirely in $C$. Then we have $M|_p \equiv (\lambda y.M_1)M_2 \rightarrow M_1[y := M_2]$ and $N|_p \equiv (\lambda y.N_1)N_2 \rightarrow N_1[y := N_2] \equiv O|_p$ with $M_1 \rightsquigarrow_{C|_{p11}} N_1$, $M_2 \rightsquigarrow_{C|_{p2}} N_2$ by Lem. 9. Hence $M_1[y := M_2] \rightsquigarrow_{C'} N_1[y := N_2]$ for some context $C'$ by Lem. 11. Let $M' \equiv M[M_1[y := M_2]]_p$ then $M \rightarrow M' \rightsquigarrow_{C[C']_p} O$.

If $\rho$ is below $C$, then it is contained in one of the $\boldsymbol{x}$-vectors $B_i$ and 'absorbed' by $\rightsquigarrow$, that is, $M \rightsquigarrow_C O$. Finally if $\rho$ is neither in $C$ nor below $C$, then $C|_p \equiv [\,]C'$. Then $M|_p$ is an $\boldsymbol{x}$-vector since $C \blacktriangleleft M$ and therefore $M|_{p1}$ is an $\boldsymbol{x}$-vector. Hence $C[[\,]]_p \blacktriangleleft M$ and $\rightsquigarrow_C \subseteq \rightsquigarrow_{C[[\,]]_p}$ by Lem. 10. Observe that $\rho$ is below $C[[\,]]_p$, a case that we have already considered, $M \rightsquigarrow_{C[[\,]]_p} O$. □

**Theorem 13 (RuS).** *If* $M^* \twoheadrightarrow N$, *then* $M \twoheadrightarrow C[B_1, \ldots, B_n] \rightsquigarrow_C N$ *for some context* $C$ *and* $\boldsymbol{x}$*-vectors* $B_1, \ldots, B_n$. *See Fig. 2.*

*Proof.* Follows from $M \rightsquigarrow M^* \twoheadrightarrow N$ and an application of Lem. 12. □

## 5 Undefinability Proofs

In this section we use reduction under substitution to give new proofs of some well-known consequences of Berry's Sequentiality Theorem.

Given $\boldsymbol{x} = x_1, \ldots, x_m$, we define the following notions relative to this choice of variables, that are assumed to be free.

**Definition 14.** An occurrence of $x_i$ in $M$ is *leading* if $M$ contains no $x_j$-vector of the form $x_j\boldsymbol{P}$ such that the occurrence of $x_i$ is in $\boldsymbol{P}$. A variable $x_i$ is *leading* if it has a leading occurence. $LV(M)$ denotes the set of leading variables in $M$.

**Lemma 15.** *For terms* $M$, $N$ *we have*

*(i) If at least one of the variables* $x_1, \ldots, x_m$ *occurs in* $M$, *then* $LV(M) \neq \varnothing$.
*(ii) If* $C \blacktriangleleft M$ *and* $M \equiv C[y_1\boldsymbol{P_1}, \ldots, y_n\boldsymbol{P_n}]$, *then* $LV(M) \subseteq \{y_1, \ldots, y_n\}$.
*(iii) If* $M \twoheadrightarrow N$, *then* $LV(N) \subseteq LV(M)$.

*Proof.* (i) Take an outermost occurrence of $x_i\boldsymbol{P}$, then $x_i \in LV(M)$.
(ii) Directly from the definition together with the fact that $x_1, \ldots, x_n \notin FV(C)$.
(iii) Note that if a variable $x_i$ is leading in a term $M[y := N]$ then it must have been leading in $M$ or $N$ and hence in $(\lambda y.M)N$. The claim follows by closure under contexts and induction on the reduction length. □

We start by showing the undefinability of Gustave's function.

**Theorem 16.** *There is no lambda term* $G$ *such that:*

$$G\,01x = x \qquad G\,1x0 = x \qquad G\,x01 = x$$

*Proof.* We employ RuS with $M \equiv Gxyz$ and $\boldsymbol{x} = x, y, z$. We have $G\,01\Omega = \Omega$ and $G\,01\Omega \twoheadrightarrow \Omega$ by confluence. By RuS there exists $N_z$ with $M \twoheadrightarrow N_z \rightsquigarrow \Omega$. If $z$ is leading variable in $N_z$, then $N_z \equiv z$, otherwise every $\rightsquigarrow$-reduct of $N_z$ would contain $\Omega$ at a non-root position. But if $N_z \equiv z$ then we would have $G\,1x0 \twoheadrightarrow 0$. Hence $z \notin LV(N_z)$ and likewise there exist $N_x$ and $N_y$ with $M \twoheadrightarrow N_x$, $M \twoheadrightarrow N_y$ and $x \notin LV(N_x)$, $y \notin LV(N_y)$. By confluence $N_x$, $N_y$ and $N_z$ have a common reduct $N$ with $LV(N) = \varnothing$ and then by Lem. 15 none of the variables $x, y, z$ occur in $N$. Therefore we obtain $\forall \boldsymbol{L} : M[\boldsymbol{x} := \boldsymbol{L}] \twoheadrightarrow N[\boldsymbol{x} := \boldsymbol{L}] \equiv N$, and hence $x = G\,01x = N = G\,01y = y$, contradicting consistency of the $\lambda$-calculus. □

*Remark 1.* For a variant of Gustave's function where $x$ is replaced by $Z$ ranging over all closed terms the proof stays valid. For a variant where the right-hand sides are closed terms $A$, $B$, $C$ we refer to the Perpendicular Lines Theorem (Thm. 22).

It is interesting to note that Thm. 16 is obtained in [BKOV99] by a different argument, involving an analysis of residuals along head reductions. Their Lemma 5.2, on the undefinability of a general form of the $G$ of Thm. 16, can be proved by our method in the same way as Thm. 16. The undefinability of the other two variants of $G$ mentioned in this remark are not covered by Lemma 5.2 in [BKOV99].

Before continuing we state a few lemmas that capture the common essence of the following undefinability results. Fig. 3 illustrates Lem. 17 and 18 applied to "parallel or" (Por). If $x_i$ is substituted by $\Omega$ in $M$ and $M$ rewrites to a normal form, then $x_i$ cannot have been leading in $M$. If such a reduction exists for every $x_i$ then all $M[\boldsymbol{x} := \boldsymbol{L}]$ (and hence all reducts) are convertible for arbitrary $\boldsymbol{L}$.

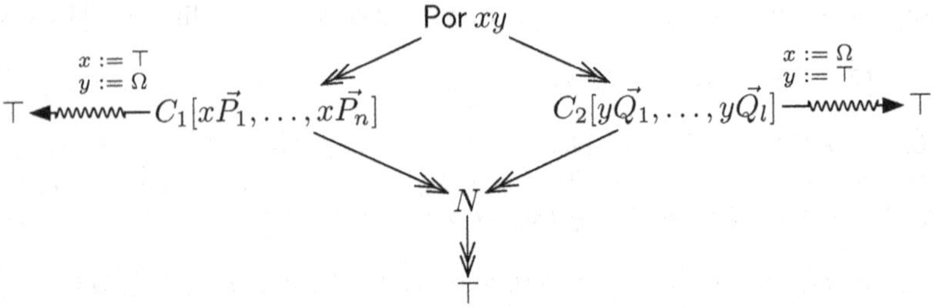


**Fig. 3.** Lem. 17 and Lem. 18 at the example of "parallel or"

**Lemma 17.** *If for $i = 1, \ldots, m$ there exist $N_i$ with $M = N_i$ and $x_i \notin LV(N_i)$, then there exists an $N$ such that $LV(N) = \varnothing$ and $\forall \boldsymbol{L}$: $M[\boldsymbol{x} := \boldsymbol{L}] \twoheadrightarrow N$.*

*Proof.* By confluence the terms $M, N_1, \ldots, N_m$ have a common reduct $N$ and by Lem. 15(iii) we have $LV(N) \subseteq LV(N_1) \cap \ldots \cap LV(N_m) = \varnothing$. So by Lem. 15(i) the variables $x_1, \ldots, x_m$ do not occur in $N$. Hence for arbitrary $\boldsymbol{L}$ we have $M[\boldsymbol{x} := \boldsymbol{L}] \twoheadrightarrow N[\boldsymbol{x} := \boldsymbol{L}] \equiv N$. □

**Lemma 18.** *If for $i = 1, \ldots, m$ there are normal forms $N_i$ and $\boldsymbol{L}^{\boldsymbol{i}}$ with $L_i^i \equiv \Omega$ such that $M[\boldsymbol{x} := \boldsymbol{L}^{\boldsymbol{i}}] \twoheadrightarrow N_i$, then $N_1 \equiv \ldots \equiv N_m$ and $\forall \boldsymbol{L}$: $M[\boldsymbol{x} := \boldsymbol{L}] \twoheadrightarrow N_1$.*

*Proof.* Let $i \in \{1, \ldots, m\}$ arbitrary. An application of RuS to $M[\boldsymbol{x} := \boldsymbol{L}^{\boldsymbol{i}}] \twoheadrightarrow N_i$ yields that there exist a term $N_i'$, a context $C$ and $\boldsymbol{x}$-vectors $B_1, \ldots, B_n$ such that $M \twoheadrightarrow N_i' \equiv C[B_1, \ldots, B_n] \rightsquigarrow_C N_i$. None of the $B_j$'s can be an $x_i$-vector. For suppose it were, then every reduct of $B_j^*$ and hence also $N_i$ would contain $\Omega$, contradicting $N_i$ being a normal form. We conclude by Lem. 15(ii) that $x_i \notin LV(N_i')$. Since $i$ was arbitrary, by Lem. 17 there exists an $N$ such that $\forall \boldsymbol{L} : M[\boldsymbol{x} := \boldsymbol{L}] \twoheadrightarrow N$. Hence $N_1 = \ldots = N_m$. By confluence and the fact that all $N_i$ are normal forms we get $M \twoheadrightarrow N_1 \equiv \ldots \equiv N_m$. □

### 5.1 Undefinability of "Parallel or"

We can now show the undefinability of "parallel or".

**Theorem 19.** *There are no lambda terms* Por *and normal forms* $\top \not\equiv \bot$ *s.t.:*

$$\mathsf{Por}\,\top x = \top \qquad \mathsf{Por}\,x\top = \top \qquad \mathsf{Por}\,\bot\bot = \bot$$

*Proof.* Assume that Por exists, we consider $M \equiv \mathsf{Por}\,xy$ with $\boldsymbol{x} = x, y$. Since $\mathsf{Por}\,\Omega\top \twoheadrightarrow \top$ and $\mathsf{Por}\,\top\Omega \twoheadrightarrow \top$, we get $\forall L : \mathsf{Por}\,xy[\boldsymbol{x} := \boldsymbol{L}] \twoheadrightarrow \top$ by Lem. 18. Then in particular $F\bot\bot \twoheadrightarrow \top \not\equiv \bot$ contradicting the assumption. □

The following is a variant of "parallel or" from [Bar84], that is also undefinable.

**Theorem 20.** *There is no lambda term $F$ s.t. for arbitrary closed $M, N$:*

$FMN = I$ *if $M$ or $N$ is solvable*
$FMN = \Omega$ *otherwise*

*Proof.* Assuming there is such an $F$, we consider $M \equiv Fxy$ with $\boldsymbol{x} = x, y$. Since $F\Omega I \twoheadrightarrow I$ and $FI\Omega \twoheadrightarrow I$, we get $\forall L : Fxy[\boldsymbol{x} := \boldsymbol{L}] \twoheadrightarrow I$ by Lem. 18. Then in particular $F\Omega\Omega \twoheadrightarrow I \not\equiv \Omega$ contradicting the assumption. □

## 6 Extension to Context Filling

We extend reduction under multiple substitution to context filling; the difference being that variables of the arguments might get bound.

**Corollary 1.** *Let $C[L_1, \ldots, L_m] \twoheadrightarrow N$. Let $\boldsymbol{x} = x_1, \ldots, x_m$ be fresh variables. For $i = 1, \ldots, m$ let $\boldsymbol{y_i}$ be a vector consisting of all variables that are bound at the i-th hole of C. Then there are a context $D$, $\boldsymbol{x}$-vectors $B_1, \ldots, B_n$ and terms $A_1, \ldots, A_n$, such that:*

$$\begin{array}{lcl}
C[x_1\boldsymbol{y_1}, \ldots, x_m\boldsymbol{y_m}] & \twoheadrightarrow & D[B_1, \ldots, B_n] \\
 & & \quad\mapsto \cdots \mapsto \quad [x_1 := \lambda\boldsymbol{y_1}.L_1, \ldots, x_m := \lambda\boldsymbol{y_m}.L_m] \\
 & & D[B_1^*, \ldots, B_n^*] \\
 & & \quad\twoheadrightarrow \cdots \twoheadrightarrow \\
 & & D[A_1, \ldots, A_n] \equiv N
\end{array}$$

*Proof.* Let $\sigma$ be shorthand for $[x_1 := \lambda\boldsymbol{y_1}.L_1, \ldots, x_m := \lambda\boldsymbol{y_m}.L_m]$ and we define $M \equiv C[x_1\boldsymbol{y_1}, \ldots, x_m\boldsymbol{y_m}]$. Then clearly $M\sigma \twoheadrightarrow C[L_1, \ldots, L_m] \twoheadrightarrow N$. As a consequence of Reduction under Multiple Substitution there exist a context $D$ and $\boldsymbol{x}$-vectors $B_1,\ldots,B_n$ such that: $M \twoheadrightarrow D[B_1, \ldots, B_n] \rightsquigarrow_C N$. □

**Lemma 21.** *If for $i = 1, \ldots, m$ there are a term $N_i$ with $z \notin N_i$ and terms $\boldsymbol{L^i}$ with $L^i_i \equiv z$ such that $C[\boldsymbol{L^i}] = N_i$, then $\forall \boldsymbol{L}$: $C[\boldsymbol{L}] = N_1 = \ldots = N_m$.*

*Proof.* After an application of Thm. 1 the proof continues analogously to the proof of Lem. 18, from $z \notin N_i$ follows that no $B_j$ is an $x_i$-vector. □

### 6.1 Perpendicular Lines Theorem

The Perpendicular Lines Theorem is a result from [Bar84], Ch. 14, stated there in terms of Böhm equivalence, together with a suggestion to extend it to $\beta$-equality. In [BS99] a counterexample is given to PPL with respect to $\beta$-equality, which, however, concerns the variant where the equations are only required to hold for substitutions of a closed term for the variable $z$. They added a suggestion to try to use [Bar84], Exercise 15.4.8, for the open variant. Indeed, it turns out that we can use RuS to prove the following theorem.

**Theorem 22 (PPL).** *Assume that for lambda terms $M_{ij}, N_i$ with $z \notin N_i$:*

$$\begin{aligned} C[M_{11}, M_{12}, \ldots, M_{2n-1}, z] &= N_1 \\ C[M_{21}, M_{22}, \ldots, z, M_{2n}] &= N_2 \\ \vdots \quad & \ \ \vdots \ \vdots \\ C[z, M_{n2}, M_{n3}, \ldots, M_{nn}] &= N_n \end{aligned}$$

*Then*

- $N_1 = N_2 = \ldots = N_n = N$
- *For all* $Z_1, \ldots, Z_n$ *we have* $C[\boldsymbol{Z}] = N$

*Proof.* Follows from an application of Lem. 21 to the above equations. □

## 7 Sequentiality

Berry's Sequentiality Theorem (BST) is about Böhm trees and these can be obtained as infinite normal forms with respect to $\beta$-reduction extended with the Böhm reduction rules. To be able to deal with this we will have to adapt the Reduction-under-Substitution Lemma to this extended notion of reduction. But before doing so, we formulate a strictly finitary sequentiality result for $\beta$-reduction alone. We give the version for multiple substitution, but a functional and a context-filling version can be straightforwardly derived.

**Theorem 23.** *Let $M[\boldsymbol{x} := \boldsymbol{L}] \twoheadrightarrow N$ with $\boldsymbol{x} = x_1, \ldots x_m$. Then $N$ can be written as $N \equiv C[A_1, \ldots, A_n]$ in such a way that:*

1. *The prefix $C$ is independent of the substitution, that is, for any $\mathbf{P}{=}P_1, \ldots P_m$ we have $M[\boldsymbol{x} := \boldsymbol{P}] \twoheadrightarrow C[\ldots]$.*
2. *Each $A_i$ depends on exactly one of the substituted terms $L_j$ in the sense that at the position of $A_i$ any term $B$ can be realized by an appropiate replacement $Q$ of $L_j$, regardless of the choice of the other substituted terms. That is,*
$$(\forall i)(\exists j)(\forall B)(\exists Q)(\forall \boldsymbol{P})\colon P_j \equiv Q \Rightarrow M[\boldsymbol{x} := \boldsymbol{P}] \twoheadrightarrow C[\ldots, B, \ldots]$$
*where $1 \leq i \leq n$, $1 \leq j \leq m$ and with $B$ at position $i$ of $C$.*

*Proof.* This is an immediate consequence of RuS, Thm. 6. If $B_i$ is an $\boldsymbol{x}$-vector $x_j K_1 \ldots K_k$ then $Q$ can be chosen as $\lambda y_1 \cdots y_k.B$. □

Undefinability results like the ones mentioned earlier can also be obtained by applying this new sequentiality theorem.

Now we turn to BST. We consider $\Lambda(\bot)$, the $\lambda$-calculus enriched with the constant $\bot$ (bottom). The Böhm rewrite relation $\rightarrow_{\beta\bot} = \rightarrow_\beta \cup \rightarrow_\bot$ on $\Lambda(\bot)$ consists of $\beta$-reduction $\rightarrow_\beta$ together with $\rightarrow_\bot$ defined by:

$$\bot M \rightarrow \bot \qquad \lambda y.\bot \rightarrow \bot \qquad u \rightarrow \bot \text{ if } u \text{ is an unsolvable}$$

For $\twoheadrightarrow_{\beta\bot}$ we have to adapt the definition of $\boldsymbol{x}$-vector. Let $\boldsymbol{x} = x_1, \ldots, x_m$. The set of *$\boldsymbol{x}$-clusters* is inductively defined as follows:

- $x_1, \ldots, x_m$ are $\boldsymbol{x}$-clusters
- if $B$ is an $\boldsymbol{x}$-cluster and $M \in \Lambda(\bot)$ a term, then $BM$ is an $\boldsymbol{x}$-cluster
- if $B$ is an $\boldsymbol{x}$-cluster and $y \not\equiv x_1, \ldots, y \not\equiv x_m$, then $\lambda y.B$ is an $\boldsymbol{x}$-cluster

Note that for every $x$-cluster $B$ we have $B[x := \bot] \twoheadrightarrow_{\beta\bot} \bot$. We tacitly assume that all occurrences of the variables $x_1, \ldots, x_m$ are free.

We adapt the notation from Def. 8 to Böhm reduction by exchanging $\twoheadrightarrow_{\beta\bot}$, $\rightsquigarrow_{\beta\bot}$, $\blacktriangleleft_{\beta\bot}$ and $\boldsymbol{x}$-clusters for $\twoheadrightarrow$, $\rightsquigarrow$, $\blacktriangleleft$ and $\boldsymbol{x}$-vectors, respectively. Likewise we obtain lemmas $9^\bot$-$11^\bot$ for $\twoheadrightarrow_{\beta\bot}$ identical to Lem. 9-11 for $\twoheadrightarrow$. In order to lift Lem. 12 to Böhm reduction, the proof has to be adopted and extended.

**Lemma 24.** $\rightsquigarrow \cdot \twoheadrightarrow_{\beta\bot} \subseteq \twoheadrightarrow_{\beta\bot} \cdot \rightsquigarrow$

*Proof.* By induction $\rightsquigarrow_{\beta\bot} \cdot \rightarrow_{\beta\bot} \subseteq \twoheadrightarrow_{\beta\bot} \cdot \rightsquigarrow_{\beta\bot}$ suffices. Let $M \rightsquigarrow_{\beta\bot} N \rightarrow_{\beta\bot} O$, then there are a context $C$, $\boldsymbol{x}$-clusters $B_i$ and terms $A_i$ such that: $M \equiv C[B_1, \ldots, B_n]$, $B_i \mapsto B_i^* \twoheadrightarrow_{\beta\bot} A_i$ for all $1 \leq i \leq n$, $N \equiv C[A_1, \ldots, A_n]$, and a step $\rho : N \rightarrow_{\beta\bot} O$ at position $p$. The case of $\beta$-steps $\rho$ is analogous to the proof of Lem. 12.

Hence let $\rho$ be a $\bot$-step according to one of the three $\bot$-rules: (i) $\rho : \bot M \rightarrow \bot$, (ii) $\rho : \lambda y.\bot \rightarrow \bot$, or (iii) $\rho : u \rightarrow \bot$ if $u$ is an unsolvable.

If $\rho$ is below $C$, then it is contained in one of the $\boldsymbol{x}$-clusters $B_i$ and 'absorbed' by $\rightsquigarrow_{\beta\bot}$, that is, $M \rightsquigarrow_{\beta\bot,C} O$. Therefore assume $\rho$ is not below $C$.

First we consider the $\bot$-rules (i) and (ii). If the redex pattern of $\rho$ is entirely in $C$, then we have $M \rightarrow_{\beta\bot} M[\bot]_p \rightsquigarrow_{\beta\bot,C[\bot]_p} O$. Otherwise $\rho$ is neither in $C$ nor below $C$, that is, $\rho$ is overlapping in-between. Then in case of (i) $C|_p \equiv [\,]C'$ and (ii) $C|_p \equiv \lambda y.[\,]$. In both cases it follows that $M|_p$ is an $\boldsymbol{x}$-cluster since $M|_{p1}$ is an $\boldsymbol{x}$-cluster. Then we are done since $C[[\,]]_p \blacktriangleleft_{\beta\bot} M$, $\rightsquigarrow_{\beta\bot,C} \subseteq \rightsquigarrow_{\beta\bot,C[[\,]]_p}$ by Lem. $10^\bot$ and $\rho$ is now below $C[[\,]]_p$.

The remaining case is $\bot$-rule (iii) with redex position in $C$. Note that if $U \twoheadrightarrow U'$ and $U'$ is unsolvable then $U$ is unsolvable. Thus $M^*|_p$ is unsolvable; either $M|_p$ is unsolvable, then $M \rightarrow_{\beta\bot} M[\bot]_p \rightsquigarrow_{\beta\bot,C[\bot]_p} O$, or the head reduction sequence $M|_p \twoheadrightarrow M'$ yields a term $M'$ having an $x_i$ as head. Then $M'$ is an $\boldsymbol{x}$-cluster with $M' \rightsquigarrow_{\beta\bot,[\,]} \bot$, hence $M \twoheadrightarrow M[M']_p \rightsquigarrow_{\beta\bot,C[[\,]]_p} O$. □

**Theorem 25.** *Let $M[\boldsymbol{x} := \boldsymbol{L}] \twoheadrightarrow_{\beta\bot} N$. Then there exist an $n$-hole context $C$, $\boldsymbol{x}$-clusters $B_1, \ldots, B_n$ and terms $A_1, \ldots, A_n$, such that $M \twoheadrightarrow_{\beta\bot} C[B_1, \ldots, B_n]$, $B_i[\boldsymbol{x} := \boldsymbol{L}] \twoheadrightarrow_{\beta\bot} A_i$ for all $1 \leq i \leq n$, and $N \equiv C[A_1, \ldots, A_n]$.*

*Proof.* The statement of the theorem is equivalent to $M^* \twoheadrightarrow_{\beta\bot} N \Rightarrow M \twoheadrightarrow_{\beta\bot} \cdot \rightsquigarrow_{\beta\bot} N$, which follows from $M \rightsquigarrow_{\beta\bot} M^* \twoheadrightarrow_{\beta\bot} N$ and an application of Lem. 24.

**Theorem 26.** *Let $M[\boldsymbol{x} := \bot, \dots, \bot] \twoheadrightarrow_{\beta\bot} N$ with $\boldsymbol{x} = x_1, \dots x_m$. Then $N$ can be written as $N \equiv C[A_1, \dots, A_n]$ in such a way that:*

1. *The prefix $C$ is independent of the substitution, that is, for any $\boldsymbol{P}$=$P_1, \dots P_m$ we have $M[\boldsymbol{x} := \boldsymbol{P}] \twoheadrightarrow C[\dots]$.*
2. *Each $A_i$ depends on exactly one of the substituted terms $\bot$ in the sense that:*
   - *A refinement of the corresponding $\bot$ to a free variable will give rise to a reduction to $C[\dots, A_i', \dots]$ where $A_i' \not\twoheadrightarrow_{\beta\bot} \bot$, regardless of the choice of the other substituted terms. That is,*
     $$(\forall i)(\exists j)(\forall \boldsymbol{P})\colon P_j \equiv x \Rightarrow M[\boldsymbol{x} := \boldsymbol{P}] \twoheadrightarrow_{\beta\bot} C[\dots, A_i', \dots],\ A_i' \not\twoheadrightarrow_{\beta\bot} \bot$$
   - *At the position of $A_i$ a $\bot$ can be realized regardless of the choice of the other substituted terms. That is,*
     $$(\forall i)(\exists j)(\forall \boldsymbol{P})\colon P_j \equiv \bot \Rightarrow M[\boldsymbol{x} := \boldsymbol{P}] \twoheadrightarrow_{\beta\bot} C[\dots, \bot, \dots]$$

*Proof.* This is an immediate consequence of Thm. 25, noting that an $x_i$-cluster with $\bot$ substituted for $x_i$ rewrites to $\bot$. □

Berry's Sequentiality Theorem can be derived as a corollary of Thm. 26 in the following way.

Let $M \equiv C[\bot, \dots, \bot]$ and let $B$ be the Böhm tree of $M$. For arbitrary depths $d \in \mathbb{N}$ there exists a reduction $M \twoheadrightarrow_{\beta\bot} N$ such that $N$ is in $\twoheadrightarrow_{\beta\bot}$-normal form up to depth $d$; then $N$ coincides with $B$ up to depth $d$. An application of Thm. 26 to $M \equiv (C[x_1, \dots, x_m])[\boldsymbol{x} := \bot, \dots, \bot] \twoheadrightarrow_{\beta\bot} N$ yields $N \equiv D[A_1, \dots, A_n]$. For all $A_i$ above depth $d$ we have $A_i \equiv \bot$, since they are $\boldsymbol{x}$-clusters with $\bot$ substituted for the leading variable in normal form. Now a $\bot$ in the Böhm tree $B$ above depth $d$ is either (a) one of the $A_i$ or (b) it is in the context $D$. In case (b) the $\bot$ is independent from all substituted $\bot$'s, and will be at this position in the Böhm tree of every $C[\boldsymbol{L}]$ for arbitrary $\boldsymbol{L}$. In case (a) the $A_i$ depends on exactly one of the substituted $\bot$'s from the input. If this $\bot$ is refined to a free variable, then the Böhm tree will no longer have a $\bot$ at this position. On the other hand, refining all other $\bot$'s from $M$ will not affect the $\bot$ in the Böhm tree at this position.

## 8 Concluding Remarks

On intuitive grounds it seems plausible that there is an "inverse" correspondence of Barendregt's Lemma and the reported properties of reduction under substitution with the notions of tracing and origin tracking, and especially with the prefix property, see [BKV00]. This relation was already indicated in [BKV00] and, with the SqBL in the place of BL, also in [Oos97] and [Ter03], Sec. 8.6. It would be interesting to investigate this correspondence in more detail and to compare the techniques of dynamic labelling used in tracing and origin tracking with the special underlining techniques that were employed in [Bar72] and [Bar74].

It seems likely that reduction under substitution can contribute to a better understanding of sequentiality, a direction that merits further investigation. The same holds for the connection with work on stability, semi-standardization and factorization, see e.g. [GK94], [Mel97, Mel98] and [Ter03], Ch. 8.

Although we didn't need it in order to obtain a sequentiality result concerning the, potententially infinite, Böhm tree of the output, it might be possible to prove also an infinitary version of RuS. This is an objective of further investigation.

## Acknowledgements

We would like to thank Jan Willem Klop, Vincent van Oostrom, Ariya Isihara and Clemens Grabmayer for stimulating conversations on the subject matter of this paper and helpful comments.

## References

[Bar72] Barendregt, H.P.: Non-definability of $\delta$ (unpublished manuscript)(1972)

[Bar74] Barendregt, H.P.: Pairing without conventional restraints. Zeitschrift für mathematische Logik und Grundlagen der Mathematik 20, 289–306 (1974)

[Bar84] Barendregt, H.P.: The Lambda Calculus: Its Syntax and Semantics, 2nd revised edn. Studies in Logic and the Foundations of Mathematics, vol. 103. North-Holland, Amsterdam (1984)

[Ber78] Berry, G.: Séquentialité de l'évaluation formelle des $\lambda$-expressions. In: Robinet, B. (ed.) Program Transformations, Troisiéme Colloque International sur la Programmation, pp. 65–78. Dunod (1978)

[Ber79] Berry, G.: Modèles complètement adéquats et stables des $\lambda$-calculs typés. PhD thesis, Université de Paris 7 (1979)

[BKOV99] Byun, S., Kennaway, J.R., van Oostrom, V., de Vries, F.-J.: Separability and translatability of sequential term rewrite systems into the lambda calculus (unpublished) (1999)

[BKV00] Bethke, I., Klop, J.W., de Vrijer, R.C.: Descendants and origins in term rewriting. Information and Computation 159, 59–124 (2000)

[BS99] Barendregt, H.P., Statman, R.: Applications of Plotkin-terms: partitions and morphisms for closed terms. J. Funct. Prog. 9, 565–575 (1999)

[Daa80] van Daalen, D.T.: The Language Theory of Automath. PhD thesis, Technische Universiteit Eindhoven. Large parts of this thesis, including the treatment of RuS, have been reproduced. In: [NGdV94] (1980)

[GK94] Glauert, J.R.W., Khasidashvili, Z.: Relative normalization in orthogonal expression reduction systems. In: Conditional Term Rewriting Systems, pp. 144–165 (1994)

[Lév75] Lévy, J.-J.: An algebraic interpretation of the $\lambda\beta K$-calculus and a labelled $\lambda$-calculus. In: Böhm, C. (ed.) Lambda-Calculus and Computer Science Theory. LNCS, vol. 37, pp. 147–165. Springer, Heidelberg (1975)

[Mel97] Melliès, P.-A.: Axiomatic rewriting theory III: A factorisation theorem in rewriting theory. In: Moggi, E., Rosolini, G. (eds.) CTCS 1997. LNCS, vol. 1290, pp. 46–68. Springer, Heidelberg (1997)

[Mel98] Melliès, P.-A.: Axiomatic rewriting theory IV: A stability theorem in rewriting theory. In: Proc. of the 14th Annual Symposium on Logic in Computer Science, LICS 1998, pp. 287–298. IEEE Computer Society Press, Los Alamitos (1998)

[NGV94] Nederpelt, R.P., Geuvers, J.H., de Vrijer, R.C.: Selected Papers on Automath. Studies in Logic and the Foundations of Mathematics, vol. 133. North-Holland, Amsterdam (1994)

[Oos97] van Oostrom, V.: Finite family developments. In: Comon, H. (ed.) RTA 1997. LNCS, vol. 1232, pp. 308–322. Springer, Heidelberg (1997)

[Ter03] Terese: Term Rewriting Systems. Cambridge University Press, Cambridge (2003)

[Vri87] de Vrijer, R.C.: Surjective Pairing and Strong Normalization: Two Themes in Lambda Calculus. PhD thesis, Universiteit van Amsterdam (January 1987)

[Vri07] de Vrijer, R.C.: Barendregt's lemma. In: Barendsen, Geuvers, Capretta, Niqui (eds.) Reflections on Type Theory, Lambda Calculus, and the Mind, pp. 275–284. Radboud University Nijmegen (2007)

# Normalization of Infinite Terms

Hans Zantema[1,2]

[1] Department of Computer Science, TU Eindhoven, P.O. Box 513, 5600 MB Eindhoven, The Netherlands
[2] Institute for Computing and Information Sciences, Radboud University Nijmegen, P.O. Box 9010, 6500 GL Nijmegen, The Netherlands
H.Zantema@tue.nl

**Abstract.** We investigate the property $\mathsf{SN}^\infty$ being the natural concept related to termination when considering term rewriting applied to infinite terms. It turns out that this property can be fully characterized by a variant of monotone algebras equipped with a metric. A fruitful special case is obtained when the algebra is finite and the required metric properties are obtained for free. It turns out that the matrix method can be applied to find proofs of $\mathsf{SN}^\infty$ based on these observations. In this way $\mathsf{SN}^\infty$ can be proved fully automatically for some interesting examples related to combinatory logic.

## 1 Introduction

Rewriting terms is a natural underlying concept in programming. Some programs generate infinite data represented by infinite terms. Moreover, many recursive specifications like $t = f(t)$ have interpretations as infinite terms but not as finite terms. So as argued in [7,6,10] it is natural to consider rewriting infinite terms and investigate its behavior.

For infinite terms the usual notion of termination, defined to be the non-existence of an infinite reduction, does not make sense. As soon as terms $t, u$ satisfying $t \to u$ exist and there is at least one symbol of arity $> 1$, then we can make an infinite term in which the term $t$ occurs infinitely often, allowing an infinite reduction in which these occurrences of $t$ are reduced to $u$ one by one.

Instead the natural notion is convergence of infinite reductions, meaning that every fixed position is affected only finitely often. This is explained in more detail in [10]; the abstract notion already goes back to [5]. As in [10,5] this notion is denoted by $\mathsf{SN}^\infty$.

In this paper we further investigate this notion. First in Section 2 we define infinite terms and investigate $\mathsf{SN}^\omega$ being the variant of $\mathsf{SN}^\infty$ restricted to reductions of length $\omega$. It turns out that $\mathsf{SN}^\omega$ is equivalent to the property that every $\omega$-reduction has only finitely many root steps. Next in Section 3 we give an if-and-only-if-characterization of $\mathsf{SN}^\omega$ by means of weakly monotone algebra provided with a metric. In Section 4 we extend $\mathsf{SN}^\omega$ to $\mathsf{SN}^\infty$ by also involving transfinite reduction lengths. It turns out that $\mathsf{SN}^\infty$ and $\mathsf{SN}^\omega$ are equivalent if all left hand sides of the rules are both linear and finite, exactly the same condition

A. Voronkov (Ed.): RTA 2008, LNCS 5117, pp. 441–455, 2008.

as required for decidability whether a rule is applicable. In Section 5 we see how the metric properties are obtained for free in case the monotone algebra is finite. Finally, in Section 7 we see how such finite monotone algebras proving $\mathsf{SN}^\infty$ can be found by matrix interpretations. In this way existing implementations for proving termination by matrix interpretations based on SAT solving can be reused for automatically proving $\mathsf{SN}^\infty$. We give an example of such a proof found for the rule for the combinator $\delta$ in combinatory logic.

## 2 Preliminaries

Intuitively, a term (both finite and infinite) is defined by saying which symbol is on which position. Here a *position* $p \in \mathbf{N}^*$ is a sequence of natural numbers. In order to be a proper term, some requirements have to be satisfied as indicated in the following definition. Here we write $\perp$ for undefined, and every symbol $f$ has an arity $\mathsf{ar}(f) \in \mathbf{N}$.

**Definition 1.** *A (possibly infinite)* term *over a signature $\Sigma$ is defined to be a map $t : \mathbf{N}^* \to \Sigma \cup \{\perp\}$ such that*

- *the root $t(\epsilon)$ of the term $t$ is defined, so $t(\epsilon) \in \Sigma$, and*
- *for all $p \in \mathbf{N}^*$ and all $i \in \mathbf{N}$ we have*

$$t(pi) \in \Sigma \iff t(p) \in \Sigma \wedge 1 \leq i \leq \mathsf{ar}(t(p)).$$

*We write $\mathsf{T}^\infty(\Sigma)$ for the set of all terms over $\Sigma$.*

An alternative equivalent definition of $\mathsf{T}^\infty(\Sigma)$ can be given based on co-algebra, but for the results in this paper we do not need this co-algebraic view.

A position $p \in \mathbf{N}^*$ satisfying $t(p) \in \Sigma$ is called a *position of* $t$. The usual notion of finite term coincides with a term in this setting having finitely many positions.

For $f \in \Sigma$ with $\mathsf{ar}(f) = n$ and $n$ terms $t_1, \ldots, t_n$ we write $f(t_1, \ldots, t_n)$ for the term $t$ defined by $t(\epsilon) = f$, $t(ip) = t_i(p)$ for every $p \in \mathbf{N}^*$ and $i = 1, \ldots, n$, and $t(ip) = \perp$ if $i \notin \{1, \ldots, n\}$.

On terms there is a natural *metric* in which terms are closer in the metric if their difference is deeper. More precisely, the distance $d(t, u)$ between two distinct terms $t, u$ is defined to be $2^{-k}$, where $k$ is the least number such that there is a position $p \in \mathbf{N}^*$ of length $k$ such that $t(p) \neq u(p)$. For equal terms terms the distance is defined to be 0.

Generalizing rewriting over finite terms, we define an infinitary *term rewrite system* (TRS) to be a set of rewrite rules. Here a *rewrite rule* $\ell \to r$ is a pair $(\ell, r) \in \mathsf{T}^\infty(\Sigma \cup \mathcal{X})^2$, where $\mathcal{X}$ is a fresh set of symbols called *variables*, all having arity 0. For effectively computing rewrite steps it is natural to require that $\ell$ is finite and linear, but for the general concept there is no need for these restrictions. In earlier texts like [6,8] finiteness was included and linearity was excluded as part of the definition.

We define a *substitution* to be a map $\sigma : \mathcal{X} \to \mathsf{T}^\infty(\Sigma \cup \mathcal{X})$. For a term $t$ the term $t\sigma$ is obtained from $t$ by replacing every occurrence of a variable $x$ by $\sigma(x)$, more precisely, for $p \in \mathbf{N}^*$ we have

$$t\sigma(p) = \begin{cases} t(p) & \text{if } t(p) \in \Sigma \\ \sigma(t(p_0))(p1) & \text{if } p = p_0 p_1 \text{ and } t(p_0) \in \mathcal{X} \\ \bot & \text{otherwise.} \end{cases}$$

For a TRS $R$, two terms $t, u$ and $p \in \mathbf{N}^*$ satisfying $t(p) \in \Sigma$ we write $t \to_{R,p} u$ if there is a rule $\ell \to r$ in $R$ and a substitution $\sigma$ such that

- $t(pq) = (\ell\sigma)(q)$ and $u(pq) = (r\sigma)(q)$ for every $q \in \mathbf{N}^*$, and
- $t(q) = u(q)$ for every $q \in \mathbf{N}^*$ not having $p$ as a prefix.

Alternatively an equivalent inductive definition of $\to_{R,p}$ can be given:

- $\ell\sigma \to_{R,\epsilon} r\sigma$ for every rule $\ell \to r$ in $R$ and every substitution $\sigma$,
- if $t \to_{R,p} u$ then $f(\ldots, t, \ldots) \to_{R,ip} f(\ldots, u, \ldots)$ for every symbol $f$, where $t$ and $u$ are in the $i$–th position of $f$ and the other arguments of $f$ remain unchanged.

The step $t \to_{R,p} u$ is called a *rewrite step at position* $p$; we write $t \to_R u$ if a position $p$ in $t$ exists such that $t \to_{R,p} u$. A step $t \to_{R,\epsilon} u$ at position $\epsilon$ is called a *root step.*

As mentioned in the introduction, instead of termination it is more natural to consider $\mathsf{SN}^\infty$, representing convergence of all infinite reductions with respect to the metric $d$. In this section and the next we focus on the variant $\mathsf{SN}^\omega$ in which the infinite reductions do not have length exceeding $\omega$, shortly indicated as $\omega$-reductions. One reason for doing so is that this is simpler: for the $\omega$ version we do not need convergence requirements for smaller limit ordinals. In Section 4 we extend our results to reduction lengths of arbitrary ordinals. In particular we will see that $\mathsf{SN}^\omega$ and $\mathsf{SN}^\infty$ coincide if all left hand sides of the rules are both finite and linear. Since these conditions are also required for decidability of whether a rule is applicable, these are very natural conditions in the setting of infinitary rewriting, making it reasonable to focus on $\mathsf{SN}^\omega$.

A characterizing property of $\mathsf{SN}^\omega$ is that for every $\omega$-reduction and every position after a finite initial part of the reduction this particular position is not affected any more. Here by 'affected' we refer to the redex position, independent whether the symbol on this position changes or not. In [6] this is called *strong convergence.* Since this characterization does not refer to the metric, and is consistent with [10], we choose this as the definition of $\mathsf{SN}^\omega$.

**Definition 2.** *A TRS $R$ is defined to satisfy* $\mathsf{SN}^\omega(R)$ *if for every $\omega$-reduction* $t_1 \to_{R,p_1} t_2 \to_{R,p_2} t_3 \to_{R,p_3} \cdots$ *and for every* $p \in \mathbf{N}^*$ *there are only finitely many $i$ such that $p_i$ is a prefix of $p$.*

The notions termination and $\mathsf{SN}^\omega$ are incomparable as the following two examples show. In the examples we use $x, y, z$ for variables.

*Example 1.* For a unary symbol $f$ we define the infinite term $t$ by $t(1^n) = f$ for all $n \geq 0$, and undefined otherwise. This term $t$ satisfies $t = f(t)$. As a consequence, the terminating TRS $R$ consisting of the single rule $f(x) \to x$ does not satisfy $\mathsf{SN}^\omega(R)$ since $t \to_{R,\epsilon} t$.

*Example 2.* The non-terminating TRS $R$ consisting of the single rule $f(x) \to g(f(x))$ satisfies $\mathsf{SN}^\omega(R)$. Intuitively this is because every step forces next redexes to be deeper; a formal proof easily follows from Theorem 5 later in this paper, by choosing $1 > \bot$ and $[f](x) = 1$, $[g](x) = \bot$ for all $x$.

*Example 3.* In [10] it was claimed that the non-terminating TRS $R$ from combinatory logic consisting of the single S-rule $a(a(a(S,x),y),z) \to a(a(x,z),a(y,z))$ satisfies $\mathsf{SN}^\omega(R)$, both for finite and infinite terms. However, this is not true when allowing infinite terms. Define the infinite terms $t, u$ by $t = a(S,t)$ and $u = a(u,u)$. Then we have

$$a(a(a(S,t),u),u) \to_{R,\epsilon} a(a(t,u),a(u,u)) = a(a(a(S,t),u),u).$$

Just like for termination it is straightforward to define a local version of $\mathsf{SN}^\omega$ depending on an initial term of a reduction. However, for termination practically all techniques only apply for the global version requiring termination for all initial terms. Since we want to link $\mathsf{SN}^\omega$ with termination techniques, we do not elaborate on local $\mathsf{SN}^\omega$.

The following theorem is a straightforward extension of a similar well-known result for finite terms.

**Theorem 1.** *For a TRS $R$ the property $\mathsf{SN}^\omega(R)$ holds if and only if every $\omega$-reduction contains at most finitely many root steps.*

*Proof.* The 'only if'-part is trivial. For the 'if'-part we assume that every $\omega$-reduction contains at most finitely many root steps. We have to prove that for every $p \in \mathbf{N}^*$ and for every $\omega$-reduction $t_1 \to_{R,p_1} t_2 \to_{R,p_2} t_3 \to_{R,p_3} \cdots$ there are only finitely many $i$ such that $p_i$ is a prefix of $p$. We do this by induction on the length $k$ of $p$. Given such a reduction, by the assumption we know it contains only finitely many roots steps. So there is an $N$ such that $p_i$ is non-empty for all $i > N$. So there is $f \in \Sigma$ of arity, say, $n$, such that for all $i > N$ we can write $t_i = f(t_{i,1},\ldots,t_{i,n})$, where for all $i > N$ and all $j = 1,\ldots,n$ we have $t_{i,j} = t_{i+1,j}$ or $t_{i,j} \to_{R,q} t_{i+1,j}$ where $p_i = jq$. For values $j$ where the latter case occurs infinitely often, by the induction hypothesis we conclude that for only finitely many of these steps the length of $q$ is $< k$. The total number of remaining steps in the reduction is finite. Hence for only finitely many values of $i$ the length of $p_i$ is $\leq k$, from which the claim easily follows. □

## 3 Monotone Algebras

A standard way to prove termination is by interpreting terms in a well-founded order. It turns out that the notion of *monotone algebra* ([11,12]) gives an if-and-only-if-characterization of termination. Now we wonder whether something similar exists for $\mathsf{SN}^\omega$. A crucial difference with earlier settings is that now we allow infinite terms that should be interpreted in the monotone algebras. One way to deal with this is the following. An infinite term can be seen as the limit of a sequence of finite terms, where each finite term is obtained by truncating the original infinite term. Now to have a well-defined interpretation of the infinite term we require that the sequence of interpretations of the finite terms has a limit. One way to formally deal with limits makes use of a *metric*, i.e., a symmetric map $d : A \times A \to \mathbf{R}_{\geq 0}$ satisfying the triangle inequality and $d(a, b) = 0$ if and only if $a = b$. We assume that $d$ is a metric on $A$.

**Definition 3.** *A sequence $a_1, a_2, \ldots$ in $A$ is said to* converge *with limit $a$ if for every $\epsilon > 0$ there exists $N$ such that $d(a, a_i) < \epsilon$ for every $i > N$. In this case we write $\lim_{i\to\infty} a_i = a$.*

*A map $f : A^n \to A$ is called* continuous *if for any $n$ converging sequences $a_{i,1}, a_{i,2}, \ldots$ for $i = 1, \ldots, n$ the sequence $f(a_{1,1}, \ldots, a_{n,1}), f(a_{1,2}, \ldots, a_{n,2}), \ldots$ also converges and $\lim_{i\to\infty} f(a_{1,i}, \ldots, a_{n,i}) = f(\lim_{i\to\infty} a_{1,i}, \ldots, \lim_{i\to\infty} a_{n,i})$.*

We will characterize $\mathsf{SN}^\omega$ in terms of a variant of weakly monotone algebras. So first we recall the notion of weakly monotone algebras from [2,3].

**Definition 4.** *A $\Sigma$-algebra $(A, [\cdot])$ is defined to consist of a non-empty set $A$, and for every $f \in \Sigma$ a function $[f] : A^n \to A$, where $n$ is the arity of $f$. This function $[f]$ is called the* interpretation *of $f$.*

*An operation $[f] : A^n \to A$ is* monotone *with respect to a binary relation $\to$ on $A$ if for all $a_i, b_i \in A$ for $i = 1, \ldots, n$ with $a_i \to b_i$ for some $i$ and $a_j = b_j$ for all $j \neq i$ we have $[f](a_1, \ldots, a_n) \to [f](b_1, \ldots, b_n)$.*

*A* weakly monotone $\Sigma$-algebra $(A, [\cdot], >, \gtrsim)$ *is a $\Sigma$-algebra $(A, [\cdot])$ equipped with two relations $>, \gtrsim$ on $A$ such that*

- *$>$ is well-founded;*
- *$> \cdot \gtrsim \subseteq > \subseteq \gtrsim$;*
- *for every $f \in \Sigma$ the operation $[f]$ is monotone with respect to $\gtrsim$.*

For a $\Sigma$-algebra $(A, [\cdot])$ and a map $\alpha : \mathcal{X} \to A$ the evaluation $[t, \alpha]$ of a finite term $t$ is defined inductively by

$$[x, \alpha] = \alpha(x), \quad [f(t_1, \ldots, t_n), \alpha] = [f]([t_1, \alpha], \ldots, [t_n, \alpha])$$

for $f \in \Sigma$ and $x \in \mathcal{X}$. For a ground term $t$ (not containing variables) $[t, \alpha]$ does not depend on $\alpha$ and is shortly written as $[t]$.

In order to deal with $\mathsf{SN}^\omega$ infinite terms should also be interpreted. We choose to consider this as the limit of the finite terms obtained by truncation. In order to define truncation we fix a constant $c \in \Sigma$; if $\Sigma$ does not contain constants then we add it. Write $|p|$ for the length of a string $p$. For a (possibly infinite) term $t$ its *truncation* $\mathsf{trunc}(t, n)$ on level $n$ is defined by

$$\mathsf{trunc}(t, n)(p) = \begin{cases} t(p) & \text{if } |p| < n \\ c & \text{if } |p| = n \text{ and } t(p) \in \Sigma \\ \bot & \text{otherwise.} \end{cases}$$

**Theorem 2.** *Let $R$ be a TRS over $\Sigma$ of which both $\ell$ and $r$ are finite for all rules $\ell \to r$. Then $\mathsf{SN}^\omega(R)$ holds if and only if there is a weakly monotone $\Sigma$-algebra $(A, [\cdot], >, \gtrsim)$ provided with a metric $d$ for which*

- *$[\ell, \alpha] > [r, \alpha]$ for every rule $\ell \to r$ in $R$ and every $\alpha : \mathcal{X} \to A$, and*
- *for every $f \in \Sigma$ the operation $[f]$ is continuous, and*
- *for every infinite ground term $t$ the sequence $[\mathsf{trunc}(t, n)]$ converges for $n \to \infty$.*

Before proving Theorem 2 we give two lemmas. We say that a $\Sigma$-algebra $(A, [\cdot])$ provided with a metric $d$ is a *metric algebra* if

- for every $f \in \Sigma$ the operation $[f]$ is continuous, and
- for every infinite ground term $t$ the sequence $[\mathsf{trunc}(t, n)]$ converges.

For a ground term $t$ we write $[t] = \lim_{n\to\infty}[\mathsf{trunc}(t, n)]$; note that for finite ground terms $t$ we now have two equivalent definitions of $[t]$.

**Lemma 1.** *For a metric algebra $(A, [\cdot], d)$, a symbol $f \in \Sigma$ of arity $k$ and ground terms $t_1, \ldots, t_k$ we have $[f(t_1, \ldots, t_k)] = [f]([t_1], \ldots, [t_k])$.*

*Proof.* Using the definition of $[\cdot]$ and continuity we obtain

$$\begin{aligned}
[f(t_1,\ldots,t_k)] &= \lim_{n\to\infty}[\mathsf{trunc}(f(t_1,\ldots,t_k), n)]\\
&= \lim_{n\to\infty}[\mathsf{trunc}(f(t_1,\ldots,t_k), n+1)]\\
&= \lim_{n\to\infty}[f(\mathsf{trunc}(t_1, n),\ldots,\mathsf{trunc}(t_k, n))]\\
&= \lim_{n\to\infty}[f]([\mathsf{trunc}(t_1, n)],\ldots,[\mathsf{trunc}(t_k, n)])\\
&= [f](\lim_{n\to\infty}[\mathsf{trunc}(t_1, n)],\ldots,\lim_{n\to\infty}[\mathsf{trunc}(t_k, n)])\\
&= [f]([t_1],\ldots,[t_k]).
\end{aligned}$$

□

**Lemma 2.** *For a metric algebra $(A, [\cdot], d)$, a finite term $t$ and a ground substitution $\sigma$ we have $[t\sigma] = [t, \alpha]$ for $\alpha : \mathcal{X} \to A$ defined by $\alpha(x) = [x\sigma]$.*

*Proof.* We apply induction on $t$. If $t \in \mathcal{X}$ then $[t\sigma] = \alpha(t) = [t, \alpha]$. In the remaining case we have $t = f(t_1, \ldots, t_k)$ for some $f \in \Sigma$ of arity $k \geq 0$. Then we obtain

$$\begin{aligned}
[t\sigma] &= [f(t_1\sigma,\ldots,t_k\sigma)]\\
&= [f]([t_1\sigma],\ldots,[t_k\sigma]) \quad \text{(by Lemma 1)}\\
&= [f]([t_1,\alpha],\ldots,[t_k,\alpha]) \quad \text{(by the induction hypothesis)}\\
&= [f(t_1,\ldots,t_k),\alpha]\\
&= [t,\alpha]
\end{aligned}$$

□

Now we are ready to give the proof of Theorem 2.

*Proof.* (Theorem 2)
For the 'only if'-part we assume $\mathsf{SN}^\omega(R)$ and have to construct an appropriate algebra. We choose $A = \mathsf{T}^\infty(\Sigma)$, and $[f](t_1, \ldots, t_n) = f(t_1, \ldots, t_n)$. As the metric we define $d(t, u)$ to be $2^{-k}$ for $t \neq u$, where $k$ is the least number such

that there is a position $p$ such that $|p| = k$ and $t(p) \neq u(p)$, and $d(t,t) = 0$. We define $> = \to_R^* \cdot \to_{R,\epsilon} \cdot \to_R^*$ and $\gtrsim = \to_R^*$. Now we prove all requirements. Assume $>$ admits an infinite decreasing sequence. By definition this is an $\to_R$-$\omega$-reduction containing infinitely many $\to_{R,\epsilon}$ steps, contradicting $\mathsf{SN}^\omega(R)$ by Theorem 1. So $>$ is well-founded. The property $> \cdot \gtrsim \subseteq > \subseteq \gtrsim$ holds by definition. Monotonicity of $[f]$ with respect to $\gtrsim$ is immediate from the definition of rewriting. So we checked the conditions of a weakly monotone algebra; we still have to check the three conditions of the theorem. If $\alpha : \mathcal{X} \to A$ then using our definition of $A = \mathsf{T}^\infty(\Sigma)$ we obtain $[\ell, \alpha] = \ell\alpha \to_{R,\epsilon} r\alpha = [r, \alpha]$, so $[\ell, \alpha] > [r, \alpha]$. From the definition of the metric it is easy to check that a sequence of terms $t_i$ converges to a term $t$ if and only if for every $n$ there exists $N$ such that $\mathsf{trunc}(t_i, n) = \mathsf{trunc}(t, n)$ for all $i > N$. From this observation both continuity of $[f]$ and convergence of $[\mathsf{trunc}(t, n)] = \mathsf{trunc}(t, n)$ easily follows.

Conversely for the 'if'-part we assume a weakly monotone algebra $(A, [\cdot], >, \gtrsim)$ with the given properties, and we have to prove $\mathsf{SN}^\omega(R)$. Assume not, then by Theorem 1 there is an $\omega$-reduction $t_1 \to_R t_2 \to_R t_3 \to_R \cdots$ containing infinitely many root steps. By applying a ground substitution on the reduction, we may assume that all $t_i$ are ground terms. For a root step $t \to_{R,\epsilon} u$ we have that $t = \ell\sigma$ and $u = r\sigma$ for some rule $\ell \to r$ and some ground substitution $\sigma$. Defining $\alpha : \mathcal{X} \to A$ by $\alpha(x) = [x\sigma]$ we obtain by Lemma 2:

$$[t] = [\ell\sigma] = [\ell, \alpha] > [r, \alpha] = [r\sigma] = [u].$$

Since $> \subseteq \gtrsim$ we conclude that $[t] \gtrsim [u]$ for every root step $t \to_{R,\epsilon} u$. Since $[f]$ is monotone with respect to $\gtrsim$ for every $f \in \Sigma$, we conclude from the inductive definition of $\to_R$ that $[t] \gtrsim [u]$ for every step $t \to_R u$. So in the $\omega$-reduction $t_1 \to_R t_2 \to_R t_3 \to_R \cdots$ we have $[t_i] > [t_{i+1}]$ for all infinitely many roots steps $t_i \to_{R,\epsilon} t_{i+1}$, and $[t_i] \gtrsim [t_{i+1}]$ for all other steps $t_i \to_R t_{i+1}$. Repeatedly applying $> \cdot \gtrsim \subseteq >$ removes all '$\gtrsim$'-steps in this sequence after the first '$>$' giving rise to an infinite decreasing $>$-sequence, contradicting well-foundedness of $>$. □

## 4 Transfinite Reductions

Until now we focused on $\mathsf{SN}^\omega$ regarding infinite reductions of length $\omega$. In this section we investigate what happens if we allow transfinite reductions: infinite reductions of higher ordinals. For any ordinal $\alpha$ an infinite $\alpha$-sequence $\{t_\beta\}_{\beta<\alpha}$ of terms is a *reduction* if $t_\beta \to_R t_{\beta+1}$ for every $\beta < \alpha$. Such a reduction is called *strongly continuous* if for every limit ordinal $\beta < \alpha$ it holds that the reduction depth of step $t_\gamma \to_R t_{\gamma+1}$ tends to infinity if $\gamma$ approaches $\beta$ from below. A reduction is called *strongly convergent* if the same holds for every limit ordinal $\beta \leq \alpha$. Note that every $\omega$-reduction is strongly continuous since no limit ordinal $\beta < \omega$ exists. As in [10] $R$ is defined to satisfy $\mathsf{SN}^\infty$ if in every strongly continuous reduction every (finite) position is affected only finitely often; stated equivalently: every strongly continuous reduction is strongly convergent. Similar to Theorem 1 we obtain

$\mathsf{SN}^\infty$ holds if and only if every strongly continuous reduction contains only finitely many root steps.

*Example 4.* Let $R$ be the TRS consisting of the three rules

$$a \to s(a), \quad b \to s(b), \quad f(x,x) \to f(a,b).$$

Then $R$ allows a strongly continuous reduction of length $\omega \times \omega$ having infinitely many root steps by repeating

$$f(a,b) \to^\omega f(s^\omega, b) \to^\omega f(s^\omega, s^\omega) \to f(a,b).$$

So $R$ does not satisfy $\mathsf{SN}^\infty$.

Next assume $R$ allows a reduction of length $\omega$ having infinitely many root steps. Then between two consecutive root steps there are only finitely many steps. Since an $s$-symbol at the root never can be removed by rewriting, both these root steps should be applications of the last rule:

$$f(t,t) \to_\epsilon f(a,b) \to^* f(u,u) \to_\epsilon f(a,b).$$

Since $\to^*$ represents finitely many steps $u$ should be both of the shape $s^n(a)$ and $s^m(b)$, contradiction.

So $R$ satisfies $\mathsf{SN}^\omega$.

Similar to Theorem 2 we obtain the following theorem, in which the fourth item appears to achieve $[t,\alpha] \gtrsim [u,\alpha]$ for $t$ rewriting to $u$ in infinitely many steps.

**Theorem 3.** *Let $R$ be a TRS over $\Sigma$ for which both $\ell$ and $r$ are finite for all rules $\ell \to r$. Then $\mathsf{SN}^\infty(R)$ holds if and only if there is a weakly monotone $\Sigma$-algebra $(A, [\cdot], >, \gtrsim)$ provided with a metric $d$ for which*

- *$[\ell,\alpha] > [r,\alpha]$ for every rule $\ell \to r$ in $R$ and every $\alpha : \mathcal{X} \to A$, and*
- *for every $f \in \Sigma$ the operation $[f]$ is continuous, and*
- *for every infinite ground term $t$ the sequence $[\mathsf{trunc}(t,n)]$ converges for $n \to \infty$, and*
- *if $a_i \in A$ for $i = 1,2,3,\ldots$ and $a_i \gtrsim a_{i+1}$ for all $i$ and $a_i$ converges for $i \to \infty$, then $a_1 \gtrsim \lim_{i\to\infty} a_i$.*

*Proof.* (sketch)
The proof is a modification of the proof of Theorem 2. For the 'only if'-part we again choose $A = \mathsf{T}^\infty(\Sigma)$, and $[f](t_1,\ldots,t_n) = f(t_1,\ldots,t_n)$, and the same metric. We choose $> \, = \, \to_R^\# \cdot \to_{R,\epsilon} \cdot \to_R^\#$ and $\gtrsim \, = \, \to_R^\#$, where $\to_R^\#$ is defined by $t \to_R^\# u$ if and only if there is a strongly convergent reduction from $t$ to $u$ of length $\alpha$ for any ordinal $\alpha$.

For the 'if'-part we observe that in a strongly continuous reduction containing infinitely many root steps, the reduction between two consecutive root steps is of the shape $t \to_R^\# u$. Here we need the last condition to conclude $[t] \gtrsim [u]$, for the rest the proof is the same as for Theorem 2. □

As was pointed out by Clemens Grabmayer for proving equivalence of $\mathsf{SN}^\infty(R)$ and $\mathsf{SN}^\omega(R)$ in case all left hand sides of $R$ are linear and finite we need the following known variant of the compression lemma (Theorem 12.7.1 in [6]). The proof can be given along the same lines as the proof of the original compression lemma given there. For any ordinal $\alpha$ we write $t \to^\alpha u$ if there is a strongly convergent reduction from $t$ to $u$ of length $\alpha$.

**Lemma 3.** *Let $R$ be a TRS over $\Sigma$ of which $\ell$ is both linear and finite for all rules $\ell \to r$. Let $t \to^\alpha u$ be any infinite strongly convergent reduction containing at least one root step. Then there is a finite number $n$, an ordinal $\beta \leq \omega$ and two terms $t', t''$ such that $t \to^n t' \to_\epsilon t'' \to^\beta u$.*

**Theorem 4.** *Let $R$ be a TRS over $\Sigma$ of which $\ell$ is both linear and finite for all rules $\ell \to r$. Then $\mathsf{SN}^\infty(R)$ and $\mathsf{SN}^\omega(R)$ are equivalent.*

*Proof.* The direction $\mathsf{SN}^\infty(R) \Rightarrow \mathsf{SN}^\omega(R)$ is trivial. For the other direction assume that $\mathsf{SN}^\omega(R)$ holds and $\mathsf{SN}^\infty(R)$ does not hold. Then there is a strongly continuous reduction

$$t_1 \to^{\alpha_1} u_1 \to_\epsilon t_2 \to^{\alpha_2} u_2 \to_\epsilon t_3 \to^{\alpha_3} u_3 \to_\epsilon \cdots$$

Applying Lemma 3 to $t_1 \to^{\alpha_1} u_1 \to_\epsilon t_2$ yields $t_1 \to^{n_1} t_1' \to_\epsilon t_1'' \to^\beta t_2$. Next we apply Lemma 3 to $t_1'' \to^\beta t_2 \to^{\alpha_2} u_2 \to_\epsilon t_3$ yielding $t_1'' \to^{n_2} t_2' \to_\epsilon t_2'' \to^\beta t_3$. Repeating this process yields the $\omega$-reduction

$$t_1 \to^{n_1} t_1' \to_\epsilon t_1'' \to^{n_2} t_2' \to_\epsilon t_2'' \to^{n_3} t_3' \to_\epsilon t_3'' \cdots$$

containing infinitely many root steps, contradicting $\mathsf{SN}^\omega(R)$. □

The proof of this theorem is essentially the same as the proof of Lemma 5.2 in [9].

## 5 Finite Algebras

In the original monotone algebra approach for proving termination of rewriting finite terms, all monotone algebras are infinite since finite algebras do not allow non-trivial strictly monotone operations. Surprisingly, for proving $\mathsf{SN}^\infty$ by our adjusted monotone algebra approach, finite algebras are useful and often provide non-trivial proofs.

The first observation is that in finite algebras satisfying some mild conditions on $>$ and $\gtrsim$ all requirements involving the metric are obtained for free.

**Theorem 5.** *Let $R$ be a TRS over $\Sigma$ of which both $\ell$ and $r$ are finite for all rules $\ell \to r$. Let $(A, [\cdot], >, \gtrsim)$ be a weakly monotone $\Sigma$-algebra satisfying*

- *$A$ is finite,*
- *$\gtrsim$ is transitive,*
- *if $a \gtrsim b$ for $a, b \in A$ then either $a > b$ or $a = b$,*

- $\bot \in A$, *and* $a \gtrsim \bot$ *for all* $a \in A$, *and*
- $[\ell, \alpha] > [r, \alpha]$ *for every rule* $\ell \to r$ *in* $R$ *and every* $\alpha : \mathcal{X} \to A$.

*Then* $\mathsf{SN}^\infty(R)$ *holds.*

*Proof.* On the algebra $(A, [\cdot], >, \gtrsim)$ we define the discrete metric defined by $d(a, a) = 0$ for all $a \in A$ and $d(a, b) = 1$ for all $a, b \in A, a \neq b$. For this metric we will prove that

- for every $f \in \Sigma$ the operation $[f]$ is continuous, and
- for every infinite ground term $t$ the sequence $[\mathsf{trunc}(t, n)]$ converges for $n \to \infty$, and
- if $a_i \in A$ for $i = 1, 2, 3, \ldots$ and $a_i \gtrsim a_{i+1}$ for all $i$ and $a_i$ converges for $i \to \infty$, then $a_1 \gtrsim \lim_{i\to\infty} a_i$.

Then by Theorem 3 we have proved the theorem.

For proving the first item choose $f \in \Sigma$ of arity $k$ and $k$ converging sequences $a_{i,1}, a_{i,2}, \ldots$ for $i = 1, \ldots, k$. Due to the definition of the discrete metric we obtain that for every $i$ there is some $N_i$ such that $a_{i,n} = a_{i,m}$ for every $n, m > N_i$. Let $N$ be the maximum of $N_1, \ldots, N_k$. Then $a_{i,n} = a_{i,m}$ for all $n, m > N$ and all $i$. Hence the sequence $f(a_{1,1}, \ldots, a_{k,1}), f(a_{1,2}, \ldots, a_{k,2}), \ldots$ converges too and

$$\lim_{i\to\infty} f(a_{1,i}, \ldots, a_{k,i}) = f(\lim_{i\to\infty} a_{1,i}, \ldots, \lim_{i\to\infty} a_{k,i}).$$

Hence $[f]$ is continuous.

For proving the second item we need a constant $c \in \Sigma$ for the definition of $\mathsf{trunc}$. We choose this constant $c$ to be a fresh one, add it to $\Sigma$ and define $[c] = \bot$. Now we prove by induction on $n$ that $[\mathsf{trunc}(t, n+1)] \gtrsim [\mathsf{trunc}(t, n)]$ for every $n \in \mathbf{N}$. For $n = 0$ this holds since $[\mathsf{trunc}(t, 0)] = [c] = \bot$ and $a \gtrsim \bot$ for all $a \in A$. For the induction step write $t = f(t_1, \ldots, t_k)$. Then using monotonicity of $[f]$ and the induction hypothesis we obtain

$$\begin{aligned}
[\mathsf{trunc}(t, n+1)] &= [f(\mathsf{trunc}(t_1, n), \ldots, \mathsf{trunc}(t_k, n))] \\
&= [f]([\mathsf{trunc}(t_1, n)], \ldots, [\mathsf{trunc}(t_k, n)]) \\
&\gtrsim [f]([\mathsf{trunc}(t_1, n-1)], \ldots, [\mathsf{trunc}(t_k, n-1)]) \\
&= [f(\mathsf{trunc}(t_1, n-1), \ldots, \mathsf{trunc}(t_k, n-1))] \\
&= [\mathsf{trunc}(t, n)].
\end{aligned}$$

Due to the condition of the theorem for every $n \in \mathbf{N}$ we either have $[\mathsf{trunc}(t, n+1)] > [\mathsf{trunc}(t, n)]$ or $[\mathsf{trunc}(t, n+1)] = [\mathsf{trunc}(t, n)]$. Assume that '>' happens infinitely often. Then due to finiteness of $A$ we obtain a cycle $[\mathsf{trunc}(t, n)] > \cdots > [\mathsf{trunc}(t, m)] = [\mathsf{trunc}(t, n)]$, contradicting well-foundedness. Hence $[\mathsf{trunc}(t, n+1)] > [\mathsf{trunc}(t, n)]$ occurs only finitely often while for all other $n$ we have $[\mathsf{trunc}(t, n+1)] = [\mathsf{trunc}(t, n)]$, hence the sequence $[\mathsf{trunc}(t, n)]$ converges for $n \to \infty$.

It remains to prove $a_1 \gtrsim \lim_{i\to\infty} a_i$ if $a_i \gtrsim a_{i+1}$ for all $i$ and $a_i$ converges for $i \to \infty$. Since $A$ is finite we conclude that $\lim_{i\to\infty} a_i = a_n$ for some $n$, the rest follows since $\gtrsim$ is transitive. □

*Example 5.* The fixpoint combinator $Y$ is a constant. Combined with the binary application operator $a$ there is only one rewrite rule $a(Y,x) \to a(x, a(Y,x))$. Choose $A = \{0,1,2,3\}$ with $>$ being the usual order and $\gtrsim = \geq$, and $[Y] = 3$, and

$$\begin{array}{ll} [a](3,3) = 2, & \\ [a](3,x) = 1 & \text{for } x < 3, \\ [a](x,y) = 0 & \text{for } x < 3 \text{ and all } y. \end{array}$$

All requirements of Theorem 5 are easily checked, proving $\mathsf{SN}^\infty$ of this rewrite rule.

## 6 Marking Symbols

When playing around with this kind of interpretation we observe that strict inequality $[\ell,\alpha] > [r,\alpha]$ is often quite a strong requirement. In fact it is only required for root steps; for non-root steps it is sufficient to require $[\ell,\alpha] \gtrsim [r,\alpha]$. Therefore we now borrow a syntactic trick from the dependency pair method [1] to be able to distinguish between these two kinds of steps: we mark root symbols of the left hand sides and right hand sides of the rules. This only works if all right hand sides have a root symbol: the TRS should be *non-collapsing*, i.e., no right hand side of a rule is a single variable. This is not a serious restriction: for any collapsing rule $C[x,\ldots,x] \to x$ one easily builds an infinite reduction starting in $t$ defined by $t = C[t,\ldots,t]$ containing infinitely many root steps. So all TRSs satisfying $\mathsf{SN}^\infty$ are non-collapsing.

For a non-collapsing TRS $R$ over $\Sigma$ we extend $\Sigma$ to $\Sigma_\#$ by adding a fresh symbol $f_\#$ for every $f \in \Sigma$ occurring as the root symbol of a right hand side or left hand side of a rule in $R$. For all such $f$ we define $\mathsf{ar}(f_\#) = \mathsf{ar}(f)$. For a non-variable term $t = f(t_1,\ldots,t_n)$ over $\Sigma$ we define $t_\# = f_\#(t_1,\ldots,t_n)$. We define

$$R_\# \;=\; \{\ell_\# \to r_\# \mid \ell \to r \in R\},$$

so $R_\#$ is obtained from $R$ by marking the root symbols of all left hand sides and right hand sides.

The corresponding monotone algebra may be *two-sorted* now, as in [2,3]. More precisely, the algebra consists of two sets $A$ and $A_\#$. On the set $A$ we have a relation $\gtrsim$ and on the set $A_\#$ we have a well-founded relation $>$. For every $f \in \Sigma$ of arity $n$ we have $[f] : A^n \to A$ as before, monotone with respect to $\gtrsim$. For the fresh symbols $f_\#$ we have $[f_\#] : A^n \to A_\#$, with the monotonicity requirement

> if for all $a_i, b_i \in A$ for $i = 1,\ldots,n$ with $a_i \gtrsim b_i$ for some $i$ and $a_j = b_j$ for all $j \neq i$ we have $[f_\#](a_1,\ldots,a_n) \geq [f_\#](b_1,\ldots,b_n)$, where $\geq$ is the union of $>$ and $=$.

For a term $t$ over $\Sigma$ and $\alpha : \mathcal{X} \to A$ we keep the definition of $[t,\alpha] \in A$ as before, and we have the obvious definition of $[t_\#,\alpha] \in A_\#$.

**Theorem 6.** *For proving $\mathsf{SN}^\omega$ or $\mathsf{SN}^\infty$ for a non-collapsing TRS by Theorem 2, Theorem 3 or Theorem 5, the signature $\Sigma$ may be extended to $\Sigma_\#$ and the algebra may be two-sorted as presented above, while the requirement*

– $[\ell, \alpha] > [r, \alpha]$ *for every rule* $\ell \to r$ *in* $R$ *and every* $\alpha : \mathcal{X} \to A$

*is replaced by*

– $[\ell, \alpha] > [r, \alpha]$ *for every rule* $\ell \to r$ *in* $R_\#$ *and every* $\alpha : \mathcal{X} \to A$,
– $[\ell, \alpha] \gtrsim [r, \alpha]$ *for every rule* $\ell \to r$ *in* $R$ *and every* $\alpha : \mathcal{X} \to A$.

*Proof.* In the 'if'-part of the proof of Theorem 2 we replace every term $t$ in the assumed infinite reduction by $t_\#$. In this way the $R$-reduction is transformed to an $R \cup R_\#$-reduction in which all $R_\#$-steps are at the root and all $R$-steps are below the root. Since the $R$-steps are not root steps the corresponding requirement weakened to $[\ell, \alpha] \gtrsim [r, \alpha]$. The rest of the proofs of the three theorems remain unchanged. □

*Example 6.* As in Example 5 we consider the single rewrite rule

$$a(Y, x) \to a(x, a(Y, x)).$$

Choose $A = \{0, 1\}$ and $A_\# = \{0, 1, 2\}$ with $>$ and $\gtrsim = \geq$ as usual, and

$$\begin{array}{lll} [Y] & = 1, & \\ [a](x, y) & = 0 & \text{for all } x, y, \\ [a_\#](x, y) & = x + y & \text{for all } x, y. \end{array}$$

We have

$$[a(Y, x), \alpha] = 0 = [a(x, a(Y, x)), \alpha],$$
$$[a_\#(Y, x), \alpha] = \alpha(x) + 1 > \alpha(x) = [a_\#(x, a(Y, x)), \alpha].$$

All other requirements of Theorem 5 and Theorem 6 are easily checked too, proving $\mathsf{SN}^\infty$.

*Example 7.* The TRS consisting of the four rules

$$a(f(a(x))) \to f(x),\ f(b) \to g(b),\ g(x) \to a(g(a(x))),\ g(a(x)) \to f(x)$$

satisfies $\mathsf{SN}^\infty$ but this can not be proved by Theorem 5, not even with its marked version as in Theorem 6. Assume it can, then there exists $n > k$ satisfying $[a^n(b)] = [a^k(b)]$. For every $m$ we have a reduction $a^m(f(a^m(b))) \to^+ a^m(f(a^{m-1}(b)))$ containing a root step. Applying this $n - k$ times and marking yields

$$[a_\# a^{n-1} f a^n(b)] > [a_\# a^{n-1} f a^k(b)] = [a_\# a^{n-1} f a^n(b)],$$

contradiction.

## 7 Matrix Interpretations

An interesting instance of monotone algebras is given by matrix interpretations [2,3], in which the algebra elements are vectors over the natural numbers. It turns out that for an interesting class of matrix interpretations the set of vectors involved can be restricted to a finite algebra which by Theorem 5 and Theorem 6 applies for proving $\mathsf{SN}^\infty$.

Fix a dimension $d$. We choose $A = \mathbf{N}^d$. The relation $\gtrsim$ on $A$ is defined by

$$(v_1, \ldots, v_d) \gtrsim (u_1, \ldots, u_d) \iff v_i \geq u_i \text{ for } i = 1, 2, \ldots, d.$$

We choose $A_\# = \mathbf{N}$ with the usual order $>$. For the interpretation $[f]$ of a symbol $f \in \Sigma$ of arity $n \geq 0$ we choose $n$ matrices $F_1, F_2, \ldots, F_n$ over $\mathbf{N}$, each of size $d \times d$, and a vector $\boldsymbol{f} \in \mathbf{N}^d$, and define

$$[f](\boldsymbol{v_1}, \ldots, \boldsymbol{v_n}) = F_1\boldsymbol{v_1} + \cdots + F_n\boldsymbol{v_n} + \boldsymbol{f}$$

for all $\boldsymbol{v_1}, \ldots, \boldsymbol{v_n} \in A$. One easily checks that $[f]$ is monotone with respect to $\gtrsim$. For the interpretation $[f_\#]$ of a marked symbol $f_\#$ corresponding to $f$ of arity $n \geq 0$ we define

$$[f_\#](\boldsymbol{v_1}, \ldots, \boldsymbol{v_n}) = \boldsymbol{f_1}\boldsymbol{v_1} + \cdots + \boldsymbol{f_n}\boldsymbol{v_n} + c_f$$

for $n$ row vectors $\boldsymbol{f_1}, \ldots, \boldsymbol{f_n}$ over $\mathbf{N}$ of size $d$, and a constant $c_f \in \mathbf{N}$. Here $\boldsymbol{f_i}\boldsymbol{v_i}$ denotes the inner product, corresponding to matrix multiplication of a row vector by a column vector.

A square matrix $F$ is called *strict upper triangular* if $F_{ij} = 0$ for $i, j$ satisfying $i \geq j$. So the main diagonal and everything below is zero.

**Theorem 7.** *Let $R$ be a non-collapsing TRS over a finite signature $\Sigma$. Let an interpretation in the above setting be given satisfying*

- *All matrices $F_i$ are strict upper triangular,*
- *$[\ell, \alpha] > [r, \alpha]$ for every rule $\ell \to r$ in $R_\#$ and every $\alpha : \mathcal{X} \to A$,*
- *$[\ell, \alpha] \gtrsim [r, \alpha]$ for every rule $\ell \to r$ in $R$ and every $\alpha : \mathcal{X} \to A$.*

*Then* $\mathsf{SN}^\infty(R)$.

*Proof.* (sketch) To define $A$ we start by the zero vector $= \bot$ and $[c]$ for all constants $c$, and close this set by application of $[f]$ for all $f \in \Sigma$. For strict-upper-triangular matrices $B_1, \ldots, B_d$, where $d$ is the dimension, one easily proves that $B_1 \times \cdots \times B_d$ is the zero matrix. As a consequence, for a ground term $t$ the value of $[t]$ does not depend on symbols in the term $t$ deeper than $d$. Since the signature is finite we conclude that there are only finitely many vectors in $\mathbf{N}^d$ that can be written as $[t]$ for a ground term $t$. As a consequence, the resulting closure $A$ is a finite set of vectors. Choose $A_\#$ to be the finite set of numbers obtained by applying the operations $[f_\#]$ on these finitely many vectors. In this way we have a corresponding algebra on which Theorem 5 and Theorem 6 apply for proving $\mathsf{SN}^\infty$. □

Strict-upper-triangularity is essential: the single rule $f(f(x)) \to f(x)$ does not satisfy $\mathsf{SN}^\infty$, while $[f](x) = x + 1$, $[f_\#](x) = x$ satisfies all requirements of Theorem 7 in dimension $d = 1$ except for strict-upper-triangularity of $F_1 = (1)$. Note that in dimension 1 this requirement states that all $[f]$ are constant.

*Example 8.* We consider the combinator $\delta$ of which the behavior is given by the single rewrite rule

$$a(a(\delta, x), y) \rightarrow a(y, a(x, y)).$$

This combinator is used as a building block for constructing new fixpoint combinators: if $Y$ is a fixpoint combinator then by $Y\delta x \rightarrow \delta(Y\delta)x \rightarrow x(Y\delta x)$ we see that $Y\delta$ is a fixpoint combinator too. To get a feeling for this rule, write $d_1 = \delta$ and $d_i = a(\delta, d_{i-1})$ for $i > 0$, omit the application $a$ as usual, and observe that the term $d_2 d_2 = a(a(\delta, \delta), a(\delta, \delta))$ admits an infinite reduction

$$d_2 d_2 \rightarrow d_2 d_3 \rightarrow d_3 d_4 \rightarrow d_4(d_2 d_4) \rightarrow (d_2 d_4)(d_3(d_2 d_4)) \rightarrow \cdots$$

for which the first four steps are all root steps. For applying Theorem 7 we choose $d = 2$ and

$$[\delta] = \begin{pmatrix} 0 \\ 1 \end{pmatrix}, \quad [a](\boldsymbol{x}, \boldsymbol{y}) = \begin{pmatrix} 0 & 1 \\ 0 & 0 \end{pmatrix} \cdot \boldsymbol{x} + \begin{pmatrix} 0 & 1 \\ 0 & 0 \end{pmatrix} \cdot \boldsymbol{y},$$

$$[a_{\#}](\boldsymbol{x}, \boldsymbol{y}) = \begin{pmatrix} 1 & 0 \end{pmatrix} \cdot \boldsymbol{x} + \begin{pmatrix} 1 & 1 \end{pmatrix} \cdot \boldsymbol{y}.$$

For $\alpha(x) = (x_1, x_2)$ and $\alpha(y) = (y_1, y_2)$ we have

$$[a(a(\delta, x), y), \alpha] = \begin{pmatrix} y_2 \\ 0 \end{pmatrix} = [a(y, a(x, y)), \alpha],$$

and

$$[a_{\#}(a(\delta, x), y), \alpha] = 1 + x_2 + y_1 + y_2 > y_1 + x_2 + y_2 = [a_{\#}(y, a(x, y)), \alpha],$$

proving $\mathsf{SN}^{\infty}$.

It turns out that the implicit finite monotone algebra used in this proof consists of

$$A = \{\begin{pmatrix} 0 \\ 0 \end{pmatrix}, \begin{pmatrix} 0 \\ 1 \end{pmatrix}, \begin{pmatrix} 1 \\ 0 \end{pmatrix}, \begin{pmatrix} 2 \\ 0 \end{pmatrix}\}, \quad A_{\#} = \{0, 1, 2, 3, 4\}.$$

In particular, $|A_{\#}| - 1$ is an upper bound on the number of root steps that can occur in any reduction. As we observed that $a(a(\delta, \delta), a(\delta, \delta))$ admits an infinite reduction starting by four root steps, in this example this upper bound is sharp.

To find this type of proofs for $\mathsf{SN}^{\infty}$ we have to find values of the matrix entries in such a way that the requirements hold. This requires only a minor modification of the existing search engines for finding matrix interpretations based on SAT solving: only the strict-upper-triangularity requirement has to be added. In fact the above proof for the $\delta$ combinator was found by entering the corresponding dependency pair problem in the web interface of AProVE [4] and inspecting that the matrix interpretation found by AProVE satisfies the extra strict-upper-triangularity requirement.

# 8 Conclusions

The property $\mathsf{SN}^\infty$ is the natural concept related to termination when considering rewriting infinite terms. We gave a full characterization by weakly monotone algebras equipped with a metric. We applied this by modifying the matrix method to find proofs of $\mathsf{SN}^\infty$ via finite weakly monotone algebras. This provides the first method to prove $\mathsf{SN}^\infty$ automatically. Unfortunately, comparing this to other methods fails since other methods do not yet exist.

**Acknowledgments.** I want to thank Jan Willem Klop, Clemens Grabmayer, Vincent van Oostrom and Fer-Jan de Vries for valuable discussions on this topic. Moreover, I want to thank the anonymous reviewers for fruitful remarks by which the paper has improved substantially.

# References

1. Arts, T., Giesl, J.: Termination of term rewriting using dependency pairs. Theoretical Computer Science 236, 133–178 (2000)
2. Endrullis, J., Waldmann, J., Zantema, H.: Matrix interpretations for proving termination of term rewriting. In: Furbach, U., Shankar, N. (eds.) IJCAR 2006. LNCS (LNAI), vol. 4130, pp. 574–588. Springer, Heidelberg (2006)
3. Endrullis, J., Waldmann, J., Zantema, H.: Matrix interpretations for proving termination of term rewriting. Journal of Automated Reasoning (to appear, 2008)
4. Giesl, J., et al.: Automated program verification environment (AProVE), `http://aprove.informatik.rwth-aachen.de/`
5. Kennaway, J.R.: On transfinite abstract reduction systems. Computer Science Reports CS-R9205. CWI Amsterdam (1992)
6. Kennaway, R., de Vries, F.-J.: Infinitary rewriting. In: Term Rewriting Systems, by Terese, pp. 668–711. Cambridge University Press, Cambridge (2003)
7. Kennaway, R., Klop, J.W., Sleep, M.R., de Vries, F.-J.: Transfinite reductions in orthogonal term rewriting systems. Information and Computation 119(1), 18–38 (1995)
8. Kennaway, R., Severi, P., Sleep, R., de Vries, F.-J.: Infinitary rewriting: From syntax to semantics. In: Middeldorp, A., van Oostrom, V., van Raamsdonk, F., de Vrijer, R. (eds.) Processes, Terms and Cycles: Steps on the Road to Infinity: Essays Dedicated to Jan Willem Klop on the Occasion of His 60th Birthday. LNCS, vol. 3838, pp. 148–172. Springer, Heidelberg (2005)
9. Ketema, J.: On normalisation of infinitary combinatory reduction systems. In: Voronkov, A. (ed.) Proceedings of the 19th Conference on Rewriting Techniques and Applications (RTA). LNCS. Springer, Heidelberg (2008)
10. Klop, J.W., de Vrijer, R.C.: Infinitary normalization. In: We Will Show Them! Essays in Honour of Dov Gabbay, vol. 2, pp. 169–192. College Publications (2005)
11. Zantema, H.: Termination of term rewriting: Interpretation and type elimination. Journal of Symbolic Computation 17, 23–50 (1994)
12. Zantema, H.: Termination. In: Term Rewriting Systems, by Terese, pp. 181–259. Cambridge University Press, Cambridge (2003)

# Author Index